Advanced Ceramic Technologies & Products

The Ceramic Society of Japan

Advanced Ceramic Technologies & Products

Springer

The Ceramic Society of Japan
2-22-17 Hyakunin-cho
Shinjuku-ku, Tokyo 169-0073
Japan

ISBN 978-4-431-53913-1 ISBN 978-4-431-54108-0 (eBook)
DOI 10.1007/978-4-431-54108-0
Springer Tokyo Heidelberg New York Dordrecht London

Library of Congress Control Number: 2012945732

Printed on acid-free paper

Springer is part of Springer Science+Business Media (www.springer.com)

Foreword

On the occasion of the publication of *Advanced Ceramic Technologies &Products.*

I would like to express my deep respect to everyone associated with the publication of *Advanced Ceramic Technologies & Products* in English. As an engineer having been involved in the ceramics industry, I sincerely hope that this practical book will be helpful to everyone involved in developing ceramic products now and well into the future.

Since the latter part of the twentieth century, various industries—such as electronics and information processing, transportation and aerospace, construction and factory building, and biomedicine—have developed and expanded globally. Through these developments, people's lives around the world have been enriched and become more convenient. The social climate in which addressing individual diversified needs and values easily is becoming a reality. I think that the appetite of human beings for such technologies will further increase the complexity and sophistication of those technologies. Ceramics have attracted significant attention as one of the important materials that supported the development of these industries with its excellent thermal, mechanical, and electrical characteristics and electronic properties, which conventional materials do not have.

Since 1959, when I established the Kyocera Corporation, I have engaged in the development, manufacture, and sales of many ceramic products. Shortly after World War II, the field of ceramics was referred to as "kiln work." Many techniques depended on a worker's intuition and experience, and were handed down from craftsman to craftsman for generations. Around this time, in Europe and the United States, purified, mineral-rich clays, such as alumina, forsterite, and steatite, began to see use in special industrial component applications for electronics and automobiles. In Japan, we largely depended on imports from Europe and the United States for the vast majority of our modern products such as automobiles, television sets, and other electric appliances, as well as the cutting-edge technologies employed at production facilities. This was around the time when I graduated from college and started my career as an engineer. I recall the days I spent feverishly working on developing new industrial ceramics. Unlike today, information was hard to come by back then, and it was not easy to access overseas research papers. I used to go the college library to transcribe research papers and use them as reference guides.

In order for fine ceramics to be used by our customers, we had to explain what fine ceramics were, because they were generally unknown and unaccepted at that time. We would listen to customer requirements, then, according to purpose and usage, choose materials and make a prototype. After many trials and evaluations by the customer, we could finally develop a product that satisfied them. I believe accumulating such tedious efforts formed today's robust ceramics industry in Japan.

Now, the study of ceramics is well-established and excellent products have been developed using state-of-the-art computer simulation technology and cutting-edge analytical equipment. These newly developed products must continuously satisfy the advanced requirements and cost demands of today's complex society. Throughout the ages, as technology continued to develop, it was almost like a fate that engineers had to bear to succeed in the next stage of development. The development history of more than 100 impeccable, or so called "sharp," products, listed in this book describe the field of application, development background, usage, technical specifications, and manufacturing processes. I believe this will provide guidelines to many engineers working in product development now and in the future. It will surely become a valuable book summarizing the history of ceramics product development.

Lastly, I would like to express again my deep respect to everyone involved in this book.

Kyoto, Japan
June 2012

Dr. Kazuo Inamori
Founder and Chairman Emeritus
Kyocera Corporation

Preface

During the twentieth century and the following decade, industrial ceramics have made significant contributions to various fields, including electronics, communications, transports, aeronautics, astronautics, environment, and energy. Many of these new ceramic products have made tremendous contributions to our society. It may not be an exaggeration to say that ceramic products in fact enhance the quality of our daily lives. Japan is one of the most advanced countries in the world with regard to materials science and related technology. Over the years, Japan has spawned many world-class-level ceramics manufacturers that have provided many highly functional, high-quality ceramic products and exported them worldwide. This would not have been possible without the fundamental and important research contributions from universities and national laboratories as well as by industry's utilizing continuous product improvements resulting from advanced technology development and the enthusiasm of engineers.

This book is a collection of ceramic products that have played active roles in various fields of industry and describes in detail how different ceramics are used in society. In addition, we have included behind-the-scene stories on materials development and detailed illustrations of the products that have reached the world-class level in manufacturing. There is probably not a single book that has categorized ceramics from this point of view. By reading this book, the reader will learn the significance that ceramics play in each field, and realize how ceramics continue to play important roles in improving the quality of our lives.

Even at this very moment, many engineers in industry are working hard to develop new and innovative products. A proverb states, "Look at the past to predict the future". It is our desire that this book describing the past and current product technologies can somehow serve to guide engineers working in the manufacturing sector to strive even further for the benefit of us all. Past achievements are not a guarantee for future success, but ceramics technology and product development remains strong.

This book fundamentally consists of revised articles from a series first published as "Ceramics Archives" in the journal *Ceramics* and includes material from a book on fine ceramics published by the Ceramic Society of Japan in Japanese and then translated into English. Many people have been involved in completing this book. In particular, we thank the committee members of the "Ceramics Archives", the authors of the original article, and those who have given advice and support in coordinating the English translation. We deeply thank the people listed in the Acknowledgements section.

For readers who are interested in learning more about Japanese ceramic products and technology, please see the Ceramic Society of Japan website http://www.ceramic.or.jp/index.html.

March 2012

Editior-in-Chief
Yoshihiko Imanaka

Editors
Yoshikazu Suzuki, Toru S. Suzuki,
Kiyoshi Hirao, Tetsuo Tsuchiya,
Hajime Nagata, Jeffrey S. Cross

Acknowledgments

The editors thank the following people for their support and assistance in the publication of this book.

Committee Members of "Ceramics Archives"

Toshiyuki Abe, Tsuneo Hiraide, Junichi Iwasawa, Yuji Katsuda, Hiroaki Kinoshita, Takayuki Komatsu, Masahiro Miyauchi, Hidetoshi Mizutani, Atsushi Omote, Shoko Suyama, Shigeki Takeda, Hiroshi Tamura, Masayuki Tsuda, Masami Uzawa, Hiroaki Yanagita, and Toyohiko Yano.

Contributors (Original Authors)

Minoru Akaishi, Kenji Akiyama, Hideyasu Ando, Tamotsu Asai, Kenji Awamoto, Shunka Cho, Takashi Eshita, Katsumi Fujimoto, Takuzo Fujimoto, Junzo Fukuzawa, Yasuo Goto, Kazuhide Gotou, Koji Hakamazuka, Masateru Hattori, Hiroshi Hayashi, Tsuneo Hiraide, Yasuhiko Hirayama, Shinichiro Hirota, Youichi Hoshi, Tomoyasu Ichiki, Kenji Ikegami, Takayasu Ikegami, Yoshihiko Imanaka, Takamasa Ishigaki, Mamoru Ishii, Masato Ishizaki, Atsuo Ito, Shinichiro Ito, Hitoshi Iwaya, Nobuo Iyi, Shusaku Kakita, Yasushi Katanuma, Yoshiaki Katsuyama, Seiko Kawashima, Kazuhiro Kido, Kazunori Kijima, Shigeyuki Kimura, Hirohiko Kishikawa, Tomoharu Kitabayashi, Satoru Kitahara, Mitsuaki Kitano, Shigeo Kittaka, Takashi Kodama, Kazunori Koga, Yutaka Komorida, Masao Kondo, Noriaki Kondo, Hiroshi Kurihara, Kei Maeda, Hiroshi Maiwa, Yutaka Maruyama, Kiyoyuki Masuzawa, Takao Matsumoto, Takeshi Matsumoto, Yasuhide Matsuo, Kenjiro Mihara, Toshiyuki Mima, Tetsuo Minaai, Takefumi Mitsuhashi, Jun Miyaji, Yoshinori Mochizuki, Hiroshi Mori, Shinichi Mukoyama, Toshiaki Murakami, Youichi Muraki, Takayuki Naba, Masayuki Nagai, Takehiko Nakajima, Osamu Nakano, Yuji Nakano, Masayuki Ninomiya, Michio Nishi, Tsuneaki Ohashi, Hiroshi Ohnishi, Akira Okada, Kiyoshi Okada, Takayuki Okamoto, Masanori Ookawara, Akira Oomae, Shoichi Ozawa, Chiharu Sakaki, Hiroyuki Sango, Takayoshi Sasaki, Manabu Sato, Yoshio Satoh, Masatoshi Seki, Tohru Sekino, Shinji Senda, Shigeyo Sonoda, Hitoshi Sumiya, Akihito Suzuki, Yoshihiro Suzuki, Takahiro Tabei, Mitsuaki Tada, Koichiro Takahashi, Koji Takahashi, Kyoji Takakura, Masatoshi Takao, Toshikazu Takeda, Masahiro Takizawa, Shohei Tamaki, Hiroyuki Tanabe, Chikao Tanaka, Hidehiko Tanaka, Tsutomu Tatekawa, Satoshi Teramura, Makoto Togawa, Junji Tominaga, Katsuhiko Tsuno, Hideyoshi Tsuruta, Masaru Ueno, Motohiro Umehara, Kazuhiro Urashima, Masahiro Wakida, Osamu Watanabe,

Yoshiyuki Watanabe, Katsuhisa Yabuta, Naohito Yamada, Yohachi (John) Yamashita, Hiroshi Yamauchi, Hitoshi Yokoi, Hideki Yokoyama, Yasushi Yokoyama, Shigeyoshi Yoshida, and Ryuichi Yoshimoto.

Executive Advisors

Junichi Hojo, Junichi Inagaki, Satoru Inoue, Shinichi Inui, Tetsuya Kameyama, Koichi Niihara, and Akira Okada.

Contents

Part II Information Technology and Electroceramics

Part I

Introduction

1 Introduction to Ceramics

Definition of Ceramics

The term "ceramics" represents both man-made and manufactured non-metallic inorganic solid materials. The name comes from keramos (clay used by potters, objects made of clay), which originated from keramikos (of clay, of pottery) or keramikos in Greek. Heat treatment has been typically used for ceramic "manufacturing" since ancient times. Therefore, ceramics is the term used to express ceramic products such as pottery and porcelain products made by high-temperature processing in furnaces. The term also indicates technologies and the science utilized to manufacture these products. The majority of potteries, porcelains, refractories, cement and glass have been made from silicate-based natural materials since early times, and these products are sometimes called traditional ceramics or classical ceramics. The terms "traditional" or "classical" do not necessarily indicate that the technologies are outdated. Both traditional ceramics and classical ceramics have been evolving with social needs and the incorporation of the latest technologies.

Since the mid-twentieth century, various kinds of ceramics have been produced using processed raw materials featuring strictly controlled chemical compositions instead of using natural materials. These ceramics are characterized by their thermal, mechanical, electromagnetic optics and biological functions. These new ceramics are sometimes called "technical", "fine", "new" or "advanced" ceramics so as to be distinct from traditional ceramics.

Classification of Ceramics by Use

Ceramics are classified by use as in Table 1.1.

Pottery and porcelain products are generic terms used to refer to the so-called "fired wares" made by mixing, forming and firing clay or stone powders. Tableware, tiles and sanitary wares are classified as pottery and porcelain products and are essential for daily life.

Refractories, machine components and cutting tools are classified as high-temperature structural materials and take advantage of the hardness, thermal resistance and abrasion resistance of ceramics. Among them, refractories feature high thermal insulation properties and resist high temperatures from several hundreds to over a thousand degrees Celsius and higher. A large amount of refractories are used in iron-manufacturing furnaces essential for steel production. Refractories are also used in waste incinerators, cement calcination furnaces, electronics industry, reactors and in reusable aerospace craft such as the space shuttle to maintain high temperatures and to protect machines and human bodies from high temperatures. The machine components include ceramic glow plugs for diesel engines, turbocharger rotors and other engine parts, parts for gas turbine engines, heat exchangers and mechanical seals. These components are slightly different from refractories, because high dimensional accuracy, in addition to high strength and thermal resistance is required of these components. Ceramics used as high temperature structural materials feature hardness and high abrasion resistance and are also used as materials for cutting tools, grinding tools and bearings.

Gypsum is made from natural gypsum or synthetic gypsum and used in products for dental treatment as well as to produce industrial models, blasters, molds and boards. Lime is made from limestone and processed lime is used as mortar, wall materials, filler for rubber and plastic, and construction materials. Clay and limestone are the main raw materials of cement, but side products such as iron and steel slag and refuse incineration ash as well as wastes are also used as raw materials in cement. The applications of cement include construction materials, refractories and dental materials.

Enamel is a compound material made by burning inorganic glass-based glaze onto a metal surface and it features mechanical strength of metals, beauty of glass and chemical stability of glass. Enamel is used widely in kitchenware, bath

Y. Imanaka et al. (eds.), *The Ceramic Society of Japan, Advanced Ceramic Technologies & Products*,
DOI 10.1007/978-4-431-54108-0_1, © Springer Japan 2012

Table 1.1 Classification of ceramics by use

Pottery and porcelain products	High-temperature structural materials
Gypsum, lime, cement	Enamel
Glass, optical material	Electric, electronic and magnetic materials
Bioceramics	

tubs, chemical industry equipment such as reaction cans, brewing tanks, building materials, art products, etc.

Glass is used to make window glass and containers and constitutes the ceramics most frequently seen in daily life. Various types of glass featuring various compositions are also used as optical materials such as lenses and prisms, glass fibers and optical fibers used in composite materials (such as FRP), picture tubes, flat panel displays, hard disk substrates and circuit substrates. Glass is amorphous, but fine crystals (β-spodumene, quarts, mica, etc.) precipitate inside when it is heat-treated in a strictly controlled temperature, depending on the type of glass. Glasses that have crystals precipitated inside are called glass ceramics. Glass ceramics feature higher strength, higher corrosion resistance and lower thermal expansion properties compared with glass. They are used as ceiling boards of electromagnetic cooking devices, turn tables of microwave ovens and exterior walls of buildings.

Ceramics featuring electrical, electronic and magnetic properties are called electroceramics. Electrical and magnetic properties of ceramics usually change substantially due to slight differences in composition and crystal structure. On the other hand, if appropriate compositions and crystal structures are selected, nearly all of the electromagnetic properties, such as insulation properties, electrical conductivity, semiconductive properties, magnetic properties, and superconductive properties, are achieved. Many of the ceramics are high quality insulators and used to manufacture cable insulators, while some of them feature high-temperature superconductivity at the liquid nitrogen temperature. Electrical resistance rises sharply above a certain temperature in some types of semiconductor ceramics. These ceramics are used as smart ceramic heaters because they control the electric current at temperatures higher than a certain level and effective for preventing fires due to overheating. Many ceramic parts featuring a variety of characteristics are incorporated in electronic appliances such as PCs and flat-screen televisions. It is not an exaggeration to say that compact and high performance mobile phones, etc. have been achieved thanks to light and small high-performance electroceramics.

Some of the ceramics feature extremely high biocompatibility and are used in bioceramic applications. Bioceramics are roughly classified into three types: materials that are used inside bodies, materials that are used outside bodies in contact with mucous membranes and skins and materials that are used without direct contact with the human body. The three types are represented by ceramic artificial bones, ceramic crowns and column fillers for high performance liquid chromatographies. They have been incorporated into products used in medical, pharmaceutical and biochemical fields.

Literature

1. The Ceramic Society of Japan (ed) (2002) Handbook of ceramic engineering, 2nd edn. Gihodo Shuppan, Tokyo [in Japanese]
2. The Ceramic Society of Japan (ed) (1997) Ceramics dictionary, 2nd edn. Maruzen, Tokyo [in Japanese]

Classification of Ceramics 2

Y. Imanaka et al. (eds.), *The Ceramic Society of Japan, Advanced Ceramic Technologies & Products*,
DOI 10.1007/978-4-431-54108-0_2, © Springer Japan 2012

Monolithic Ceramics (Single Crystal, Sintered Body, Glass, etc.) 2.1

Ceramics are non-metallic inorganic solids. Ceramics are classified into "monolithic ceramics" composed of a single chemical compound and "composite ceramics" composed of multiple chemical compounds.

Monolithic ceramics that are typically composed of a single chemical compound are explained below. Crystals represented by precious stones such as diamonds, potteries and porcelains are frequently produced by high temperature firing processes (Note 2.1) and glasses are classified as monolithic ceramics (Fig. 2.1). This classification is based on atomic arrangement and structure, and it is intended for the understanding of ceramics.

2.1.1 Crystalline Materials and Amorphous Materials

Ceramics consist of atoms that are bonded together. Ceramics can be roughly classified into two types depending on the arrangement of atoms that constitute the particular substance.

1. Crystalline solid: a solid in which atoms are arranged periodically and in a certain order throughout the material.
2. Amorphous (non-crystalline solid): a solid in which atomic arrangement does not have long-range order.

According to the examples described above, precious stones, potteries and porcelains are classified as "crystalline solids" and glass is classified as "amorphous." Amorphous solids have homogeneous qualities, but the atomic arrangement is irregular, unlike crystalline solids in which atoms are arranged regularly or periodic (Fig. 2.2). However, the arrangements are regular at certain distances depending upon the atoms, but this range is limited (this type of arrangement is called "short-range order.")

It is actually very difficult to tell if a ceramic is classified as crystalline solid or glass. For example, some watch faces are covered by a transparent alumina crystal instead of glass because alumina is resistant to scratches, and it is difficult to distinguish between the two materials using the naked eye. Is there any method in which to distinguish between crystalline solids from amorphous solids? Generally, X-rays, which have wavelengths much shorter than normal light, is utilized for this determination. If the atoms are arranged regularly in repeating atomic planes, reflected X-rays interact with each other, causing a strong reflection facing toward a certain direction. Strong reflections are not observed in amorphous solids because the atoms are arranged irregularly.

2.1.2 Crystalline Solid: Single Crystal and Polycrystals

Monolithic ceramics are divided into "crystalline solids" and "amorphous solids," as explained above. "Crystalline solids" can be subdivided into "single crystals" and "polycrystals (more commonly known as polycrystalline materials)" (Fig. 2.2).

1. Single crystals: solids in which atoms are arranged periodically from one end of the material to the other end and where grain boundaries are not present.
2. Polycrystals: a crystalline solid consisting of many grains, where the orientation of one grain is usually different from that of the adjacent grain.

In single crystals, atoms are arranged periodically throughout the solid, but the properties may vary depending on the direction of the arrangement (anisotropy). "Cleavage," or cleaving uniformly in a particular crystalline direction, is observed in a number of single crystals such as in diamonds.

Note 2.1 Technically speaking, potteries and porcelains usually contain several compounds and glass. But they were explained in this manner because ceramic materials for electronic parts, etc. are made from a single compound, although they are usually made by a similar method using a refined material.

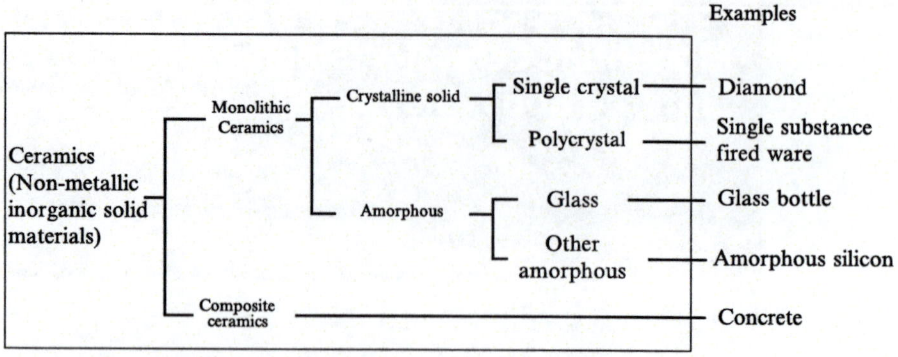

Fig. 2.1 Classification of ceramics

Fig. 2.2 Structure of ceramics (schematic)

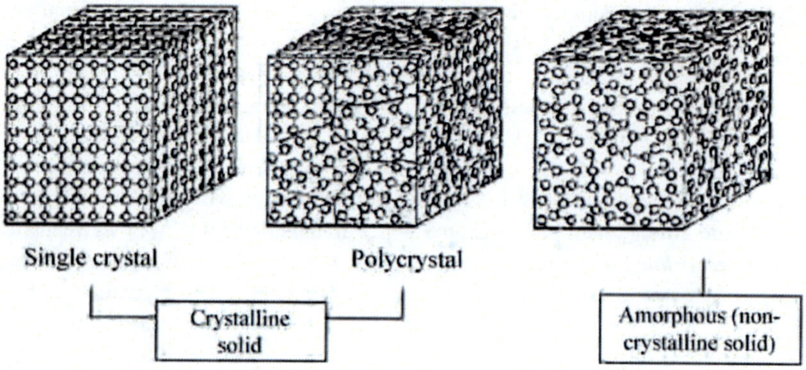

This results when cleavage along atomic planes or crystal faces is possible because atomic bond strength is weak and density of bonds is low. It is well known that calcite tends to crack or cleave in three directions and that the fractured faces are flat planes. Mica also has pronounced cleavage and cracks into sheets.

Polycrystals are usually manufactured by sintering the materials in high-temperature furnaces. Polycrystals manufactured using this method are called "sintered bodies." Pottery and porcelain products are examples. Even if one type of compound is used as the raw material, boundaries with irregular structures are created between grains (grain boundaries). Therefore, sintered bodies exhibit properties not observed in single crystals. The grains are oriented in various directions, and in general, the directions do not determine the properties. Fractured surfaces of polycrystals are jagged and exhibit granular or semi-dull luster.

2.1.3 Amorphous: Glass and Other Amorphous Solids

"Amorphous" materials can also be divided into two types: (1) glass and (2) solids other than glass. Glass is normally made by melting glass material containing silicate at high temperatures and quenching it. It becomes a liquid with high viscosity below its solidification temperature. Hence, crystals are not formed (supercooled condition), and the viscosity increases as the temperature decreases. The material becomes a glass near the glass transition temperature, after which the free movement of atoms is no longer possible. The glass at this stage is considered to be a "rigid liquid." It is essentially in a thermodynamically unstable state and energetically favors crystallization at room temperature. We can use glass without worry because the time before it actually crystallizes is extremely long. In other words, materials that exhibit a "glass transition state" are glasses. Amorphous solids other than glasses are the same as glasses with respect to the homogeneous atomic arrangement without periodicity or regularity in a wide range. However, they are manufactured without the process of super-cooling or passing through the glass transition.

As for the cleavage of crystals, glass is homogeneous and the structure is non-directional (expressed as "isotropic"). Therefore, it fractures into various jagged shapes or shells, not flat objects (Note 2.2). Because glass is isotropic, it exhibits no difference with regards to the transmission of light in each direction in terms of optical properties.

Note 2.2 Some of the crystal solids such as garnet fracture into jagged shapes.

Composite Materials 2.2

2.2.1 Definition of Composite Materials

Composite materials, unlike monolithic ceramics explained in the previous section, are made by blending or combining multiple chemical compounds, metals and polymers. Each of the substances that constitute composite materials is often expressed as a "phase." Composite materials contain one or more phases that can be clearly distinguished from the matrix. The matrix is mixed with a variety of metals and polymers as well as ceramics. Earthenwares, potteries and porcelains, and mud walls that have been used in Japanese houses since early times are in a broad sense composite materials. Composite materials are also observed in a number of living organisms such as wood (composite of cellulose and lignin) and shells (composite of calcium carbonate and protein) as well as artifacts. Many of the composite materials created in nature have rational structures and high performance. Efforts are being made to utilize them in the designing and development of artificial materials.

Composite materials in a narrow sense are advanced materials made by combining multiple industrially-produced high-purity raw materials. Composite materials are represented by ferroconcrete, but advanced composite materials such as cutting tools, carbon fiber-reinforced carbon (C/C composite), fiber-reinforced plastic (FRP) have also been commercialized.

2.2.2 Structural Characteristics

Functions of composite materials can be controlled by adjusting the configuration and arrangement of dispersed phases. The dispersed phases include continuous fibers, short fibers, needle-like crystals (whiskers), platelet crystals and grains. In order to achieve the desired or high performance properties, these phases need to be homogeneously dispersed. Continuous fibers and whiskers featuring shape anisotropy are sometimes dispersed two- or three-dimensionally by facing them toward one direction or arranging them like a fabric. Fiber-reinforced ceramic composite materials, in which the expansion and growth of cracks within matrix material is blocked by continuous fibers (reinforcing materials), have extremely high toughness and high reliability in terms of dynamics.

Meanwhile, with respect to nanocomposite materials, in which the composite structure is controlled at the nanometer level, research on physical property characterization started in the 1980s, when ultrafine metal particles, etc. were added to the inorganic matrixes of ceramics, etc. Later, mechanical properties of structural ceramics were improved, which led to the development of materials with new properties and functions such as superplasticity and machinability. Ceramic nanocomposite materials are produced by dispersing nano-sized ceramic or metal particles as the second phase in the matrix of single crystal grains and by intentionally causing structural defects (stacking faults, dislocation, etc.) in order to improve the properties. The materials can also be used to enforce grain boundaries. They are classified into intragranular-type, intergranular-type, intra/intergranular-type, nano/nano-type, etc., as shown in Fig. 2.3.

In addition to particle dispersed composite materials explained as an example, nanocomposite materials made by utilizing nano-level grain-boundary phases as well as those made by adding carbon nanotubes, have been studied. Typical structural ceramics such as alumina and silicon nitride as well as dielectric materials and magnetic materials are used as the matrix of nanocomposite materials.

2.2.3 Functional Characteristics

The improvement of characteristics and the appearance of new functions through compositing are important and essential for composite materials. One of the aims of using composite

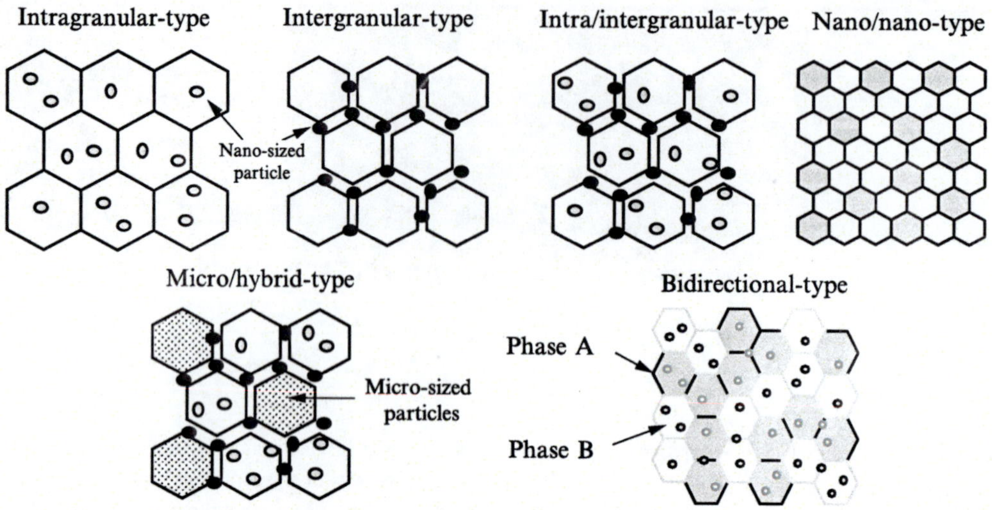

Fig. 2.3 Classification of nanocomposite materials by microstructure

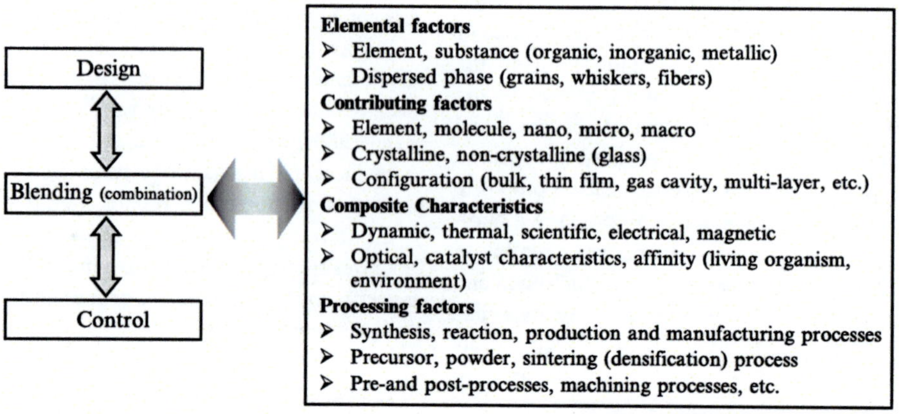

Fig. 2.4 Design guide for function-conscious ceramic matrix composite materials

materials is to improve the mechanical properties of the matrix. For example, the materials made by dispersing silicon carbide (SiC) in alumina (Al_2O_3) feature higher hardness, toughness and abrasion resistance than alumina and are used as cutting tools.

Meanwhile, composite materials with a focus on properties other than mechanical properties have also been developed. For example, ceramic composites that can be processed by electrical discharge have been developed by adding a certain amount of conductive substances such as nitride or carbide to ceramic materials, which are generally insulators (electrical discharge machining allows for the cutting into intended shapes). Electronic ceramics such as varistors and laminated capacitors also belong to composite materials in terms of their structures and functions.

2.2.4 Future Composite Materials and Their Issues

Materials that have multiple superior functionalities will be produced by the introduction of composites of various structural scales into ceramics. In order to realize materials of this kind, in addition to the combination of constituting factors such as types and configurations of materials (substances), what is needed is the design and control of multiple factors such as structural factors that include sizes and configurations, characteristic factors including mechanical and electromagnetic properties and processing factors involving material production, as shown in Fig. 2.4. For the creation of composite materials, advanced knowledge and understanding of technologies involving physics, chemistry and biochemistry will be required, in addition to conventional science and technologies relating to ceramic materials. Meanwhile, it is important to give due consideration to the environment and recyclability of micro- and nano-level composite materials.

Literature

1. Sakka S (1983) Science of glass amorphous material. Uchida Rokakuho, Tokyo [in Japanese] (2.1)
2. The Ceramic Society of Japan (ed) (2003) Introduction to ceramics science. Ceramic Society of Japan [in Japanese] (2.1)
3. Kagawa Y, Hatta H (1990) Ceramic matrix composite material. Agne Shofusha, Tokyo (2.2)
4. Niihara K (1991) J Ceram Soc Jpn 99:974–982 (2.2)
5. Sekino T (2001) Mater Integr 14(1):23–28 (2.2)

Raw Materials of Ceramics

3

A majority of the elements listed in the periodic table are used to manufacture ceramics. A variety of ceramics are manufactured using these raw materials. Raw materials of ceramics are classified into natural materials, minerals with nearly no processing or refinement either by physical or chemical methods being needed, and artificial raw materials (also called synthetic raw materials), which are produced by extracting needed components from natural minerals. Natural raw materials are used to manufacture traditional ceramics such as potteries, porcelains, refractories, cement and glass, while artificial materials are used to manufacture "fine ceramics" (or advanced ceramics) intended for electronics, photonics, etc.

Natural Raw Materials

Data on major natural raw materials are listed in Table 3.1. As the column of major application indicates, the majority of materials are intended for mass production. Self-sufficiency rates in the table also indicate that Japan relies largely on natural resources from overseas. Import trading partners are concentrated in the USA, China, Australia, and other specific countries, and issues involving stable supply routes and the stabilization of prices have yet to be resolved.

Artificial Materials

Data on major artificial raw materials are listed in Table 3.2. Manufacturing methods of oxide-based raw materials differ greatly from those of nonoxide-based raw materials such as nitride, carbide and boride. Oxide-based materials are manufactured mainly by (1) the solid-state reaction method where mixed powders react at high temperatures, (2) the precipitation method where raw materials are produced from solutions containing intended components through precipitation reactions caused by concentration control, pH change, etc. and (3) the sol–gel method where fine particles are produced by the hydrolysis of organic metals such as alkoxides. In general, the grain diameter of raw material powders becomes smaller and purity and quality become higher in the order of (1)–(3). However, production costs increase in the same manner. Suitable manufacturing methods are selected depending on the application, but method (1) is widely used following advancements in crushing technologies.

Meanwhile, nitride, carbide and boride are important raw materials for the production of nonoxide-based raw materials. The materials are manufactured mainly by (1) the direct method where unblended powder reacts at a high temperature and in a controlled atmosphere, (2) the reduction method where oxide powder reacts at a high temperature while being reduced using carbon and (3) the gas-phase method where the object material is synthesized by a reaction involving gas-phase components at a high temperature. Thermal decomposition of synthesized imide compounds or amide compounds has been incorporated by the industry into the production of silicon nitride. Method (3) is suitable for the production of nano particles, but surface treatment is required because the particles are unstable due to oxidation and hydrolysis reactions on the particle surface. Raw material powders are normally equiaxial. However, highly anisotropic materials containing plate or needle crystals are also being produced. Raw materials containing needle crystals and fibrous crystals, which are used to manufacture composite materials containing ceramics, are also used to manufacture composite materials containing plastic, metal, etc. A variety of single crystals (whiskers), polycrystals and noncrystalline fibers, are manufactured by methods including the extension of melt fused at a high temperature and the sol–gel method where fibrous polymer precursors are produced from solutions.

Y. Imanaka et al. (eds.), *The Ceramic Society of Japan, Advanced Ceramic Technologies & Products*,
DOI 10.1007/978-4-431-54108-0_3, © Springer Japan 2012

Table 3.1 Data on major natural raw materials for ceramics produced in Japan

Name of raw material	Major mineral components	Major applications	Self-sufficiency rate[a]	Major exporters
Silica stone, silica sand	Quartz [SiO_2]	Sheet glass, potteries and porcelains, semiconductors	○	Australia, India
Kaolin	Kaolinite [$Al_4(OH)8(Si_4O_{10})$]	Powder for papermaking, potteries and porcelains, glass fibers	Δ	USA, New Zealand
Bauxite	Gibbsite [$Al(OH)_3$]	Ceramics, refractories, abrasives	×	Australia, Brazil
Talc	Talc [$Mg_3(OH)_2(Si_4O_{10})$]	Cordierite ceramics	×	China, Australia
Magnesia	Magnesite [$MgCO_3$]	Refractories	×	China, North Korea
Lime stone	Calcite [$CaCO_3$]	Cement, powders for construction and papermaking	◎	–
Zirconium mineral	Zircon [$ZrSiO_4$]	Ceramics, refractories, casting sand	×	Australia, South Africa
Titanium mineral	Ilmenite [$FeTiO_3$]	Pigments, electronic materials, photocatalysts	×	Australia, Canada
Rare earth mineral	Bastnaesite [$La(CO_3)F$]	Optical glass, magnets, abrasives, phosphors	×	China, USA
Borate mineral	Colemanite [$Ca_2B_6O11{\cdot}5H_2O$]	Glass fibers, potteries and porcelains	×	Turkey, USA
Graphite	Graphite [C]	Refractory, carbon materials, lubricants	×	China, Korea

◎ –100%; ○ >50%; Δ <50%; × –0%

[a]Self-sufficiency rate in Japan

Table 3.2 Data on major artificial raw materials for ceramics produced in Japan

Name of raw material	Chemical formula	Major manufacturing method	Purity[a] (N)	Grain diameter (μm)	Major applications
Silica	SiO_2	Gas-phase method	1–4	~0.01	Fillers, absorbents, semiconductors, glass
Alumina	Al_2O_3	Precipitation method	2–4	0.1–40	Cutting tools, machine parts, abrasives, IC substrates Catalyst supports
Zirconia	ZrO_2	Precipitation method	2–3	0.3–5	Ceramics, pigments, refractories
Magnesia	MgO	Precipitation method	1–3	0.01–150	Refractories, ceramics
Barium titanate	$BaTiO_3$	Solid-state reaction method	2–3	0.1–1.5	Electronic materials
Mullite	$Al_6Si_2O_{13}$	Sol–gel method	1–3	~1	Fireproof materials, high-temperature jigs
Spinel	$MgA_{12}O_4$	Solid-state reaction method	1–3	~0.1	Refractories
Apatite	$Ca_5(PO_4)_3(OH)$	Precipitation method	1–3	0.1–10	Materials for living organisms, fillers
Silicon nitride	Si_3N_4	Direct nitriding	1–2	0.3–10	Machine parts, structural materials, slide materials, tools
Aluminum nitride	AlN	Reduction nitriding method	1–2	~1	High thermal conductivity substrates, heat radiation materials
Boron nitride	BN	Reduction nitriding method	1–2	0.1–200	Solid lubricants, superhard materials
Silicon carbide	SiC	Reduction carbonizing method	1–2	0.03–5	Refractories, heating elements, jigs for semiconductor
Tungsten carbide	WC	Direct carbonization method	1–3	0.7–10	Superhard materials
Boron carbide	B_4C	Reduction carbonizing method	1–2	0.8–40	Reactor materials
Lanthanum	LaB_6	Reduction method	1–2	~1	Thermal electron emission materials

[a]a = number of nines, for example 2 = 99%, symbol ~ indicates approximately

Literature

1. The Ceramic Society of Japan (ed) (2002) Ceramic engineering handbook, 2nd edn [advanced]. Gihodo, Tokyo, pp 16–129 [in Japanese]
2. The Ceramic Society of Japan (ed) (1992) Advanced ceramics handbook. Ohmsha, Tokyo, pp 121–12 [in Japanese]
3. Otsuga N, Okada K (1985) Ceramics 20:964–968 [in Japanese]
4. Supervised by Saito H (1956) Utilization of fine ceramics, vol 1. Taiga Shuppan, Tokyo, pp 65–76 [in Japanese]

4 Synthesis of Ceramics

Y. Imanaka et al. (eds.), *The Ceramic Society of Japan, Advanced Ceramic Technologies & Products*,
DOI 10.1007/978-4-431-54108-0_4, © Springer Japan 2012

Sintering 4.1

Ceramics are originally defined as sintered bodies of inorganic compounds. Examples are pottery and porcelain that have been manufactured since early times and have a long history. It is difficult to process minerals into various shapes. However, if crushed mineral powders are formed into a desired shape and sintered by heating, products having high strength and functions required for practical use can be manufactured. Previously, crushed natural minerals were used as raw materials, but high-purity powders processed by chemical methods are currently being used. Sintering is essential for the production of ceramic materials, regardless of raw powder manufacturing methods.

4.1.1 Sintering Phenomenon

Figure 4.1 shows pictures of a formed body (a), a preliminary sintered body (b) and a sintered body (c), and scanning electron microscope (SEM) pictures of each body. In (a), powder is tightly packed. In (b), the contact area between grains is larger and there are less voids. In (c), pores have been removed completely. Picture (c) indicates that many grains are tightly assembled within ceramic materials. It also indicates that the size of grains grow in line with densification.

4.1.1.1 Three Stages of Sintering

Microstructures of sintered bodies change drastically during the sintering process. It is difficult to explain the entire process using one model. Therefore, modeling is normally done by dividing the process into three stages, the initial, intermediate and late stages.

1. Initial stage (Fig. 4.2a). The contact area of grains (grain boundary) increases rapidly to a level of 20% of the cross sectional area of the grain (this phenomenon is called neck growth). The grain boundaries do not shift because necks are constricted.
2. Intermediate stage (Fig. 4.2b). The pores are located along the ridges of grains in a cylindrical shape, and the sintered density ranges from 60% to 95% of the theoretical density (Note 4.1). Necks are less constricted, and grain boundaries migrate toward the center of small grains, making small grains smaller and large grains larger (this phenomenon is called grain growth).
3. Late stage. The pores isolated on corners of grains disappear and the sintered density becomes 95% or higher.

4.1.1.2 Driving Force of Sintering

Atoms on the surface of liquids and solids which are unbound on the gas side have higher energy than internal atoms which are totally bonded. Surface tension is generated on the surface for canceling the excess energy state. The surface of water in a capillary tube is higher than the surface of water in the surrounding area because surface tension γ acts on the concaved water surface in the capillary, generating upward pressure (this is called capillary force).

Outward pressure is generated on the surface of necks, which are concaved as Fig. 4.3 indicates. If the curvature radius of a neck surface is r_s, the pressure acting on surface P is expressed as γ/r_s ($2\gamma/r_g$ in spherical surface). The pressure acting on the water surface with a curvature radius of 0.5 mm is low at 3 g/cm^2. Surface tension is negligible. However, it is not uncommon that powders used for sintering are of 50 nm in radius. The surface tension is about ten times that of water, and extremely high pressures reaching 200 kg/cm^2 are generated on the surface of necks. Sintering is promoted utilizing high pressure as the driving force.

Note 4.1 Theoretical density is the density of a sintered body without any cavities nor voids. Single crystals of ceramics typically have theoretical density.

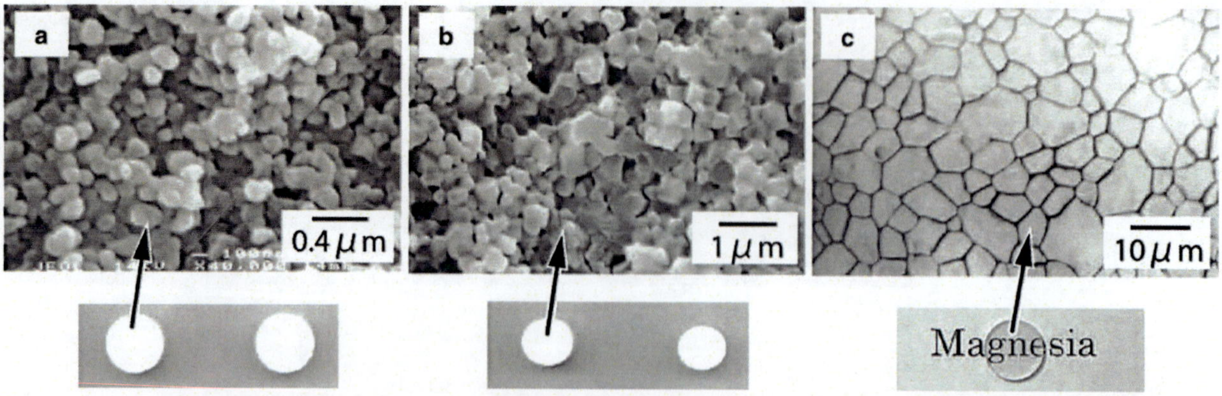

Fig. 4.1 Stages in microstructure during sintering of a formed body (**a**), a preliminary sintered body (**b**) and a sintered body (**c**). (**a**) and (**b**) are the fractured surface and (**c**) shows the polished surface with SEM pictures. Voids (called gas pores) become smaller following the sintering process. The pores disappear completely and the sintered body becomes transparent if the sintering is optimized. The grain size increases (grain growth) as densification is promoted

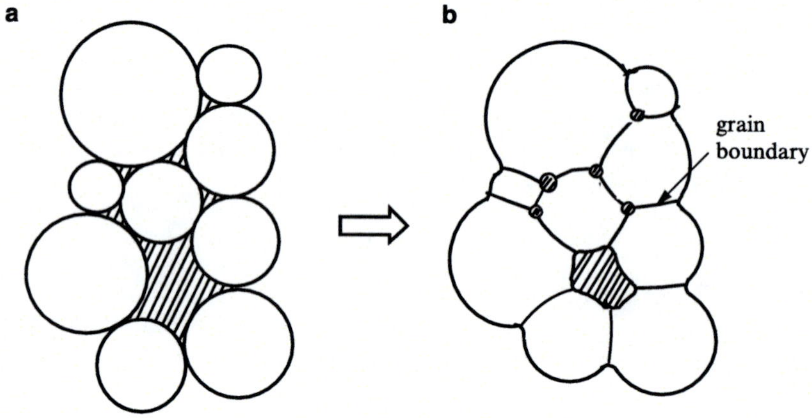

Fig. 4.2 Sintering process. (**a**) Shows the initial stage of sintering, while (**b**) shows the intermediate stage of sintering. The *shaded areas* indicate voids. In the intermediate stage, grains become large enough and grain growth is induced by the migration of grain boundaries. Since it is difficult to remove large voids, powders that do not generate large voids in green bodies need to be used in order to create dense sintered bodies

Figure 4.2 shows that the surface area decreased while the grain boundary area increased by sintering. In other words, sintering is the phenomenon where a part of the free energy stored on the particle surface is stored as the grain boundary free energy, while the remaining energy is used to diffuse the atoms around the necks to induce neck growth and densification. On the other hand, grain growth is a mass transfer phenomena where grain boundary free energy is consumed (and grain boundary area is reduced).

4.1.2 Diffusion Types

When impurities of different valences exist, atomic defects (lattice point defects without atoms that are supposed to exist, atoms between lattice points where atoms are not supposed to exist, hereinafter referred to as defects) are introduced to maintain electro-neutrality. Defects may also be caused by thermal energy. With respect to the former defects, the amount of defects is proportional to the amount of impurities, while the amount remains the same regardless of the temperature change. As for the latter defects, the amount of defects increases exponentially with the temperature.

Atoms jump from one lattice point to the adjacent defect site for diffusion. The amount of atom diffusion is proportional to the product of the defect amount and jump frequency. A large energy is needed for the jumping of atoms, and therefore, diffusion at room temperature is negligible even if the substance contains a large amount of impurities. Jumping frequency also increases exponentially with the temperature, and therefore, when the temperature exceeds 1/2 of the melting point of the substance, phenomena caused by diffusion, such as sintering and grain growth, are observed. Naturally, powders containing a large amount of impurities are sintered more quickly because they contain a larger amount of defects.

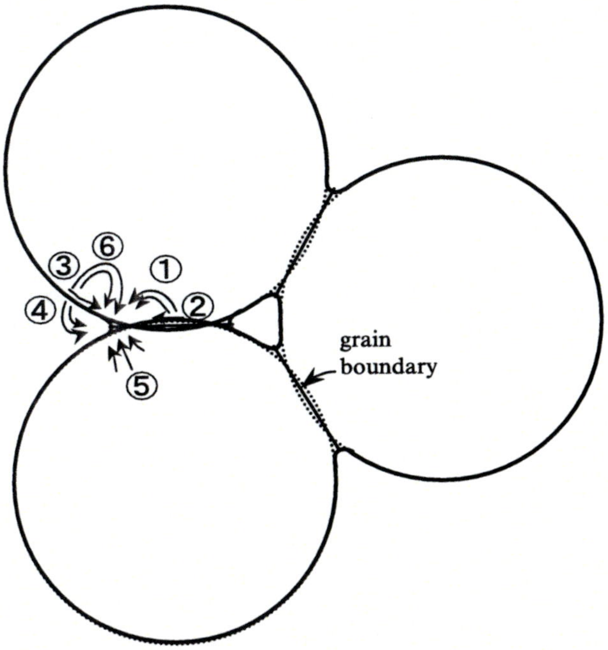

Fig. 4.3 Mass transfer routes and mechanisms. (*1*) Volume diffusion mechanism, (*2*) grain boundary diffusion mechanism, (*3*) surface diffusion mechanism, (*4*) evaporation/condensation mechanism, (*5*) fluidization mechanism, (*6*) volume diffusion mechanism

4.1.3 Sintering Mechanism

The mechanism of mass transfer that promotes sintering is called the sintering mechanism. At present, five mechanisms—volume diffusion, grain boundary diffusion, surface diffusion, evaporation/condensation and fluidization—are known. These mechanisms are closely related to the mass transfer route. Figure 4.3 shows the routes (1) and (6) volume diffusion mechanism, (2) grain boundary diffusion mechanism, (3) surface diffusion mechanism, (4) evaporation/condensation mechanism and (5) fluidization mechanism. Among the routes, (3), (4) and (6) are the routes through which atoms are diffused from grain surface to the neck surface. In this mass transfer process, the neck grows but center-to-center distance between grains remains the same. Therefore, no densification occurs. The strength of sintered bodies is not improved substantially. These routes are not suitable for the production of sintered bodies, excluding those used for catalyst supports and filters, which take advantage of porosity. In other routes, the mass is transferred from grain boundaries and inside the grains to the neck surface, promoting neck growth and densification at the same time. The fluidization is observed in sintering substances of low viscosity such as glass. In general ceramics, sintering is promoted via the volume diffusion mechanism or the boundary diffusion mechanism.

4.1.4 Sintering Speed Formula

Sintering speed is expressed by the neck growth speed or the interparticle shrinkage speed. In model experiments, large grains are used so that neck diameters and interparticle shrinkage rates can be measured. However, powders for practical use are sized 1 μm or less and are difficult to measure. Therefore, the reduction speed of specific surface areas or the shrinkage speed of green pellets is normally measured instead. The speed formula most frequently used in textbooks is introduced below.

$$\frac{\Delta L}{L_0} = \left(k_1 \frac{\gamma D t}{r^q} \right)^n \tag{4.1}$$

k_1 is the proportional constant, ΔL is the shrinkage of green pellet, L_0 is the thickness of green pellet, γ is the average diameter of grains, D is the diffusion coefficient, t is the sintering time, and n and q are the constants that change depending on the sintering mechanism. $n=2$ in the volume diffusion mechanism, while $n=3$ in the grain boundary mechanism. $q=3$ in the volume diffusion mechanism, while $q=4$ in the grain boundary mechanism. Formula (4.1) indicates that the speed becomes faster by 10^3–10^4 times when the grain diameter is reduced to 1/10. High-purity raw materials are required for high performance materials. The drop in diffusion coefficients due to high purity can be compensated for by the microparticulation of raw powders. Because of these reasons, fine raw powders are used in the production of fine ceramics.

4.1.5 Actual Sintering

4.1.5.1 Blending of Powders for Sintering

Sinterability largely depends on the powder manufacturing conditions. Densification at low temperatures leads to low production costs and realizes fine microstructure with superior functions. Research has been promoted for production of high-sinterability powders. Generally speaking, powders with fine spherical particles that are evenly arranged with limited voids feature high sinterability.

4.1.5.2 Utilization of Additives

Atoms with valences different from those of atoms of the major component generate a large volume of defects. Therefore, if a compound containing such atoms is added, mass transfer is enhanced extremely and the sinterability can be improved. Additives that improve sinterability without resulting in the deterioration of material properties are generally used in the industrial field.

4.1.5.3 Sintering Method

1. Normal sintering. Formed bodies made by the compacting of powder are sintered under atmospheric pressure and at high temperature.
2. Pressure sintering. Densification is significantly accelerated if pressure is applied during sintering. In the hot-press method, densification is accelerated by pressing powders into dies using punches. In the HIP method, powder in capsules or sintered bodies with isolated gas cavities are densified by applying hydrostatic pressure generated by gas pressure at high temperatures.
3. Reaction sintering. Adopted for the sintering of non-oxide materials. Silicon nitride (Si_3N_4), for example, is densified by means of the reaction between Si and N.

Single Crystal Synthesis Methods 4.2

Atoms and molecules are sometimes arranged in a crystal with high periodicity. If the periodic crystal structure is observed throughout the specimen, it is called a single crystal. Strictly speaking, the cyclic patterns in single crystals are not perfect, and contain foreign substances or are partially deformed, as well as the composition is sometimes inhomogeneous. These imperfections are called crystal defects and the quality of single crystals depends on the amount of defects.

Single crystal synthesis methods are roughly classified into (1) the solution-growth method where crystalline elements are diluted with a solvent to deposit crystals, (2) the melt-growth method where crystal elements are heated, melted and cooled to crystallize the melt and (3) the vapor-phase growth method where reactant gases react in the gas phase and deposit a solid on to initially formed seed crystals that grow into a single crystal (Note 4.2). There are many other methods for single crystal synthesis using various processes. A suitable method should be selected based upon considerations for the quality of materials as well as the intended size, shape and quality of products.

If salt (NaCl) is dissolved in water in a beaker and the water is evaporated gently, numerous single crystals sized approximately 1 mm^3 are seen at the bottom of the beaker (one example of a solution-growth method). Structures of single crystals of this size can be examined with X-rays. The salt used in optical circuits of infrared prisms, for example, needs to be larger with higher homogeneity and higher quality. In order to create single crystals with higher quality, the raw material is placed in a cylindrical container. The container is heated in a furnace to melt the material and then cooled down, moving it slowly within the furnace so as to obtain crystals. The area of high quality is cut out from the solid crystal so as to obtain a single crystal. This growing method is called the Bridgman method (refer to Table 4.1).

Growing methods, applications and functions of ceramic single crystal materials that have been commercialized are listed in Table 4.1. Among them, quartz, used widely in instruments, communication devices, ultrasonic generators, optical circuit devices, etc., has the largest market. Quartz is grown in a sealed high-pressure container under high temperature and high pressure (approximately 400°C, 1,000 kg/cm^2). Quartz raw material is placed at the bottom of a large vertically-long container (with a capacity of 1,000 L) and seed crystals are placed at the top inside the container. The container is filled with an alkali solution and heated from the outside using a heater. The pressure inside the container increases, and the raw material blends into the alkali solution at the high-temperature bottom area when the temperature at the top area is controlled and lower than the temperature at the bottom area. Solubility decreases in the low-temperature top area, inducing deposition on the seed crystals and promoting crystallization. This method is called the hydrothermal synthesis method. Crystals grow at a speed of 0.5 mm to several mm per day (it varies depending on the direction of growth) and growth takes more than 1 month.

LN ($LiNbO_3$) and LT ($LiTaO_3$) have also been commercialized for use as surface acoustic wave (SAW) devices of TVs, videos and mobile phones. These crystals are grown by the Czochralski method (Fig. 4.4). In this method, a crucible containing raw material is located in the center of a heater to melt the material by heating. Seed crystals are brought into contact with the surface of the melt (the surface temperature is controlled to the level that does not melt the seed crystal). When the seed crystal is turned in the temperature-controlled melt and the top is pulled up, cylindrical single crystal is grown as shown in the figure. The speed of the growth is several mm

Note 4.2 The vapor-phase growth method is mainly used to grow thin film crystals. Refer to Sect. 4.7. Thin film.

Table 4.1 Typical ceramic single crystal materials that have been commercialized

Type of crystal	Growing method	Applications	Utilized functions and effects
Quartz (SiO_2)	Hydrothermal synthesis method	Oscillators, optical circuit materials	Piezoelectricity, translucency
LN, LT	Czochralski method	SAW elements, light modulation devices, frequency conversion devices	Piezoelectricity, acousto-optical effect, electro-optical effect, nonlinear optical effect
KDP family	Aqueous solution method	Acoustic elements, optical circuit materials, light modulation devices	Piezoelectricity, translucency, electro-optical effect, nonlinear optical effect
Al_2O_3	Verneuil method Czochralski method	Bearings, ornaments, window parts	High hardness, luminescence, uniformity
YAG	Czochralski method	Solid lasers	Luminescence
NaCl family	Bridgman method	Optical circuit materials	Translucency, uniformity
BGO	Czochralski method	Scintillators	Luminescence
NaI	Bridgman method	Scintillators	Luminescence
KTP family	Flux method, hydrothermal method	Wavelength conversion elements	Nonlinear optical effect
LBO, CLBO	Flux method	Wavelength conversion elements	Nonlinear optical effect, ultraviolet translucency
CaF_2	Bridgman method	Optical circuit elements	Translucency

Abbreviations: Chemical formula *LN* $LiNbO_3$, *LT* $LiTaO_3$, *KDP* KH_2PO_4, *YAG* $Y_3Al_5O_{12}$, *BGO* $Bi_4Ge_3O_{12}$, *KTP* $KTiOPO_4$, *LBO* LiB_3O_5, *CLBO* $CsLiB_6O_{10}$

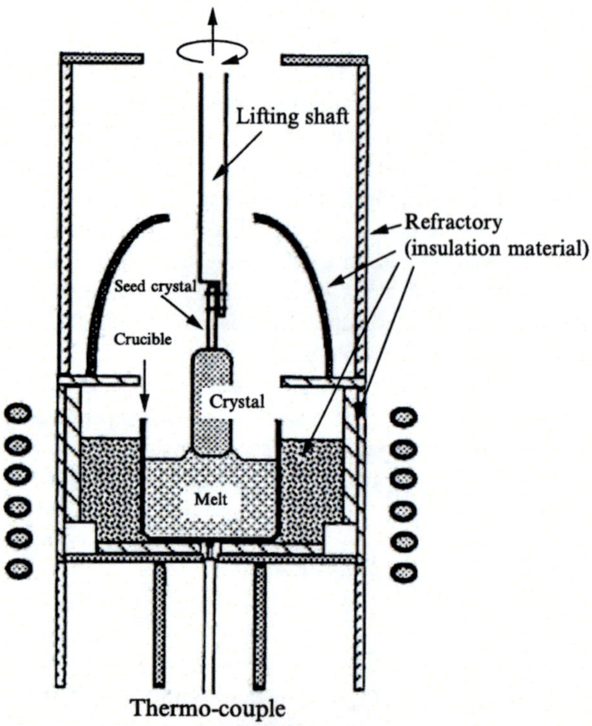

Fig. 4.4 Outline of Czochralski method

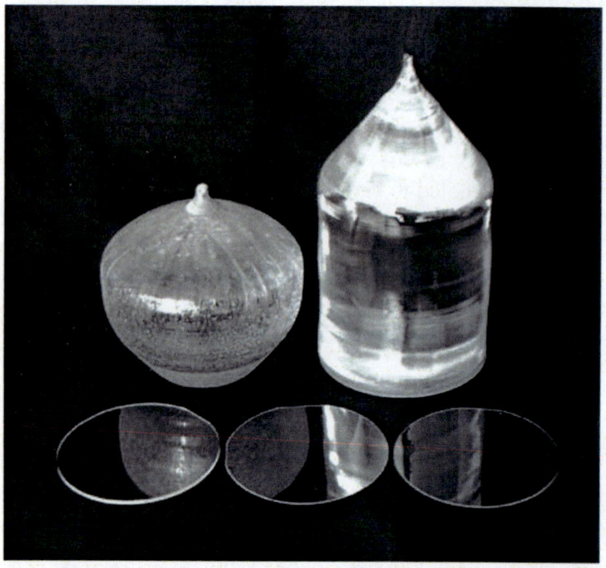

Fig. 4.5 LN (*right*) and LT (*left*) single crystal, and 3-in. substrate grown by the Czochralski method

per hour, which is much faster than the solution-growth method. Figure 4.5 shows LN single crystal with a diameter of a 3-in. (Note 4.3) and LT single crystal with a diameter of a 4-in. in which nonstoichiometric defects are controlled as well as substrates made by grinding the crystals. This method is most commonly adopted for growing crystals from melt. Large silicon single crystal intended for semiconductors is also grown with this method.

Sapphire (Al_2O_3), a typical ceramic single crystal material, is grown by various methods. The materials used as bearings and window parts do not have to be very high in quality and are grown by the Verneuil method. In this method, a container filled with raw powder is vibrated so as to release

Note 4.3 1 in. = 2.54 cm. Diameter of single crystal grown by the Czochralski method is traditionally expressed in inches.

the raw material from the bottom of the container. The raw powder melts while it passes through an oxy-hydrogen flame. The melt deposits on top of a seed crystal and crystallization is promoted. This method does not require a container for melt, and a large single crystal can be produced at a low cost, although the quality is not high. In the case when materials of high uniformity and active ions are needed for solid lasers with high quality, these materials are grown with the Czochralski method or the heat exchange method in which the melt is slowly cooled down in a container.

Single crystal synthesis technologies have been developed actively in order to utilize various characteristics and functions of ceramic single crystals. However, the materials that have been commercialized and have a certain level of market share are limited. It is not always true that commercialized materials are superior in terms of characteristics. Availability of materials depends on whether the material, with large size and high-quality, can be offered at a low price. There are many materials, among those with superior characteristics, which have not been commercialized because of the difficulty in growing them. Development of single-crystal growth technologies is extremely important for the application of new materials, and breakthroughs in growing technologies continue to be sought after.

Glass Synthesis 4.3

The majority of the glass is synthesized by the fusion method. Solid raw materials are melted at a temperature above its melting point and the liquid phase is cooled to create the glass using the fusion method. The raw materials do not have to be solids, and the glass can be synthesized by using raw materials of a gas or liquid phase (gas-phase method, liquid-phase method). These methods are utilized to synthesize high-purity glass with special functions (Fig. 4.6).

4.3.1 Fusion Method

Glass does not crystallize in the cooling process and is an amorphous solid material produced by fusion. Fusion, results when solid materials are transformed physically and chemically by heating to create homogeneous melts, is the most important part of the glass manufacturing process using solid materials. The process of the fusion method is roughly divided into four stages—the blending of glass materials, fusion at high-temperature, hot forming of the initial shape and annealing (during which stress is removed). Purified synthetic oxides, carbonates, hydroxides, natural minerals, etc. are used as the raw materials. Each of the raw materials is weighed before mixing to achieve the specified composition. The blended material is made by mixing and stirring is melted uniformly at a specified temperature (1,500–1,600°C) so as to create a melt. This melt is kept at a high temperature until it is completely free from bubbles. The melt is formed into the intended shape and then is annealed in the period during which residual stress is removed.

The glass manufacturing process is repeated using crucibles using the batch method, while the materials are melted continuously in a tank furnace using a continuous method. The most distinctive feature of glass is its transparency, so inclusions (bubbles, pebbles, lines) must be avoided. Therefore, the inclusion removal technology is important.

4.3.1.1 Batch Method

Fusion in crucibles (batch method) is employed for the limited production of diversified types of glass materials such as optical glass. Reaction containers (crucibles) are required to be made of materials that do not generate inclusions such as bubbles during the high-temperature melting process. Platinum or platinum alloy is the most suitable crucible material, but high alumina content refractories are normally used because of lower costs. For the production of optical glass, the melt is stirred as highly homogeneous glass melt is required.

4.3.1.2 Continuous Method

General glass products are manufactured in large volumes by using tank furnaces by melting the raw materials continuously at high temperatures of 1,500–1,600°C, utilizing the heat generated by the combustion of gas or heavy oil. Joule heat is also used as the heat source in the electrical melting process. In industrial production, processes from melting to the forming of the initial shape are performed during one operation. In other words, the glass melt is formed while it is at a high temperature and has high fluidity. The forming methods are diversified. Glass bottles are made by the blowing method where air is blown into a lump of molten glass (gob), TV glass panels are made by the press forming method, glass tubes of fluorescent lamps, etc. are made by the tube pulling method and glass sheets are made by the float method.

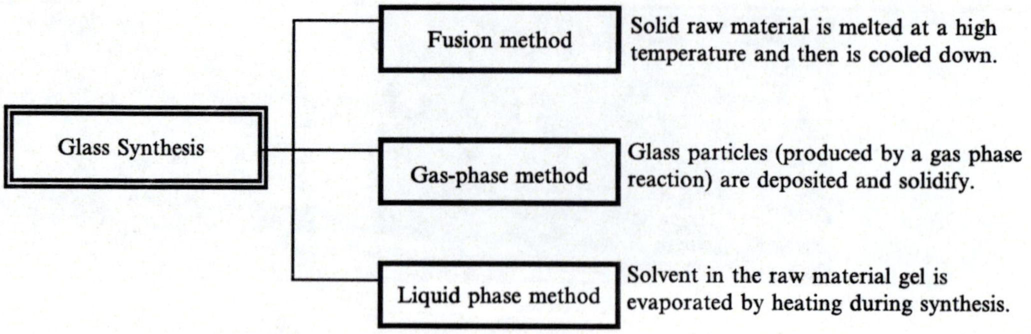

Fig. 4.6 Glass synthesis methods

4.3.2 Gas-Phase Method

Synthesized silica glass is manufactured by the gas-phase method using silicon tetrachloride ($SiCl_4$), a gas, as the starting material. Synthesis reactions in the gas-phase method involve the generation of glass particles (soot) via the thermo-oxidative reaction of metal halides: $SiCl_4+O_2=SiO_2+2Cl_2$ and dehydration reaction that occurs during transparentization through additional heating: $SiOH+Cl_2=SiOCl+HCl$.

If the product is allowed to contain water, silicon tetrachloride is evaporated in the gas-phase, which is introduced to oxy-hydrogen flames in order to synthesize soot by means of flame hydrolysis. The soot is fused and deposited on a rotating target to produce clear glass. High-purity glass is produced by this method because the raw material is in the gas phase and containers are not used for melting. Anhydrous silica glass, prepared to avoid the absorption in the infrared range, is synthesized by a reaction with oxygen only, using a high-frequency plasma torch as the heat source so as to prevent intrusion of the OH group. The gas-phase method is used to synthesize silica glass-based fibers and a variety of thin films.

4.3.3 Liquid Phase Method

In the sol–gel method, glass is made from a room-temperature liquid material, without utilizing the melt cooling process. A sol, having fine colloid particles suspended in liquid, is turned into a gel with no fluidity by flocculating the colloid particles. The glass is produced by evaporating the liquid, followed by heat treatment and firing.

The sol–gel method is classified roughly into two methods. In one of the methods, the sol, the starting material, is adjusted and poly-condensation reaction is induced by the hydrolysis of organic metal compounds such as alkoxide in an appropriate solvent. In the other method, a gelatinizing agent is added to non-crystalline silica powder, which is obtained by flame hydrolysis of silicon tetrachloride, dispersed in a suitable solvent.

When metallic alkoxide is used as the starting material, silicon methoxide $[(CH_3O)_4Si]$ or silicon ethoxide $[(CH_3CH_2O)_4Si]$ is used as the raw material for silicon. As for alkali, alkaline earth, alumina, etc. are used as the modifier oxide, the solutions are acetates, nitrates, etc. The sol is placed in a container, sealed and kept at a temperature below the boiling point of the solvent, it gradually changes into a gel with a gradual increase in viscosity. The liquid trapped in the voids within the porous gel is evaporated and dried. Then firing is performed to create a glass lump of a specified shape and without gas cavities. If the sol is coated on glass or ceramic substrates or spun directly on fibers before its fluidity is completely lost, thin film and fibrous glass will be created after going through the drying and firing processes. Firing to create glass from the gel is performed at a temperature near the transition point of the intended glass, which means the glass can be produced at a much lower temperature compared with the conventional fusion method.

Plasma 4.4

4.4.1 Plasma Material Science

Solid substances change to liquid and then to gas when they are heated. The solid state, liquid state and gas state are called the three states of matter. If the gas is heated further, some atoms will become ionized by forming positive ions due to electrons separation from atoms. The gas-like substance in which electrons co-exist with their dissociated positive ions was named "plasma" by Langmuir. Plasma is called the fourth state of matter because of this reason (Fig. 4.7).

Plasma is an electromagnetic fluid composed of electrons and ions (charged particles) and molecules and atoms (neutral particles). However, plasma flames generally contain the same number of positive and negative charges and are electrically neutral. Nuclear fusion reactions are observed in plasma exceeding 100 million°C. Plasma is expected to be applicable to energy generation (nuclear power generation utilizing nuclear fusion) and element conversion (producing gold out of water), but these technologies have yet to be developed.

In a plasma at a temperature of approximately 10,000°C, one or two electrons are separated from atoms and molecules. Advanced technologies are in the process of commercialization to take advantage of plasma processes. Plasmas are classified roughly into the "non-equilibrium plasma" and the "equilibrium plasma." Non-equilibrium plasma features electron temperatures of several tens of thousand degrees Celsius, while the temperature of ions and molecules are less than 1,000°C (also called low-temperature plasma). This type of plasma is utilized in semiconductor device processing because of the chemically active radicals. It may be generated using low electrical power and compact devices, but reacted substances are limited in amount. It is not effective for heating. However, it is effective for promotion of reactions in low temperatures. In equilibrium plasma, regional temperatures of electrons, ions and neutral particles are nearly equal at about 10,000°C. This type of plasma is also called high-temperature plasma, thermal plasma or chemical plasma. It is chemically active and the heat can also be utilized, but it requires large amounts of power and large-scale equipment. The plasma is generated at high pressure, making it possible to handle large amounts of a substance.

Characteristics of plasma reactions and processes are as follows:

1. Changes substances to an active, exited state (high energy state that promotes reactions).
2. Various chemical reactions can be accelerated.
3. Ultra-high temperatures can be utilized.
4. High enthalpy (the state where effective heating energy is high).
5. Generated light, heat and ions can be utilized.
6. Quenching of substances is possible.

These plasma characteristics are often utilized in sintering, surface treatment, etching (surface grinding), CVD processes, etc. and are expected to be effective for the creation of new substances and the production of new materials.

4.4.2 Plasma Sintering

Sintering, the core technology of ceramic manufacturing processes has been studied and advanced by many ceramists. New technologies such as reaction sintering and the HIP method have been introduced for actual use, but they are still under development. Plasma sintering is expected to develop further into a new technology for ceramic processing (Fig. 4.8).

The process of producing compressed formed bodies (so-called "green bodies") prior to sintering is the same as the process in the normal sintering method, but firing and sintering in plasma flames is unique with this method. In addition to the utilization of the high temperatures and high enthalpy of plasma, utilization of its chemically active state, charged particles (electrons, ions) in plasma, quick heating/annealing properties are anticipated. Plasma sintering has advantages such as (1) sintering of substances that are difficult to be sintered (low sinterability substances), (2) short-time sintering, (3) a wide range of

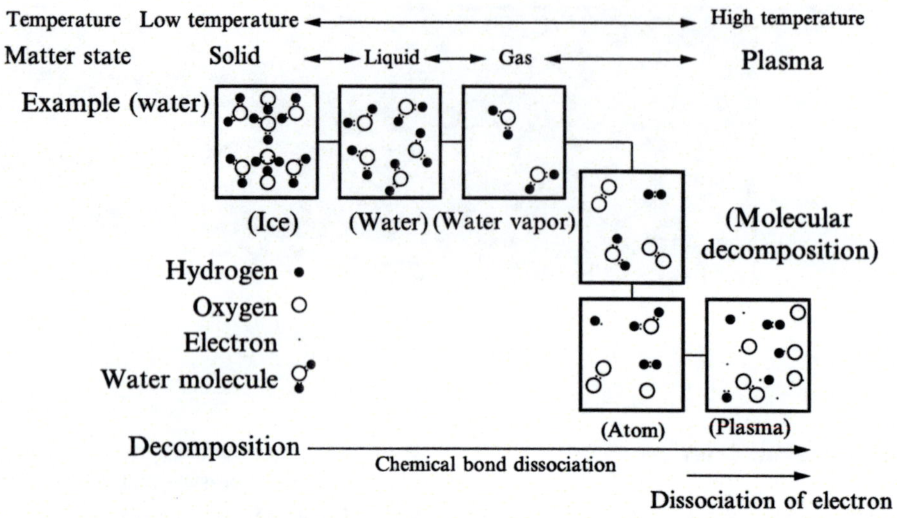

Fig. 4.7 Changes in temperature and matter state

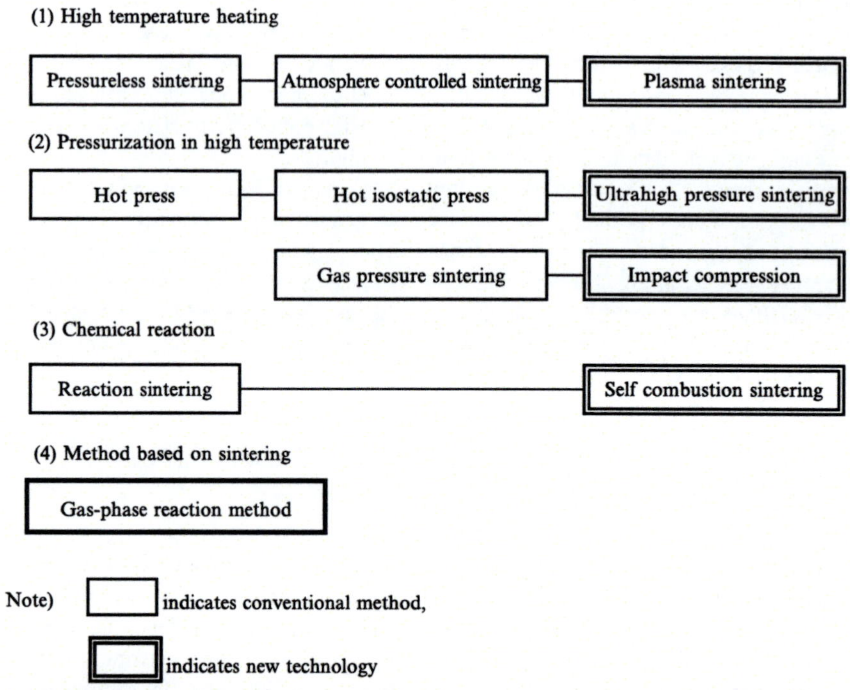

Fig. 4.8 Sintering technology

densified compositions and (4) control of grain growth. However, there are some issues including evaporation and the generation of cracks, requiring fine control of plasma. Therefore, know-how in plasma processing is essential. Low-temperature plasma is easy to handle, but determination of proper conditions for the densification of low sinterability substances is difficult.

For example, defining the actual temperature is somewhat complicated. The plasma temperature is high while the specimen temperature is low in comparison, thus, there is a so-called "heat barrier" between plasma and specimens. Temperatures of electrons, ions, molecules and atoms, etc. in the plasma flame vary and are complicated.

4.4.3 Detailed Examples of Plasma Processes

Semiconductor films are manufactured by plasma etching and plasma CVD. The majority of metallic powders such as W, Mo, Mn, Ag, Si, Cr and Pd are turned into ultrafine particles in a hydrogen plasma. Ceramic ultrafine particles of SiC, Si_3N_4, AlN, WC, Al_2O_3, TiN, ZrN, MgO and ZrO_2 are also being manufactured. Furthermore, a mixture of these powders as well as a mixture made by adding sintering agents of carbon (C) and boron (B) to SiC have also been synthesized. Diamond can be synthesized in the non-equilibrium range by using plasma. Coating films of WC–Co-based cermet, WC, TC, etc. have also been manufactured by plasma spraying.

Plasma sintering has been successful in Al_2O_3, BeO, HfO_2, UO_2, SiC, ZrO_2, etc. With respect to silicon nitride (Si_3N_4), which has is likely to dissociation at high temperature, a sintered body with a theoretical density has not yet been produced.

4.5 Ultrahigh Pressure Synthesis

The research field on ultrahigh pressure covers a wide range from design and development of ultrahigh pressure apparatus for materials synthesis to physical property measurement under ultrahigh pressure. Ultrahigh pressure is a pressure of 1 GPa or higher applied by a solid pressure medium (substance that transmits the pressure).

4.5.1 Generating Method of Ultrahigh Pressure

Ultrahigh pressure is divided into two methods: the static ultrahigh pressure method and the dynamic ultrahigh pressure method. In the former method, a mechanical press is used to generate ultrahigh pressure in a high pressure container. An ultrahigh pressure state can be created in a large space for a long time. In the latter method, ultrahigh pressure states exceeding 100 GPa can be generated for a short time of μs order by means of the impact of high-speed flying objects, etc. Substances such as diamond are synthesized by the static and dynamic methods. Ultrahigh pressure synthesis using the static ultrahigh pressure method is explained below.

Among the various devices used in the static ultrahigh pressure generation method, the belt-type ultrahigh pressure device is most frequently used for the production of substances and for research on materials synthesis as a relatively large sample volume is available for specimens. Figure 4.9 shows an example. The high pressure container of a belt-type device is composed of a pair of cone-shaped anvils made of WC–Co (tungsten carbide–cobalt) and a cylinder. When a specimen or a gasket is placed in the cylinder, and pressure is applied to the area between the anvils, ultrahigh pressure is generated on the specimen and a graphite heater heats the specimen.

4.5.2 Synthesis of Ceramics by Ultrahigh Pressure

Oxide materials synthesized by the ultrahigh pressure synthesis method feature a high density phase, which is thermodynamically stable at high pressures, and a crystalline phase, which needs to be kept in a high oxygen pressure atmosphere for the stability of the materials. Super-hard materials such as diamond and cubic boron nitride (cBN), a number of inorganic compounds and deep earth minerals are synthesized by the ultrahigh pressure synthesis method. In particular, diamond and cubic boron nitride are produced in large amounts with this method. The ultrahigh pressure synthesis method has been contributing to studies in various fields and has also been attracting attention from the industrial field. Examples of high pressure synthesis are illustrated below.

An ultrahigh pressure synthesis method of diamond was first discovered by a researcher at GE Inc. in the USA. With this method, graphite accompanied by metal catalysts such as iron, cobalt and nickel are used to synthesize diamond under a thermodynamically stable condition at 5 GPa and approximately 1,500°C. In some cases, diamond was synthesized directly from graphite without using any metal catalyst.

Figure 4.10 shows the state of carbon and diamond synthesis conditions. Diamond synthesis incorporating nonmetallic or inorganic compound, shown in the figure, is useful for analysis of natural diamond components and is important from an academic perspective. In terms of the materials science, the method has also achieved interesting results, represented by the diamond sintered body, which incorporates carbonate as the sintering agent and features high thermal resistance.

In recent years, active efforts have been made to synthesize not only diamond but also superconductive substances by the

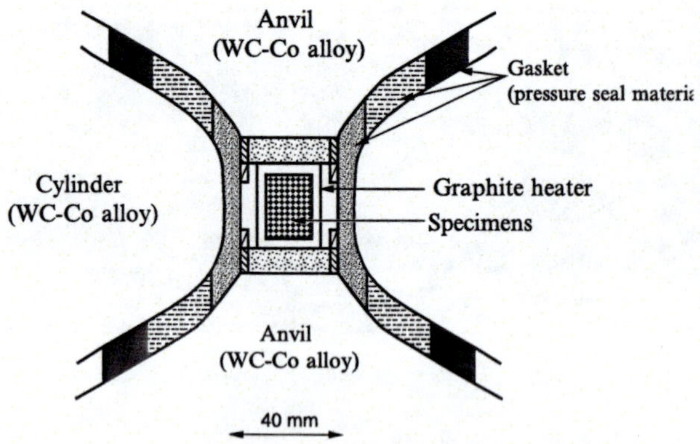

Fig. 4.9 Example of belt-type ultrahigh pressure device

ultrahigh pressure synthesis method. Synthesis was believed to be impossible under normal pressure that has been realized by changing temperatures and compositions (a two-dimensional matter search and synthesis method). The ultrahigh pressure synthesis method is the three-dimensional matter search and synthesis method involving the pressure axis. It is expected to play a very important role in the discovery of new materials.

Fig. 4.10 State of carbon and diamond synthesis

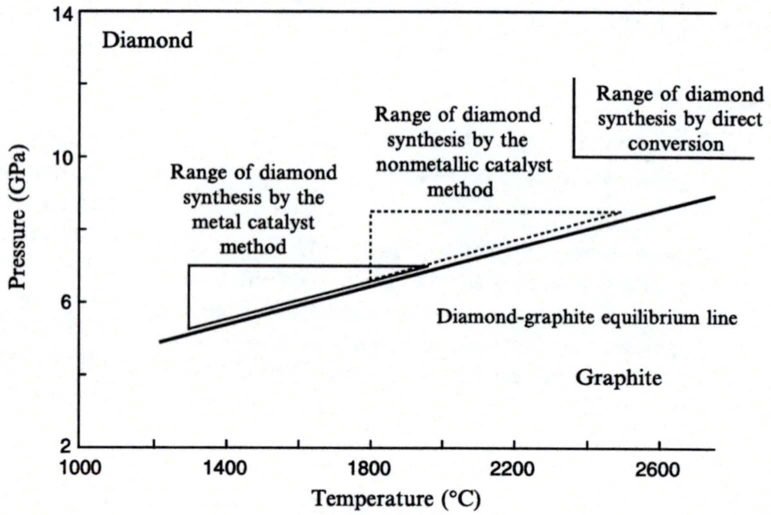

Soft Chemical Synthesis 4.6

Soft chemical synthesis is a method where the intended product is synthesized through moderate chemical reactions under temperatures ranging from room temperature to several hundred degrees Celsius. The synthesis operation is a reaction process or the combination of reactions and heat treatment as shown below (Fig. 4.11). The solid-state synthesis method, the general ceramic synthesis method, involves one step, where raw powders are mixed and crushed to create the intended composition. This is then calcined at a temperature of approximately 1,000°C. Meanwhile, in soft chemical synthesis, the composition and structure are partially modified and converted toward the intended substance through sequential and multistage synthesis operations by means of taking advantage of low-temperature chemical reactions. Therefore, the desired substance is fabricated at the nanometer level. This synthesis method is also important from the perspective of nanotechnology.

While the solid-state synthesis method involves thermal equilibrium, soft chemical synthesis is a non-equilibrium process, which usually enables synthesis of metastable phases that cannot be attained with normal methods. Starting materials used in soft chemical synthesis must be active in chemical reactions in low temperatures. Compounds with layer structure or tunnel structure and featuring host–guest reaction are normally used.

4.6.1 Ion Exchange

When a compound composed of an electrically-charged host framework such as clay mineral and ions that exist in the vacant spaces of the host framework interacts with a solution containing different type of ions or dissolved salt, the ions in the voids can be replaced. This reaction allows for the introduction of heterogeneous substances into voids of nano-level crystal lattices, making it possible to synthesize an organic/inorganic nanocomposite simply by using organic ions.

4.6.2 Intercalation/Deintercalation

The reaction for insertion of a guest chemical in the vacant space of crystal lattice of the host compound is called intercalation and the reversed reaction is called deintercalation. The reaction is accompanied by the reduction or oxidation of the host compound. Insertion reaction of donor-type guests such as alkali metals in graphite layers is an example of the intercalation reaction accompanied by reduction or oxidation. The guest substance that acts as the oxidant or the reducing agent is reacted in the liquid phase or the gas phase, or the reaction is induced electrochemically using the host compound as the electrode. Note that, Li-ion secondary batteries used as a power source in mobile phones and PCs incorporate intercalation/deintercalation of Li ions between graphite and $LiCoO_2$.

4.6.3 Pillaring

When multinuclear hydroxide ions are introduced between layers of a layered host substance with ion-exchange properties and calcined at around 500°C, hydroxide ions are dehydrated and condensed, acting like pillars that support the layer structure. The voids between the pillars function like fine pores in zeolite, converting the layered substance to a madreporic body with a high specific surface area.

4.6.4 Templating

The method used to crystallize inorganic components under hydrothermal condition, etc. around a template (mold), using organic molecules or the molecular assembly as the template. A substance with porous structure is obtained by heating so as to remove the template.

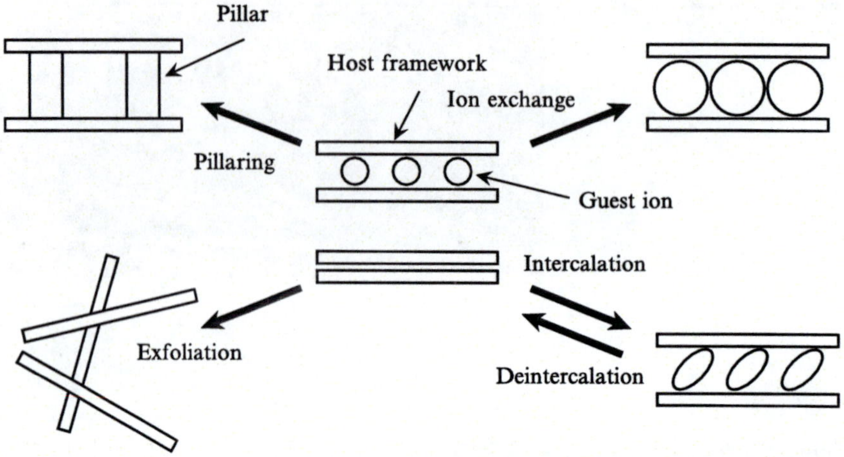

Fig. 4.11 Soft chemical synthesis process

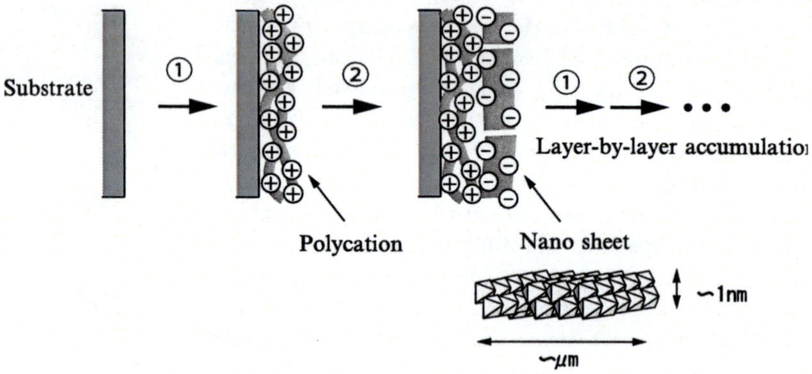

Fig. 4.12 Conceptual diagram of LBL accumulation. ① shows reaction in polycation solution immersion and ② shows reaction in nano sheet colloid solution immersion

4.6.5 Exfoliation

By reacting an appropriate guest with a layered host compound under proper conditions, extensive swelling is induced and the layered host compound is sometimes exfoliated to single layers. It is known that ions with hydration energy such as Li and Na, and bulky quaternary ammonium ions are effective for inducing single-layer exfoliation. The layer is about 1 nm in thickness, while the width is normally in the micron range. The layer is regarded as a new type of nano substance (nano sheet). The nano sheet is in the form of a colloid, which can be used as the starting material of soft chemical synthesis. If the colloid is dispersed in a polymer to create a nanocomposite, mechanical strength and gas barrier properties can be added. By utilizing the property where layer structure is recovered due to recondensation caused by the addition of electrolytes, various applications are possible. For example, it is used to make composites with guest substances that cannot be introduced between layers easily by ion exchange or intercalation, and to build of multilayer nano thin films utilizing LBL accumulation as explained below.

4.6.6 LBL (Layer-by-Layer) Accumulation

A method to build a high-order nanostructure through the effective control of interaction among ions, ligands, clusters, colloid particles, nano substances, etc. to array and assemble these substances on solid–liquid interfaces by molecular layer processing. The most popular one is the layer-by-layer (LBL) method, where two types of substances having charges opposite from each other are repeatedly absorbed on the substrate by monolayer and in a self-assembly manner. LBL accumulation utilizing a nano sheet created by exfoliation and an organic polymer is explained in Fig. 4.12 by a schematic. When a substrate with clean surface is immersed in a cationic polymer solution under appropriate conditions, a monolayer of polymer is absorbed within the surface of the substrate. When the entire surface is covered, absorption of excess polymer in the solution is inhibited due to electrostatic repulsion and the reaction is stopped after coating by the monolayer. After washing, the nano sheet is immersed in a colloidal solution. The nano sheet, having a negative charge, is absorbed into the surface of the substrate, which is stopped due to self-assembly. This process involves charge inversion on the substrate surface and can be theoretically repeated endlessly. In this method, the substances are assembled layer by layer at the nano level. The controllability of the nanostructure is equivalent to that of the molecular beam epitaxy (MBE) method.

Thin Film Deposition 4.7

Thin films are produced by depositing atoms, molecules or ions on a substrate from either a gas phase or liquid phase. Manufacturing methods of ceramic thin films are roughly classified into gas-phase or liquid-phase methods. This is shown in Table 4.2.

The most popular among the gas-phase thin film manufacturing methods are vacuum evaporation method, laser abrasion method, sputtering method and CVD (chemical vapor deposition) method.

In the vacuum evaporation deposition method, films are deposited at a relatively low pressure. Measures are required for effective reaction of metal elements with nonmetal elements. Reaction is normally promoted with the aid of active plasmas or radicals.

In the laser abrasion method, films are produced by irradiating a solid material (which is called a target) with pulsed laser beams. The irradiated area of the target is gasified and plasmas are generated with laser beams. It is a superior method that creates thin films with limited deviation from the composition of the target at a comparatively low temperature. It is however not appropriate for the production of films on large substrates.

In the sputtering method, atoms or molecules are ejected from the surface of a target by incident ions to transport the ejected atoms to the substrate. This is shown in Fig. 4.13. Ceramics are used as the target in many cases, while in some cases metals are sputtered and metal atoms can be reacted with oxygen or nitrogen gas to produce a ceramic film. The sputtering method produces film's with a composition similar to the ceramic target and high energy of incoming particles produces films with superior adhesion properties. High productivity is achieved with this method because targets do not have to be replaced frequently, allowing for continuous operation. This method allows for film production on large substrates.

In the CVD method, gasified raw material compounds are made to chemically react by means of thermal decomposition, oxidation, or reduction in the gas phase to create a thin film with intended chemical composition on the substrate. Although structures and characteristics of the created thin film vary depending on physical and chemical properties of the raw material compounds, and complicated compositions are difficult to control with this method, it has advantages such as high step coverage and the capability to create high quality thin films on large substrates.

The chemical solution deposition method utilizes a liquid phase, which is a solution containing metallic salts or metal-organic compounds in a solvent that is coated on the substrate by the spin coating method or other techniques, which is then thermally treated to produce a ceramic thin film. Among the chemical solution deposition methods, the method where thin films are made from a viscous precursor solution is called the sol–gel method. The composition of the precursor solution is readily transferred to the thin film, allowing easy control of the composition, and films can be formed on large substrates using this method. The electrophoretic deposition method is a solution based method that is suitable for producing films on substrates with complicated shapes.

Ceramic thin films are used in the electronics field for sensors, surface acoustic wave filters, bulk acoustic wave filters, transparent electrodes, memory capacitors, etc. and the optical field for nonreflective, abrasion resistant coating, etc. Functions, materials and applications of popular ceramic thin films, including those under development, are listed in Table 4.3.

The thin films are produced by depositing atoms of molecules on substrates, which is unique to thin film manufacturing processes. The characteristics described below are attributable to these unique processes. Target compounds are normally prepared at process temperatures lower than the sintering temperature of ceramics. Materials that are thermodynamically unstable can also be synthesized. Non-equilibrium phases are sometimes observed, and materials that are not originally soluble in one another can be mixed forcibly by simultaneous sputtering or multilayer formation. Furthermore, compounds with structures and characteristics not observed in sintered ceramics, such as with epitaxial growth, where

Table 4.2 Manufacturing methods of ceramic thin films

Reaction media	Principle of film formation technique	Manufacturing method
Gas phase	Physical vapor deposition (PVD)	Vacuum evaporation deposition
		Laser abrasion
		Sputtering
	Chemical vapor deposition (CVD)	CVD
Liquid phase	Chemical solution	Sol–gel
		Metal organic deposition (MOD)
	Electrochemistry	Electrophoretic deposition

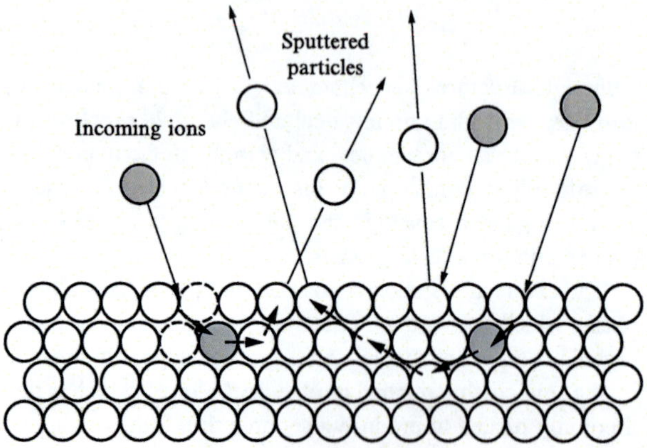

Fig. 4.13 Principle of the sputtering method

the growth is based on the atomic order of the substrate, and oriented growth, where a certain crystalline ordering is selectively grown on the surface of the substrate, is possible.

Table 4.3 Typical applications of ceramic thin films

Function	Material	Usage
Dielectric	SiO_2–Si_3N_4, Hf–Si–O–N	Capacitor
Piezoelectric	ZnO, AlN	Filters (SAW, BAW)
Pyroelectric	(Pb, La)TiO_3	Infrared ray sensor
Ferroelectric	Pb(Zr,Ti)O_3, $SrBi_2Ta_2O_9$	Nonvolatile memory
Electroconductive	SnO_2	Gas sensor
	SiC	Thermistor
	TiN	Barrier metal for IC
Electroconductive, transparency	In_2O_3–SnO_2, ZnO:Al	Transparent conducting layer of display
Ion conductive	$LiCoO_2$, $LiMn_2O_4$	Battery
Ferromagnetic	γ-Fe_2O_3	Magnetic recording media
Electroluminescence (EL)	ZnS:Mn	Light-emitting element of display
Others	TiO_2, Ta_2O_5, SnO_2, SiO_2	Nonreflective, optical coating
	SiC, TiN, ZrN	Corrosion resistance, abrasion resistance coating

Thin films are generally subjected to the stress from the substrate. The stress may cause poor adhesion with the substrate and cracking in thicker films, but materials with characteristics not observed in bulk products can be produced if the stress is properly controlled. The significance of thin film manufacturing technologies as well as its use in advanced film manufacturing devices will continue to increase responding to the needs for lower operating voltages and high integration density in electronic devices.

Powders and Fine Particles 4.8

Ceramics are normally used as sintered bodies made from raw materials in the form of powders, and thus, powders play important roles in ceramic manufacturing. Production of homogeneous sintered bodies made of high-purity powders have lead to the significant advancement of traditional ceramics (potteries and porcelains, refractories) known as "fine ceramics." Manufacturing methods, separation methods and particle size analysis methods of various sizes of fine particles are shown in Fig. 4.14. Powders with a grain diameter of 0.1 to several μm are generally used as raw materials for ceramics. However, demands for nano-size powders have been increasing recently. Responding to this trend, the significance of powder manufacturing has been increasing. Powders are manufactured by liquid-phase manufacturing methods such as the precipitation method, sol–gel method and hydrothermal reaction method as well as gas-phase methods that incorporate gas-phase reaction or evaporation–condensation performed by heating with reactors, flames, plasmas and lasers, in addition to grinding and sorting of fine particles. The need for spherical particles with small and homogeneous grain diameters for improvement of sintering properties following the sintering theory promoted the development of new synthesis methods. Furthermore, not only simple spherical particles but also particles with crystalline habits and fine particles with anisotropy such as rods and whiskers can be synthesized now by utilizing the liquid-phase or gas-phase method and controlling the speed and density of nucleation fundamental to particle generation.

One of the major achievements of the new synthesis methods is realization of nanoparticles with a grain diameter of several tens of nm or smaller. Nanoparticles exhibit peculiar physical phenomena. An increase in magnetic susceptibility, a decrease in the melting point, an increase in light absorption, an increase in transition temperature to superconductive states, improvement in thermal conductivity at ultralow temperature, were observed in ceramic materials made of nanoparticles. In chemical applications, catalyst characteristics have been improved by the utilization of nanoparticles.

Clusters represented by the fullerene (C_{60}) related compound have been attracting attention recently. Following the discovery of nanotubes and nanofibers derived from cluster structures, the range of applications has been further expanded. The synthesis of clusters that are smaller than nano particles has started only recently. However, clusters are expected to be applicable to a wider range of fields because of their novel characteristics, including their superconductive property, which is the first potential application, and is of interest in order nano devices.

The fine particles explained above were synthesized basically for the synthesis of high-purity and homogeneous powders. In addition, it is also important to introduce heterogeneous properties to powder particles to create complex compounds with added functionality. However, the composite powder needs to be macroscopically homogeneous. This is a big issue to be solved in the synthesis of composite particles.

Measurement and evaluation of particles are also important for the utilization of fine particles. For measurement of grain diameters, the most basic property of particles, interaction with X-rays, and light, varies depending on the grain diameter and several methods need to be used for accurate measurements as shown in Fig. 4.14. If the measured particles overlap, grain diameter distribution changes depending on the measuring method. Therefore, establishment of precise measurement/evaluation methods is one of the important issues to be solved for the utilization of fine particles. Grain boundary control of polycrystalline sintered bodies is another important characteristic in all ceramic materials. Because of this reason, raw powders should be synthesized with active grain boundary control of core–shell structure composite particles and nanoparticle coated composite particles. Furthermore, development of fine particles that play a role for physical and chemical interaction are required in order to realize functionalization of the powder surface.

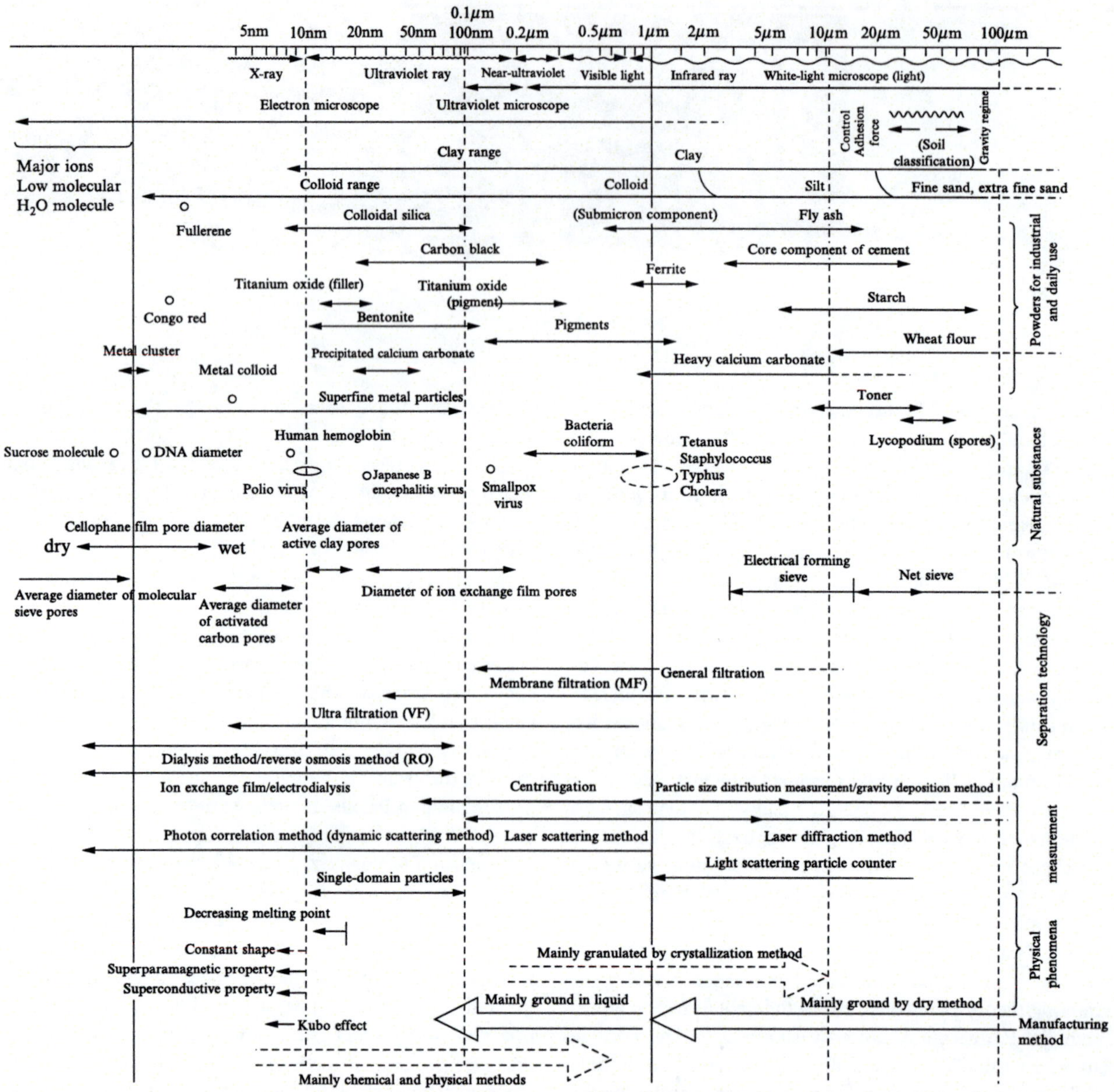

Fig. 4.14 Characteristics by fine particle size [reference: Jinbo M et al. (1991) Powders – functions and applications, the Japanese Standards Association, pp 16–17]

Literature

1. Ashbey MF (1974) A first report of sintering diagrams. Acta Metall 22:275–289 **(4.1)**
2. Editing Committee of Crystal Engineering Handbook (1971) Crystal engineering handbook. Kyoritsu Shuppan, Tokyo [in Japanese] **(4.2)**
3. Nakamura K, Nakata I (eds) (1981) Experimental physics course 13. In: Preparation and processing of specimen. Kyoritsu Shuppan, Tokyo [in Japanese] **(4.2)**
4. Fukunaga O, Sosetsu K (1979) (Chemistry review). In: The Chemical Society of Japan (ed) Tokyo, Ultrahigh pressure and chemistry, no. 22. pp 9–30 [in Japanese] **(4.5)**
5. Sawaoka A, Kondo K, Fujiwara S, Kusakabe M, Sosetsu K (1979) (Chemistry review). In: The Chemical Society of Japan (ed) Tokyo, Ultrahigh pressure and chemistry, vol 22. pp 31–45 [in Japanese] **(4.5)**
6. Hall HT (1960) Rev Sci Instrum 31:125 **(4.5)**
7. Yamaoka N, Akaishi M, Kanda H, Osawa S, Taniguchi N, Sei H, Fukunaga O (1992) Pres Tech 30:249 [in Japanese] **(4.5)**
8. Bovenkerk HP, Bundy FP, Hall HT, Strong HM, Wentorf RH Jr (1959) Nature 184:1094 **(4.5)**
9. Bundy FP (1963) J Chem Phys 38:631 **(4.5)**
10. Akaishi M, Kanda H, Yamaoka S (1990) J Cryst Growth 104:578 **(4.5)**
11. Akaishi M, Kanda H, Yamaoka S (1992) J Hard Mater 3:75 **(4.5)**
12. Muromachi E, Buturi O (1995) Appl Phys 64:325 [in Japanese] **(4.5)**
13. Jinbo M, Igasaki F, Tsubaki J, Higashitani K, Masuda H, Yamamoto H (eds) (1991) Powders – functions and applications, the Japanese Standards Association, pp 16–17, Figure 1.1.1, the Japanese Standards Association, Tokyo [in Japanese] **(4.8)**

5 Characteristics of Ceramics

Y. Imanaka et al. (eds.), *The Ceramic Society of Japan, Advanced Ceramic Technologies & Products*,
DOI 10.1007/978-4-431-54108-0_5, © Springer Japan 2012

5.1 Mechanical Properties

5.1.1 Hardness and Elasticity

Hardness of ceramics is normally expressed by the size of the indent made by pressing a diamond indenter against the ceramic surface. If the indenter has a square pyramidal shape, the Vickers hardness is expressed as P/S, where the load is P and the surface area of the indent is S, and the unit is GPa. The Knoop hardness is measured by a rhombic-based pyramidal indenter, but the Vickers hardness is more popular. When a solid is stressed (force/area), the solid is deformed in accordance with Hooke's law, which is expressed as (stress) = (elasticity) × (deformation). Strictly speaking, an isotropic solid has two independent elastic moduli. For example, if tensile stress is applied to a bar, it elongates in the same direction as the applied force. The elasticity (Young's modulus, E) is obtained by dividing the stress by the deformation (elongation/length). The other is Poisson's ratio (ν) which is the ratio of shrinkage in the direction normal to the direction of the elongation (Fig. 5.1). The shear modulus (or modulus of rigidity) and the bulk modulus of elasticity that affect shearing stress and volume compression, respectively, are other elastic moduli, but they can be calculated using the Young's modulus and the Poisson's ratio. The elastic coefficient of a material is calculated by measuring the deformation by bending tests and tensile tests or by measuring sonic speed of a material. The values obtained are called the static (isothermal) elasticity and dynamic (adiabatic) elasticity, respectively. E and ν of ceramics are 100–400 GPa and about 0.2, respectively.

5.1.2 Strength

The theoretical strength of a material is expressed by the force required to separate the atoms that constitute the material, which is approximately 1/10 of its elasticity. However, the actual strength is about 1/1,000 of the elasticity and the strength of ceramics intended for structural materials ranges from 400 to 800 MPa. The actual strength of materials is lower than the expected values because the stress is concentrated at defects (cracks) inside the material. The strength decreases as the size of the defect increases (Note 5.1). Defect sizes are statistically distributed in materials. Therefore, the strength of a material varies and is evaluated statistically (Weibull distribution).

The strength of a material is expressed accurately by the mean value or the Weibull distribution function obtained by testing many samples to failure. The strength is classified into bending strength, tensile strength, and compressive strength, which are measured by different methods. Figure 5.2 shows a method known as the three-point bending test.

5.1.3 Fracture Toughness

Materials contain defects (cracks) and the material fracture is caused by the stress concentrated at the tip of a crack (Fig. 5.3). The value of the stress concentration is expressed by the stress intensity factor K (Note 5.2). If the stress and the crack are small and K is also small, the crack extends very slowly. However, the crack starts to extend rapidly, leading to breakage, when K reaches a certain value. The value at this point is the critical stress intensity factor or the fracture toughness value K_C, which expresses the toughness (or fragility) of the material. Stress σ at this point is the fracture strength (strength) of the material. Strength depends on the length and shape of the crack as well as the fracture toughness value and is not an intrinsic characteristic of the material. Ceramics have low tensile strength (mode I) and

Note 5.1 According to Griffith, strength $\sigma=(2\gamma E/\pi\alpha)^{1/2}$, γ: surface energy, E: Young's modulus, α: length of crack.

Note 5.2 Expressed by stress intensity factor $K=Y\sigma(\mathrm{p}\alpha)^{1/2}$, Y: shape factor, σ: external stress, α: length of crack.

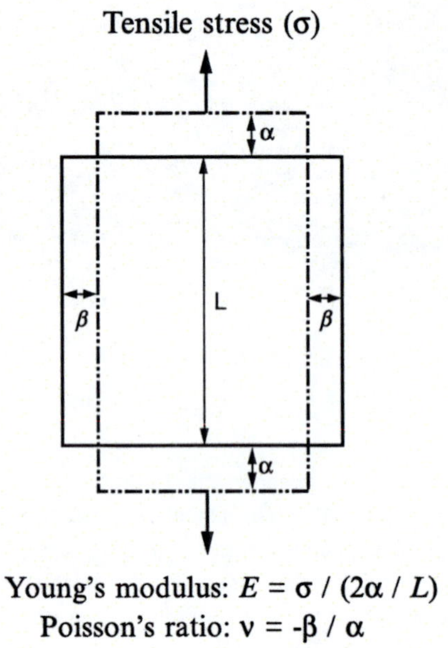

Fig. 5.1 Young's modulus and Poisson's ratio

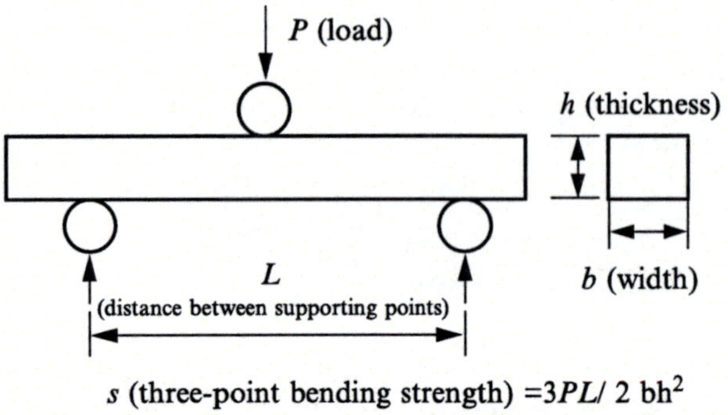

Fig. 5.2 Bending strength measurement by the three-point bending test

are normally evaluated by the fracture toughness value K_{IC} in mode I. K_{IC} of ceramics typically vary from 3 to 10 MPa $m^{1/2}$, and fracture tests are performed by using specimens with a crack.

5.1.4 Other Mechanical Properties

In addition to the above-mentioned basic properties, the fatigue strength, creep, abrasion resistance and corrosion resistance are important mechanical properties related to the strength of ceramics. Fatigue is the phenomenon where a crack extends gradually when loaded below the fracture strength, eventually leading to fracture. The service life is predicted by estimating that the rupture time that is proportional to stress or K^n. Creep is the deformation of materials under high temperatures. The rate of deformation is proportional to (diffusion coefficient) × (stress)^n/(grain diameter)^m. Abrasion resistance is a combination of mechanical properties and chemical properties and becomes higher when the hardness and the fracture toughness value increases.

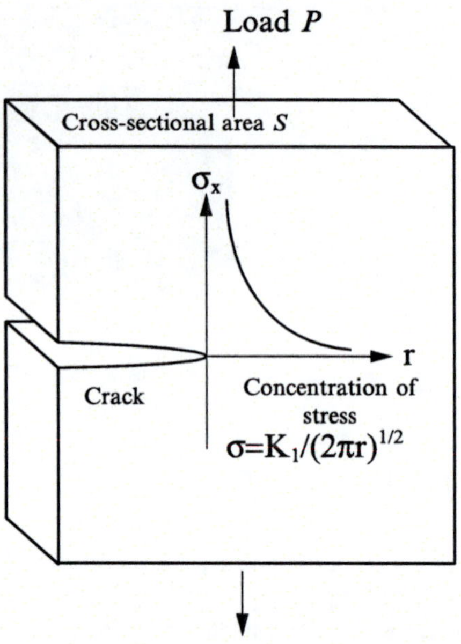

Fig. 5.3 Concentration of stress at a crack

5.1.5 Evaluation of Mechanical Properties

Ceramics have higher hardness and elasticity compared with metallic and organic materials. They are resistant to abrasion, light weight and show limited deterioration in mechanical properties at high temperatures. The major disadvantage of ceramics is extremely high fragility. This indicates that the strength of ceramics depends on defects, but low fragility undermines their reliability with respect to strength. The key to utilization of ceramics is taking countermeasures to overcome their fragility. Mechanical properties of ceramics are evaluated by measuring the basic properties explained above at temperatures ranging from room temperature to high temperatures.

Thermal Properties 5.2

Physical properties (solid-state characteristics) that indicate thermal absorption and loss characteristics of materials include specific heat capacity and thermal expansion, while properties related to heat transfer and insulation are expressed by thermal conductivity and thermal diffusivity.

5.2.1 Specific Heat Capacity (Specific Heat)

When a substance with a mass of m absorbs energy in the form of heat written as Q and the temperature increases by ΔT, $Q/(\mathrm{m}\cdot\Delta T)$ is the specific heat capacity (specific heat). If m is a molecular quantity in moles, $Q/(\mathrm{m}\cdot\Delta T)$ then it is the molar heat capacity (molar specific heat capacity). Heat capacities are classified into the molar heat capacity at constant volume C_v and the molar heat capacity at a constant pressure C_p, and $C_p-C_v=TV\beta/K_T$ (T: absolute temperature, V: mole volume, β: coefficient of thermal expansion, K_T: isothermal compressibility). C is a fundamental thermodynamical quantity. Thermal vibration (phonon is quantized) occurs in atoms of solid bodies when the temperature increases, and Q is equivalent to the amount of increase in thermal vibration energy (internal energy) of atoms. Temperature change of C is proportional to the cube of the absolute temperature (Debye model) at low temperatures, but it changes toward a certain value that does not depend on the substance ($C=3R$ in monatomic solid, R: gas constant; known as the Dulong–Petit law, 3nR where n refers to an atomic solid; Kopp–Neumann law) at high temperatures. The summation rules of specific heat capacities of oxides are applicable in an approximate sense to multiple oxides over a wide range of temperatures while not being limited to high temperatures. In actual materials, thermal changes due to property changes such as thermal vibration, defect formation, electron thermal excitation and phase transition properties affect the change in heat capacity, and therefore, precise measurements and analyses are performed.

5.2.2 Thermal Expansion

The expansion of substances responding to temperature change is characterized by the coefficient of thermal expansion β. When an interatomic potential develops in a material becomes asymmetric, thermal expansion occurs responding to the temperature increase. As Fig. 5.4 indicates, covalently bonded crystals of diamond and SiC have strong bonds and low thermal expansion. Meanwhile, thermal expansion is high in ionic crystals such as MgO, which has a high asymmetric nature, and metals with low bonding energy. However, some ionically bonded crystals exhibit extremely low or negative thermal expansion. Silica glass, for example, exhibits low thermal expansion because the SiO_4 tetrahedrons, structural units of the glass network, fill the gaps in the network during the process of displacement. Cordierite shows positive or negative coefficients of thermal expansion, depending on the crystalline axis (Fig. 5.4). In the case of a cordierite sintered body, changes in the two are balanced, exhibiting low thermal expansion. It is believed that the low thermal expansion of aluminum titanate sintered bodies is attributable to the cracks generated near the grain boundaries due to thermal expansion anisotropy. Low thermal expansion materials (thermal shock resistant materials) are essential for honeycombs and so on, that are used in environments with sudden temperature changes. When various components consisting of difference materials are subject to changes in temperature, design specifics need to be considered due to differences in thermal expansion.

5.2.3 Thermal Conductivity

Thermal conductivity κ, which indicates the amount of heat transferred in a material, is expressed as $\kappa=c\alpha\rho$ (α is thermal diffusivity, ρ is density) according to the heat conduction equation. For convenience, thermal conductivity is normally

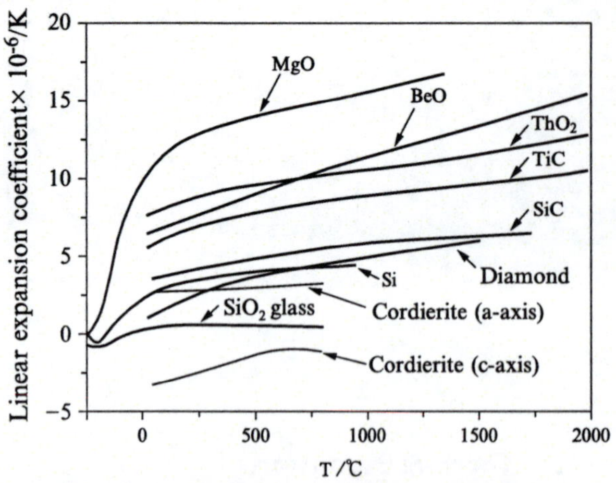

Fig. 5.4 Coefficients of thermal expansion of ceramics versus temperature

obtained from thermal diffusivity (speed of heat diffusion). The thermal conductivity of solids ranges from 0.01 to 3,000 W/mK (at ambient temperature), depending on various factors explained below. Phonons, conduction electrons and photons are the carriers that transfer heat in solids. As for metals in which conduction electrons play the role of the carriers, κ_e is proportional to electrical conductivity, which is expressed as $\kappa_e = L\sigma T$ (L: Lorenz number, Wiedemann–Franz law). However, many ceramics show electrical insulating properties and therefore are controlled by lattice thermal conduction κ_L. Solids that satisfy the following conditions generally have high thermal conductivity: (1) strong chemical bond, (2) high atomic packing density, (3) high symmetry and (4) are composed of light elements. Diamond, which exhibits the highest thermal conductivity, fully satisfies all of these conditions. The degree of thermal conductivity and the temperature change are expressed as $\kappa_L = (1/3)\,\omega l_L c$ (ω: group velocity of phonons) in relation to the mean free path l_L of phonons, and thermal diffusivity α is proportional to l_L and ω.

l_L of crystals is constant at low temperature because of the size effect and the temperature change is as rapid as the change in thermal capacity. Meanwhile, the thermal conductivity decreases as l_L decreases exponentially responding to an increase in scattering of phonons at high temperature (higher than Debye temperature specified by Debye model). The thermal conductivity reaches a maximum value at a temperature in-between (normally lower than room temperature). At high temperature, collision of phonons against one another is proportional to T and l_L is inversely proportional to the temperature, which means κ_L is inversely proportional to T. Meanwhile, in case of glass, l_L responds to temperature changes like the thermal capacity even if the temperature is low because l_L is limited to a certain degree by the distance between atoms.

l_L generally becomes shorter due to decreased thermal transfer caused by structural factors (defects) such as gas cavities, grain boundaries, transitions and impurities inside ceramics, and therefore, thermal conductivity reflects the structural sensitivity of ceramics. In sintered bodies of high thermal conductivity, the apparent thermal conductivity changes by digits responding to a change in porosity, and control of porosity is important for sintered bodies. Materials with high porosity are suitable for thermal insulation, while single crystals and sintered bodies with nearly theoretical density are suitable for IC substrates, which need to have high thermal conductivity. In translucent ceramics, heat transfer κ_p by photons responding to radiation is proportional to T^3 at temperatures above 1,000°C, and the thermal conductivity increases rapidly responding to an increase in temperature. Internal thermal radiation is not negligible in opaque materials with gas cavities and voids.

Electrical Properties, Dielectric, Pyroelectric and Piezoelectric Properties 5.3

5.3.1 Dielectrics: Paraelectrics and Ferroelectrics

Some ceramics conduct electricity, while others do not. Ceramics that do not conduct electricity are called electrical insulators. When an insulator is placed in an electric field, positive and negative charges are pulled away from their original positions when opposite electrical charges are applied, displacing the charge symmetry, resulting in positive and negative charges being induced, respectively, on each side of the insulator (Note 5.3). Because of this phenomenon, insulators are sometimes called dielectrics.

Insulators are classified into paraelectrics, ferroelectrics, piezoelectrics and pyroelectrics. Electrical charge in dielectrics generally return to their original condition when external voltage is removed. Dielectrics that satisfy this condition are called paraelectrics. Meanwhile, dielectrics that remain polarized (residual polarization) even after removal of external applied voltage are called ferroelectrics. Piezoelectrics materials that generate voltage when the crystals are stressed by a force, while pyroelectrics generate voltage when the crystals are applied with heat. The classification is not as strict as classification of plants. For example, ferroelectrics also have piezoelectric and pyroelectric properties. Refer to Fig. 5.5 for an easy-to-understand illustration.

Insulators (dielectrics) that are believed to be unresponsive to electricity exhibit very interesting electrical properties as explained above. Dielectric bodies that are used in various electrical fields because of their polarization can be flexibly controlled by voltage. For example, due to the property where polarization is aligned (dipoles) for the storage of charges, they are used as capacitors, e.g. barium titanate, etc. which exhibit high capacitance that is used in many electrical products.

Note 5.3 This phenomenon is called polarization, and electrical dipoles are generated under this condition. Electrical dipoles indicate a pair of positive and negative electric charges separated by a certain distance.

5.3.2 Pyroelectrics

Many types of sensors are made from ceramics. The sensors that can detect faint infrared rays released from human skin apart from the sensors by about 1 m are also made from ceramics. The sensors utilize the properties of pyroelectrics. The principle is explained below (Fig. 5.6).

Ferroelectrics are polarized when a DC voltage above a threshold value is applied, generating positive charges on the surface of one surface and negative charges on the other. The charges on the surface remain even after external voltage application is removed. Floating electrical charges such as dust and electrons in the air adhere to the surface to neutralize the charges on the surface (Fig. 5.6a). The amount of electrical charge remaining on the surface is called residual polarization (Coulomb/m^2) which is the amount of charge per unit area (Coulomb). When infrared pulses are applied to a ferroelectric material, the polarization changes resulting from thermal expansion or contraction and results in a change in the amount of electrical charge on the surface (Fig. 5.6b). If an external resistor (*R*) is connected to a point between the two electrodes, electrical charge flows through the *R*. The current is called the pyroelectric current and the sensor is activated upon detection of the voltage (Fig. 5.6b). The dotted circles indicate electrical charges that disappear following application of infrared rays).

This can be expressed as shown in Fig. 5.7 from a different point of view. Residual polarization $P_s=0$ at Curie point T_c (temperature at which the property changes from ferroelectric to paraelectric). When infrared rays are applied to elements at room temperature T_A, the temperature rises slightly (ΔT=0.01°C, for example) to T_B, and P_S drops from A to B. The surface charges equivalent to $\Delta P_S=P_S(T_A)-P_S(T_B)$ flow through the external resistor *R*.

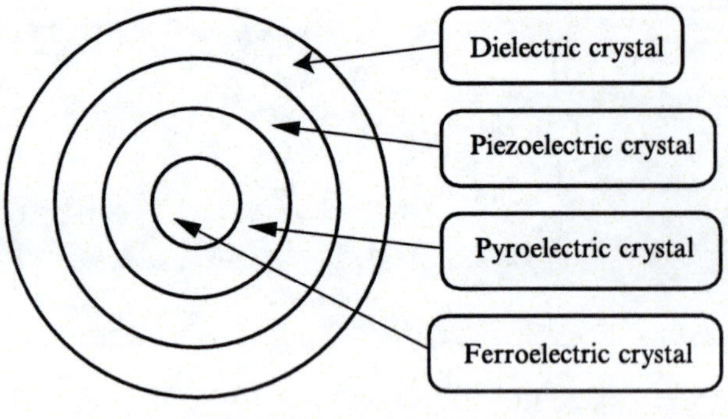

Fig. 5.5 Classifications of dielectrics by electrical properties

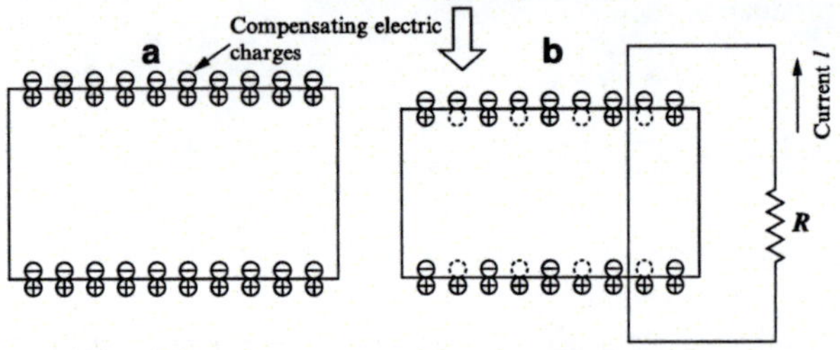

Fig. 5.6 Explanation of pyroelectric effect and piezoelectric effect. ⊕: positive charge, ⊖: negative charge, :( ): points at which electrical charges disappear, *R* external resistor, *I* pyroelectric current or piezoelectric current

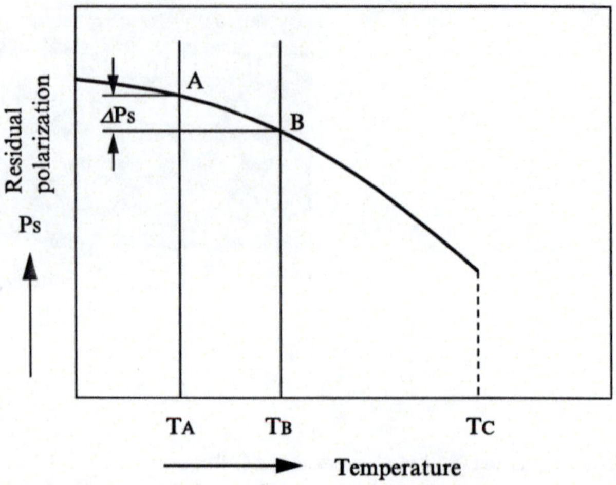

Fig. 5.7 Residual polarization versus temperature

5.3.3 Piezoelectrics

Piezoelectricity is a phenomenon where voltage is generated by a crystalline material when an external stress (pressure or tensile force) is applied to it. The phenomenon where the material expands or shrinks when voltage is applied to it is called the piezoelectric effects. The former is also called the direct piezoelectric effect, while the latter is called the converse piezoelectric effect. Piezoelectric ceramics featuring the piezoelectric effect are also used widely as electronic materials. They are used as ultrasonic transducers, ignition elements, microphones, and piezoelectric transformers.

The principle is explained in Fig. 5.6 in an easy-to-understand illustration. Figure 5.6a shows the surface polarized by external voltage. When pressure is applied vertically on the top and bottom, the polarization changes and some of the charges in the surface layer disappear (Fig. 5.6b). If an external resistor (*R*) is connected to a point between the two electrodes, electric charges having no counterparts flow through the *R*. This is called the direct piezoelectric effect.

Electrical Properties, Electronic Conductivity, Ionic Conductivity, Mixed Conductivity

5.4

Materials that have electrical conductivity are classified by the type of the electrical charge carrier. Electrical charge carriers such as electrons, ions or electrons plus ions vary in electronic conductors, ionic conductors and mixed conductors, respectively. Electronic conductors are the most important among the three types because they have the widest variety of applications. Research on solid-state properties has been performed with a focus on electronic conductors, and a number of theoretical analyses have been carried out. Applications of ionic conductors are limited unlike electron conductors, but basic research has been emphasized because they have diversified structures and properties. With respect to mixed conductors in which both electrons and ions move, they can be used as electrodes and catalysts. Therefore, basic research as well as applied research has been promoted. Electronic conductivity, ionic conductivity and mixed conductivity are explained below in that order.

5.4.1 Electronic Conductivity

Electronic conductivity is explained using the energy band structure. Electrons, which are much smaller than atoms, have the properties of particles and waves in isolated atoms. Therefore, behavior of electrons is controlled by the quantum mechanics, and the energy values are discontinuous (quantization of energy). However, if the atoms come closer and interaction is induced, the energy value of electrons deviates from the value under the isolated condition little by little because the same quantum number is prohibited by the Pauli exclusion principle. As a result, the energy values become extremely wide in a number of atoms in materials, although the number of electrons is limited. Therefore, the electron energies, which are discontinuous in isolated atoms, can be virtually viewed as continuous energies. Electron energy is referred to as energy bands as they become quasi-continuous within a certain range.

Active interaction is observed between atoms in solids such as ceramics because the distance between atoms is small. The number of atoms is extremely large, excluding ultrafine particles and ultrathin films, and solid-state properties can be discussed on the basis of energy. The energy range of electrons is called the allowed band. Even if the number of electrons is huge, there is an energy range where electrons are not allowed to exist. This range is called the forbidden band. Electrons occupy the energies step by step starting from lowest energy levels, if a disturbance is caused by thermal energy is neglected. The allowed band in which all of the energy levels are totally filled with electrons is called the filled band. The band with the highest energy is called the valence band. The allowed band that has the highest energy and which is partially or totally empty is called the conduction band. The minimum energy value in the band gap between the valence band and the conduction band is called the energy gap or the band gap. The energy band varies in a complicated manner depending on the direction in some crystals, and it is not always easy to estimate the minimum value. However, errors will be limited if the mean value of the maximum energy in the valence band and minimum energy in the conduction band are adopted.

Electrical conduction of ceramics can be explained using the energy band shown in Fig. 5.8. When ceramics are coated with electrodes and an electric field is applied, electron energy is increased. However, even if the electrons receive energy when they are in the valence band, transition to a different state is prohibited because all of the energy levels are occupied.

As shown in Fig. 5.8a, if the energy gap is large, the electrons cannot jump into the conduction band even if thermal and optical energy is supplied externally, which means no electron contributes to electrical conduction, and electrical conductivity is not exhibited even if an electrical field is applied. Therefore, the material is an insulator and examples are MgO, Al_2O_3 and SiO_2.

Meanwhile, if the energy gap is small, when the electrons receive external energy they can jump from the valence band

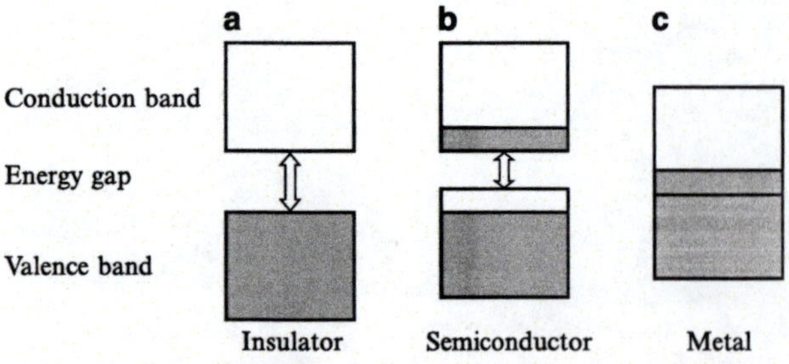

Fig. 5.8 Structure diagram of energy bands (*shaded areas* indicate the areas occupied by electrons)

to the conduction band, as shown in Fig. 5.8b. The electrons jump into the conduction band are shifted by the energy from the electric field, carrying electrical charges and contributing to electrical conduction. These materials are called semiconductors because they exhibit electrical conductivity lower than metals and higher than insulators and examples are ZnO, Fe_2O_3 and TiO_2. The number of electrons that jump into the conduction band and the number of holes (electron holes) in the valence band that remain are equal. If the electrons have higher mobility, electrons with negative charges carry the current and the material is called an n-type (n stands for negative) semiconductor. If the holes in the valence band have higher mobility, the holes that have positive charge transport the current, and the material is called the p-type (p stands for positive) semiconductor. If the valence band and the conduction band overlap or part of the allowable band is occupied from the beginning, as shown in Fig. 5.8c, electrons can be moved by small amounts of energy. Metals have this type of band structure. Some of the ceramics also have this band structure and metallic conduction, although the number is limited. The examples are TiO, CrO_2 and ReO_3.

5.4.2 Ionic Conductivity

There are many ceramics with highly ionic bonding, and hence, ions may move when an external electrical field is applied to these ceramics. Figure 5.9 shows the change in the internal energy caused by the movement of ions from point a to b in the crystal lattice (in this case, free energy should be used instead of the internal energy). The energy is low in areas near the lattice points, but the energy becomes higher in areas between the lattices in principle because a large amount of repulsion energy is generated due to the overlapping of electron clouds. The ions need to have high thermal energy and move beyond the large energy mountain (energy barrier) between the lattices for ionic conduction by electric field application. If two ions are located adjacent to each other, the space cannot be occupied by two ions. Therefore, the probability of transfer is low unless there is no ion from the beginning or lattice cavities are formed. Conditions for achieving ionic conduction can be decided with considerations to these points.

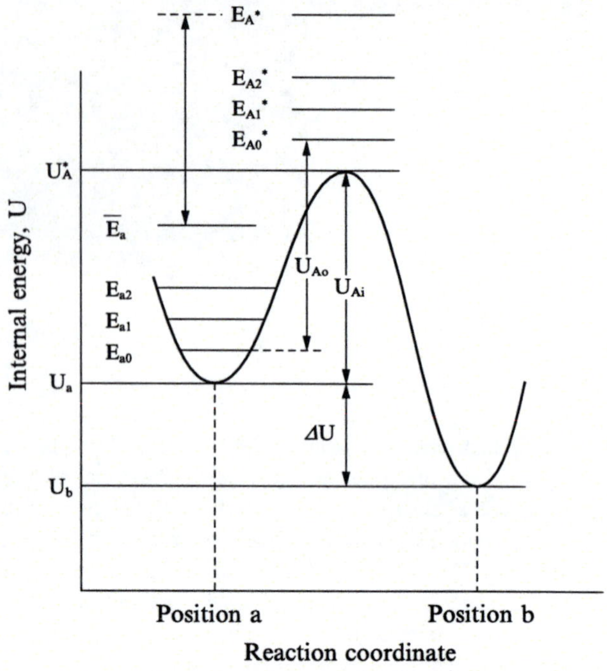

Fig. 5.9 Internal energy change caused by ionic transfer (vibration energy is shown above potential energy)

The first condition is that there are many atomic positions that are not occupied by ions inside the atomic lattice (there are many vacancies between atoms even in pure crystals) and mobile ions can to move via these positions. The second condition is that the number of atomic positions that can be occupied by ions exceeds the number of ions in the crystal, and the difference in potential energy between them is not large. The third condition is that the lattice defect concentration of mobile ions can be increased with the external introduction of lattice defects by solid-solution using additives. The fourth condition is that the vacancies are larger than those in crystals and mobile ions exist because atoms and ions are disorganized like those in glass that was frozen at a state close to the liquid state.

Typical ionic conductors are listed in Table 5.1. Positive ionic conductors are represented by H^+, Li^+, Na^+, K^+, Ag^+ and Cu^+ and negative ion conductors are represented by F^-, Cl^- and O_2^-. Among them, H^+, Li^+ and O_2^- are attracting attention in terms of application.

Table 5.1 Typical ionic conductors

Mobile ions	Ionic conductors	Conductivity (S cm^{-1})
Ag^+	α-AγI	2×10^0 (200°C)
	$RbAg_4I_5$	2.7×10^{-1} (25°C)
	$AgI–Ag_2O–B_2O_3$ (glass)	1×10^{-2} (25°C)
Cu^+	$Rb_4Cu_{16}I_7Cl_{13}$	3.4×10^{-1} (25°C)
	$CuI–Cu_2MoO_4–Cu_3PO_4$ (Glass)	4×10^{-3} (25°C)
	$CuBr–Cu_2MoO_4–CuPO_3$ (glass)	2×10^{-3} (25°C)
Li^+	Li_3N	1.2×10^{-3} (25°C)
	$Li1_4Zn(GeO_4)_4$ (LISICON)	1.3×10^{-1} (300°C)
	$Li_{1.3}Ti_{1.7}Al_{0.3}P_3O_{12}$	8×10^{-4} (25°C)
	$LiI–Li_2S–SiS_2$ (glass)	2×10^{-3} (25°C)
Na^+	$Na_2O{\cdot}11Al_2O_3$ (β-alumina)	1.3×10^{-1} (300°C)
	$NaZr_2P_2SiO_{12}$ (NASICON)	3×10^{-1} (300°C)
H^+	$HU_2PO_4{\cdot}4H_2O$	7×10^0 (100°C)
	$H_3(PMo_{12}O_{40}){\cdot}29H_2O$	2×10^{-2} (25°C)
	$SrCe_{0.95}Yb_{0.05}O_{3-\alpha}$	8×10^{-3} (800°C)
	$BaCe_{0.9}Nd_{0.1}O_{3-\alpha}$	2.2×10^{-2} (800°C)
O_2^-	$Zr_{0.85}Ca_{0.15}O_{1.85}$	2×10^{-3} (800°C)
	$Zr_{0.82}Y_{0.18}O_{1.91}$	2×10^{-2} (800°C)
	$Bi_{1.5}Y_{0.5}O_3$	3×10^{-1} (800°C)
F^-	CaF_2	3×10^{-6} (300°C)
	β-PbF_2	1×10^{-3} (200°C)
	β-$PbSnF_4$	8×10^{-2} (200°C)

5.4.3 Mixed Conductivity

It is not unusual that electronic conduction and ionic conduction occur at the same time at high temperatures or at low oxygen partial pressures. However, it is rare that the rate of contribution (transport number) to the total electrical conductivity is high for both electrons and ions and controllable. CeO_2 is an oxide ion conductor, but it has high ionic conductivity and the transport number reaches 0.5 in some areas as electrons are generated due to the following reaction, which is expressed by the Kröger–Vink notation under low oxygen partial pressure.

$$O_O^x \rightarrow 1/2O_2 + V_O + 2e^-$$

Some compounds with a perovskite structure exhibit electronic conductivity and oxide ion conductivity at the same time. Compounds made by adding Sr to $LaCoO_3$ or $LaNiO_3$ exhibit electronic conductivity and oxide ion conductivity. These compounds can be used in electrochemical devices such as electrodes in fuel cells.

Magnetic Properties 5.5

Materials are divided into two groups, that is, those with a permanent magnetic dipole and those without a permanent magnetic dipole. In the case of a permanent magnetic dipole, atoms themselves have magnetic moments even if there is no magnetic field (it is easy to understand that each of the atoms is a small magnet). Each of the electrons that spin around the nucleus has a magnetic moment, and magnetic properties exhibited by transition metal elements and rare earth metallic elements are induced by the orbital motion of electrons that spin around nuclei and the magnetic moments of the electrons themselves (electrons are magnets of the smallest size).

Materials having electrons with permanent magnetic dipoles are classified into four types, as shown in Fig. 5.10, depending on the magnetic interaction of their permanent magnetic dipoles. They are "paramagnetic" in which magnetic interaction is extremely small without an externally applied magnetic field and dipoles are oriented in random directions, "ferromagnetic" in which magnetization directions are the same due to the force that parallelizes adjacent dipoles, "antiferromagnetic" in which magnetization arrangement is anti-parallel with one another and "ferrimagnetic" in which the magnetization arrangement is the same as antiferromagnetic substances but the magnetization is strong because the magnitudes of magnetization are different and the magnetization is equivalent to the difference in these magnitudes. Among the magnetic materials, ferromagnetic materials and ferrimagnetic materials have been used widely as permanent magnets, magnetic recording media, and transformers.

In ferromagnetic materials and ferrimagnetic materials, dipoles are arranged in parallel or anti-parallel at temperatures below the characteristic temperature T_c called the Curie temperature, but the dipoles are not aligned at temperatures above the Curie temperature and the materials exhibit paramagnetic properties. As for antiferromagnetic materials, dipoles are in ordered arrangement at temperatures below characteristic temperature T_N (called the Néel temperature). The ordered arrangement disappears at temperatures above the Néel temperature. At temperatures T above the Curie temperature, the magnetic susceptibility, χ, becomes

$$\chi = \frac{C}{T - \Theta}$$

C: Curie constant specific to each substance, Θ: Curie temperature

which is called as Curie–Weiss law.

Meanwhile, ferromagnetic materials and ferrimagnetic materials exhibit strong spontaneous magnetization at temperatures below the Curie temperature and magnetic hysteresis characteristics shown in Fig. 5.11. These magnetic materials have magnetization properties as indicated in the figure because of the following reasons. Under the condition A (demagnetized state) in Fig. 5.11, the magnetic body is composed of microscopic regions called magnetic domains, where magnetic dipoles are aligned in the same direction, and the magnetic domains are separated by thin domain walls. Each of the magnetic domains is magnetized without the application of a magnetic field as magnetic dipoles are aligned in the same direction. However, the direction varies depending on the magnetic domain and the whole magnetic body is in the demagnetized state without magnetization. If a magnetic field is applied from this state, the rate of magnetic domains that are aligned in the same direction as the magnetic field increases (condition B). Moreover, when the magnetic field is strengthened, all of the magnetic domains face the same direction as the magnetic field (condition C), inducing magnetization rotation. All of the magnetic domains in the magnetic body are then parallel to the magnetic field, reaching magnetization saturation (condition D). If the magnetic field is weakened by this state, magnetic domains facing different directions are generated due to magnetization rotation and the volume increases (condition E), but the magnetized state does not return to its original conduction before magnetization, and residual magnetization, the magnetization component in the direction of the applied magnetic field, remains (condition F). If a magnetic field is applied in the

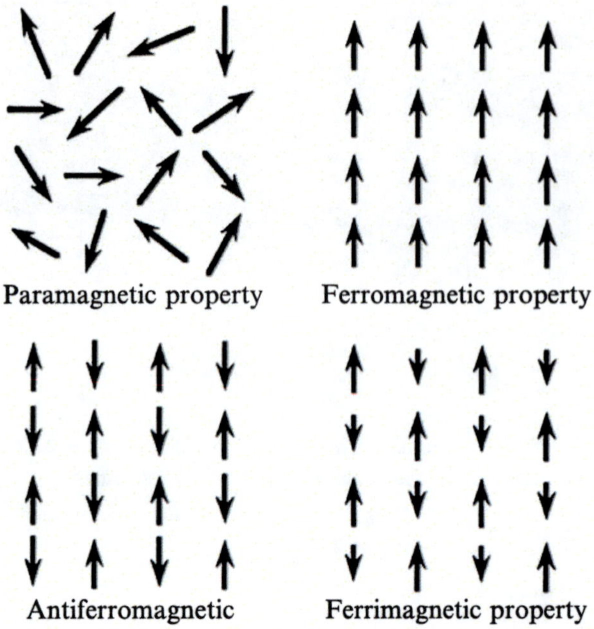

Fig. 5.10 Arrangement of permanent magnetic dipoles in each magnetic property

opposite direction from this state, the total magnetization drops rapidly (condition G) and the apparent coercive force becomes zero (condition H). If the magnetic field is in the opposite direction is strengthened, the ratio of magnetic domains which face the same direction as the magnetic field increases, causing magnetization rotation, and the magnetic domains align in the same direction as the magnetic field (condition I). These magnetic properties of ferromagnetic materials are utilized in a variety of fields.

Materials that have a small coercive force and are capable of changing magnetization directions responding to weak magnetic fields are called soft magnetic materials and are used in transformers, inductors, and magnetic heads. These soft magnetic materials are used as such due to properties that allow for the easy transmission of magnetic flux (high permeability). Meanwhile, magnetic materials that require the application of a strong magnetic field for the changing of magnetization direction are called hard magnetic materials. These materials have high coercive force and residual magnetization are used as permanent magnets. The performance of permanent magnets is expressed by the product of magnetization B, which is generally obtained from demagnetization properties, and magnetic field H (maximum magnetic energy product). Permanent

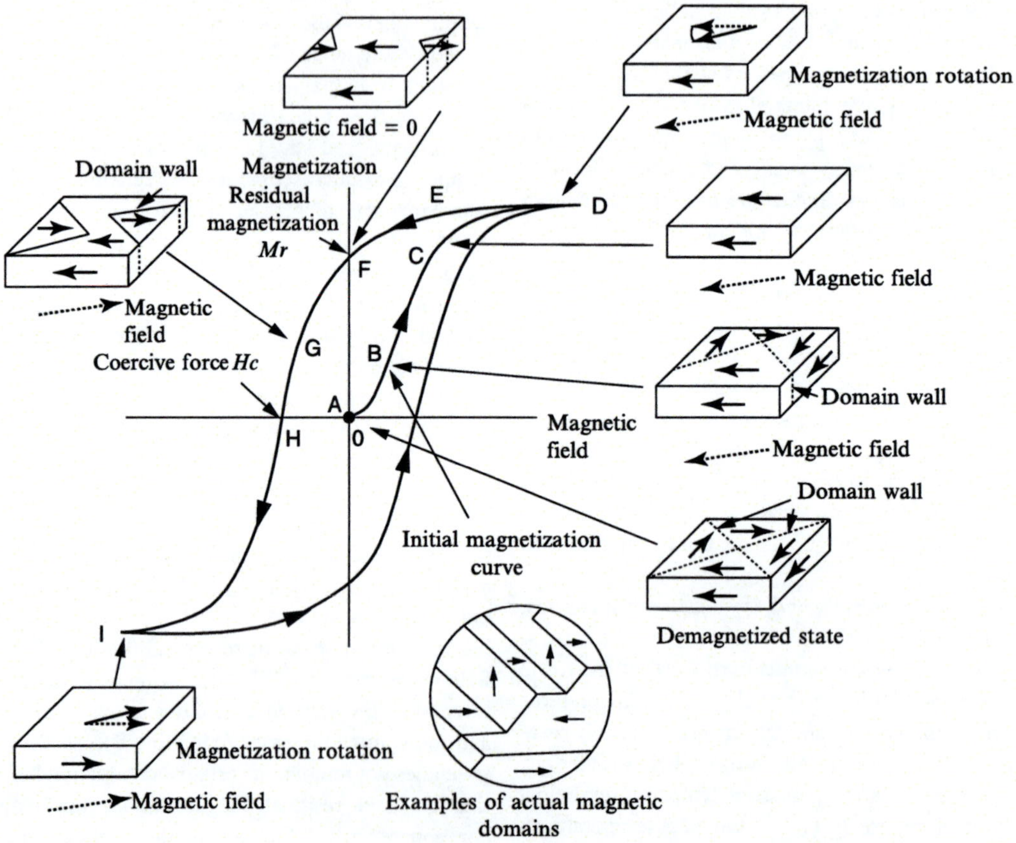

Fig. 5.11 Models showing magnetization curves and magnetized states of magnetic materials

magnet materials such as Nd–Fe–B have maximum magnetic energy products such as high as 450 kJ/m^3. Performance improvement of permanent magnet materials is directly related to miniaturization and performance improvement in motors.

Magnetic recording technologies that utilize the magnetic materials have progressed remarkably. They are capable of retaining residual magnetization state stably, realizing a recording density of 100 Gbit/in.^{2} (bit length 30 nm, track width 200 nm). Media created as a result of a number of efforts are used to realize high-density recording. For example, magnetic fine particles with a uniform size of 6 nm (with large magnetic anisotropy energy) are arranged in one direction.

Phenomena such as magnetostriction, where magnetic materials expand or contract depending on the magnetization direction, and the magnetoresistance effect, where electrical resistance changes, are observed. These phenomena are utilized in the generation of ultrasonic waves and magnetic field detection by actuators and magnetic heads.

The phenomena where light waves' polarization is transmitted or reflected by interaction of magnetic spin or magnetic waves in magnetic materials rotate are known as the Faraday effect or the Kerr effect. These characteristics are utilized in nonreciprocal circuit elements of isolators and circulators, which are the major application targets of magnetic materials. In optical magnetic recording, information is read based upon the detection of magnetization, which utilizes the Kerr effect.

In recent years, development of electronic devices that utilize electrical conduction that is dependent on the spin, which is called spin electronics (spintronics), is attracting attention. The technologies were made possible following the realization of distinguished control depending on spinning state (↑ spin and ↓ spin) through the miniaturization of devices. Magnetic heads for reading, which utilize the giant magnetoresistance effect (GMR) and tunnel magneto-resistance elements, along with magnetoresistive random access memory (MRAM), have already been commercialized.

As explained above, magnetic properties of magnetic materials are utilized directly in some cases, while interactions between magnetism and magnetic waves (or lights), between magnetism and current and between magnetism and elasticity are utilized in others. Antiferromagnetic materials have been attracting attention in recent years as materials that control magnetic properties of ferromagnetic layers by the magnetic interaction between thin magnetic layers. Their importance will further increase in line with the expected progress of spin electronics.

5.6 Optical Properties

Many ceramics transmit near-ultraviolet, visible and near-infrared light resulting in their incorporation in various applications. Transmission, absorption and reflection are basic interactions of light with a material and they vary depending on the wavelength of the light. Crystals produce characteristic colors such as red and green depending on the difference in absorption of visible light wavelengths.

5.6.1 Refraction of Light

Refractive index is an important element that expresses the nature of crystals that transmit light. Refraction is a phenomenon where the light is bent when it enters it from a certain medium (such as vacuum or air) into a crystal. The relationship between the refractive index of the medium n_i and that of the crystal *nr* is expressed as below:

$$n_i / n_r = \sin\theta_r / \sin\theta_i \qquad (5.1)$$

θ_i in the figure is the incident angle and θ_r is the refraction angle. On the other hand, when the light enters into vacuum or air from a crystal, a phenomenon called total reflection, where the light is totally reflected, is sometimes observed. This phenomenon occurs when sin θ_r in the formula above is 1 (θ_r=90°) (Fig. 5.12).

Reflection of light is always observed at the boundary of media with different refractive indexes. Light is composed of waves and can be weakened or strengthened by the superposition of waves. This is called interference of light. The interference effect and the reflection phenomenon of light are utilized in anti-reflection and high-reflection coatings. Reflected light is cancelled or strengthened by creating multilayer thin films, in which refraction changes layer by layer by controlling the thickness, and inducing interference of light at each layer boundary.

5.6.2 Anisotropic Crystals and Refractive Index

Light waves progress in a wave-like manner from a source. The amplitude of the wave oscillation varies in natural light. In linearly polarized light, the orientation of the oscillation is uniform (Fig. 5.13). Eye glasses that filter out the light reflected from the surfaces such as water or glass are made of polarized glass, which is based on the fact that reflected light contains a large amount of horizontal polarization.

In some crystals, properties related to light interactions depend on the oscillation direction of light. These crystals are called "anisotropic" crystals. The refractive index of these crystals changes depending on the oscillating direction of the transmitted light. Anisotropic crystals exhibit complex optical phenomena due to interaction of traveling direction of light, vibrating direction of light and crystal orientation. One of them is the birefringence. Birefringence is well known as the phenomenon where double vision of letters and figures occurs when they are viewed through crystals such as calcite. This phenomenon occurs because the crystal has different values of refraction depending on the direction of the crystallographic planes (Fig. 5.14). Polarizing elements are used to generate linear polarization and takes advantage of birefringence.

5.6.3 Wavelength and Refractive Index of Light

Refractive index changes depending on the wavelength of light. This phenomenon is called dispersion. The refractive index generally is higher as the light wavelength shortens. In some of the crystals that have anisotropic refractive index, dependency of absorption coefficient on wavelength changes

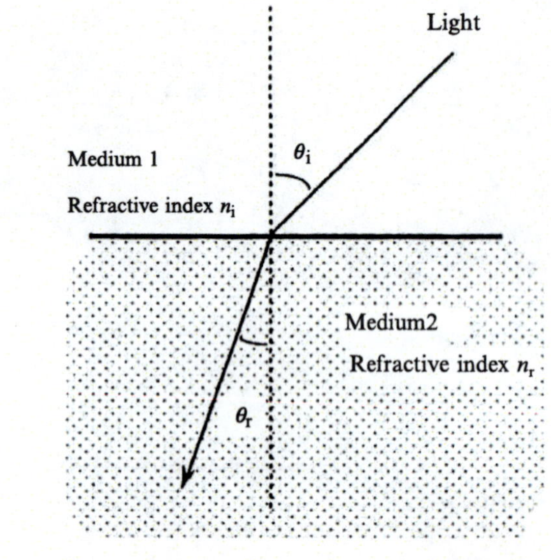

Fig. 5.12 Refraction of light

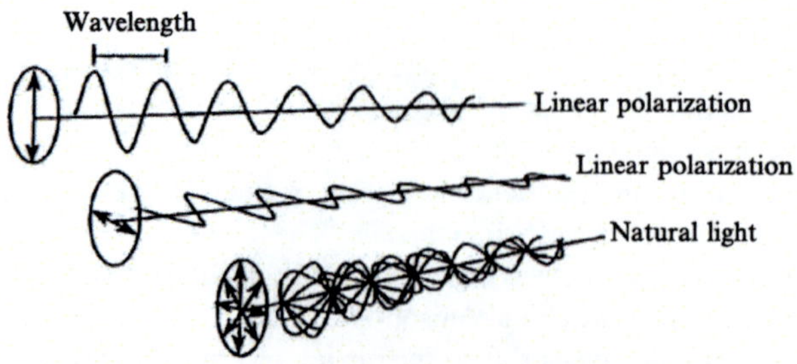

Fig. 5.13 Differences between natural light and linear polarization

depending on the vibration direction of light. Colors of these crystals change depending on the viewing direction due to the linear polarization of light. This is called dichroism.

5.6.4 Magnetic Field, Electric Field and Refractive Index

Refractive indexes of dielectric crystals change when an electric field is applied to them. Their optical properties then become more complex. The phenomena are referred to as the Pockels effect or the Kerr effect and the refractive index varies depending on the direction of the electric field applied to the crystal. The phenomenon where the direction of light is changed by the electric field application is utilized in optical shutters, etc.

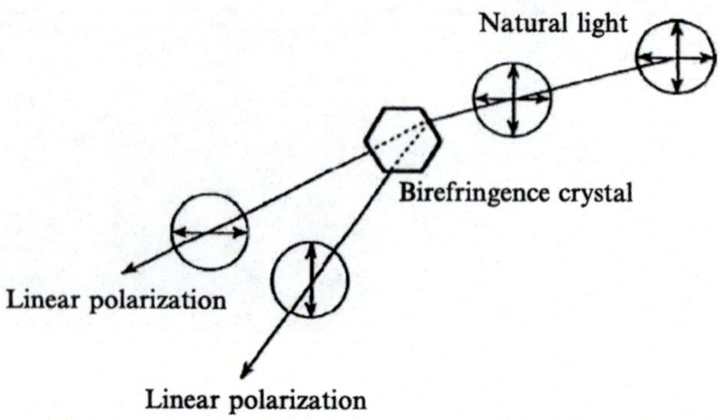

Fig. 5.14 Birefringence phenomenon

The photorefractive effect is observed in some unique crystals. It is a combined effect that involves the generation of electrical conductivity and a change in the refractive index, which are caused by the increase in the mobility of electrons following irradiation by light. If the crystal that is irradiated with light is uneven in terms of thickness, the mobility of electrons inside the crystal will be uneven, inducing charge distribution. Refractive index distribution that varies depending on the thickness of the radiated light is then observed in the crystal. The distribution is maintained for a certain period of time after irradiation ceases. Furthermore, if light is radiated, the light that was originally radiated can be regenerated. This phenomenon indicates that images are stored in the crystal and is expected to be applicable to elements for pattern storage.

Magnetic crystals also exhibit complex optical properties responding to interaction with a magnetic field. The phenomenon where the oscillating direction of the light wave that passes through a crystal in a magnetic field rotates in the direction of the magnetic field is called the "Faraday effect," while the phenomenon where birefringence is observed in the light that passes in the direction normal to the magnetic field is called the "magnetic Kerr effect."

5.6.5 Nonlinear Optical Phenomenon

Light is a kind of electromagnetic wave and electrons around the atoms of a crystal are vibrated by the electromagnetic field of light when the light passes through the crystal. As a result, the electrons emit light at various wavelengths in addition to the light wavelength of the incident light. Harmonics with frequencies two, three and more times of the incident light are important components. The ratio of harmonic component generation is expressed by the "nonlinear susceptibility."

5.6.6 Other Optical Properties

The properties explained above are passive, but some ceramics emit light or fluoresce when they are irradiated by light or electrons. The phenomena are observed when electrons of light emitting ions, provided with energy, return to their original condition (called excitation/relaxation process of an electronic state). In the case of fluorescent lamps, ultraviolet rays generated from mercury vapor excite electrons in certain elements of ceramic particles coated on the internal surface of the fluorescent lamp, and visible light is generated during the process of relaxation. With respect to laser crystals, electrons absorb excitation light of a certain wavelength and are stored in a high-energy state, and the electrons emit light of a uniform wavelength when they are released all at once.

Other optical properties include photochromism where colors are changed by radiation of light, color center formation and rotatory power of crystals.

Literature

1. Nishida T, Yasuda E (1986) Dynamic property evaluation of ceramics. Nikkan Kogyo Shimbun, Tokyo [in Japanese] **(5.1)**
2. Nakazawa H, Kobayashi H (1976) Strength of solid bodies. Kyoritsu Shuppan, Tokyo [in Japanese] **(5.1)**
3. Okamura H (1976) Introduction to linear fracture mechanics. Baifukan [in Japanese] **(5.1)**
4. Nakamura T (1984) Ceramics and heat. Gihodo Shuppan, Tokyo, pp 1–101 [in Japanese] **(5.2)**
5. The Japan Society of Thermophysical Properties (ed) (1990) Thermophysical property handbook. Yokendo, Tokyo [in Japanese] **(5.2)**
6. Kittel C (2005) Introduction to solid state physics, 8th edn. **(5.2)**
7. Sakka S (ed) Dictionary for understanding of ceramics. Agne, Tokyo, p 337 [in Japanese] **(5.3)**
8. Mook NK (1982) 12 new material encyclopedia. Nikkan Kogyo Shimbun, Tokyo, p 88 [in Japanese] **(5.3)**
9. The Chemical Society of Japan (ed) (1984) Designing of functional ceramics. Japan Scientific Societies Press, Tokyo, pp 25, 76 [in Japanese] **(5.3)**
10. Kreuer KD (2002) Chem Phys Chem 3:771 **(5.4)**
11. Supervised by Tamura H, Sato M (2003) Next-generation lithium secondary battery, NTS, Tokyo, p 312 [in Japanese] **(5.4)**
12. Chikazumi S. Physics of magnetic materials, Shokabo [in Japanese] **(5.5)**
13. The Laser Society of Japan (ed) (1982) Laser handbook. Ohmsha, Tokyo [in Japanese] **(5.6)**

Part II

Information Technology and Electroceramics

Portable Mobile Terminals and Information Appliances

6

Mobile communication service began in December 1979, in Japan, ahead of the United States and European countries. The automobile telephone service provided by the Nippon Telegraph and Telephone Public Corporation (currently NTT), targeting the 23 wards of Tokyo, was the first service of its kind. The mobile communication service originally adopted was the FDMA system (Note 6.1), an analog system that utilizes the 800 MHz range, but was transformed later to adopt the TDMA system (Note 6.2), a second-generation digital system that can effectively utilize higher frequencies, responding to an increase in the number of subscribers. Today, the TDMA system is giving way to the CDMA system (Note 6.3), a third-generation system that uses higher frequency effectively and enables high speed data transmission.

In this section, various kinds of passive components will be introduced used in mobile phones, the discrete ceramic parts used in information appliances, and various active components such as piezoelectric gyros as well as passive components used in mobile phones, including the SAW (Surface Acoustic Wave) filter (Sect. 6.1), the TCXO (Temperature Compensated X'tal Oscillator) (Sect. 6.3), the dielectric filter (Sect. 6.4), the isolator (Sect. 6.6), the chip capacitor (Sect. 6.7) and the noise reduction sheet (Sect. 6.12) will be discussed. In order to show the outline of functions of these parts, a block diagram of the W-CDMA system is shown in Fig. 6.1. The illustration in Fig. 6.2 shows a circuit substrate equipped with these parts.

In Fig. 6.1, a dielectric duplexer, a combination of a receiving filter and a transmission filter, is placed directly below the antenna. Weak radio waves received by the antenna are amplified by a low noise amplifier (LNA) via a dielectric filter on the receiving side. The SAW filter placed in the subsequent stage of the LNA supports the dielectric duplexer by supplementing its attenuation performance. It also functions as a band-pass filter that attenuates signals leaking out of the mixer in the subsequent stage.

In Fig. 6.1, the received frequency is converted by a two-step down converter to a low frequency that can be processed by the baseband filter. The received frequencies in the GHz band are converted to the initial intermediate frequencies in the 100 MHz range by mixing with frequencies from a local oscillator in the initial stage by means of a mixer, followed by a conversion to lower frequencies, the secondary intermediate frequencies, in the secondary stage. The SAW filter positioned between the initial stage mixer and the secondary stage mixer is a 100 MHz bandpass filter that sorts out the initial intermediate frequencies only. Frequency signals transmitted to the local oscillator need to be temperature-stable, and the Temperature Compensated X'tal Oscillator (TCXO) is used to supply temperature-stable reference signals. The transmit signals moves in the opposite direction through the line at the bottom of Fig. 6.1. The isolator, which is mounted to prevent part of the transmit signals from being reflected by the antenna port and returning to the power amplifier, is necessary to protect the power amplifier by reducing its output load fluctuations. The LNA and mixer on the receiving side as well as the power amplifier and mixer on the transmitting side, for example, of the block diagram, are, respectively, composed of a single IC chip. A large number of chip capacitors are mounted around these IC chips for decoupling and smoothing. Meanwhile, the noise reduction sheets are mounted randomly on the substrate so as to effectively reduce noise.

Note 6.1 *FDMA: Frequency Division Multiple Access.* A communication method in which a channel is divided into multiple bands and each band is allotted to individual channels, allowing multiple users to communicate simultaneously.

Note 6.2 *TDMA: Time Division Multiple Access.* A communication method in which one channel is divided into different time slots and allotted to multiple users. In this method, frequencies can be utilized effectively because one channel can be shared by multiple users.

Note 6.3 *CDMA: Code Division Multiple Access.* A communication method in which individual codes are used to distinguish users who share the same channel. The voice data, etc. of users is coded by spectrum spread, featuring high efficiency in frequency utilization and eavesdrop resistant properties.

Y. Imanaka et al. (eds.), *The Ceramic Society of Japan, Advanced Ceramic Technologies & Products*,
DOI 10.1007/978-4-431-54108-0_6, © Springer Japan 2012

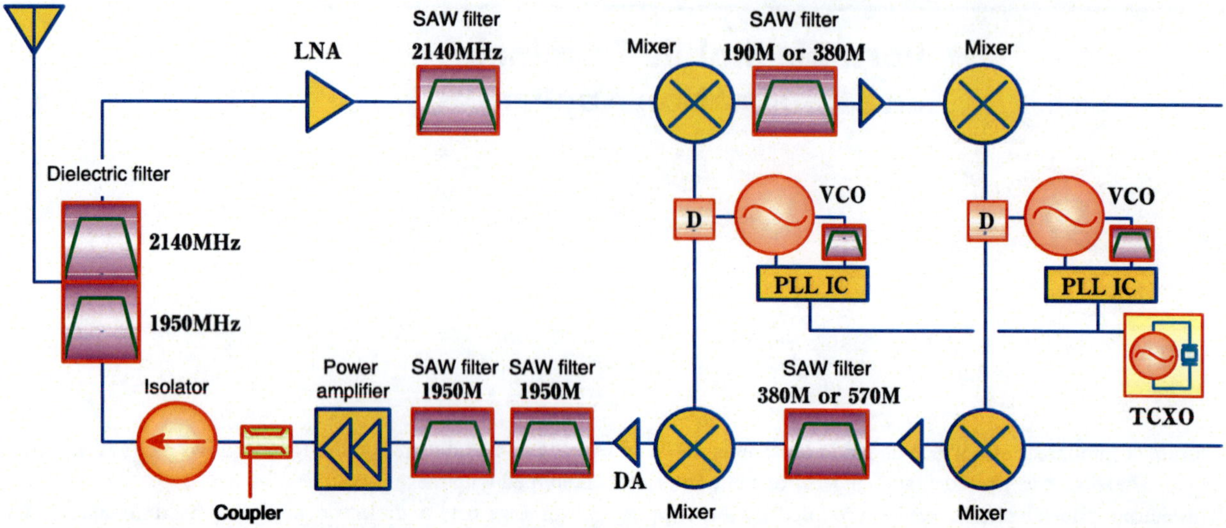

Fig. 6.1 Example of mobile phone block diagram

Details of individual ceramic parts that are used in devices such as information appliances are also explained in this section. The multilayer chip LC filter (Sect. 6.5) is a multilayer part, as the name indicates, created by integrating an inductor and a capacitor. It was developed by employing the world's first technology in which a dielectric ceramic, screen-printed with magnetic ferrite and silver conductor were printed, sintered, and then annealed using co-firing technology (1983). Multilayer LC filters were used in various electronic devices as signal-processing filters in the 1980s, when they were first developed. The demand for multilayer LC filters as signal-processing devices declined following recent trends in electronic devices towards higher frequencies and high-performance digitalization. They are currently used in a wide variety of electronic devices for EMC (Electro-Magnetic Compatibility).

The FeRAM (Ferroelectric Random Access Memory) (Sect. 6.2) is currently used in IC cards, electronic tags, and other equipment as a nonvolatile memory that utilizes ferroelectric PZT polarization. It is expected to be developed and progressed as a type of memory for energy-saving quick-on computers in which data is not lost when the power is turned off and the operation is resumed quickly when the power is turned back on.

Ceramics are used in many home appliances such as in grain insulation type barrier layer capacitors (Sect. 6.8) and cylindrical ceramic capacitors/(surface mount) winding inductors (Sect. 6.9). These ceramic capacitors are currently represented by multilayer ceramic condensers. However, barrier layer capacitors and cylindrical ceramic capacitors, featuring stable dielectric characteristics and superior frequency characteristics, are widely used in many home appliance products. The winding inductor is the most basic inductor, manufactured by winding copper wire, as the name indicates. The characteristics of these inductors are controlled by the shape of the ferrite core material and the total number of copper coils, and therefore, the stabilization of material properties as well as an improvement in dimension accuracy after sintering are required. Many home appliances have been refined in accordance with the needs of customers through additions that include miniaturization and improvements in functionality. Responding to this trend, chip-type components that satisfy the needs for miniaturization and high-density packing are being developed, targeting nearly all electronic components (Sect. 6.10). In developing chip-type electronic components, cofiring with the electrode material is a necessity, and therefore the processing technology for temperature and atmosphere control during cofiring is important. Along with the mass production of SMDs (surface mount devices), such know-how is fully utilized for the development of electronic components for electronic devices represented by mobile phones and PCs.

A variety of appliances, such as ceramic fan heaters and bathroom driers, utilize heat. The PTC (positive temperature coefficient) thermistor utilizes the resistance temperature characteristic of ceramics, in which the resistance drastically changes at the Curie point. It is widely used for temperature detection, temperature compensation and overheat detection. The PTC thermistor (Sect. 6.11), with unique volt–ampere and ampere-time characteristics, is also adopted for use in many electronic devices.

Gyro sensors incorporating piezoelectric ceramics (Sect. 6.13) are mainly used in navigation systems for detecting

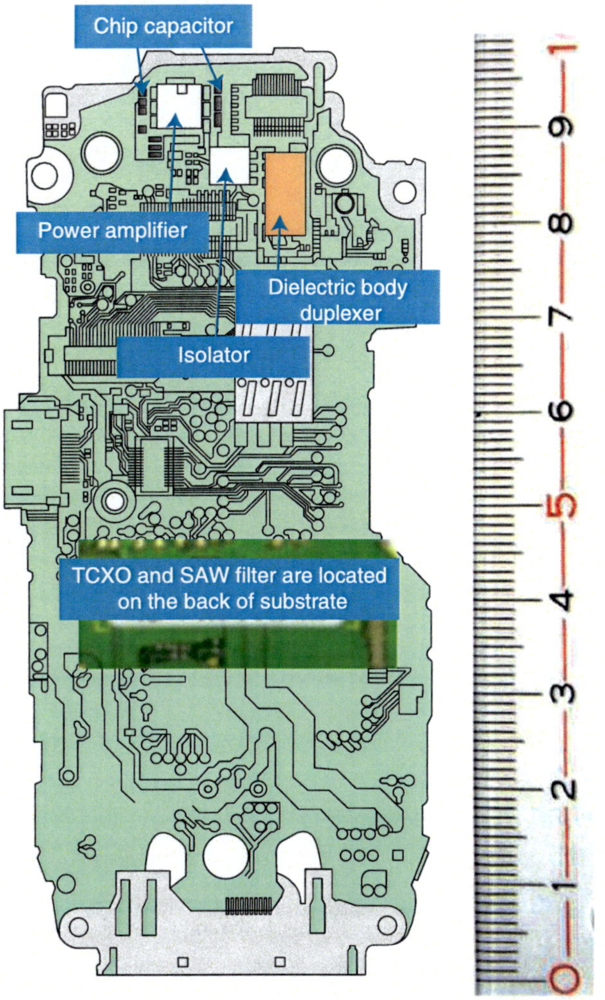

Fig. 6.2 Illustration showing a circuit substrate for mobile phone

driving path and in cameras for detecting destabilization. Gyro sensors that incorporate ceramics are effective in miniaturization and cost reduction. The gyro sensors detect angular velocity, etc. through voltage conversion of the Coriolis force generated by moving and rotating parts that are mounted with piezoelectric driving elements. In respect to the structure, a number of types such as the tuning fork type and the bimorph type have been planned and developed. Recently, gyro sensors incorporating silicon MEMS (Micro Electro Mechanical Systems) technology have been attracting attention, promoting the compartmentalization of sensors.

The ultrasonic motor (Sect. 6.14) is an actuator that utilizes a piezoelectric element to convert electric energy into mechanical energy. It is superior to magnetic motors because of its many advantages such as a high torque in low speed, quietness, and an absence of magnetic noise. The ultrasonic motor, which was first used in the early 1990s to drive an auto-focus lens for single-lens reflex cameras, is currently used in a variety of devices. Cost reduction and the utilization of lead-free materials are expected to be promoted in the future.

The multilayer ceramic speaker (Sect. 6.15) was developed from an oscillator manufactured by using the ceramic piezoelectric transducer made of PZT. It has been used since the mid 1970s in all types of electronic devices for such purposes as generating the beeping tones of AV devices and telephones as well as the alarm tones of automobiles. The structure was maximized to realize high sound pressure levels and high definition sound by changing the dielectric body to a multilayer structure so as to enhance the oscillation energy. It is expected to be incorporated into portable devices such as mobile phones, which will be further miniaturized and improved in functionality.

SAW Filters (1989)* 6.1

A variety of radio frequency (RF) filters, which suppress interfering waves and transmit electrical signals, are used in mobile phones. The RF filters require high performance and low signal-loss to effectively utilize the input waves. At the same time, they need to be compact, thin and low cost for use in mobile phones. The Surface Acoustic Wave (SAW) filter satisfies all of these requirements and is currently used in almost all mobile phones. In order to electrically use the SAW, interdigitated electrodes are used for signal propagation and detection. Electrodes such as Al films are commonly employed on a piezoelectric single crystal such as $LiTaO_3$, which effectively converts electrical and mechanical vibrations, by using semiconductor lithography based integration technology (Note 6.4), hermetically sealing it in a ceramic package for use.

6.1.1 Background of Development

Mobile phones have been rapidly spreading throughout the world since 1989, when the first pocket-size mobile phone was released. This rapid growth was supported by the miniaturization and functional improvements in the components as well as the longer battery life. SAW filters are used in mobile phones as RF filters and the intermediate frequency (IF) filters (Note 6.5). They have been widely used as a key component in the miniaturization of mobile phones because they are compact and thin, as shown on the right of Fig. 6.3. The RF filters in mobile phones are used to cut unwanted frequency components or noise from the assigned transmission frequency. The transmission signal needs to be utilized efficiently, and the SAW filter, featuring a low-loss and rapid cutoff performance, is suitable for use in mobile phones. Approximately 4–5 SAW filters are used in a single mobile phone at present.

6.1.2 Characteristics

6.1.2.1 Characteristics of the Device

Elastic waves are converted into a mechanical wave that propagates along the surface of a solid during their propagation. The SAW filter is a passive component that effectively utilizes this physical phenomenon. As shown in Fig. 6.4, interdigitated electrodes, having an electrode finger cycle of λ, are formed by the patterning method using a thin film of a light metal such as Al on the surface of the piezoelectric substrate. The SAW is excited and received by employing the piezoelectric effect of the substrate, enabling electrical input, mechanical wave propagate and electrical output. Through the series of operations of SAW excitation and reception by the interdigitated electrodes, only the frequency components f_o expressed in the formula below are selected, which results in the filtering function.

$$f_0 = V / \lambda$$

The V indicates the acoustic velocity of the piezoelectric substrate. The SAW wavelength λ is extremely short and about 100-thousandths of the wavelength of magnetic waves. Therefore, compact, thin filters with sharp skirt characteristics can be created by utilizing mechanical vibrations generated by the piezoelectric single crystal thus resulting in high Q

*The number in parenthesis indicates the year that the product was first commercialized.

Note 6.4 Lithography is a type of semiconductor manufacturing technology. A technology to form fine patterns on a wafer by a series of processes including film formation, resist coating, exposure, development, etching and resist removal. This technology was developed for semiconductor production, but is also used in other fields today.

Note 6.5 IF stands for intermediate frequency. For wireless transmission and reception of signals, signals in the baseband need to be modulated to enhance the frequency to the radio frequency. As an intermediate step in this process, the signals are often converted to an intermediate frequency. The channel signals alone are filtered by the IF filter.

Fig. 6.3 SAW filter for mobile phones. The photograph on the *left* shows a piezoelectric substrate, on which the SAW filter is formed by patterning a Al film using lithography. Shown on the *right* are the final products mounted in a ceramic package. These products, exhibit high performance, compact and thin structure and low price, are used as key devices in mobile phones

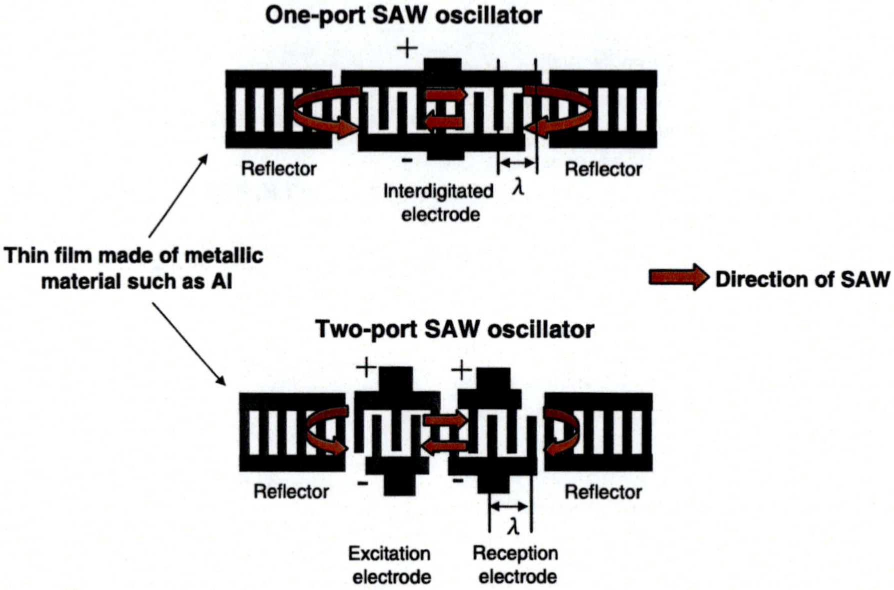

Fig. 6.4 SAW filter electrode structure of used as RF filter. Patterns are formed on the piezoelectric crystal substrate by using Al or such like. Interdigitated electrodes are designed based on a structure incorporating oscillators, which provide excellent filtering performance with low loss

values (Note 6.6). The structure of electrodes used in a SAW filter that is used as an RF filter is shown in Fig. 6.4 as an example. The structure of the IF filter is substantially different from that of the RF filter, but it will not be explained here. As shown in the figure, the RF filter, featuring a resonate structure that enables effective utilization of the SAW energy, has low-loss and sharp cutoff characteristics. The figure shows one-port type and two-port type resonators.

Note 6.6 The Q value is a dimensionless parameter that describes how under-damped an oscillator or resonator is and it characterizes a resonator's bandwidth relative to its center frequency.

The one-port resonator is constructed in the shape of a ladder to create bandpass characteristics, while the two-port resonator has bandpass characteristics inherently in its design. These two types have advantages and disadvantages and are chosen depending on their intended purpose.

6.1.2.2 Manufacturing Method

The SAW filter is manufactured by simply forming interdigitated patterned electrodes of Al or Al alloy film on a piezoelectric single crystal made of materials such as $LiTaO_3$ (lithium tantalate), $LiNbO_3$ (lithium niobate) and crystal. The piezoelectric single crystal and a wafer, which is made by slicing the piezoelectric single crystal and forming interdigitated patterns of Al or such like on it, are shown on the left in Fig. 6.3. Four-inch wafers are currently the mainstream. The SAW filter chip is as small as 1×1 mm, and therefore several thousands of chips can be produced from a 4-inch substrate. The interdigitated electrode, featuring cycles of 2–5 μm and a minimum pattern width (resolution) of 0.5–1.2 μm, can be manufactured using general semiconductor manufacturing equipment. Because they can be mass-produced using semiconductor manufacturing equipment, the products can be offered at competitive prices. As the final step, the chips are sealed hermetically in a compact ceramic package so as to prevent them reacting with ambient air that contains moisture and oxygen. Figure 6.5 shows a cross-sectional view of the chips in a ceramic package, and a photograph of a package that is actually offered as a product is shown on the right side in Fig. 6.3. The package is the surface-mounted leadless type.

6.1.2.3 Performance

The SAW filter has excellent performance in a range of frequencies, with a center frequency f_0 ranging from 800 MHz to 2.5 GHz, a bandpass width ranging from 2% to 3% in fractional bandwidth ($\Delta f/f_0$), an insertion loss of 2–3 dB and an out-of-band suppression of 30–60 dB. A transmission power up to 1 W can be achieved by using a special electrode material. For these reasons, filters that meet almost any specification can be produced.

6.1.3 Future Prospects

Billions of filters are currently produced throughout the world owing to the spread of mobile phones. The number of filters used in one unit has been increasing in line with the shift to multiband components, and needs for miniaturization and cost reduction remain. The filters have evolved from a wire-mounted type to a flip-mounted type that incorporates bumps, as shown in Fig. 6.5. For miniaturization, filters that are almost equivalent to the chip in size, with less wasted space, as shown in Fig. 6.6, have been developed recently. SAW filters with higher performance and others in which the wafers are sealed as one unit for scaling and cost reduction will be developed in the future.

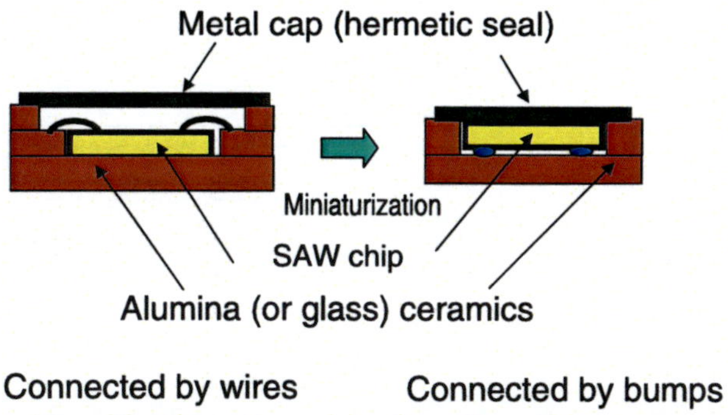

Fig. 6.5 Structure of SAW filter package. The ceramic package is hermetically sealed because the SAW chip is susceptible to ambient temperature. Recently, the wire bonding method has been replaced by the solder bump-connection method due to miniaturization

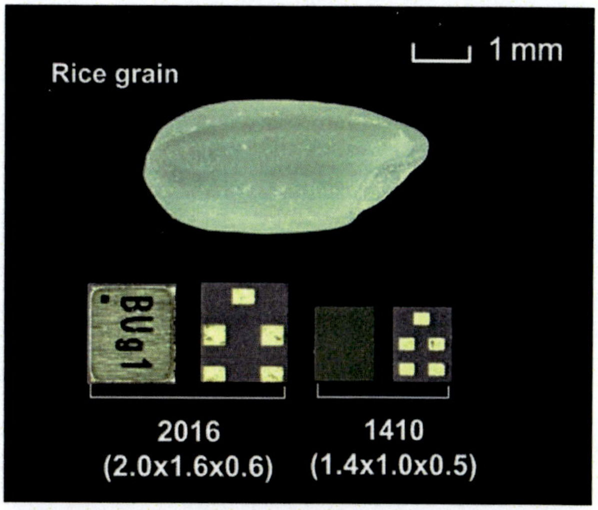

Fig. 6.6 Scaling of SAW filter. Miniaturization and cost reduction of SAW filters have been achieved, and a filter smaller than a rice grain was developed recently. This miniaturization is expected to continue

FeRAM (1999) 6.2

In ferroelectrics (Fig. 6.7), an internal polarization initially generated by applying an external electric field exists even after the electric field is removed. Ferroelectric memory (FeRAM: Ferroelectric Random Access Memory) is produced by utilizing this characteristic. FeRAM not only retains memory even after the power is turned off (nonvolatility) but also allows high-speed writing of information, making the frequency of rewriting extremely high. FeRAM (Fig. 6.8) is used in IC cards, and electronic tags among other things, taking advantage of these superior characteristics.

6.2.1 Background of Development

Ferroelectrics were first developed for use in semiconductor memories as early as in the 1950s but reliability issues impeded its commercialization until the 1980s when further development (Note 6.7) by a U.S. venture firm lead to a commercial product. It was about 10 years later, that a Japanese company started large-scale mass production of FeRAM. Currently, over one billion FeRAM chips are currently in the market after more two decades of mass production.

6.2.2 Characteristics

6.2.2.1 Comparison of Characteristics with Leading Semiconductor Memories

A comparison of FeRAM characteristics with various leading semiconductor memories is shown in Table 6.1. Nonvolatile memories, such as Flash memory and EEPROM, require high voltage and are slow to write data, and the data rewriting frequency is comparatively slow in these memories. Compared with these nonvolatile memories, FeRAM allows high-speed data writing, requires lower voltage for rewriting and is fast in rewriting. For these reasons, EEPROM, SRAM and DRAM, which have each been used for their differing memory characteristics, can be replaced by a single FeRAM.

6.2.2.2 Application Field of the Product

FeRAM has been used for data retention during power-off because it is nonvolatile and its data rewriting frequency is much higher than that of FLASH and EEPROM. Previously, the battery-mounted SRAM (battery backup SRAM) was used for this purpose. Utilization of FeRAM eliminates the need to use batteries and the cost of mounting batteries. Figure 6.8 shows pictures of 1 Mbit and 256 Kbit FeRAM chips used for this purpose.

Other products that take advantage of FeRAM characteristics include IC cards (Fig. 6.9), electronic tags and mobile phones. These products need to be compact and lightweight, and therefore, it is beneficial for them to use simple-structured circuits. FeRAM can be used both as a nonvolatile memory and a high-speed processing memory, making it possible to simplify the circuit structure. FeRAM consumes less power because it does not require the booster circuit that EEPROM and FLASH memory require, realizing a longer communication distance in RFID applications.

The usage of electronic money that utilize credit card with IC chips and mobile phones has been rapidly increasing. Network security is indispensable for the usage of these functions. To assure security, data needs to be protected with an advanced encryption technology. FeRAM is nonvolatile and at the same time is capable of rewriting data at high speeds. Therefore, devices incorporating FeRAM can easily carry out high-level encryption processing with enhanced security.

Note 6.7 FeRAM or F-RAM (also called FRAM) is a ferroelectric memory.

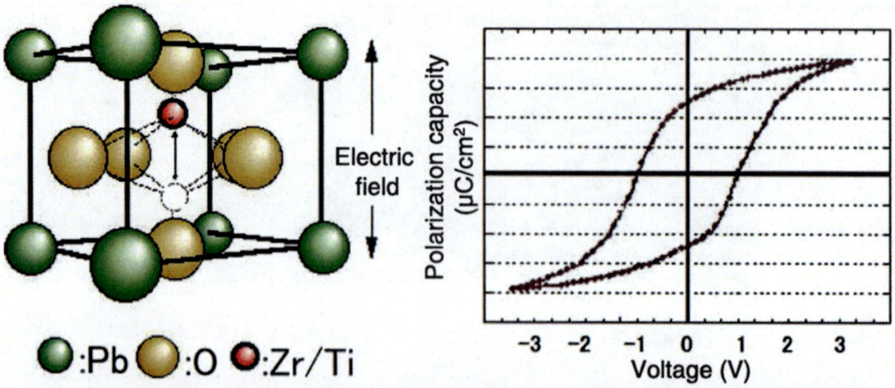

Fig. 6.7 Crystal structure and polarization principle of lead zirconate titanate (PZT) (*left*), and crystallization/voltage characteristics of ferroelectric (*right*). (1) The polarization occurs when an electric field is applied (the Zr/Ti element has two occupancy sites and moves to one of the sites depending on the direction of electric field application). (2) Polarization remains even after the external electric field is no longer applied. (3) The two atomic sites can be used to store data such as "0" and "1" data, making it nonvolatile

Table 6.1 Comparison of characteristics of types of semiconductor memory

	FeRAM	FLASH	EEPROM	SRAM	DRAM
Type of memory	Nonvolatile	Nonvolatile	Nonvolatile	Volatile	Volatile
Write time (write cycle)	50–180 ns	1 s	10 ms	10–70 ns	60 ns
Read time (access cycle)	50–180 ns	90 ns	200 ns	10–70 ns	60 ns
Write voltage	1.8 V	14 V	14 V	1.8 V	1.8 V
Rewrite frequency	10 billion to 1 trillion cycles	100,000 cycles	100,000 cycles	Infinitely often	Infinitely often

ns nano second, *FLASH* batch-erase type nonvolatile memory, *EEPROM* nonvolatile memory, *SRAM* on-demand writing, high-speed memory, *DRAM* on-demand writing, high-integration memory

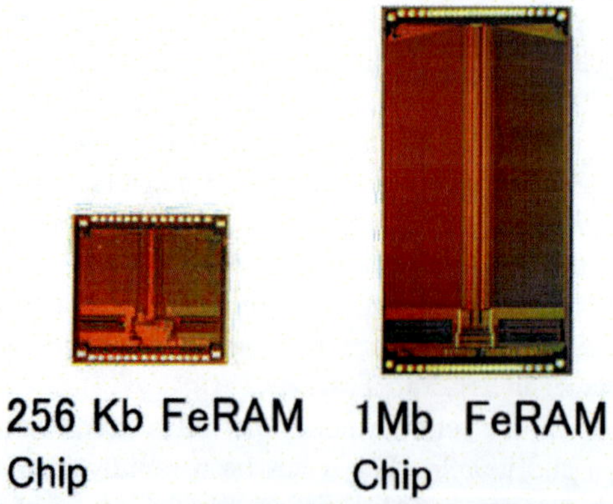

Fig. 6.8 Pictures of mass-produced FeRAM chips. Both of the chips use the PZT ferroelectric capacitor. They are widely used for data retention during power off, etc.

6.2.2.3 Production Process

The semiconductor production process generally involves integration of various circuit components such as a transistor, resistor and condenser to an Si substrate (called the bulk process) and the process of connecting elements with aluminum wires (called the wiring process) (Fig. 6.10). For the production of FeRAM, the process of building the ferroelectric capacitor is added between the semiconductor component process and the metallization process. Therefore, FeRAM production process has been adapted to manufacturing process of traditional semiconductors, making it comparatively easy to produce.

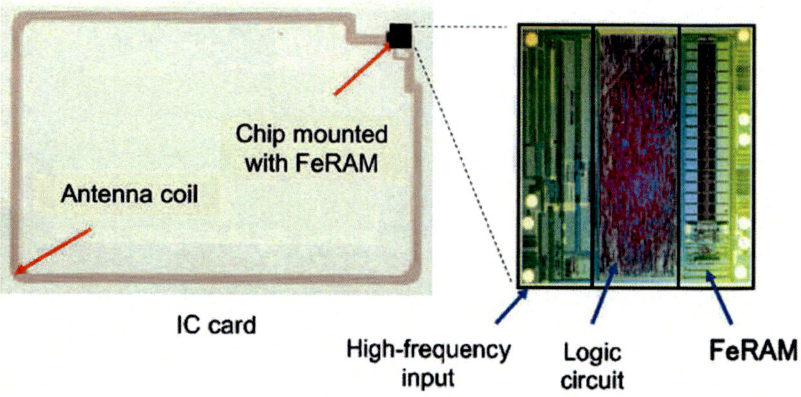

Fig. 6.9 Noncontact IC card and FeRAM-mounted chip. In the noncontact IC card, the power and signals are supplied from an antenna wire placed around the edge of the card to the semiconductor chip, which logically processes the information and outputs the information through the antenna. The FeRAM, a nonvolatile memory featuring a fast rewriting speed and low power consumption, has been used widely in IC cards in recent years

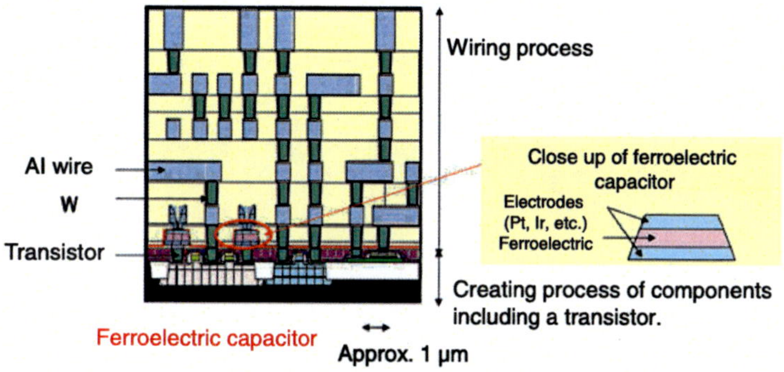

Fig. 6.10 Cross section structure of FeRAM. In order to manufacture the FeRAM, components such as a transistor are created on the Si substrate first, and then the ferroelectric capacitor is created. The components are electrically connected by Al wires, etc. to produce the FeRAM

Currently, ferroelectrics made of lead zirconate titanate [PZT: Pb(Zr, Ti)O_3] (Note 6.8) or bismuth-layered oxide (represented by SBT: $SrBi_2Ta_2O_9$) are mainly used in FeRAM. The former is high in polarization capacity, while the latter is high in rewriting frequency, although the amount of stored charge or polarization capacity is low. The capacitor thin film is produced by the sputtering method when PZT is used, and produced by the chemical solution deposition method when STB is used. The organic chemical vapor deposition method, which can create capacitors with thinner films as well as 3D structures, was developed in recent years to replace the above mentioned deposition methods.

6.2.2.4 *Reliability*

Data retention characteristics are vital to nonvolatile memory. FeRAMs at the early stage of development were insufficient in data rewriting and reading performance. However, they were improved substantially to attain a retention performance sufficient for practical use through modification of the circuit and process optimization of the ferroelectric thin layers.

The rewriting frequency is also important for FeRAM. According to the specifications in current product catalogs, the frequency is approximately 10 billion cycles (10^{10} cycles), which is much less than volatile types of memory such as DRAM. However, some FeRAMs that incorporate improved electrodes and the organic chemical vapor deposition method have endurance exceeding 10^{14} read/write cycles making them comparable to non-volatile memories such as DRAM or SRAM.

6.2.3 Future Prospects

Overall FeRAM has many excellent characteristics that had never been observed in a traditional semiconductor memory. However, it has yet to fully penetrate the market, hindered by a lack of applications that utilize its unique advantages, and an immature design technology base that lacks sufficient understanding of FRAM characteristics, physics and degradation mechanisms. In addition, miniaturization of the FeRAM capacitor are needed to produce higher density memory devices requested by the marketplace. It is expected that products that effectively utilize the unique characteristics of the FeRAM will need to be developed before FeRAM will become a mainstream memory.

Note 6.8 EU countries began enforcing the RoHS regulations, which basically prohibits the use of lead in electrical and electronic devices, from July 1, 2006. PZT-based materials contain around 60 wt% of Pb as a main constituent element, but the materials are allowed to be used at present because the RoHS indicates that "lead used in electronic ceramic parts is excluded from the subjects of the RoHS order." The RoHS order is reviewed every 4 years and a great amount of research and development is being conducted on lead-free materials.

TCXO (Temperature Compensated X'tal Oscillator) (1994) 6.3

Temperature Compensated X'tal Oscillator (TCXO) is a quartz oscillator that was designed to provide extremely stable operation by incorporating a temperature compensation circuit that compensates for frequency-temperature characteristics peculiar to the quartz oscillator. The temperature compensation circuit was incorporated because temperature dependency is the most influential factors for frequency stability. The TCXO stably operates at frequencies over a wide range of temperature fluctuations and is used as a reference signal source for time in electrical devices and communication devices.

6.3.1 Background of Development

In Japan, mobile phones, which were previously available only on a rental basis, started to be offered for sale in 1994. As a result, they came into widespread use, and the demand for TCXO greatly expanded. The TCXO was 0.4 cm^3 in size in those days, but has since shrunk to 0.008 cm^3 in line with improvements in performance, diversification of functions, miniaturization, and a reduction in the weight of mobile phone units. Figure 6.11 shows the progress in the miniaturization of the TCXO.

6.3.2 Characteristics

6.3.2.1 TCXO Types and Specifications

Crystals with Q-values of several million, provides extremely high oscillation stability compared with LC resonance circuits, which only exhibit Q-values in the range of hundreds. TCXOs can be largely classified into two types, the "analog type" and the "LSI type (Note 6.9)". This classification was made based on the components mounted on the TCXO and the temperature compensation method adopted by the TCXO.

Both the analog type and the LSI type utilize control signals arising from the inverse temperature characteristics peculiar to the quartz oscillator to cancel the temperature characteristics, realizing stable frequency-temperature characteristics. In the analog type, the quartz temperature characteristics are compensated for by components such as the thermistor, condenser, resistor, and transistor. An example of the analog type is shown in Fig. 6.12. In the LSI type, the quartz temperature characteristics are compensated for by a single-chip LSI equipped with a memory element. A circuit block of the LSI type is shown in Fig. 6.13.

The analog type has advantages such as low phase noise and low power consumption. However, it is comprised of many elements, making it difficult to miniaturize, and its temperature compensation range is −30 °C to 80 °C, which is narrower than that used in LSIs. The LSI type, which is composed of the LSI and quartz oscillator only, can be miniaturized and made thinner. It has several disadvantages including inferior phase noise characteristics (Note 6.10) and a high equipment cost, which leads to higher product prices.

The LSI type is in a mainstream component at present, perhaps due to its preferred use in compact devices. However, the analog type is utilized when priority is placed on phase noise. Table 6.2 shows the standard electrical specifications of the analog type and the LSI type. In the LSI type, which incorporates a stabilized power supply, fluctuations in the oscillation frequency and output amplitude caused by power fluctuation are extremely low. The LSI type deteriorates about 10 dBc/Hz (per 1 kHz offset) compared with the analog type, due to the noise component of the control voltage applied to the variable capacitance that controls the frequency and the noise component of the oscillation circuit.(Fig. 6.14)

Note 6.9 LSI = Large Scale Integration

Note 6.10 Noise factors at output signals around oscillation frequencies. Communication performance is influenced by the characteristics of this factor.

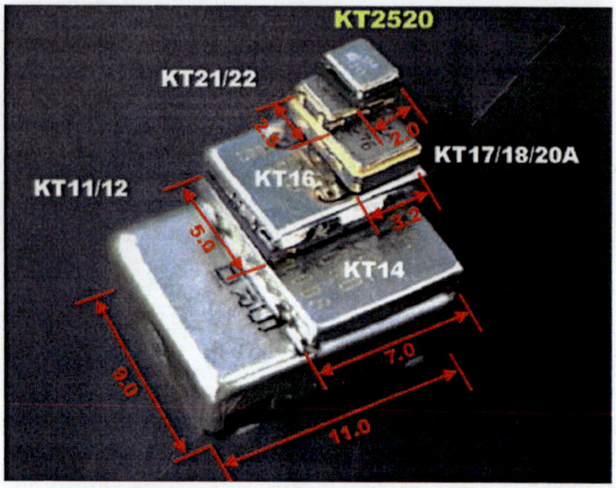

Fig. 6.11 Transitions in TCXO sizes. The TCXO continues to be shrunk responding to an increase in functionality and a miniaturization of mobile phones. In the process of downsizing, the length of the long side of the next TCXO product was shortened approximately to the length of the short side of the previous product

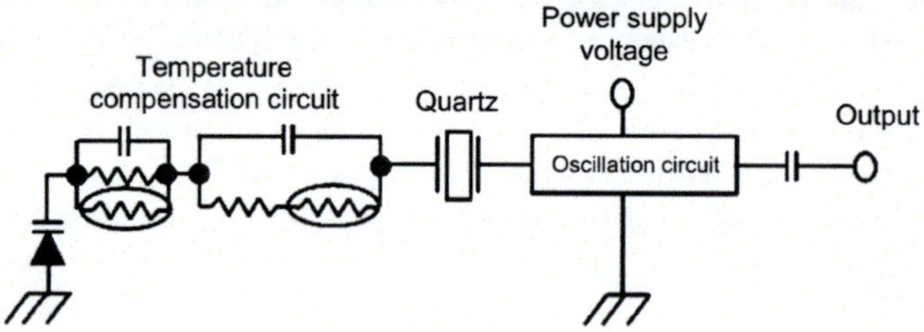

Fig. 6.12 Example of an analog circuit. The circuit is composed of a resister, a condenser, a thermistor, and a transistor. It directly changes the load capacity of the quartz oscillator and varies oscillation frequency with temperature compensation

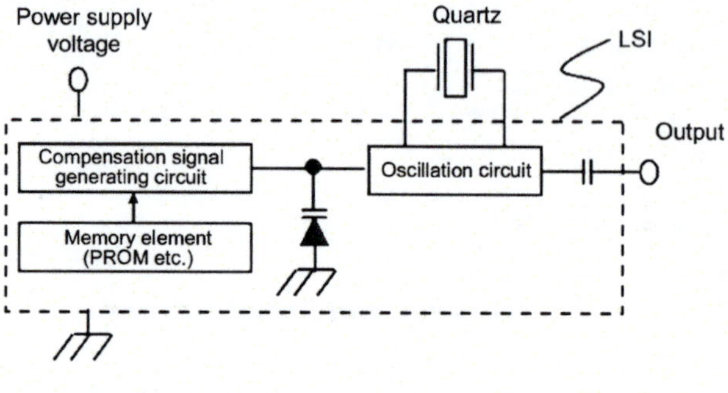

Fig. 6.13 Circuit block of the LSI type. It uses a single chip IC mounted with a temperature compensation signal, an oscillation circuit, a memory element, and a temperature sensor. The circuit supplies a temperature compensation signal voltage to the variable capacitance diode and indirectly varies the frequency for temperature compensation

Fig. 6.14 Comparison of phase noise characteristics between analog type and LSI type. The analog type, which does not use CMOS, has superior phase noise characteristics

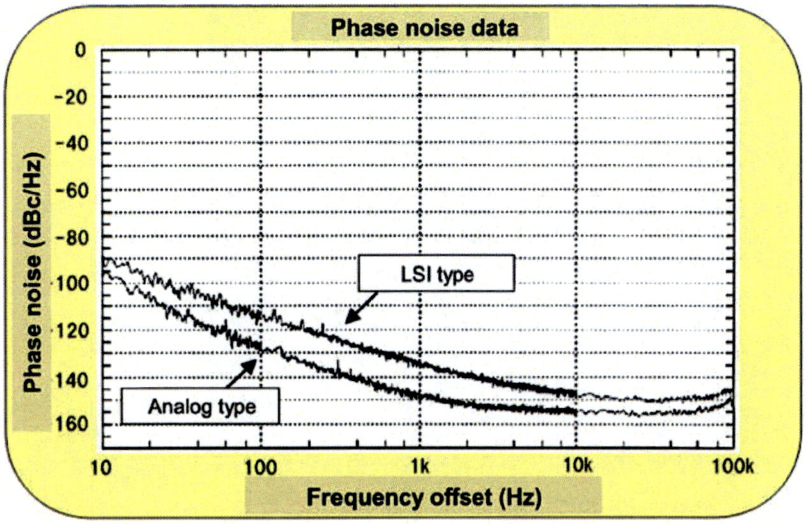

Table 6.2 Standard electrical specifications of analog type and LSI type

Item	Analog type	LSI type
Power supply voltage (V)	2.7±0.1	2.6±0.3
Output frequency (MHz)	19,200, 19,680, 19,800, 26,000	12,600, 14,400, 19,200, 19,680, 19,800, 26,000
Service temperature range (°C)	−30 to +80 °C	−30 to +85 °C
Storage temperature (°C)	−40 to +85 °C	−40 to +85 °C
Initial deviation of frequency (ppm)[a]	±2.5 (max)	±1.5 (max)
Frequency stability (ppm)[b]	±2.0/±2.5 (max)	±2.0 (max)
Secular change of frequency (ppm/year)	±1.0 (max)	±1.0 (max)
Output voltage[c]	0.8 (min)	0.8 (min)
Consumed current (mA)	1.2 (max)	1.2 (max)
Harmonic ratio (dBc)	−0.8 (max)	−0.8 (max)
Voltage control range (ppm)[d]	(Ex.) ±8.0 to ±12.0	(Ex.) ±8.0 to ±12.0
Phase noise characteristics (dBc/Hz)[e]	−145 (max)	−135 (max)

Comparison of the analog type and the LSI type TCXO. The LSI type is superior in frequency deviation, while it is inferior in phase noise, compared with the analog type

[a]Include two reflows
[b]Within specified temperature range
[c]CL: 10 kΩ//10 pF
[d]$V_c = 1.5 \pm 1.0$ V
[e]1 kHz offset, 19.2 MHz output frequency

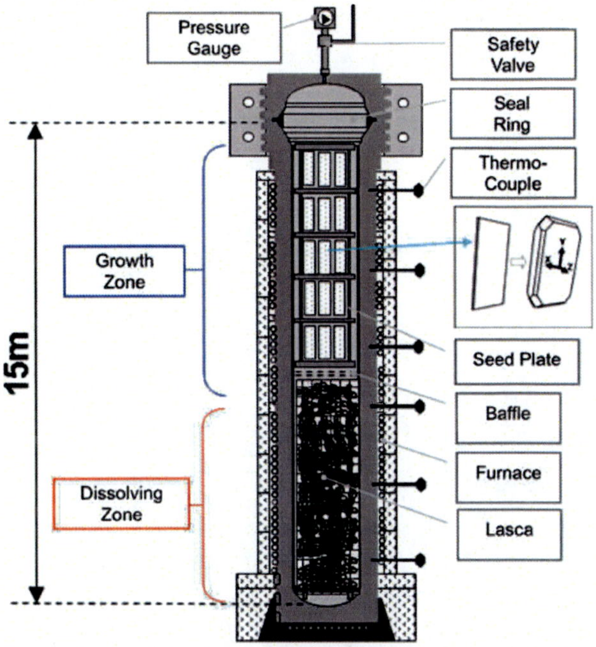

Fig. 6.15 Autoclave. A high-temperature and high-pressure furnace for growing synthetic quartz at 350 °C and 1,000 atm

6.3.2.2 Manufacturing Method

Synthetic quartz crystal is produced by an autoclave (Note 6.11) (Fig. 6.15). The quartz oscillator is created by processing a wafer produced through the cutting of synthetic quartz (Fig. 6.16) into a specified shape and attaching electrodes on the wafer. Figure 6.17 shows a quartz oscillator mounted on the TCXO. The characteristics change depending on the cutting angle, but the AT cut (Note 6.12) (Fig. 6.18), which features preferable frequency-temperature characteristics, is generally adopted.

Note 6.11 A high-temperature and high-pressure furnace for growing synthetic quartz at approximately 350 °C and 1,000 atm.

Note 6.12 A table cut: is a cutting direction which is most frequently adopted in MHz channels. It features a cubic curb with an inflection point at around 25 °C.

Fig. 6.16 Synthetic quartz. Quartz industrially-produced by autoclave. Lasca (small quartz piece), which is dissolved in sodium hydroxide solution, is recrystallized on the seed plate, producing high-purity artificial quartz

Fig. 6.17 Quartz mounted in TCXO. In order to produce the quartz oscillator, the quartz is processed to a wafer, and then the wafer is mounted with electrodes

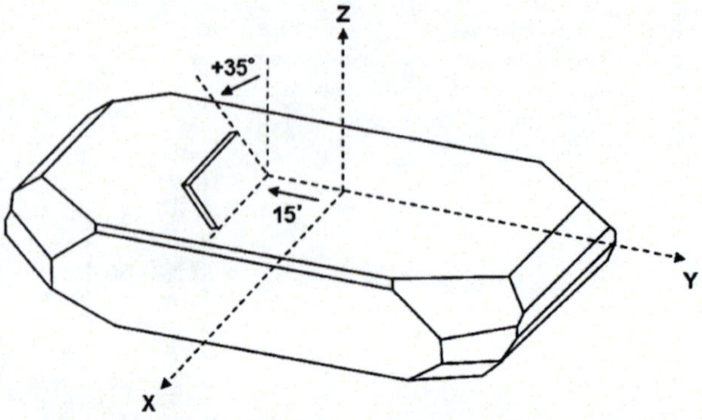

Fig. 6.18 AT cut (+35° 15) synthetic quartz. This cutting angle is most frequently used for MHz channels. It features a cubic curb with an inflection point at around 25 °C

6.3.3 Future Prospects

The growth rate of mobile phone subscriptions has declined when compared with explosive growth rates of the past, but it still remains at approximately 10% per year, covering all age groups throughout the world. In addition, the functions and performance of cameras, e-mail, TVs, etc. continue to be enhanced, and the devices are becoming more compact, lightweight and thin. Because of these changes, greater performance is required of TCXO. Meanwhile, further miniaturization seems unlikely because the quartz already is approaching limits to miniaturization. Therefore, it is necessary to ameliorate the production process of the quartz oscillator and the semiconductor, reduce the number of TCXO components, increase the packaging density and examine the component materials. It appears the elemental technologies such as the micro soldering technology used in micro wiring of the substrate, as well as IC bare chip assembling technology, which will be used in the future, need to be re-examined.

Dielectric Filter for Microwaves (1979) 6.4

Filters are used, in general, to remove unnecessary substances of various sizes from a signal or a stream. For example, a filter in a coffee pot removes the coffee beans from the extracted coffee liquid. Meanwhile, filters in electronic circuits remove unnecessary signals (radio waves) and takes out unnecessary signals. Microwaves ranging from 800 MHz to 2 GHz are used in mobile phones, and the dielectric filter lets through the frequency bands necessary for transmission and reception only, blocking unnecessary frequency bands. Dielectric filters are used as filters (wave filters) for communication devices primarily represented by mobile phones. The dielectric ceramic features a high permittivity and its utilization allows the miniaturization of filter components, contributing greatly to a reduction in the size and weight of communication devices. Figure 6.19 shows the mounting position of a dielectric filter used in a mobile phone.

6.4.1 Background of Development

In line with the rapid development of mobile communication devices represented by mobile phones, electronic components used in communication devices have been required to be more compact and lightweight. The dielectric filter was developed in response to these requests. The history of the dielectric filter is symbolized by transitions in size, and strenuous development efforts have been made in miniaturization, involving the technologies on materials, processing and design. The transition in the shrinking of the antenna duplexer, which is used in automobile phones and mobile phone units, is shown in Fig. 6.20.

6.4.2 Characteristics

6.4.2.1 Types and Specifications of Products

Communication devices need to be mounted with a filter (wave filter) that allows the passage of required signals (electromagnetic waves) only. The dielectric filter is one of the filters intended for microwaves. The dielectric filter is free from mechanical vibration, unlike filters that incorporate a piezoelectric element, and the microwaves themselves resonate inside the dielectric resonator. The length of microwaves are reduced to $1/\sqrt{\varepsilon_r}$ (ε_r: relative permittivity) of the free space status inside the dielectric body, making it possible to reduce the size of the resonance system. The dielectric filter is widely used in communication devices because of its compact size and excellent temperature characteristics.

The main units of automobile phones at the time of its first commercialization were so large that they took up the majority of trunk space. A smaller filter was highly needed to downsize the main unit, and the metal case type dielectric filter shown in Fig. 6.21 was developed in response to this need. Following the adoption of this type of filter, the main unit was shrunk to about 1/8 of its original size, making a huge contribution to miniaturization and the spread of automobile phones. In those days, $MgTiO_3$–$CaTiO_3$ ceramic ($\varepsilon_r = 21$) was primarily used.

In later years, mainstream mobile communication systems shifted to mobile phones, and filters were required to be further shrunk and surface-mountable on printed substrates. The surface-mounted dielectric filter shown in Fig. 6.22 was developed in response to these needs. High permittivity

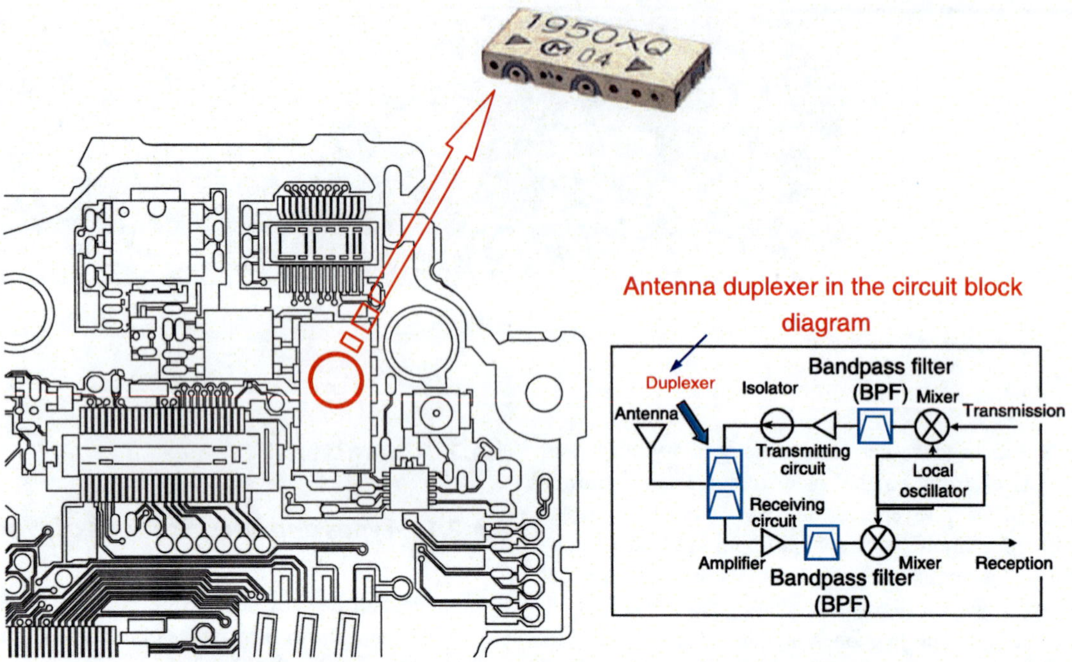

Fig. 6.19 Dielectric filter location in mobile phone. In a W-CDMA (Note 6.13) terminal, it is mainly used as an antenna duplexer. It functions as a branch circuit and one antenna can be shared by the transmitting filter and the receiving filter

materials, mainly BaOPbO–Nd_2O_3–TiO_2 ceramics (ε_r = 92), were used for downsizing.

With the improvements in mobile phone functions, further downsizing and weight reduction were required of dielectric filters. The block-integrated dielectric filter shown in Fig. 6.23 was developed responding to these requests. It was produced by utilizing advanced material and processing technologies, realizing a W-CDMA duplexer dimensioned 5.4 × 5.0 × 1.7 mm. It mainly uses $Ba(Sm,Nd)_2Ti_4O_{12}$ ceramic (ε_r = 82) and $Ba_2Ti_9O_{20}$–$BaSm_2Ti_4O_{12}$ ceramic (ε_r = 47).

6.4.2.2 *Production Method*

Dielectric filters are basically made by forming silver or copper electrodes on the surface of dielectric ceramics. The colored areas in Figs. 6.21, 6.22, and 6.23 are the electrodes. As shown in Fig. 6.24, they are generally produced by forming synthetic ceramic powder made by the solid-phase method into specified forms. After calcination, electrodes are formed and a tuning is carried out to attain the target filter characteristics. High technologies are required for forming and tuning operations because the hole diameter of the coaxial resonator keeps getting smaller in line with reductions in the product's size and height.

6.4.3 Future Prospects

The dielectric filter was introduced above as one of the ceramics that contributed to the downsizing of mobile phones. It is hoped that research and development will be continued, in terms of material, processing and design so as to take advantage of the characteristics of dielectrics, resulting in widespread use of the products in communication devices.

Note 6.13 The communication method for the third generation mobile phones (3G) developed by NTT DoCoMo, Ericsson, etc. It enables a high-speed data communication at 384 kbps and real-time communications by moving images and voices.

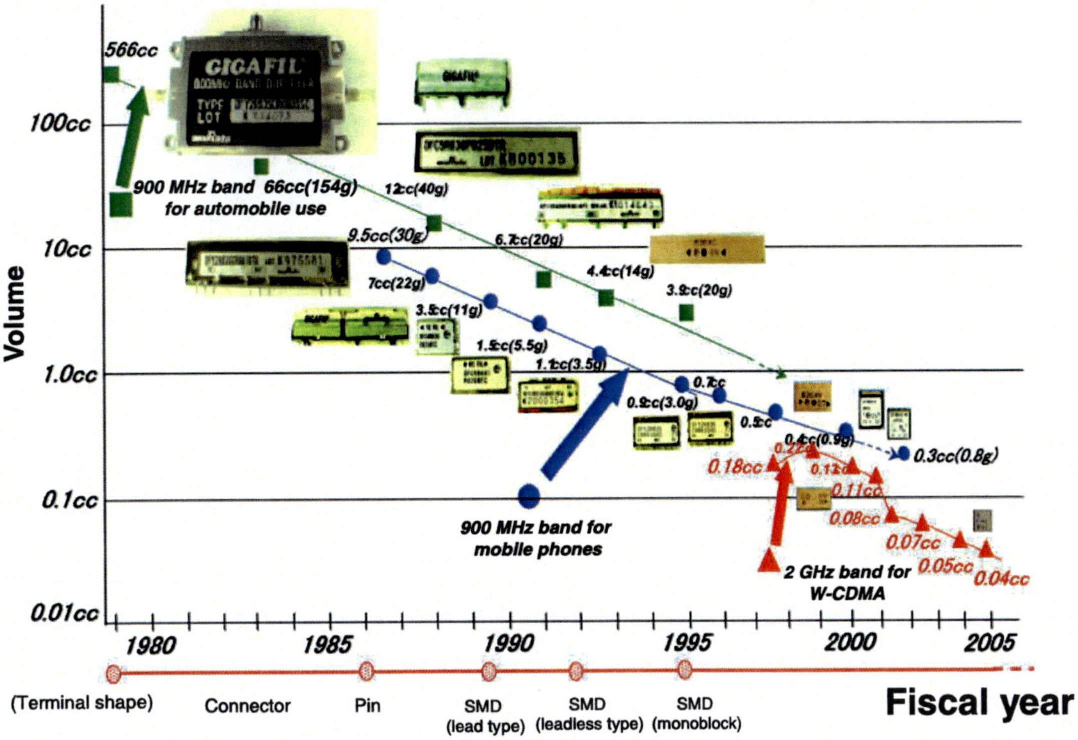

Fig. 6.20 Transition in the miniaturization of the dielectric filter (antenna duplexer). Following the shift to third-generation systems (W-CDMA), the 2 GHz band is being used instead of the 900 MHz band, but the volume was reduced to less than 1/1,000th of its pioneers

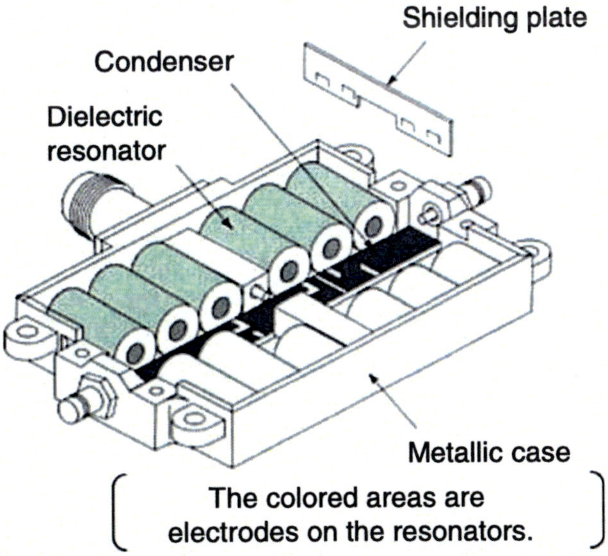

Fig. 6.21 Metallic case-type dielectric filter. Multiple dielectric resonators are placed, facing each other, in a metallic case. A shielding plate is placed between the dielectric resonators to separate them, connecting the resonators by the condenser

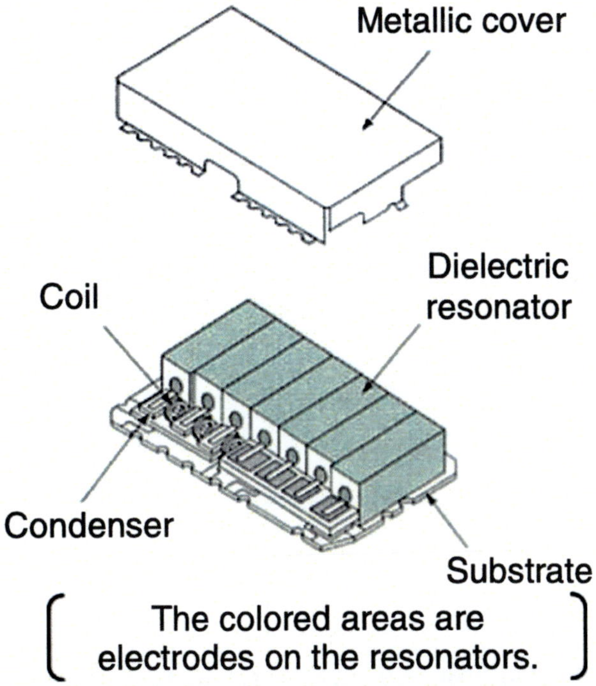

Fig. 6.22 Surface-mounted dielectric filter. Multiple dielectric resonators and connecting parts such as condensers and coils are placed on a substrate, upon which surface mounting electrodes are formed. All of the components are covered with a metallic cover

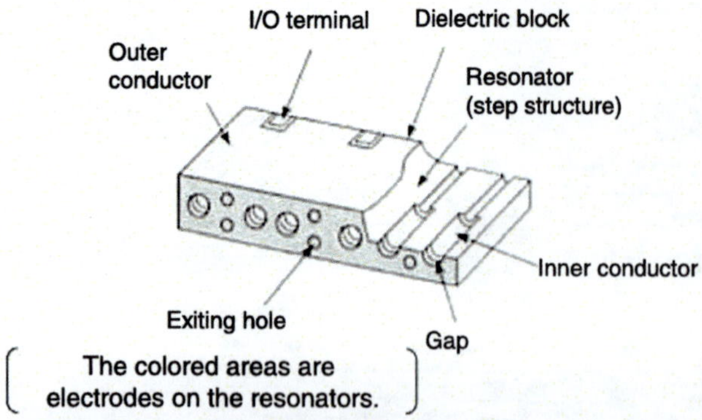

Fig. 6.23 Block-integrated dielectric filter (partial cross section). It is composed of a single ceramic block and does not use any other components such as condensers, coils or a metallic cover

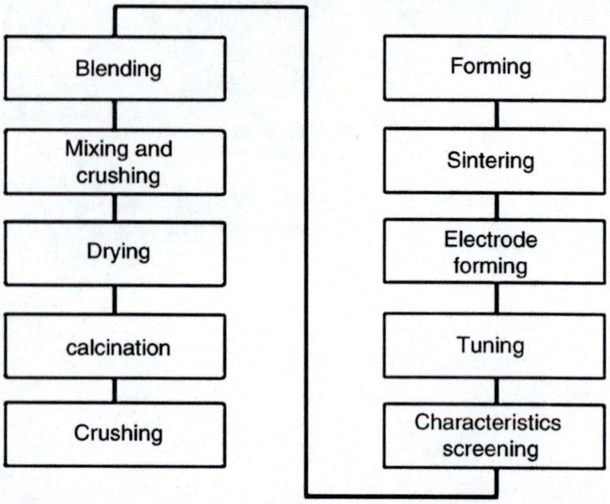

Fig. 6.24 Dielectric filter production process. Materials such as $BaCO_3$, Nd_2O_3 and TiO_2 are blended and crushed, and then calcined for CO_2 component removal and provisional synthesis. After these operations, binder components, selected depending on the forming method, are mixed to create a form conforming to the specified shape

Multilayer Chip LC Filter (1983) 6.5

As electronic devices become increasingly smaller and lighter, smaller electronic components are also needed. However, the simple miniaturization of single-function parts such as the inductor and condenser causes circuit board mounting problems. Under these circumstances, chip composite technologies have been developed that break the existing barriers for the realization of high density mounting. The multilayer chip LC filter (Note 6.14) was created by integrating an inductor and a condenser. Products such as the signal processing LC filter and the EMC (electro magnet compatibility) (Note 6.15) LC filter are among the multilayer chip LC filters. They were created by laminating into a unit and simultaneously calcining the magnetic ferrite used in the inductor and the dielectric ceramic used in the condenser. The product was first incorporated in video cameras in early 1983. Even today, they are widely used in a variety of digital electronic devices such as mobile phones and automobile audio equipment.

6.5.1 Background of Development

The downsizing of passive components, including the chip condenser, chip resistor and chip inductor, has been proceeding at a rapid pace, and 0402 products (0.4×0.2 mm) have already been mass-produced. Various issues are emerging following this rapid downsizing, in terms of mounting technologies. These issues include the problems in handling small parts by an automatic mounting machine, the limits in increasing density through adjustments of conductor patterns on printed circuit boards or through the coordination of distance between parts, the mounting accuracy of parts and the reliability of soldered areas. These issues can no longer be solved by the conventional method of downsizing components. The chip composite technologies have been developed as a tool to break existing barriers to realize high density mounting. Multiple passive components with different functions are integrated into one chip by using chip composition technology. The multilayer chip LC filter was created by integrating an inductor and a condenser. Figure 6.25 shows the image of a typical multilayer chip LC filter. Actual products include the signal processing LC filter and the EMC LC filter. Outlines of these products are indicated in Table 6.3.

Note 6.14 A circuit composed of an inductor (coil) and a condenser. It removes electrical noise and picks up signals within a designed frequency band.

Note 6.15 Electronic devices, which use electrical signals, emit electromagnetic noise. EMC (electro magnet compatibility) means "no emission of electromagnetic noise from electronic devices" and "no influence of external electromagnetic noise." Components that realize this property are called EMC countermeasure parts or noise suppression parts.

6.5.2 Characteristics

6.5.2.1 *Types and Specifications of Products*

Analog video cameras, video decks, and TVs that were commercialized from the mid 1980s used the LC filter to process video and sound signals. In video cameras, which were required to be compact and lightweight, signal processing LC filters were mainly used because the LC filters in video cameras, in particular, required smaller dimensions. The lineup of products is shown in Fig. 6.26. However, following the digitalization of electronic devices, semiconductor ICs are now used for signal processing and LC filters are no longer used.

In the 1990s, the digitalization of electronic devices proceeded at a rapid pace. The issues arising from digitalization were addressed by employing EMC, and the EMC LC filter (Note 6.16) which began to be used to counter this. The EMC LC filter has a wide lineup, because it varies depending on the circuit composed of an inductor and a condenser as well as a constant value. The lineup of the EMC LC filters is shown in Fig. 6.27. In order to respond to the needs for enhanced high density mounting, products that incorporate multiple EMC filters (arrays) are also offered. Advantages of

Note 6.16 The EMC LC filter is sometimes called a three-terminal filter because it has three terminals, namely the input terminal, output terminal and GND terminal.

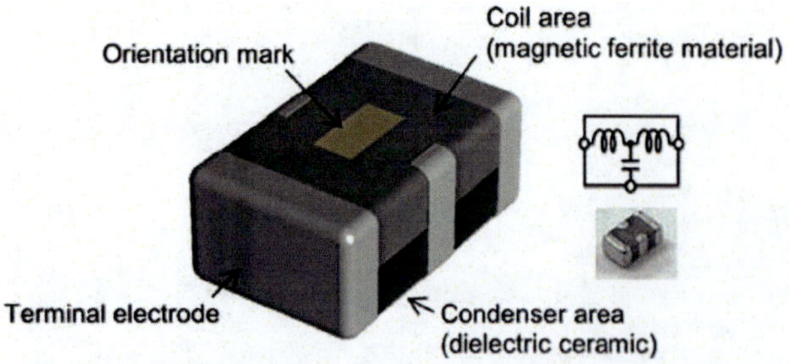

Fig. 6.25 Multilayer chip LC filter (EMC LC filter). The multilayer LC filter developed in 1983. It was realized by the world's first technology that allows the integrated lamination and co-firing of the magnetic ferrite and dielectric ceramic

Table 6.3 Types of multilayer LC filters

Type		Signal processing LC filter	EMC LC filter
Used material	Coil area	Magnetic ferrite	Magnetic ferrite
	Condenser area	Magnetic ceramic	Magnetic ceramic
	Internal conductor	Ag	Ag
Application		Analog video cameras	Digital electronic devices

Signal processing LC filters were produced in the 1980s. Currently, the EMC LC filter is in the mainstream

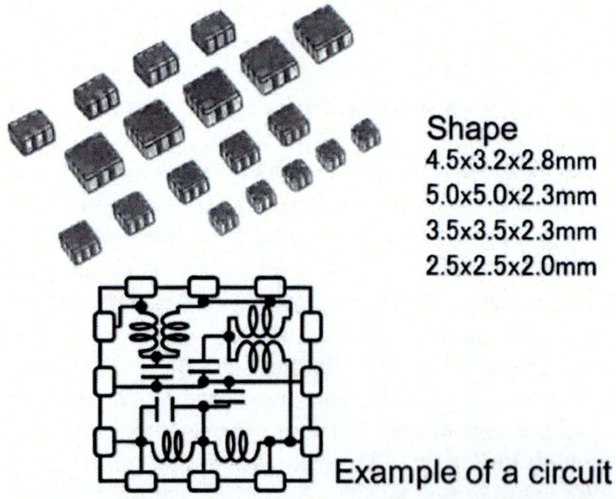

Fig. 6.26 Signal processing LC filter. The signal processing IC filter was mainly used in analog video cameras. It processed electrical signals ranging from several MHz to several tens of MHz

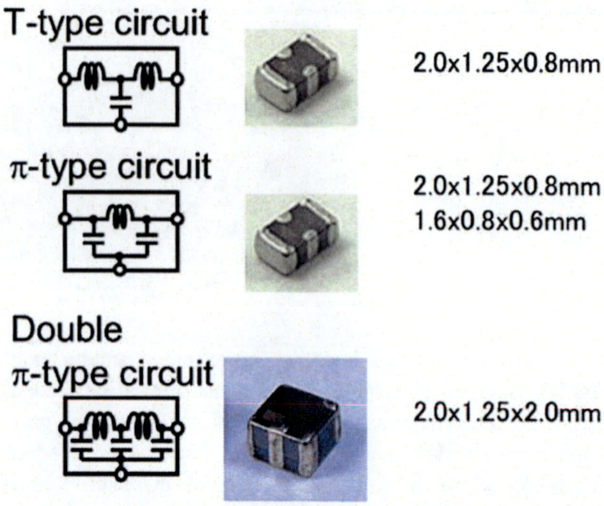

Fig. 6.27 EMC LC filter. In the 1990s, electronic devices were digitalized at a rapid pace. The EMC LC filter (three-terminal filter) began to be used as a countermeasure for EMC, which needed to be addressed for digitalization

the EMC filter array are indicated in Fig. 6.28, and the major products are shown in Fig. 6.29.

6.5.2.2 Manufacturing Method

The technology for creating a circuit inside a laminated body through integrated lamination and the simultaneous calcination of materials, namely low-temperature sintered ferrite and low-temperature sintered dielectric ceramic, was developed in 1983. In order to build LC composite parts into a single chip, ferrite used in the inductor and dielectric ceramic used in the condenser need to be integrated. The properties of ferrite are completely different from those of dielectric ceramic, and therefore, various problems will arise, in terms of reliability, if they are simply laminated together. The two materials with completely different properties are integrally laminated and simultaneously fired with high accuracy control in the calcination shrinkage ratio and thermal expansion coefficients of ferrite and dielectric ceramic, incorporating advanced lamination and integration technologies to significantly improve pattern accuracy. The material used for the internal electrodes is silver (Ag), which has high electrical conductivity. In the first step, ferrite and dielectric ceramic that can be sintered at a temperature lower than the melting point of silver (960 °C) were developed. Ni–Cu–Zn-based soft ferrite was used as the material for the magnetic body, while titanium oxide-based ceramic was

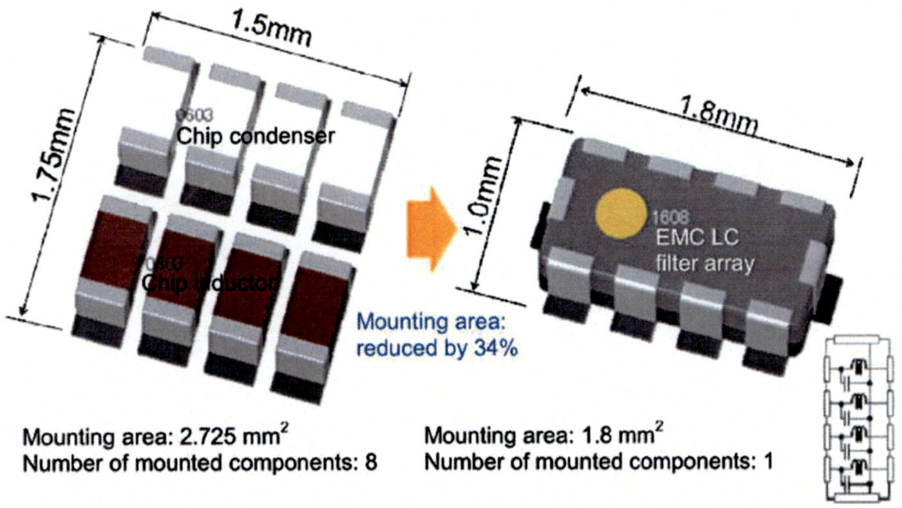

Fig. 6.28 Advantages of an EMC LC filter array. The EMC LC filter array is created by combining multiple EMC LC filters. In this example, eight units of coils and condensers are combined

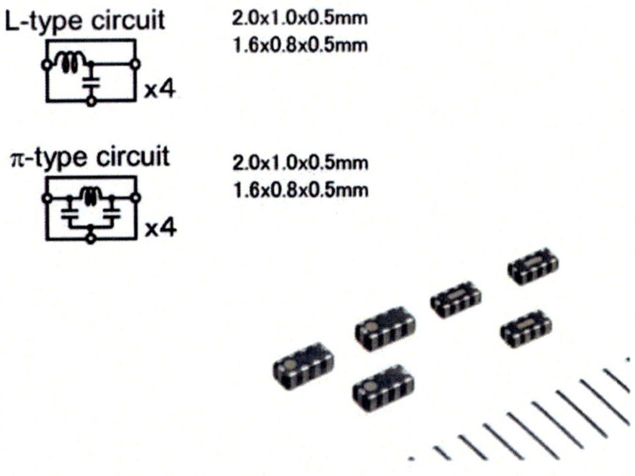

Fig. 6.29 EMC LC filter array. The EMC LC filter array is incorporated in various digital electronic devices

used as the material for the dielectric ceramic. The method for lowering the calcination temperature of each material was developed on the basis of the material compositional design technology and grain diameter control technology. The outline of the production process is shown in Fig. 6.30.

6.5.3 Future Prospects

Multilayer chip LC filters continued to evolve based on the technologies developed in the early 1980s and through constant development of materials and processing technologies thereafter. Continued progress is expected, responding to the demands of the market for miniaturization, weight reduction and functional improvements.

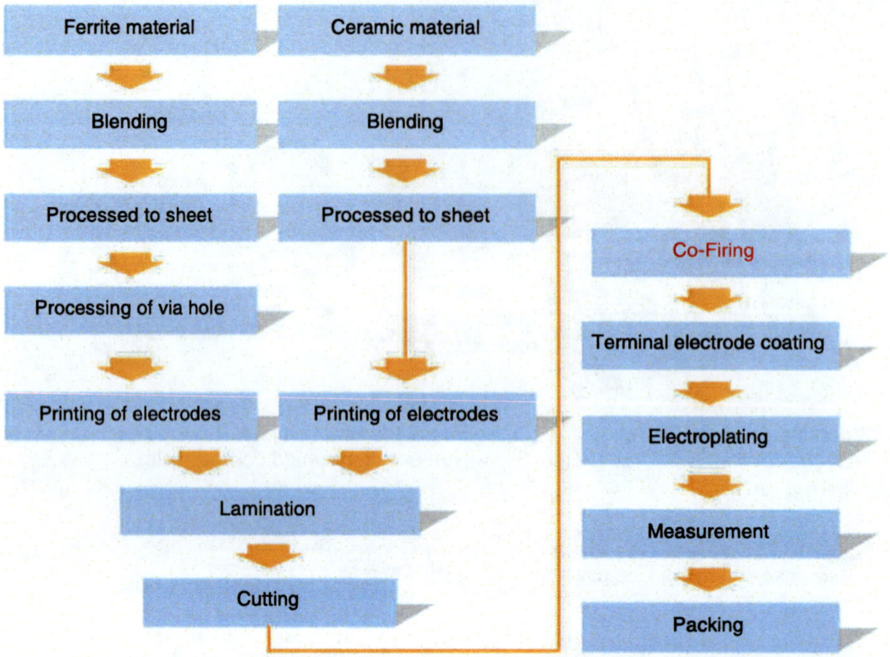

Fig. 6.30 Manufacturing process. In respect to the coil area composed of ferrite material, the internal electrodes need to be positioned helically. Therefore, via holes are opened on the sheet for vertical continuity. The via holes are usually formed by laser irradiation

6.6 Isolator (1960s)

An isolator (Fig. 6.31) is an irreversible circuit component that utilizes the gyromagnetic phenomena (Note 6.17) in microwave ferrite. Portable terminals that began to be used at the end of the 1970s had a tendency to be unstable in operation due to the influence of the waves reflected from the antenna on the power amplifier (transmitted power amplifier) or due to the generation of intermodulation distortion (IMD) (Note 6.18) caused by the intrusion of unnecessary radio waves into the power amplifier via the antenna. Isolators, featuring their irreversible property, transmit forward signals, but do not transmit reverse signals. Because of this property, isolators have been widely used to protect power amplifiers from antenna load fluctuations.

6.6.1 Background of Development

The utilization of ferrite for microwaves started in the 1950s in the United States. Originally, spinel ferrite was used. However, since rare-earth garnet ferrite was discovered in 1956 in France and the material was confirmed to dramatically improve magnetic loss, rare-earth garnet ferrite has come into general use. At the same time, a portable communication system was started from the automobile phone service provided by the Nippon Telegraph and Telephone Public Corporation (currently NTT) in Japan in 1979. Later, portable shoulder phones were introduced, followed by the rapid spread of handy-type mobile phones, which were released in 1987.

In mobile phones, problems such as a drop in power efficiency of the power amplifier and the generation of unnecessary signals such as adjacent channel interfering waves occur when the antenna impedance fluctuates due to the influence of surrounding objects such as the human body and part of the output waves are reflected by the antenna and returned to the power amplifier. Isolators, which let forward signals through and block reverse signals, can reduce load fluctuations and are effective for antenna impedance fluctuations. The isolator is generally mounted on the RF portion (Note 6.19) inside a mobile phone. The circuit block diagram of a common RF portion is shown in Fig. 6.32.

6.6.2 Characteristics

6.6.2.1 *Structure of the Product*

Basic operations of the isolators are explained in Fig. 6.33a, three-port circulator. Figure 6.33b shows a circulator, which is connected with a terminating resistance on one of its ports. It functions as an isolator, preventing reflected waves from the load side from being transmitted to the port (1) on the transmitting side.

Distributed parameter isolators, which include waveguide and junction types, were originally used. However, distributed parameter isolators had to use a ferrite isolate with a size of several centimeters in diameter for mobile phones, which use communication frequencies ranging from 800 MHz to 2 GHz, because the size of the ferrite to be incorporated is inversely proportional to the frequency.

Note 6.17 The phenomena where gyroscopic movement (also called precession) occurs in the magnetic moment applied with an AC magnetic field. The magnetic resonance is generated when the frequency of the AC magnetic field becomes identical to the natural frequency of the magnetic moment precession.

Note 6.18 The phenomena where signals with frequency components containing the sum or difference of an equal multiple of the signals ($\omega = n\omega_1 \pm m\omega_2$) are generated when signals containing different frequency components, ω_1 and ω_2, are input to the transmission system, causing a suppression of reception sensitivity.

Note 6.19 The portion where high frequency signals used for wireless communication is processed.

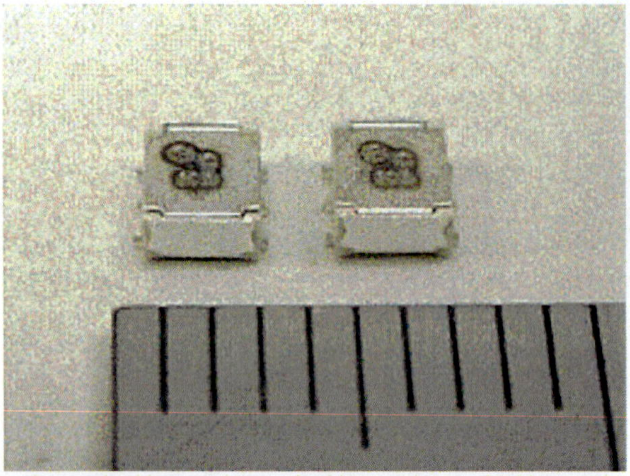

Fig. 6.31 Image of lumped parameter isolators. In line with the miniaturization and weight reduction of mobile phones, isolators were also required to be smaller and lighter (3.2 mm H × 2.5 mm W × 1.2 mm H)

Therefore, it was difficult to miniaturize these components. For this reason, lumped parameter isolators, which can be miniaturized, were used as the transmitting circuit of mobile phones. The basic configuration of the lump parameter isolator is shown in Fig. 6.34. Electrically-isolated center electrodes are placed on top of each other and at an angle of 120° against each other on a ferrite applied with a DC field.

6.6.2.2 Characteristics of Ceramics

The microwave ferrite used in isolators is required to have properties that include:

(1) Small magnetic loss, i.e., a small material constant ΔH (ferromagnetic resonance half width) (Note 6.20).
(2) A temperature coefficient of saturated magnetization that is close to the temperature coefficient of the biasing magnet.

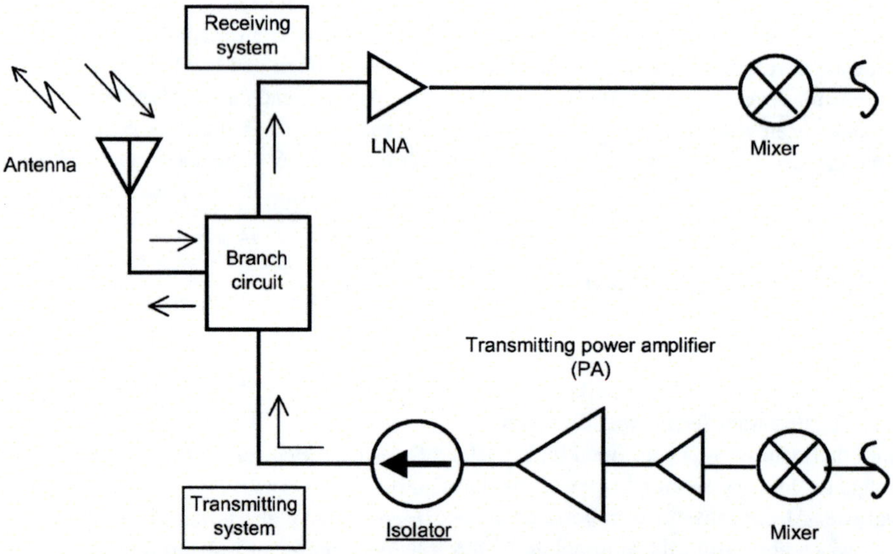

Fig. 6.32 Block circuit diagram of the RF portion of mobile phones. Isolators are mainly used in the RF portion inside mobile phones, with the aim of protecting the power amplifier (transmitting power amplifier) from antenna load fluctuations. Isolators prevent drops in power efficiency of power amplifiers by avoiding antenna load fluctuations, which results in reducing consumption current and minimizing the drop in talk time of mobile phones

Note 6.20 An index that indicates the magnetic loss of ferrite in microwave frequency bands. When magnetic resonance occurs in microwaves, the energy in high frequency magnetic fields is absorbed, causing magnetic loss. Magnetic loss reaches maximum value at or around the resonance frequency. The lumped parameter isolators are operated at frequencies far from the resonance frequency, and the influence of magnetic loss is smaller in ferrites with a smaller ferromagnetic resonance half width.

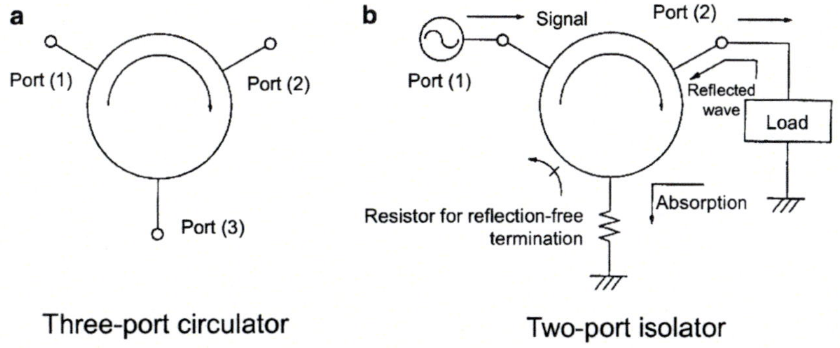

Fig. 6.33 Basic operations of isolators. (**a**) The signal input to port (1) is transmitted in the direction indicated by the *arrow* only and is output to port (2). The same can be said about ports (2) and (3) as well as ports (3) and (1). (**b**) When a signal is transmitted from port (1) to port (2) and a reflected wave is returned from the load side, the reflected wave is transmitted in the arrowed direction and absorbed by the termination resistor

With respect to (1), the calorific value of the ferrite needs to be minimized by reducing the insertion loss, because high-frequency signals with a high power of approximately 1 W are let through. As for (2), the reasons are as follows. In the case of a ferrite isolator with an external magnetic field is applied with a high-frequency magnetic field that crosses the external magnetic field at right angles, the circular polarized permeability $\mu\pm$ of the ferrite is affected by saturated magnetization (Ms) and the internal DC field (H_{in}) of the ferrite (refer to Fig. 6.34). Therefore, electrical properties relative to temperature changes need to be stabilized by making the temperature coefficient of the residual magnetic flux density of the magnet and the temperature coefficient of the microwave ferrite identical.

Because of the reasons explained above, YIG (yttrium iron garnet) is generally used to produce the microwave ferrite used in isolators, because YIG features a small ΔH and the temperature coefficient of saturated magnetization can be controlled by selecting an appropriate composition ratio.

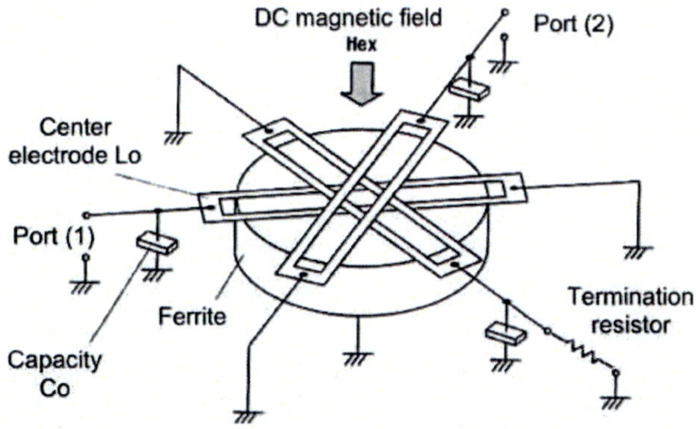

Fig. 6.34 Basic configuration of lumped parameter isolators. When an external DC magnetic field H_{ex} is applied and a high-frequency magnetic field is applied at right angles to H_{ex}, the magnetic moment Ms of the microwave ferrite starts precession. Circular polarized wave permeabilities, $\mu+$ and $\mu-$, are expressed in the following formulas. $\mu+ = 1 + |\gamma| Ms/(\mu_0|\gamma|H_{in} - \omega)$ positive circularly polarized wave permeability. $\mu- = 1 + |\gamma| Ms/(\mu_0|\gamma|H_{in} + \omega)$ negative circularly polarized wave permeability. Ms: saturated magnetization of ferrite, γ: gyromagnetic constant, μ_0: permeability in vacuum, H_{in}: internal DC magnetic field, ω: angular frequency of high frequency magnetic field. When the high frequency magnetic field is a positive circularly-polarized wave (clockwise rotation), the rotation direction is the same as the magnetic moment precession and substantial magnetization is observed. On the other hand, when the high frequency magnetic field is a negative circularly-polarized wave (counterclockwise rotation), the rotation direction is opposite the magnetic moment precession and the magnetization is minimal. Phase displacement is caused by the directional difference of polarized waves, curving the direction of microwaves

6.6.2.3 Manufacturing Method

The basic manufacturing process of the microwave ferrite is as follows. Raw materials such as Fe_2O_3 and Y_2O_5 are weighed, mixed and calcined, and are later ball milled and granulated through the addition of a binder. The powders are

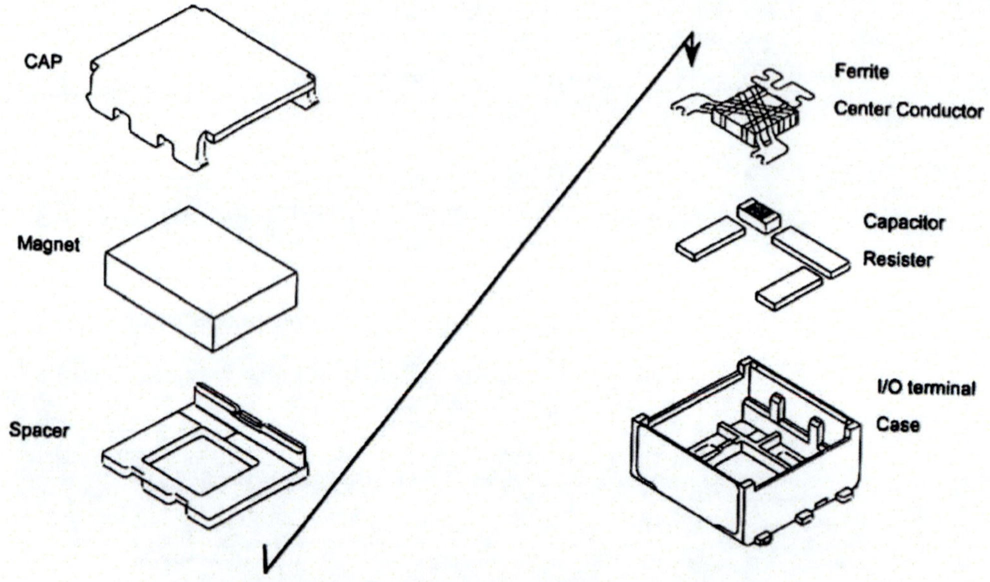

Fig. 6.35 Exploded view of a compact isolator intended for mobile phones. The isolator is composed of a permanent magnet for bias magnetic field application, a microwave ferrite on which a center electrode is formed, a capacitor for impedance matching, a terminating resistor, and an external case

then pressed by a press machine, cast to the specified shape and sintered at high temperatures to produce a sintered compact used as a microwave ferrite. The isolator is produced by assembling the microwave ferrite as shown in Fig. 6.35.

6.6.3 Future Prospects

The number of components and power usage of mobile phones have been increasing due to the enhancement of transmission speed and an increase in functionality. In addition, frequency bands used by mobile phones are planned to be increased from two bands to three bands, leading to a further increase of components. Under these circumstances, further performance improvement of the isolators for the RF portion of mobile phones will be required to contribute to the miniaturization, thinning, weight reduction and consumption current reduction of mobile phones. It is expected that further improvements will be made to reduce the size and thickness of isolators as well as to enhance performance, through reconsiderations of the basic configuration, incorporation of high precision ferrite electrode forming technology and utilization of the properties of ceramic multilayer substrates for structural design.

Chip Capacitor (1979) 6.7

The chip capacitor is an important passive component used for the decoupling of analog and digital circuits as well as for the smoothing of disturbances from power supply circuitry in compact portable devices such as mobile phones. The chip capacitor is used in a variety of modules in mobile phones and the number of capacitors per mobile phone has been increasing. Figure 6.36 shows the locations of chip capacitors in a mobile phone.

6.7.1 Background of Development

Demand is increasing for the further shrinking of electronic components as mobile devices are increasingly becoming smaller, thinner, and more enhanced in terms of function and performance. In respect to chip capacitors, miniaturization and capacitance improvements have also been advancing. The transition of chip capacitor usage rate by size is shown in Fig. 6.37. Currently, the 1005 size (1.0×0.5 mm) has been the most frequently used, and the usage rate of ultracompact size products sized 0603 (0.6×0.3 mm) or smaller is increasing.

6.7.2 Characteristics

6.7.2.1 *Characteristics and Specifications of Products*

Chip capacitors are used in a variety of circuits for mobile phones and function to reduce signal noise, continuously supply stable current (decoupling), and smooth the signals in order to bring them closer to direct current (smoothing). There are a variety of types in chip capacitors, including ceramic, aluminum electrolysis, tantalum electrolysis and film types. Among them, the ceramic type is most widely used in mobile devices. This is because the ceramic capacitor is compact and features a wide capacitance range of 0.5 pF to 100 μF, nonpolar properties, high reliability attributable to the high insulation resistance, a high breakdown voltage and high heat resistance, and low equivalent series resistance.

Figure 6.38 shows an image of the multilayer structure of a ceramic type capacitor. Dielectric bodies and internal electrodes are piled up alternately and the internal electrodes are alternately connected with external electrodes. The main material of the dielectric body is barium titanate ($BaTiO_3$). The relative permittivity of $BaTiO_3$ is approximately 1,000 at normal temperature, but it is 10,000 or higher at the Curie temperature of 120°C. In actual use, the material composition is adjusted to shift the Curie temperature and reduce the permittivity dependency on temperature. Examples include replacing Ba site with alkaline-earth elements or rare-earth elements, and replacing Ti sites with Zr or Sn.

Figure 6.39 shows the transitions of the dielectric layer thickness and electrostatic capacitance (volumetric capacitance) of chip capacitors. In 1980, the dielectric thickness was 20 μm and about 60 of them were layered in total. In 2005, a capacitor incorporating 500 dielectric layers with thicknesses of 1 μm was commercialized, increasing the volume–capacitance ratio by 400 times. Figure 6.40 shows a cross-sectional picture of a capacitor incorporating 1 μm dielectric layers. Measures have been taken to reduce the thickness of the dielectric layer and increase the number of dielectric layers. However, when the dielectric layer is made thinner, the electric field intensity applied to the dielectric layer increases. In order to reduce the thickness without lowering the reliability, technologies for improving breakdown voltage and reducing the deterioration of the insulation resistance are needed. Responding to these needs, a number of elemental technologies, including sheet forming technologies, manufacturing and printing technologies of Ni powder for internal electrodes, technologies for better powder material dispersion, and technologies for forming high accuracy layers have progressed, in addition to improvements in dielectric materials through the fine structure control of ceramics and fine ceramic grain processing.

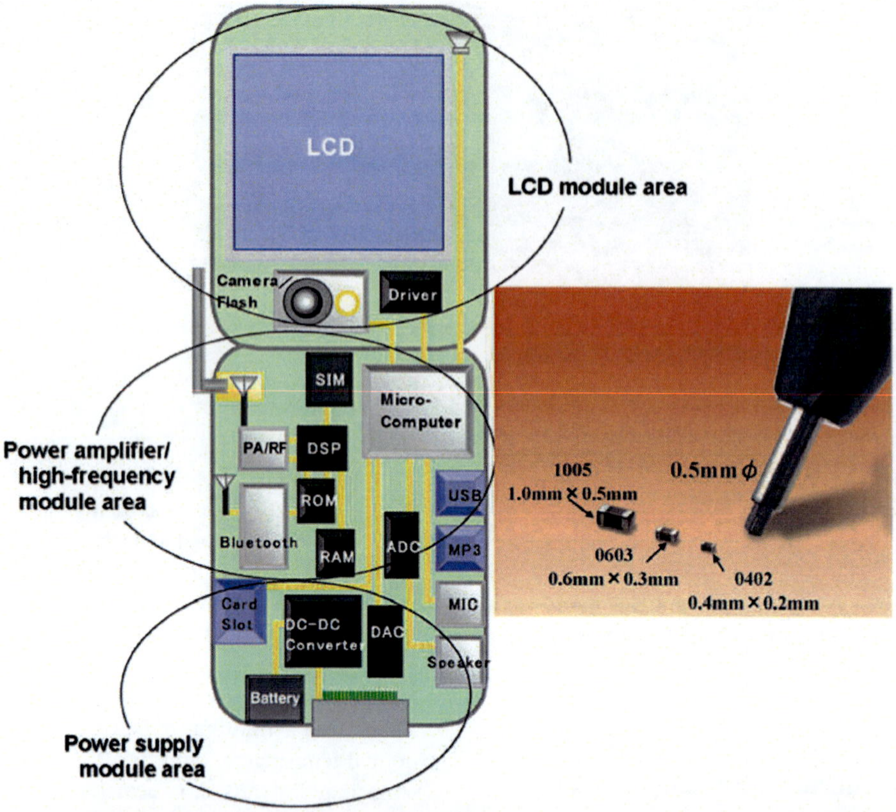

Fig. 6.36 Locations of chip capacitors in a mobile phone. The circuits of mobile phones are largely classified into four parts: the LCD module, power amplifier module, high-frequency module, and power supply module. The chip capacitor is used in all of the circuits. The chips are *cubic* in shape, and some of them are smaller than the tip of a mechanical pencil

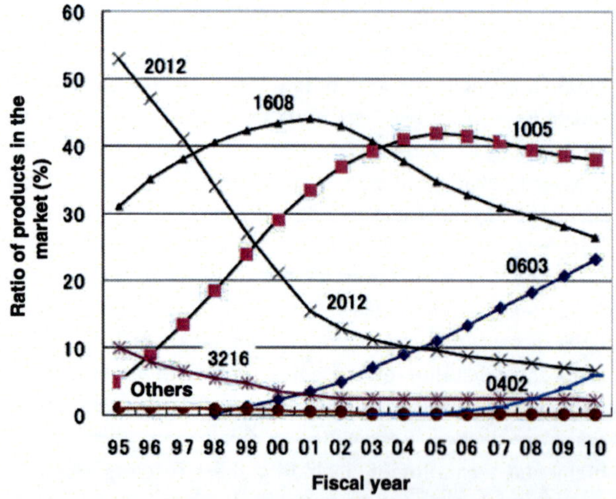

Fig. 6.37 Ratios of chip capacitors by size. The numbers in the figure indicate chip sizes. 3216, 2012, 1608, 0603, and 0402, respectively, mean 3.2 mm × 1.6 mm, 2.0 mm × 1.25 mm, 1.6 mm × 0.8 mm, and 0.4 mm × 0.2 mm. Chip capacitors were miniaturized in response to the reduced size and improved functionality of mobile devices, as well as through advances in mounting technologies. The smallest of the currently commercialized products is 0402 (0.4 mm × 0.2 mm)

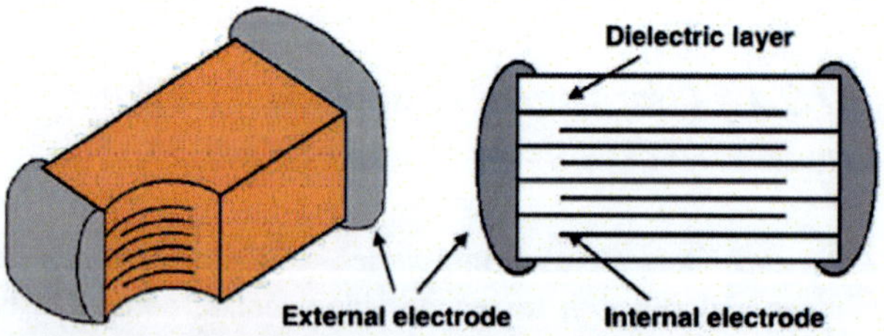

Fig. 6.38 Outline of chip capacitor structure. A chip capacitor is composed of dielectric layers, electrode layers, and external electrode layers. In general, $BaTiO_3$ ceramics with a high relative permittivity are used for the dielectric layers, Ni for the internal electrodes, and Cu for the external electrodes. The external electrode is sometimes coated with two layers of metallic plating, Ni and Sn, for solder mounting

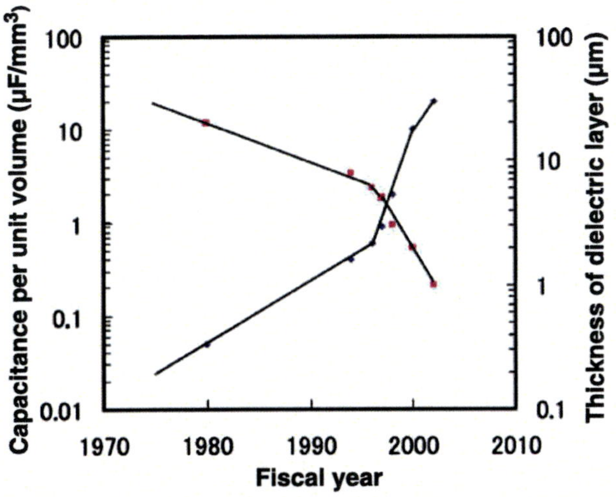

Fig. 6.39 Transition in thickness of dielectric layer and capacitance per unit volume of chip capacitor. Capacitance per unit volume, *C/V* [F/m³] is obtained by the formula below. $C/V = C/(NSd) = \varepsilon_r \varepsilon_0 / d^2$ In the formula: *C*: capacitance, *V*: volume of dielectric layer, *N*: number of dielectric layers, ε_r: relative permittivity of dielectric body, ε_0: permittivity in vacuum, *S*: area of electrode, *d*: distance between electrodes. Capacitance *C/V* is proportional to the relative permittivity of the dielectric body and the area of the electrode *S*, and is inversely proportional to the distance between electrodes *d*. In order to attain a large value, materials of a high relative permittivity need to be used to create dielectric layers, and the layers need to be as thin as possible

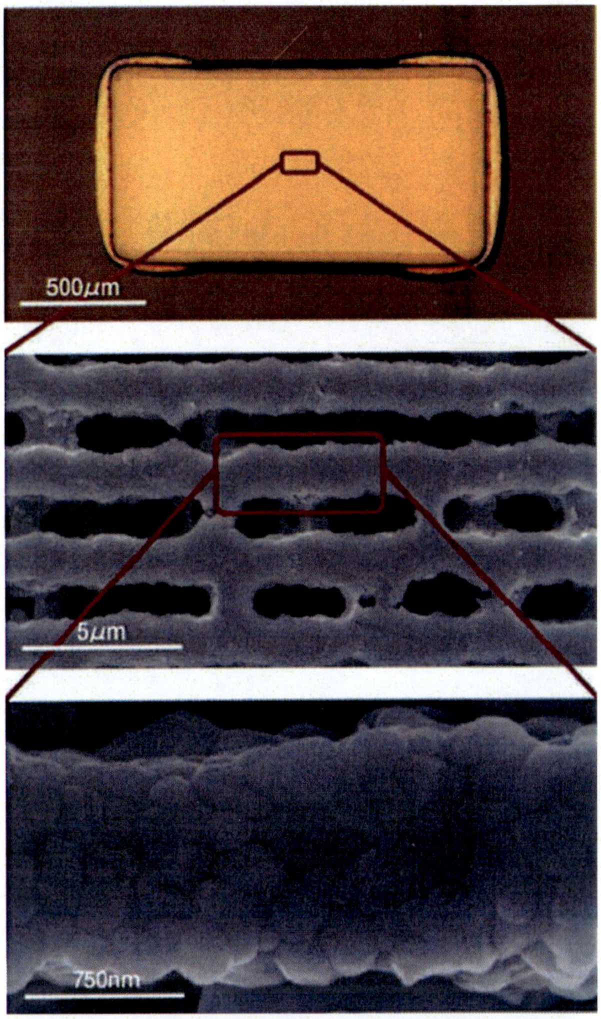

Fig. 6.40 Cross-sectional view of a capacitor incorporating 1 μm dielectric layers. Cross-sectional view of a 1608 (1.6 mm × 0.8 mm) B characteristic (Note 6.21) 10 μF chip capacitor. Electrodes and dielectric layers are layered in an orderly manner and cannot be identified by the naked eye. As seen in the SEM picture at the *bottom*, about five small ceramic grains, sized 100–200 nm, are formed between the electrodes

6.7.2.2 Manufacturing Method

Figure 6.41 shows the general manufacturing method of chip capacitors. In the dielectric body manufacturing process, high-purity materials are mixed for even dispersion and chemically reacted through calcination (thermal treatment). In the green sheet manufacturing process, the sheets must be several μm in thickness, have little variation in thickness, and be free of defects. Chip processing consists of a number of steps such as, a thin and homogeneous metallic film of Ni is formed on the green sheet, and then press bonded and formed with careful attention paid to the displacement of electrodes. Then, the formed body is diced into chips, which are co-fired in the electric furnace at 1,000–1,350°C. A laminated body incorporating ceramics and internal electrodes is created through these processes. In the measurement and packing process, electrical properties such as the capacitance value, breakdown voltage, and insulation resistance are measured and appearance check screenings are carried out, covering all of the products. The products are taped, packed in a bulk case, checked for a final time, put in an exterior package, and delivered.

Note 6.21 A value, specified by the JIS standard, that expresses the capacitance change rate against temperature. B characteristics means the capacitance change rate is ±10% or less at temperatures −55°C to 85°C. The value is expressed by the maximum value and the minimum value of the capacitance change rate, against the capacitance change rate at the reference temperature, 20°C, in the applicable temperature range.

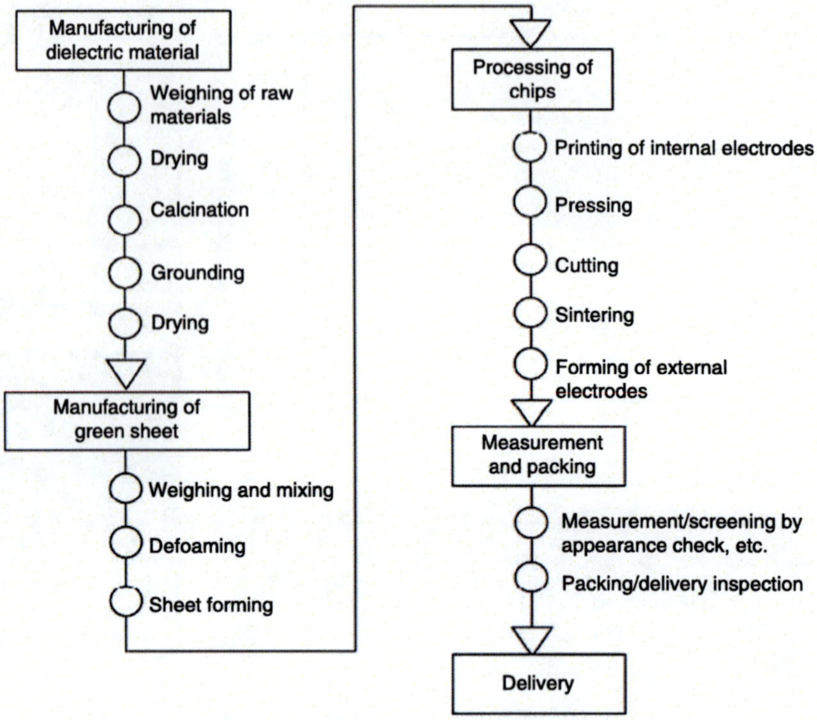

Fig. 6.41 Manufacturing process of chip capacitors. Dispersing agents, organic binders and solvents are added to the raw material for dielectric body to disperse the raw material and create the slurry, which is thinly extended on the carrier film by coating methods using doctor blades and die coaters to create the green sheet. The green sheet is screen-printed with internal electrodes in a paste form, then press-bonded and calcined after being cut into chips. If Ni is used for the internal electrodes, sintering has to be carried out in a reducing atmosphere to prevent the oxidization of electrodes, requiring sintering in a reducing atmosphere at an oxygen partial pressure of 10^{-5}–10^{-8} MPa

6.7.3 Future Prospects

As a result of progress in miniaturization and functional improvements in mobile devices, chip capacitors will be required to be smaller in size and higher in performance. Mobile devices will continue to be smaller in size and higher in performance, and this performance will be improved through the coordination of material designs that allow thinner layers and an increase in the number of layers, as well as improvements in process and product designs.

6.8 Grain Boundary Insulation Type Barrier Layer Capacitor (1973) (Production of Disk Type Was Discontinued in 2004)

Grain boundary insulation type barrier layer capacitors (Fig. 6.42) have an insulating substance evenly segregated on the grain boundary area of a semiconductor ceramic and utilize a hybrid function of the grains and the grain boundary. The product, which has excellent electrical properties, is normally cylindrical in shape and is used as a compact and high capacitance product. Some of them are disk-shaped and others have irregular shapes.

6.8.1 Background of Development

Electrical components are now universally required to be small-sized, high in functionality and reliability, and low-priced in response to progress in the high density mounting technology realized by enhanced surface mounting technologies. For capacitor components, requests for smaller size, larger capacitance and lower price are also on the rise. Multilayer ceramic capacitors and semiconductor ceramic capacitors have emerged, responding to these needs. The semiconductor ceramic capacitors are largely classified into two types, the grain boundary insulation type (Note 6.22) and the surface reoxidation type, and chosen depending on the required characteristics (Table 6.4). They are incorporated in communication devices including mobile phones and consumer products including tuners, DVD recorders, TVs and automobile audio equipment. Figure 6.43 shows the location of the grain boundary insulation type capacitor mounted in a tuner.

6.8.2 Characteristics

6.8.2.1 *Characteristics of Ceramics*

The grain boundary insulation type barrier layer capacitor features dielectric characteristics observed in the ceramic-based material and the grain boundary diffusion material (Note 6.23). Ceramics such as $SrTiO_3$, (Sr, Ca)TiO_3, $BaTiO_3$, (Ba, Sr)TiO_3 can be used as base materials. In particular, $SrTiO_3$ and (Sr, Ca)TiO_3 have stable dielectric characteristics and are appropriate for use as base materials. $SrTiO_3$ can be turned into a semiconductor by utilizing valence control method and forced reduction method at the same time. Another requirement for semiconductor ceramics is controlled grain size. The grain sizes that are several tens of μm are controlled in a large part by utilizing process expertise, but the factors that decide the sizes also include the grain size of the $SrTiO_3$ powder, the Sr/Ti ratio and sintering additives such as SiO_2 and Al_2O_3. This makes it necessary to control these factors. During thermal treatment the additives diffuse into the grain boundary which controls the property of the semiconductor grains adjacent to the grain boundary utilizing its dielectric characteristics. The most effective substance among the materials is Bi_2O_3. Other substances are also used, such as PbO and B_2O_3, which control the diffusion to the grain boundary, as well as CuO and MnO, which are believed to compensate for the atomization of $SrTiO_3$ in the vicinity of the grain boundary and supply oxygen to the grain boundary. The additives diffuse in the grain boundary during the secondary sintering. Figure 6.44 shows a TEM image of the triple point of the sample, which was slow-cooled after being treated at the secondary sintering temperature for diffusion of the diffusion material in the grain boundary. The composition

Note 6.22 The grain boundary of semiconductor-type ceramics is controlled by the dielectric body. The capacitance, breakdown voltage, and insulation resistance are determined by the permittivity and thickness of the dielectric body as well as the size and number of the grains.

Note 6.23 The additive diffuses in the semiconductor grain boundary to create an insulation layer. The additive, a paste of oxide such as Bi, Pb, Cu and Mn, is printed or dip-coated on the ceramic.

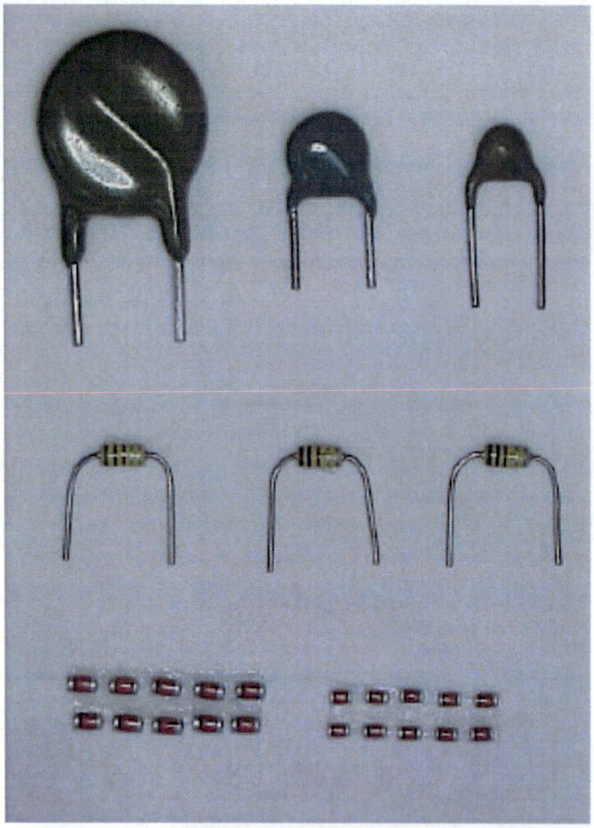

Fig. 6.42 Grain boundary insulation type barrier layer capacitors. They come in two shapes, a *disk shape* and *cylinder shape*, and are selected depending upon their application

Table 6.4 Structure and characteristics of the two types of semiconductor capacitors

Reoxidation type	Grain boundary insulation type
Utilizes dielectric body, diffused from surface and controlled	Utilizes dielectric body in grains diffused from surface and controlled
Atomic value control type, atomic value compensation	Atomic value control type + forced reduction type, insulation material diffusion and atomic value compensation
• Capacitance is determined by permittivity and thickness of dielectric body • Breakdown voltage and insulation resistance depend on dielectric layer	• Capacitance is determined by permittivity and thickness of dielectric body, as well as diameter and number of grains • Breakdown voltage and insulation resistance corresponding to dielectric body and number of grains • High humidity resistance
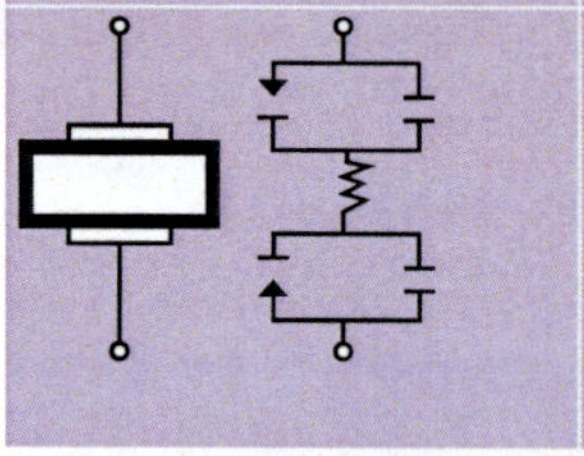	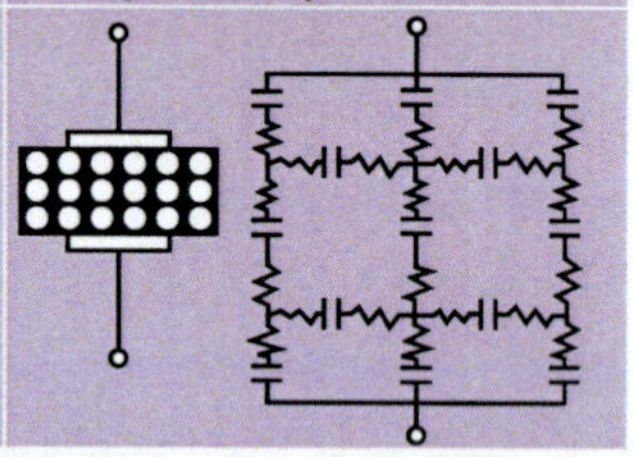

The semiconductor capacitor is a capacitor that has a capacitive insulation layer formed on the grain boundary, which is the outer or inner surface of a semiconductor ceramic. They are classified into surface reoxidation types and grain boundary insulation types

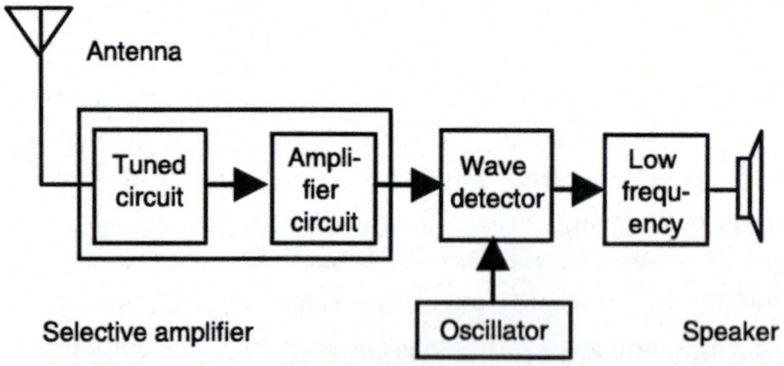

Fig. 6.43 Location of a grain boundary insulation type barrier layer capacitors in a tuner. Products with high electrostatic capacitance are used as a coupling capacitor or a bypass capacitor in amplifier circuits

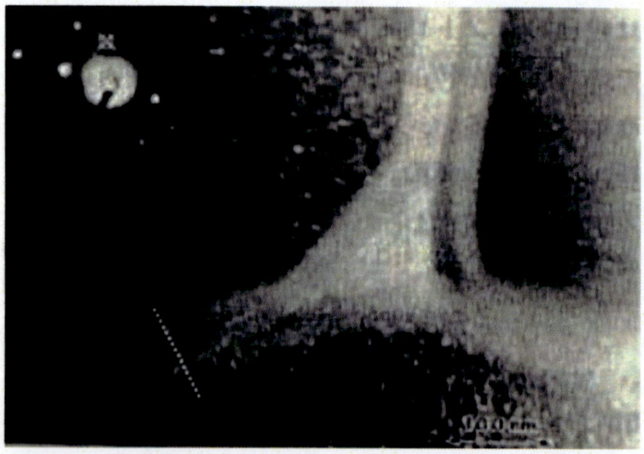

Fig. 6.44 TEM dark-field image of sample after heat diffusion of metal oxide and slow cooling. The liquid phase is present at the triple point only, and is not observed in the grain boundary area (*dotted line*)

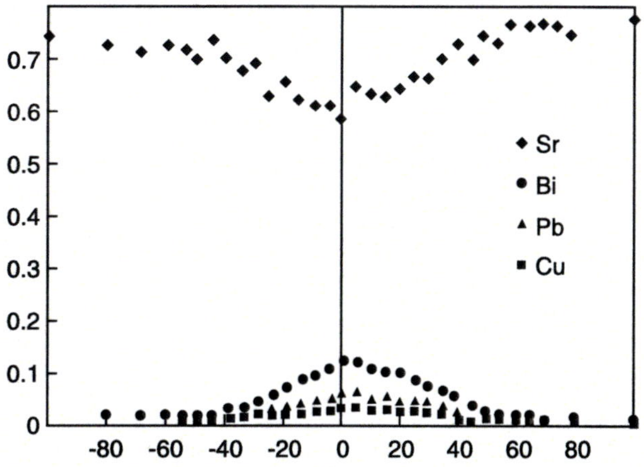

Fig. 6.45 Composition change of the area in the vicinity of grain boundary for a sample in which metal oxide was heat treated resulting in diffusion observed by STEM chemical analysis. It is observed that the diffused material diffuses from the grain boundary into the grains

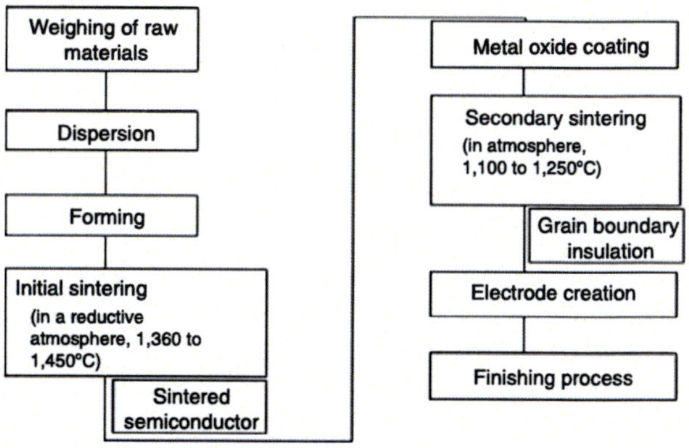

Fig. 6.46 The manufacturing process of grain boundary insulation type barrier layer capacitor. Ceramic powder is wet-dispersed, and the dried powder is mixed with a binder to create a paste. The paste is formed, dried and calcined

changes the diffusion material additive in the dotted area of Fig. 6.44, obtained by STEM analysis and indicated by distance from the grain boundary, is shown in Fig. 6.45.

6.8.2.2 Manufacturing Method

The manufacturing process of the basic grain boundary insulation type barrier layer capacitor is shown in Fig. 6.46. Trace additives, metal oxides such as Dy, Nd, Y and Nb, are mixed into the raw material. The mixed material goes through a wet dispersion process, followed by a drying process. The dried power is mixed with an organic binder and extrusion molded to a cylindrical shape or formed into a green sheet, followed by a drying process. The initial sintering is carried out in a reducing atmosphere in order to create a semiconductor-like ceramic with grains grown to an appropriate size. During the secondary sintering in air, the grain boundary becomes insulating by coating the semiconductor ceramic with metal oxide and its diffusion. The grain boundary insulation type barrier layer capacitor is manufactured with this method.

6.8.3 Future Prospects

Grain boundary insulation type barrier layer capacitors are gradually being replaced by multilayer ceramic capacitors as a result of rapid progress in capacitance improvement and the miniaturization of ceramic capacitors. However, grain boundary insulation type capacitors, with their stable dielectric characteristics and excellent frequency characteristics, will continue to play an important role.

Wound Chip Inductor (Mid 1980s) 6.9

Wound chip inductors are mainly used in filter circuits and switching circuits. Surface mounting mold type inductors were developed as replacements for conventional lead-type inductors in the mid 1980s in response to the miniaturization and performance improvements of home appliances. Later, in the late 1990s, they evolved into square chip inductors, following the spread of portable digital appliances such as digital cameras and digital video cameras. Square chip inductors have enabled further device miniaturization and performance improvement. Even today, lineups of inductors are increasing so as to meet the changing needs of devices.

6.9.1 Background of Development (Table 6.5)

6.9.1.1 Mid 1980s

Electronic devices are increasingly being miniaturized following the advancement of TVs and videos for consumer use, leading to a need for high density and multilayer mounting substrates. As a result of these advances, lead-type electronic components began to be replaced by surface mounting type components. In line with this trend, the LE series, which was manufactured by housing a lead type component in a resin mold, was developed for surface mounting, and came into widespread use.

6.9.1.2 Late 1990s

Following the spread of portable digital appliances such as digital cameras and digital video cameras, surface mounting components were required to be smaller in size with higher performance. In line with this trend, the resin mold type explained above evolved to the LB series, which was manufactured by forming electrodes directly on the ferrite core in order to reduce the dead space, and then housing it in a resin mold.

During this period, the performance of large appliances such as TVs and video players also improved, which accelerated the incorporation of surface mounting methods, and the products began to be used widely in DC/DC switching converters in addition to the conventional application to signal filters and power supply filters. To meet requirements, a wide variety of products featuring various shapes and characteristics were developed.

6.9.2 Characteristics

6.9.2.1 Product Characteristics

6.9.2.1.1 Mold Type

The lead-type inductor was originally made by winding a wire around a cylindrical core and coating the outer surface. The mold type was made by removing the lead from the cylindrical core and creating external terminals on a frame, housing them entirely in a resin mold (Fig. 6.47).

6.9.2.1.2 Square Chip Type

Silver electrodes were formed directly on a square core for miniaturization and product performance improvement, and the winding area alone was housed in a resin mold in order to reduce dead space as much as possible. Because of this structure, the winding area was enlarged, allowing the reduction of resistance and substantial performance improvement (Fig. 6.47).

Table 6.5 Progress of products since the 1980s

Structure	1980s	1990s	2000s
Lead type			
Mold type			
Square chip type			

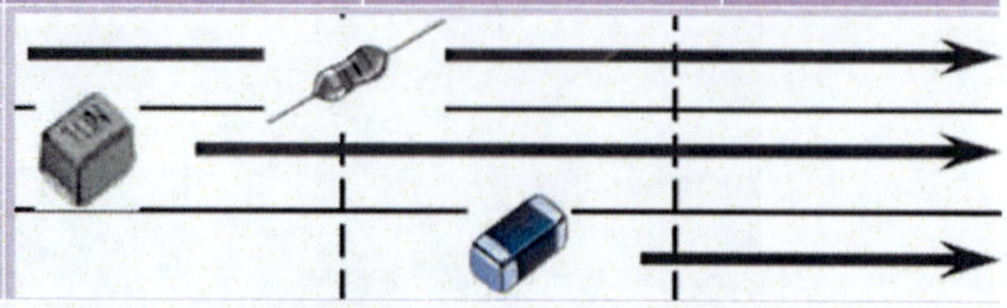

In the mid 1980s the surface mounting method began to be employed and mold type products were developed as a result of progress made in consumer appliances. Square chip type products, which are smaller in size and higher in performance, were developed following the expansion of portable digital devices in the late 1990s

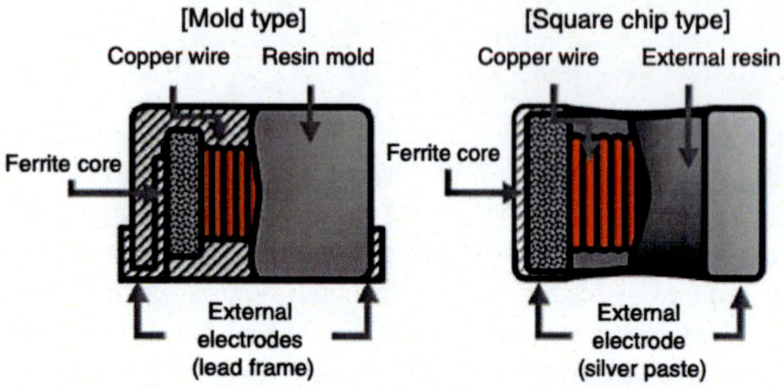

Fig. 6.47 Comparison of internal structure. Unlike conventional mold type products, the resin mold houses the winding area only in square chip type products, realizing effective use of that internal area and enabling miniaturization and performance improvement

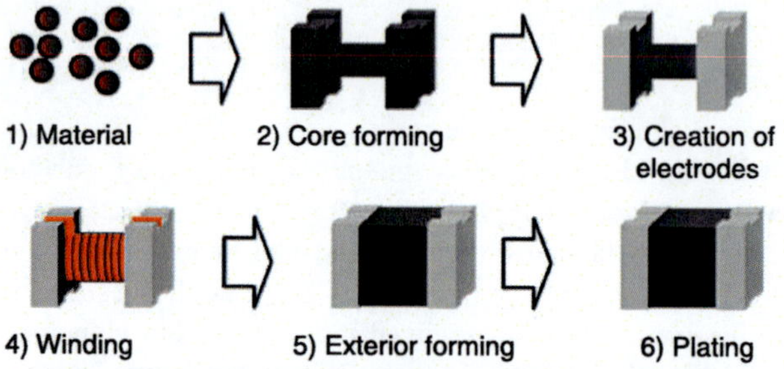

Fig. 6.48 Flow of basic manufacturing process. The characteristics are controlled for various applications by changing the materials used and dimensions

6.9.2.2 Manufacturing Method

The basic manufacturing process is shown in Fig. 6.48.

1. Ni–Zn ferrite material powder (Table 6.6)
2. Formed to a shape of square drum and sintered
3. Ag electrodes are formed on the ends of the core
4. A copper wire is wound around the core axis
5. The winding is housed in a resin molding
6. Completed by plating the electrodes

The characteristics are controlled by adjusting the materials and size to meet various requirements.

6.9.2.3 Lineup of Products (Figs. 6.49 and 6.50)

In addition to conventional series, the current lineup includes products of various sizes and shapes including special low-height shapes for use in filters and converters. The lineup features a variety of characteristics such as low resistance and large current.

Table 6.6 Physical values of Ni–Zn ferrite materials

Name	Material	AC initial permeability	Effective saturation magnetic flux density	Curie temperature
		μ_{iac}	B_{ms} (mT)	T_c (°C)
Material a	Ni–Cu–Zn ferrite	600	410	>210
Material b	Ni–Cu–Zn ferrite	240	370	>220

Material a is for high current, while material b is for normal to low resistance

AC initial permeability (μ_{iac}): permeability measured in minute AC magnetic fields

Effective saturation magnetic flux density (B_{ms}): magnetic flux observed when nearly saturated external magnetic field is applied

Curie temperature: the temperature at which the ferromagnetic property is converted to a paramagnetic property

6.9.3 Future Prospects

The need for compact and low-height products that feature low inductance will increase in switching devices, in response to the improvement in the speed of ICs. Meanwhile, the need for products that feature narrow tolerance (Note 6.24) will increase in filters. Further improvement and expansion of the lineup will be needed to meet these requirements.

Note 6.24 Means to narrow the guaranteed tolerance in the L value specified by the JIS standard, in which the tolerance is expressed as J (±5%), K (±10%), M (±20%), N (±30%), etc.

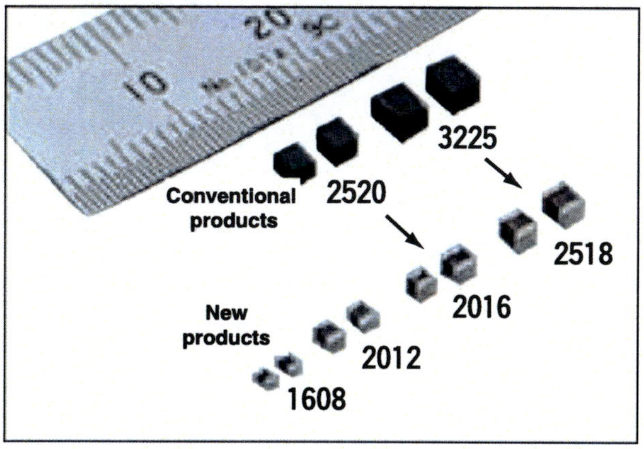

Fig. 6.49 Mold type products and square chip type products. The square chip type, which has less dead space than the mold type, is smaller in size yet realizes the same performance

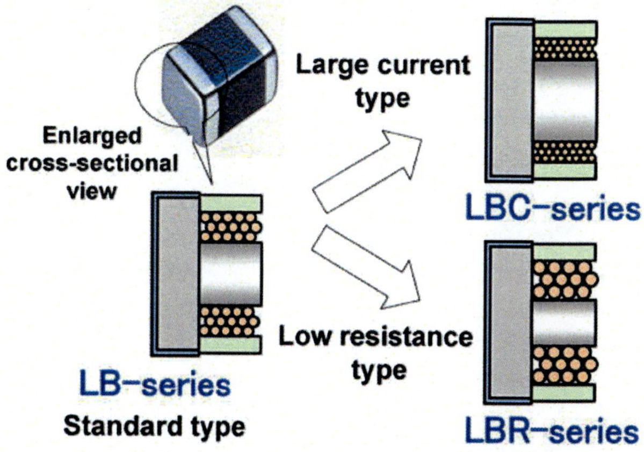

Fig. 6.50 Characteristics of internal structures. The characteristics are controlled by adjusting ferrite core dimensions and wire sizes to meet requirements. (1) The LBC series: core diameter was increased to enhance the degree of flux saturation under current application, so as to guarantee the L value in high current. (2) The LBR series: core diameter was reduced and a thick wire was used to keep the resistance low

Multilayer Chip Inductor (1980) 6.10

An inductor (coil) (Fig. 6.51) is an impedance element that utilizes the action of electromagnetic force generated by applying current to conducting wires wound around cores of ferrite, ceramic, etc. Because of this principle, the structure is more complex than other passive components such as capacitors and resistors, delaying its adoption for use in surface mounted devices (SMD).

An ultra compact SMD using no winding wires at all was realized by a multilayer chip inductor through the incorporation of multilayer integration technology, which enables micron-order lamination of ferrite and conductor materials. The multilayer chip inductor uses no mechanical windings at all, and therefore, can be miniaturized by refining the layer patterns, which is a major advantage and effective for reducing both size and weight.

It was first incorporated in portable radios and headphone stereos in the early 1980s, and currently sees widespread use in devices such as in the high-frequency circuits of mobile phones.

6.10.1 Background of Development

In the late 1970s, a portable radio that could be stored in a dress shirt pocket was commercialized. Surface mounting technology, a new mounting technology that was used only in electronic devices for industrial and military use, began to be widely used in consumer electronic devices, stimulated by the release of this portable radio. In surface mounting technology (Note 6.25), a mounting technology incorporating surface mounted devices (SMD) (Note 6.26), electronic components are soldered directly on a conducting pattern called the Land, which is created on a printed substrate or circuit board. Because through holes are not required in this method, component mounting density is enhanced, contributing to the drastic miniaturization of substrates and electronic devices.

Among the passive components, surface mounting technology was first applied to capacitor and resistors. With respect to inductors, the application of surface mounting technology was delayed because of its complicated structure, which was designed to utilize the action of electromagnetic force generated by current application on a conducting body wound around cores of ferrite, etc.

An inductor incorporating the laminating method (multilayer chip inductor) was developed in 1980 as a measure for application of the surface mounting method on inductors. Since that time, the multilayer chip inductor has been used in many electronic devices.

Note 6.25 A technology to mount electronic components on a printed substrate that is coated with solder paste by using an automatic mounting device. The substrate is then heated to a temperature exceeding the melting point of solder to mount the electronic component to the substrate.

Note 6.26 Devices on which surface mounting technology is applicable.

6.10.2 Characteristics

6.10.2.1 Features and Specifications of Products

According to their usage, multilayer chip inductors can largely be classified into two types (Table 6.7). The first type consists of a "multilayer ferrite coil," which uses a magnetic ferrite material and is operated in ranges from several MHz to a hundred MHz. It is used in the choke coil, the matching circuit, and the filter circuit, as well as in the resonance circuit in combination with a capacitor. Its magnetic shield structure allows high-density mounting. The inductor, featuring a high Q value, is highly effective for the miniaturization and weight reduction of devices. They come in three shapes, 1005 (1.0×0.5 mm), 1608 (1.6×0.8 mm) and 2012 (2.0×1.25 mm).

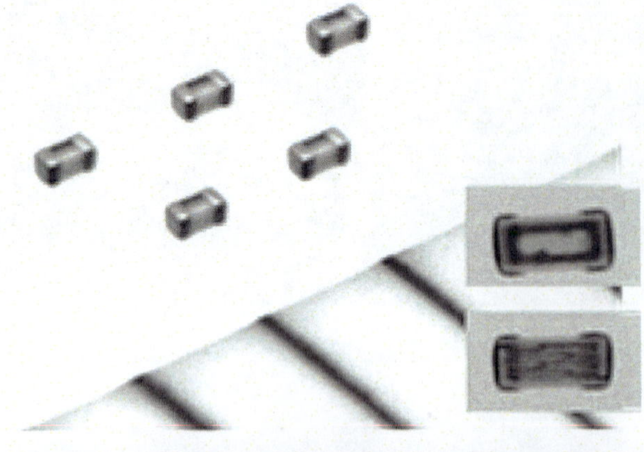

Fig. 6.51 The multilayer chip inductor (example: MLG0402). One scale represents 1 mm. An inductor incorporating the lamination method (multilayer chip inductor) was developed in 1980 as an application method of the surface mounting method on inductors. Its original size is 3216 (3.2 mm × 1.6 mm), but products as small as 0402 (0.4 mm × 0.2 mm) are mass-produced at present

Table 6.7 Types of multilayer chip inductors

Name of product	Multilayer ferrite coil	Multilayer ceramic coil
Image		
Material used	High-Q magnetic ferrite	Low-loss dielectric ceramic
Internal conductor	Ag	Ag
Main usage	General electronic devices	Mobile phones, etc.

Multilayer chip inductors are classified largely into two types according to the materials used. These types are the multilayer ferrite coil, which uses a magnetic ferrite material and is operated in the range of several MHz to a hundred MHz, and the multilayer ceramic coil, which uses dielectric ceramic and operates at high frequencies exceeding several hundred MHz

The advantages of the multilayer ferrite coil include its magnetic shield structure and a low direct-current resistance (RDC). It is a magnetic shield type (closed magnetic path) because the electrodes are sealed inside the magnetic ferrite. This means that the crosstalk (Note 6.27) among components is limited, making the product appropriate for high-density mounting.

The other type is a "multilayer ceramic coil (high-frequency coil)" used in the high-frequency circuits of mobile phones. It is operated at high frequencies around several hundred MHz. A dielectric ceramic material, which has high high-frequency properties, is used as the core material, because the frequency is very high and ferrite magnetic material cannot be used. A miniature multilayer chip inductor was realized by creating a three-dimensional internal conductor inside the dielectric ceramic material. The miniaturization of chips has been rapidly accelerated in response to reductions in the weight and size of mobile phones, and a miniature product dimensioned at 0402 (0.4×0.2 mm) has already been mass-produced. Figure 6.51 is a picture of the 0402 multilayer chip inductor. However, the 0603 product, which features an inductance value ranging from 1 to 100 nH in the E24 series (Note 6.28) supports lead-free soldering, and is still in the mainstream.

Note 6.27 A phenomenon where a signal track picks up the signal of another track or leakage flux. This phenomenon is expressed as crosstalk.

Note 6.28 The values of passive components such as resistors, capacitors and inductors are expressed not by whole numbers such as 1–3 but by standard figures obtained by dividing the numbers from one to ten by a geometric series, which is specified by the international standards (JISC 5063 in Japan). The values are called the E series. The E24 series are the series obtained by dividing the numbers from 1 to 10 by 24 and are very small. The series from E3 to E6, the E12 series and the E24 series are common and frequently used.

The differences in the electrical properties of the multilayer ferrite coil and the multilayer ceramic coil, as well as product lineups, are shown in Fig. 6.52 and Table 6.8.

6.10.2.2 Manufacturing Method

The manufacturing method of multilayer chip inductor is shown in Fig. 6.53. It is similar to the manufacturing method of multilayer chip capacitors, but uses different via hole processing.

The internal electrodes of the multilayer chip inductor need to be three-dimensionally structured in a spiral form. Therefore, via holes are opened in the sheets for vertical conductivity. The via holes are created mainly by laser radiation. Silver, Ag, is generally used as the internal electrode material. Ag, with its high electric conductivity, is capable of suppressing the copper loss, improving the Q value. The melting point of Ag is 962°C, and therefore, the sintering temperature has to be less than 900°C to use Ag as the internal electrode material. The ferrite material and ceramic material that can be sintered at this temperature were developed for this reason.

6.10.3 Future Prospects

An increasing number of multilayer ferrite coils are used today as current-responsive choke coils in DC–DC converters. Development efforts for ferrite materials and proposals for current-resistant multilayer structures have been made to

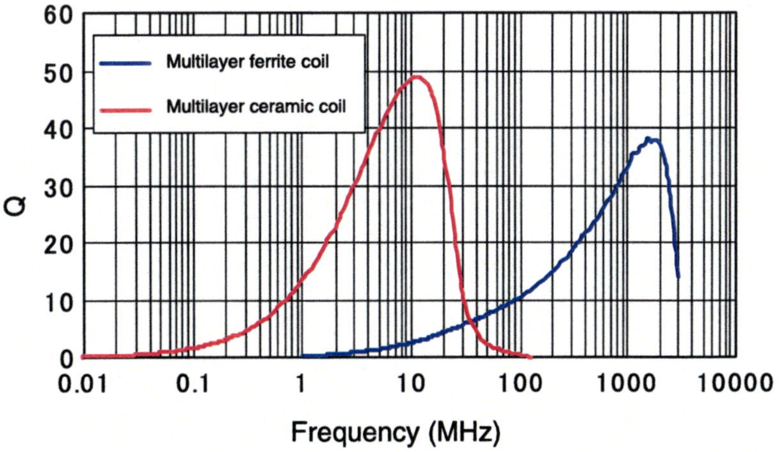

Fig. 6.52 Electrical characteristics (Q-frequency characteristics). The multilayer ferrite coil and multilayer ceramic coil are compared by Q-frequency characteristics. Q (quality factor) indicates coil quality and the loss becomes smaller as the value increases. Its relationship with tan δ, which indicates capacitor quality, is expressed as $Q = 1/\tan\delta$

Table 6.8 Lineups of multilayer chip inductors

Shape	Dimensions	Inductance (μH)	Tolerance
Lineup of multilayer ferrite coils			
1005	1.0 mm × 0.5 mm × 0.5 mm	0.1–2.2	±10%
1608	1.6 mm × 0.8 mm × 0.8 mm	0.047–33	±5%, ±10%
2012	2.0 mm × 1.25 mm × 0.85 mm or 1.25 mm	0.047–100	±5%, ±10%
Lineup of multilayer ceramic coils			
0402	0.4 mm × 0.2 mm × 0.2 mm	1.0–12	±0.2 nH, ±0.3 nH, ±5%
0603	0.6 mm × 0.3 mm × 0.3 mm	0.3–100	±0.2 nH, ±0.3 nH, ±5%
1005	1.0 mm × 0.5 mm × 0.5 mm	0.6–390	±0.2 nH, ±0.3 nH, ±5%
1608	1.6 mm × 0.8 mm × 0.8 mm	1.0–270	±0.3 nH, ±0.5 nH, ±5%

Lineups of multilayer ferrite coils and multilayer ceramic coils are shown above. The multilayer ceramic coils that are used in mobile phones have been miniaturized at a rapid pace

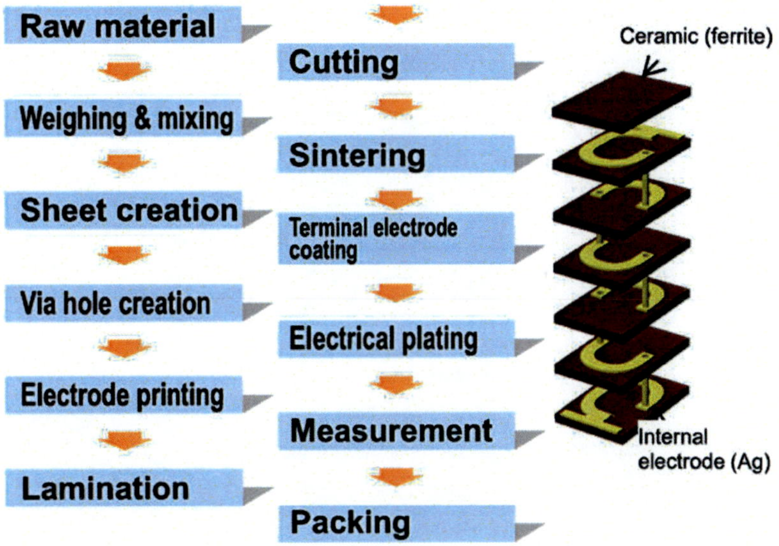

Fig. 6.53 Manufacturing process. The internal electrodes of a multilayer chip inductor need to be three-dimensionally structured in a spiral form. Therefore, via holes are opened in the sheets for vertical conductivity. The via holes are created mainly through laser radiation

meet current requirements, and the product will continue to be used for this purpose.

With respect to the multilayer ceramic coil, a small inductance is generated inside the module substrate of a high-frequency module intended for mobile phones (the majority of which incorporate the multilayer ceramic coil) by using the LTCC substrate. The trend of mounting the inductor inside the substrate will continue to accelerate. However, the multilayer ceramic coil will also continue to be used in large coils featuring high inductance and in devices that feature high Q characteristics and narrow tolerance. It is anticipated that development will continue to find measures for devices with high performance and narrow tolerance.

PTC Thermistors (1959) 6.11

The name thermistor comes from *thermally sensitive resistor*, and they are classified into two types: one with a positive temperature coefficient (PTC) and one with a negative temperature coefficient (NTC). The PTC thermistor is produced by adding a small amount of a rare earth element to $BaTiO_3$, which is known as a material for condensers. The element is characterized by its extremely large change in temperature vs. resistance characteristics, i.e., low electrical resistance at room temperature and a rapid increase in the resistance at and above the Curie temperature (Note 6.29). A variety of home appliances that utilize the characteristics of this element have been commercialized. They are widely used for a variety of purposes, including demagnetizing color TVs, starting refrigerator motors and protecting electronic circuits by way of overcurrent protection and overheat protection. Thermistors are expected to continue to be utilized in a variety of electric and electronic devices.

6.11.1 Background of Development

The PTC thermistor was discovered by Haayman in 1952. Then, the basic mechanism of the PCT thermistor was summarized by Saburi in 1959. The major properties of the PTC thermistor are explained in Fig. 6.54. Later, the PTC thermistor was mass produced in Japan in 1961, for the first time in the world. After that time, its applications expanded to various products. In the early days, it was commercialized as a constant-temperature heating element featuring static characteristics and was used in heaters. Initially, the PTC thermistor was used in electronic heating pads, electronic *kotatsu* (heated tables) and electronic rice cookers. Then, it was applied to ceramic fan heaters. After that, it began to be used for demagnetization (Note 6.30) as well as for starting motors, taking advantage of its dynamic characteristics and in color TVs and refrigerators. Today, substrate-mounted chip PTC thermistors are used in applications such as detecting the overheating of MOS-FET (Note 6.31) in the power supply circuit mounted on LSIs (large scale integration) and protecting USB (universal serial bus) circuits against excessive current.

6.11.2 Characteristics

The PTC thermistor, featuring a positive temperature coefficient, is characterized by three fundamental characteristics which vary with its operation: electrical resistance-temperature variation, shown in Fig. 6.54, static or steady-state operation (volt–ampere characteristics), and dynamic operation (current–time characteristics). They are shown in Fig. 6.55. It is also influenced by its heat capacity due to its structure and dimensions, heat emission coefficient of the surrounding media, thermal time constant and permissible power.

Characteristics of products that utilize these fundamental properties are explained below.

6.11.2.1 Utilization of Resistance-Temperature Characteristic

The PTC resistance-temperature characteristic is utilized for temperature detection. Electrical resistance variations caused by changes in ambient temperature are converted to electrical signals for temperature detection, temperature compensation

Note 6.29 The temperature at which the electrical resistance starts to increase rapidly. The Curie point is the temperature at which the electrical resistance value doubles compared with the electrical resistance value at 25 °C.

Note 6.30 Cathode-ray type picture tubes of color TVs are magnetized, causing permanent color shading, due to geomagnetism and the external magnetic field. AC current is used to demagnetize the tubes.

Note 6.31 Metal Oxide Semiconductor Field Effect Transistor.

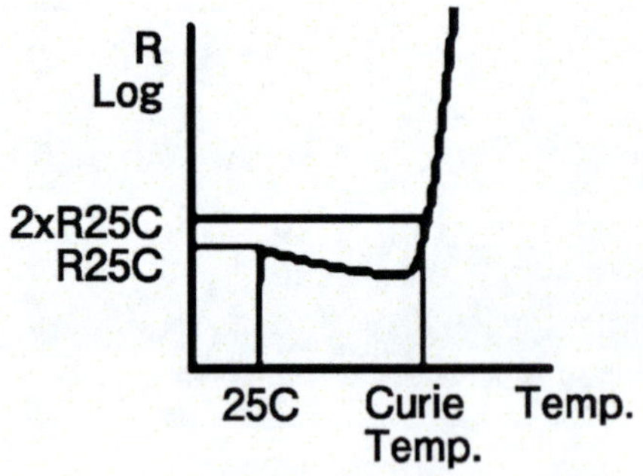

Fig. 6.54 Electrical resistance temperature characteristics. The figure shows the typical characteristics of a PTC thermistor. The *vertical axis* R indicates the electrical resistance value. R25C is the electrical resistance value at 25 °C, and 2×R25C is twice the electrical resistance value at 25 °C. The Curie temperature can be changed by adding Sr, or Pb to control the operating temperature

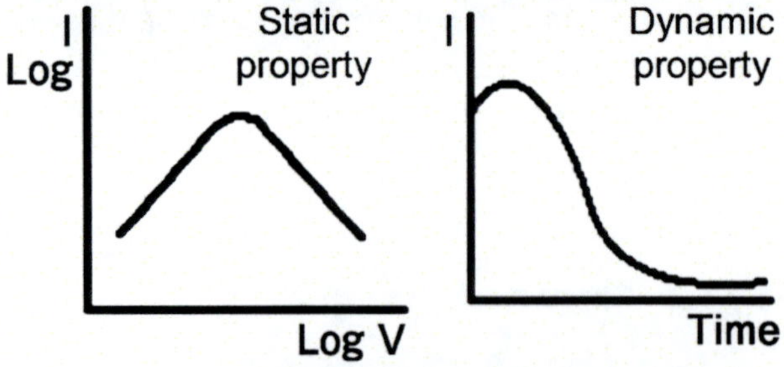

Fig. 6.55 Static or steady-state and dynamic operation. The static property expresses the volt–ampere characteristics and the electrical resistance remains constant in the area to the left of the local maximum point, and the electrical power remains constant in the area to the left of the local maximum point. Dynamic property expresses the current-time characteristics, indicating that the electrical resistance value increases due to self-heating after voltage application, attenuating the current

and overheat detection. Lead-frame type packages were conventionally used for this purpose, but they are being replaced by surface mount chip-type packages in response to market trends toward compact, thin and low-height devices, and the size has been reduced to 1005 (Note 6.32). Figure 6.56 shows examples of circuits used for overheat detection.

6.11.2.2 *Utilization of Steady-State Operation*

Steady-state operation of the PTC is the condition where the applied voltage and the current are stable that is where internal heat generation is kept in equilibrium with external heat dissipation. If the operating point is set above the local maximum point of current, the current is kept constant and the PTC thermistor temperature adjusts to the new equilibrium. Therefore, the steady-state condition utilizes constant-temperature heat generators and heaters. The PTC heater does not require an ON–OFF controller, unlike nichrome heaters, and is capable of retaining a constant temperature without fluctuations. The operating point remains below the local maximum point of the current under steady-state. Therefore, the property is also utilized to protect against overcurrent because the current in a circuit can be limited by setting the circuit so that the operating point is shifted to an area above the local maximum point of current when an abnormality occurs in the electrical circuit, causing excessive current. When it is used to protect against overcurrent, the function is automatically restored when the abnormality disappears, and this is one of the major characteristics of the PTC thermistor.

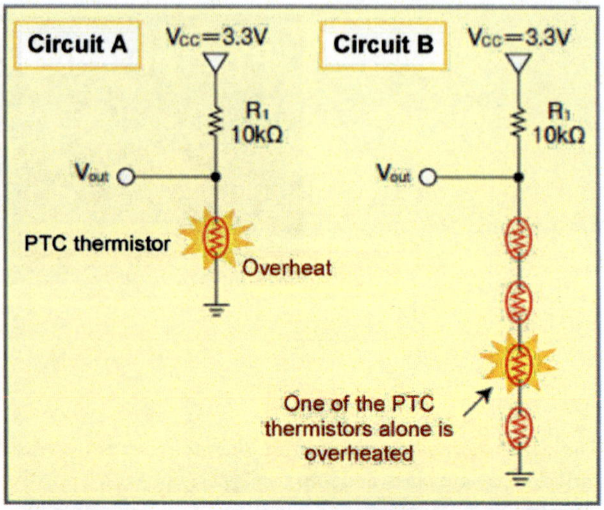

Fig. 6.56 Example of a circuit used for overheat detection. Overheat is detected at one point in circuit A, while it is detected at multiple points in circuit B. Abnormal overheating at multiple points can be detected by tandemly arranging PTC thermistors in the detection circuit

6.11.2.3 *Utilization of Dynamic Characteristics*

Operation of the PTC dynamic mode (current–time characteristics) causes the electrical resistance value to drop and the current to decrease over time due to the self-heating of the element, while a large current flows immediately after voltage application, when the electrical resistance value is low. This property is utilized in TVs for demagnetization as well as for starting motors.

Note 6.32 Method of expressing chip resistors sizes. In this case, it is 1.0 mm L×0.5 mm W.

6.11.2.4 Manufacturing Method

The PTC thermistor is manufactured by adding a small amount of dopants such as rare earth elements for creating semiconductors and a slight amount of property improvement agent such as Mn to its major components, $BaCO_3$ and TiO_2 powders. A separate dopant (Sr or Pb) is then added for the desired Curie temperature and the powder is then lightly calcined, a binder is added, and it is cast, calcined with electrodes. The PTC thermistor usually incorporates ohmic Ag electrodes or alternatively Zn in Ni plating or base Ni plating. It does not use the typical Ag electrodes used in general ceramic electronic components because these are n-type semiconductors and the PTC requires ohmic contact electrodes.

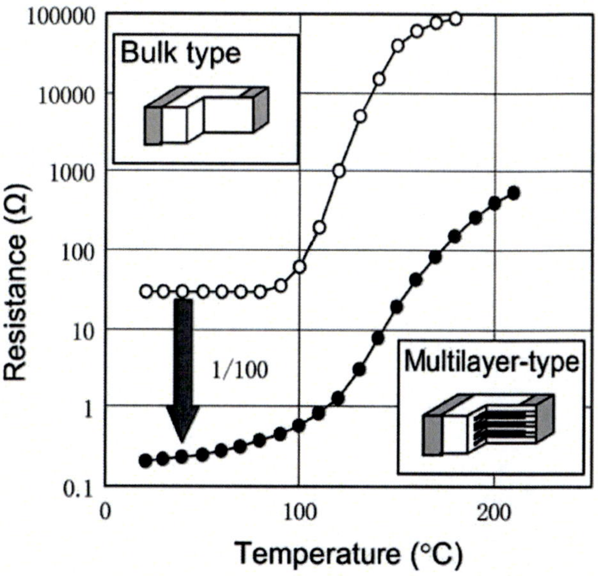

Fig. 6.57 Reduction of electrical resistance by adopting a multilayer type PTC. Electrical resistance temperature characteristics of the conventional bulk-type and the multilayer type, both incorporating PTC thermistor chips sized 2012, are compared in the figure. The minimum electrical resistance value of the multilayer type is one-hundredth of the minimum electrical resistance value of the conventional bulk-type

6.11.3 Future Prospects

Recently, a multilayer PTC thermistor (Fig. 6.57), a new PTC thermistor featuring one-hundredth the electrical resistance of conventional bulk-type elements, was commercialized. Demands for a lower electrical resistance have been strong in the PTC thermistor market, but it was difficult to calcine PTC ceramics without oxidizing the internal base metal electrodes. This is because a base metal such as Ni has to be used to form ohmic contact (Note 6.33) electrodes, while the PTC property is lost when processed under a reducing atmosphere. To solve these conflicting issues, research on calcination technologies were developed, and a multilayer PTC thermistor that uses Ni for the internal electrodes was commercialized in 2003, solving these longstanding issues. A photograph of a PTC is shown in Fig. 6.58. The electrical resistance of the PTC thermistor will further decrease in the future through enhanced scaling, creating thinner layers and improving PTCs and the PTC thermistor market will steadily expand.

In general, PTCs contain a small amount of Pb, which is added to increase the Curie temperature. Research has been undertaken to develop a technology to prepare a lead-free device, responding to the recent trend toward reduction of environmental burden. A technology for totally removing Pb from PTC thermistors is likely to be developed in the near future. PTC thermistors have been in the market for more than half a century, but active research and development still continues to produce new products even today. PTC thermistors are expected to continue to be used in home appliances, electronic devices, and other products.

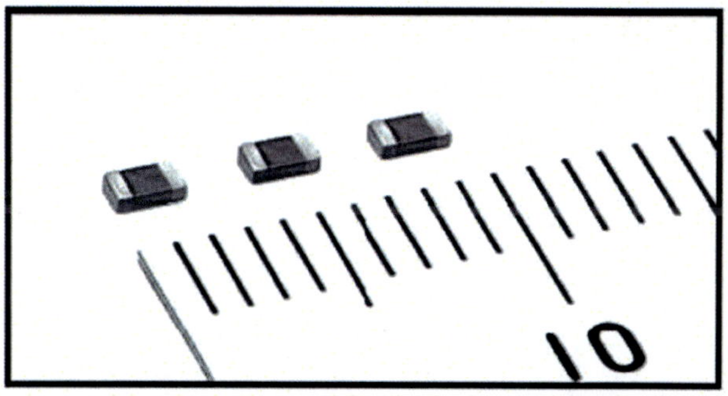

Fig. 6.58 World' first multilayer PTC thermistor. Photographs of multilayer PTC thermistors, 2012 (2.0 mm × 1.2 mm), R25 = 0.2 Ω. They are produced by reducing atmosphere calcination and reoxidation, and using Ni for the internal electrodes

Note 6.33 Resistive contact in which the electrical resistance value remains constant regardless of the electrical current direction and voltage height. Voltage V is proportional to current I with R as a coefficient.

Noise Suppression Sheet (1995) 6.12

Noise suppression sheets are made of composite materials in which thin and flat metal flakes are dispersed and aligned in one direction and in high density in polymers. These magnetic sheets feature high flexibility and workability. They are characterized by a high electrical resistance and large magnetic loss in the UHF band (300 MHz to 3 GHz), and therefore, can absorb (and convert to heat) high frequency noise without side effects. Inside mobile electronic devices such as mobile phones and digital cameras, where many components are mounted in high density, various magnetic interferences can occur that are hard to anticipate at the designing stage, frequently cause external noise radiation and internal malfunctions. Noise suppression sheets are widely incorporated, especially in compact digital electronic devices, as an easy-to-use measure against problems caused by high frequency magnetic fields.

6.12.1 Background of Development

Mobile phones these days are equipped with a variety of functions, including high-pixel digital cameras and RFID (radio frequency identification) (Note 6.34) units for electronic money functionality, and have been evolving remarkably as personal IT terminals. Mobile phones contain many densely mounted with various electronic components to support these functions, resulting in the frequent occurrence of intra-system electro-magnetic interference, which is a major obstacle in realizing both high speed and reliable operation of all functions in conformity with the original design. To overcome this serious problem of high-frequency electromagnetic interference, a composite structure noise suppression sheet (Busteraid®) made by dispersing flaked soft magnetic metal powder in polymers was proposed and commercialized.

The noise suppression sheet, featuring high electric resistance and loss properties based on the frequency selectivity attributable to magnetic resonance, is capable of effectively suppressing high-frequency noise without side effects such as the deterioration of signal waves. Therefore, the sheet rapidly gained popularity as an easy-to-use solution and the International Electrotechnical Commission (IEC) (Note 6.35) has been working on standardization of the product. With respect to portable electronic devices, the suppression of electrostatic discharge (ESD) and enhancement of RFID communication distance are also big issues. The capacitance of the noise suppression sheet is highly effective as a solution to the ESD issue, while the magnetic shield properties of the sheet provide a good solution for the RFID issue.

6.12.2 Characteristics

6.12.2.1 Structure of the Noise Suppression Sheet

Figure 6.59 shows how the signals and noise are transmitted when the microstrip transmission line is mounted with a noise suppression sheet. When signals containing noise enter or incident the line, a slight reflection of signals/noise occurs directly under the noise suppression sheet (the reflection ratio is called S_{11}). In general noise control products, the transmission of noise is suppressed by minimizing the reflection of signals (low-frequency components) and reflecting the noise

Note 6.34 A mechanism for identifying humans and identifying and controlling objects through the use of a micro wireless chip. It was originally developed for product identification and control technology to replace barcodes used in the retail industry, but has also been utilized widely for the introduction of IT and automation to society.

Note 6.35 An international organization that coordinates and creates standards for the fields of electricity, electronics, communication and atomic energy. It was established in 1906, and has been responsible for the electric/electronic sector of ISO since 1947. It is headquartered in Geneva, Switzerland.

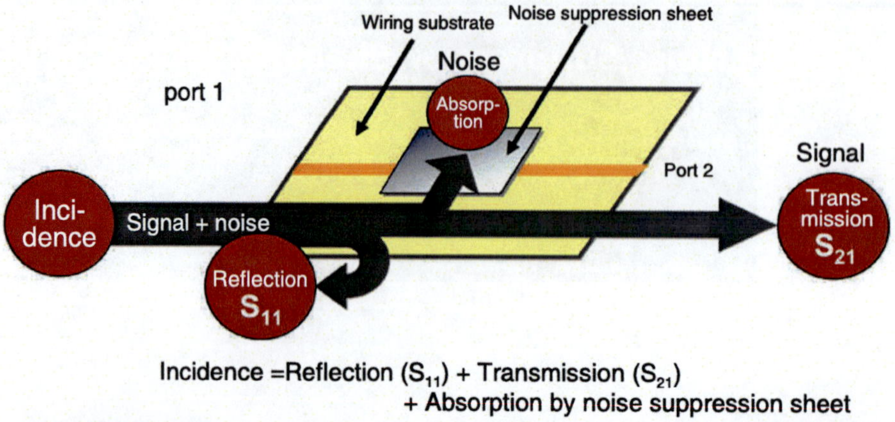

Fig. 6.59 Transmission characteristics changed by noise suppression sheet. When signals (input) containing a noise component incident a circuit mounted with a noise suppression sheet (NSS), a slight reflection (S_{11}) occurs directly below the NSS, due to impedance mismatching. The majority of the remaining noise components (i.e. high-frequency components) are absorbed (converted to heat) due to the high-frequency magnetic loss of the NSS, while the majority of signal components are transmitted without being absorbed (S_{21})

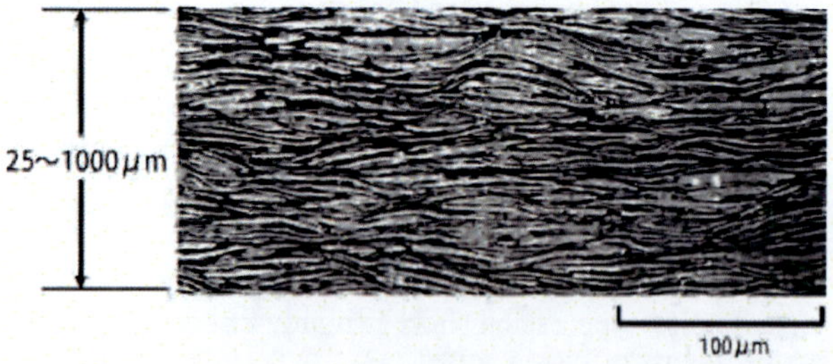

Fig. 6.60 Cross sectional SEM image of a noise suppression sheet. The noise suppression sheet is made by densely arranging flat-shaped soft magnetic metal powders with a high aspect ratio and aligning the same direction in polymers. This structure reduces demagnetizing fields ($N_d \cdot M_s$) and displacement currents, the factors that drop effective magnetic permeability (μ_{eff}) in the in-plate direction of the sheet, realizing a large μ_{eff} and high electric resistance in the high-frequency region

(high-frequency components) only, by taking advantage of the difference in frequencies between signals and noise. With this method, however, the reflected noise can cause other electromagnetic interference. With the noise suppression sheet, signals and noise enter the transmission line with minimal reflection, and the high-frequency current, i.e. the noise component, is absorbed (converting current to heat) through the magnetic loss of the noise reduction sheet placed directly above it. The noise reduction sheet functions as an absorption lowpass filter. Therefore, the noise suppression sheet is required to have (1) high electrical resistance that prevents reflection and (2) a large amount of resonance-type loss components (μ'', ε'', ρ) that speed up the cutoff operation of the lowpass filter. Figure 6.60 is a cross sectional SEM image of the noise suppression sheet. It is of a composite structure, in which soft magnetic metal (Fe–Si–Al alloy) powders, in the form of flakes as thin as approximately 1 μm, are densely dispersed and aligned the same direction, in polymers on the sheet. Because of the two-dimensional shapes of the powder and the sheet, the in-plane direction demagnetizing field on the sheet is substantially reduced, and the imaginary part permeability μ'', a loss term, is enhanced, and the generation of overcurrent that inhibits the frequency property of permeability is blocked, realizing a large permeability and resonance-type permeability distribution in the UHF bands as well as an absorption-type noise solution with limited side effects.

Radio frequency identification systems (RFID) that utilize 13.56 MHz are widely used, and mobile phones are equipped with RFID tags, providing the phone with mobile wallet and credit functions. In using these functions, if the device mounted with an RFID tag has a high conductive property, a reverse magnetic field is generated by the eddy current, severely deteriorating the transmission distance. This problem is prevented by utilizing the magnetic shield property of the noise suppression sheet to shunt the high-frequency magnetic field generated from the antenna. Figure 6.61 shows the frequency characteristics of the complex magnetic permeability of the noise suppression sheet. If the thickness of a magnetic material is thinner than the depth from the surface (Note 6.36), resonance of magnetic spins occur at specified frequencies, reducing the real part magnetic permeability

Note 6.36 The depth where a electromagnetic field that incidents a material is attenuated to 1/e (≈1/2.718≈ −8.7 dB). It is expressed by 1/ , in which the magnetic permeability of the material is μ and the electric conductivity is σ.

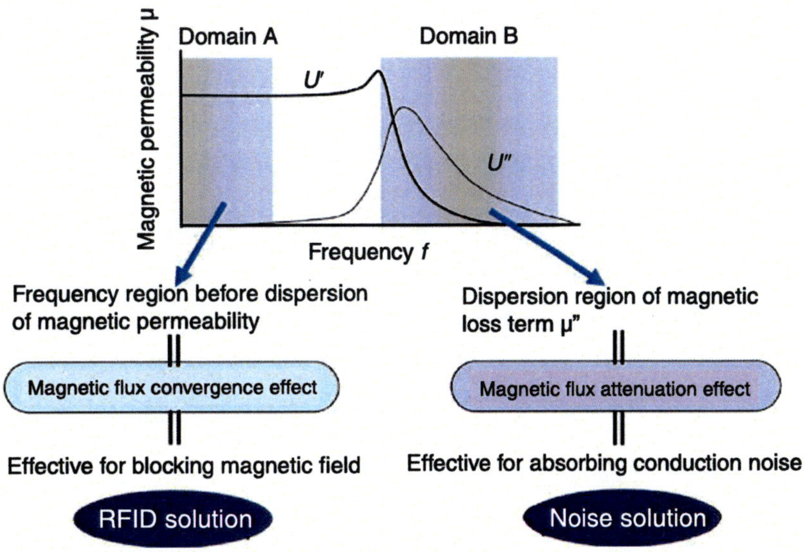

Fig. 6.61 Two functions of the noise suppression sheet. A reversal of magnetization cannot follow the frequency when the frequency of external magnetic field becomes high. This state is expressed as a function of frequency, separating the magnetic permeabilities into the actual part (μ′) and the imaginary part (μ″), in Fig. 6.61. For suppressing high-frequency noise, the domain (domain B) where the imaginary magnetic permeability (μ″), the magnetic loss term, is large, is utilized. Meanwhile, the domain (domain A) where the actual magnetic permeability (μ′) is large and the imaginary magnetic permeability (μ″) is small is used for blocking of the magnetic field such as improving the transmission/reception distance of the RFID

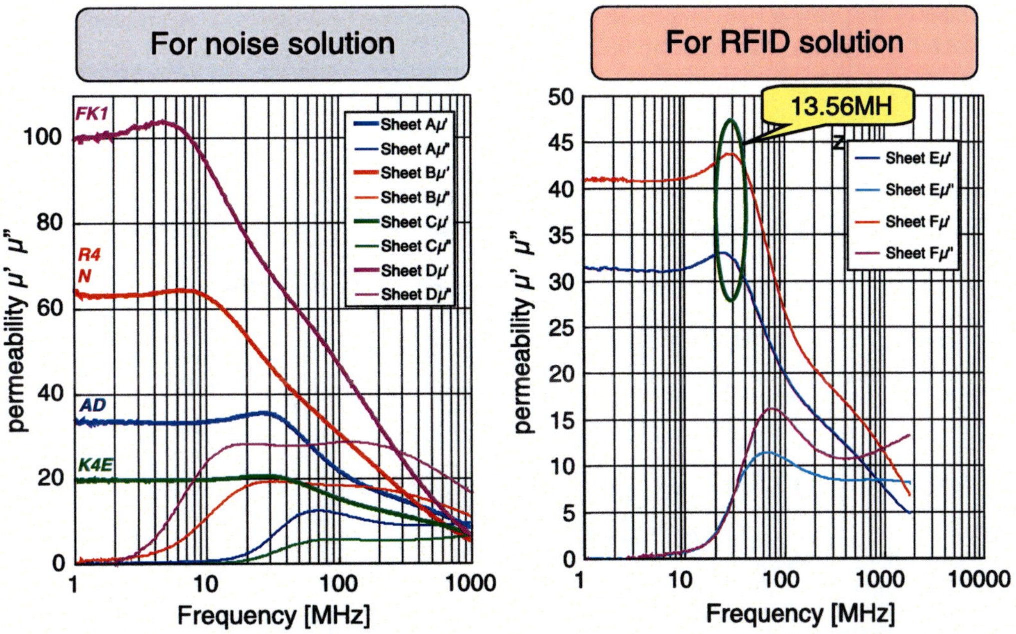

Fig. 6.62 Examples of noise suppression sheet permeability properties. As indicated in Fig. 6.61, the frequency property requirements of magnetic permeability vary depending on the usage of the noise suppression sheet. Figure 6.62 shows the frequency properties of magnetic permeabilities of two noise suppression sheets, one intended for a noise solution and the other for an RFID solution

(μ') and generating imaginary part magnetic permeability (μ''). In this process, the frequency domain (domain A) before the start of μ'' generation indicates the convergence of magnetic flux (magnetic shield effect), and domain B, where μ'' is large, indicates an attenuation of magnetic flux. The magnetic shielding function is required for the improvement of transmission/reception sensitivity of the RFID tag, requiring a large actual part magnetic permeability μ' in the carrier frequency domain and the smallest possible μ'' at the same time. Figure 6.62 shows the frequency characteristics of various types of noise suppression sheet.

6.12.2.2 Mechanism and Effects of the Noise Suppression Sheet

The most important effect of the noise suppression sheet is the conduction noise suppressing effect. The frequency characteristics shown in Fig. 6.59, which explains the ratio of "absorption," are also important. In Fig. 6.63, the frequency characteristics are expressed by absorption ratios, which are indicated by the indexes $P_{\text{loss}}/P_{\text{in}}$ (=incident power − reflected power − transmitted power). Conduction noise is absorbed through the separation and absorption of signal/noise frequency domains by equivalent resistance with frequency properties. The absorption ratio $P_{\text{loss}}/P_{\text{in}}$ is proportional to the product of the magnetic loss term μ'' and the frequency f, therefore the noise suppression sheet needs to have resonance-type magnetic permeability. The compound magnetic material made by placing flat metallic powders in polymers is very effective for the "absorption" of high-frequency conductive noise. The noise suppression sheet is applicable to electronic device components that need to be controlled for suppression of noise radiation and signal integrity deterioration generated by high-frequency magnetic interference.

6.12.3 Future Prospects

Issues involving radiation noise and signal quality deterioration are on the rise and solutions to the issues are becoming more important than ever. Today, it is impossible to attain the planned operating quality without solving the issues of high-frequency magnetic interference. Various magnetic interference problems are arising now in minute electronic circuits such as semiconductor components, and the requirement for the thickness of the noise suppression sheet is about to reach 10 μm. Under these circumstances, ferrite-plated film (Fig. 6.64) and nano-granular magnetic thin film are attracting attention as materials for solving electromagnetic interference. The ferrite-plated film (Busterferrix®), invented by Professor Abe and his colleagues at the Tokyo Institute of Technology, is a new type of electromagnetic interference solution material that does not contain nonmagnetic inclusion

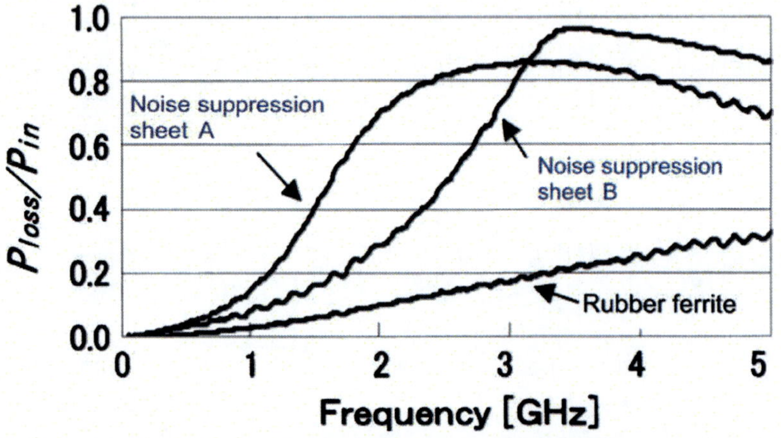

Fig. 6.63 Conduction noise absorption (P_{loss}/P_{in}) property of noise suppression sheets. The frequency characteristics of conduction noise absorption property (P_{loss}/P_{in}: ratio of "absorption" indicated in Fig. 6.59), under the condition that the noise suppression sheet is mounted on the wiring substrate as shown in Fig. 6.59. The noise suppression sheet features a higher absorption rate and sharper change in absorption amount in response to an increase in frequency, compared with rubber ferrite made by dispersing spinel-type ferrite powders on a rubber material

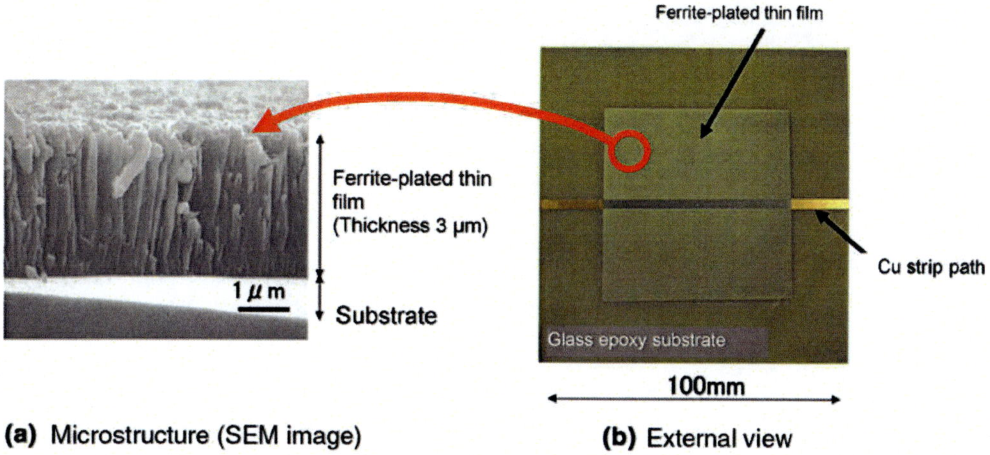

Fig. 6.64 Ferrite-plated thin film formed directly on a glass epoxy printed wiring substrate. (**b**) Shows the appearance of the ferrite plated thin film, which is attracting attention as the next-generation noise suppression material that will replace the noise suppression sheet, directly formed on glass epoxy substrate, a typical wiring substrate. (**a**) Shows the cross sectional SEM picture. The ferrite-plated thin film, featuring a high effective magnetic permeability μ_{eff}, can be formed directly on the wiring substrate and the film that is as thin as several μm provides a sufficient noise reduction effect

such as binders. It features high electric resistance and large resonance-type magnetic loss, making it possible to form a film directly on wiring substrates. With a width as thin as 3 μm, it has an excellent conductive noise absorption property and is applicable for practical use, enabling the suppression of high-frequency electromagnetic interference inside semiconductor components and multilayer wiring substrates.

Piezoelectric Gyro (1965) 6.13

Gyro sensors have been actively used in recent years in car navigation systems to detect the driving path and in cameras for shake detection. For detecting angular velocity, which is a physical quantity, by gyro sensors, the resonance type that utilizes the resonance vibration of solids is currently believed to be more advantageous in terms of cost and shape, aside from high accuracy types such as optical fiber. The piezoelectric gyros, which effectively utilize piezoelectric/inverse piezoelectric effects, are widely used for camera shake detection because they are becoming less expensive and feature high sensitivity. They have come to be incorporated in input devices such as the mice and remote controls of PCs and game consoles. Meanwhile, the electrostatically-driven resonance type has been gaining ground in the market, following advancement in Si deep-etching technology. Each type has found its own market, depending on its advantages.

6.13.1 Background of Development

The principle of the vibrating gyroscope was already explained in a 1943 U.S. Patent. The first piezoelectric gyro, the "Vyro," underwent full-scale commercialization by GE in the 1960s. Therefore, the principle of piezoelectric gyros is usually explained based on this GE model, which utilizes a quadrangular prism-shaped tuning fork in Japan, Imano, Tomikawa et al. have been actively researching it since the 1980s.

When a rotation is applied to a vibrating surface at right angles to the surface, an elliptic trajectory is traced out due to the influence of the Coriolis force (Note 6.37). As shown in Fig. 6.65, the free resonance in the quadrangular prism oscillates as if the central portion is a pendulum with its pivot point on the node. When a rotation is applied in the axial direction of the quadrangular prism, it starts an elliptic motion, due to the lateral displacement stress applied on the vertical vibration. If the frequency amplitude of the vibration is controlled to remain constant, the magnitude of the lateral stress applied on the ellipse is proportional to the input angular velocity (angular variation per second: deg/s). When a quadrangular prism with a piezoelectric element laminated on it is applied with a voltage, they expand and contract, vibrating like a bimetal (Note 6.38). Meanwhile, the piezoelectric element laminated on the side is stressed and generates a voltage, which is proportional to the Coriolis force. The gyroscope is designed to improve sensitivity by effectively using the piezoelectric property of reversibility and generating large amplitude in resonant condition.

Some of the piezoelectric gyros incorporate quadrangular/triangular prisms and cylinders, as explained above, and have both the longitudinal direction and the latitudinal direction on one surface, while others have an independent driving piece and a detecting piece, in the same manner as a tuning fork (Fig. 6.66). Recently, products that incorporate quartz and piezoelectric thin film have also been commercialized.

6.13.2 Characteristics

6.13.2.1 *Applications and Performance*

In the case of operating vehicles, where the center of rotation changes every second and is sometimes located outside the vehicle, the point of angle measurement is an unknown. The shaking of video cameras is attributable to the rotational

Note 6.37 Objects that move in rotating systems are subjected to forces vertical to both the axial direction and the moving direction and the forces are proportional to the movement speed. The force is called the Coriolis force, which is what creates the whirl in a typhoon.

Note 6.38 Made by joining two metal plates with different coefficients of thermal expansion (CTE). When the temperature changes, the two metal plates try to maintain balance and are flexed due to the difference in volume change arising from the difference of CTE. It is used as an easy-to-use temperature switch.

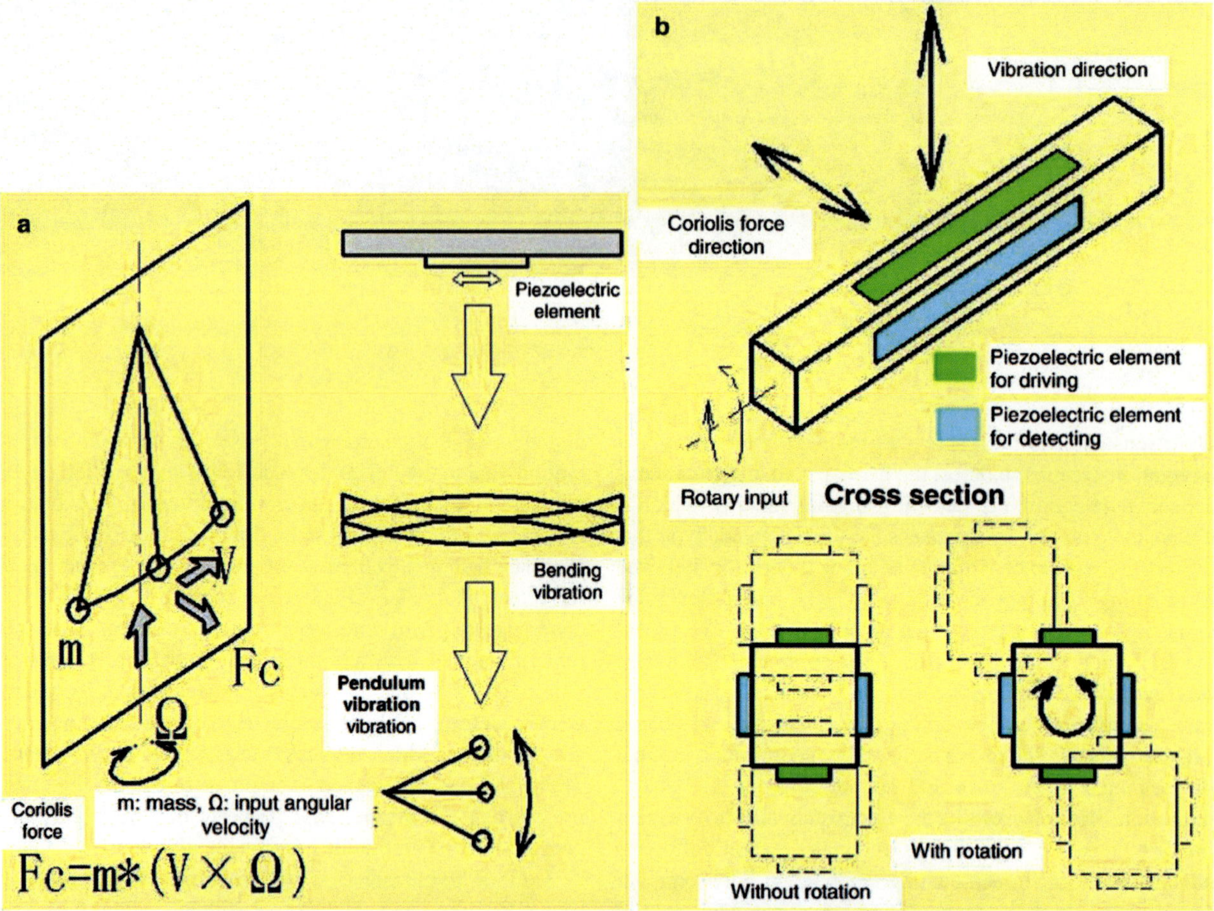

Fig. 6.65 (**a**) The Coriolis force. When a rotation is applied to a vibrating surface of a pendulum at right angles to the surface, a driving force is generated outside the vibrating surface. The driving force is called the Coriolis force and is proportional to the weight of mass point and vibration velocity. Meanwhile, if the plates are mounted with a piezoelectric vibrator and applied with a voltage, they expand and contract, generating flexural vibration. This means that an electrical pendulum can be formed by using piezoelectric elements. (**b**) Principle of detection. A vibrating quadrangular prism can be created by using piezoelectric elements. If a piezoelectric element is laid out at right angles to the vibrating surface, a voltage proportional to the stress induced by the Coriolis force is output. This voltage, if controlled to a certain mass and a certain vibration velocity, is proportional to the input angular velocity

motion having a pivot point on the shoulder or the wrist, however, it is unrealistic to mount the angle meter on joints. For measuring these rotational motions, angular velocity sensors that give the same value regardless of their location in a vehicle are essential.

For shake correction that captures the variation of AC current (peak of sensitivity at 10 Hz), miniaturization and cost reduction are required to respond to the miniaturization of devices, as well as a high-speed response. Meanwhile, for the measurement of motions where the angular variation from the initial direction matters, such as the measurement carried out in automobile navigation systems, the process of angular velocity integration is needed. This requires that the center point of right and left angular variations remains at the reference potential. Deviation from the reference voltage, i.e. the offset, is caused by temperature change, etc., but displacements are undesirable because they are added as angular errors in the integration process. Therefore, gyros intended for this usage are required to have an especially stable offset performance, and a memory is sometimes used for numerical correction.

6.13.2.2 Manufacturing Process and Circuit System

Among the various types of piezoelectric gyros that have been commercialized, the image and structure of a product incorporating a bimorph tuning fork that is most widely used for shake detection are explained in Fig. 6.67. A cast metal

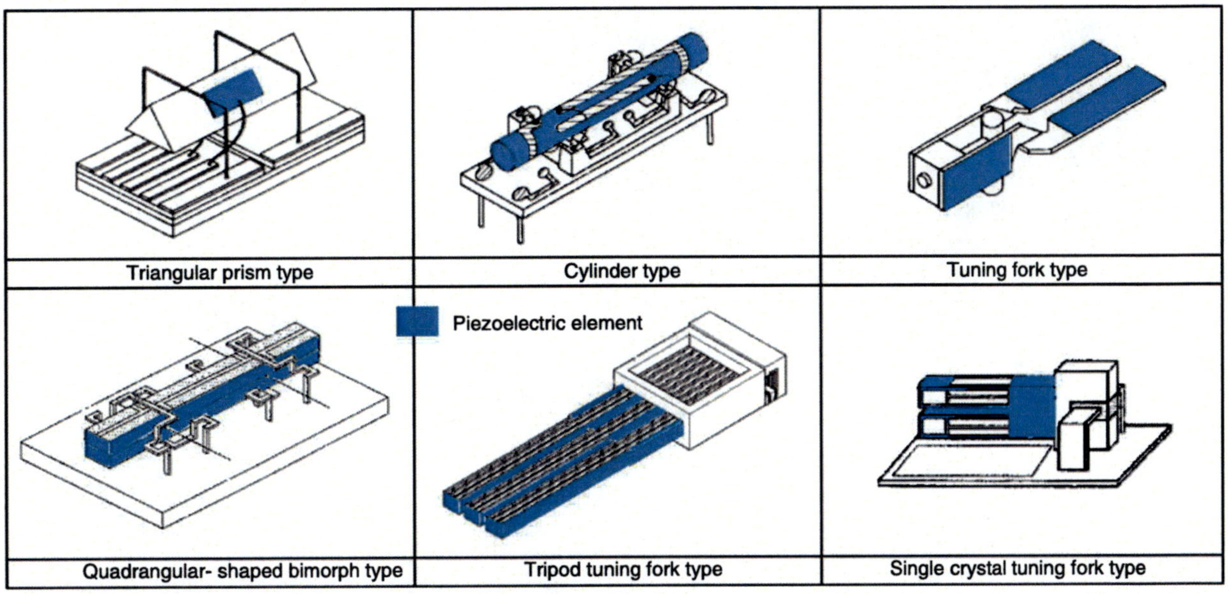

Fig. 6.66 Commercialized piezoelectric gyros. These are roughly classified into the stick type vibrator (tuning fork type), which are triangular prism, quadrangular prism and cylindrical in shape, and the tuning fork type, which varies in materials and shapes. Adapted from the Patent Gazette of Japan

part called a supporting pin is mounted on a substrate equipped with an ASIC (Application Specific Integrated Circuit) (Note 6.39). The bimorph vibrator is sandwiched from above and below and the support pin also functions to dispense from the electrode.

As shown in Fig. 6.68, the bimorph element is cut out of a wafer, which was made by laminating piezoelectric elements in opposite polarization directions from each other. If voltage is applied under this condition, the upper element contracts when the lower element expands, and the reverse phenomenon occurs if the reversed voltage is applied, exciting flexural resonance through AC signals. In addition, if the upper electrode is divided in half and the Coriolis force generated by rotation is applied to the halves, the expansion/contraction stress on the electrodes divided by the groove become opposite from each other, generating potential difference between the right and left electrodes. A voltage proportional to the Coriolis signal is generated by synchronously detecting (Note 6.40) the differential signals on the basis of the oscillating signal. Then, the voltage is transduced to direct current and output.

Note 6.39 ASIC stands for Application Specific Integrated Circuit, and is intended for special applications. They are large in size because they normally incorporate universal circuits which are sometimes unnecessary. Their characteristics are fully utilized because the application is limited, but they cannot be used for other purposes, meaning that they are unsuitable for small quantity production.

6.13.3 Future Prospects

The piezoelectric gyro is designed to detect minute forces as small as 1.01G, in terms of acceleration, by effectively utilizing the reversible piezoelectric property. However, as the accuracy of angular velocity detection improves, a more stable offset property is required. The MEMS (Micro Electro-Mechanical System) (Note 6.41) gyro, which uses silicon featuring greater material stability, has been attracting attention. It is anticipated that applications of vibration-type gyros will continue to expand through selection by cost and performance.

Note 6.40 A technology to pick up voltage and current values at a specified AC signal phase point. Various external noises are randomly generated, but signals buried in the noise can be effectively taken out by placing a focus on a certain frequency at a certain time point, taking advantage of periodicity in signals.

Note 6.41 MEMS stands for Micro Electro-Mechanical System. In a narrow definition, it is a mechanism for using static electricity to oscillate various shapes of resonators that are made by the microfabrication of silicon. In recent years, systems incorporating crystal, piezoelectric thin films, and other materials have come to be included in this category.

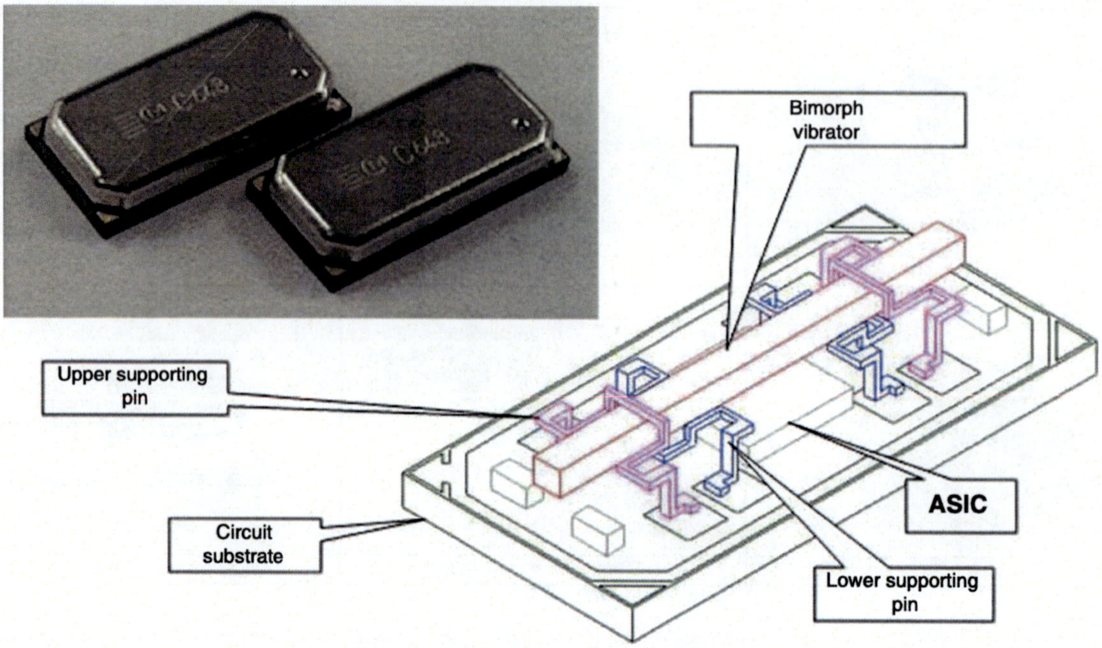

Fig. 6.67 Image and structure of bimorph piezoelectric gyroscope. ICs for driving and detection are mounted on the substrate, and electrical signals are exchanged between the ICs and bimorph elements via the three-dimensional supporting pins. Because of this three-dimensional structure, vibrations in the driving and detecting directions can resonate freely without being obstructed

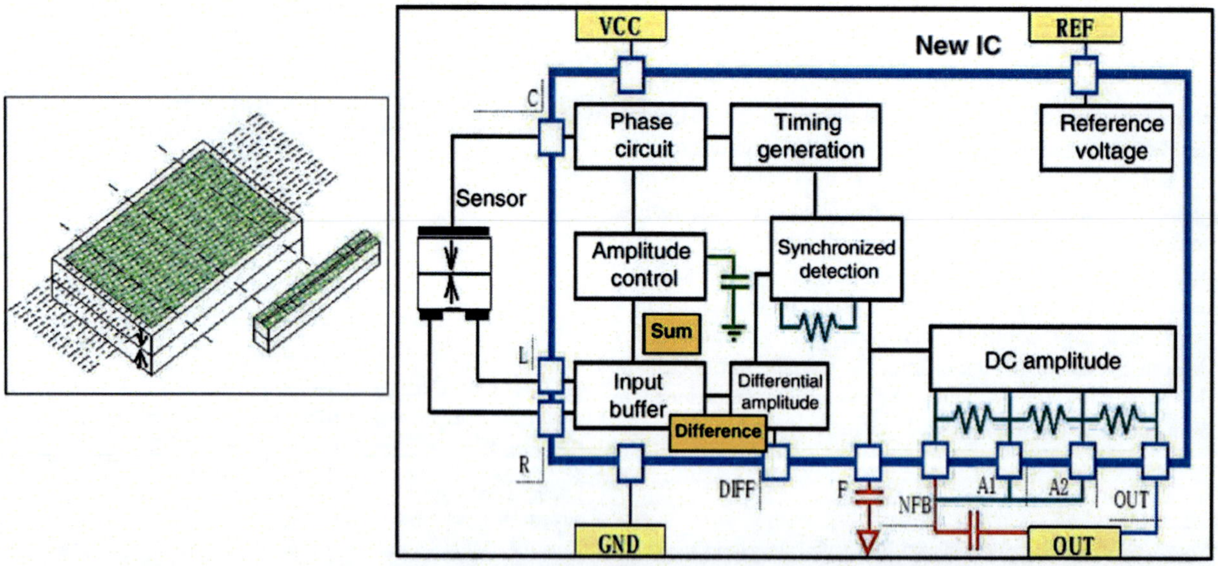

Fig. 6.68 Processing method. Bimorphs are cut out by a diamond blade, and several tens of bimorphs can be cut from one sheet of ceramic wafer. They can be driven and used for detection without any further processing because electrodes are formed at the same time. Signals are processed by the ASIC, Coriolis force signals only are extracted through the oscillation loop and the synchronized detection

Ultrasonic Motor (1986) 6.14

The ultrasonic motor is broadly interpreted as a piezoelectric actuator (Note 6.42) that utilizes a piezoelectric element (Note 6.43) to convert electrical energy to mechanical energy. The piezoelectric element generates ultrasonic waves (Note 6.44) in a vibrator (stator), driving a movable body (rotor or slider) via friction force. A rotation type ultrasonic motor was commercialized for the first time in 1986 in Japan. Compared with electromagnetic motors, ultrasonic motors are characterized by high torque with low-speed, high start/stop responsiveness, the retention of torque during shutdown periods, quietness, and no generation of magnetic noise. Meanwhile, although the degree of design freedom is comparatively high in ultrasonic motors, which incorporate simple-shaped piezoelectric elements and have a simple mechanism, there are some aspects, such as wear due to vibration and friction, that make the design of these motors difficult. A number of companies and universities have carried out research and development of ultrasonic motors and proposed a variety of types. Some of them have been commercialized.

6.14.1 Background of Development

Traditionally, electromagnetic motors are used in a wide variety of devices. However, new types of actuators are needed to realize miniaturization, accuracy enhancement and new functions. In particular, the piezoelectric actuator, which incorporates a piezoelectric element that converts electrical energy to mechanical energy, has been attracting attention and is expected to be applicable to a variety of instruments including optical, precision, information, medical, transportation, residential, and industrial instruments.

6.14.2 Characteristics

6.14.2.1 *Characteristics of the Ultrasonic Motor and Piezoelectric Element*

Figures 6.69 and 6.70 show rotation type ultrasonic motors commercialized in early 1990 for driving the autofocus (AF) lens of single-lens reflex cameras. Examples of actual applications of ultrasonic motors are introduced below with a focus on the piezoelectric element.

Figure 6.69 shows a ring-shaped ultrasonic motor, which is composed of a ring-shaped stator and a rotor, as shown in Fig. 6.69a. The stator was created by laminating a piezoelectric element made of piezoelectric ceramic on a ring-shaped metal. The piezoelectric element features an electrode pattern composed of Phase A and Phase B as shown in Fig. 6.69b so as to position different poling directions alternately along the circumference through poling treatment (Note 6.45). When Phases A and B of this piezoelectric element are, respectively, applied with high-frequency voltages with phases deviated by 90°, the stator goes into resonance (Note 6.46) conditions under a predetermined frequency. As a result, traveling waves caused by bending (flexural) vibration are generated in the stator as shown in Fig. 6.69c, and the rotor, pressurized by the stator, is rotated by the friction force.

Note 6.42 An actuator that utilizes a piezoelectric element to convert electric energy to mechanical displacement or force.

Note 6.43 An element made by creating electrodes on a shaped piezoelectric material.

Note 6.44 Sound waves and elastic waves above the audible frequency range, at 20 kHz or higher.

Note 6.45 The process of generating piezoelectricity through the application of DC voltage on a piezoelectric material for domain rearrangement.

Note 6.46 The phenomenon where the amplitude of a vibrator rapidly increases when the character vibration frequency of the vibrator matches the external vibration frequency.

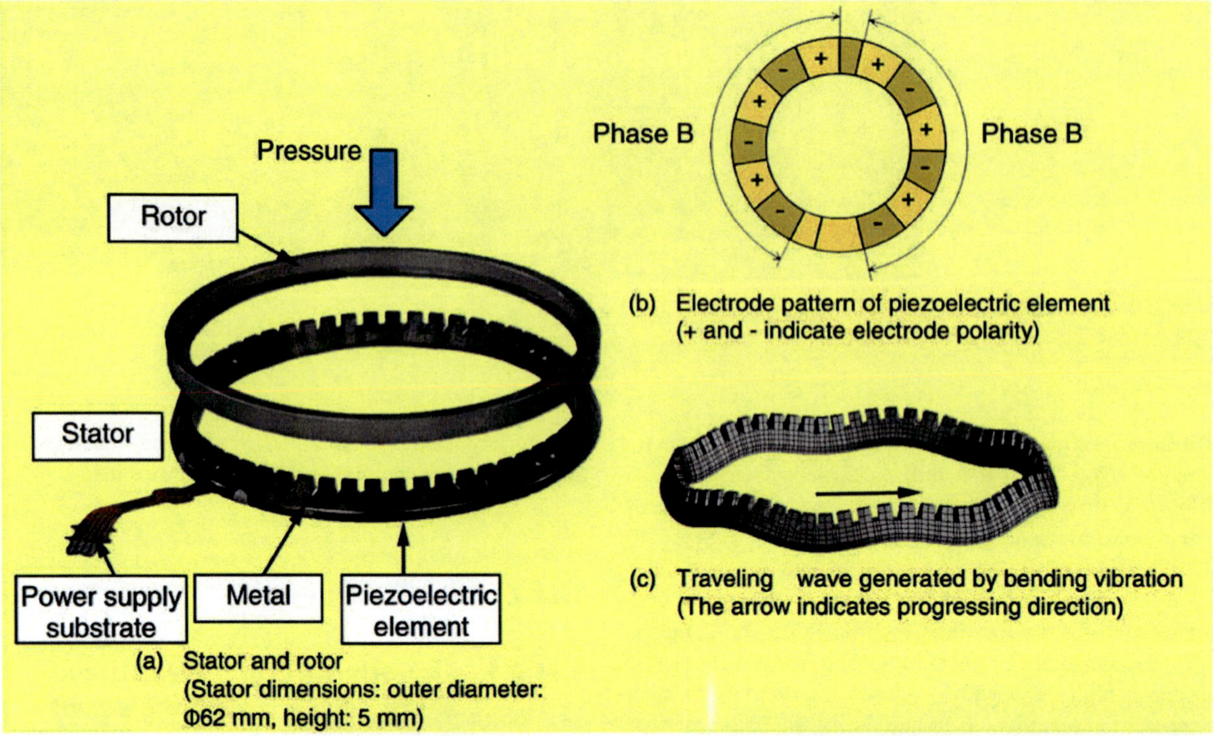

Fig. 6.69 Ring-shaped ultrasonic motor. (**a**) Stator and rotor: the stator is created by laminating a piezoelectric element on a ring-shaped metal. The rotor is in contact with stator by pressure. (**b**) Electrode patterns of a piezoelectric element: the marks indicate poling direction. High-frequency voltages with phases deviated by 90° are applied, respectively, to Phase A and Phase B. (**c**) Traveling wave: the amplitude is magnified in the figure. Traveling waves that rotate along the circumference are generated by the bending vibration

Figure 6.70 shows an example of a bar-shaped ultrasonic motor, which is composed of a stator and a rotor, with a piezoelectric element connected by a metal part, as shown in Fig. 6.70a. The piezoelectric ceramic element is disk-shaped as shown in Fig. 6.70b and the right and the left sides have different poling directions. Phase A and Phase B, respectively, have two ceramic elements placed in opposing directions. When Phase A and B of this piezoelectric element are, respectively, applied with high-frequency voltages with phases deviated by 90°, the stator goes into resonance (Note 6.46) conditions under a predetermined frequency. As a result, the upper part of the stator circumnutates due to the bending vibrations generated in two mutually-perpendicular directions of the stator as shown in Fig. 6.70c, and the rotor, pressurized by the stator, is rotated by a frictional force. The friction surfaces of the stator and the rotor are treated with a wear-resistant hard plating or a surface-hardening process, for both the ring-shape and the bar-shape. In addition, multilayer piezoelectric elements exclusive to motors are used in current bar-shaped ultrasonic motors instead of a single piezoelectric plate as explained later.

Figure 6.71 shows inner views of replacement lenses for single-lens reflex cameras, in which an ultrasonic motor, either the one shown in Fig. 6.69 or the one in Fig. 6.70, is assembled, depending on the shape. Products that incorporate the ultrasonic motor for driving AF lens, which is introduced here, are characterized by high-speed start/stop performance as well as quiet operation. They are only a part of many examples and various types of ultrasonic motors with various characteristics that have been proposed and commercialized by a number of companies and universities. The types vary from linear-types (vs. rotation-types), standing wave-types (vs. traveling wave-types) to nonresonant-types (vs. resonant-types), and some of them are flat-shaped, not ring or disk-shaped. They are generally characterized by high torque with low-speed, high start/stop responsiveness and controllability, the retention of torque during the shutdown period, and quietness, as well as a property that keeps them free from being affected by magnetic fields and affecting others by magnetic fields. Meanwhile, although the degree of design freedom is comparatively high in ultrasonic motors, which have a simple mechanism, there are some aspects, such as the wear due to vibration and friction, that make the designing of ultrasonic motors difficult. Consideration is needed for the power supplies, drive circuits, control technologies, etc. that satisfy the needs of each of these motors.

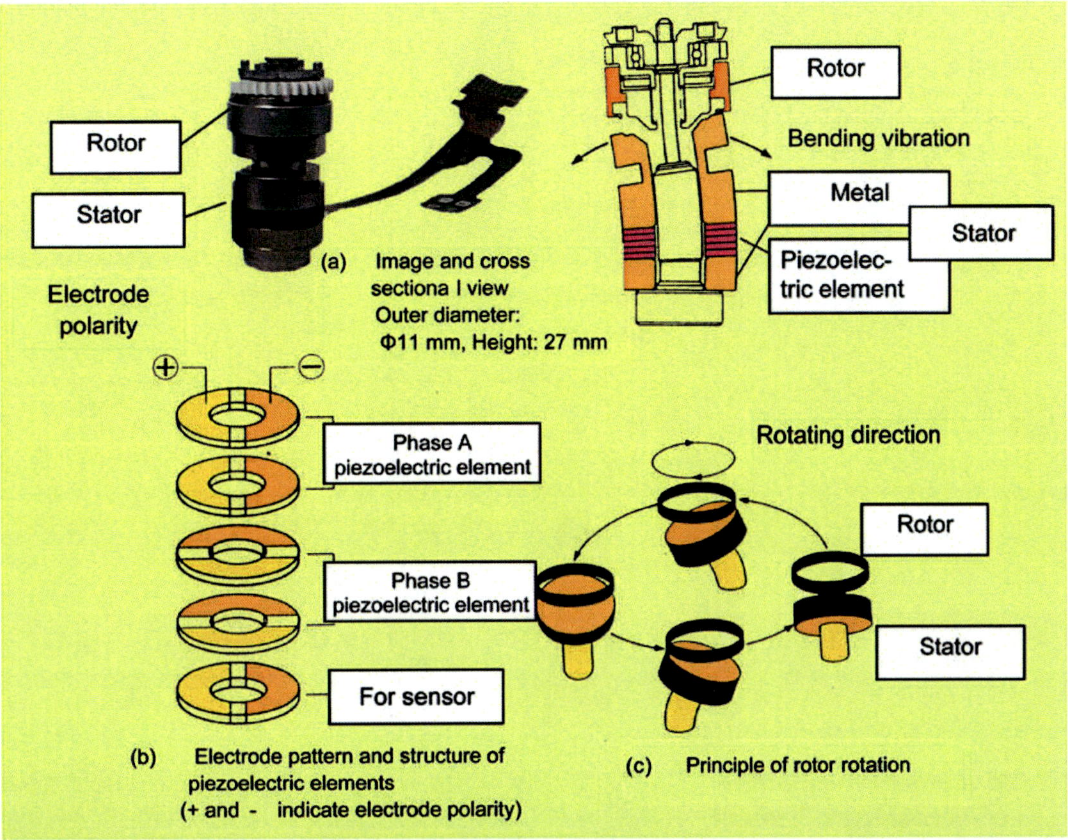

Fig. 6.70 Bar-shaped ultrasonic motor. (**a**) The image and cross sectional view: the stator is created by connecting a piezoelectric element to a metal. The rotor is in contact with the stator by pressure. (**b**) Electrode patterns of a piezoelectric element: the marks indicate poling direction. The upper portion of the stator circumnutates when high-frequency voltages with phases deviated by 90° are applied, respectively, to Phase A and Phase B. (**c**) Principle of rotor rotation: the circumnutating movement and the rotation of the rotor, which is pressurized for contact

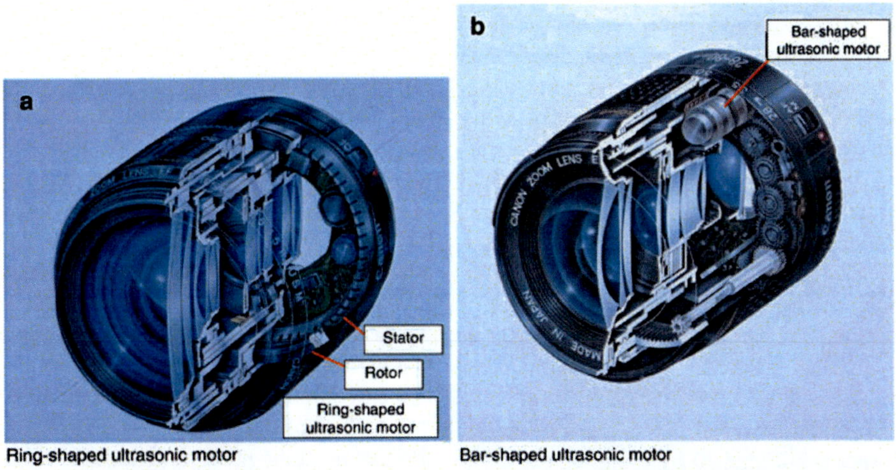

Fig. 6.71 Ultrasonic motors inside interchangeable lenses. (**a**) The ring-shaped motor is mounted outside of the lens group. The simple structure allows for easier assembly and direct driving. (**b**) The bar-shaped motor is mounted outside of the lens group. It can be assembled regardless of the lens size

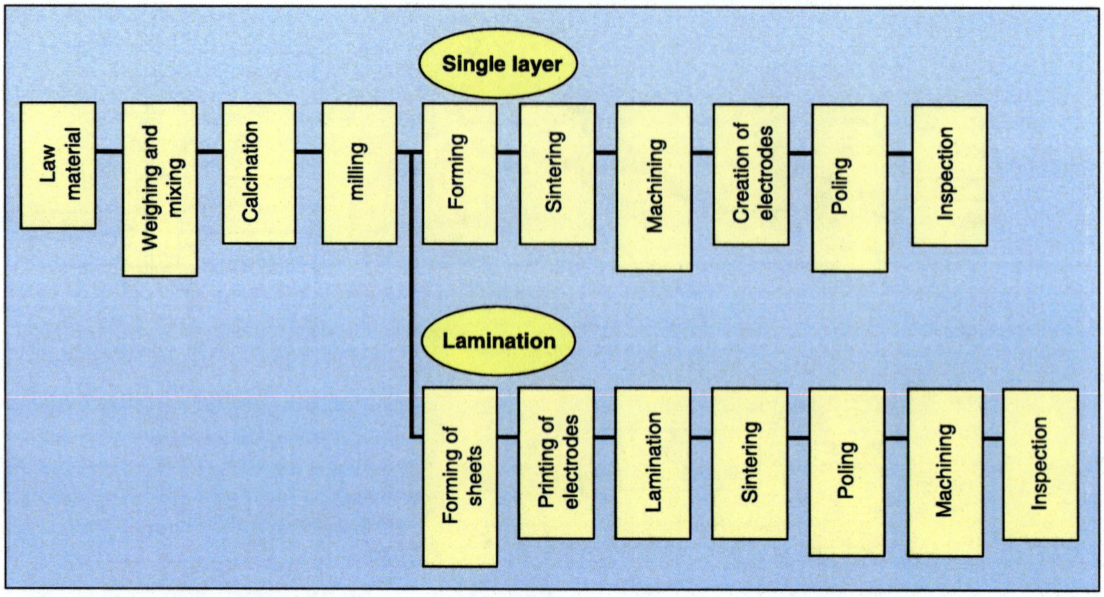

Fig. 6.72 Manufacturing process of piezoelectric elements (single layer and multilayer). Electrodes are created and poled after forming and sintering to produce the single layer type. Meanwhile, electrodes are printed on green sheets, which are then laminated and sintered before the poling process, to produce the multilayer type

6.14.2.2 Manufacturing Method of Piezoelectric Element

Piezoelectric elements used in ultrasonic motors have comparatively simple shapes such as plate and disk shapes. Figure 6.72 shows the outline of the manufacturing process of a piezoelectric element. Piezoelectric elements are normally made of piezoelectric ceramic, a three-component material created by adding a doping element to lead zirconate titanate. The powder is calcined, milled and then sintered to produce the piezoelectric ceramic. The single plate is formed by dry pressing or extrusion and sintered, after which electrodes are burned into the surface of silver or made by deposition of gold or nickel for poling. Piezoelectric ceramics featuring low vibration damping, a large mechanical quality factor Q_m and a comparatively large electromechanical coupling coefficient k and piezoelectric constant d are suitable for ultrasonic motors. They are also required to be mechanically strong and hard to crack. Recently, piezoelectric ceramics featuring low loss and low heat generation under high load application have been studied to improve output of power.

Low-voltage operation was realized in the bar-shaped ultrasonic motor shown in Fig. 6.70 by using the exclusive multilayer piezoelectric element shown in Fig. 6.73. It incorporates an electrode pattern and a structure suitable for driving a bar-shaped ultrasonic motor. For example, the electrode pattern inside the multilayer is divided into four parts. Multilayer piezoelectric elements are manufactured by laminating a number of green sheets, upon which internal electrodes are printed, and then sintering and poling them, as shown in Fig. 6.72.

6.14.3 Future Prospects

Twenty years have past since the ultrasonic motor was first commercialized. It is expected that various types of ultrasonic motors that allow full utilization of the characteristics of various devices and create new functions will be developed in the future. The production volume of ultrasonic motors intended to drive the camera AF-lens introduced above has been increasing and developments will continue for cost reduction, miniaturization and increasing output power so as to commercialize enhanced products. It is hoped that in line with the progress of ultrasonic motors, the development of piezoelectric element production technologies that enable cost reduction and the realization of characteristics that meet the needs of ultrasonic motors. Further development of environmentally-conscious lead-free piezoelectric materials is also expected.

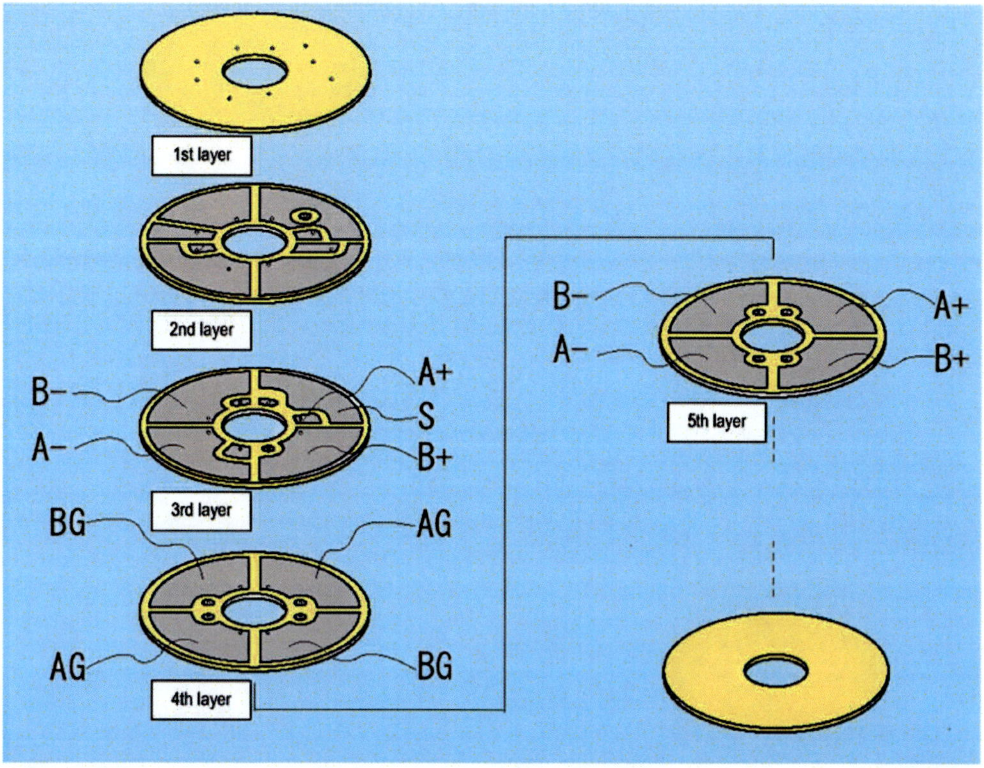

Fig. 6.73 Multilayer piezoelectric element intended for a bar-shaped ultrasonic motor. The figure shows the electrode pattern and structure of each layer. The layers below the 4th/5th layer have the same electrode pattern as the 4th/5th layer, and the power is supplied to each of the electrodes from the surface of the 1st layer via a through hole. The electrode patterns are basically divided into four portions and consist of Phase A (a piezoelectric layer between electrodes A+ and AG, and a piezoelectric layer between electrodes A− and AG) and Phase B (a piezoelectric layer between electrodes B+ and BG, and a piezoelectric layer between electrodes B− and BG). They are poling so that poling direction are different between A+ and A− and between B+ and B−. Because of this configuration, two-directional bending vibrations that cross each other at right angles are generated. The piezoelectric layer between Electrode S on the 3rd layer and Electrode AG on the 4th layer is used as a control sensor

Multilayer Ceramic Speaker (1999) 6.15

The ceramic sounding body has a long history and various improvements and enhancements were made by a number of engineers from the late 1960s to the late 1970s. In those days, high sound pressure was required for the ringers of home telephone sets. The sound pressure of a ceramic sounding body depends on the displacement amount of bending vibration, if the size of the diaphragm remains the same. Therefore, engineers in those days researched ways to improve the displacement amount. For example, they creatively attempted to maximize the Young modulus for diaphragms and the supporting method and geometric dimensional ratio of diaphragm and lead zirconate titanate (PZT), as well as to remove and effectively utilize air resistance. However, none of these efforts led to a substantive improvement. As of 2007, the products are accepted in the market as low-price, mature sounding bodies.

6.15.1 Background of Development

In the past several years, mobile devices such as mobile phones have gone through remarkable miniaturization and weight reduction. In particular, the contents business based on high-speed communication, in particular, has expanded to the distribution business of high quality music and images, and it is easy to imagine that the widespread use of terrestrial digital waves will create a new market for mobile phones with the keyword "One seg" with regards to TV broadcasting. Meanwhile, mobile phones lighter than 100 g have been widely used, and it is not an exaggeration to say that the miniaturization, increase of functions and improvement in performance in mobile devices have been achieved by the evolution of electronic devices.

Compact electromagnetic speakers have been widely used since around 1980. Among others, the invention of Nd–Fe–B magnet, a rare-earth magnet, changed not only the characteristics but also the market of electromagnetic speakers (Note 6.47). Currently, electromagnetic speakers are incorporated in the majority of all mobile phones.

The multilayer ceramic speaker (Fig. 6.74) is an ultraslim speaker developed on the basis of ceramic lamination technology, as a next-generation speaker that functions as an interface between mobile devices and humans. The thin and light speaker, with characteristics such as high sound pressure, high sound quality, and low power consumption, which were not achieved by conventional ceramic speakers, has been incorporated in devices such as mobile phones and digital still cameras since 2000.

6.15.2 Characteristics

Figure 6.75 shows the basic principle of the lead zirconate titanate piezoelectric ceramic (hereinafter called PZT). When voltage is applied between the electrodes, which are made of silver, etc. and created on the upper face and the lower face of a PZT that is sintered to a thin plate, the PZT expands and contracts in the longitudinal and thickness directions. The amounts of displacement in the longitudinal and thickness directions are expressed by Δt and Δl, respectively, which depend on voltage V and piezoelectric constants d_{33} and d_{31}. The amount of displacement is approximately 0.1 μm in the thickness direction when a voltage of 150 V is applied to a plate with a thickness of 100 μm. The PZT is an electromechanical transducer that generates a minute displacement and is mainly used in the minute positioning mechanism of semiconductor steppers and as a compact motor. When this element is used as an acoustic device, it is difficult to generate air vibration for sufficient sound pressure, because the displacement amount is minimal. As a solution, displacement

Note 6.47 A speaker in which the diaphragm is vibrated by supplying current to a coil, using Fleming's rules.

enhancing mechanisms including the one shown in Fig. 6.76 were suggested in the past. Structures of current piezoelectric sounding bodies (Note 6.48) are based on them.

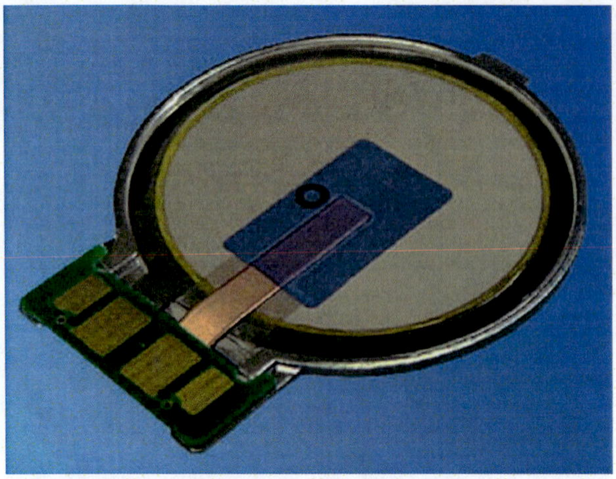

Fig. 6.74 A multilayer ceramic speaker. Piezoelectric ceramic speakers are superior to electromagnetic speakers in terms of the features such as low power consumption, thin structure and non-magnetic properties. However, they are disadvantageous in terms of sound pressure and sound quality. We have successfully developed an ultraslim ceramic speaker with improved sound pressure, maintaining its advantages

During the period when remarkable inventions were made in the field of magnet materials, new progress was made in ceramic technology, which resulted in "ceramic lamination technology," supported by the large capacitance multilayer ceramic capacitor. The multilayer ceramic speaker explained below successfully achieved sound pressure and sound quality that had been totally impossible in conventional ceramic speakers by applying this core technology to a conventional ceramic sounding body. The formula (6.1) expresses the vibration energy of the multilayer ceramic speaker. The sound pressure increases as the vibration energy increases.

$$energy \approx \frac{1}{2}\,\frac{d_{31}{}^{2}\cdot V^{2}\cdot D^{2}}{S_{31}}\cdot\frac{n}{t} \qquad (6.1)$$

(general area) (new technology area)

d_{31}: piezoelectric strain constant in the transverse direction, s_{31}: elastic compliance constant in the transverse direction, V: applied voltage, D: diameter of piezoelectric ceramic, t: thickness of ceramic sheet, n: number of laminated ceramic sheets.

The vibration energy of a multilayer ceramic speaker depends on the applied voltage and the PZT diameter. However, a large voltage application is difficult for devices such as mobile phones and PDA that are powered by batteries. And usage of large ceramics runs contrary to miniaturization efforts. Therefore, it is difficult to simply apply the

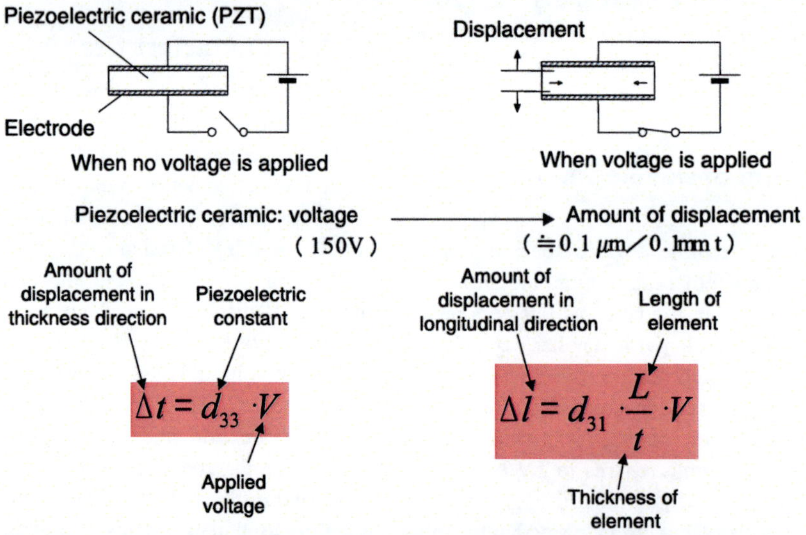

Fig. 6.75 Basic operation of piezoelectric ceramics. Piezoelectric ceramics are electromechanical transduction elements and the amount of deformation is defined by the direction of the electrical field

Note 6.48 A sounding body in which the sound is produced by bending vibrations generated by a pair of metal plates laminated with a piezoelectric ceramic, which is distorted by the application of an electric field.

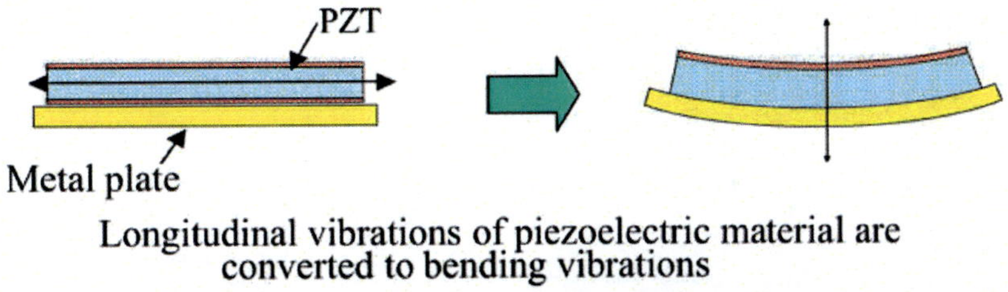

Fig. 6.76 Operating principle of a piezoelectric sounding body. The piezoelectric sounding body utilizes the lateral deformation of piezoelectric ceramics. Unimorph and bimorph structures are generally incorporated. In the unimorph structure, PZT (ceramic) is attached only on one side of the metal plate, while PZT is placed on both sides of the metal plate in the bimorph structure

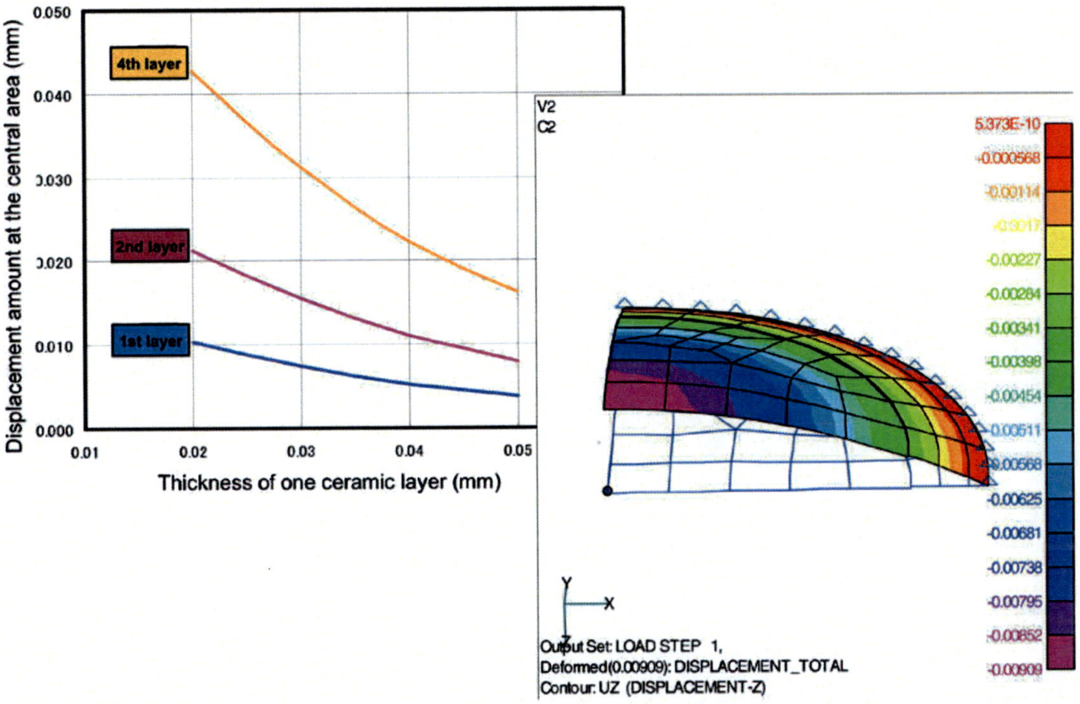

Fig. 6.77 Example of displacement magnitude using the finite element method for sounding body. The sound pressure of a sounding body depends on the displacement amount of bending vibrations. Therefore, the finite element method is very effective for analyzing the displacement magnitude. The figure shows a static analysis only, but the characteristics of the sounding body were designed by carrying out other analyses including a frequency dependency analysis of the displacement amount and an analysis of high-order vibration modes

general formula (6.1) to mobile devices; that has been blocking the application of ceramic speakers to mobile devices. The author and colleagues have added a multilayer structure, a new technology, to the vibration energy formula (the new technology area in the formula). The vibrating energy in the formula depends on the thickness of the ceramic and the number of laminated layers. This means that the vibrating energy can be increased by increasing the layers of thin ceramic sheets. However, the analysis is based on the radial direction of the ceramics only and the vibration in the axial direction need to be changed into bending vibrations in actual operation. Therefore, if only the number of laminated layers n in the formula (6.1) is increased, the ceramics will be thicker, resulting in a suppression of bending displacement. In respect to the number of laminated layers n and the thickness of the ceramic sheet t, they were optimized by utilizing unique multilayer reliability technology accumulated through the research and development of multilayer capacitors and designing a vibrator based on the general finite element method (refer to Fig. 6.77).

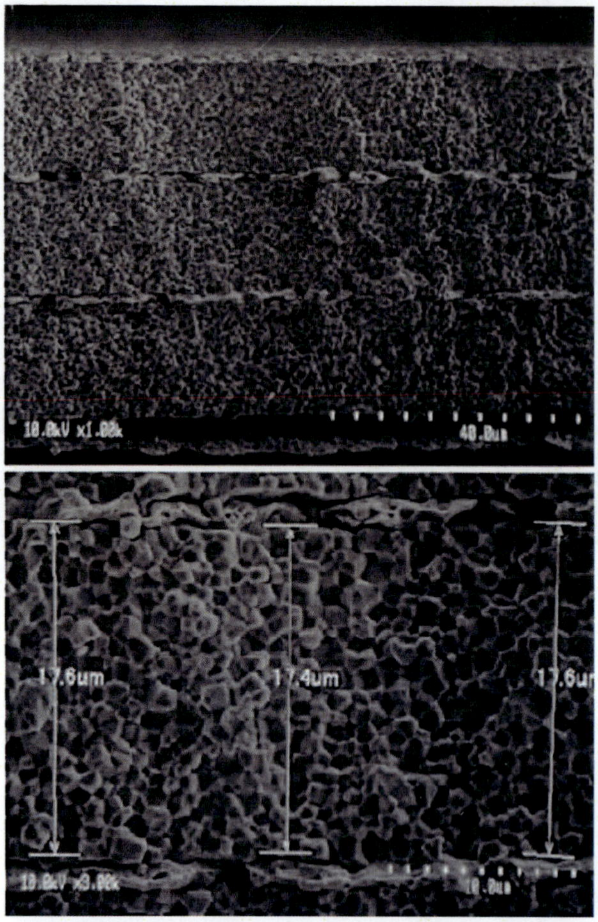

Fig. 6.78 Cross sectional view of a multilayer element. This ceramic was produced by simultaneously co-firing internal electrodes and green bodies, which enabled the multilayer lamination of thin sheets

The next-generation ceramic speaker, the multilayer ceramic speaker, was created through the fusion of dynamic design, material technology, lamination technology, and the coordination of technologies for vibration analysis, acoustic analysis, and others.

Figure 6.78 shows a cross sectional view of the element used in the multilayer ceramic speaker that we developed. It is composed of three layers of 18 μm sintered ceramic, with incorporated Ag/Pd electrodes. The via hole method, which is generally incorporated in multilayer inductors, was used for electrical wiring.

Major characteristics of multilayer ceramic speakers are as follows:

1. Ultraslim structure of 0.7 mm
2. Ultralight structure of 0.4 g or less
3. Low power consumption, 1/5–2/3 of the electromagnetic type
4. High sound pressure and sound quality in a wide range of bands
5. Does not generate magnetism and does not influence magnetic cards such as cash cards
6. Realizes acoustic properties that are insusceptible to back cavity (Note 6.49).

Note 6.49 The back cavity refers to the back-side volume of audio devices. The ideal condition is indicated as infinite volume, and therefore, larger volumes are ideal for taking advantage of audio device properties.

Lithium-Ion Batteries (1991) 6.16

The lithium-ion secondary battery, a new type of battery commercialized in 1991, has been used widely as the power source of IT equipment such as mobile phones and laptop PCs. The global market size is currently about 1 trillion yen (13 billion US $). The batteries are beginning to be used in electric cars and are expected to play important roles in resolving issues related to natural resource usage, the environment and energy. Functional ceramics called lithium transition metal oxides ($LiCoO_2$, etc.) are used as the cathode material in these lithium-ion secondary batteries. It is important that lithium ions diffuse freely in the crystal structure of the cathode material. Utilizing ceramic material technologies, various diffusion pathways are being designed and a series of new cathode materials are being created.

6.16.1 Background of Development

Lithium-ion secondary batteries are "non-aqueous secondary batteries that use carbon as the anode active material and lithium transition metal oxide ($LiCoO_2$) as the cathode." The operating principle is shown in Fig. 6.79. Lithium ions desorb from the cathode, $LiCoO_2$, and diffuse to the carbon anode during charging. Because of this electrochemical reaction, electrons flow from the cathode to the anode. The reversed reaction occurs during discharging. The new batteries are fundamentally different from conventional secondary batteries in that they involve no chemical reaction and involve ions and electrons only. The lithium-ion battery has been used widely as the power source in IT equipment such as mobile phones and laptop PCs. The global market size is currently about 1 trillion yen (13 billion US $). Furthermore, the batteries are beginning to be applied to electric vehicles and energy storage systems.

6.16.2 Characteristics

6.16.2.1 *Characteristics of Products and Constituent Materials*

Lithium-ion secondary batteries have the following characteristics: (1) electromotive force (4.2 V), (2) small and light, (3) large current discharge, (4) low self-discharge rate, (5) charge–discharge efficiency of nearly 100%, (6) contains no toxic substance. These characteristics are attributable to non-aqueous electrolytes (Note 6.50). Figure 6.80 indicates that energy density (Note 6.51) of the batteries is overwhelmingly higher than that of nickel-cadmium secondary batteries and nickel-hydrogen secondary batteries that use aqueous electrolytes.

Batteries are generally composed of the cathode material, anode material, electrolytes and the separator. Other materials such as the electric collector and binder are also used.

Major constituent materials used to manufacture lithium-ion secondary batteries are listed in Table 6.9. Cylindrical and square lithium-ion secondary batteries are manufactured using these materials. Cylindrical products are mainly used in laptop PCs and square products are used in mobile phones.

As shown in Fig. 6.81, $LiCoO_2$, the cathode material, is coated on both sides of an aluminum foil, which is about 15 μm in thickness. The binder connects the $LiCoO_2$ powders and bonds them on the electric collector. In respect to the anode, carbon, the anode material, is coated on both sides of a copper foil, which is about 15 μm in thickness.

Note 6.50 While electrolysis occurs at 1.5 V in aqueous electrolyte, the electrolyte incorporating organic solvent has a voltage resistance of 4 V or higher.

Note 6.51 Amount of electric energy that can be stored per unit volume or weight of a battery.

6.16.2.2 Lithium-Ion Secondary Batteries and Ceramic Materials

A lithium transition metal oxide compound is used as the cathode material of lithium-ion secondary batteries. It is important for cathode materials of secondary batteries that lithium ions diffuse freely in the crystal structure of the cathode material. Diffusion pathways inside crystals are classified into three types, one-dimensional, two-dimensional and three-dimensional types. Typical positive-electrode materials that are currently known are explained in accordance with these classifications.

– Layered rocksalt cathode material

A group of compounds with two-dimensional lithium-ion diffusion pathways represented by $LiCoO_2$, which is currently used most widely as the positive-electrode material of lithium-ion secondary batteries. Other compounds belonging to the group of this crystal structure include $LiNiO_2$ and $LiMnO_2$. However, they cannot be used independently due to their battery characteristics. They have been processed into composite oxides such as $Li(Ni_xAl_{1-x})O_2$ and $Li(Ni_{1/3}Mn_{1/3}Co_{1/3})O_2$ for actual use.

– Spinel cathode material

A group of compounds with three-dimensional lithium-ion diffusion pathways represented by spinel type $LiMn_2O_4$. The electrical discharge capacity is smaller compared with layered rocksalt cathode materials. However, the materials are prospective positive-electrode materials for medium and large-sized lithium-ion secondary batteries that are expected to expand in the future because they are inexpensive and safe.

– Olivine cathode material

Group of compounds with one-dimensional lithium-ion diffusion pathways represented by olivine type $LiFePO_4$. The lithium ion diffusion pathways are one directional, which is a disadvantage as cathode materials. However, the disadvantage has been overcome by the use of nanoparticles. Utilization of the materials as a cathode in the lithium battery will be realized in the near future. The electromotive force is low at about 3.5 V and the energy density cannot be enhanced. However, commercialization of this product has begun because of its stable battery characteristics.

Cathode Anode

$$LiCoO_2 + C \underset{\text{Discharging}}{\overset{\text{Charging}}{\rightleftharpoons}} Li_{1-x}CoO_2 + Li_xC$$

Fig. 6.79 Equations of lithium-ion secondary battery reaction

6.16.3 Future Prospects

Secondary battery technologies are expected to become important for solving major social issues such as effective natural resource utilization, the environment and energy. Technologies for ceramic materials, in particular, will play important roles in future technologies of cathode materials

Table 6.9 Major materials used to manufacture lithium-ion secondary batteries

	Constituent materials
Cathode materials	$LiCoO_2$, $LiNiO_2$, $LiMn_2O_4$
Anode materials	Carbon (graphite, hard carbon)
Electrolyte solvent	Ethylene carbonate, propylene carbonate, dimethyl carbonate, diethyl carbonate
Electrolyte salt	$LiPF_6$, $LiBF_4$
Separator	Porous polyethylene film
Binder	Polyvinylidene fluoride, SB latex
Cathode	Electric collector made of aluminum foil (10–25 μm)
Anode	Electric collector made of copper foil (10–25 μm)

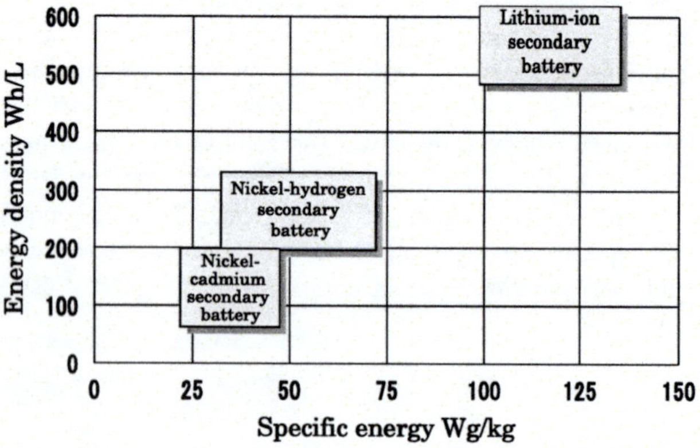

Fig. 6.80 Comparison of energy density of secondary batteries

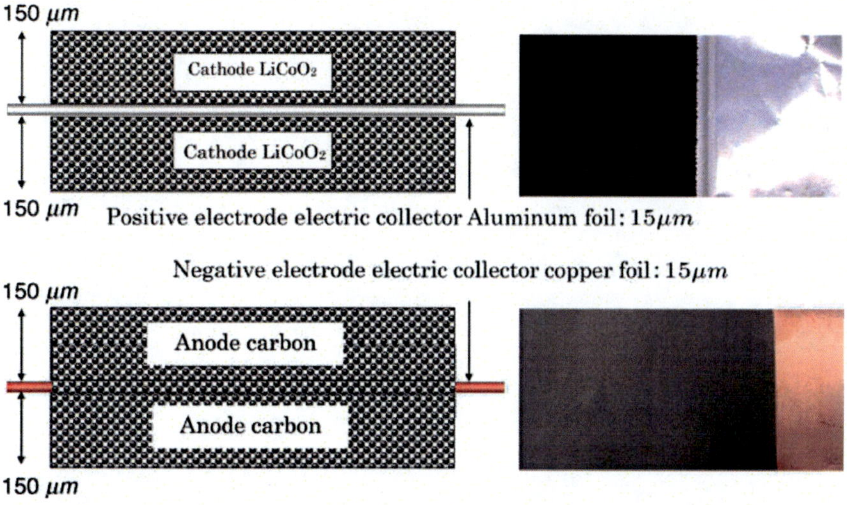

Fig. 6.81 Structure of positive and negative electrodes of lithium-ion secondary batteries

intended for secondary batteries. The designing of lithium ion diffusion pathways, above all, has unlimited potential. It is expected that new cathode materials will be developed one after another in the future.

Literature

1. Sato Y (2001) RF band filter for mobile phones. J Inst Electron Inform Commun Eng 84(11):782–789 **(6.1)**
2. Yamazaki T et al (1997) IEDM Tech Dig 613 **(6.2)**
3. Nakamura T et al (1995) Integr Ferroelectr 9:179 **(6.2)**
4. Kawashima S et al (2001) Proceedings of Symposia of VLSI Technology and VLSI Circuits C12-3 **(6.2)**
5. Horii Y et al (2002) IEDM Tech Dig 539 **(6.2)**
6. Konishi Y (1965) J Inst Telecommun Eng 48:899–908 **(6.6)**
7. Hashimoto T (1977) Microwave ferrite and application technologies. Sogo Denshi Shuppan, Tokyo, pp 74–109 **(6.6)**
8. Murata Manufacturing Co., Ltd. (ed) (2003) Fundamentals and applications of ceramic condensers. Ohmsha, Tokyo **(6.7)**
9. Iguchi Y, Saito T, Yamaoka N (1992) Ceramics 27(8):758–764 **(6.8)**
10. Saburi O (1959) J Am Ceram Soc 14(9):1159–1173 **(6.11)**
11. Saburi O (1961) J Am Ceram Soc 44(2):54–63 **(6.11)**
12. Niimi H (2005) Monthly J Funct Mater 25(May issue):28–34 **(6.11)**
13. Kand A, Tashiro S, Igarashi H (1994) J Appl Phys 33:5431–5434 **(6.11)**
14. Aoto W, Takeda H, Nishida T, Okamura S, Iwasaki T, Shimada T, Terao K (1994) Proceedings of the annual meeting of the Ceramic Society of Japan 220 **(6.11)**
15. Sato M, Yoshida E, Sugawara E, Shimada H (1996) J Magn Soc Jpn 20:4214–4216 **(6.12)**
16. Yoshida E, Kondo K, Ono H (2006) Nikkei Electron. (Nikkei Business Publication, Inc.) 918:119–126 **(6.12)**
17. Yoshida S et al (2001) IEEE Trans Magn 37:2401–2403 **(6.12)**
18. Abe M (2001) Sci Ind 75(8):342–344 **(6.12)**
19. Konno M, Sugawara S, Kudo S (1995–2011) Piezoelectric-type vibrating gyroscope angular velocity sensor. J Inst Electron Inform Commun Eng C-1, J78-C-1(11):547–556 **(6.13)**
20. Sugawara S, Tomikawa Y (1999–2007) Feature article on vibrating gyroscope – commercialization of 'elastic wave devices' accelerated. J Acoust Soc Jpn 55(7):496–503 **(6.13)**
21. Maeno T (2003) J Robotic Soc Jpn 21(1):10–14 **(6.14)**
22. the Solid Actuator Research Group, the Japan Technology Transfer Association (ed) (1994) New actuator handbook for precision control of actuator. Fuji Techno System, Tokyo, pp 825–1008 **(6.14)**
23. Editorial Committee for Micromachine Technology List (2003) List of micromachine technologies. The Industrial Technology Service Center, pp 468–471 **(6.14)**
24. Kataoka K (2002) Electron Technol 44(4):14–18 (Nikkan Kogyo Shimbun, Ltd.) **(6.14)**
25. Maruyama H, Kojima N, Okumura I (2000) 2000 motor technology symposium. The Japan Management Association, pp B2-2-1–B2-2-8 **(6.14)**
26. Yoshino A, Otsuka K, Nakajima T, Koyama A, Nakajyo S (2000) J Chem Soc Jpn, Chem Ind Chem (Nippon Kagaku Kaishi) 8:523–553 (in Japanese) **(6.16)**

Computer

7

Computers have been finding their ways into everyday life and contributing to affluence, safety and personal convenience, as well as responding to progress towards a information society, IT society or ubiquitous society. Computers, which have been supporting social changes, began to change the social structure in the 1980s, when integrated semiconductors were developed and the speed of computers was rapidly enhanced. Current computers and high-speed servers contain many small ceramic capacitors mounted on circuit boards, but their shells are mainly made of resin and metallic materials. Previously, when priority was also placed in areas where components were mounted as well as semiconductors, ceramics were sometimes used as the base material of the main circuit board. A number of electronic ceramics were introduced in response to the ceramics boom from the mid 1980s to the early 1990s. In this chapter, the ceramics intended for computers are explained in detail.

The multilayer ceramic circuit substrate (Sect. 7.1), a high-speed, high-density circuit substrate intended for supercomputers, was developed in the 1980 and was commercialized in the early 1990s. Ceramic substrates are superior to resin substrates because of the low thermal expansion property and circuit design of workability, which enabled a circuit substrate incorporated with a long micro wiring. Later, in line with the progress in semiconductor technologies, the mounting method was required to be changed and ceramic substrates are no longer used as circuit substrates. However, the technologies are currently utilized in the LTCC (low temperature co-fired ceramics) intended for high-frequency devices and vehicles.

The ceramic package (Sect. 7.2) is indispensable, even today, for computer packaging. They are now competing in the market with resin packages that have been commercialized responding to the needs for low cost and weight reduction. Ceramic packages are far superior to resin packages in the high-frequency range, and aluminum-based ceramic packages are used in optical communication and high-frequency devices, playing a major role in these fields.

The hybrid IC (Sect. 7.3), which was created by mounting a compact discrete circuit component on a ceramic circuit substrate, was fully commercialized in the 1990s and was applied in compact video cameras. The hybrid ICs are currently used in various devices including mobile phones.

Aluminum nitride (Sect. 7.4) and high thermally conductive silicon carbide (Sect. 7.5) attracted attention as materials for heat transmission in computer ICs, and were used to produce radiator boards, heat sinks and LSI packages. Aluminum nitride is a material that is difficult sinter. A large amount of research and development were carried out in a careful manner to find optimal sintering additives for realizing high thermal conductivity and translucency, believing that the key to bulk production is the sintering aid. It was finally decided that rare-earth oxides are extremely effective for high thermal conductivity. Meanwhile, silicon carbide is a semiconductor by nature and was believed to be difficult to create insulating substrates out of the material. However, the material was provided with an insulation property and high thermal conductivity following the discovery of adding beryllium oxide as a sintering additive. Although it was classified as a structural material, this discovery attracted a great deal of attention as serendipity (running into unexpected good luck) in the ceramics industry. This material is widely used in the CD, DVD and optical communication fields as a heat sink for optical elements such as semiconductor lasers, and is indispensable for supporting the components with high power consumption.

The single-crystal sapphire substrate features thermal, mechanical and optical characteristics, but its applications were limited due to the substrates high price. The cost was reduced following the technological advancement of recent years and they began to be used widely as substrates in blue and while LEDs (light emitting diodes) and high power devices. The LCD projectors had radiation problems arising from enhanced luminance and downsizing. Transparency, thermal resistance property and high thermal conductivity (so as to release the heat generated from the lamp quickly) were required of the polarizing plate, which is used to project clear images. The polarizer retention board is one of the examples in which the characteristics of the single-crystal sapphire meet the requirements of a specific component (Sect. 7.6).

Y. Imanaka et al. (eds.), *The Ceramic Society of Japan, Advanced Ceramic Technologies & Products*,
DOI 10.1007/978-4-431-54108-0_7, © Springer Japan 2012

Multilayer Ceramic Circuit Substrate (1990–1995)* 7.1

Multilayer ceramic circuit substrates (Figs. 7.1 and 7.2) were created in the early 1990s, when Japanese and American computer manufacturers competed for higher performance in supercomputers (Note 7.1) in response to needs for higher speed wiring substrates. These substrates were used in supercomputers for scientific computation and large-scale general purpose computers. The material properties (low-thermal expansion) and manufacturing process properties (via hole forming, interconnect process properties) of ceramic are superior to resin, which was used in conventional substrates. These advantages led to the commercialization of large-scale, multilayer high-speed ceramic circuit substrates featuring high-density LSI (large scale integration) packaging. Currently, this technology is used in high-frequency functional components, substrates and packages.

7.1.1 Background of Development

During the period from the 1980s to the early 1990s, computer manufacturers in Japan and the U.S. competed with each other to improve the performance of large-scale supercomputers (Fig. 7.3), and computing speed was enhanced remarkably during this time.

In order to improve signal speed, LSIs and circuit substrates capable of processing and transmitting signals at a high speed are required. When logic and computation processing is carried out inside a single LSI, the speed depends on the LSI itself. However, when signals are transmitted via wirings on a circuit substrate and computed in a different LSI, the speed of the system is hampered by the time required for transmission of the signals via the wiring. Furthermore, if signals are transmitted via an LSI on a different substrate, the speed is delayed by the cable that connects the substrates.

For these reasons, the following conditions stimulated high speed circuit substrate development during this period.

1. Increased LSI packaging density
   Reduce LSI packaging pitch and length of wiring to shorten signal transmission time on a substrate.
2. Increased layers and an enlarged substrate
   Integrate wiring in a substrate to increase the number of LSIs loaded on a single substrate.
3 Use of substrates with low dielectric constant
   Use of substrates with low dielectric constant surrounding signal wires so as to reduce the signal transmission time per unit length of wiring.

   Furthermore, the substrate was required to have high heat dissipation performance, because the amount of heat generated by the components increased due to enhanced LSI component speed and development of the large-scale integration method.

7.1.2 Characteristics

7.1.2.1 Ceramic Material Technology

The majority of the previous circuit substrates were printed circuit boards comprised of polyimide resin. Ceramic circuit substrates were developed as replacements for resin substrates because ceramic is superior to resin in terms of function and manufacturing method with respect to the above-mentioned circuit substrates requirements.

1. In order to realize high density packaging, it is necessary to establish highly reliable interconnects between the substrate and LSI. Therefore, the thermal expansion properties of the substrate material need to be close to that of silicone (thermal expansion coefficient: 3.6×10^{-6}/°C).

*The number in parenthesis indicates the year that the product was first commercialized.

Note 7.1 Computers that perform and process complicated, large-volume numerical calculations such as the analysis of phenomena, and prediction and designing in advanced scientific fields including material science, atomic energy, meteorological phenomena, environment and aerospace.

Fig. 7.1 Image of a multilayer ceramic circuit substrate (Bar = 50 mm). Slurry, a mixture of ceramic powder, organic binder and solvent, is tape-cast to create a green sheet, on which via holes and wiring patterns are screen-printed to create a thick film conductor. Following this, thermal compression stacked layers are sintered at high temperature to produce the substrate

Fig. 7.3 Image of Supercomputer. Supercomputers are used in a variety of fields for scientific computation, specifically, fluid analysis, structural analysis, nuclear fusion, molecular chemistry, image processing, resource exploration, weather prediction, energy and economic analysis, and utilization forms are open to users

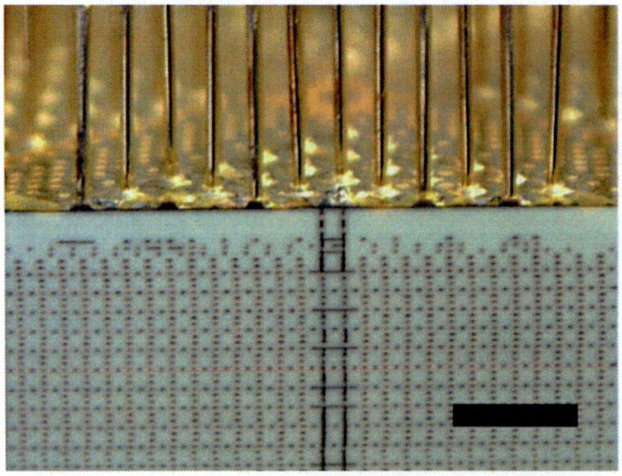

Fig. 7.2 Cross section of multilayer ceramic circuit substrate (Bar = 5 mm); 80 μm wiring patterns and 50 μm via hole patterns formed between insulating layers, each of which is 200 μm in thickness, are three-dimensionally connected. The substrate achieved a packaging density of 24 LSI/100 cm^2, which is about five times that of conventional resin printed circuit boards

The thermal expansion coefficient of resin is around 10^{-5}/°C, while that of ceramic is in the 10^{-6}/°C order, meaning that ceramic is more appropriate for realizing high density packaging.

2. In manufacturing conventional resin substrates, insulation resin layers were laminated before drilling via holes, followed by a plating process to establish continuity between the layers. Under this manufacturing method, drills tend to be flexed while opening via holes when the substrate layers exceed a certain number, resulting in lower accuracy of hole positions. In addition, design criteria of substrates, including the number of layers and wiring dimensions, are decided depending on the pitch of drilled holes in this manufacturing process of resin substrates, which influences the realization of the aforementioned high density packaging

   Meanwhile, in ceramic substrates, via holes for interlayer continuity can be formed on arbitrary locations of each insulation layer and insulation layers with via holes can be laminated in one operation to create a multilayer, realizing a super multilayer without difficulty.

3. The relationship between signal propagation delay time T_{pd} and the dielectric constant (ε) of a substrate material is as follows:

$$T_{pd} = \sqrt{\varepsilon} / C \quad C: \text{velocity of light}$$

For the realization of high speed transmission, it is desirable that the dielectric constant of the substrate material be is low. The dielectric constant of ceramic is 5 or higher, while that of resin is 4 or lower in general. In respect to this material potential, resins are superior to ceramics. However, as explained below, the effect of high density packaging was more remarkable for achieving high speed transmission.

In addition, characteristics of ceramics themselves, were improved by shifting the material from alumina to glass ceramics.

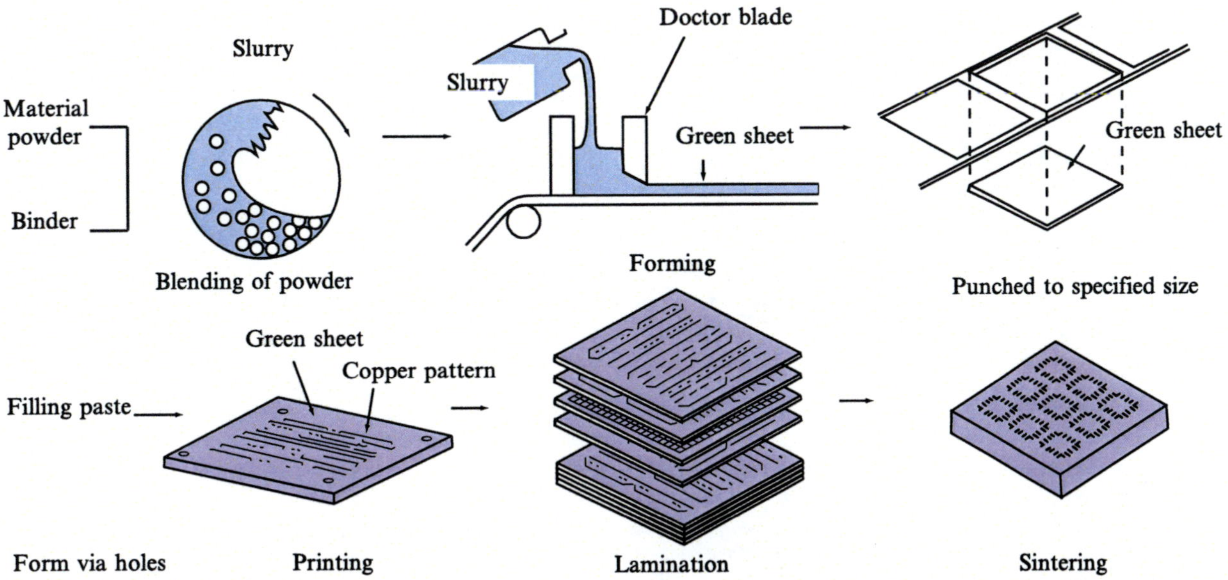

Fig. 7.4 Manufacturing process of multilayer ceramic circuit substrates. Slurry, a mixture of ceramic powder, organic binder and solvent, is tape-cast to form a green sheet, on which via holes and wiring patterns are screen-printed to create a thick film conductor. Following this, thermal compression stacked layers are sintered at high temperature to produce the substrate

7.1.2.2 Manufacturing Process

The basic manufacturing process is shown in Fig. 7.4. In the first process, ceramic powder and organic binder are mixed to produce a paint-like slurry. This slurry is tape cast by the doctor–blade method to create a ceramic raw sheet (green sheet) that is as flexible as paper. Then via holes for inter-layer conduction and wiring patterns are screen-printed on the green sheet using a conductive paste. A number of printed green sheets are piled up and laminated by applying heat and pressure (an organic binder in the printed green sheets plays the role of an adhesive that bonds the layers). After the organic binder in the green sheet is burned out, the conductor metal and ceramic are co-fired to create a multilayer ceramic substrate.

7.1.2.3 Performance and Specifications of Product

Because an alumina/glass composite (Fig. 7.5), which has higher dielectric constant compared with resin, was used as the substrate material, transmission speed became slower. The substrate wiring delay time was 83–85 ps/cm in the ceramic circuit substrate, while that of resin-based printed substrate was 70–75 ps/cm. However, the LSI density was two times higher than that of resin substrates, which resulted in a reduction of wiring length between LSIs, achieving a reduction in the total delay time by half. In addition, the transmission speed was improved by approximately 20% when compared with alumina substrates (100 ps/cm), the first generation of ceramic circuit substrates. Copper with low electrical resistance (DC resistance: 100 mΩ/cm) was used as wiring, which enabled the mounting of as many as 4,000 wirings (total length: 1 km) (Table 7.1).

7.1.3 Future Prospects

The configurations of printed circuit boards are forced to respond to changes in the performance of active elements mounted on substrates. This circuit substrate was created to maximize the performance of bipolar transistor (Note 7.2) elements, which boasted the highest performance in its time. Later, the mainstream shifted from bipolar transistors to CMOS (Note 7.3) and clock speed and concurrent processing

Note 7.2 Transistor elements control the flow of a small number of carriers by application of current. They involve electrons and positive holes for operation, and therefore, are bipolar. In general, they consume a lot of electricity but feature high speed operation.

Note 7.3 CMOS stands for complementary metal oxide semiconductor. CMOS is an IC in which Metal Oxide Semiconductors for both the P and N channels are formed on one chip. It is connected and structured so that the channels supplement operations of each other. While the current is applied to one channel, the other channel receives no current, reducing power consumption.

Fig. 7.5 Microstructure of alumina/glass composite (Bar=5 μm). Ceramics that feature low-temperature sintering ability, low dielectric constant and low thermal expansion coefficient were developed by utilizing the characteristics of glass, namely, high-temperature fluidity, dielectric properties and thermal expansion properties. The dielectric constant was reduced by half compared with products made of alumina, and transmission speed was enhanced by approximately 20%

Table 7.1 Example of characteristics of multilayer circuit substrate for supercomputer (Fujitsu VP2000)

External dimensions		245 mm × 245 mm × 13 (thickness) mm
Number of layers		61 (incl. 36 signal layers)
Number of mounted LSIs		144 at max.
Base material	Dielectric constant	5.7
	Thermal expansion coefficient	4×10^{-6}/°C
Conductor material	Pattern width	95 μm
	Pattern length inside substrate	1 km

Note: the product performance and specifications described above are for the circuit substrate developed for the Fujitsu VP2000 supercomputer, which was manufactured and commercialized using an alumina/glass composite as an insulation material and copper as conductive material

performance in computers were further improved. Therefore, this substrate is not used as a circuit substrate intended for large-scale computers. Currently, they are supplied to the market in a variety of forms such as electronic parts, substrates and packages for mobile phones and PCs, and the basic technology continues to be utilized even today.

Ceramic Package (Mid 1960s) 7.2

The ceramic package is the "container" in which silicon ICs (integrated circuit) or chips, etc. are mounted for attachment to printed circuit boards which is the core of electrical appliances. The package also protects the IC from the environment. It has been used since the mid 1960s. The ceramic package mounted with a microprocessor (Note 7.4) for calculator operation was commercialized in 1971. Later, as computer usage increased due to growing use of personal computers, the performance of IC chips increased and the ceramic packages used to mount IC chips also developed into highly functional multilayer components with a complicated internal structure (Fig. 7.6). The ceramic package is now used in a wide variety of devices, and it plays a major role in communication devices such as mobile phones. Responding to demands for a smaller SAW (surface acoustic wave) filter (Note 7.5), which is used as a frequency filter, and a smaller crystal oscillator (Note 7.6), which is used as a reference frequency generator, metal packages have shifted to being surface mounts on ceramic packages (Fig. 7.7), and continue to develop.

7.2.1 Background of Development

High reliability is a requirement for LSI (large scale integration) circuits. As the transistor numbers increase in the silicon die, the number of input/output terminals increase, but the size of the package for mounting the silicon die must remain compact. When the silicon die is enlarged for higher performance, the amount of heat generation tends to increase, requiring the package material to have a thermal expansion coefficient close to that of the silicon die, as well as having high thermal conductivity. At the same time, mobile phone SAW filters and crystal oscillators, which are currently mounted in metal packages, need to be surface-mounted on smaller packages in correspondence to the miniaturization of communication devices, and packages that are capable of protecting SAW and crystal chips from both external environment change and dropping impacts are required.

Note 7.4 Manufactured by mounting a computer's central processing unit on one or more integrated circuits.

Note 7.5 A frequency filter that utilizes a surface acoustic wave as called SAW filter.

Note 7.6 A crystal chip, which consists of a crystal which has been integrated into a device. Its resonance frequency depends on the crystal's piezoelectric characteristics.

7.2.2 Characteristics

7.2.2.1 Role of the Package

The package is required to have a variety of functions, including (1) full utilization and complete functionality of the electrical properties of the chips, (2) easy mounting and handling of dies and chips, (3) protection of dies and chips from the external environment and (4) high thermal conductivity of heat generated from within the die.

The package at the same time must be scalable to meet future requirements: (a) increase in the number of input/output terminals per unit area, (b) three-dimensional (3D) high-density internal wiring, (c) higher frequencies and high speed electrical signals, (d) compactness and thin structure, (e) metal cap and metal pin brazing capability and (f) enhancement of heat release property.

7.2.2.2 Alumina Package

The alumina package is a widely used for ceramic packages and explained below in detail. Alumina (Al_2O_3) has a high Young's modulus and is strong enough to be used in small-sized and thin 3D structures. In addition, brazed input/output

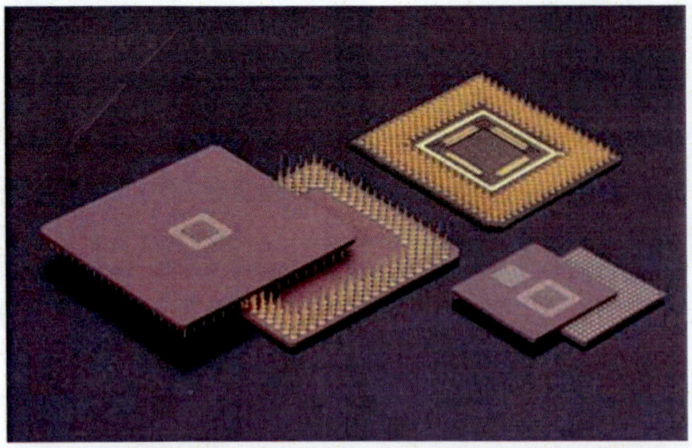

Fig. 7.6 Ceramic package for LSI. Silicon chips for LSI are mounted in a cavity and the C4 (Note 7.7) portion of a ceramic package. The ceramic package is equipped with a number of metal pins and connection terminals for electrical connection with circuit substrates, responding to the recent increase of electrical wiring terminals

Fig. 7.7 Ceramic packages for communication devices. SAW and crystal chips are strongly affected by external environment factors such as moisture and need to be fully protected by the ceramic package. Although demands for miniaturization are strong, both toughness that resists damage by dropping and a sealing performance of metal caps are required at the same time

Table 7.2 Examples of alumina ceramic characteristics

	92% alumina	FR5 (reference)
Dielectric constant (@1 MHz)	9.8	4.6
Dielectric loss (10^{-4} @ 1 MHz)	3	80
Resistivity (Ω cm)	$>10^4$	$>10^4$
Breakdown voltage (kV/mm)	15	–
Thermal expansion coefficient (10^{-6}/K)	6.8	14 (x–y directions)
Thermal conductivity (W/(mK))	17	0.5

(I/O) pins and metal caps for sealing must firmly adhere to the alumina package by soldering. It also features high corrosion resistance properties against acid and alkali, when exposed to a wide range of selection for chemicals used in the plating process.

Because of the high electrical insulation property of the alumina, a number of thin ceramic sheets can be laminated to create a 3D high density wiring structure, which is extremely advantageous in miniaturizing the package. The thermal expansion coefficient of alumina is smaller than metallic and organic materials by a factor of 10 and is close to that of silicon, therefore, the material is especially suitable for loading the large silicon dies used in LSI circuits. In addition, alumina features a low dielectric loss, which is advantageous in increasing the frequencies of I/O signals and exchanging high-speed signals (Table 7.2).

7.2.2.3 Manufacturing Process

The manufacturing method of multilayer ceramic packages incorporating alumina materials is explained below (Fig. 7.8). A high-purity alumina powder is used as the main raw material, and alkali components such as Na and impurities that deteriorate electrical insulation performance are fully removed from the raw material. As an auxiliary material, additives such as MgO are used to control sinterability.

These raw materials are blended to form a specified composition, then mixed together with organic binders and organic solvents to produce the slurry, in which the raw materials are uniformly dispersed. The slurry is cast to a form a resin tape (on a carrier tape) through a slit between a metal plate called the doctor blade and the carrier tape. Then the slurry goes through a drying process, and a row of ceramic sheet (green sheet) of specified thickness and flexibility is formed on the carrier tape. Characteristic impedance matching as well as the design of an electrical circuit that ensures characteristics such as high signal transmission properties and low crosstalk are performed on the ceramic package to fully utilize the electrical properties of the functional chips mounted in it. In order to create complex three-dimensional internal wiring structures, some of the packages are fabricated by stacking several tens of sheets. Each of the green sheets is printed with specifically designed via holes and wiring patterns, etc. by screen printing, using molybdenum (or other material) as the electrode material.

Note 7.7 Controlled collapse chip connection abbreviated C4. A chip connection method incorporating a height-controlled solder, used for mounting silicon chips in a package.

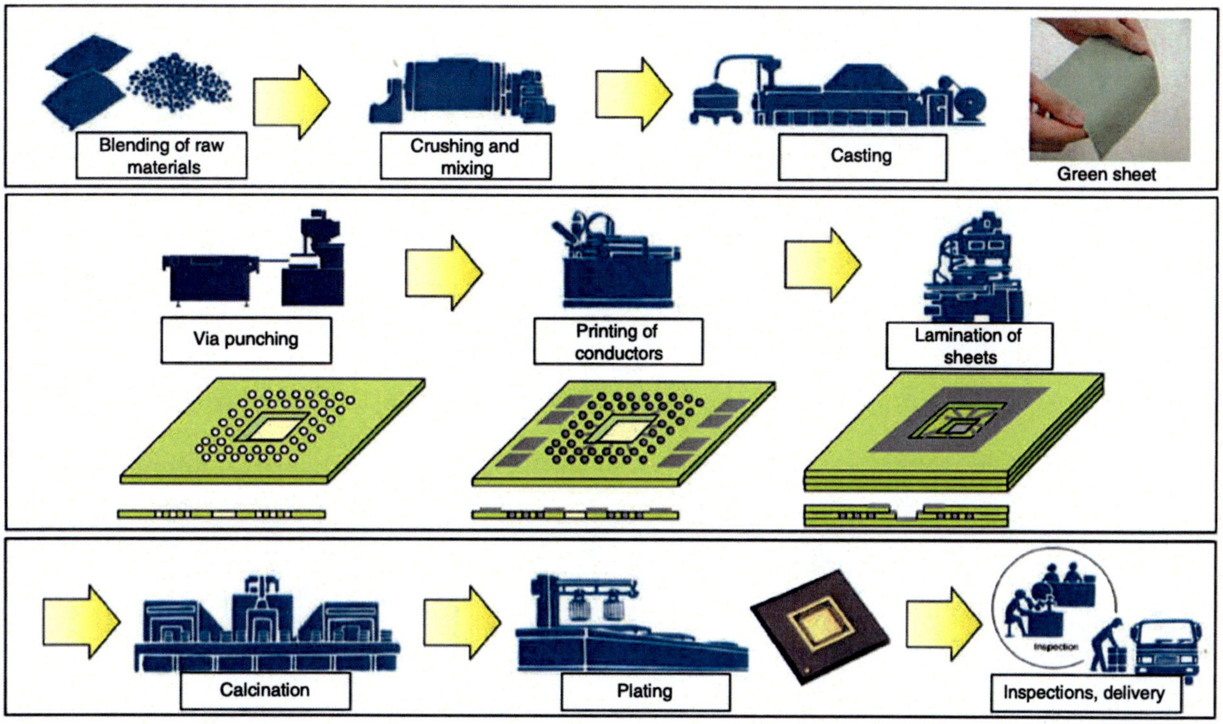

Fig. 7.8 Manufacturing process of a ceramic multilayer package. Ceramic raw powders are crushed and mixed together with organic binders and organic solvents, which are cast to create a green sheet. Via holes are created and conductor patterns are printed on the green sheet, in accordance with the ceramic package design. It is then laminated, calcined and plated to produce the finished product

The green sheets, printed with electrodes, are piled up in the order based on the circuit design, and then laminated by vacuum thermal compression bonding. The laminated body goes through an organic binder removal process, and then the ceramic and electrodes are co-fired in a highly atmosphere controlled furnace to create a sintered body featuring a complex internal structure and high dimensional accuracy. Surface electrodes and metal pins are plated after brazing the metal pins for I/O terminals in some cases, depending on the design. The product is completed after going through several inspection processes.

7.2.3 Future Prospects

The ceramic package has been evolving over the past 40 years since it was first commercialized. The ceramic materials used to produce the packages range from alumina to aluminum nitride having high thermal conductivity and low temperature co-fired ceramic (LTCC) that allows the utilization of high conductivity conductors such as silver and copper. The ceramic packages are selected depending on the application. As a replacement for the ceramic package, a low-priced plastic package is also used. However, ceramic packages are superior in terms of strength, toughness, low thermal expansion and designability, and will continue to evolve and be widely used in devices that require high reliability due to their robustness.

7.3 Hybrid IC Incorporating Low Temperature Co-Fired Ceramic Multilayer Substrate (1990–1992)

Japanese home appliances manufacturers competed in the development and commercialization of compact and portable video camera recorders during the 1990s. The manufacturers incorporated many types of modules, including the hybrid IC, in order to increase the density of circuits inside the package, aiming to realize compact products. During that period, when hybrid ICs incorporating thick film substrates were in the mainstream, hybrid ICs incorporating newly-commercialized low temperature co-fired ceramic multilayer substrates were developed. This development contributed to a dramatic reduction of circuit board space and the improvement of the component mounting density, which led to the miniaturization of video camera recorders.

7.3.1 Background of Development

As a device for recording motion pictures, 8 mm movie cameras were once conventionally used by a limited number of people for many years. Following progress in the development of video technology and the charge coupled device (CCD) sensors, which converts visible images to electrical signals, video camera recorders were released in Japan in the second half of 1980. These motion picture recording devices can be operated easily by consumers and spread rapidly. However, many of the devices in those days used VHS tapes as the recording media, and the large devices were inconvenient for users to carry. Device developers made substantial efforts to create small video cameras that were easy to carry and took the following measures in the 1990s.

1. Shrinking the recording media
   a. Miniaturizing of VHS (Note 7.8) video tapes
   b. Compact 8 mm videotapes
2. Scaling down the electrical circuits
   a. Miniaturizing the ICs
   b. Improvement of high density mounting technology for electrical circuit substrates
   c. Incorporation of the hybrid IC with finer pitch electrical circuits

The devices developed through these measures made video camera recorders easier to carry. Ceramic technology, in particular, contributed to the miniaturization of electrical circuits by realizing the ultracompact hybrid IC, which was manufactured by using a low temperature co-fired ceramic multilayer substrate that was introduced to the market at the time. Figure 7.9 shows a schematic view of the hybrid IC incorporating a low temperature co-fired ceramic multilayer substrate.

7.3.2 Characteristics

7.3.2.1 *Low Temperature Co-Fired Ceramic Multilayer Substrate*

In those days, thick film ceramic substrates and glass epoxy resin substrates were mainly used as hybrid IC substrates. Both types of substrates had problems in terms of structure

Note 7.8 VHS stands for the Video Home System, a video recording/playing system developed by JVC in 1976. The system became the de facto standard for home videos. During the same period, Sony Corporation developed the Betamax system. VHS and Betamax competed in an intense format war for about 10 years, resulting in victory for VHS. The VHS system is being replaced by recording media such as DVD and HDD.

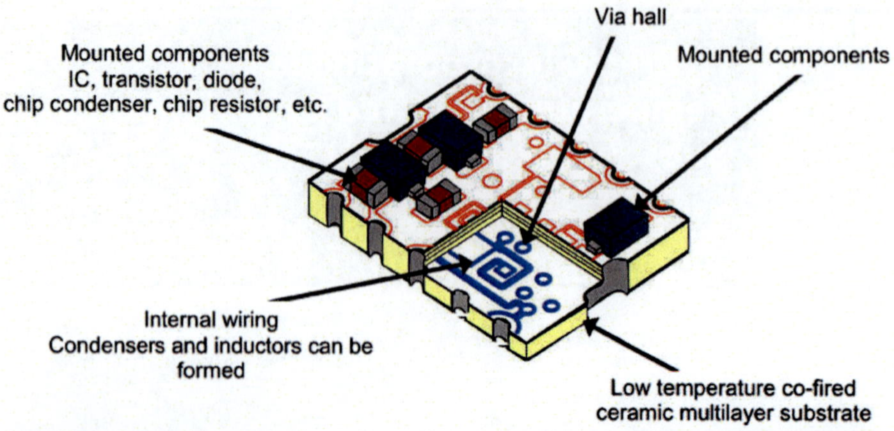

Fig. 7.9 Schematic view of the hybrid IC incorporating the low temperature co-fired ceramic multilayer substrate. By taking advantage of the multilayer ceramic structure, capacity inside the substrate was secured and the inductor was created inside the substrate, greatly contributing to the miniaturization of hybrid ICs

Table 7.3 Comparison of the substrate wiring rules or dimensions

	Wiring rule	
Name of substrates	Line/space	Via hole diameter
Low temperature co-fired ceramic multilayer substrate	75 μm/75 μm	100 μm
Thick film ceramic substrate	100 μm/100 μm	150 μm (one layer only)
Glass epoxy resin substrate	100 μm/100 μm	200 μm

For circuit substrates at the time, the line-space wiring (design) rule was set at 100 μm/100 μm and via hole diameter at 200 μm Φ due to practical reasons concerning mass-production. The values of 75 μm/75 μm and 100 μm Φ attained by the low temperature co-fired ceramic multilayer substrate were breakthroughs that enabled the shrinking of circuit substrates

and reliability, making it difficult to realize smaller sizes and high density, resulting in a maximum component mounting density of 50% or lower. In contrast, the low temperature co-fired ceramic multilayer substrate, which can accommodate the wiring surface in layers and have the component-mounting site formed on the surface layers only, improves the component mounting density by 70%. The three key points in this difference are as follows:

1. Advanced design and wiring technology for finer line-and-space pitch
2. Difference in via hole diameter
   Differences in values are shown in Table 7.3. Figure 7.10 shows the size of the same electrical circuit made by using the glass epoxy resin substrate IC of those days for comparison. The area was reduced by about 60% over the area of products from that time (the product size is 35 mm×20 mm). Figure 7.11 shows the product that was actually mounted in the limited space of an 8 mm video camera recorder.
3. Manufacturing process
   The manufacturing process of this product is shown in Fig. 7.12. First, solder paste is printed on the low temperature co-fired ceramic multilayer substrate (70 mm×70 mm sheet), which is assembled to form a unit of six sheets. Then, the components including the IC, transistor, chip condenser, and chip resistor are mounted at a high speed (0.1 s/pcs) through the use of an automatic mounting machine. The substrate is heated in a reflow furnace to a temperature exceeding the solder melting temperature (approximately 190°C) to melt the solder for the metal binding of the electrode to the pad on the substrate and the terminal electrodes of each component. Then, flux residues generated in the soldering process are washed away, followed by visual inspection, dimension verification, and an electrical quality assurance process.

7.3.3 Future Prospects

The hybrid IC that incorporates the low temperature co-fired ceramic multilayer substrate began to be fully utilized in early 1990. The hybrid IC currently has a huge share in the market for high-frequency modules used to miniaturize high-frequency circuits in mobile phones, etc., taking advantage

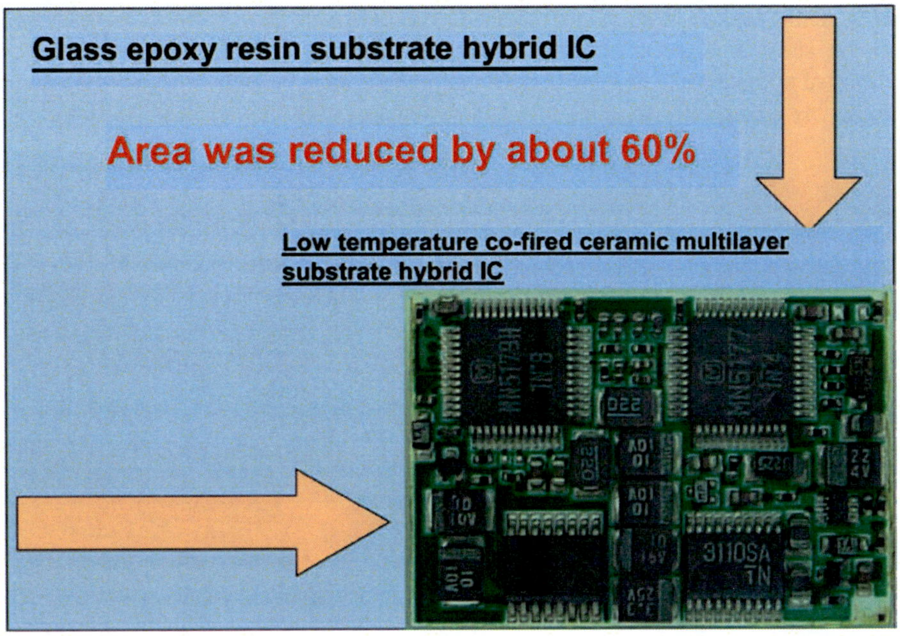

Fig. 7.10 Comparison of the areas of glass epoxy resin substrate hybrid IC and low temperature co-fired ceramic multilayer substrate hybrid IC. The size of the hybrid IC was substantially reduced by 60% in area, due to the difference in the structure and the design rules and more advanced semiconductor processing

Fig. 7.11 Product mounted in an 8 mm video camera recorder. The large CCD sensor control circuit was miniaturized to an ultra-compact size, making it possible to mount needed circuits inside the limited finder space, which contributed to the miniaturization of the device

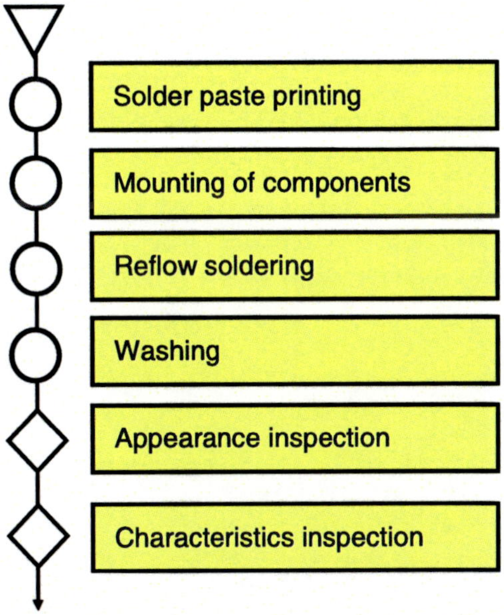

Fig. 7.12 Manufacturing process of low temperature co-fired ceramic multilayer substrate hybrid IC. Solder paste is printed on the low temperature co-fired ceramic multilayer substrate, which is assembled to form a unit of six sheets. The components are then mounted by using an automatic mounting machine. The substrate is heated in a reflow furnace to melt the solder for metal binding. Then, flux residues are washed away, followed by an appearance inspection and a characteristics inspection before completion of the product

of its ability to use low conductor resistance materials such as Ag and Cu as electrode materials. The hybrid IC is expected to expand into new markets with new applications by adding high performance, high frequency modules, such as Bluetooth modules and a wireless LAN for Near Field Communication (Note 7.9). A high-frequency module of interest that incorporates metamaterials as an antenna will be introduced in the future.

Note 7.9 A system for short distance communication within about 100 m, including Bluetooth, wireless LAN, UWB, etc. Wireless LAN has a variety of standards such as the IEEE802.11 series, and the IEEE802.11 g (54 Mbps) type using 2.4 GHz frequencies. It is currently in the mainstream in Japan.

7.4 Aluminum Nitride Substrate for Semiconductor Device (1985)

Semiconductor devices requiring high heat dissipation performance are incorporated in a variety of devices ranging from inverters that can control a large amount of power to commercially available DVD devices. Aluminum nitride, whose thermal conductivity is higher than that of metallic aluminum, is used as semiconductor device mounting boards on which metallic circuits are formed in advance by thick copper plates or Au thin film depending on the usage. This technology has greatly contributed to the device performance improvement. Improvements in semiconductor performance will continue to accelerate and the utility of aluminum nitride is expected to further increase.

7.4.1 Background of Development

Heat generated by semiconductors is a major reason for property degradation or cause of failures in semiconductors. Device reliability has been steadily improving together with increased functionality and greater integration. For these reasons, priority is placed on heat dissipation performance and thermal expansion properties, as well as electrical insulation properties, in the selection of mounting and circuit boards.

The power module, which is the heart of the inverter (Note 7.10) used for power control, is required to have high heat dissipation performance, as well as a thermal expansion coefficient that matches Si. In the case of a large Si chip that is directly mounted on a board, it results in a increase in the power usage and greater integration as well as a need for higher reliability.

Furthermore, semiconductor lasers (Note 7.11) are incorporated into DVD devices, which have been contributing to the recent remarkable improvements in the characteristics of PCs and visual devices. The laser itself tends to be affected by heat and stress attributable to heat generated during operation. Therefore, the substrate used in these laser devices is required to have a high thermal dissipation property as well as a thermal expansion coefficient equivalent to that of the laser chip (Note 7.12) (made of GaAs, etc.), responding to improvement in the performance of the devices. Figure 7.13 shows the structure of a semiconductor laser.

7.4.2 Characteristics

7.4.2.1 *Characteristics of Aluminum Nitride Circuit Substrate*

Aluminum nitride (AlN) is a nitrogen containing material that does not exist in nature and has a theoretical high thermal conductivity of 320 W/(m K). Its thermal conductivity as a commercial product is lower than this theoretical value, but a thermal property exceeding that of metallic aluminum can be attained by decreasing the defects in the AlN grains through microstructure control. In order to use aluminum nitride as a semiconductor mounting board, metallic circuits have to be formed on top of it. Various metallic circuit forming methods were developed for a variety of applications. Table 7.4 shows the characteristics of aluminum nitride and Fig. 7.14 shows an example of the AlN circuit substrate forming method. The important point in forming the circuit involves how the metals are jointed to the nitride material.

Note 7.10 A device that arbitrarily converts the power-supply frequency applied to the motor so as to control the motor speed freely, continuously and effectively.

Note 7.11 A laser oscillation module incorporating a compound semiconductor, which includes the GaAs type and the GaN type, mounted on optical pickups of CDs and DVDs. The module is composed of a compound semiconductor laser chip, heat sink, submount, lead, etc.

Note 7.12 A compound semiconductor laser chip mounted in a semiconductor laser module. GaAS crystal is used for lasers from the infrared laser to the red laser and GaN crystal is used for the blue laser.

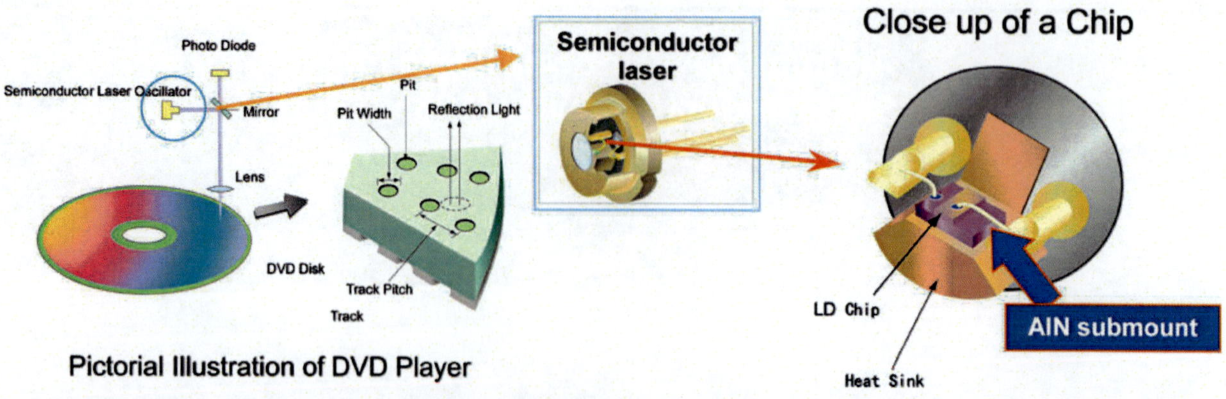

Fig. 7.13 Structure of semiconductor laser. The figure shows the structure of a semiconductor laser mounted with aluminum nitride (AlN). The characteristics of the semiconductor laser is improved by positioning the AlN, which has a high thermal conductivity and a thermal expansion coefficient close to that of the laser chip, directly below the chip

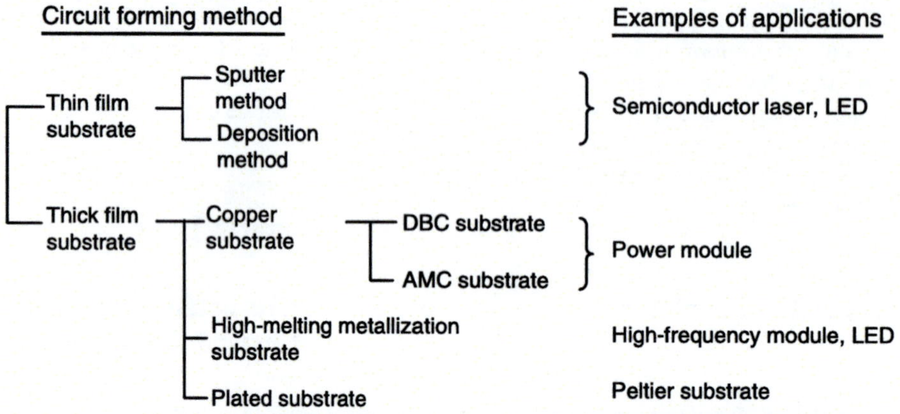

Fig. 7.14 Example of AlN substrate circuit forming method. The figure shows the typical circuit formation method of the AlN substrate as well as its applications. It is expected that they will be applied not only to semiconductor lasers and power modules but also to the LED and Peltier substrates of high-frequency modules

Table 7.4 Examples of aluminum nitride characteristics

Item	Unit	Al_2O_3	AlN		
Bending strength (at 3 points)	MPa	320	350	330	330
Fracture toughness	MPa $m^{1/2}$	3.5	3.3	3.0	3.0
Coefficient of thermal expansion	$\times 10^{-5} \times 10^{-5}$ 1/K	7.3	4.6	4.6	4.6
Thermal conductivity	W/(mK)	21	170	200	230
Dielectric strength	kV/mm	14	14	14	14

Values expressing the major characteristics of aluminum nitride (AlN) are shown in the table. Compared with alumina (Al_2O_3), which is frequently used as a semiconductor mounting substrate, the thermal conductivity is remarkably high

7.4.2.2 Metalized Substrate with Thick Copper

The power module mentioned above incorporates a substrate on which a thick copper plate ranging from 0.2 to 0.5 mm in thickness is directly bonded. It is a simple structure with a thick copper plate directly bonded on AlN, and it features high heat dissipation characteristics. The thermal expansion coefficient of the copper circuit is equivalent to that of AlN, and therefore, a large Si chip can be directly joined with the circuit, making it possible to simplify the power module structure needed for matching different thermal expansion properties. Figure 7.15 shows the image of the substrate incorporating a copper plate and structures of the power modules are compared in Fig. 7.16.

The copper circuit is bonded by two methods: the direct bonding method in which a Cu plate is directly joined with AlN, and the active metal method in which a brazing filler material added with active metals such as Ti and Zr is used as an intermediate material. The former is called the direct bonding copper (DBC®) substrate, while the latter is called the active metal brazing copper (AMC) substrate. The DBC substrate utilizes the Cu–O eutectic liquid phase generated by using a small amount of oxide contained in Cu as a bonding material. The Cu–O eutectic liquid phase rarely wets a nonoxide materials, and therefore, a thin oxide layer is formed on the surface of the AlN to enable the bonding.

Meanwhile, the AMC substrate usually uses Ti as an active metal and contains Ag and Cu in general. Ti preferentially diffuses on to the surface of AlN to form TiN, bonding the Cu plate via the Ag–Cu alloy.

Figure 7.17 shows cross sectional views of the DBC method and the AMC method.

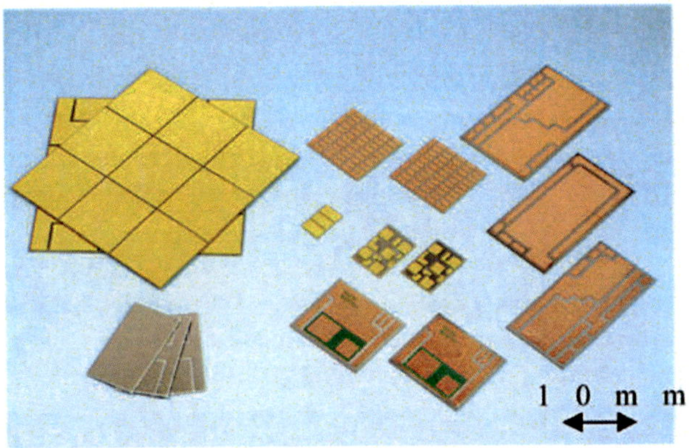

Fig. 7.15 Image of metalized substrate with thick copper. The figure shows the images of circuit substrates on which thick copper plates are bonded. The copper plates are plated with Ni or Au in some cases

7.4.2.3 Thin Film Substrate

AlN circuit substrates used as submount substrates (Note 7.13) of semiconductor lasers are generally manufactured by forming metal thin films on the surface of AlN by deposition or sputtering. The metal films are usually three layered structures composed of Ti, Pt and Au. Each layer has the following function:

1. Ti: reacts with AlN to form nitride, and functions as a contact layer with ceramic.
2. Pt: prevents diffusion as an intermediate barrier.
3. Au: used as a surface layer because Au-Sn eutectic liquid solder is used for mounting chips and Au wires are bonded.

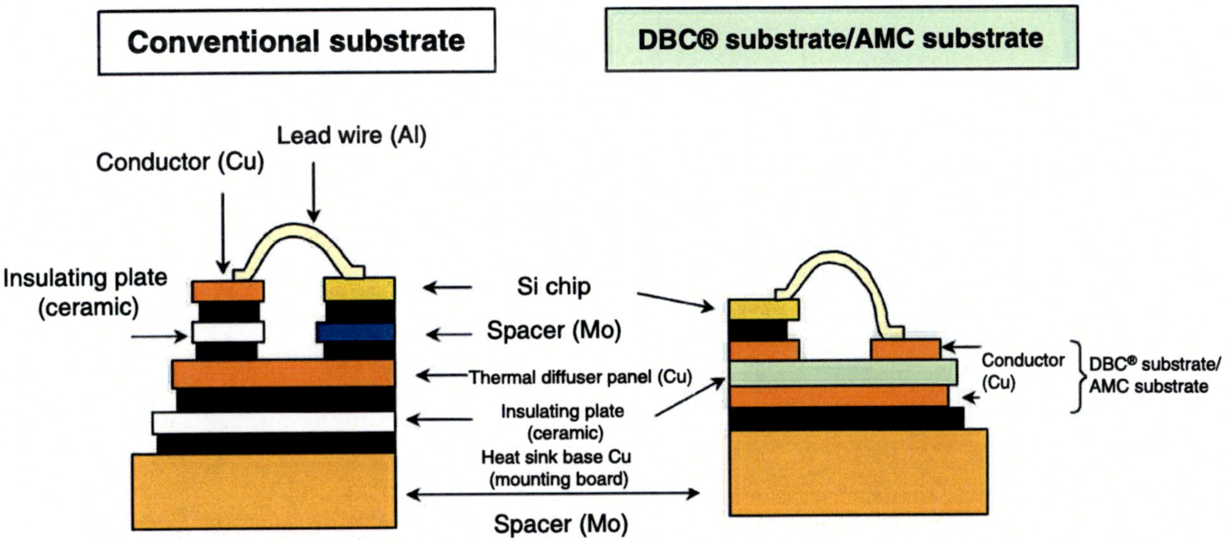

Fig. 7.16 Comparison of power module structures. By using substrates incorporating copper plates (DBC® substrate/AMC substrate) the power module structures were simplified when compared with conventional substrates, reducing the mounting cost and heat resistance

Note 7.13 A submount substrate is mounted between the laser chip and the metallic heat sink of a semiconductor laser module. It is used to distribute the heat generated from the laser chip uniformly and to reduce the stress caused by the difference in thermal expansion mismatch between the chip and the metallic heat sink.

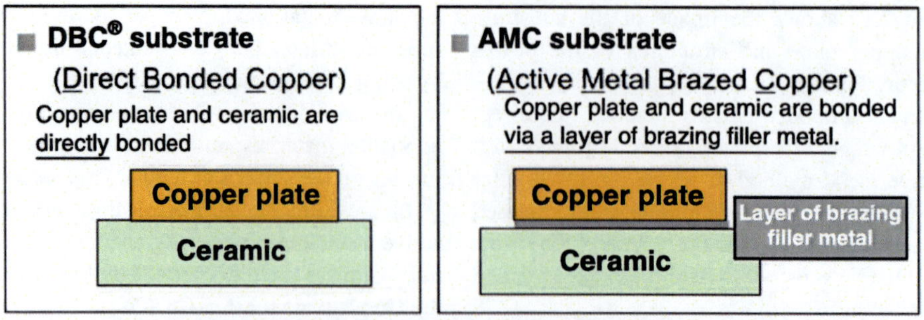

Fig. 7.17 Cross-sectional views of substrates bonded by the DBC method and the AMC method. The figures show cross-sectional views of the boards bonded with copper plates with two different methods

7.4.3 Future Prospects

Aluminum nitride substrates utilization are expected to grow in line with the rapid improvement in the performance of semiconductor products. Meanwhile, the stumbling block that prevents the utilization of aluminum nitride in semiconductor devices, which continue to drop in price, is the production cost of aluminum nitride. Price reduction is difficult because the amount of aluminum nitride raw material in the market is limited and also because its production process requires special equipment, unlike oxide ceramics. A breakthrough in reducing production costs, as well as expansion of applications will be needed for the expansion of this material.

Silicon Carbide for High Thermal Conductivity Substrate (1985)

7.5

A high thermal conductivity (Note 7.14) ceramic substrate was first commercialized in the early 1980s in response to the change from ICs into VLSIs (Note 7.15), which started in the second half of 1970, as a countermeasure to remove the heat generated from ICs during operation. The amount of heat per unit area generated from a VLSI is higher than that of an electrical heater, and a need for ceramic materials that have a heat conductivity equivalent to that of metallic aluminum and related materials was established. Because of this situation, high thermal conductivity SiC substrates were commercialized as IC package mounting boards. Their application has decreased due to a change in computer mounting technology and a change in ICs from bipolar to Complementary Metal Oxide Semiconductor (CMOS). They are currently used as mounting boards for optical devices of semiconductor lasers (Note 7.16) used in communication and information technology.

7.5.1 Background of Development

In the second half of the 1970s, efforts were made to shrink bipolar devices and create VLSIs in order to increase the scale and processing speed of computers. Following these changes, the heat generated from the LSI became extremely high, which led to the need for a new cooling technology as well as utilization of ceramics that have high thermal conductivity equivalent to metals for use as IC package materials. Under these circumstances, focus was placed on silicon carbide (SiC) and aluminum nitride (AlN), which both have high theoretical thermal conductivities. Following this, the SiC substrate, a ceramic substrate having a thermal conductivity higher than AlN substrate, was developed.

7.5.2 Characteristics

7.5.2.1 *Characteristics of Materials*

Ceramics represented by alumina have high strength, good insulation properties, and reliability, but they also have low thermal conductivity, and thermal expansion property that is extremely different from that of the Si used in devices. In contrast, beryllium oxide (beryllia, BeO) has high thermal conductivity comparable to metal, but was used in only limited fields due to its toxicity. Silicon carbide (SiC) developed in the early 1980s is a substrate having high thermal conductivity, excellent electrical insulation and thermal expansion coefficient equivalent to Si, and can be used in versatile conditions.

7.5.2.2 *Manufacturing Method*

The basic manufacturing process of SiC ceramics is explained below. In the initial process, SiC powder, organic binder, and sintering additives are mixed to create a uniformly-mixed slurry, which is processed into granular powder via the spray dry method. Then, the powder is evenly filled in a die and press-formed. The press-formed body, piled up with separators, is sintered in high temperature vacuum state (2,100°C) while kept under pressure (hot pressing method, HP method).

Note 7.14 The figure of merit is a measure of a materials heat conducting capability. In general, ceramics have low thermal conductivity, i.e., 17 W/(m K) in the case of alumina. The development of ceramics having thermal conductivity equivalent to metals such as A1 [234 W/(m K)] and Cu [395 W/(m K)] had been desired. Materials of high thermal conductivity are needed because the power consumption of semiconductors is approximately 0.5 W/mm^2 or higher.

Note 7.15 VLSI, is an abbreviation for Very Large Scale Integrated Circuit.

Note 7.16 The semiconductor laser is the heart of devices such as CD players, DVD players, optical pickups, and signal transmission devices for optical communication.

The sintered substrate is surface polished to adjust the thickness, then cut into specified sizes. The metallic wiring films, etc. are formed on the surface of the substrate by the thick film method/thin film method, for such uses as an LSI mounting package substrate or a wiring substrate.

Table 7.5 Comparison of characteristics of ceramic substrates

	SiC	Al_2O_3	BeO	Si
Thermal conductivity (W/(mK))	300	17	260	135
Thermal expansion coefficient (10^{-6}/°C)	3.7	7.2	7.5	3.5
Dielectric property (RT, 1 MHz)	45	8.5	6.7	11.9

7.5.2.3 Product Performance

Characteristics of ceramics are shown below, in comparison with the alumina substrate.

Now, the substrates will be compared under mounted conditions. By incorporating SiC with high thermal conductivity as the substrate material, the thermal resistance was reduced to 55% of the value attained by conventional alumina substrates (using the same shape and structure). Table 7.5 shows images of the SiC substrate and the alumina substrate mounted on large-scale computer LSIs. The conventionally used alumina package, shown on the left side, has a unique structure that was designed to combine the thermal conductivities of the metals to attain sufficient thermal resistance. Meanwhile, the SiC package incorporates high thermal conductivity SiC with a thermal conductivity that is equivalent to metals, is structured simply, and can attain the same thermal resistance.

7.5.3 Future Prospects

Utilization of high thermal conductivity SiC in computers in the method shown in Fig. 7.18 was discontinued in the second half of 1980 due to the introduction of CMOS LSIs

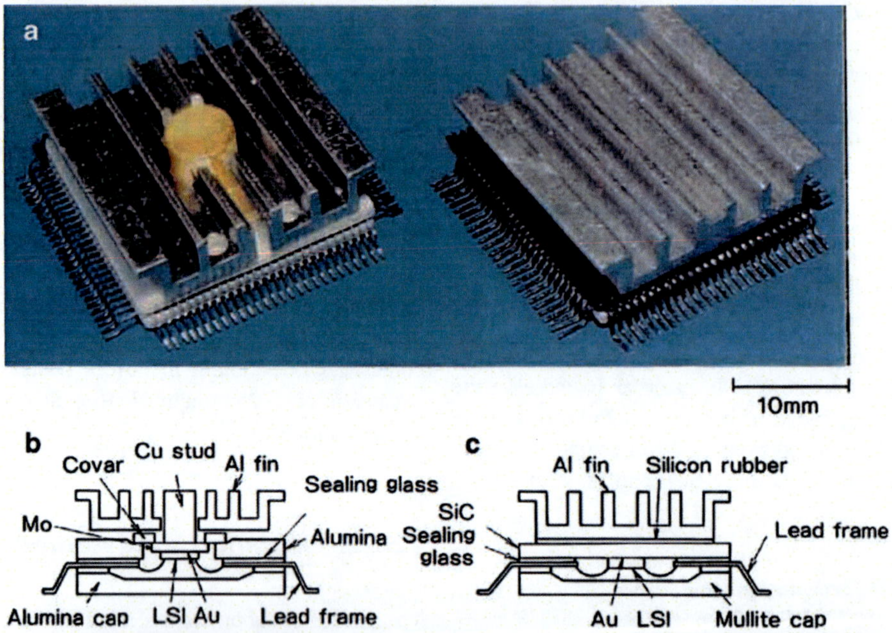

Fig. 7.18 LSI package for large computers incorporating high thermal conductivity SiC ceramics. The conventionally used alumina package (image on the left side of photo **a** and **b**) has a unique structure that was designed to combine metals so as to attain proper thermal resistance. Meanwhile, the SiC package (image on the right side of photo **a** and **c**), incorporating high thermal conductivity SiC, is structured simply and can attain proper thermal resistance

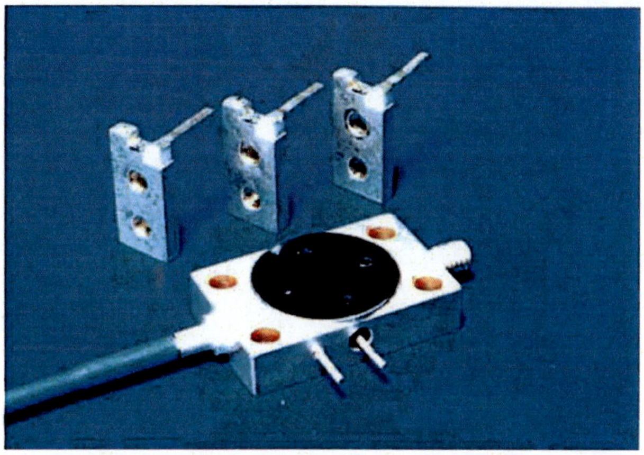

Fig. 7.19 Semiconductor laser for optical communication incorporating high thermal conductivity SiC heat sink. Because the power consumption of semiconductor lasers is approximately 0.5 W/mm^2 or higher, ceramics featuring high thermal conductivity as well as thermal expansion properties equivalent to silicon are needed for incorporation in laser based devices

and multi chip modules. Since that time, it has been supplied for use in the heat sinks of optical elements such as semiconductors lasers, as shown in Fig. 7.19. High thermal conductivity SiC will continue to be used and supplied as a material for components such as optical communication devices, and optical storage devices, among others.

Single Crystal Sapphire Substrate (1995) 7.6

Single crystal sapphire, an aluminum oxide corundum (Note 7.17) crystalline material composed of natural aluminum and oxygen, is widely known as a precious stone. The pure crystal is transparent and colorless, but becomes blue (blue sapphire) if the aluminum oxide contains impurities such as Ti and Fe, and red (ruby) if the impurity is Cr.

This single crystal sapphire, displays excellent mechanical and thermal properties as well as chemical stability and light permeability, and is incorporated as a material for a variety of industrial products, including the substrate for GaN crystal growth used in blue and white LEDs (light emitting diodes) and the polarizer retention substrate used in LCD (liquid crystal display) projectors.

7.6.1 Background of Development

Utilization of single crystal sapphires was limited to certain fields because it was previously difficult to grow and process high quality crystals, although the single crystal sapphire, with its excellent mechanical and thermal properties, chemical stability, and light permeability. There were a wide variety of devices and optical elements that would use sapphire substrates if the substrates were inexpensive. Because of advances in crystal growth and processing technologies in recent years, it is now possible to produce large, high quality single crystal sapphires with diameters >100 mm. Following the developments of these technologies, the crystal is currently incorporated as a material for a variety of industrial products, including the substrate for GaN crystal growth used in blue and white LEDs (light emitting diodes) and the polarizer retention substrate used in LCD projectors. It has also begun to be used in devices operated at high speeds and frequencies, which incorporate silicon on sapphire (SOS) (Note 7.18) made by growing silicon on sapphire. Its applications are expected to expand further, including application to devices that are mounted with large-scale CMOS (Complementary Metal Oxide Semiconductor) circuits.

7.6.2 Characteristics

7.6.2.1 *Material Property*

Sapphire is an α-alumina (α-Al_2O_3) crystal and is similar to a hexagonal system (rhombohedral system to be precise) as shown in Fig. 7.20. Its properties are shown in Table 7.6. The major properties are as follows:

1. Flat and smooth surface [after chemical mechanical polishing (CMP) Ra ≤2 Å]
2. High optical transparency in a wide range of wavelengths (200–5,000 nm)
3. High temperature melting point(2,053°C)
4. High thermal conductivity (42 times higher than that of silica glass)
5. High chemical resistance

7.6.2.2 *Manufacturing Method*

Single crystal sapphire was first synthesized in 1876 by French chemist Fremi using lead-based flux in a joint experiment

Note 7.17 Generic name of corundum crystalline precious stones grown from aluminum oxide crystals. The pure crystal is transparent and colorless, but crystals containing dopants or impurities have a characteristic color.

Note 7.18 A type of fully-depleted SOI (Silicon On Insulator). SOS stands for Silicon On Sapphire. Single crystal silicon is grown on a single crystal sapphire substrate and a CMOS circuit is formed on the substrate. SOS features higher speed, higher high-frequency properties and lower power consumption, compared with general silicon substrates.

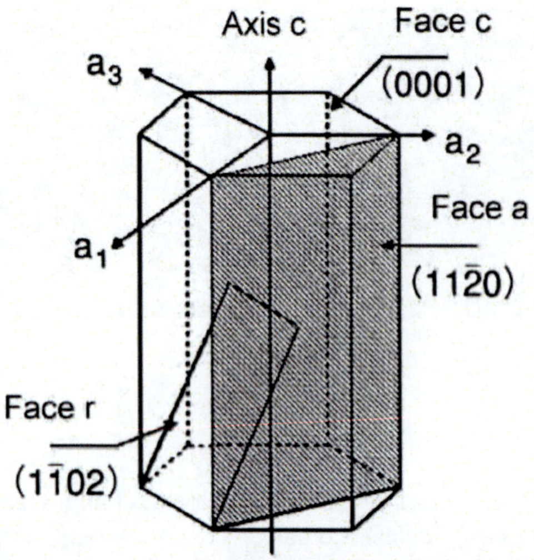

Fig. 7.20 Unit cell of sapphire single crystal. Sapphire is an α-alumina (α-Al_2O_3) crystal and is approximately a hexagonal system (a rhombohedral system to be precise)

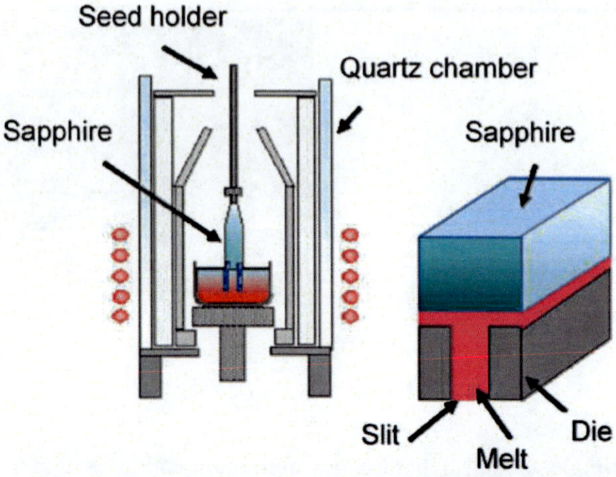

Fig. 7.21 Schematic view of the EFG method (enlarged conceptual diagram of die is on the *right side*). The crystal is grown by using a die placed inside a melt crucible, which regulates the supply of melt and the shape of crystal. The melt inside the crucible moves upward through the slit on the die by capillary action, gets crystallized, and is pulled upward. The cross-sectional shape of the grown crystal is regulated by the shape of the *upper end* of the die, creating crystals with the same approximate size

Table 7.6 Material characteristics of sapphire

Physical properties		Electrical properties	
Crystal system	a=4.763 Å	Electrical resistance	1×10^{14} Ω m (normal temperature)
	c=13.003 Å		1×10^{11} Ω m (500°C)
Density	3.97×10^3 kg/m^3	Permittivity	11.5 (//Axis C)
Tensile strength	2,250 MPa		9.3 (⊥Axis C)
Thermal properties			Both (10^3–10^{10} Hz, 25°C)
Thermal expansion coefficient	5.3×10^{-6}/k (//Axis C)	Dielectric strength	48×10^3 kV/m (60 Hz)
	4.6×10^{-6}/k (⊥Axis C)	Optical properties	
Thermal conductivity	42 W/m K (normal temperature)	Refractive index	no=1.768
Specific heat	75 KJ/kg K (normal temperature)		ne=1.760

with Feil. It is believed to be the first crystal that was artificially synthesized. Later, sapphire came to be produced at low cost using the Verneuil method, a method for the recrystallization of synthetic gems. However, large-sized ingots could not be grown using this method and the crystallinity was inferior, limiting its range of application. Later, the Czochralski method began to be employed to grow synthetic sapphire, with the aim of improving crystallinity. However, the material was not grown in bulk using this method and the manufacturing cost was high. Following the development of the EFG (edge-defined film-fed growth) method, a method that makes it possible to grow inexpensive materials with high crystal quality, single crystal sapphire began to be widely used as an industrial material.

The EFG method (Note 7.19) was reported by Labelle and his colleagues in 1971, leading to the development of sapphire crystal fiber growth technology, followed by the developments of growth technologies for ribbon and cylindrical sapphire crystals. A diagram of the EFG method is shown in Fig. 7.21.

The crystal is grown by using a die placed inside a melt crucible, which regulates the supply of melt and the shape of crystal. The melt inside the crucible moves upward through

Note 7.19 One of the single crystal growth methods. It stands for edge-defined film-fed growth. It is characterized by the easy controllability of crystal orientation, high growth speed, and controllability of the cross-sectional shape.

the slit on the die by capillary action, crystallizes, and is pulled upward. The cross-sectional shape of the grown crystal is regulated by the shape of the upper end of the die, creating crystals of the same approximate size. Therefore, crystals of various cross-sectional shapes, such as a circle, quadrangle and tube, can be grown by simply adjusting the shapes of the slit and the upper end of the die. In addition, multiple crystals can be grown in one die by making changes to the die.

7.6.2.3 Applications to Blue and White LEDs

Technologies for growing crystal on GaN-based III–V group compound semiconductors progressed at a remarkable pace during the past few years, leading to the commercialization of green, blue and white LEDs, etc. LED light sources are characterized by speedy on/off operation, a long working lifetime that is 50–100 times longer than that of incandescent light bulbs, and a low power consumption that is 1/3–1/15 lower than that of conventional light bulbs. Because of these characteristics, LED applications have been rapidly increasing and they include mobile phones, LCD backlights, full-color outdoor displays, toys and general lighting devices. GaN, the main material of blue and while LEDs, can be epitaxially grown on single crystal sapphire substrates. Because the single crystal sapphire has properties suitable for epitaxial growth, including a lattice constant that is relatively close to that of GaN, high temperature stability, thermal expansion properties, and chemical stability under the GaN epitaxial growth condition (approximately 1,200°C). It can also be purchased for a comparatively low price. It is used as a standard substrate for these reasons.

Figure 7.22 shows a picture of the sapphire substrate used for thin films deposition. It is preferable that sapphire substrates be large in diameter for use in semiconductor processing equipment. The initial size was 10 × 10 mm in the early stage of development, and was increased to 50.8 mm or 2 inches in diameter in response to the start of LED mass production. Currently, substrates up to 300 mm or 12 inches in diameter are in use. In order to obtain high quality crystalline GaN films, bidirectional control, as well as axial orientation control for the reduction of variations in cutting the chips, is important. In addition, accuracy enhancement with respect to warping and flatness, which affects the uniformity of GaN films and forming of device circuits, is also important.

7.6.2.4 Polarizer Retention Substrate Used in LCD Projectors

Responding to remarkable progress in the IT (information technology) revolution, PCs began to be used frequently for preparing oral presentations, increasing the demand for compact and high luminance LCD projectors. Given these circumstances, measures for dissipating heat in polarizer films used in these types of projectors became an important issue. Polarizer retention boards are required to have a variety of characteristics including clearness, a high thermal conductivity that enables high heat radiation, a low thermal expansion property and low anisotropic nature to avoid stressing the polarizer, and a resistance to optical anisotropy. Polarizers featuring high heat radiation properties were realized by optimization of orientation and establishment of effective manufacturing methods, leading to the production of compact and high luminance LCD projectors. Figure 7.23

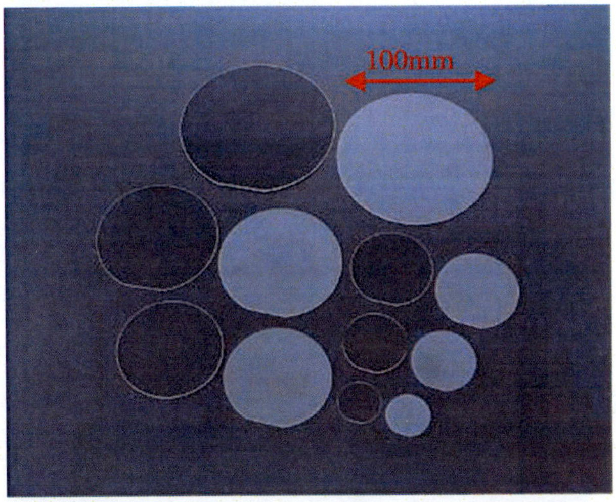

Fig. 7.22 Sapphire substrate for thin films

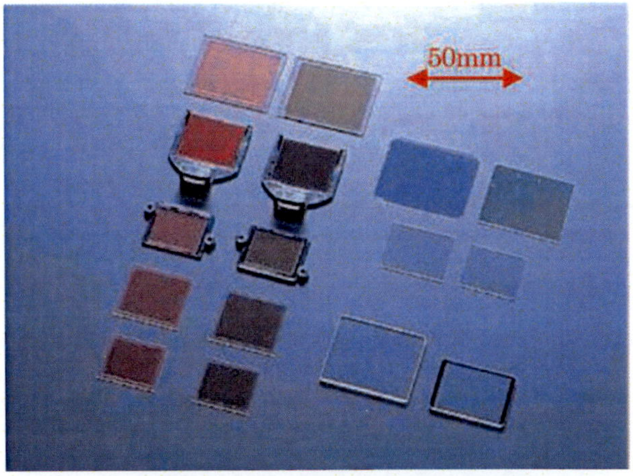

Fig. 7.23 Polarizer retention board used in LCD projectors

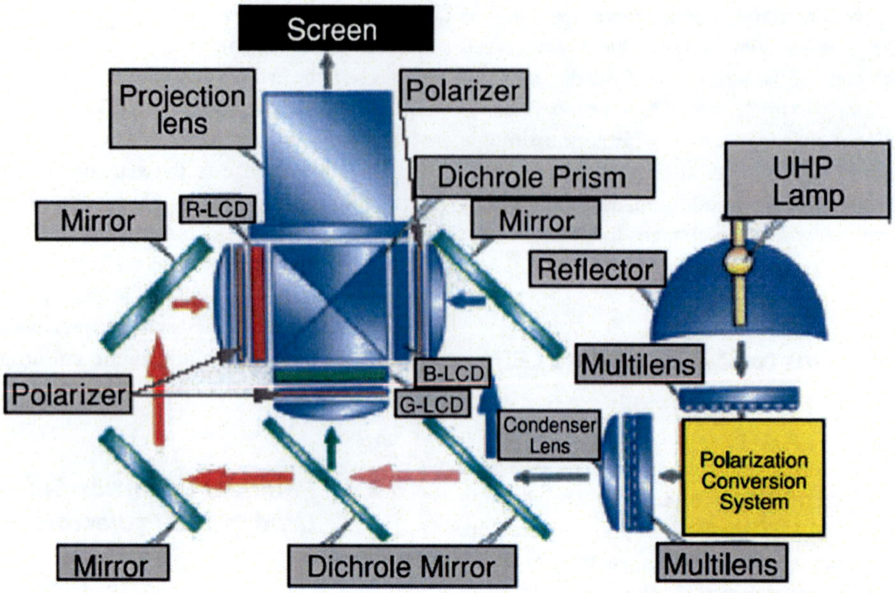

Fig. 7.24 Example of an optical engine for LCD projectors (3LCD type)

shows a picture of a polarizer retention board used in LCD projectors, and Fig. 7.24 shows a schematic drawing of an optical engine for LCD projectors. The EFG method, which allows for the efficiently growing board-shaped materials, makes it possible to effectively produce square-shaped substrates such as a polarizer retention board at a low cost.

7.6.3 Future Prospects

In line with the increasingly serious environmental concerns of recent years, greater expectations for energy-saving, environmentally-friendly LEDs have been on the rise. LED technologies will continue to progress in terms of improved luminance and luminance efficiency. The need for LED chips will greatly increase when they begin to be used in automobiles and general household/office lighting devices, and larger substrates will be needed for the effective production of these chips. In addition, they have begun to be incorporated in high-frequency devices, following active research on the SOS, which is made by forming a Si film on a single crystal sapphire substrate. Demand for sapphire substrates are expected to continue to grow.

Literature

1. Imanaka Y (2005) Multilayered low temperature co-fired ceramics (LTCC) technology. Springer, New York **(7.1)**
2. Imanaka Y (2006) Multilayer ceramic substrate. Mater Integr 19(3): 37–46 **(7.1)**
3. The Electronic Materials Manufacturers Association of Japan (ed) (1994) Advanced ceramic substrate for electronic circuit. Japan Industrial Publishing Co., Ltd., pp 29–50 **(7.2)**
4. (1995) Handbook on crystal growth. The Japanese Association of Crystal Growth, pp 634–636 **(7.6)**
5. LaBelle HE Jr (1971) Mater Res Bull 6:581 **(7.6)**
6. Chalmers B, LaBelle HE Jr, Mlavsky AI (1972) J Cryst Growth 13/14:84 **(7.6)**
7. LaBelle HE Jr (1980) J Cryst Growth 50:8 **(7.6)**

8 Office Automation Devices

Ceramics have contributed greatly to the development and growth of the information technology (IT) industry. Various ceramic products have been commercialized as products, intended for use as the functional material and/or structural component in peripheral Office Automation (OA) systems. This chapter discusses ceramic products incorporated as these vital components in printing machines, namely printers used as output devices.

In general, printers are used as output devices for producing hard copies or print outs of information transmitted from remote locations and produced locally by OA devices. Ink-jet type printers are widely used and well known. Ink-jet printers are used widely not only for office use but also in households for printing of decorative items, pictures, home business, etc. Printers incorporating piezoelectric elements have been widely commercialized as one type of ink-jet printer. The printers control precisely the ink discharge rate by means of an actuator, which utilizes the piezoelectric effect and produces vibrant high color images as a result. The ink discharging printer head (Sect. 8.1) is the heart of a printer. The piezoelectric ceramic, mounted inside the printer head, consists of a piezoelectric elements.

Although the field of application differs from that of ink-jet printers, thermal recording printers are also widely used in a variety of business applications. They are mainly used in OA devices such as fax machines and printers, as well as simplified portable printers for the printing. They are also used in logistics systems for printing data and bar codes transmitted from a variety of electronic devices within the freight industry. They have also have wide ranging applications that includes printing of letters and images produced by medical devices. This thermal print head (Sect. 8.2) was made by forming fine resistive layers that function as heating elements and electrode layers in a circuit on an alumina ceramic substrate that is covered by protective layers.

Y. Imanaka et al. (eds.), *The Ceramic Society of Japan, Advanced Ceramic Technologies & Products*,
DOI 10.1007/978-4-431-54108-0_8, © Springer Japan 2012

Inkjet Printer Head (1995)* 8.1

Inkjet printers have many advantages such as high resolution printing equivalent to that of photographs, low costs and compact sizes. They are used widely in offices and households worldwide. Print heads are classified roughly into two types, the piezoelectric (piezo) type and the bubble jet type. Piezo print heads are classified into three types, namely the flexural vibration type (Note 8.1), the longitudinal vibration type and the shear vibration type, depending on the vibration mode being utilized. In this section, the flexural vibration type print head will be introduced. Flexural vibration print heads are comparatively easy to manufacture and feature high performance. Electrodes are formed on both sides of a thin piezo element which contracts when an electric field is applied. Ink flows via channels when the diaphragm is flexed to discharge ink drops. Print heads today are capable of discharging 7 picoliter (pL) ink droplets at a frequency of 50 kHz from each of the discharge outlets. They can also discharge finer ink droplets of 2 pL.

8.1.1 Background of Development

Personal computers and the Internet have evolved and expanded remarkably in recent years. There was a theory that the paperless era would come with the increase in computer use. However, the amount of printed materials has increased in line with the increase of information following the wide spread use of computers. Inkjet printers, featuring advantages such as high resolution printing equivalent to that of photographs, low costs and compact sizes, are used more widely than other types in offices and households. In addition to single function printers, multi-function printers (printers that can also fax, scan and copy) have become available and an increasing number of users are selecting them. Besides, there are many compact printers that are exclusively used for the printing of photographs on the market. Users have been increasing rapidly as these printers do not have to be connected with personal computers and can be operated simply by the press of a button when printing photographs. In response to these needs, tremendous efforts have been made in recent years toward the improvement of inkjet head performance so that higher-speeds and higher-resolution printing may be achieved. The number of ink discharge outlets has increased substantially, increasing ink discharge amount per unit time, and the minimum size of an ink droplet that can be discharged has decreased as well.

8.1.2 Characteristics

8.1.2.1 Characteristics and Speci cations of Products

The heads of inkjet printers are classified roughly into two types, the piezoelectric type and the bubble jet type. Piezo print heads are classified into three types, namely the flexural vibration type, the longitudinal vibration type and the shear vibration type, depending on the vibration mode. In this section, the flexural vibration type ink jet head will be introduced.

The principle of discharging ink from the head is illustrated in Fig. 8.1. The actuator is formed on a diaphragm which forms the ink pressure chamber (Note 8.2). The actuator is composed of alternating layers of thin electrode films and piezoelectric films. When a voltage is applied to electrodes, an electric field is generated within the piezoelectric film. This reduces the area of the piezoelectric film and

*The number in parenthesis indicates the year that the product was first commercialized.

Note 8.1 Vibration similar to the drum face being convexed and concaved.

Note 8.2 The flow channel is connected to the ink discharge outlet and its internal pressure changes responding to the activation of the actuator.

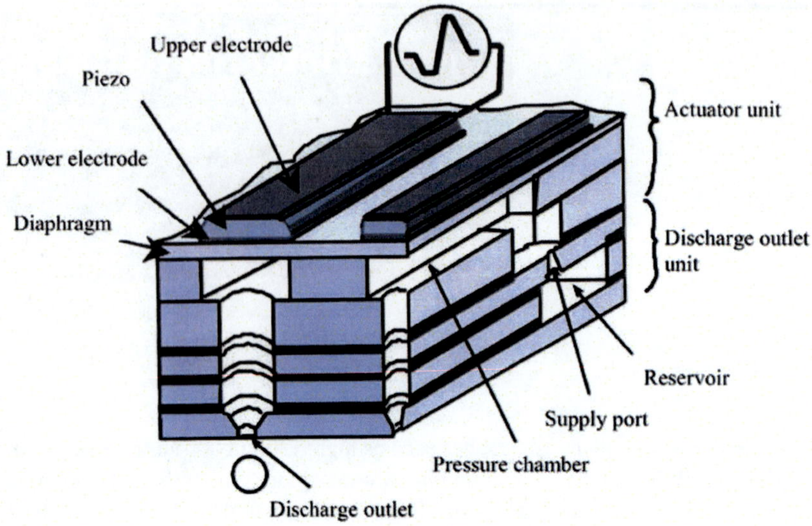

Fig. 8.1 Cross-sectional view of head. The ink flow channel is formed by punching a hole in the plate and then laminating the plates. Ink is supplied from the ink container to the reservoir, and then enters the pressure chamber via the supply port. It is applied with a driving pulse and is discharged from the discharge outlet in the form of ink drops

results in the flexing downward of the diaphragm, which in turn reduces the volume of the pressure chamber. Ink inside the pressure chamber discharges in the form of an ink droplet via the discharge outlet connected to the pressure chamber. When the applied voltage is released, the diaphragm returns to its original position, increasing the volume within the pressure chamber. As the discharge outlet is covered with ink, surface tension generates negative pressure in the pressure chamber. The ink inside the reservoir flows into the pressure chamber via the supply port in preparation for the next discharge.

Figure 8.2 shows a compact printer. This printer is mounted with the printer head shown in Fig. 8.3. This printer head incorporates a double layer piezoelectric actuator of the flexural vibration type, which is shown in Fig. 8.4. Flexural vibration heads are commercialized in two types, the single piezoelectric film layer type and the double layer type. The double layer type has higher rigidity and is effective for achieving a head with a higher drive frequency for discharging ink droplets via the discharge outlet. Typical specifications of the print head are indicated in Table 8.1. It is capable of printing L-size (89 mm × 127 mm) photographs at a speed of 35 s per sheet.

Fig. 8.2 A compact printer with a print resolution equivalent to photographs (215 mm × 152 mm × 145 mm). The printer is equipped with 4 colors with 90 discharge outlets per color. The head is composed of two actuator units, which are bonded side by side on a large discharge outlet unit

8.1.2.2 Structure and Manufacturing Method of the Head

In order to manufacture the head, an actuator unit and a discharge outlet unit, bonded with adhesive, are bonded on a case head which is produced via injection molding of a resinous material.

The actuator unit is composed of a communication plate with a communication port, a plate with a pressure chamber, a plate used as a diaphragm and an actuator. The manufacturing method is described as follows. A green sheet of the

Fig. 8.3 Photograph of the print head. The discharge outlet plate on *top* (25 mm×27 mm) and a component made by laminating multiple plates and piezo films which are bonded on the case head that is made via the injection molding of a resinous material to the print head

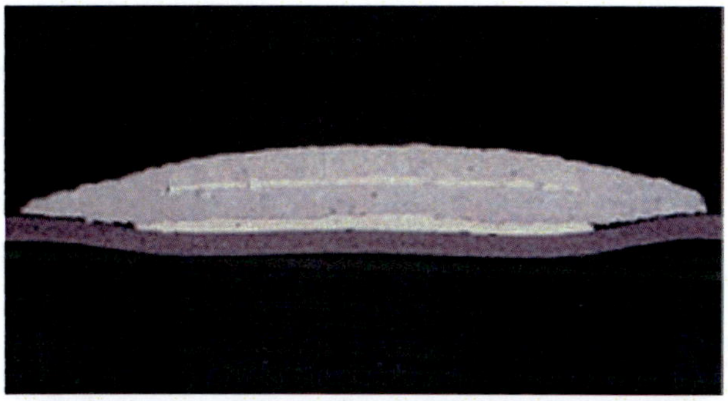

Fig. 8.4 Cross-sectional view of double layer piezo actuator. The lower electrode, lower piezo, intermediate electrode and upper electrode are formed on the diaphragm in this order. The total thickness is several tens of microns

required thickness is produced from a mixture of zirconia (ZrO_2) powder and a binder. The flow channel is formed on the green sheet using a die, and then the sheets are sintered in three layers at a high temperature. A laminated three-layer sheet featuring high sealing performance can be created without any special bonding process. Then, a paste composed mainly of platinum (Pt) is printed on the laminated

Table 8.1 Typical specifications of the print head

	Specification
Number of nozzles	90×40 rows
Density of discharge outlets	120 dpi
Piezoelectric layer	Double layer
Maximum drive frequency	43 kHz
Max. size of ink droplet	7 pL
Min. size of ink droplet	2 pL
Nozzle diameter	20 mm
Length of pressure chamber	1.1 mm

sheet. The laminated sheet is sintered and a lower electrode is formed on the sheet. Paste films for the lower piezo material [$Pb(Ti,Zr)O_3$], an intermediate electrode (Pt) and the upper piezo are printed in that order above the lower electrode and then sintered. The actuator unit is completed by printing a paste that forms the upper electrode (Au) above the upper piezoelectric and sintering the entire unit.

The discharge outlet unit is made by bonding three sheets of stainless steel by film adhesive. Prior to the bonding process, a discharge outlet is formed on the discharge outlet plate, a reservoir and a communication port are formed on the reservoir plate, and a supply port and a communication port are formed on the supply plate by etching, punching as well as by other methods.

8.1.3 Future Prospects

Inkjet printers can print images and text one by one and are more appropriate for small lot (several pages at a time) printing in comparison with methods that utilize a printing plate. A line head having the width equivalent to that of a sheet of paper has been developed and the printing speed has been improved so that they may be used instead of conventional printing machines.

They are not limited to printing images only on paper but can also print on the surface of metals and plastics. They have begun to be used in the manufacturing of interim products. For example, they are used to form circuit patterns on substrates and to add pigment for LCD panel color filters. The viscosity of discharged liquid is high in some applications, and further development of print heads capable of discharging high viscosity liquid is expected. It is believed that the piezoelectric type, which does not require the heating of liquid, can handle a wider variety of liquids than the bubble jet type print heads.

Thermal Print Head (1975) 8.2

The thermal print head is an electronic device that prints on recording media by selectively heating the media with resistors, which are lined up on an alumina ceramic substrate. It features a simple structure and high maintainability, and is very common as it is incorporated in a wide variety of printers ranging from monochrome high-speed printers to full-color photo printers.

8.2.1 Background of Development

The thermal recording system that utilizes the thermal print head expanded rapidly in the 1980s following its utilization in fax machines. At the same time, the thermal print head has been applied to printers intended for various applications. It is currently being used in many fields, including for "train and bus tickets," "shopping receipts," and "account statements from automated telling machines" to "X-ray, CT and MRI records used in the medical field." Figure 8.5 shows the appearance of the thermal print head and its major applications. In today's IT dependent society, a large amount of information is computerized, resulting in greater accessibility. Various kinds of printers have been developed and act as an "interface between humans and information," and many of these are thermal printers. Thermal print heads are now an indispensable part of our daily lives.

8.2.2 Characteristics

8.2.2.1 Structure and Characteristics of Products

Thermal print heads are classified into two types, the thin film type and the thick film type, both of which are manufactured by different methods. The two types are compared in Table 8.2. The thin film type, which is more widely used, is explained below.

Figure 8.6 shows the outline of the structure of the thermal print head. The thermal print head is composed of a wafer on which heating resistors are formed, a driver IC that supplies the electrical signal to the heating resistors, an external wiring substrate for drive signal input from the printer and a heat sink that releases heat and also functions as a support. On the wafer, a glaze layer, a barrier layer, a heating resistor layer, an electrode wiring, oxidation-resistant layer, and wear-resistant layer are stacked and processed in this order on the alumina substrate. When a current is supplied to the electrodes, the resistors are heated and the heat is transmitted to thermal recording paper via the protection film (Note 8.4), developing color for printing. The picture shows the heating resistors on the 300 dpi (Note 8.5) thermal print head seen from the thermal recording paper-side. The black diagonal line is a strand of "hair" used for the comparison of dimensions.

Figure 8.7 shows the surface temperature of the heating element under printing conditions and an example of pixels recorded on thermal paper. Although actual heating time varies depending on the specifications, it may be as short as several microseconds.

The recording media is not limited to thermal recording paper. If a fusion type thermal transfer ink ribbon (Note 8.6) is used, printing on plain paper is possible, and photographs

Note 8.4 A type of recording paper made by coating the paper with a leuco dye and an oxidizing agent. When heat is applied, the two substances are melt-mixed, forming images.

Note 8.5 A unit used to express resolution, stands for dot per inch. 300 dpi, for example, indicates 300 pixels lined up in the 1-in. (25.4 mm) distance.

Note 8.6 A thin PET base material coated with ink that is melted by heat. The base material is heated by the heat generated in the thermal print head. Then the ink is transferred to the recording paper to form images.

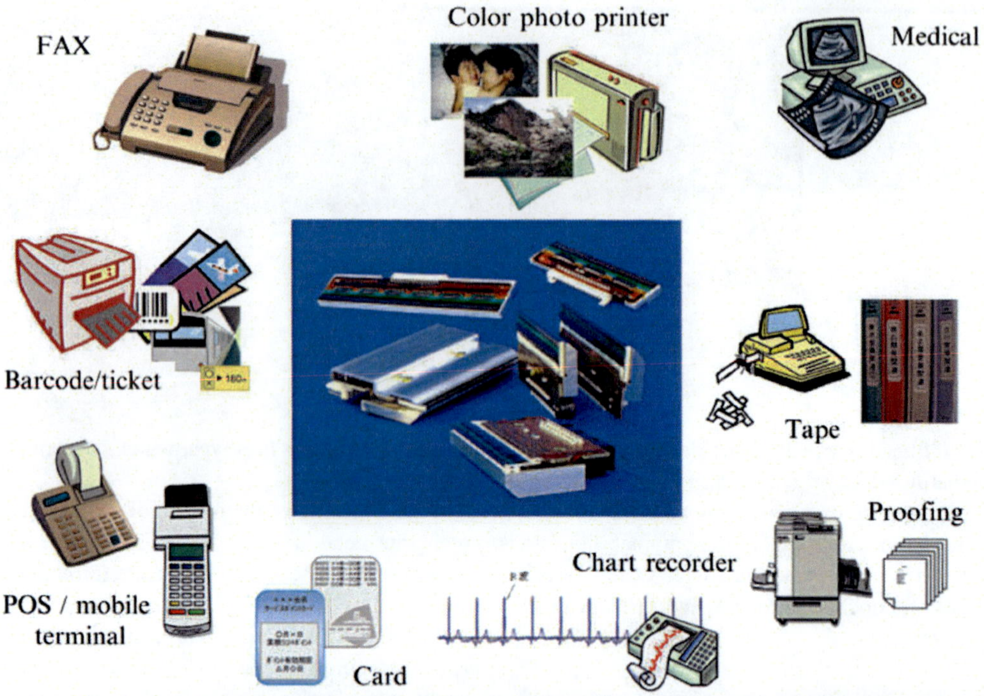

Fig. 8.5 Thermal print heads and its major markets. The production of thermal print heads intended for fax machines was started in the 1980s. The application range expanded due to the simplicity of the mechanism and high maintainability. As a result they are now indispensable in day-to-day life

Table 8.2 Comparison of the thin film type and the thick film type

	Manufacturing method	The area around the heating resistors	Characteristics of the manufacturing method
Thin film type	**Thin film of resistors is formed** ↓ **Thin film of electrodes is formed** ↓ **Patterning** ↓ **Protective film forming**	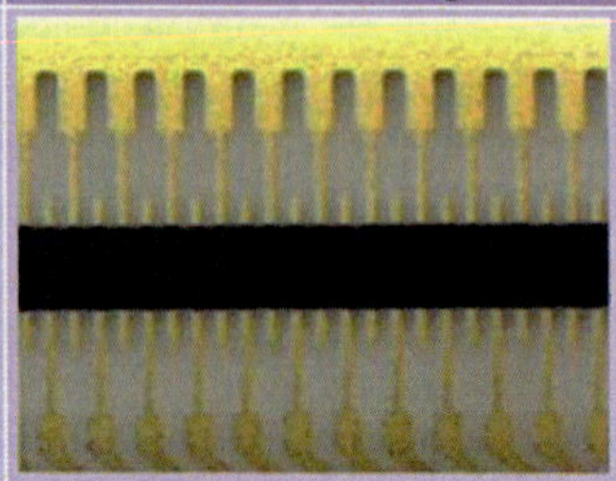	*Advantages* High density wiring can be formed Easy to equalize resistance value distribution *Disadvantages* Expensive manufacturing equipment
Thick film type	**Electrode forming** ↓ **Resistors are printed and calcined** ↓ **Protective film forming**	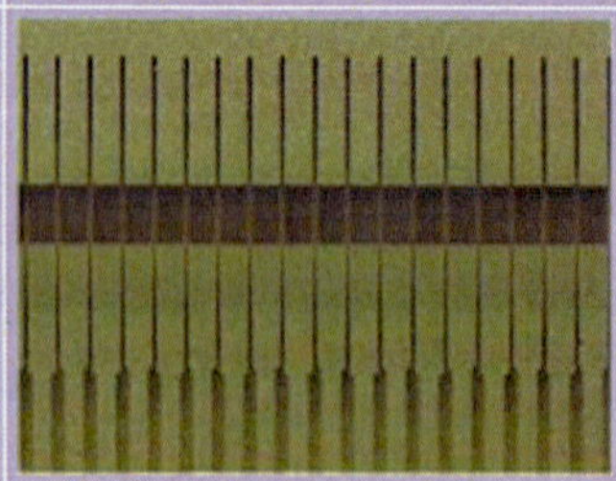	*Advantages* Inexpensive manufacturing equipment *Disadvantages* Difficult to form high-density wiring Difficult to equalize resistance value distribution

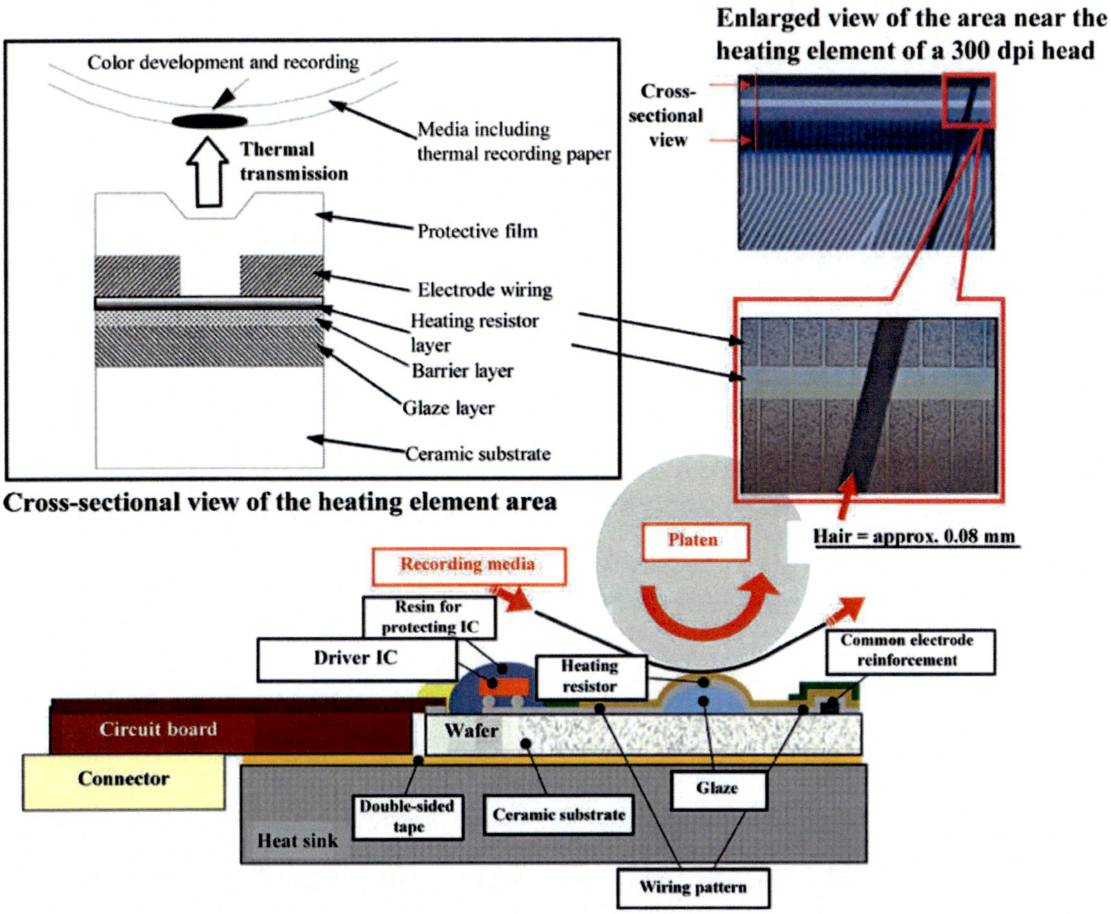

Fig. 8.6 Outline of thermal print head structure. The thermal print head is composed of a wafer, a driver IC, a circuit board and a heat sink. The heat generated by current supplied to the fine heating resistors on the wafer is transmitted to the thermal recording paper, changing color for printing

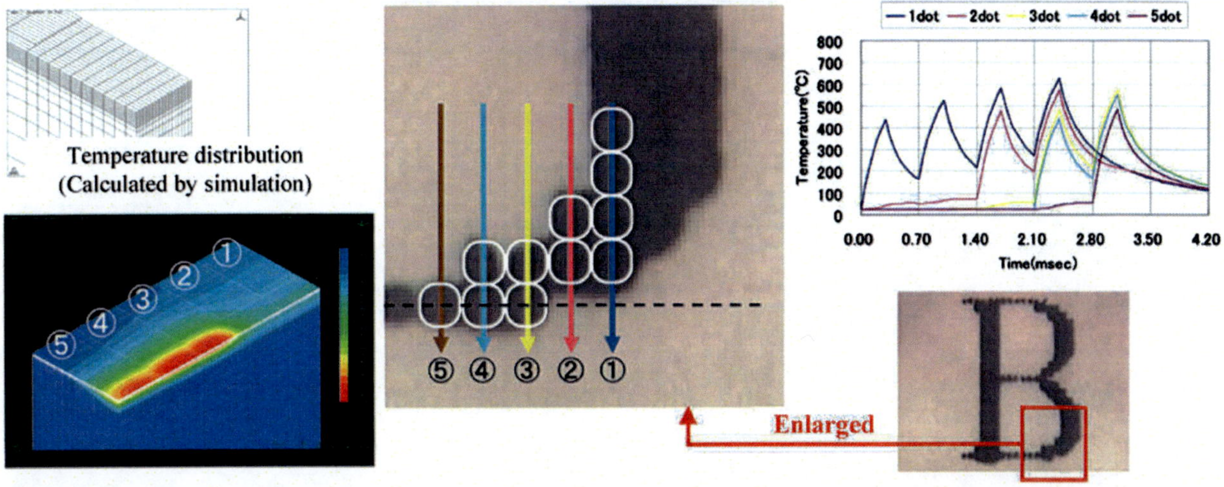

Fig. 8.7 Surface temperature of the heating element and an example of recorded pixels. The heat generation time of the heating element varies depending on the specifications. It may be as short as several microseconds (300 μs for example)

can be printed by using a sublimation thermal transfer ink ribbon (Note 8.7) with a special type of paper.

There are a variety of printing methods such as the xerographic method, the inkjet method and the thermal method. The thermal method is characterized by:

1. The possibility to produce compact and low-cost printers as the printer engine structure is comparatively simple.
2. Simple design structure that ensures the possibility of durability and high maintainability.
3. Clear printed images with no "bleeding" or "spattering" as seen in other printing methods as printing is performed by contacting between the heat source and the media.
4. Adjustable heating temperature by changing the energy applied to the heating element, which allows for a gradation expression equivalent to that of silver halide photographs.
5. Printing that can be done as soon as the power is turned ON. There is no warm-up time like that required of other methods. Due to these features, this method has been widely adopted.

8.2.2.2 Types

Thermal print heads are classified into four types, as shown in Table 8.3, depending on the location of the heating resistors.

The flat type that is the most commonly used form that has heating resistors and an IC mounted in the same plane and features high productivity. However, the recording media has to be bent because the convex portion of the driver IC is in the paper feeding pathway.

The real edge type and the corner type have lower productivity, but they support "straight pass," which does not require the bending of recording media, and are applicable to hard-to-bend media such as plastic cards. The real edge type has high glaze smoothness and achieves high print quality with no irregularity in the intermediate tone. The corner type allows high-speed printing by high-speed thermal response, as the volume of the glaze can be minimized.

The near-edge type has the heating resistors at a place away from the top of the glaze, with productivity slightly lower than that of the flat type while offering the straight pass feature. Due to its structure, a certain level of glaze height is required, making it difficult to achieve high-speed thermal response and maintain the smoothness of the glaze.

8.2.2.3 Manufacturing Method

Figure 8.8 shows the outline of the manufacturing process. A glass layer having a convex cross-sectional shape is formed on an alumina ceramic plate. A resistive layer of $TaSiO_2$ and a conductor layer of Al are formed on this substrate by the sputtering method. Then the resistive layer and the conductor layer are processed to specified patterns, forming multiple heating resistors on the substrate. A protective film layer, consisting of Si_3N_4 is formed to protect the heating resistors. Next, a driver IC for the selective heating of the thermal

Table 8.3 Types of thermal print heads

	Flat type	Real-edge type	Corner type	Near-edge type
Structure	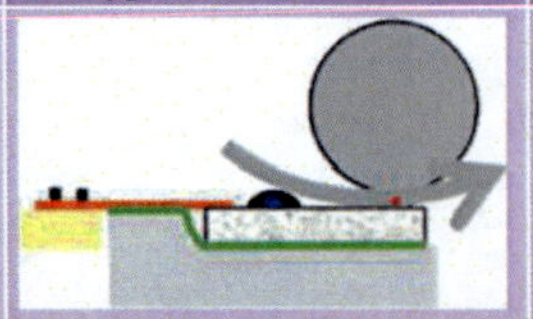	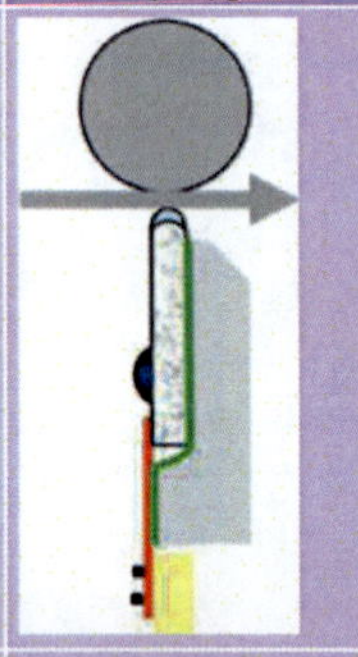	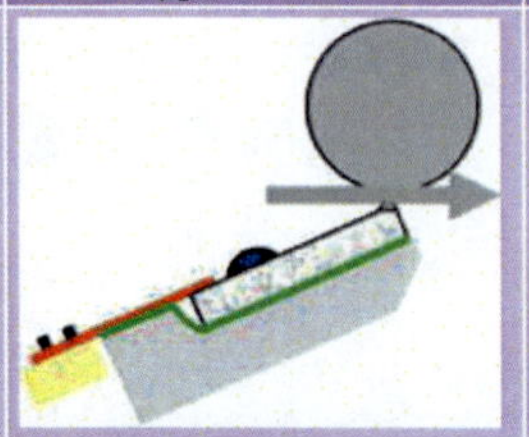	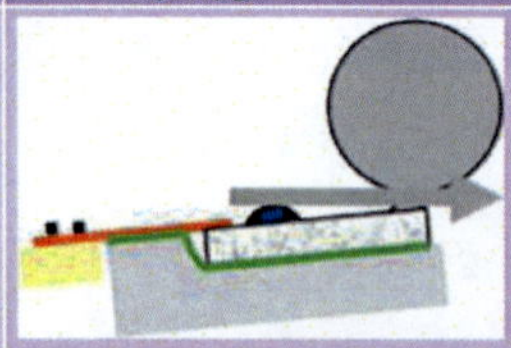
Advantages	Most widely used High productivity and inexpensive	Supports straight pass Space saving	Supports straight pass High-speed printing	Supports straight pass Price equivalent to flat type head
Disadvantages	Does not allow straight pass	More expensive than the flat type	More expensive than the flat type	Not suitable for high-speed printing

Note 8.7 A ribbon made by coating a thin PET base material with a sublimation dye and used in combination with a special paper coated with resin that settles the dye. Pristine gradation is achieved as the amount of transferred dye is proportional to the heat generated on the thermal print head.

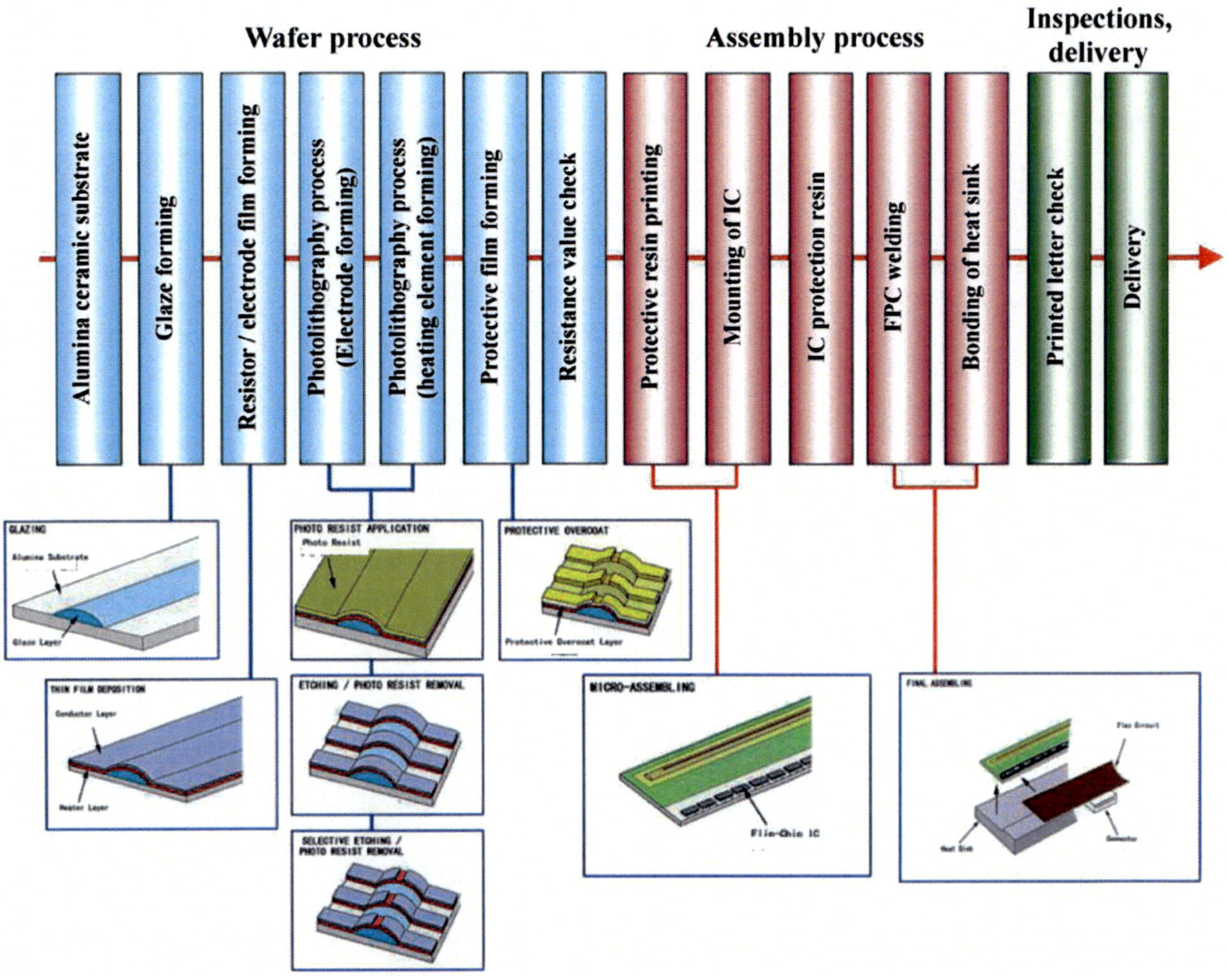

Fig. 8.8 Manufacturing process of thermal print head. The forming of the thin film on the glazed ceramic substrate is followed by both an assembly process and an inspection process

resistors is mounted and an external wiring substrate used for external drive signal input is connected before completing the thermal head by bonding the substrate on a heat sink, which disperses the residual heat and functions as a support.

8.2.3 Future Prospects

Efforts are being made to improve the performance of printers, including the "improvement of printing speed" and the "improvement of printed image quality," aiming to make life more convenient. For the improvement of printer performance, technological improvements of thermal print heads are essential, and various efforts covering a wide range of areas, including research on materials and the electric signal control method are being made. Under the advanced technologies of today, the following features are expected to be commercialized very soon.

1. Printing speed: on-demand printing at 1,000 mm/s will be realized for "date code printer applications (Note 8.8)."
2. Print quality: the printing of pictures at 16.77 million colors × 900 dpi will be achieved soon. The range of applications is expected to expand further in the future.

Note 8.8 This printer is used to print serial numbers and production dates on products. The thermal print method is employed for high-speed printing on food packages, etc.

Literature

1. Nikkei Electronics (July 11, 1977), pp 54–67, Japan **(8.2)**
2. Sublimation thermal transfer recording technology. Triceps Co. Ltd, (1988) pp 61–84, Japan **(8.2)**

9 Displays

The display business is a global industry that has been indispensable in terms of their role in the industrial development, an important role which ceramics will continue to play in the future as well. In this section, ceramics used in display devices will be discussed.

The history of the display industry in Japan coincides with the history of TV sets. The display industry has been evolving together with TV broadcasting since 1953 when NHK Tokyo Television began full-time broadcasting. During this period, TVs evolved from black-and-white broadcasting to color broadcasting, to satellite broadcasting, to hi-vision broadcasting and then to terrestrial digital broadcasting. Since the beginning of TV broadcasting, research and development has continued to improve definition, screen size, flatness, etc. of the CRT (Cathode Ray Tube), the main product of the display industry.

In a CRT display, electron beams ejected from an electron gun are bent by electromagnets (called deflection yoke) on the side of the device and illuminate the phosphor on the front side. Ceramics that feature high heat resistance and insulation properties are indispensable for the CRT even today and are used in picture tube glass, phosphors, multiform glass (electron gun supporting bar) (Sect. 9.6) around high-voltage electron guns (which become high in temperature), cathode tubes, holders, as well as in other items.

In the second half of the 1990s, following the release of the LCD (liquid crystal display) and the PDP (plasma display panel) (Sect. 9.1), a new TV market offering TVs equipped with a FPD (flat panel display), featuring thin and large screen, has been created. Price reduction was achieved in the twenty-first century and the market is currently continuing to expand. The history of LCD application within TVs has been unexpectedly short. Since the application of LCD in calculator screens (in 1973), efforts have been made towards the development of liquid crystal materials, driving systems and color displays. Various issues involving view angles and moving image displays have been solved, thus resulting in the expansion of the FPD industry. LCDs require back lights that function as light sources as they do not emit light. Ceramics are mainly used in the back light phosphors and the piezoelectric transformers of the back light driving units. Piezoelectric transformers (Sect. 9.9), in particular, are essential for the miniaturization and price reduction of back lights. Also, alkali-free glass is being developed for use in LCD displays (Sect. 9.4) for organic liquid crystals.

PDP emits light and its application in FPD has been long anticipated. However, it was used only in information displays for a long time due to the single orange color and the difficulty of pixel refining. In the 1980s, research on color PDP led by NHK of Japan became active and Japanese electric appliance manufacturers released large-screen PDP TVs one after another in the 1990s. The price reduction has occurred rapidly upon entering the twenty-first century and the LCD market, as well as the FPD market have both expanded. Explanations in this chapter cover the PDP and application of ceramics to PDP display glass (Sect. 9.4), phosphors (Sect. 9.2) and ribs (partitions) (Sect. 9.3).

Transformers are also electronic parts essential for home appliances. In Japan, electricity is supplied to homes at a voltage of 100 and 200 V. The voltage is transformed by power transformers in home appliances to the required operating voltage. They are also used to transform signals by intermediate frequency transformers, matching transformers, pulse transformers, as well as others. Ferrites are used in transformer cores (Sect. 9.7) and a variety of shapes are available depending on the application. Transformers are essential for TVs and many major home appliances. The flyback transformer and the deflection yoke transformer (Sect. 9.5) are the two major transformers used in picture tubes. The flyback transformer is used as a power transformer that supplies a high voltage (10 kV) to electron guns, while the deflection yoke transformer is used as an electromagnet that bends electron beams ejected from the electron gun. Small power supply transformers for switching (Sect. 9.8), which incorporate a ferrite core, are used in the power supply of various electronic devices. The range of their application has been expanding following technological innovations resulting in miniaturization.

Y. Imanaka et al. (eds.), *The Ceramic Society of Japan, Advanced Ceramic Technologies & Products*,
DOI 10.1007/978-4-431-54108-0_9, © Springer Japan 2012

Color PDP (1990)* 9.1

In recent years, plasma TVs that feature thin, large screens that display high definition images have become common place. The color plasma display panel (PDP) was commercialized after many years of research and development in Japan. The color PDP has more than one million mini fluorescent lamps between two glass panels which are lit in three primary colors; red, green and blue, in order to display images and text. The glass panels are equipped with horizontal and vertical electrodes as well as partitions that are called barrier ribs which are made of glass powder ceramics. Electrical discharge occurs in the spaces partitioned at 0.1 mm intervals. The phosphor coated on the inner surface of the ribs is illuminated by ultraviolet rays emitted by plasma resulting from the electrical discharge. The PDP offers bright, sharp, powerful images and its simple structure allows for the manufacturing of large-sized screens. The market for it has therefore been rapidly expanding due to growing demand for large display monitors in public spaces and TVs for personal use (Fig. 9.1).

9.1.1 Background of Development

Cathode-ray picture tubes were used to display images for many years after the beginning of television broadcasting. However, the volume and weight of large displays that incorporate picture tubes were a limitation. Developments of new displays that replace conventional displays were focused on thin and slim displays. These new displays are formed on thin panels such as glass panels and are called flat panel displays.

Although the operating principle of plasma displays that represent thin and large displays was invented in the U.S. in the 1960s (Note 9.1), the commercialization of current color PDPs was realized by means of many years of research and development in Japan. Fujitsu developed the basic structure for the world's first 21-in. full color PDP in 1992, followed by the development of technologies used in large, high-definition displays as well as technologies used in the mass production of PDPs by manufacturers. This has led to the development of the current generation of PDPs.

9.1.2 Characteristics

9.1.2.1 Structure and Function of Products

Figure 9.2 shows (a) panel structure and (b) luminescent cell structure of a color PDP. The PDP panel is basically composed of two glass panels on which longitudinal and latitudinal electrodes, partitions called barrier ribs and phosphor containing three primary colors, are laminated with a space of approximately 0.1 mm between the two panels. Each of the spaces that are blocked by the longitudinal and latitudinal electrodes and the partitions act as a luminescent cell.

When voltage is applied to the area between display electrode couples on the front panel, an electrical discharge occurs on the surface of the electrodes (called surface discharge), ionizing the internal discharged gas into a plasma. Ultraviolet rays generated from the plasma stimulate the phosphor coated on the back panel, generating colored light illumination in the direction of reflection (called the reflection type). The surface of display electrodes are covered with insulation glass called a dielectric layer. When electrical discharge occurs, electrical charge (electrons, ions) accumulated on the surface impedes the electrical discharge after a period of time. When the polarity of the applied voltage is

*The number in parenthesis indicates the year that the product was first commercialized.

Note 9.1 Bitzer, Slottow and Willson at the University of Illinois invented electrical discharge cells made by covering the surface of a metallic electrode with insulation glass to provide the capacitance. The structure designed to control electrical discharge by capacitance is the basic principle of the AC type PDP.

a 42-inch color PDP **b** PDP monitor for point-of-sale advertisement

Fig. 9.1 Color PDP and its applications. PDPs are used as monitors for public display viewed by multiple people at the same time as they are thin, large and have wide viewing angles. They have begun to be widely used as home-use plasma TVs due to their performance with regard to reproducing high resolution images

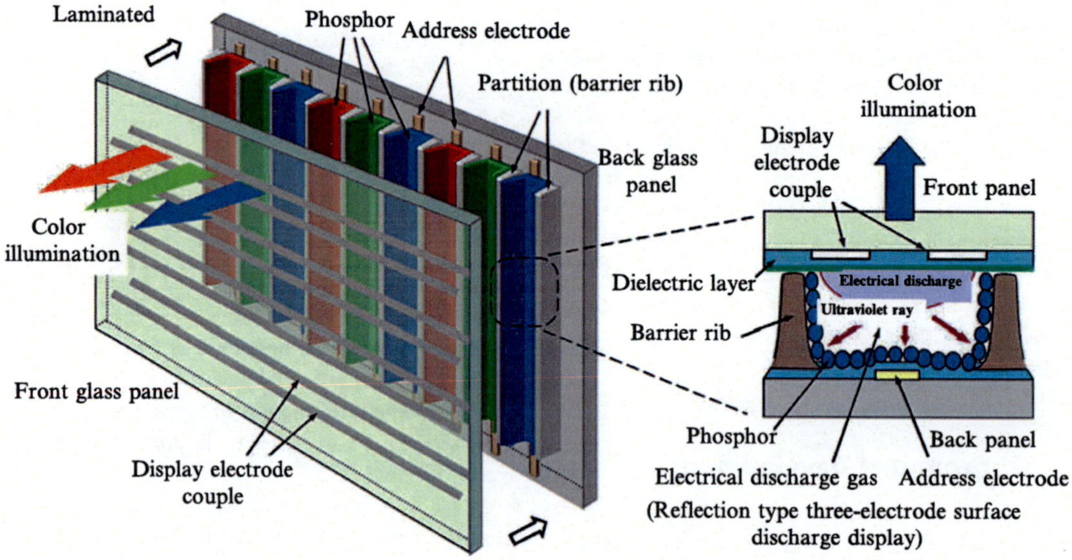

a **Panel structure** **b** **Luminescent cell structure**

Fig. 9.2 Color PDP structure. When the front glass panel mounted with latitudinal electrode couples is laminated with the back glass panel mounted with longitudinal electrodes, partitions and phosphors, the luminescent cells, spaces surrounded by crossing electrodes and partitions, are created

reversed, the accumulated electrical charge acts as an igniter and causes an electrical discharge, realizing a continuous pulse discharge and bright illumination through alternating polarity voltage pulses. The addressed cell electrode on the back panel is applied with a voltage at the time of selecting luminescent cells to be turned ON, causing a small-scale ignition discharge. This PDP is called the three-electrode surface discharge display because its operation is controlled by address electrodes and electrode couples composed of two display electrodes.

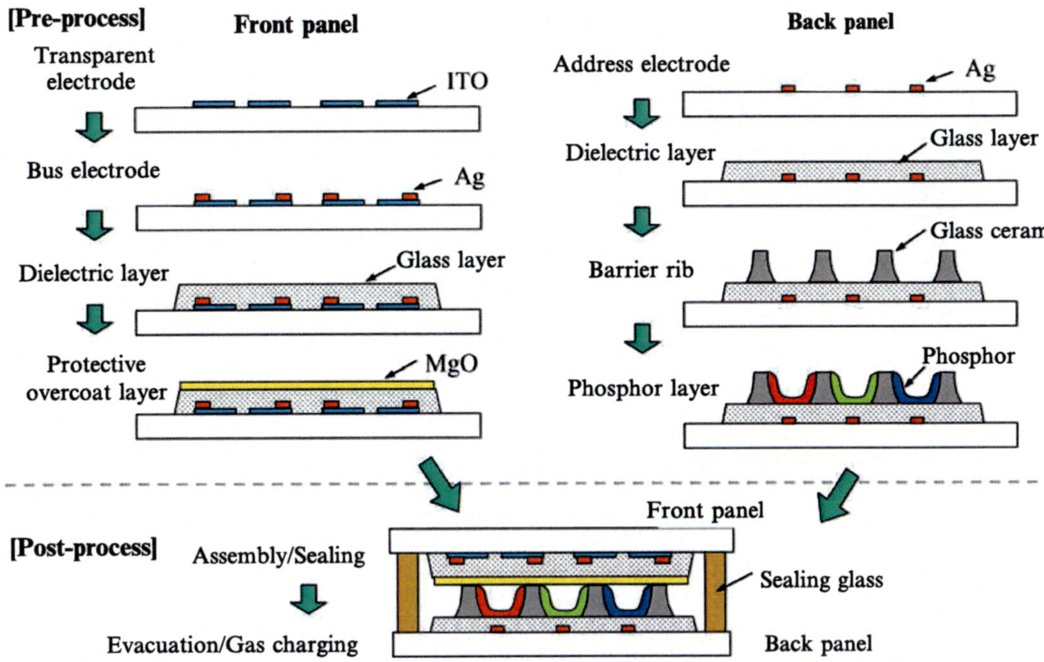

Fig. 9.3 Manufacturing process of PDP panel. In the pre-process, electrodes, dielectric layers are formed by printing, deposition and calcination, respectively, on the glass substrates of the front panel and the back panel. In the post-process, the two substrates are laminated and electrical discharge gas is charged

9.1.2.2 Manufacturing Method

Figure 9.3 shows the manufacturing method of the color PDP. In the manufacturing process of the front panel, indium tin oxide tin-doped indium oxide (ITO) is deposited on the glass substrate and formed into clear display electrode couples. Thin bus electrodes are formed to prevent voltage drop because clear electrodes are high in resistance. Bus electrodes, made of silver paste, are formed by printing or other methods. Glass powder paste of a low melting point is formed by printing or pasting a sheet on the entire surface of the display electrodes, which is then heated to a temperature of approximately 600°C to create a clear glass dielectric layer. Then, an MgO protective overcoat layer (Note 9.2) is formed by the vacuum evaporation method. In the back panel manufacturing process, address electrodes and a dielectric layer are formed on the glass substrate first. Then the glass substrate is coated with rib paste, a mixture of glass with a low melting point, resin and solvent, which is processed into shape by sandblasting after resist patterning. It is then calcined to form glass partitions. Red, blue and green phosphor pastes are applied, respectively, in this order through the screen printing method and dried, forming a layer by calcination. The processed front panel and back panel are laminated by using sealing glass material, and then the space is vacuumized and filled with a mixture of neon (Ne) and xenon (Xe) gas (Note 9.3).

As explained in the above, the panel is manufactured with the incorporation of popular printing and thin film technologies that allows for high mass production. The manufacturing process is short due to the simple structure, making it easy to manufacture large panels.

Barrier ribs are the structural elements that largely affect the display performance of PDPs. Figure 9.4 shows an example of the barrier rib forming method. High-accuracy processing technologies have been developed and ribs can be formed into various shapes. The ribs are formed by the sandblast method, in which particles are blasted under high pressure to create a etched pattern, or the photosensitive paste method,

Note 9.2 MgO (magnesium oxide), a protective film coating featuring high thermal and impact resistance, is known as a material in fire resistant bricks. It is also an important structural element featuring smooth electron emission and the lowering of electrical discharge voltage.

Note 9.3 Ne gas, which lowers the voltage at the start of electrical discharge, is mixed with Xe gas (4–10%), which generates ultraviolet rays. As the ratio of Xe gas is increased, the amount of ultraviolet rays increases, improving the luminance. However, the voltage at the start of electrical discharge increases. Therefore, the gas composition ratio is optimized while taking into consideration various factors.

a Box-type barrier rib **b** Serpentine barrier rib

Fig. 9.4 Example of "barrier rib (partition)" forming method. The types of barrier ribs include the stripe type, the box type, which prevents interference by electrical discharge in the neighboring space, and the serpentine type, in which the cells are circular, achieving high luminance efficiency

in which the pattern is formed by using glass paste mixed with photo polymerization resin (Note 9.4) that hardens in response to light.

9.1.3 Future Prospects

The display performance of new products released in 1995 and 2005 is shown in Table 9.1. Since the development of the world's first 42-in. color PDP in 1995, various technologies have been developed. A model supporting the so-called full spec high definition was released in 2005. Remarkable progress was achieved during this 10 year period and it is expected that technological developments and improvement of luminance efficiency with a focus on the reduction of power consumption and cost, will continue and products will evolve further.

Note 9.4 A photosensitive organic material, that hardens through polymerization caused by energy absorption of ultraviolet rays. Materials that effectively expose and harden thick films have been developed for use as paste in PDP ribs.

Table 9.1 Example of progress made in PDP display performance over 10 years

	42-in. W-VGA developed in 1995	42-in. Full-HD developed in 2005
Luminance	300 cd/m^2	1,000 cd/m^2
Contrast	400:1	3,000:1
Resolution	852×480 pixels	1,920×1,080 pixels

The display performance of the world's first 42-in. color PDP (*left*) was insufficient for use in TVs, but it was improved remarkably during a 10 year period. Satisfactory display performance was realized via the full spec high definition (HD) PDP, which has the highest resolution among displays for high vision TVs

Phosphors (1950) 9.2

Phosphors are generally powders with a grain diameter of several microns and each of the grains is close to a single crystal. A small amount of an activating agent is added to the host crystal in the form of a solid solution to change the composition and to produce various colors of "light." Although phosphors are used in various fields, the phosphors introduced in this section are those used in displays (Fig. 9.5).

9.2.1 Phosphors for CRT (Cathode Ray Tube)

Development of phosphors became active in line with the expansion of picture tube TVs (since the 1950s). The current composition of phosphor was established in the 1970s. From this period onward, the majority of blue and red phosphors were tinted with pigments, aiming to enhance contrast and reduce reflection of outside light. Development after that period focused on optimization of the coating properties of fluorescent films for actual use rather than on the luminescent properties of crystalline grains (Fig. 9.6; Table 9.2).

9.2.2 Phosphors for CCFL (Cold Cathode Fluorescent Lamp)

In liquid crystal displays (LCD), images are produced by a backlight combined with a liquid crystal shutter and color images are created by adding color filters to them. A CCFL is normally used as the backlight (Fig. 9.7).

Therefore, the quality of the LCD color displays depends on the product of CCFL spectral distribution and color filter spectral transmittance. Blue, green and red phosphors mixed with an appropriate ratio are used in the CCFL to create the intended white color. Although phosphors developed for lighting focused on light flux, color rendering properties, mercurial resistance (Note 9.5) (Table 9.3), advanced LCD TV specifications emphasized color reproducibility. Therefore, CCFLs with spectral distribution that matches the spectral properties of color filters are needed. The color reproducibility can be expanded by improving green and red phosphors (Fig. 9.8).

9.2.3 Phosphors for Plasma Displays (PDP)

PDPs, like LCDs, are rapidly becoming more popular as flat panel displays (FPD). Phosphors used in PDPs require high discharge stability in microscopic cell spaces in addition to high luminescence intensity by ultraviolet rays in vacuums, color reproducibility and resistance to plasma emissions (Fig. 9.9).

Composition of the phosphor that is most widely used currently was proposed in the early 1980s. Later, blue phosphors featuring long service life and high luminance, green phosphors featuring high afterglow and electrical properties and red phosphors featuring high color purity were developed, contributing to the progress of PDPs (Fig. 9.10).

9.2.4 Phosphors for Other Displays

Phosphors are also used in fluorescent display tubes, FEDs (Note 9.6) and LEDs (Note 9.7). They will continue to evolve as materials essential for displays.

Note 9.5 Absorption of mercury by phosphors is believed to be one of the factors that shortens the service life of lamps.

Note 9.6 Stands for Field Emission Display. Displays that have small electrodes (electron emission source) as many as the number of pixels laid out on a glass substrate. Each of the electrodes irradiates electrons over the phosphor on a glass substrate, which faces the electrodes across a space of several mm, for illumination. SED (Surface-conduction Electron-emitter Display) is also included.

Note 9.7 Stands for Light Emitting Diode. Elements that incorporate a semiconductor that generates light when electric current is applied. By using the LED alone or an assembly of LEDs and phosphors, full-color displays are made possible.

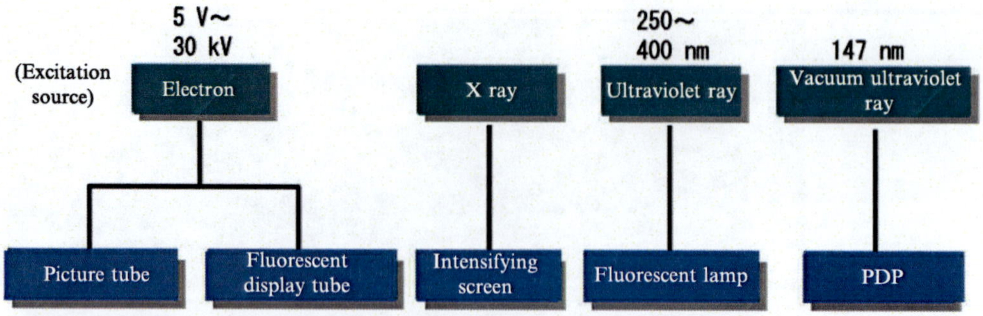

Fig. 9.5 Usages of major phosphors by excitation source. Phosphors are used for various industrial purposes by selecting phosphors that match excitation sources

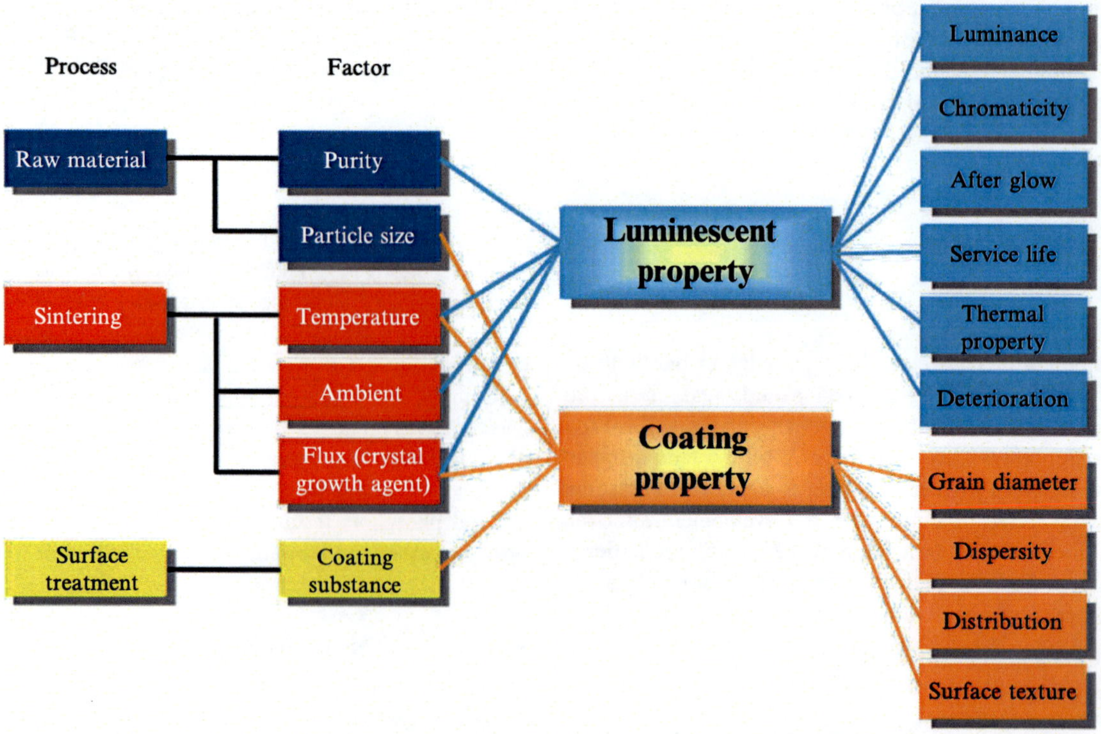

Fig. 9.6 Factors that control characteristics of phosphors. Properties required of phosphors are divided into luminescent properties and coating properties. Factors that affect the properties and manufacturing conditions are indicated

Table 9.2 Phosphors for CRT: compositions and chromaticity (CIE chromaticity) of phosphors (three colors) for CRT are listed below

	Composition formula	Chromatic coordinate x	Chromatic coordinate y
Blue	ZnS:Ag+pigment	0.146	0.061
Green	ZnS:Cu, Al	0.282	0.620
Red	Y_2O_2S:Eu+pigment	0.656	0.332

Fig. 9.7 Mechanism of LCD and the roles of CCFL. The liquid crystal shutter determines the portions to be transmitted and the white light from CCFL is divided into three primary colors by the filter and transmitted. Various colors are reproduced by adjusting the three primary colors

Table 9.3 Typical phosphors for CCFL: compositions and chromaticity (CIE chromaticity) of phosphors (three colors) for CCFL are listed below

Color	Composition formula	Chromatic coordinate x	Chromatic coordinate y
Blue	$BaMgAl_{10}O_{17}$:Eu	0.145	0.066
Green	$LaPO_4$:Ce, Tb	0.361	0.574
Red	Y_2O_3:Eu	0.648	0.347

	Composition formula	Luminescence peak nm	Relative luminance
Green	$LaPO_4$:Ce,Tb	543	100
	$BaMgAl_{10}O_{17}$:Eu,Mn	515	105
Red	Y_2O_3:Eu	611	100
	YVO_4:Eu	619	98
	$3.5MgO\cdot0.5MgF_2\cdot GeO_2$:Mn	658	34

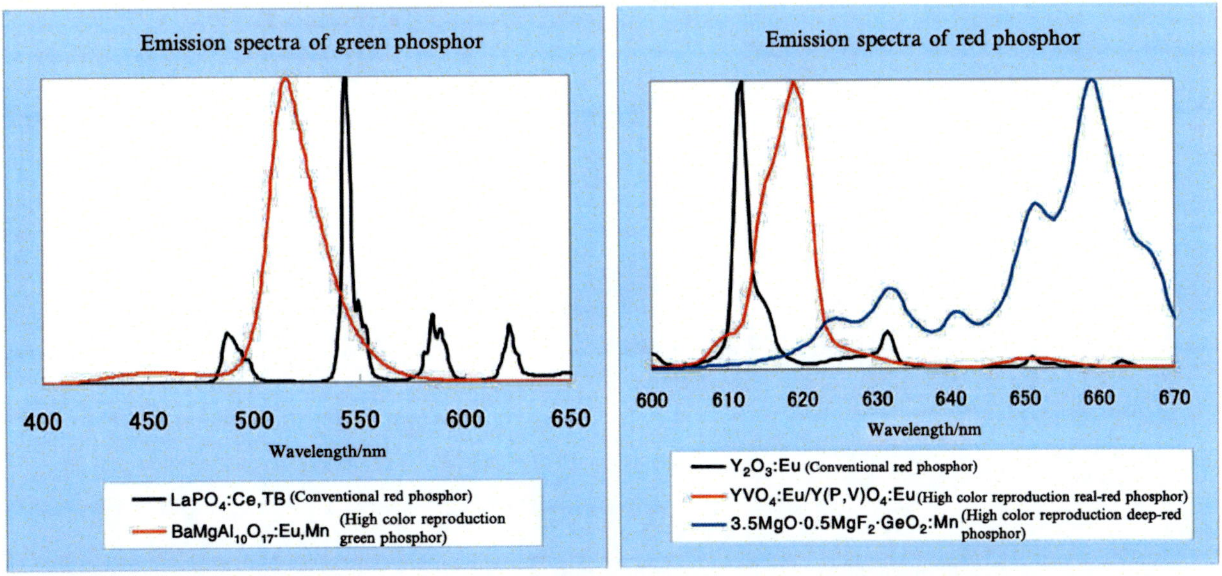

Fig. 9.8 High color reproduction green and red phosphors for CCFLs. General green and red phosphors for CCFLs are compared with high color reproduction phosphors

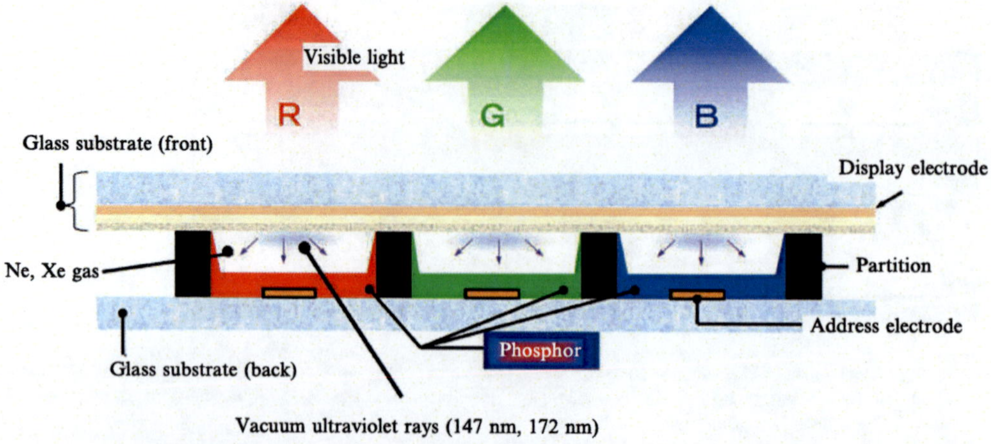

Fig. 9.9 Mechanism of PDP and roles of phosphors. Vacuum ultraviolet rays are generated by electrical discharge to produce colors from blue, green and red phosphors corresponding to pixels

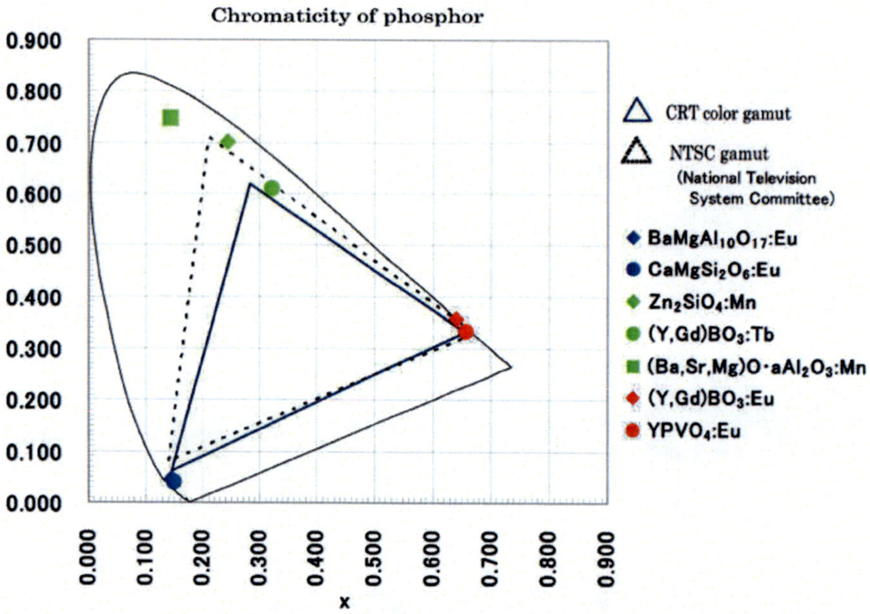

Fig. 9.10 Comparison of the color reproduction range between PDP phosphors and CRT phosphors. Typical PDP phosphors are indicated by color. The color reproduction ranges are normally wider than those of CRT phosphors

9.3 PDP Rib (1980)

A plasma display panel (PDP) is a display device that utilizes a gas discharge for light emission. Ribs, in general used for full-color PDPs, have three roles. They form a discharge space of approximately 0.1 mm between two glass plates that are major components of the panel, they provide the light emitting area created by phosphor coating on the internal wall, as well as separating each display pixel (dot) in terms of electricity and emission color (Fig. 9.11).

9.3.1 Background of Development

The DC-driven PDP was invented in 1954 and the AC type was invented in 1966, both in the U.S. The display was originally produced in a single color, e.g. neon orange, and the application was limited to information displays for industrial usage. Later, in the 1980s, they were developed for use in PC terminals, with the AC-type full-color panel being commercialized in 1992. PDPs are suitable for large image displays because they are self-emissive and have easily produced structures. Due to these reasons, PDPs spread rapidly, mainly for use in large flat TVs, and they currently outperform LCDs in the market for large sizes that exceed 40 in. Rib materials and forming processes have been changing as a result of the need for large screens, high resolution and low prices. Large plasma TVs measuring 40 in. or more (large screen with a diagonal size of 1 m or more) in the current market, feature more than two million pixels and are formed by ribs. Height variations of ribs are limited to several μm or less. The product accuracy in PDPs is among the highest in consumer ceramic parts.

9.3.2 Characteristics

9.3.2.1 Rib Materials and Forming Method

A PDP is made by laminating two glass substrates at a temperature of approximately 400°C so as to create an electrical discharge space inside. Therefore, rib materials are required to have thermal resistance of 400°C or higher. Ribs are formed at a temperature of 500–600°C, which is lower than the temperature at which glass substrates are distorted. Ribs are mainly composed of two inorganic materials—the binding component (glass frit mainly consisting of lead oxide, which softens under the forming temperature) and the structural component (ceramic filler such as alumina). In order to form the rib on the panel, these inorganic materials are turned into a powder, which is mixed with an organic component consisting of resin and solvent in order to create a viscous ink (paste). The organic component, selected according to the forming method, is volatilized or decomposed and then removed due to drying and sintering during the process of forming on the panel. The processing methods of ribs are shown in Fig. 9.12. The methods are classified into two types; the additive method in which the materials are formed on selected area, and the photolithography method, in which the materials are applied on the entire area and the materials in the discharge space are removed, leaving the ribs. In earlier years, the former screen-printing method, in which ribs were directly formed, was adopted and fine ribs were created as in the method shown in Fig. 9.13. However, later the sandblasting method (refer to Fig. 9.13), photosensitive material method and chemical etching method became mainstream

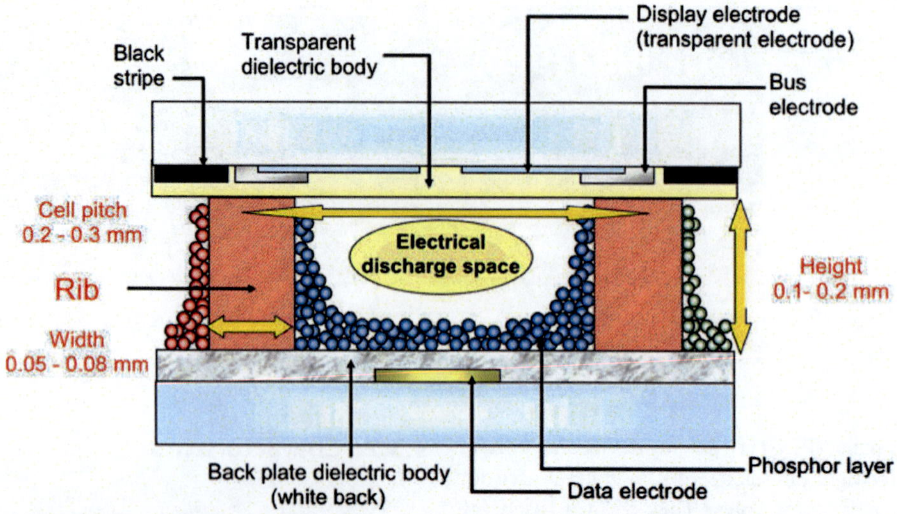

Fig. 9.11 Structure of a PDP cell. PDP ribs have three roles. They form a discharge space of approximately 0.1 mm between two glass plates which are major components of the panel, they provide the light emitting area created by phosphor application on the internal wall, and they separate each display pixel (*dot*) in terms of electricity and emission color

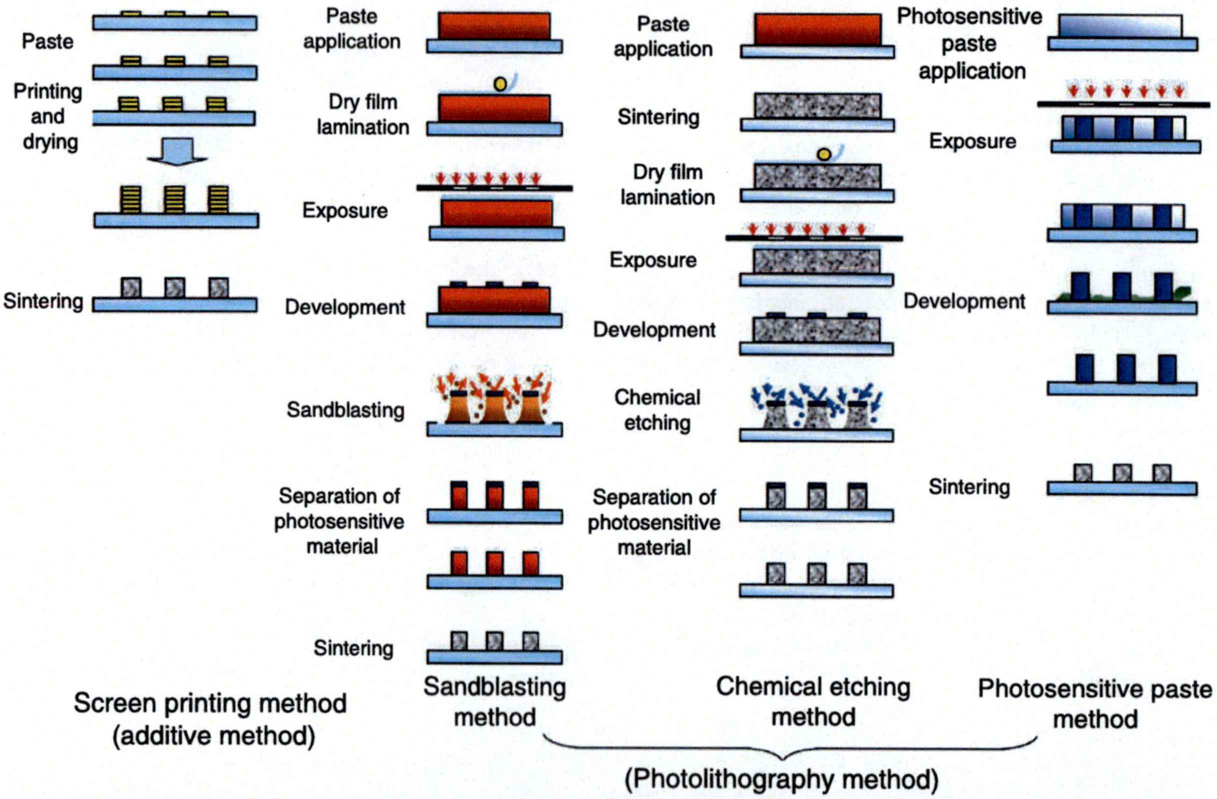

Fig. 9.12 PDP rib forming processes. The figure shows the processes involved in forming PDP ribs. They were conventionally formed by the screen printing method. However, the photolithography method is currently the mainstream. In the photolithography method, effort is being made to develop ribs featuring high aspect ratios in the height direction

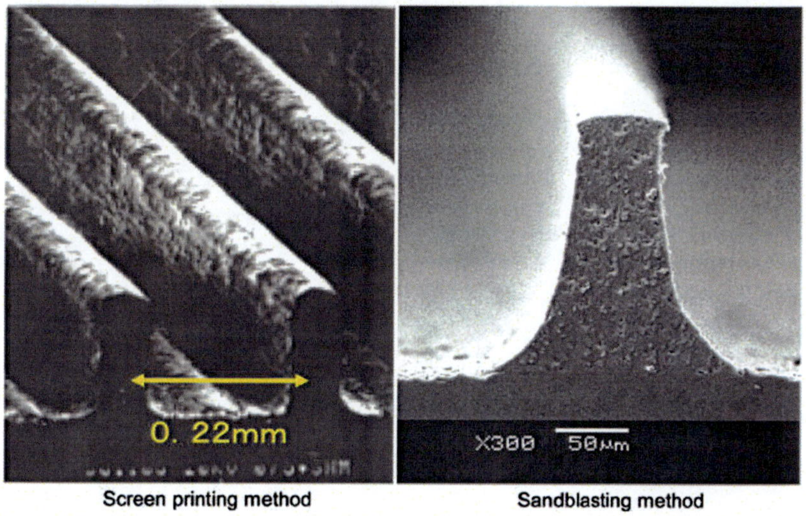

Fig. 9.13 PDP ribs formed by the screen printing method and the sandblasting method. High definition ribs formed by the screen printing method featuring a rib width of 0.05 mm, pitched 0.22 mm, have been realized. However, this method was replaced by the photolithography method, represented by the sandblasting method, due to problems with positional accuracy and productivity

due to their high productivity and reproducibility with fine scale patterning. In the sandblasting method, areas are removed by blasting it with a fine abrasive. In the case of the photoresist material method, the paste containing photosensitive components is exposed in order to harden it, and unnecessary areas are developed with an alkali solution to form the pattern. In these two methods, sintering is performed after pattern forming to create the PDP ribs. In the etching method, the whole film is sintered and unnecessary areas are removed by acid to form the ribs.

9.3.2.2 Environmental Measures

Conventional PDP ribs, as is the case with other structural materials, were made of $PbO–SiO_2–B_2O_3$-based glass frit composed mainly of lead oxide, which melts at 500–600°C and is an electrical insulator and chemical stability. However, it has been pointed out that disposed of lead glass may develop into a harmful substance. This classification is made for materials that are classified as an environmentally hazardous substances and is listed as an item in which usage and emission is to be controlled by the 2006 Restriction of Hazardous Substances (RoHS) directive. Therefore, PDP rib materials need to be replaced by lead-free materials as soon as possible. Bismuth-based glass ($BiO_2–ZnOSiO_2–B_2O$) and zinc-based glass containing alkali ($ZnO–SiO_2–B_2O_3–RO$) are possible alternatives, but the former is high priced due to the scarcity of natural reserves. Meanwhile, the latter is chemically unstable, releasing hydroxyl groups (OH) within the ribs into the electrical discharge space, leading to the degradation of PDP characteristics as well as service life. The development of zinc-alkali-based glass materials has advanced and forming conditions have also been improved upon. Prototypes of lead-free PDPs have also been produced and thus the commercialization of lead-free PDPs is expected in the near future.

9.3.3 Future Prospects

There has been high demand for PDP ribs, leading to rapid advancements of the materials and forming processes. The development of stable and accurate micro-fabrication technology is required in order to achieve high resolution and low power consumption. It is also important to lower the costs of materials and processes and reduce the amount of energy consumed by the PDPs during operation.

9.4 Display Glass (Approximately 1980)

Display devices, used to display information and images, have seen advancement with today's rapidly developing information-based society. Flat panel displays, represented by liquid crystal displays and plasma display panels; have become a familiar part of daily life. Glass, which is clear and easily processed into panels with a high degree of flatness, is essential for these display devices. Glass materials, developed exclusively and, respectively, for liquid crystal displays and plasma displays, are used to produce display glass substrates.

9.4.1 Background of Development

In today's advanced information-based society, information is digitalized for recording with various media and for transmitting via communication technologies such as broadcasting and the Internet. People obtain the information in the form of sounds, texts and images, among which visual information is the most prominent. Therefore, display devices for information displays are indispensable within the current information-based society.

TV receivers, which represent information display terminals for home-use, used conventional cathode-ray tubes for decades. However, recently liquid crystal displays and plasma displays for flat-screen TVs have been become popular due to the trend towards larger screens with space-saving displays. Liquid crystal displays, used as PC monitors and mobile device displays, are now used daily. The substrate glass, developed exclusively for such purposes, has a composition that is different from that of normal window glass (soda-lime glass) used to produce flat panel displays.

9.4.2 Characteristics

9.4.2.1 Characteristics and Specifications of Products

Glass can be formed into a large flat panel easily and is essential for creating flat panel displays. In both liquid crystal displays and plasma displays, pixels are formed between two glass substrates to display the image. Major characteristics of each substrate glass are indicated in Table 9.4, using products produced by Asahi Glass Co., Ltd. as examples. Characteristics of the glass and displays are explained below (Note 9.8):

1. Liquid crystal displays
   The structure of a liquid crystal display (transmission type) is shown in Fig. 9.14. The screen image is displayed by switching each of the pixels ON and OFF with the liquid crystal filled between the two substrates. In the transmission type, the light from the backlight passes through the panel and is seen via the front face (Note 9.8). The RGB colors are created by the color filter formed on the front substrate glass. In respect to the driving method, the switching is performed by the Si-based thin film transistors (TFT) laminated on the glass substrate (Note 9.9). TFT is affected by alkaline components, and therefore, alkali-free glass is used as the glass substrate. The thermal expansion coefficient is $30–40 \times 10^{-7}/°C$, which is smaller by more than 50% compared with soda-lime glass (approx. $85 \times 10^{-7}/°C$), while the strain point is higher than 600°C, featuring high thermal resistance.
2. Plasma display panel (PDP) (Note 9.10)

Note 9.8 There is also a reflective liquid crystal display, in which the screen is displayed utilizing reflection of outside light.

Note 9.9 This method is called the active drive method. Meanwhile, passive drive liquid crystal displays are driven by horizontal and vertical electrodes that intersect each other. In the passive drive method, normal window glass (soda-lime glass) is generally used as the substrate glass.

Note 9.10 See Sects. 9.1 and 9.3 for further details on PDP technology.

Table 9.4 Characteristics of substrate glass for displays

	Glass for liquid crystal displays AN100	Glass for plasma displays PD200	Soda-lime glass
Density (g/cm³)	2.51	2.77	2.49
Thermal expansion coefficient ($\times 10^{-7}$/°C)	38	83	85
Strain point (°C)	670	570	511
Annealing point (°C)	720	620	554
Softening point (°C)	950	830	735
Young's modulus (GPa)	77	76	71

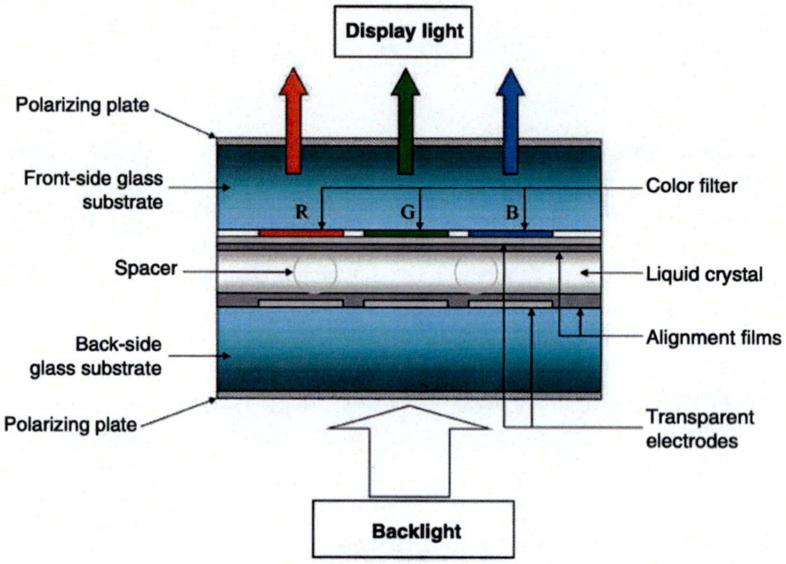

Fig. 9.14 Diagram showing the configuration of a liquid crystal panel. Liquid crystal is filled between two glass substrates and the pixels are displayed by controlling the transmission of light via switching the pixel on or off

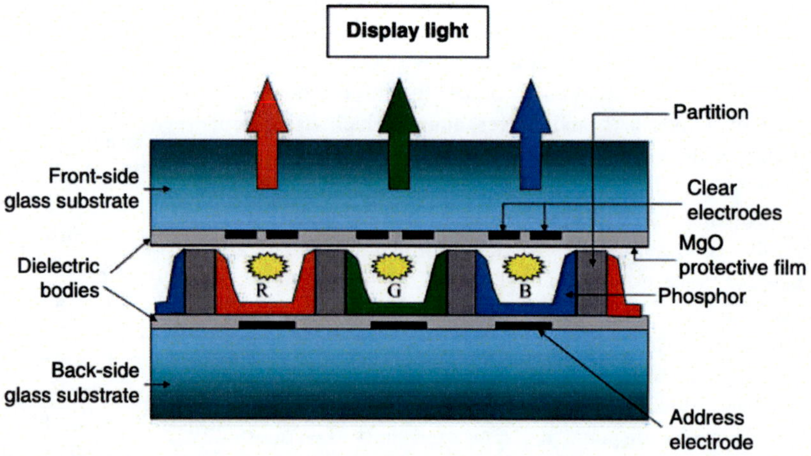

Fig. 9.15 Diagram showing the configuration of a color plasma display panel (PDP). The space between two glass substrates is divided into pixels, and the ultraviolet rays generated by plasma discharge in the spaces stimulate the phosphors in order to make them emit light

The structure of a plasma display is shown in Fig. 9.15. The display mechanism is often compared to a series of minute fluorescent lamps. Partitions that divide the plasma discharge space are formed on the backplate, and RGB phosphors are applied on the inner surface of the partitions. The front substrate is mounted with electrodes that induce plasma discharge and the electrodes are covered with a dielectric layer and an Mg protective film. Ne–Xe mixture gas is sealed inside the panel and the ultraviolet rays, generated from Xe via plasma discharge, stimulate the phosphors to make them emit light. PDPs, unlike liquid crystal displays, are light-emitting displays. The dielectric bodies, partitions, sealing materials, etc. are made of glass (frit) with a low-melting point, and the glass featuring a thermal expansion coefficient equivalent to that of soda-lime glass (approx. $85 \times 10^{-7}/°C$) is suitable as the material for the substrate, so as to match the expansion coefficient with that of the glass (frit). However, the strain point of soda-lime glass (approx. 510°C) is a problem because the maximum sintering temperature of the frit material is approximately 600°C, causing the deformation of substrates. Therefore, "glass with high strain point," with a thermal expansion coefficient equivalent to that of soda-lime glass and a strain point of 570°C or higher, has been developed and used to produce plasma display substrates.

9.4.2.2 Manufacturing Method

The substrate glass used in flat panel displays is melted in a melting kiln with the maximum temperature exceeding 1,500°C, and is formed into a plate. The float method and the fusion method are currently utilized for glass sheet formation. In the float method, the glass material is supplied to a float bath filled with molten tin, on which the glass material is formed into a plate (Fig. 9.16). This is the same method employed for forming normal window glass panels. In the fusion method, the glass material is supplied into a fusion pipe, from which the material is over-flown and formed into a plate while being pulled downward (Note 9.10) (Fig. 9.17). Currently, LCD substrates are manufactured by both the float and fusion methods, while plasma display substrates are manufactured by the float method. Both types of displays require substrate glass with an extremely high degree of flatness. The substrate size is becoming larger year by year because multiple panels are cut out of a sheet of mother glass. The standard thickness is currently 0.5–0.7 mm for liquid crystal displays and 2.8 mm for plasma displays, but thinner glass substrates are increasingly popular in order to reduce the panel thickness and weight.

9.4.3 Future Prospects

The production amount of panels has been increasing year by year, for both liquid crystal displays and plasma displays, and the market for thin-screen TVs has been growing remarkably. The shipping volume of substrate glass is expected to increase as a result of this upward trend.

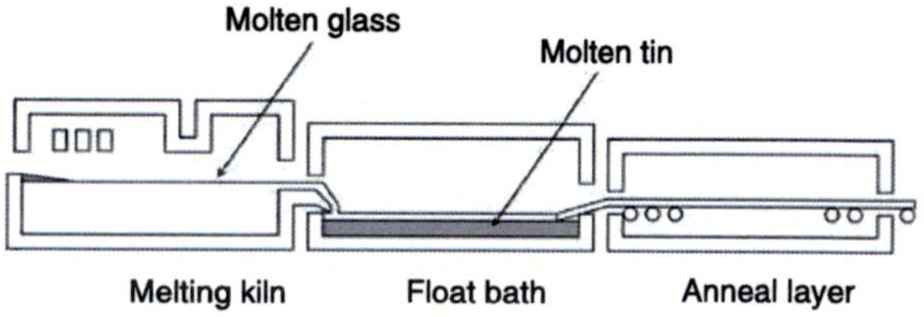

Fig. 9.16 Conceptual diagram of the float method. Melted glass in the melting kiln is floated over molten tin and formed into a plate

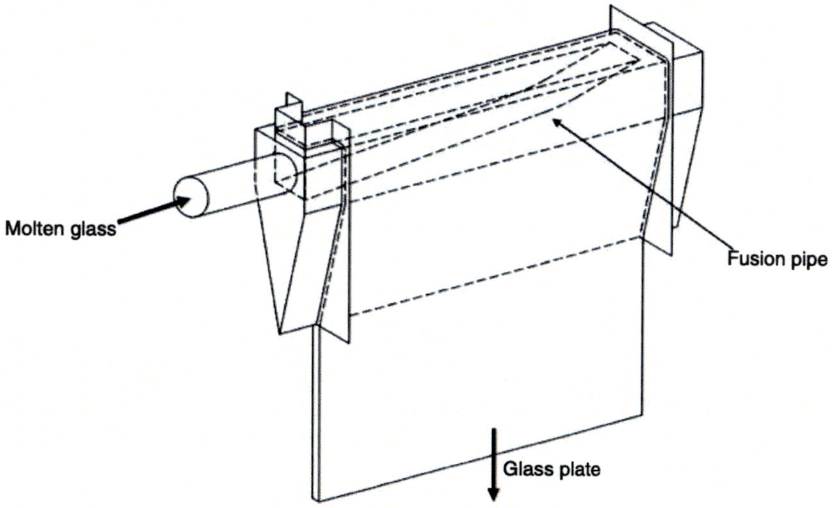

Fig. 9.17 Conceptual diagram of the fusion method (Note 9.9). Glass material that overflows from the fusion pipe is formed into a plate while being pulled downward

Ferrite Core for Deflection Yoke (1953) 9.5

Cathode-ray tubes (CRT) (Note 9.11) are being replaced by FPDs (Note 9.12). The cathode-ray tube was employed in TVs and monitors until several years ago. The deflection yoke is an essential part of the cathode-ray tube. Electron beams from an electron gun of a cathode-ray tube are deflected as specified, while they pass through a magnetic field created in the deflection yoke and reach the phosphor screen. By using ferrite cores in the deflection yoke, the magnetic flux is generated effectively and accurately. Mn–Mg–Zn-based ferrites are generally used in consideration of their high resistivity, low magnetostriction and low price, as well as for their Curie temperature, saturation magnetic flux density and core loss. Production amount of cores for deflection yokes has been decreasing following the decline in sales of cathode-ray tubes, but the production will certainly continue due to their display performance and low cost .

9.5.1 Background of Development

Mass production of cathode-ray tube TVs began in the first half of the 1950s and TVs, refrigerators and washing machines were called the "three sacred treasures." TVs spread rapidly to households during the era of high economic growth in the 1960s, and black-and-white TVs were gradually replaced by color TVs. The cathode-ray tube was also widely used as the display device for PCs when desktop PCs were the mainstream. Although PCs incorporating cathode-ray tubes are disappearing from retail stores in Japan following the rise of FPDs, PCs incorporating cathode-ray tubes far exceed other types in terms of numbers within the global market.

Note 9.11 Cathode Ray Tube (CRT) has been in use since Mr. Brown invented it in 1897.

Note 9.12 FPD, stands for Flat Panel Display, meaning a display device mounted with a thin and flat screen and represented by liquid crystal display (LCD) and plasma display (PDP).

9.5.2 Characteristics

9.5.2.1 *Principle*

In the cathode-ray tube method, light is emitted when electron beams released from the electron gun impact against the phosphor screen. The light is generated on only one central point in this method. Various images can be displayed on the phosphor screen by installing a deflection yoke (Fig. 9.18) on the electron beam orbit to deflect electron beams vertically and horizontally. The deflection yoke (Fig. 9.19) generates a horizontal magnetic field and a vertical magnetic field to deflect electron beams. Soft ferrites are mounted on the outer side of coils so as to ensure the effective and accurate generation of magnetic fields.

9.5.2.2 *Products*

Mn–Mg–Zn-based ferrites (oxide magnetic material that consists primarily of ferric oxide, Fe_2O_3) are generally used to create ferrite cores for deflection yokes. This is in consideration for magnetic properties, including high resistivity, low loss, high saturation magnetic flux density, low magnetostriction, as well as permeability and a Curie temperature of several hundreds higher than the specified temperature, and low cost. High dimensional accuracy is required for the accurate generation of magnetic fields, and especially, the cruvature on the inner surface is important. They are generally shaped like a morning glory flower so as to match the funnel curve (Note 9.13) of the cathode-ray tube. Recently, tubes which meet the need for energy saving and high

Note 9.13 The curve at the rear end of the cathode-ray tube where the deflection yoke is mounted. The name comes from its funnel like appearance.

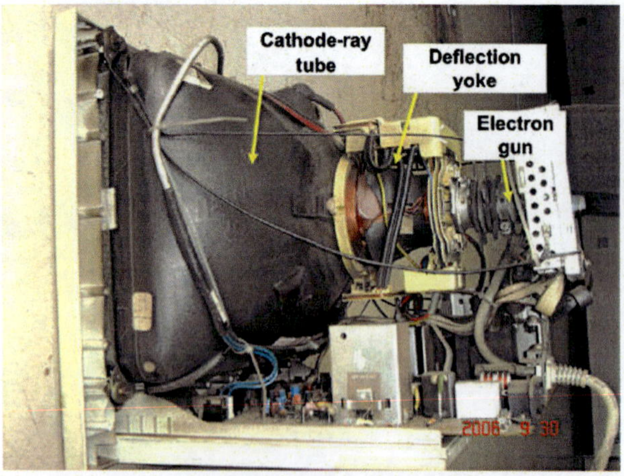

Fig. 9.18 Picture of a deflection yoke mounted on a cathode-ray tube. Display devices mounted with a cathode-ray tube need a deflection yoke. Electron beams released from the electron gun are deflected vertically and horizontally by the deflection yoke and display various images on the phosphor screen on the front side of the cathode-ray tube

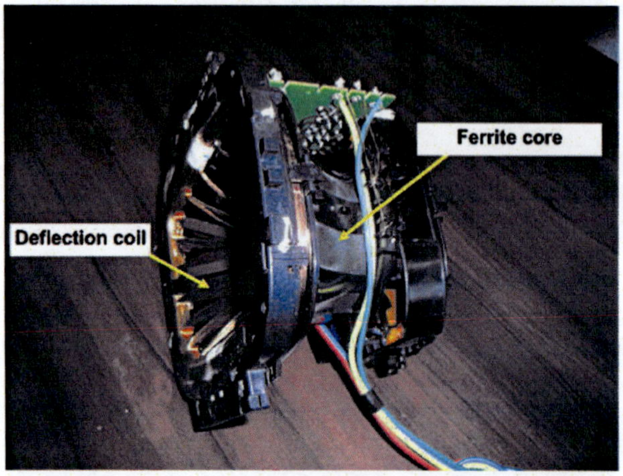

Fig. 9.19 Deflection yoke. Magnetic fields that deflect electron beams horizontally and vertically are generated by the horizontal coil and the vertical coil. The outer side of the coil is covered by the ferrite core so as to ensure the effective and precise generation of magnetic fields

efficiency are being mass produced (Fig. 9.20). In general, the product is assembled when the cores are placed on the coils and fixed by clips during the process of assembling the deflection yoke. Grooves for engaging the clips are created on the outer face of the cores.

Fig. 9.20 Ferrite based deflection yoke. They are generally shaped like a morning glory flower. Square type ferrite cores (*left bottom*) are being mass-produced with an aim to improve deflection efficiency

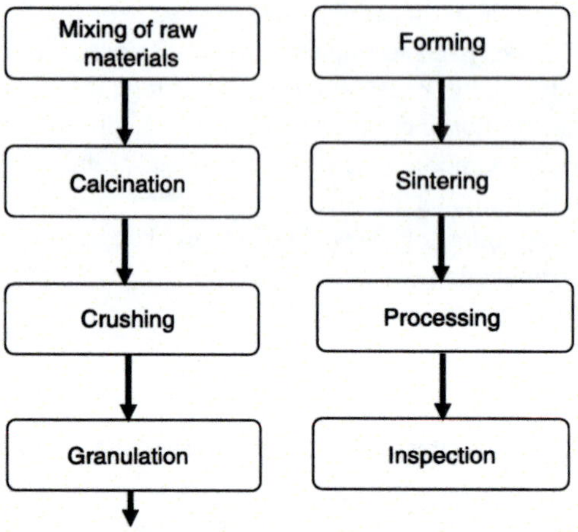

Fig. 9.21 Manufacturing process of ferrite cores. Produced by mixing primary material ferric oxide with other oxide materials, ferrite granules are calcined to produce the products

9.5.2.3 Manufacturing Method

CRTs are manufactured using the process shown in Fig. 9.21, which is also employed to produce other ferrite cores. The material production process, which is performed with the dry method or the wet method, includes the mixing of raw materials to granulation. It is very important that the granule quality is uniform. Large pressing machines ranging from several tens of tons to several hundreds of tons are used for forming.

The quality is controlled in this process, by checking whether the density of each portion of the formed ferrite core conforms to the specified value. Then, the formed body is sintered in a furnace at a temperature of one thousand and several hundred degrees. The formed body shrinks by slightly more than 10% during sintering, and therefore, careful attention is needed to ensure high dimensional accuracy and in order to prevent deformation during sintering. As one of the measures to prevent shrinkage, the formed body is hung inside a ring-shaped furnace material for sintering. With respect to cores for deflection yokes used in monitors, the inner surface is polished after sintering, as higher dimensional accuracy is required of them than of others. The products that have passed the final dimensional inspection are divided and packed. The core is divided once for easy assembly of the deflection yoke, and shallow grooves are created by the die in the process of forming, on the core for easy separation. The core is divided into two pieces by cleaving along a groove.

9.5.3 Future Prospects

The cathode-ray tube numbers will continue to decline due to growth in FPDs, but cathode-ray tubes will remain on the market and will be used as display devices for TVs and monitors due to their dynamic image display performance and cost advantage. Therefore, although the production amount will decline, a long-term demand for ferrite cores for deflection yokes will continue.

CRT Insulator (Multiform Glass) (1956) 9.6

Multiform glass is a special type of glass used to support the electron gun inside the Cathode Ray Tube (CRT) display. In the CRT, electron beams (Note 9.14) from the electron gun are directed at the phosphor coated screen to produce images. The instantaneous voltage applied to the electron gun (Notes 9.15 and 9.16) is as high as several tens of thousands of volts. The electrodes need to be supported by a material featuring low thermal expansion due to temperature changes in electrodes, because displacement of the electron gun and electrodes affects the accuracy and stability of images. Therefore, the material is required to possess high thermal insulatory properties, high breakdown voltage and low expansion properties.

9.6.1 Background of Development

CRTs are displays which are used widely in TVs and PC monitors. The electron guns used in CRTs vary in their specifications, shapes and sizes. The parts that support the electron guns are made of materials that can be formed into various shapes and feature the following characteristics.

Note 9.14 In a broad sense, electron beams are particle beams created by converging and accelerating electrons to the state of beams. A CRT has three electron beams, which, respectively, illuminates R, G and B phosphors.

Note 9.15 The operating voltage of the electron gun varies depending on the characteristics, but it is generally in the rage of 20–200 V at the cathode, 5–10 kV at the pre-focus lens and 20–30 kV at the main lens.

Note 9.16 The maximum working temperature of the CRT electron gun is approximately 1,000 K at the heater.

9.6.2 Characteristics

9.6.2.1 Characteristics and Specifications of Products

The cathode in the electron gun is heated by a heater to release thermal electrons that are accelerated by the application of high voltages, and electron gun generates three electron beams corresponding, respectively, to red, green and blue via electrodes that also play roles as the apertures and lenses. The beams are controlled to reproduce the desired colored image and the red, green and blue phosphors inside the panel generate colors responding to the electron beams. Each of the electrodes is composed of a cylindrical or plate-shaped metallic part. The electrodes are laid out along the axis of electron beams that are projected and the legs of the electrodes are supported and fixed by the multiform glass (Fig. 9.22).

The space inside the CRT is in a vacuum and the multiform glass needs to be free from residual gas release and deterioration. The multiform glass needs to have high thermal insulator properties and high breakdown voltage. In addition, it is required to feature low thermal expansion so that the distance between the electrodes remains the same regardless of temperature changes so as to ensure stable images. It is also required to feature limited viscosity change, for the improvement of accuracy in assembling of the electron gun. In order to satisfy the requirements, borosilicic acid-based glass (mainly composed of SiO_2, B_2O_3, K_2O) is used.

For the identification of electron types and production lines during the assembly of electron guns, the glass is colored by pigments. This is done without sacrificing its required specifications (the basic color is opaque white) (Fig. 9.23).

Characteristics required of the multiform glass are as follows:

1. High insulation properties against high voltages applied to the electrodes.
2. Low thermal expansion coefficient.

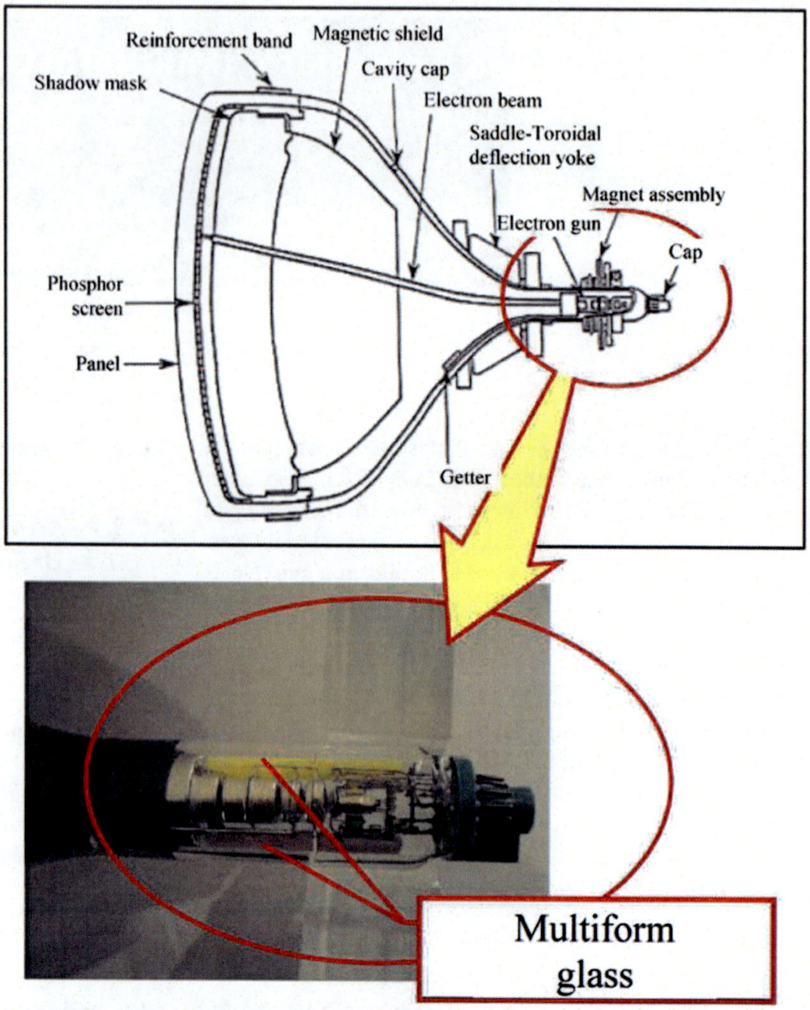

Fig. 9.22 Electron gun and multiform glass. Inside the CRT display, legs of the electron gun are supported and fixed by the multiform glass

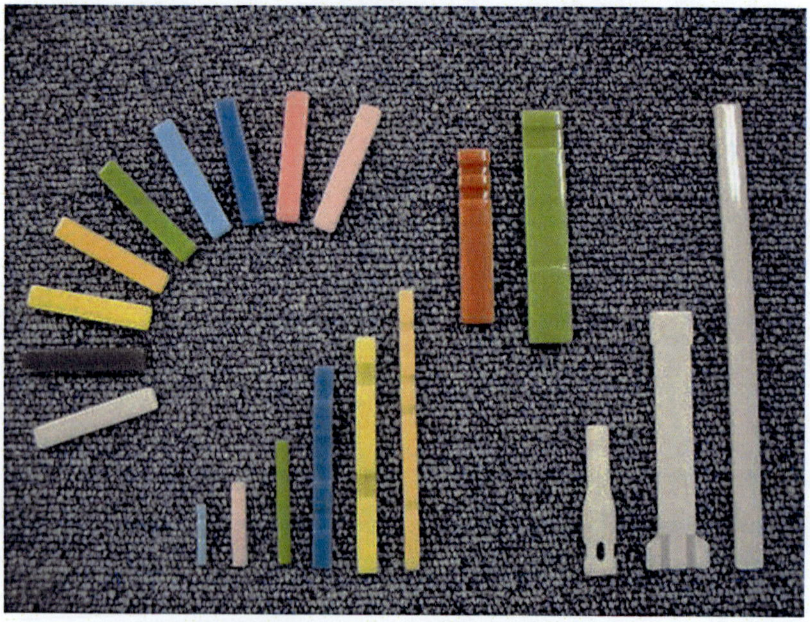

Fig. 9.23 Examples showing flexibility in shapes and color tones. Multiform glass can be formed into various shapes to meet the shapes of electron guns. The glass can be colored by pigments for the identification of electron types and production lines during assembly of electron guns, without sacrificing the required specifications

3. Limited fluctuation in viscosity responding to high-temperature heating.

During the process of mounting the electron gun onto the multiform glass, the glass surface is heated up to about 1,400°C. A viscosity that ensures appropriate operations under the temperature range is required. Improper fitting of electrode legs and the multiform glass leads to breakage.

4. Proper glass wetting with electrode metals.
5. Does not release gas in a vacuum. Free from deterioration.
6. Easy to create various shapes.
7. Can be colored by pigments.

An example of multiform glass specifications is shown in Table 9.5.

Table 9.5 Characteristics of multiform glass, example

Grain diameter during granulation	Standard: 5 μm
Appearance	Free from bubbles, scratches, cracks, chipping, burrs, deformation, adhesion of foreign objects, etc.
Volume resistivity, ρ	log $\rho = 8.9\ \Omega m$ (350°C)
Thermal expansion coefficient	27×10^{-7}/°C (300°C)
Viscosity coefficient η	Log $\eta = 6.65$ Pa·s (at softening point 810–820°C) Log $\eta = 2.4$–3.0 Pa·s (point of operation 1,400°C)
Color tone	9 color tones

For multiform glass, breakdown is prevented by controlling average grain size during the granulation process to about 5 μm in diameter and reducing the size of bubbles after forming and sintering

9.6.2.2 Manufacturing Method and Mounting of Electron gun

Multiform glass is manufactured using the sintering method, where powder material is formed and calcined instead of forming by molten glass. This is shown in Fig. 9.24. In the sintering glass method, the formed glass is ground, mixed with binders, pigments, water, etc., and is press formed using a die. This method enables forming into various shapes responding to the shapes of electron guns. Long bubbles and impurities from the die may mix in the case of molten glass that can lead to low voltage breakdown. In the sintering glass method, the average grain size is controlled to about 5 μm in diameter during the process of grinding and granulation of the molten glass prior to forming. Therefore, the size of bubbles after forming and sintering can be minimized to prevent low voltage breakdown.

Figure 9.25 shows the electron gun mounting method employed by a final assembly manufacturer.

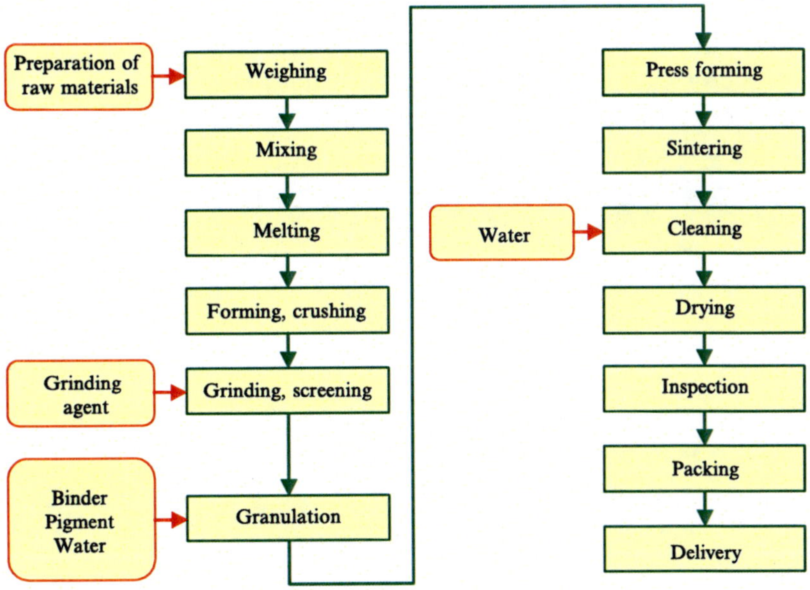

Fig. 9.24 Manufacturing process. Multiform glass is manufactured using sintered glass and not by the molten glass process. Powder material made by means of grinding and granulating molten glass is formed and sintered. Multiform glass can be formed into various shapes and the average grain size is controlled to a minimum during the process of grinding and granulation so as to prevent breakdown, etc.

The surface of the multiform glass is heated to 1,400°C before fitting the legs of electrodes.

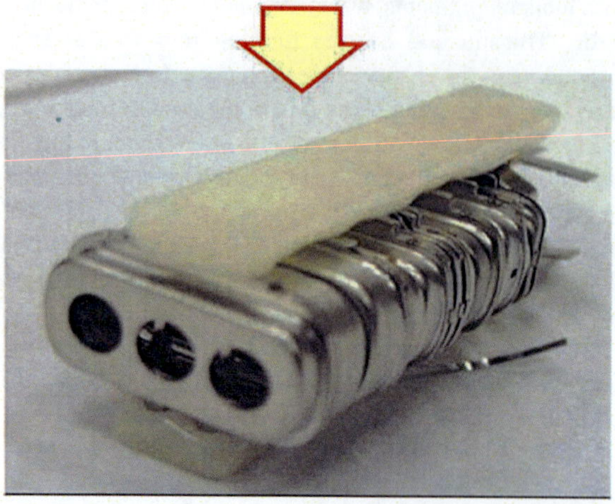

Legs of electrodes are fitted into the glass. It is maintained at a viscosity appropriate for high-temperature heating and fitting.

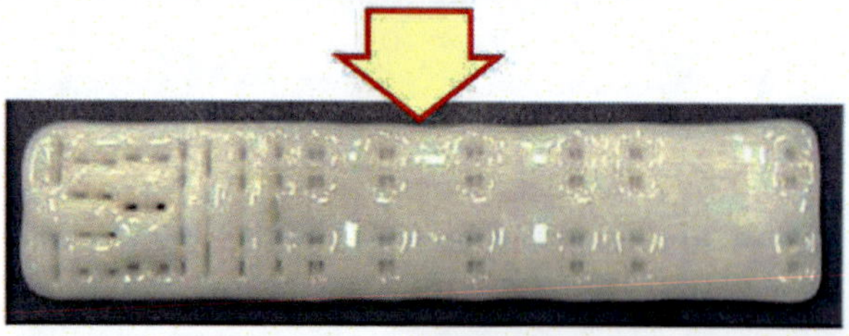

Traces of embedded legs of the electron gun are seen in the picture. The multiform glass fits properly with electrode metal pins. It does not release gas in a vacuum and is free from deterioration.

Fig. 9.25 Multiform glass and fitting of electron gun

9.7 Transformer Cores (1947)

The transformer was invented in 1831 by Faraday. They began to be used widely in electronic devices following the adoption of ferrite cores, which are magnetic ceramics. The ferrite core was invented in 1930 by Dr. Y. Kato and Dr. T. Takei (Fig. 9.26) at Tokyo Institute of Technology. Commercialization was promoted for various applications and the application of the material expanded rapidly for use in intermediate frequency transformers and in switching supply transformers.

9.7.1 Background of Development

The principle of the transformer was discovered in 1831 by the experiment of the electromagnetic induction phenomenon conducted by Faraday. The first practical transformer was designed in 1882 by Goraru of France and Gibbs of England. Two coils are wound around a core and the initial core is applied with a current to generate magnetic flux within the core. Voltage is induced on the secondary coil via electromagnetic induction by the magnetic flux. The voltage can be changed (transformed) by adjusting the turn ratio of the two coils. Transformers were originally used for power distribution. Familiar examples are the currently used pole-type transformers and the power supply voltage (200–100 V) transformers. A laminated silicon steel plate was used mainly as the core. The core material caused no problem in low frequencies at the level of commercial frequencies, but the properties deteriorated at high frequencies limiting applications. Ferrite was invented in 1930 by Dr. Yogoro Kato and Dr. Takeshi Takei at Tokyo Institute of Technology. Although soft ferrite used as the core did not serve any purpose in those days its application was sought after. Following the suspension of four-tube radio production by the second world war administration in Japan, known as GHQ in 1947, production of super heterodyne radio (Note 9.17) began and soft ferrite began to be used in large amounts as cores of intermediate frequency transformers (Fig. 9.27) that are used in high frequencies. Responding to the start of full-scale TV broadcasting in 1953, high voltage transformers (flyback transformers) (Fig. 9.28) began to be produced to light up the picture tube. Since that time, flyback transformers were substantially downsized and continued to be used in picture tube TVs. In the 1970s, they began to be used in consumer electronic devices following the commercialization of switching supplies developed for satellites by NASA. Conventional power supplies incorporated power transformers with a multi-layered silicon steel core and were large and heavy. Power transformers were substantially reduced in size and weight as the power transformer is operated at high frequencies over 10 kHz with switching supplies. Following the reduction in size and weight, switching supply transformers have been used in a wide variety of electronic devices. Transformers incorporating a ferrite core are also used in signal and measuring devices such as in impedance conversion, insulation, balance–unbalance conversion and sensing.

9.7.2 Characteristics

9.7.2.1 *Characteristics of Products*

Ferrite used as transformer cores is an iron oxide-based oxide magnetic material. The crystal structure is of a spinel type and the material is ferrimagnetic. The characteristics vary

Note 9.17 The radio system in which high-frequency signals generated on antennas by electric waves are converted to intermediate frequencies before amplification. The regenerative four-tube radio could generate intercept waves caused by oscillation. Production was discontinued after the war in response to GHQ instructions.

depending on the type and ratio of bivalent metal elements added to iron (Fe). Ferrite features high volume resistance compared with magnetic metal materials such as silicon steel and permalloy. It is widely used as a material for cores used at high frequencies as it makes it hard to generate eddy currents in the core. The ferrite used in transformers is roughly classified into two types—the Mn–Zn type and the Ni–Zn type. Mn–Zn-based ferrite features higher characteristics in terms of initial permeability, saturation magnetic flux density and core loss than does the Ni–Zn-based ferrite. However, its volume resistivity is lower when compared with the Ni–Zn-based ferrite. Therefore, the coil cannot be wound directly on the core, requiring the use of a bobbin. Cores of various shapes have been designed and manufactured for a variety of applications (Fig. 9.29). Although Ni–Zn ferrite is slightly inferior to the Mn–Zn based ferrite in characteristics, it features high volume resistance and can be looked at as an insulator. Ferrite is suitable for compact products because the coil can be wound directly around the core. It is also suitable for use at higher frequencies than Mn–Zn-based ferrite. Typical transformers incorporating a ferrite core are listed in Table 9.6. Mn–Zn ferrite is also used in signal transformers such as transmission transformers that should be free from distortion and in pulse transformers in which the connection is important.

Fig. 9.26 The world's first ferrite core. The world's first ferrite core was invented in 1930 by Dr. Y. Kato and Dr. T. Takei

9.7.2.2 Manufacturing Method

Figure 9.30 shows the manufacturing method of general ferrite cores. Iron oxide is normally used as the raw material, which is mixed to an intended composition. The characteristics

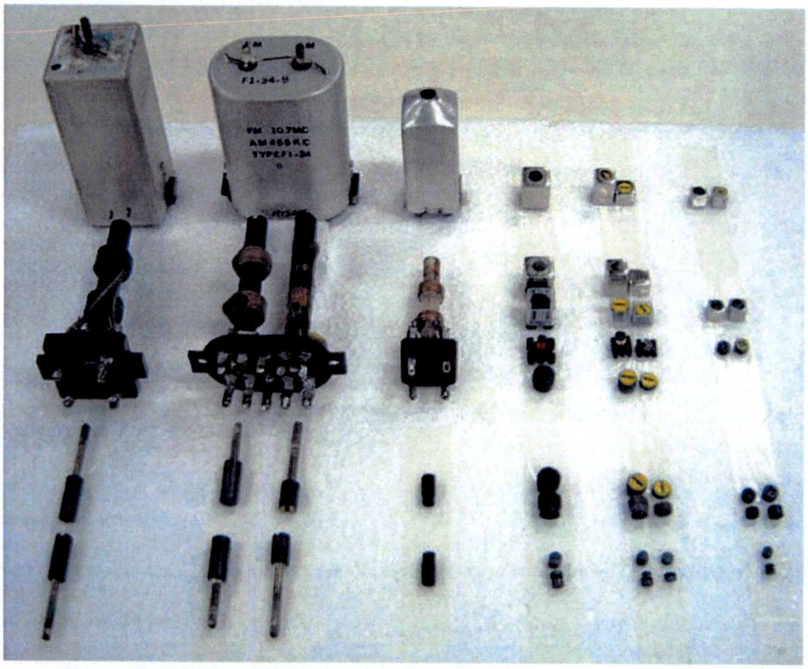

Fig. 9.27 Transition in structures of intermediate frequency transformers. Conventional radios that utilized regenerative waves could generate intercept waves caused by oscillation. However, the noise was reduced by converting the high-frequency signals generated on antennas to an intermediate frequency using intermediate frequency transformers before amplification

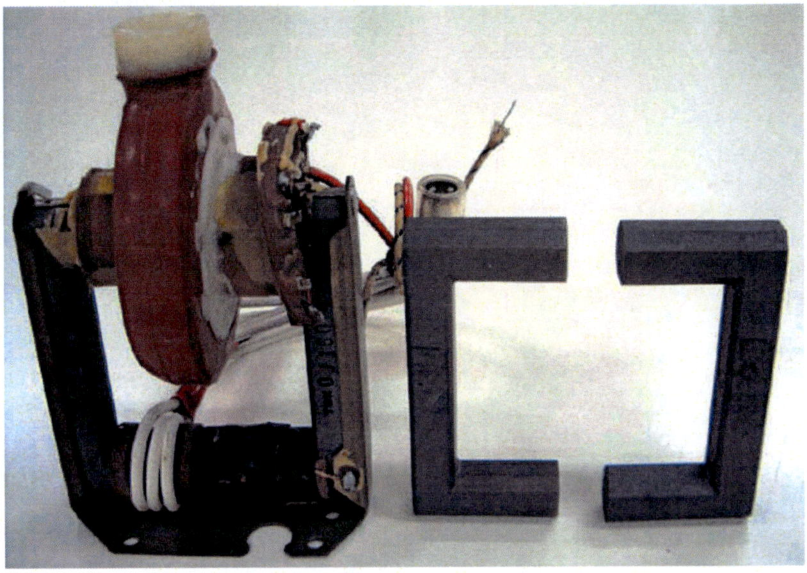

Fig. 9.28 An early version of flyback transformer and a ferrite core. A early flyback transformer and a ferrite core that were used in black-and-white TVs. They generated a high voltage of about 10 kV

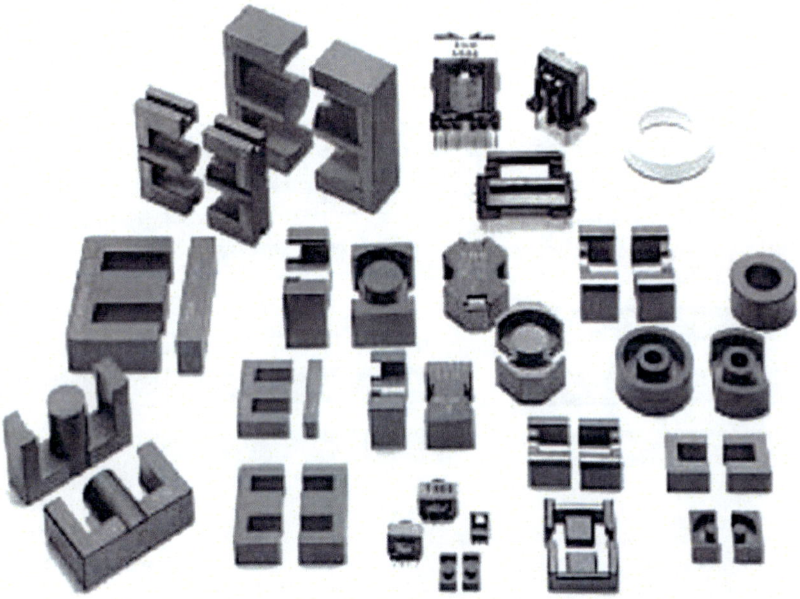

Fig. 9.29 Typical shapes of ferrite cores intended for transformers. Cores of various shapes have been designed and manufactured for a variety of transformers. Ferrite materials suitable for each transformer have been developed and commercialized

of ferrite are sensitive to compositions and impurities and attention is required for precise weighing and for the prevention of contamination. The mixed powder is temporarily calcined at 900°C to create a material close to ferrite and then the material is ground into a fine powder. Attention should be given to average particle size and particle size distribution of the ground powder as sintering properties vary depending on the grinding conditions. A binder is added to the ground powder for granulation before the powder is formed. Many of the ferrite cores are complex in shape and tend to generate cracks during the forming process. Therefore, various trials are carried out on binders and forming machines. The sintering is then performed at a temperature from 1,100°C to 1,300°C. Then if necessary, the product goes through the grinding process before completion. Various methods are also attempted with respect to the calcination conditions as the characteristics of ferrite cores change depending on the crystal state after sintering.

Table 9.6 Types and major applications of transformers that incorporate ferrite cores

Classification	Type	Major application
Power transformer	Flyback transformer	Picture tube TV, PC monitor
	Switching supply transformers	General electric appliances, AC adaptors
	Converter transformer	DC/DC converter (PC, laptop PC, HEV)
	Inverter transformer	Liquid crystal backlight (liquid crystal TV, LC monitor) Hf fluorescent lamp Inverter motor (white goods, elevators, industrial equipment)
	Current transformer	Measuring instrument, sensor
	Wireless charging	Cordless phone, electric toothbrush Carrier machine for industrial use
Signal transformer	Intermediate frequency transformers	Radio, TV
	Matching transformer	Amplifier
	Balun transformer	Antenna (balance–unbalance conversion)
	Transmission transformer	ADSL modem
	Pulse transformer	LAN terminal
	Rotary transformer	Video

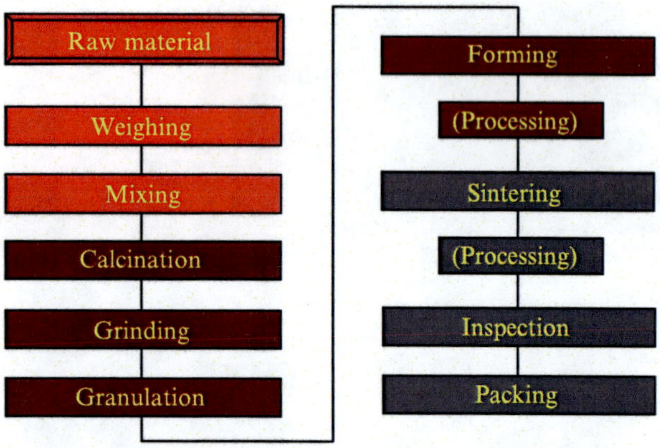

Fig. 9.30 Manufacturing method of ferrite cores. Ferrite cores are manufactured using the powder-metallurgical method, the same method as with other ceramic products. The characteristics change due to difference in composition, the mixing of impurities and difference in crystal structure making process control important

9.7.3 Future Prospects

Various efforts are being made to improve characteristics of ferrite materials intended for transformers. The characteristics continue to be enhanced and new materials are being introduced everyday. Responding to an increase in switching frequencies, the development of transformer core materials used in the MHz range has also been performed actively. Development of optimum core materials will be promoted further responding to the advance in circuit technologies and electronic parts. In the case of core shapes, they are designed to eliminate waste in terms of magnetic circuits, reduce leakage of magnetic flux and reduce size and thickness by analyzing magnetic field simulations. It is expected that the development of thin cores featuring high radiation performance will be further promoted.

Small Power Supply Transformer for Switching (1975) 9.8

Switching power supplies are widely used in consumer and industrial electronic devices because of their small size and high efficiency. Transformers used with them are important electronic devices, and the performance improvement in the transformer substantially affects the performance of the power supply. The performance of transformers is determined by the magnetic material used in them, which is normally a magnetic ferrite. The material properties have been improved in various ways. The shapes of cores have also evolved. Transformers have been downsized and the materials used to produce cores and other parts have become less expensive. Improvements benefit not only the consumer through cost reductions but also impact the environment.

9.8.1 Background of Development

Switching supplies are widely used in consumer and industrial electronic devices and have shown improved efficiencies. Responding to recent trends, demands for switching supplies with smaller sizes and higher performance have been increasing. Transformers used in them are important electronic devices along with semiconductor devices that function as switching elements. Performance improvement of transformers substantially affects the performance of power supplies. The performance of transformers is determined by the magnetic material used in them, which are normally ferrite cores. The material properties of ferrites have been improved in various ways. Figure 9.31 shows various shapes of small power supply transformers for switching. Figure 9.32 shows types of magnetic materials and products that incorporate the materials. Mn–Zn ferrite cores are normally used in power supply transformers. Among the ferrite material properties, saturation magnetic flux density (Bs) (Note 9.18) and core loss (Pcv) (Note 9.19) are important factors in downsizing the transformers. The material needs to feature high Bs and low Pcv in order to scale down the transformer size. Efforts are being made to enhance the Bs and lower the Pcv.

Note 9.18 If the magnetic field applied to a magnetic body is increased, the magnetic flux stops increasing at a certain point. This phenomenon is called magnetic saturation and the magnetic flux density at this point is expressed by the saturated magnetic flux density.

9.8.2 Characteristics

9.8.2.1 Characteristics and Specifications of Products

There has been no significant improvement or progress in the shapes of ferrite cores used in power supply transformers up until now. This means that no substantial progress in the downsizing of transformers has been made. However, substantial downsizing was recently realized following the improvement in the core shape and material. Improvements were made by changing the cross-sectional shape of the multi-output power supply transformers (widely used in home appliances) to an asymmetric one, with a focus on power supplies for transformers. In addition, high performance ferrite cores were developed and applied to the transformer. This led to a 25% volume reduction of transformers against a conventional core shaped material while maintaining equivalent output. The conventional product and the new product are compared in Fig. 9.33. The newly developed transformers have been used in home appliances such as inverter type air conditioners, DVD players and liquid crystal TVs because of their small size and low cost. Applications will be expanded to a wider range of electronic devices in the future. The downsizing of transformers contributes to the reduction of materials for all parts including the core. This contributes to cost reduction as well as environmental conservation through resource saving, production energy saving and the operating energy saving of electronic devices.

Note 9.19 If the magnetic field applied to a magnetic body is increased, magnetization inside the magnetic body can no longer catch up with the frequency of the magnetic field at a certain point. The excess magnetic field becomes the loss. This loss is called the core loss.

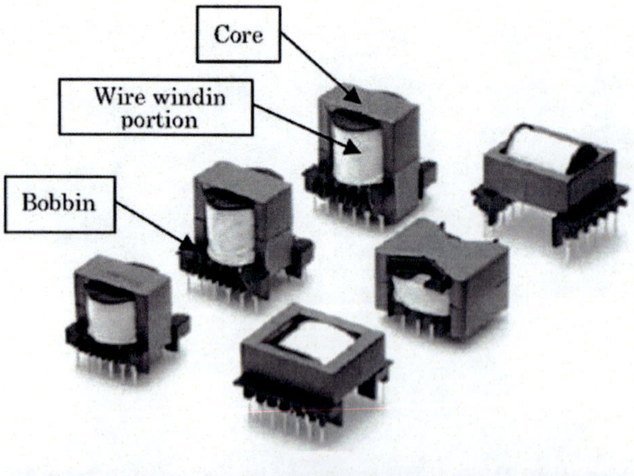

Fig. 9.31 Small power supply transformer for switching. The small power supply transformers incorporate ferrite cores. The insulator coated copper wire is wound around a bobbin, which is then assembled around a ferrite core

9.8.2.2 *Manufacturing Method*

The production flowchart of ferrite cores is shown in Fig. 9.34. Three types of major components, Fe_2O_3, MnO and ZnO, are weighed and mixed according to the intended properties. The raw materials are made to react by means of calcination so as to create the ferrite. The ferrite is ground in to a fine powder of micron-order and a binder is added to create a mixture with an appropriate fluidity. The grains fill a die of the intended shape and press formed to create cores. The press formed cores are sintered at a high temperature of 1,300°C or higher to create ferrite cores. They are ready for use in transformers after mechanical processing. An insulator coated copper wire (Note 9.20) is wound around a roller called a bobbin so as to produce a coil. The coil and ferrite core are assembled and fixed to create a transformer.

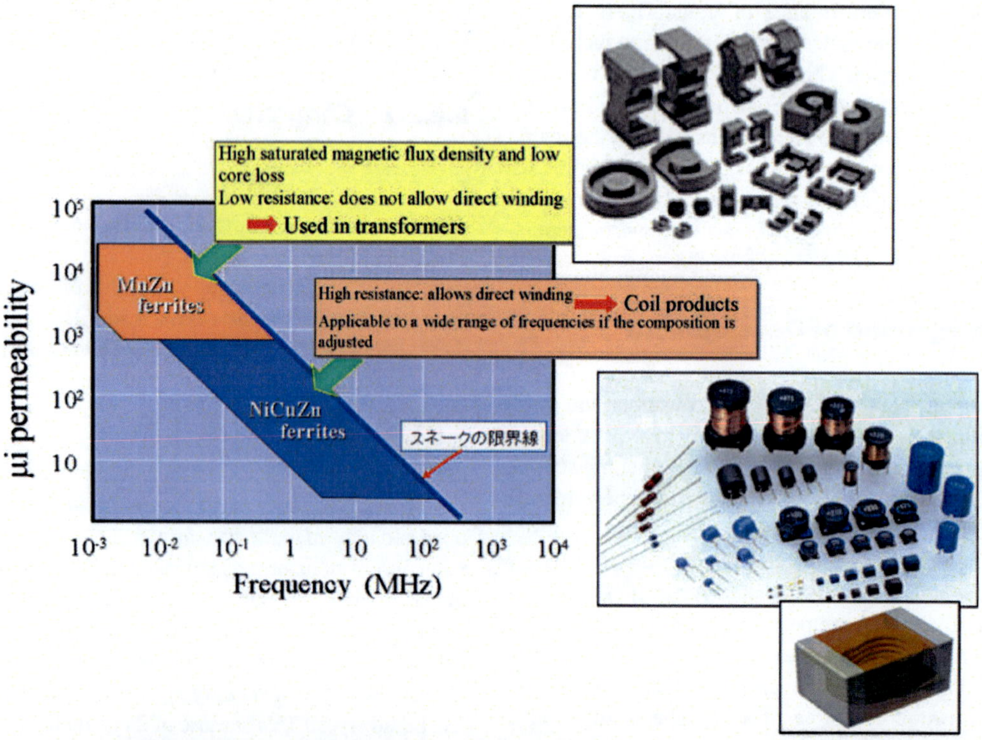

Fig. 9.32 Types of ferrite materials and applicable products. Ferrite core materials are classified into the Mn–Zn ferrite and the Ni–Zn ferrite. Mn–Zn ferrite cores are suitable for power supply transformers. However, the permeability drops abruptly in line with the sudden increase in loss at certain frequencies, which vary depending on the property of the ferrite. The limiting frequency is lower when the permeability is higher. The product of the permeability and the limiting frequency is normally constant. The line is called the Snoek's limit line. Snoek's limit line: ferrite is capable of maintaining a certain permeability and low magnetic loss at high frequencies

Note 9.20 Copper wires coated with urethane resin are used in transformers. The insulation coating is peeled off from both ends of the wire for connection with terminals after winding.

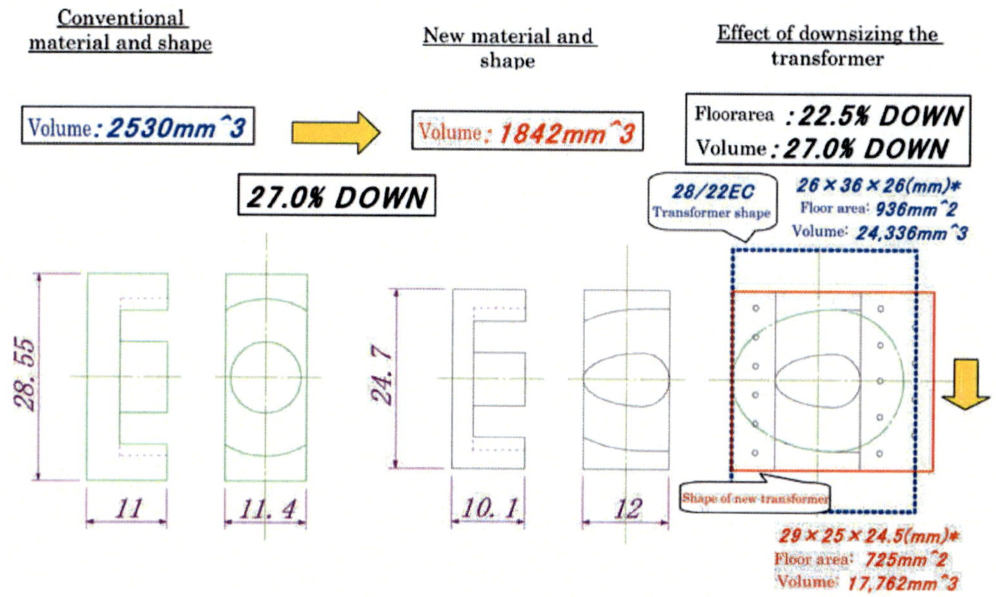

Fig. 9.33 Example of core volume reduction by core shape change. The new transformer (EEG shape), whose volume is 27% less than the conventional transformer (EER shape), while achieving equivalent performances

Fig. 9.34 Manufacturing method of Mn–Zn ferrite. Fe_2O_3, MnO and ZnO powders are mixed, calcined, press formed into intended shapes and sintered at high temperature to produce a ferrite

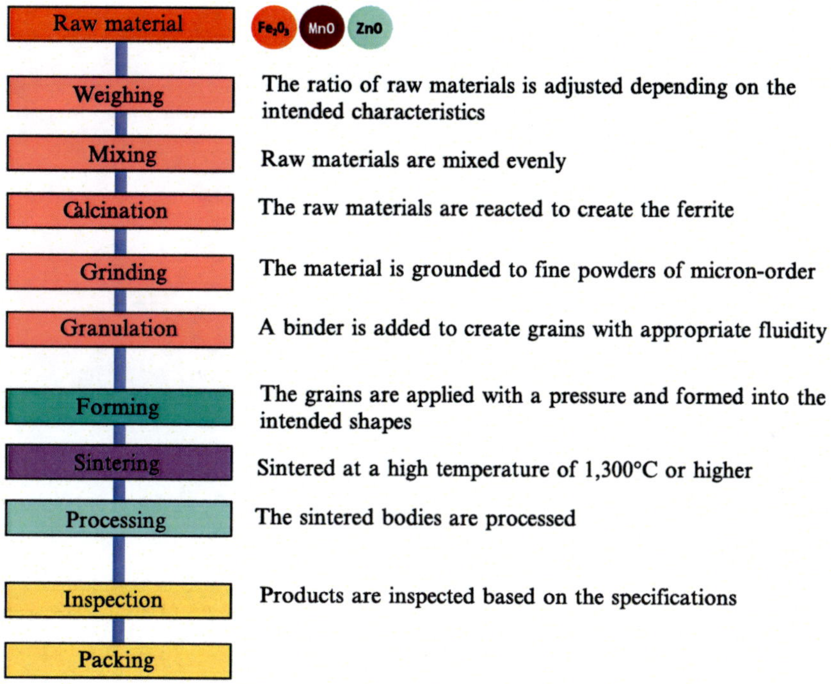

9.8.3 Future Prospects

Material performances of ferrite cores are being raised to increase the *Bs* and lower the loss. In addition to the improvement of the quality of ferrite core materials, efforts are being made to realize economical core shapes through magnetic field analysis simulations. Further scaling down will be realized by improvements in both material quality and core shapes.

Piezoelectric Transformer (1994) 9.9

Piezoelectric transformers are piezoelectric ceramics that transform an input voltage into a much higher output voltage (Fig. 9.35). The combination of the inverse piezoelectric effect and piezoelectric effect of piezoelectric ceramics was proposed by C.A. Rosen of General Electric in 1956. Piezoelectric transformers are non-flammable and have low magnetic noise. They were introduced to the market in the early 1990s. They are widely used for lighting in the LCD (liquid crystal display) backlight of laptop PCs, liquid crystal TVs that require a high voltage (Fig. 9.36).

9.9.1 Background of Development

Inverters (Note 9.21) for illumination in the LCD backlights (cold-cathode tubes) are required to be thin, light and highly efficient responding to the expansion of space-saving and energy-saving home appliances such as laptop PCs, liquid crystal TVs and liquid crystal monitors. Cold-cathode tubes intended for LCD feature several MΩ of impedance (Note 9.22) during OFF (meaning high onset voltage), while the impedance during ON is several hundred kΩ (meaning low sustaining voltage). Therefore, the transformer that turns on the cold-cathode tube is required to have the capacity to fully respond to the load fluctuations. Demands have been steadily increasing for transformers used for inverters that drive cold-cathode tubes since the piezoelectric transformer was commercialized in 1994. Coil transformers (Note 9.23) do not fully serve the purposes in terms of thickness, efficiency and magnetic noise. Meanwhile, piezoelectric transformers enable reduction in size and height, improvement in efficiency, high electrical resistance, and are also nonflammable. As a result, the output characteristics of piezoelectric transformers are suitable for illumination in cold-cathode tubes.

Note 9.21 A device that converts DC power to AC power.

Note 9.22 The amount indicating the resistance of current flow in AC circuits. The complex number is generally expressed by the Z symbol. The real part is called the resistance and the imaginary part is called the reactance. The unit is ohm (Ω).

Note 9.23 A part used to convert AC voltage by electromagnetic induction. Coils made by winding copper wires are normally used for both input and output. Therefore, they are called coil transformers.

9.9.2 Characteristics

9.9.2.1 Characteristics and Specifications of Products

Figure 9.37 shows structures of typical piezoelectric transformers (ROSEN type). Electrodes are positioned on top, bottom and one end of a rectangular ceramic plate, which is poled along the thickness and length directions, respectively. The area poled along the thickness direction is called the primary side (drive part), while the area poled along the length direction is called the secondary side (power generation part). The supplied electric energy is converted to the elastic energy by the inverse piezoelectric effect on the primary side, and the elastic energy is converted again to the electric energy by the piezoelectric effect on the secondary side. Conversion of electric energy is performed via elastic vibration, and therefore, piezoelectric transformers are used in frequencies close to the intrinsic resonance frequency (resonance frequency), which is determined by the elastic wave propagation velocity of the ceramic and the size of the piezoelectric transformer. Characteristics of piezoelectric transformers are explained below.

If the load impedance on the secondary side of the piezoelectric transformer is raised, the maximum boost-up ratio and the maximum output frequency become higher (Fig. 9.38). The load dependency of the piezoelectric transformer output is applicable for supplying high voltage to cold-cathode tubes, as mentioned above. The shapes also affect the boost-up ratio, output voltage, operating frequency and temperature rise characteristics of piezoelectric transformers. As for piezoelectric transformers incorporating the length expansion mode, the boost-up ratio increases with an

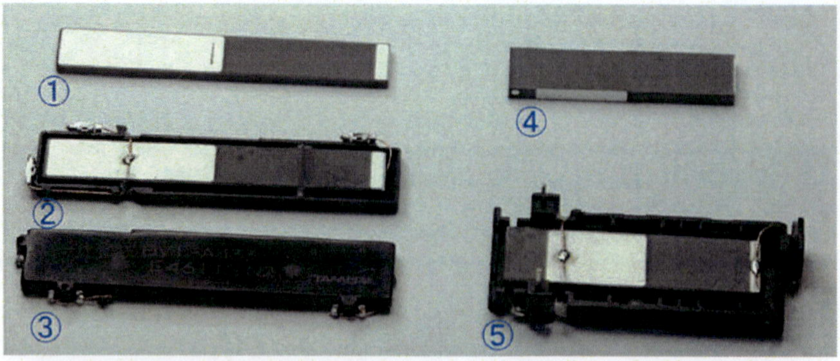

Fig. 9.35 Appearance of piezoelectric transformers. (*1*) Piezoelectric transformer unit L44×W6.5×T1.6 mm–5 W type single plate. (*2, 3*) Cased piezoelectric transformers placed in a low-height case for safety. (*4*) Multi-layered piezoelectric transformer. The primary side is laminated to improve the boost-up ratio. (*5*) Piezoelectric transformer for a TV. A case is used to ensure insulation distance, responding to the safety standard. Products with the length of 20–50 mm (frequency: approx. 30–200 kHz) are mass-produced

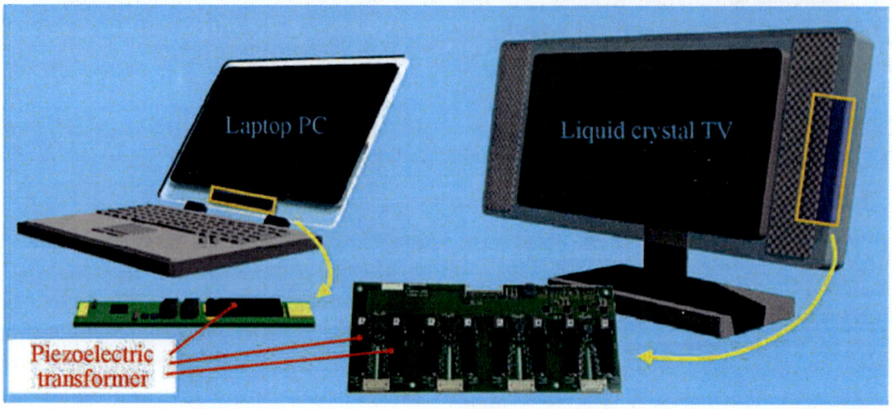

Fig. 9.36 The areas where piezoelectric transformers are used. They are mounted on liquid crystal backlight inverters (piezoelectric inverters), which are normally installed on the back side of LCDs

increase in the length/thickness ratio. The optimum length is decided with consideration for efficiency as the amount of heat generation per output increases in line with the increase in boost-up ratio. The size of piezoelectric transformers is determined by the operating frequency (half-wavelength mode/approx. 50 kHz at 30 mm), output voltage. In order to increase the boost-up ratio without changing the size, the primary side needs to be laminated so as to increase the capacitance of the primary side (laminated piezoelectric transformer).

For the driving of the piezoelectric transformer, voltage with a frequency close to the resonance frequency is applied to the primary side while a cold-cathode tube (load) is connected to the secondary side. If the driving waveform on the primary side is distorted by the overlay of harmonic overtones, the harmonic overtones result in losses, lowering the efficiency in some cases.

Table 9.7 shows specifications of a single-plate piezoelectric transformer intended for powering a 14-in. LCD backlight.

9.9.2.2 Manufacturing Method

Piezoelectric ceramics for piezoelectric transformers are required to feature a high electromechanical coupling factor, a high mechanical quality factor (Q_m), low loss due to vibration, low dielectric tangent (tanδ), high mechanical strength, high resistance to vibrating stress, and low property degradation and mechanical fatigue under long-term operation with a large amplitude. Among the materials that have been commercialized, the hard materials are made by adding three components and a variety of additives to the base material of

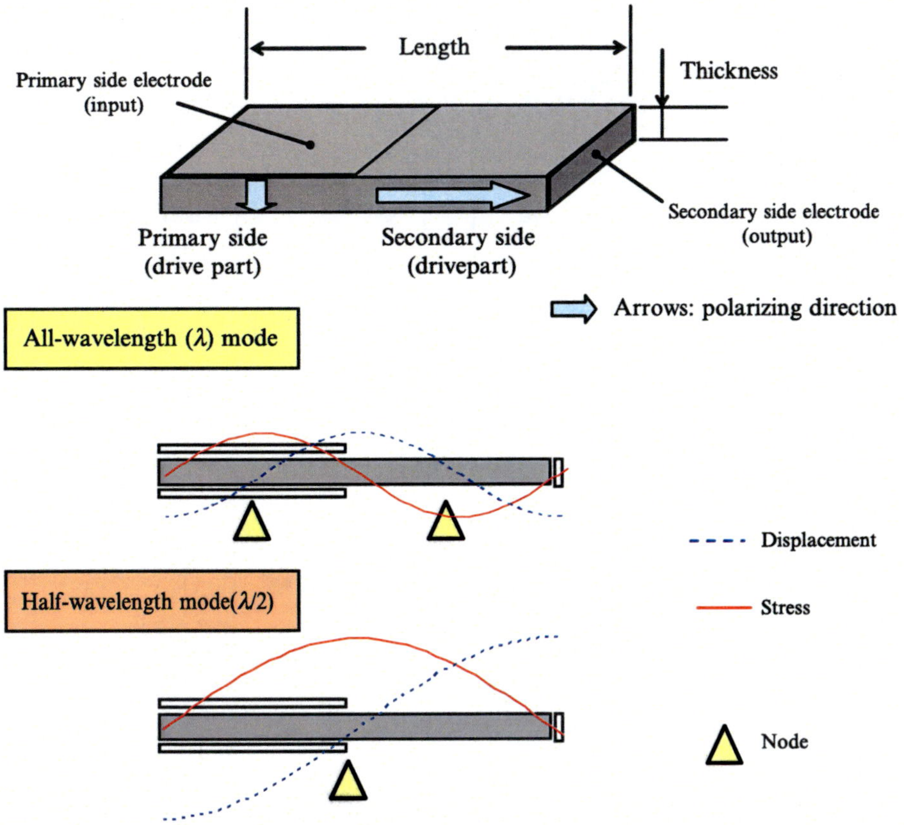

Fig. 9.37 ROSEN type piezoelectric transformer (needs to be rewritten). The types operated in full-wavelength and half-wavelength mode are called λ mode and λ/2 mode, respectively. The node (point where the maximum stress is observed), represents a change in the displacement. *Transformation ratio of γ with no load.* Without an output load and assuming there is no loss, if input/output voltage of piezoelectric transformer is expressed as V_1, V_2, respectively, and if capacitance on the primary side and the secondary side of the piezoelectric transformer is expressed as C_{01}, C_{02}, respectively, γ can be calculated from the equations below. $\frac{1}{2}C_{01}\cdot V_1^2 = \frac{1}{2}C_{02}\cdot V_2^2$ where $\gamma = \frac{V_2}{V_1} = \sqrt{\frac{C_{01}}{C_{02}}}$

Fig. 9.38 Change in characteristics of piezoelectric transformer due to load application. During actual illumination of cold-cathode tubes, frequency and voltage decrease. After the high-impedance tube is illuminated, the illuminating voltage (current) frequency changes in response to the load

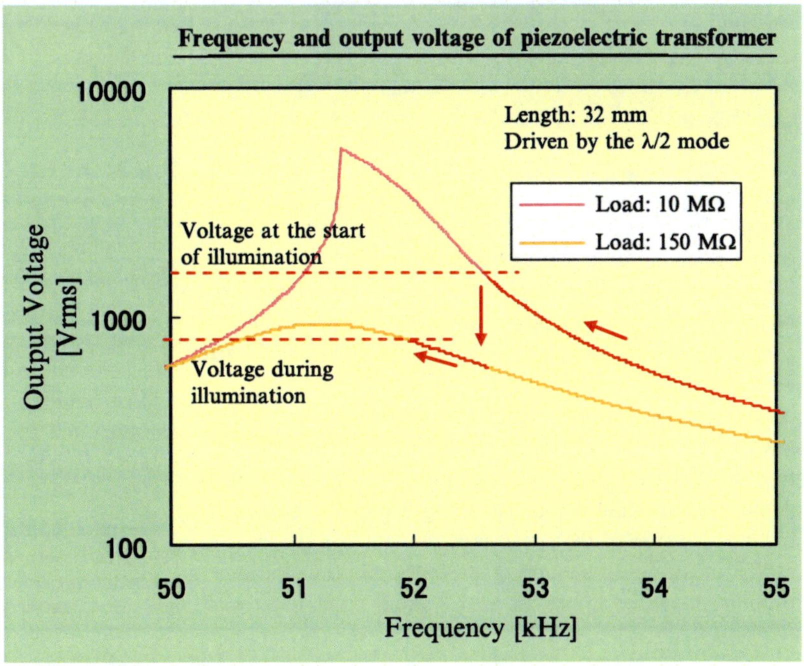

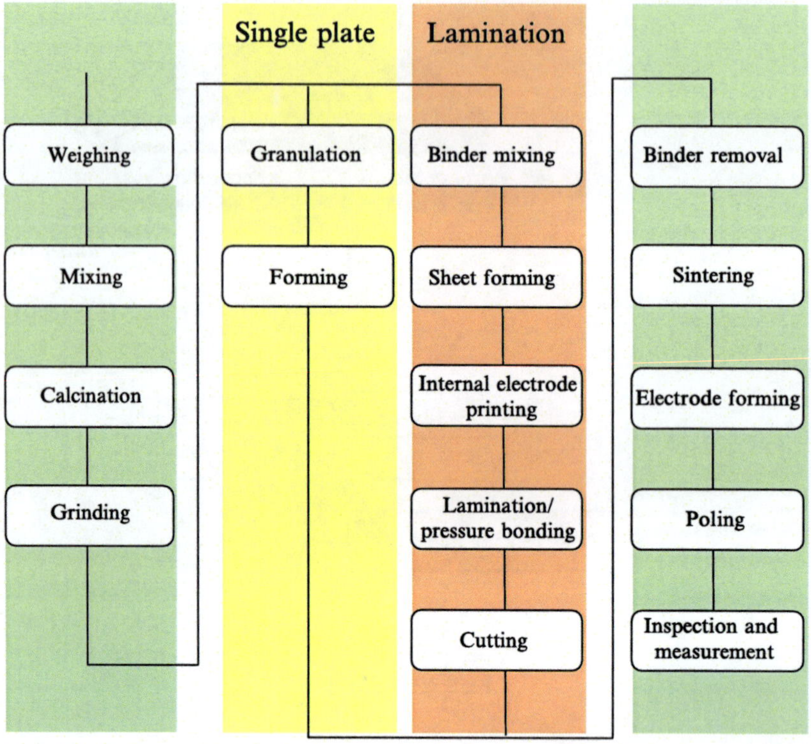

Fig. 9.39 Manufacturing process of piezoelectric transformer. Forming, calcination and electrode forming are performed in the same manner as for normal piezoelectric ceramics. The secondary side and the primary side are then poled to create piezoelectric transformers. An assembly inspection step is added if necessary

Table 9.7 Specifications of a single-plate piezoelectric transformer intended for powering a 14-in. LCD backlight

Size	L44 mm × W6.5 mm × T1.6 mm
Operating frequency	80–88 kHz
Output voltage (at the start of illumination)	–1,800 Vrms
Output current (during illumination)	–6.0 mA
Output power	–5 W
Boost-up ratio (during illumination)	–15 times

lead zirconate titanate-based ceramic. Manufacturing methods are classified into the methods for single-plate products and the methods for preparing laminated products (Fig. 9.39). When the internal electrodes and the ceramic are co-fired for multi-layered piezoelectric transformers; silver–palladium alloy, which features a high melting point and does not react with ceramics, is used to form the internal electrodes. If the ceramic material is sintered at a low temperature (950°C or lower), less expensive silver is used. The piezoelectric transformer is mounted directly on the substrate or covered with a flame-retardant case before mounting it on the substrate if required because of the power input–output wirings or as a safety measure. In mounting piezoelectric transformers, caution is required so that the area is strong enough to resist external impacts and operating vibration of the piezoelectric transformer itself. They need to be mounted so that the vibration is not impeded by wiring or other materials. For wiring, the incorporated structure resists vibration [tinsel wire (Note 9.24), conductive rubber, spring, lead wire reinforced by elastic body]. For retention, elastic bodies (silicone resin, etc.) are used to prevent the lowering of vibration efficiency and the transmission of vibration to other parts.

9.9.3 Future Prospects

Responding to the increase in the size of liquid crystal displays and the subsequent increase in output and energy-saving performance, the piezoelectric transformers featuring high efficiency, low magnetic noise, low height and high safety will continue to be widely used. Meanwhile, development will be required for materials and manufacturing processes that are more appropriate for high-power products and large panels which require illumination.

Note 9.24 Tinsel and Lead wires made by covering core wires made of resin with copper foils. They are resistant to vibration and used in products such as speakers.

Literature

1. Shinoda H, Kurimoto K (2006) J Jpn Soc Appl Phys 75:5–15 **(9.1)**
2. Material technologies for LCD, PDP and organic EL, CMC Publishing Co., Ltd. (2005) (Chapter 4.5) **(9.3)**
3. Maeda T (1994) J Jpn Soc Precis Eng 70:466–469 **(9.4)**
4. Tooley FV (1984) The handbook of glass manufacture, 3rd edn. Ashlee Pub. Co. **(9.4)**
5. Teitaro Hiraga et al (1986) Ferrite. Maruzen **(9.5)**
6. Yamane Y et al (ed) (1999) Glass engineering handbook. Asakura Publishing, p 16 **(9.6)**
7. Yamada H, Miyazawa E, Bessho K (1975) Basic magnetic engineering. Gakkensha Publishing, pp 102–122 **(9.7)**
8. Rosen CA (1956) Ceramic transformers and filters. In: Proceedings of electronic component symposium, p 205 **(9.9)**

10 Audio and Digital Information Storage

Major ceramic products that have fueled the development of audio devices are explained below. Before 1960, open reel tape deck recorders were widely used and they allowed consumers to easily enjoy high-quality music. The cassette tape recorder was introduced within Japan in the 1960s. The 1960s was the era of cassette decks. Later on, the audio market moved rapidly toward an era of diversification, where videos and PCs have also become important mediums following the miniaturization of devices as well as digitalization and the enhancement in the capacity of the recording mediums. The devices continue to evolve into mobile phones and portable music players in today's digitalized and ubiquitous computing society.

Until the 1980s, magnetic tapes (Sect. 10.1) and magnetic heads (Sect. 10.2) were widely used in audio reproduction and recording devices. The first magnetic head that was intended for tape recorders was released in the 1960s. The magnetic head is a key component for recording and reproducing sounds. The magnetic head converts electric signals from the audio circuit to magnetic signals for recording the media, while the magnetic signals from the media are converted to electrical signals for transmission to the audio circuit. The magnetic head incorporates ferrite as the core material without exception. Mn–Zn-based materials, which feature high permeability and magnetic flux, are mainly used for this purpose. Later on, in a response to the shift to optical disks, the heads have evolved into optical pickups.

Audio tapes, which were originally on open reels, utilized iron oxides such as γ hematite (γ-Fe_2O_3) and magnetite (Fe_3O_4) in magnetic powders. Later, chromium dioxide featuring high retention properties began to be used as the magnetic powder on the recording medium responding to the expansion of cassette tapes and the Hi-Fi audio boom. Magnetic tapes are being replaced by optical disks and the production of magnetic tapes has been on the decrease.

Hard disks (magnetic disks), developed by incorporating these audio technologies, are currently widely used in computers as recording medium and for storing large amounts of information. In a hard disk, alumina titanium carbide, featuring high hardness and toughness, is used in the magnetic head slider (Sect. 10.3) that writes the data. The magnetic head sliders are manufactured by using alumina titanium carbide as the substrate, and forming a magnetic circuit on the substrate by thin film processing while utilizing a precision processing technology. Optical disks are also attracting attention as one of the mediums for recording larger volumes of information. Optical disks for consumer audio applications have been commercialized and are being used as a recording medium for data, music and movies. Many types of optical disks, designed to match various types of standards, are widely used. The phase-change optical disks (Sect. 10.4), which are known as rewritable disks, incorporates a Ge–Sb–Te material, that is a type of ceramic recording material.

Y. Imanaka et al. (eds.), *The Ceramic Society of Japan, Advanced Ceramic Technologies & Products*, DOI 10.1007/978-4-431-54108-0_10, © Springer Japan 2012

Magnetic Tape (1950)* 10.1

Iron oxide was originally used as the magnetic powder used in audio tapes for audio recording. Following the rapid expansion of cassette tapes and the Hi-Fi audio boom later on, chromium dioxide, featuring higher coercive force, began to be used as a replacement for iron oxide. This was in response to demand for higher quality recording media. Because chromium dioxide was produced only overseas, cassette tapes that use cobalt modified iron oxide as the magnetic powder were introduced and they replaced chromium dioxide based tapes (including video tapes). For further improvement of recording properties, metal grain magnetic powder (metal powder) was developed. Magnetic tapes that were once extremely popular were gradually replaced by optical disks and other digital media. The production of magnetic powder continues to decrease. However, data tapes that have been improved through technological innovations for higher density continue to be used for the storage of computer data. And nonmagnetic iron oxide is used even today as the pigment for the bottom layer of the multilayered tapes.

10.1.1 Background of Development

Iron oxides such as gamma hematite (γ-Fe_2O_3) and magnetite (Fe_3O_4) iron oxides were used as the magnetic raw material for audio tapes, which were originally open-reel tapes. The compact cassette, which was developed in 1965 by Philips, expanded rapidly as the patent allowed it to be used for free. Following the Hi-Fi audio boom, higher quality recording media was required and chromium dioxide (CrO_2) featuring a higher coercive force, which was developed by DuPont of the U.S., began to be used as the magnetic powder. Because Japanese manufacturers could not produce chromium dioxide, cassette tapes that use cobalt modified iron oxide as the magnetic powder were introduced and replaced chromium dioxide tapes (including video tapes).

10.1.2 Manufacturing Method

10.1.2.1 Manufacturing Method of Magnetic Tapes

Figure 10.1 shows the manufacturing method of typical magnetic tapes. Binder resin, lubricant, abrasive and solvent are added to magnetic powder, which is milled and dispersed to produce a magnetic coating material. The material is coated thinly and evenly on the base film made of PET, and then the film undergoes magnetic field orientation while going through a drying furnace for evaporation of the solvent. The magnetic surface is then processed by the calendar machine (Note 10.1) for mirror-like finishing, creating rolls of wide magnetic tapes. The roll is cut according to a specified tape width and the tapes are wound to create cassette tapes.

10.1.2.2 Manufacturing Method of Magnetic Powder

Figure 10.2 shows the typical manufacturing method of iron oxide. Ferrous sulfate and sodium hydrate are used as raw materials for iron hydroxide, which is made to react under a certain pH and oxidation conditions so as to create needle-shaped α-FeOOH yellow crystals. The crystals are dried and reduced with hydrogen to create magnetite, which is oxidized gently in air to produce γ-Fe_2O_3. The characteristics of γ-Fe_2O_3 are affected by the α-FeOOH forming reaction and the

*The number in parenthesis indicates the year that the product was first commercialized.

Note 10.1 Equipment used to apply heat and pressure alternatively with metallic rolls and elastic resin rolls for crating mirror-like finishing of the magnetic tape surface.

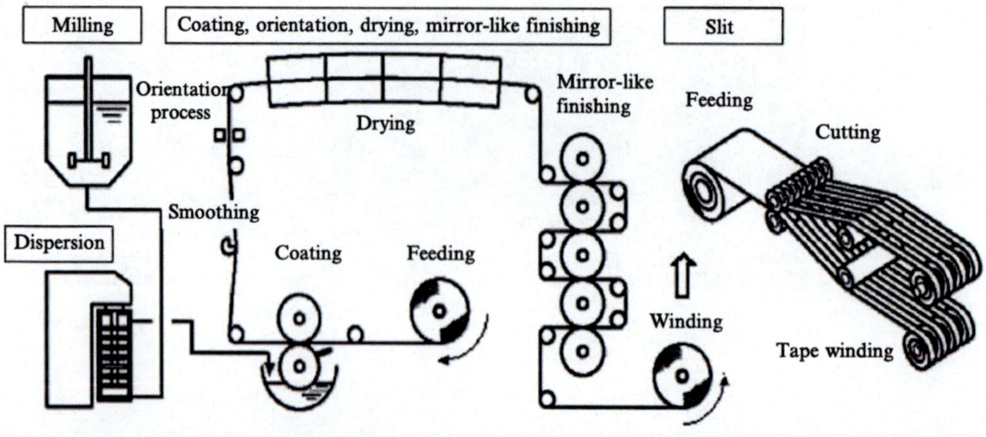

Fig. 10.1 Manufacturing method of magnetic tapes. Binder resin, lubricant, abrasive and solvent are added to magnetic powder, which is milled and dispersed to produce a magnetic coating material. The material is coated on the base film, and then the film undergoes magnetic field orientation while going through a drying furnace for evaporation of the solvent. Then it is processed for mirror-like finishing, creating rolls of wide magnetic tapes. The roll is cut according to the specified tape width

Fig. 10.2 Reaction process of iron oxide

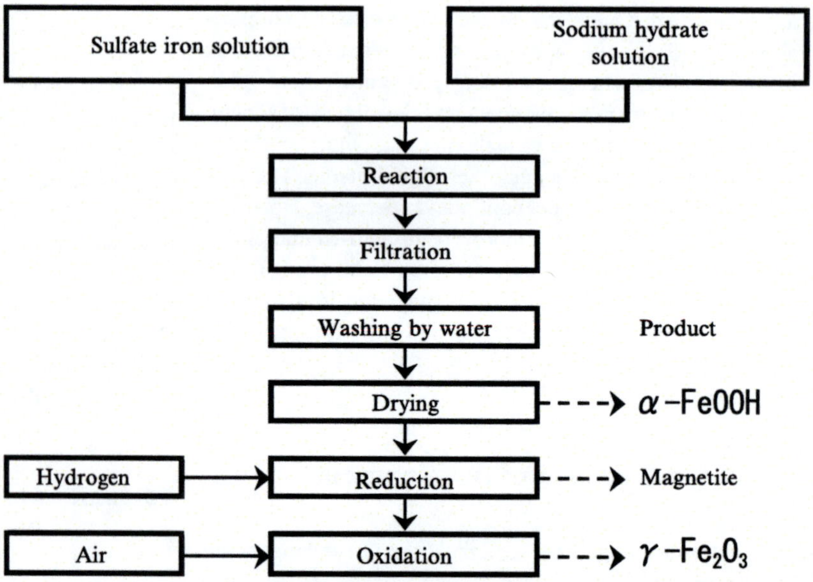

oxidizing condition to create γ-Fe_2O_3. The shape, size and particle size distribution are impacted by the reaction of α-FeOOH. The coercive force H_c is controlled during the oxidizing reaction.

However, enhancement of the coercive force was limited in iron oxide, even if the ratio of needle-shaped crystals was increased to improve shape anisotropy. The breakthrough was realized by utilizing magnetic powder of cobalt modified iron oxide, which was made by absorbing cobalt on the surface of needle-shaped iron oxide. Figure 10.3 shows an example of the production process. Iron oxide used as the core is treated with a cobalt salt reaction solution and dried to produce magnetic powder featuring higher coercive force. The magnetic property can be adjusted readily depending on the application, by the selection of the iron oxide and the adjustment of the cobalt absorption amount. The material, which can be produced at a lower cost and is safer when compared with chromium dioxide, features higher performance resulting in wider applications to video tapes and data tapes in addition to audio tapes.

Magnetic properties of a tape coated with iron oxide magnetic powder and a tape coated with cobalt modified iron oxide

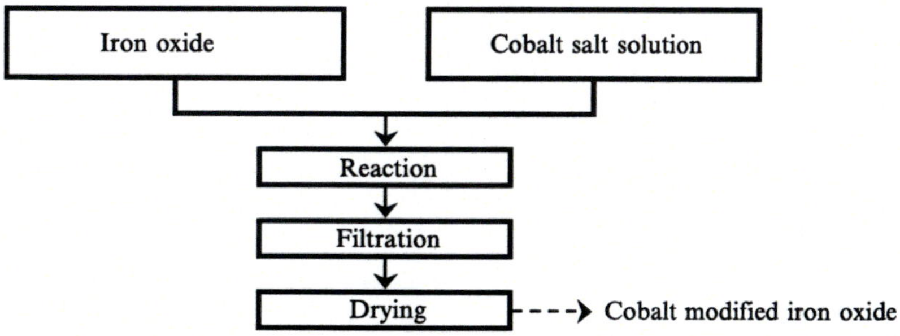

Fig. 10.3 Manufacturing method of cobalt modified iron oxide

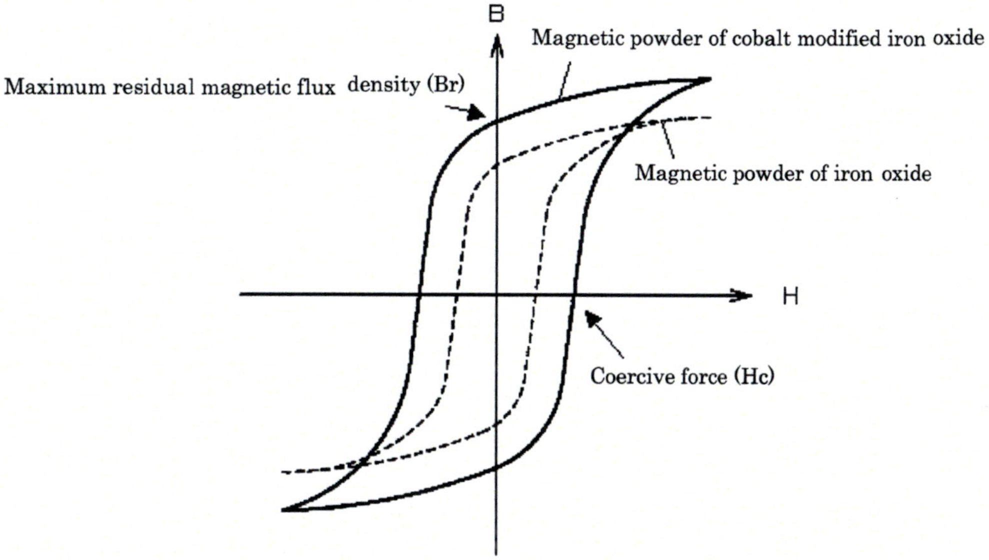

Fig. 10.4 Magnetic properties of tapes (*B–H* hysteresis loop). The tape coated with cobalt modified iron oxide has a larger loop than the tape coated with iron oxide which reduces background noise in audio tapes

magnetic powder (*B–H* curve) are compared in Fig. 10.4. The hysteresis loop shows that the tape coated with cobalt modified iron oxide is larger, meaning higher coercive force (H_c) and higher maximum residual magnetic flux density (B_r).

10.1.3 Future Prospects

For the further improvement of recording properties, magnetic powder was developed into metallic magnetic powder (metal powder). Magnetic tapes that were once extremely popular were gradually replaced by optical disks, and other silicon based memory devices. As a result production amounts have been decreasing year after year. However, data tapes that have been improved through technological innovations for higher density, including technologies for ultrafine metal magnetic powder, multilayer coating, ultrathin magnetic layers and tracking servo (Note 10.2), continue to be used for the storage of computer data. Finally, nonmagnetic iron oxide is used even today as the pigment for the bottom layer of the multilayered tapes.

Note 10.2 The number of tracks in the longitudinal direction of a tape increased to several hundred and the width of a track was reduced to μm order following improvements in capacity. The system that positions the head in the tracks on the tape utilizes pre-recorded magnetic or optical pattern signals so as to trace the narrow tracks reproducibly.

Magnetic Head (Ferrite) (1970) 10.2

The history of the magnetic recording began in 1898, when V. Poulsen of Denmark obtained a patent on the telegraphone. The magnetic head is made by bending an electrical magnet to create a gap as shown in Fig. 10.5. The magnetic flux generated in the magnetic head core has a leakage flux due to the gap. The leakage flux can be used to record, regenerate and erase signals (sound, image, data) by magnetizing the magnetic media while passing through the magnetic media. The magnetic head technology has advanced for use in consumer and industrial devices.

Permalloy, ferrite, sendust, amorphous are examples of materials used as magnetic heads. The following devices incorporate magnetic heads made of ferrite:

- HDD magnetic head ⇒ Computers and external memory units
- FDD magnetic head ⇒ Computers and external memory units (Fig. 10.6)
- VTR (image) magnetic head ⇒ VTR/audio (sound) magnetic head ⇒ Tape recorders
- Eraser head ⇒ VTR/tape recorder
- Magnetic head for cards, magnetic (streamer) head for magnetic tapes ⇒ ATM, automatic ticket gate, data recorder

Magnetic heads are explained below with a focus on floppy disk drive (FDD) magnetic heads made of ferrite.

10.2.1 Background of Development

Floppy disk (FD) magnetic heads were developed in line with the advancement of FDD technologies (Table 10.1). They were first incorporated in the large computer system 3740 by IBM in 1970. Eight-inch FDDs were used as the initial program loader and the failure diagnosis program loader in the computer system. On this occasion, hard disk drives (HDDs) that allow for the replacement of media were used, and the technologies were introduced to FDDs. The tunnel erase head is one of these types. The tunnel erase method is effective for solving the off-track problems arising from the replacement of media, which is proved by the fact that it is incorporated in the majority of current FDDs.

The 8-in. FDD, which was introduced first, was used widely as the auxiliary recording device in line with the development of small computers. Later, smaller 5.25-in. mini floppy, released in 1976 by Shugart, expanded its application in compact desktop systems such as word processors and PCs, creating a huge industry of computer peripheral devices. They developed into high-density recording media following the improvement in capacity from the 100 kB level to 1 MB. The 5.25-in. size was created by downsizing the 8-in. type, responding to the demands for small types. In 1980, SONY released the 3.5-in. micro floppy disk system featuring higher reliability. The product, responding to the needs at the time, expanded explosively and was adopted as the world standard. The 5.25-in. type was replaced by the 3.5-in. type.

The recording density of FDD, as well as other recording devices, has been improved responding to the needs for scaling down and higher capacity. 3.5-in. flexible magnetic FDDs featuring a large capacity of 100–750 MB have been commercialized.

FDDs have evolved through the improvements explained above. However, demand for FDD drives has been declining following the enlargement in HDD drive capacity and the expansion of CD-R with inexpensive recording media and compact, easy-to-handle semiconductor memory .

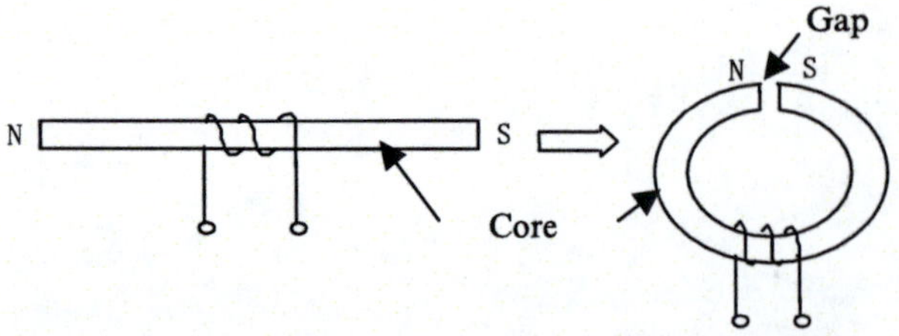

Fig. 10.5 Conceptual diagram of a magnetic head. *Left*: when a current flows through the coil wound around the bar-shaped core, the core becomes a magnet. *Right*: the bar-shaped core is circular, leaving a gap between the two ends. This is close to the basic structure of a magnetic head. When a recording media (magnetic disk, etc.) contacts the gap (*top of right-side* illustration) leaving a space smaller than the gap, the magnetic flux leaks out of the gap on one side and passes through the magnetic material of the recording media and returns to the gap on the opposite side. The recording media is magnetized (written) during this process

Table 10.1 History of FDD technology development

	1970		1976	1980
Developer	IBM		Shugart	SONY
Market demand	Intended for program/loader		Downsizing	Improvement in reliability
Size of recording media	8-in.		5.25-in.	3.5-in.
Disk diameter	200 mm		130 mm	90 mm
Coercive force	3,000e		3,000e	7,200e
Recording capacity	400 kB	800 kB	1 MB	2 MB
Track density	48TPI		96TPI	135TPI
Core type	Laminate type		Straddle type	Bulk type

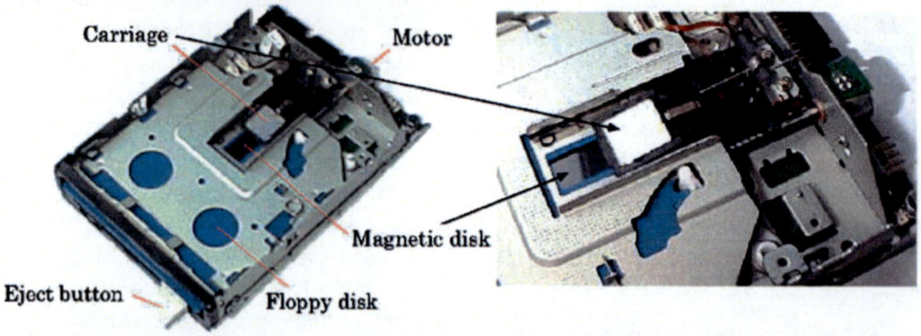

Fig. 10.6 An external memory unit (drive) which incorporates the FDD magnetic head. *Right*: the outer chassis was removed to see the entire system (drive). *Left*: the head mounting portion was enlarged. When a magnetic disk is inserted in the drive, the shutter for the magnetic disk opens. This exposes part of the magnetic (floppy) disk. The FDD magnetic head is mounted on the back of the white carriage near the magnetic disk in a manner that allows for the contact of the head and the magnetic disk

10.2.2 Characteristics

10.2.2.1 Characteristics of Products

Magnetic materials used as core materials of magnetic heads are required to feature the following characteristics:

- High saturation magnetic flux density (Note 10.3)
- High permeability (Note 10.4)
- Low coercive force
- High abrasion resistance and others

Metal permalloy was once used for magnetic heads of 8-in. and 5.25-in. FDDs, but ferrite has been used as core materials for advanced FDDs. The Mn–Zn-based ferrite and the Ni–Zn-based ferrite are mainly used. The Mn–Zn-based ferrite is made by sintering the base material, Fe_2O_3 mixed with MnO and ZnO, while the Ni–Zn-based ferrite is made by sintering the base material, Fe_2O_3 mixed with NiO and ZnO.

Figure 10.7 shows the difference in magnetic properties between the Mn–Zn-based ferrite and the Ni–Zn-based ferrite. The permeability μ in the low frequency range is higher in the Mn–Zn-based ferrite (1), while the permeability μ in the high frequency range is higher in the Ni–Zn-based ferrite (2). Due to this reason, the Ni–Zn ferrite was initially used as the core material of FD magnetic heads. Mn–Zn-based ferrite, featuring high permeability and magnetic flux density, has been in the mainstream since the improvements were

Note 10.3 The degree of magnetization is called the magnetic flux density. Saturation magnetic flux density is the index of magnetic flux density and expresses the performance of magnetic materials. The limit of magnetization is called the saturation magnetic flux density. The larger the number, the more suitable it is downsizing of the core and high-density recording.

Note 10.4 Permeability is the proportional constant μ, which is expressed as $B=\mu H$, where H is the strength of the magnetic field and B is the magnetic flux density. Magnetic flux density increases in proportion to permeability. If the permeability increases, the magnetic flux that passes through a substance increases, enhancing the strength as a magnet (improving the efficiency of the head).

made (3). Some of the materials are made by hot pressing during the sintering process so as to realize higher density and magnetic properties.

The FDD magnetic head is characterized by the configuration, which consists of the read/write head and the erase head as shown in (1), and the system that erases both sides of the data recorded by the read/write head as a measure against the displacement of tracks caused by replacement of the media. The system was incorporated to guarantee compatibility, which is essential for FDD and other recording devices that use removable media.

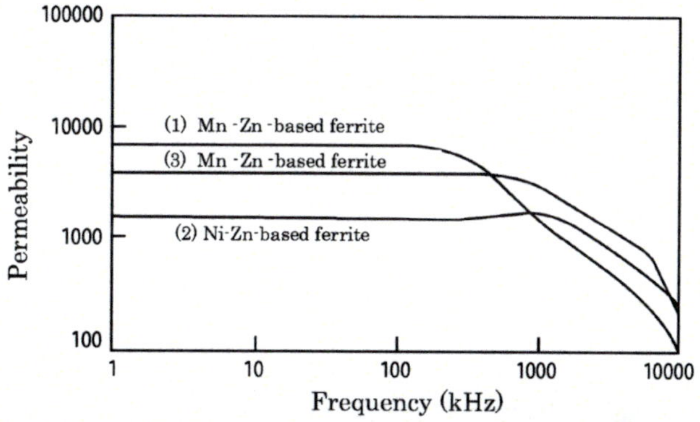

Fig. 10.7 Permeability (μ) and frequency of Mn–Zn-based ferrite and Ni–Zn-based ferrite. The Mn–Zn-based ferrite (*1*) and the Ni–Zn-based ferrite (*2*) are used as core materials of FDD magnetic heads. Initially in terms of magnetic properties, Ni–Zn-based ferrite was used (featuring high frequency properties) as opposed to Mn–Zn-based ferrite (*1*) that has higher permeability. Later, following the development of Mn–Zn-based ferrite exhibiting a frequency response equivalent to that of Ni–Zn-based ferrite, the Mn–Zn-based ferrite of type (*3*) was widely adopted

The old data that remains un-erased due to off-track during overwriting becomes noise during the time of reproduction, lowering the error rate. Therefore, the old data is erased. Both ends of the data written by the read/write head is erased by the erase head, and at the same time, the old data that remains due to off-track is erased at the time of overwriting. This head mechanism is called the tunnel erase system (Fig. 10.8).

The advanced erase system features high overwriting performance, realizing recording of higher density (short wavelength). Developments to realize even higher density magnetic recording continue. In order to achieve the goal, media with a high coercive force is needed as well as a magnetic head featuring a strong magnetic field, for writing on the media. Responding to the needs, the metal-in-gap (MIG) head was developed. It has a metallic film of high magnetic flux density near the magnetic head that causes saturation. This is because the recording magnetic field is determined by saturation magnetic flux density.

The mechanism of the entire magnetic head is explained below. The head core is bonded with a part called the slider, which is made mainly of calcium titanate. This is shown in Fig. 10.9. The slider enables stable contact between the head core and the media and also protects the head core. The head is chamfered to prevent scratching of the disk at the time of its contact with the disk. Head cores are generally made of polycrystalline ferrite. The bulk type is made by processing polycrystalline ferrite. The read/write head and the erase head are wound, respectively, with coils for electromagnetic conversion and erasing magnetic flux generation (Fig. 10.10). The disk is designed for two-sided recording, and a pair of two heads is located on both sides of the media, contacting the disk at an appropriate pressure (Fig. 10.11). The head below the disk is called the side 0 head and the head above the disk is called the side 1 disk.

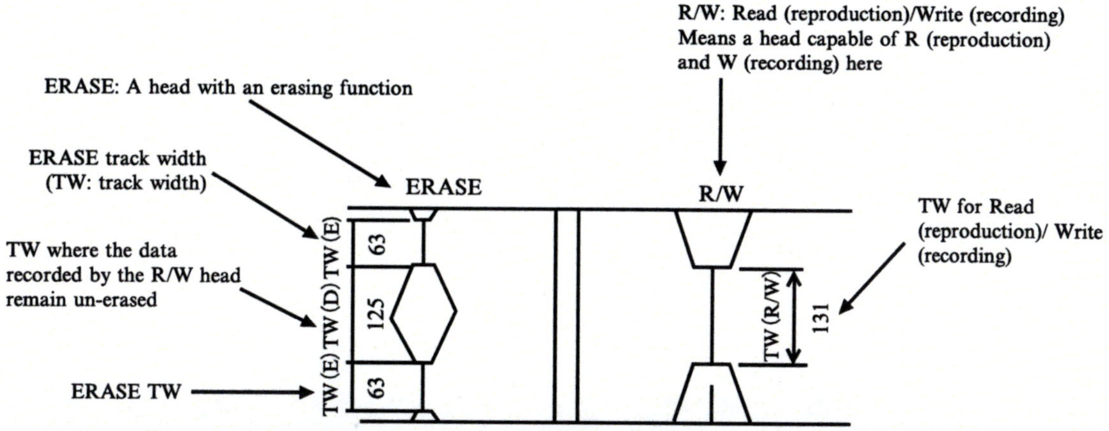

Fig. 10.8 Explanation of the tunnel erase type FDD magnetic head. The data recorded in the recording TW (R/W) 131 μm range remains un-erased in the 125 μm range after passing of the erase head, allowing necessary data to be precisely kept (due to tracking positions)

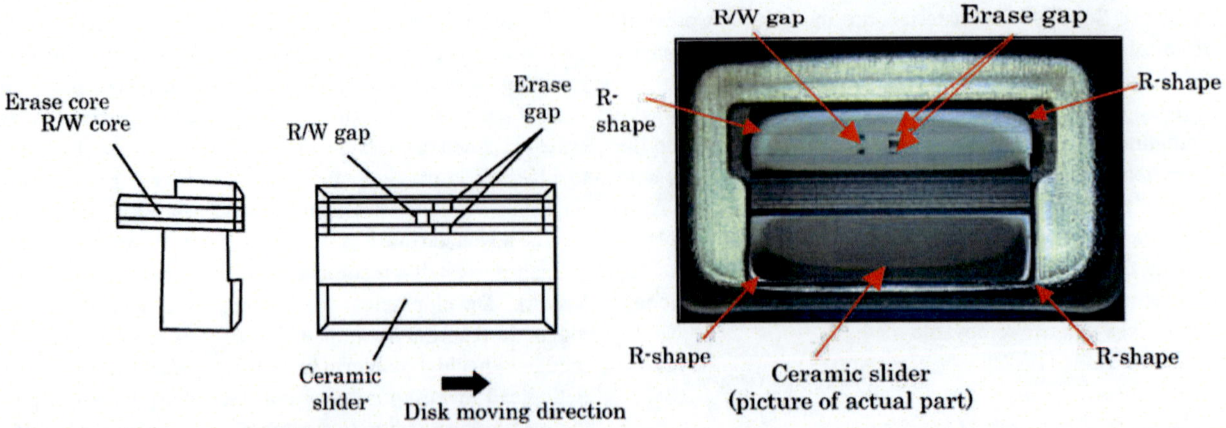

Fig. 10.9 Structure of head. The R/W core and the erase core, respectively, and independently form magnetic circuits. The cores that contact the disk are completed by bonding them on the slider and finishing. The head is R-chamfered to prevent scratching of the disk during contact with the disk

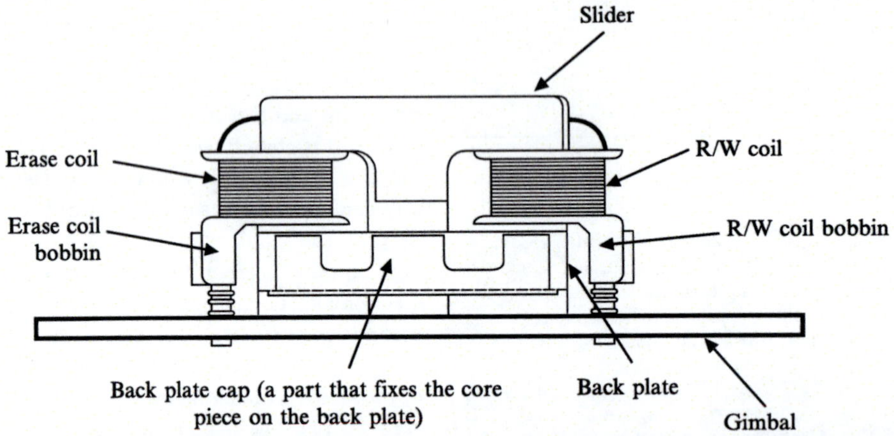

Fig. 10.10 Schematic after assembly of coils

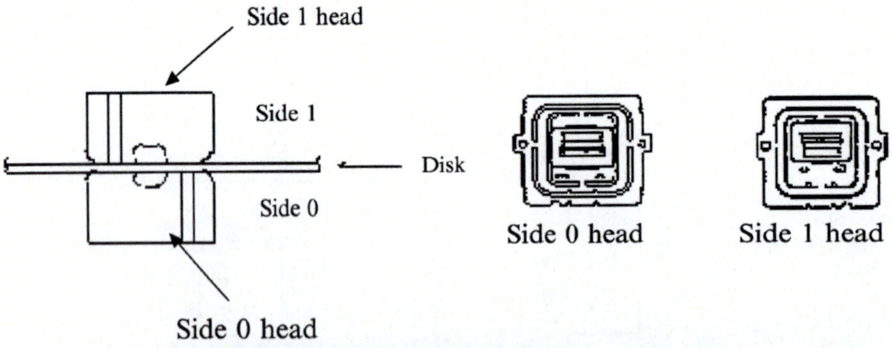

Fig. 10.11 Side 0 head and side 1 head. The side 0 head and the side 1 head are, respectively, located above and below the floppy disk (recording media) inside the floppy disk drive. When a disk is inserted in the drive, the disk shutter opens (part of the disk is exposed) and the heads contact the disk from above and below the disk as shown in Fig. 10.11. Generally, the side 0 head (fixed) does not have a function that absorbs the thickness and warping of the flexible disk. Absorption is performed by the gimbal mechanism (*springy*) of the side 1 head

The side 0 disk is generally fixed, while the side 1 head is designed to move flexibly responding to the flexible disk. The FDD magnetic head is required to feature high electromagnetic conversion properties, high resistance to external noise, low torque for prevention of damage to the disk and high collision resistance.

10.2.2.2 Manufacturing Method

Figure 10.12 shows the basic manufacturing process and Fig. 10.13 shows completed FDD magnetic heads.

Manufacturing process: gaps for read/write and erase are formed on the ferrite, the core material, by glass welding (welding process). The glass welded core block is processed to the intended thickness (Piece process). A slider is bonded on the core piece, which is processed to the intended size (semi-assembly).

Assembly and inspection process: the completed core slider is assembled with a coil, back plate, a gimbal (Note 10.5), FPC (Note 10.6), which is then delivered as a product after inspection.

10.2.2.3 Performance and Specifications of Products

Conventional 3.5-in. FDDs have a thickness of 1 in., while those intended for note PCs and external FDDs are 1/2 in. in thickness. Previously, desktop PCs incorporated the

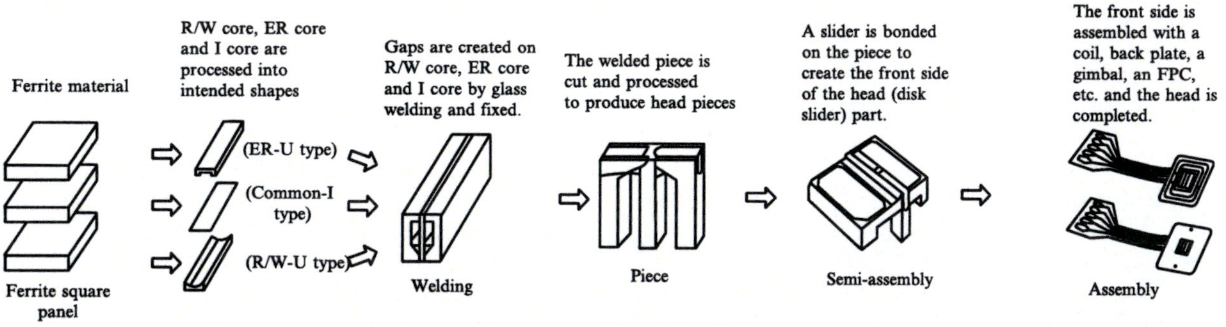

Fig. 10.12 Manufacturing method of the FDD magnetic head

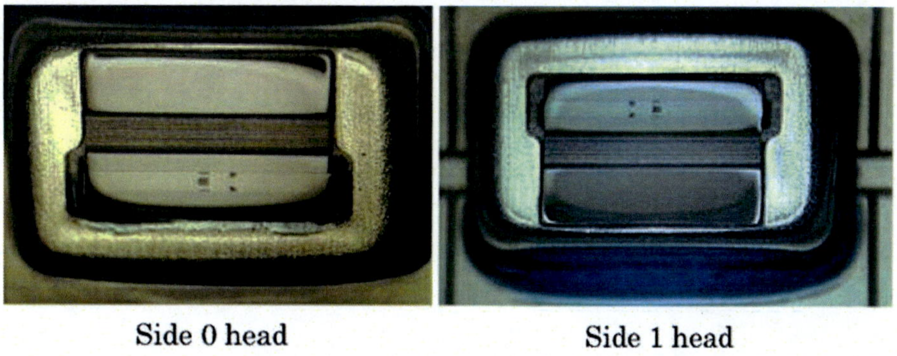

Fig. 10.13 Pictures of completed FDD magnetic heads

Note 10.5 The gimbal is a springy supporting part for the magnetic head. It is used to maintain proper contact between the magnetic head and the magnetic media (by reducing influence from thickness or warping of the magnetic media).

Note 10.6 Stands for Flexible Printed Circuit. As the name indicates, it is a flexible printed circuit.

thin-type FDD. The thin-type head was widely used. High impact resistance is required of the 1/2 in. FDD because of its applications. Therefore, the chassis (slider) of the head that supports the core chip is designed to ensure impact resistance. In addition, bonding materials featuring high bonding strength and reliability have been developed so as to ensure a flatness equivalent to glass bonding for the head sliding surfaces, realizing both high quality and a low cost. The major specifications, shape and dimensions of a magnetic head intended for 3.5-in. FDs used in the 1-in. FDD systems are listed below (Fig. 10.14) in Table 10.2.

Table 10.2 Characteristics of FDD Systems

Characteristics	Specifications
Core material	Mn–Zn-based ferrite
Head structure	Bulk type
Erase gap configuration	Tunnel erase system
Slider material	Calcium titanate ($CaTiO_3$)
Recording capacity	2 MB
Major specifications	
Write track width (mm)	0.131
Erase track width (mm)	0.071
Effective track width (mm)	0.117
Distance between gaps (mm)	0.35
Recording/reproduction coil turns (Ts)	160/160 max.
Erase coil turns (Ts)	120
Major characteristics (2 MB mode)	
Output	1 mvp-p or higher
Resolution	70 % or higher
Self-overwrite	–28 dB or less
Compatible overwrite	–21 dB or less
Asymmetry property of waveforms	200 ns or less

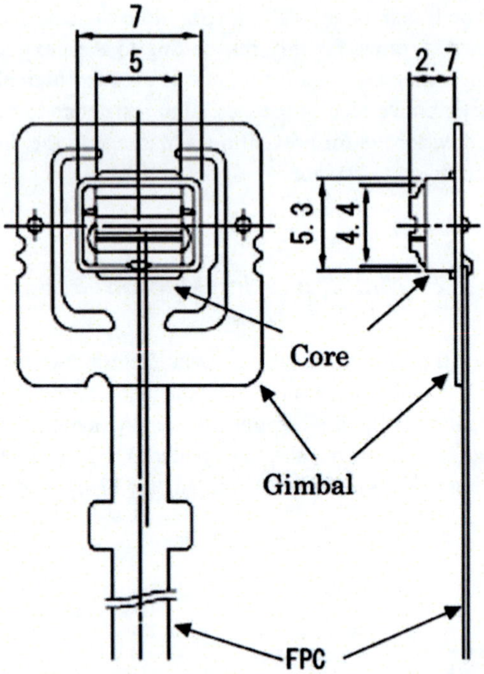

Fig. 10.14 Shape and dimensions of the completed product

Ceramic Materials for Thin Film Magnetic Head Slider (1978)

10.3

Alumina titanium carbide (ALTIC) introduced in this section is used as the thin film magnetic head slider for reading and writing in the hard disk drive (HDD), which is the major storage device of the information age. The thin film magnetic head (Note 10.7) is made by forming magnetic circuitry using semiconductor process technology on the ALTIC substrate (wafer) (Fig. 10.15). The substrate is processed further into sliders. The ALTIC substrate is required to have high density, mirror-like finishing and high dimensional accuracy. For the processing of sliders, precision processing including dry surface processing with accurate dimensional control and high reliability are required, in addition to machine processing. Responding to these requirements, the pressure sintering process was utilitized for preparing the substrate and each of the processes from the raw material to the final product is strictly controlled to maintain consistent quality.

10.3.1 Background of Development

Capacity improvement and the miniaturization of external memory devices have been pursued in the computer industry. Bulk materials such as permalloy and ferrite oxide were developed and mass produced for use as magnetic head materials. IBM developed its "Winchester Technology," an air-lift recording device incorporating a magnetic head. After the release of the 3370 type in 1978, the thin film magnetic head (Fig. 10.16) made by forming a magnetic circuit on the substrate using semiconductor manufacturing processes was used on the Al_2O_3–TiC material (ALTIC material) as the ceramic substrate.

ALTIC is made by dispersing TiC in Al_2O_3 to improve the hardness and toughness of Al_2O_3. The material was first introduced to the market by Nippon Tungsten in 1965 in the form of abrasion-resistant ceramic cutting tools, which were manufactured by utilizing hot pressing technology (Note 10.8), a high-temperature, high-pressure sintering technology. The material was widely used by the automobile industry in the U.S. and IBM selected the material as a target for evaluation. Later, the ALTIC material has been used as the standard material for substrates of thin film magnetic heads. During this period, responding to the rapid improvement of recording density and miniaturization, the ALTIC material continued to be improved to satisfy stricter specifications.

10.3.2 Characteristics

1. Characteristics and specifications of products
   Machine tools are required to have high abrasion resistance, while other properties are required of thin film magnetic head substrates. According to sources, IBM evaluated

Note 10.7 The magnetic head is used to read and write magnetic data to a hard disk. The magnetic circuit is manufactured by the semiconductor production process, while the slider is manufactured by ultraprecise processing. They need to be high in reliability and require advanced technology.

Note 10.8 The powder is placed in a die made of carbon, etc. and is sintered with an appropriate pressure applied in a uni-axial direction to ensure densification.

many items using a large number of materials. As Fig. 10.16 indicates, the ALTIC material needs to satisfy the specifications required of substrates that constitute magnetic circuits along with characteristics required for the interface with the recording media (disks). The ALTIC ceramic was selected because it satisfies all of the required specifications. Figure 10.17 shows the microstructure of Al_2O_3–TiC ceramic. TiC (gray) is dispersed in the Al_2O_3 matrix (black). Characteristics of the ALTIC-based magnetic head substrate are listed in Table 10.3.

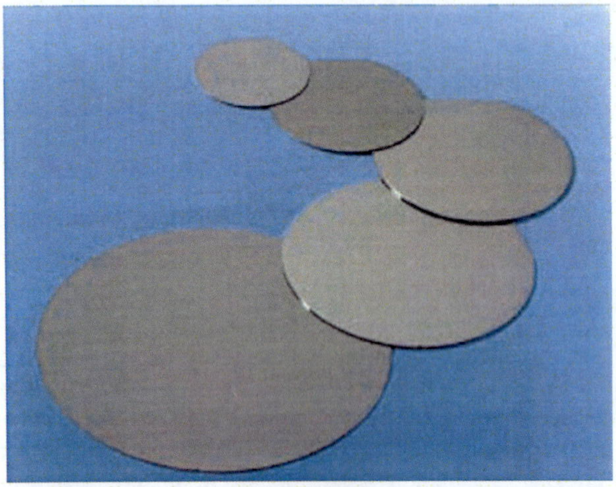

Fig. 10.15 Substrates for thin film magnetic head. The diameter was initially 2″, but larger diameters 4″ – 6″ are common today. Meanwhile, the thickness has been reduced from 4.0⇒2.8⇒2.0⇒1.2 mm, responding to miniaturization of magnetic heads

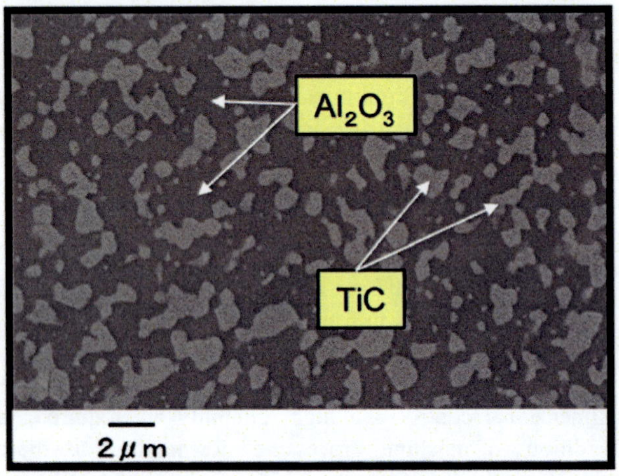

Fig 10.17 Microstructure of ALTIC. TiC (*gray*) is dispersed in the Al_2O_3 matrix (*black*)

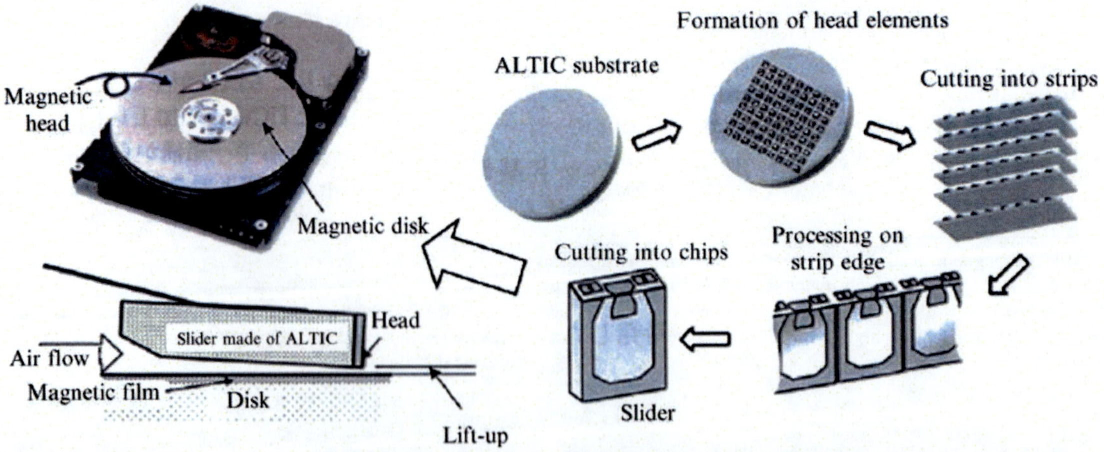

Fig. 10.16 ALTIC slider. The ALTIC substrate is also called the slider because it is a substrate on which the magnetic circuit is formed, and at the same time, it is a part that slides across the surface of a magnetic disk

Table 10.3 Mechanical and physical properties of ALTIC material

Characteristics	Unit	Al_2O_3–TiC (AC-72)	Al_2O_3
Density	g/cm^3	4.30	3.97
Specific resistance	mΩ cm	2.0	>10^{16}
Hardness	Hv	2,000	1,900
Young's modulus	GPa	390	400
Bending strength	MPa	880	800
Fracture toughness	MPa m$^{1/2}$	3.8	3.0
Thermal conductivity	W/(m K)	24	29
Thermal expansion coefficient	ppm/K	7.6	7.6
Crystal grain diameter	μm	1.0	5.0
Optical constant	n	2.24	Measured at
	k	0.45	λ=5,461 Å

(a) Requirements for substrates intended for magnetic circuit formation

- Ease of processing the substrate (mirror-like finishing, chipping resistance)
  The magnetic circuit formation is a thin film process and the surface is processed to a mirror-like finish as is the substrate. It is important that a mirror-like finish without scratches is achieved. After circuit formation, it is cut by a diamond grinder to create separate heads. The workability in cutting the substrate largely affects productivity. Chipping during cutting needs to be avoided because residual particles may affect the reliability of HDDs. Processability and chipping resistance of Al_2O_3 are substantially improved by means of dispersing TiC.
- Reduction of residual pores
  There is a need to reduce post-sintering residual pores (Note 10.9) because they may be a site that holds particles. The sinterability of the material is lowered by the presence of TiC, but residual pores are reduced by controlling the composition and utilizing the pressure sintering method.
- Reduction of internal stress
  After forming the head and the magnetic circuit on the substrate, the substrate is cut into strips (Note 10.10), as shown in Fig. 10.16. If the substrate has residual stress at the time of cutting, the rows are warped and accuracy is not achieved even after the height adjustment of magnetic poles following the cutting process. Internal stress must be reduced by process control and use of the reduction process.

(b) Requirements for head–disk interfaces

- Superior tribology of recording media
  The magnetic head lifts up as the recording disk starts to rotate. It contacts the disk when the revolution stops. The tribology characteristics at both stopping and starting positions were evaluated to see the influence on the recording media on the disk surface. Tribology was confirmed to show superior characteristics. Low lifting distance of 10 nm or less has been realized through ultra-precise processing of the slider surface, shape control and improvement in the technologies for the protective film and lubrication layer on the disk and slider surfaces.
- Superior workability in ion processing
  To increase the recording density, lowering of the lift-up height is effective. The technology has been developed actively. In line with the reduction in height of the lift-up height, the influence of difference in peripheral speeds (Note 10.11) of parts inside and outside the magnetic disk became non-negligible. With respect to the slider surface that faces the disk as shown in Fig. 10.16, the issue was solved by changing the surface to an aerodynamic shape by reactive etching, which is used in semiconductor processing, instead of conventional machine processing using grinders and abrasive grains. The processing speed and texture of the processed surface are important for

Note 10.9 A residual pore is a space or gap that remains between grains after sintering. Pores form in solid-phase sintering that does not use a liquid phase.

Note 10.10 It is cut into strips after forming the head elements, as shown in Fig. 10.16.

Note 10.11 As shown in Fig. 10.16, the slider surface is designed to prevent uneven lift-up heights of the slider, caused by differences in revolution speeds of the disk (difference in circumferential velocity) between the inner circumference and the outer circumference of the disk. The "magnetic disk" is shown in Fig. 10.16.

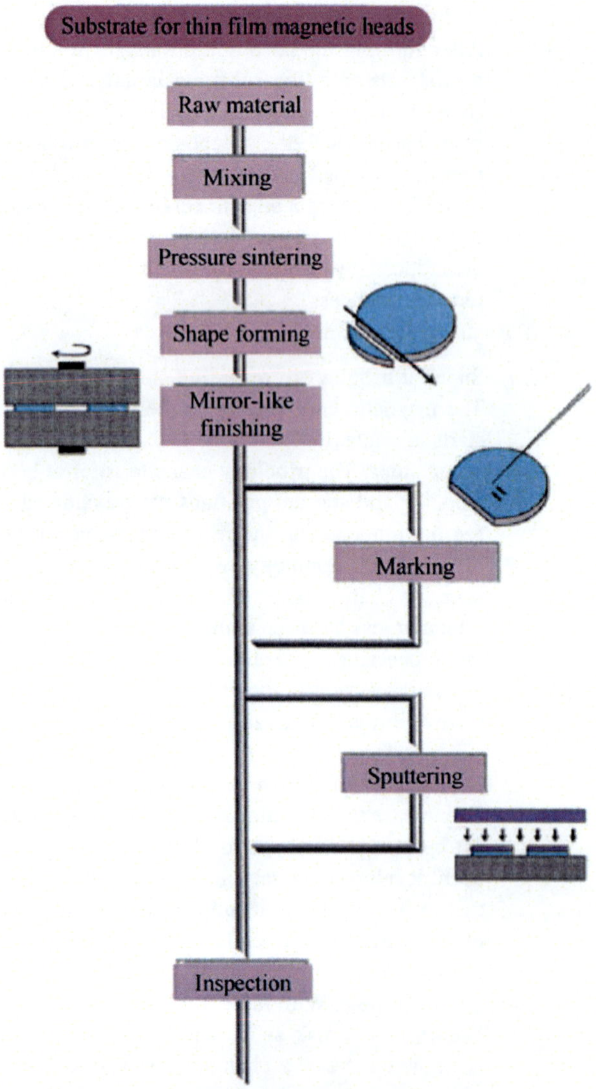

Fig. 10.18 Manufacturing process of substrate for thin film magnetic head. The ALTIC substrate is densely processed by pressure sintering and is processed in to a flat and smooth block. The substrate is marked with indications of materials and processing flow. An alumina film is formed on the substrate surface if needed

the dry process. These issues are solved by controlling the texture and composition of the ALTIC material.

2. Manufacturing method
   Carefully selected raw material powders are mixed to a specified composition, which is densely sintered by pressure sintering (Note 10.12). The sintered body is processed to a flat and smooth block to produce the magnetic head substrate (Fig. 10.18). Each of these processes needs to be strictly controlled so as to maintain stable quality. In particular, the powder mixing condition and the pressure sintering condition need to be controlled appropriately and precisely in order to create magnetic head substrates featuring homogeneous density and composition with limited residual stress.

10.3.3 Future Prospects

HDDs, which have been miniaturized and enhanced in terms of capacity by means of the increase in surface recording density explained in Fig. 10.19, are used not only as conventional recording devices of computers but also in consumer information devices such as video recorders, mobile music devices, car navigation systems and digital cameras. This has taken place even after the introduction of other recording devices such as flash memory.

The ceramic material (ALTIC) used in sliders for the thin film magnetic head will continue to improve in terms of reliability, responding to the design requirements of microfabrication sliders utilizing nano-technology, including downsizing and low lift-up height as well as counter measures to reduce particles.

Note 10.12 In this method, the powder is sintered by applying an appropriate pressure for the promotion of densification. It is usually done using a hot pressing method (HP) or the hot isostatic pressing method (HIP).

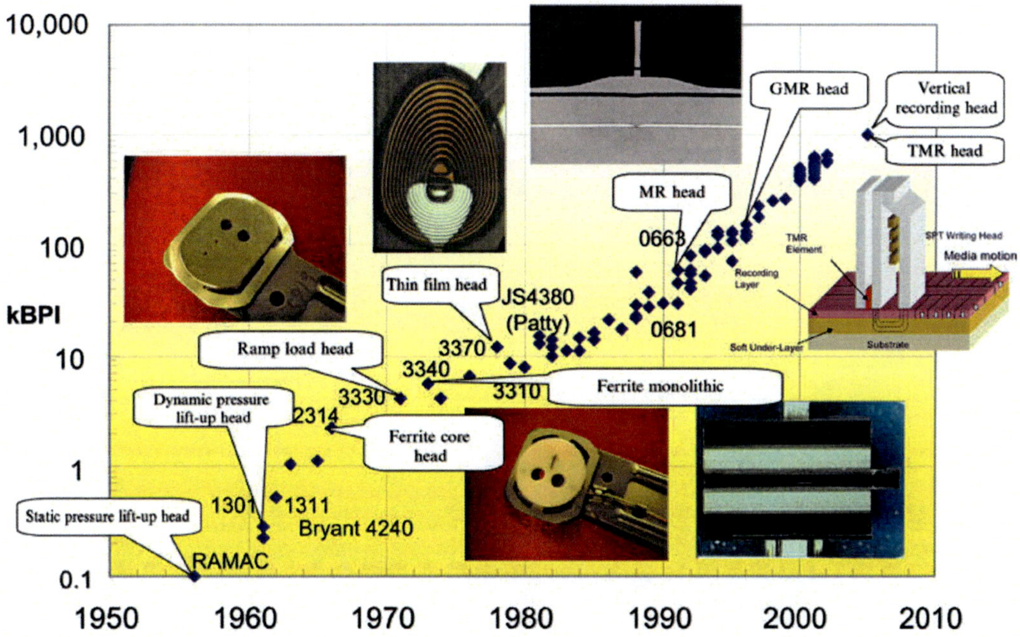

Fig. 10.19 Advancement of HDD magnetic head technologies and evolution of surface recording density. The relationship between advancement of HDD magnetic head technologies and surface recording density (*vertical axis*) versus year is shown. The ALTIC material was first used in the 3370 type thin film magnetic head released in 1978 by IBM [reference: IDEMA JAPAN, Exhibition Commemorating the 50th Anniversary of Magnetic Disk, June 2006 (approved by IDEMA JAPAN)]

Phase-Change Rewritable Optical Disk (1977)

10.4

The phase-change rewritable optical disk is one type of optical disk that allows for users to write and rewrite it in order to store information, while music CDs and movie DVDs are used to record existing contents in large amounts at factories for distribution. There are a variety of rewritable recording media on the market. As for optical disks, information is recorded utilizing the difference in optical properties between the two states (called phases) which are used to store data.

10.4.1 Characteristics

10.4.1.1 Phase-Change Type

Writing and erasing is performed by irradiating the media with a semiconductor laser beam, which is reduced to a micron in size by a lens, on a plastic disc. The cross sectional view of the disk is shown in Fig. 10.20. The writing and erasing cycles are shown in Fig. 10.21. The first phase is the "amorphous phase," which is created by increasing the temperature to the level above the melting point and then by cooling it quickly. The second phase is the "crystal phase," which is created by heating the amorphous phase by a laser to a high temperature below its melting point and then by cooling it slowly. The information is reproduced by scanning the recording material with a laser beam, which is weak enough that the temperature increase is negligible, retrieving the difference in reflectance of the two phases as signals. The name "phase change" comes from the action of the precisely power-controlled laser beam, which moves back and forth between the two phases. Figure 10.22 is a SEM image of the phases after recording. It shows that the two phases are in different states.

10.4.1.2 Recording Material

Atoms are three-dimensionally arranged neatly in a periodic pattern in the crystalline phase. The amorphous phase is a solid phase like the crystal phase, but the periodic pattern no longer exists and the atoms do not have a short-range order. Groups of materials for switching between the two phases by changing the patterns of heat application using a laser were developed by Matsushita Electric (now called Panasonic) and other companies in Japan. The Ge–Sb–Te ternary material is the most popular. A rewritable DVD with a switching speed of several tens of nano seconds is realized when utilizing this technique. The phase-change optical disk was first released in 1990. Since that time, products of several disk formats have been commercialized, as shown in Table 10.4. PD, CD±RW, DVDRAM, DVD±RW are examples. The phase-change system was later incorporated in optical disk devices such as the BD (Blu-ray disk) and HD-DVD. Ge–Sb–Te or equivalent materials are used as the recording material. The difference in composition affects rewritable frequency and rewriting time. The improvement in the recording density of optical disks is attributable to the shortening of the wavelength of the semiconductor laser used in optical recording. This is changed from the near-infrared region to red and then blue region while the recording material remains the same.

10.4.1.3 Protective Layer Material

The recording material is formed on a plastic substrate and the phase change temperature (approximately 650 °C) exceeds the disc melting temperature. To prevent thermal damage of the substrate by repeated erasing of records, thermal protection layers are formed on both sides of the recording material, as shown in the cross sectional view of Fig. 10.20. These layers, in combination with the metallic reflection coating, improve optical efficiency and control heat transfer. The typical material for the protective layer is a mixture of ZnS and SiO_2, which was developed by Matsushita Electric and is used in nearly all rewritable optical

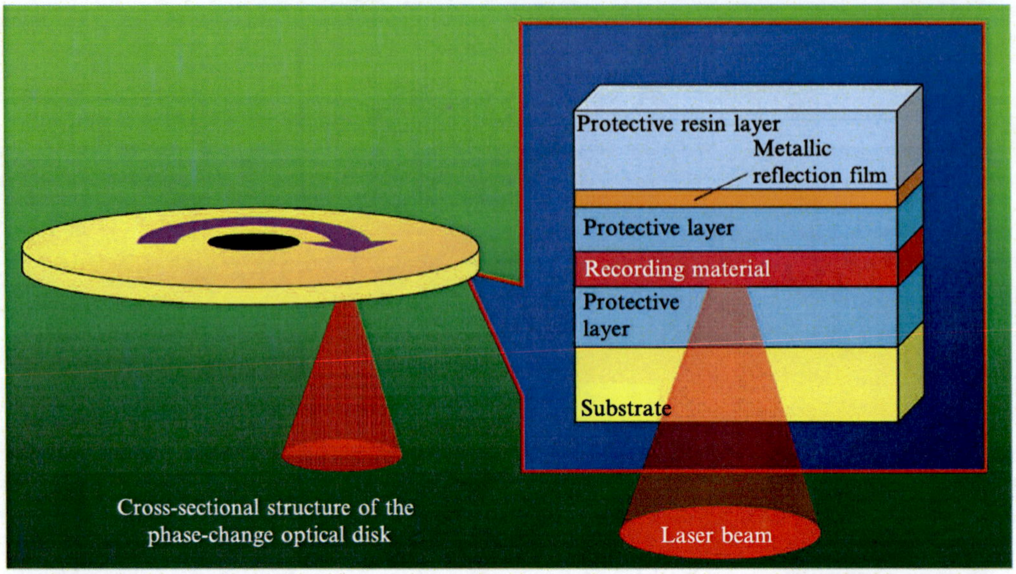

Fig. 10.20 Cross-sectional structure of the phase-change optical disk. Several functional films including the recording material are formed on the plastic substrate

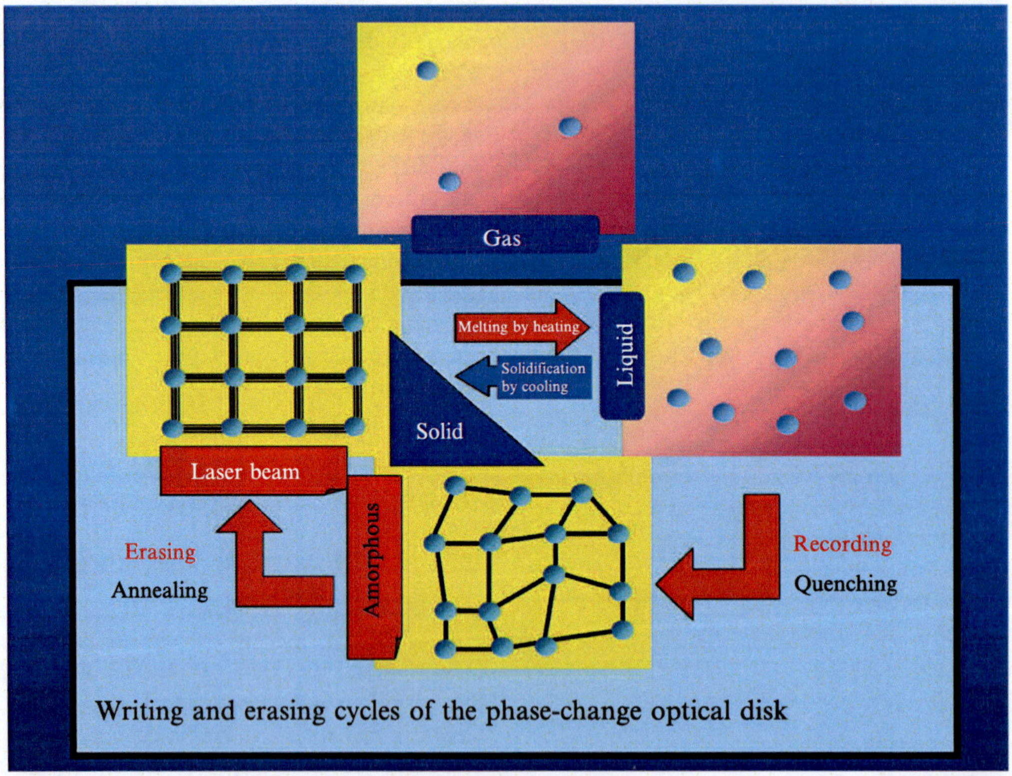

Fig. 10.21 Concept of writing and erasing cycles of the phase-change optical disk. A laser beam irradiation melts the recording material, which is quenched to create the amorphous phase for recording. For erasing, the recording material is heated until crystallization begins in the solid state

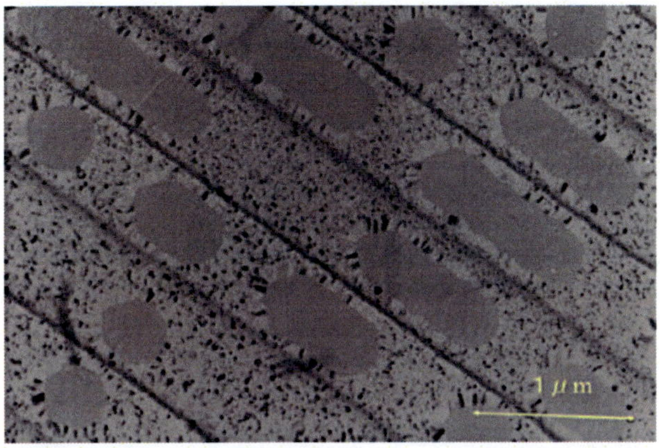

Fig. 10.22 Electron micrograph of the phase-change optical disk after recording. The areas that look comparatively homogeneous are amorphous phases, while the granular areas in the background are crystal phases. Recording of information is performed by a combination of creating amorphous tracks of various lengths

Table 10.4 Types of phase-change optical disks, recording materials and recording capacity

Disk format	Recording material	Recording capacity
PD	Ge–Sb–Te	0.65 GB
CD−RW CD+RW	Ge–In–Sb–Te Ag–In–Sb–Te	0.65 GB
DVD−RAM	Ge–Sb–Te	4.7 GB
DVD−RW DVD+RW	Ge–In–Sb–Te Ag–In–Sb–Te	4.7 GB
Bluray (BD)	Ge–Sb–Te	25 GB

There are several recording formats. The major components are antimony and tellurium for all of them. The recording capacity, rewritable frequency, erasing time are optimized by means of careful selection of the material composition

disks. The Ge–Sb–Te recording material guarantees rewriting up to several hundreds of thousands cycles due to the protective layer.

10.4.2 Future Prospects

The phase-change rewritable optical disk (Note 10.13) continues to evolve even today. Material composition remains nearly the same, but developments continue responding to the endless demands for recording media with large capacity and high rewriting speeds. One of the efforts being made is the development of a disk with multiple recording layers. The focal point of the laser beam is controlled so as to realize a disk with a recording capacity with several disks. A double-layer disk has already been realized. On the double layer disc, the recording film that is closer to the laser beam needs to be translucent. Disks with three or more recording films will be realized in the near future. Efforts have been made to irradiate the laser beam at a level below the diffraction limit by utilizing the near-field light. Phase-type materials are promising as the recording materials for other applications in the future. The recording device uses an amorphous phase (Note 10.14) that seems unstable, but proper writing time, thermal stability and high-speed rewriting have been realized by means of the careful selection of materials and precise optical irradiation. A reliable device was realized through the scientific analyses of recording materials, including analyses performed in advanced analysis facilities such as SPring-8 (Note 10.15).

Literature

1. Yamada N, Takao M, Matsunaga T (2003) Phys Condens Matter 38:357–364 **(10.4)**
2. Japanese patent No. 17788207 (1992) **(10.4)**
3. SPring-8 Research Frontier (2003) 104–105 **(10.4)**

Note 10.13 An optical disk where rewriting is performed by repeatedly moving a precisely power-controlled laser beam back and forth between the crystal phase and the amorphous phase.

Note 10.14 A solid body in which atoms are not regularly arranged. Glass is an example of an amorphous material.

Note 10.15 A high luminance X-ray analysis facility for research that uses synchrotron radiation (synchrotron light) and is located in Nishi-Harima, Hyogo Prefecture http://www.spring8.or.jp/.

11 Optical Parts and Optical Communication

Japanese companies have achieved world-class optical lens technology that is key to develop components used in digital cameras and mobile phones, and precise optical parts that enable internet optical communication. These companies hold a considerable worldwide share of the products incorporating these technologies. This field will continue to remain important for the near future, and Japan is expected to lead the world within it. In this section, typical glass and ceramic materials that are used in optical parts and optical communication are introduced.

High performance lenses are used in IT devices such as digital cameras and mobile phones equipped with cameras. Glass for optical lenses (Sect. 11.1) is introduced below. Among optical lenses, aspheric lenses, in particular, contribute to the miniaturization of lenses. Development of the aspheric lenses began in the 1980s, when an innovative technology enabled the direct press molding of high-temperature glass into a lens shape. Figure 11.5 shows a mobile phone equipped with a camera, which underwent miniaturization by incorporating an aspheric lens.

Optical parts, enablers of optical communication for data transmission such as used in the Internet, are explained below. Optical communication parts that support optical communication include the distributed refractive index lens [Gradient Index (GRIN) lens] that integrates information (Sect. 11.2), the optical fiber that transmits information (Sect. 11.3), the optical fiber amplifier that enables long-distance communication (Sect. 11.4), the optical communication ferrule that enables accurate connection (Sect. 11.5), and so on.

Optical fiber communication progressed dramatically in the 1980s and has been expanding rapidly ever since. Originally, this system was used in trunk communication networks for connecting countries, e.g., between Japan and the U.S. In recent years, optical communication has been expanding to households in the form of fiber-to-the-home (FTTH). Optical communication has characteristics for allowing for the transmission of large amounts of data and loss of information is extremely low in comparison with conventional communications that use electricity. In addition, the Dense Wavelength Division Multiplexing (DWDM) technologies that support recent broadband communication systems have been progressing. These parts are produced by Japanese companies and incorporate advanced and accurate manufacturing technology and feature extremely high reliability. These parts are introduced in detail in the following section.

Y. Imanaka et al. (eds.), *The Ceramic Society of Japan, Advanced Ceramic Technologies & Products*, DOI 10.1007/978-4-431-54108-0_11, © Springer Japan 2012

Optical Lens (Aspheric Lens; 1982)* 11.1

Optical lenses are classified into two types: spherical lenses and aspheric lenses. Spherical lenses have been produced for a long time by polishing optical glass. The mass production of glass aspheric lenses began in the 1980s following the development of the technology required to accurately press form softened glass by using precisely formed dies made of special materials. Glass aspheric lenses have contributed substantially to the downsizing of optical systems and to the improvement of their performance. This has further enhanced the resolution of optical devices such as digital cameras, digital video cameras and camera-equipped mobile phones.

11.1.1 Background of Development

Optical systems including cameras incorporate multiple spherical lenses so as to reduce aberration (Note 11.1). When using spherical lenses, spherical aberration occurs due to the displacement of the focal point between the light incident on the center and the light incident on the surrounding area, as shown in Fig. 11.1a. If aspheric lenses are used in optical systems, the number of lenses used in optical systems is reduced and performance of the systems is improved as the spherical aberration is easily removed as shown in Fig. 11.1b. The introduction of aspheric lenses had been anticipated; since the mass production of aspheric lenses via cutting and polishing methods proved to be difficult and expensive.

In the 1980s, the mass production of glass aspheric lenses was achieved following the development of the technology required to accurately press form softened glass by using precisely formed dies made of special materials, which allowed for the precise transcription of dies used in the production of high accuracy optical elements.

The lenses produced by utilizing this technology are called glass mold lenses. Introduced below are the glass aspheric lenses that are currently used in high-precision optical devices such as cameras and are indispensable for the improvement in the definition or resolution of images taken with optical devices.

11.1.2 Characteristics

11.1.2.1 Characteristics

The downsizing of optical systems and the improvement in their performance was achieved by means of the introduction of aspheric lenses into optical systems. Plastic aspheric lenses have also been produced, however glass aspheric lenses, which feature a wider range of refractive indexes and dispersion (Note 11.2) with good quality and stability to environmental changes such as temperature and humidity, have been used in components that need to be high in performance and reliability. Glass products are higher in cost than plastic products as the forming temperature of glass is higher. The two types are selected according to the application.

Following the digitalization of optical devices in recent years, the digital camera market has been expanding rapidly and the number of mobile phones equipped with high

*The number in parenthesis indicates the year that the product was first commercialized.

Note 11.1 Departure of rays from an ideal image formation points is called "aberration." They are represented by five Seidel aberrations and two types of color aberrations. Lenses are designed in consideration of glass types (types of optical glass), shapes and layouts so as to reduce aberrations.

Note 11.2 In visible light region, the glass refractive index becomes lower as the wavelength becomes longer. The refractive index and wavelength dependence (called dispersion) vary depending on the type of the optical glass and its properties are expressed simply by the refractive index n_d (d line; refractive index at 587.56 nm) and the Abbe number v_d. $v_d=(n_d-1)/(n_F-n_C)$, where n_F and n_C indicate the refractive index of the F line (486.13 nm) and the C line (656.27 nm), respectively. The v_d is smaller when the variation in refractive index due to wavelength is larger. This is interpreted as high dispersion. Please refer to Fig. 11.3.

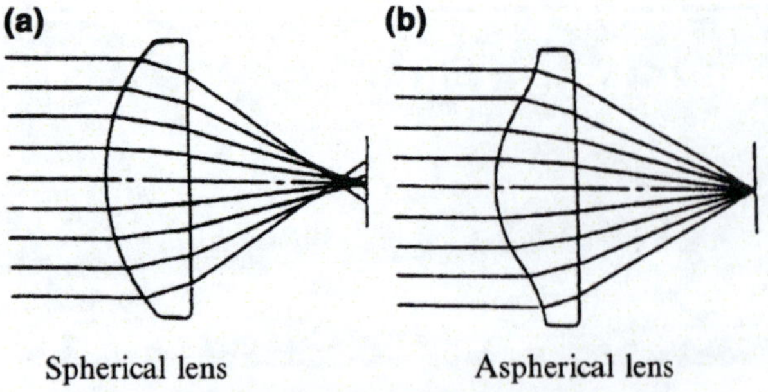

Fig. 11.1 Removal of spherical aberration using an aspheric lens. In imaging optical systems, aberrations are removed by using a combination of multiple lenses (**a**) Spherical lens (**b**) Aspherical lens

definition cameras has also rapidly increased. Glass aspheric lenses of small caliber (Fig. 11.2) are used widely in these high precision optical devices. The profile accuracy of the lens surface is 0.1–0.2 μm at its maximum. For further improvement in the performance and definition, a number of aspheric lenses in the lens system has been on the increase in recent years and optical glass of low dispersion and high refractive index and optical glass of high dispersion and high refractive index have become widely used (an example of optical glass for molding is shown in Fig. 11.3). An optical system of a digital camera is shown in Fig. 11.4. Concave meniscus lenses used as front lenses are being replaced by

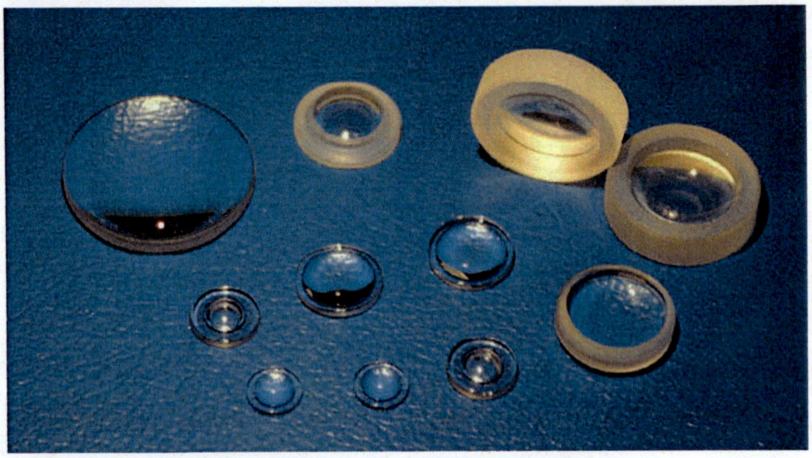

Fig. 11.2 Glass molded aspheric lenses. The majority of them have diameters of 3.5–15 mm, but there are also smaller and larger ones

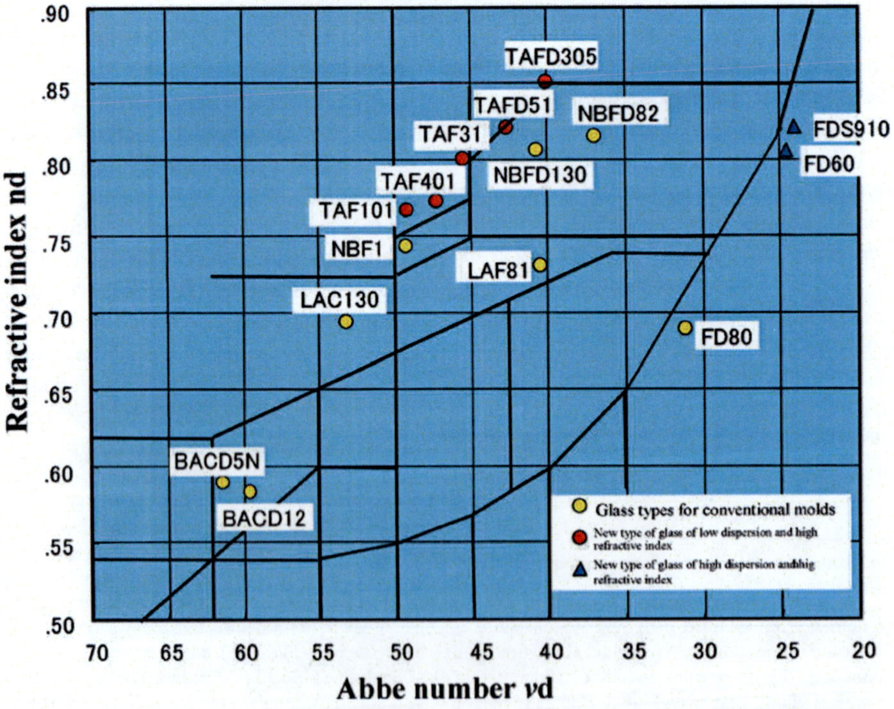

Fig. 11.3 Optical glass used in molded lenses (excerpt from HOYA's catalog). Glass featuring a high refractive index is effective in the reduction of thickness and size in lenses. Applications using high index glasses has been increasing

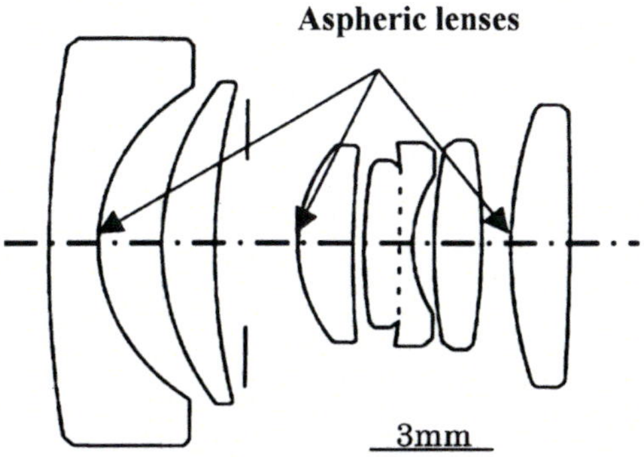

Fig. 11.4 Optical lense system in a typical digital camera. The front lens (on the *left side* of the figure) is a concave meniscus lens

Fig. 11.5 Example of application in a mobile phone equipped with a camera. High definition was achieved by incorporating glass aspheric lenses for three lenses (six faces)

aspheric lenses and optical glass of low dispersion and high refractive index are to be used in the future. Figure 11.5 shows an example of its application in a mobile phone equipped with a two million mega pixel camera. Although plastic lenses are mainly used as pickup lenses of optical discs, glass lenses will be used in the next-generation high-capacity products. Plastic lenses are also used as combination lenses in optical communication devices.

11.1.2.2 Manufacturing Methods

As for the pioneering patent of Eastman Kodak, which opened a way for forming high-accuracy lenses, glass was set in a dye made of special material (such as glassy carbon), heated up in a non-oxidizing atmosphere, and was pressed at a temperature near the glass softening point by setting the glass and the die at the same temperature. It was then cooled down to a temperature below the transition point of the glass (until the glass solidified), without releasing the pressure. This method was markedly differed from the previous lens blanks forming method in which fluid glass is pressed in low temperature dyes. Later on in Japan, a variety of production technologies have been developed and advanced.

The vital elements for producing glass mold lenses include (1) die material and surface thin film, (2) processing of an ultra-precision die, (3) glass composition and pre-form material (glass material to be formed) appropriate for forming, (4) precision forming technologies (incl. press machines, die structures), (5) measurement and evaluation of aspheric surfaces and (6) optical design technology including those for aspheric surfaces. Forming die materials are required to allow the high-accuracy processing of mirror surfaces of optical parts, to be free from change due to high temperature, to enable increased hardness and strength and to be free from the fusion of softened glass. Requirements of the dies cannot be satisfied by a single material and are therefore usually composed of the substrate and the surface film. Ceramics such as extra hard alloy and silicon carbide are widely used as the substrate, while a number of materials that include rare metal-based, carbon-based materials, nitride materials and carbide materials are widely used as the surface film. Press forming is usually performed in a non-oxidizing atmosphere so as to prevent the oxidization of dies. For example, optical glass with a comparatively low softening temperature, enabling pressing at 600°C or lower, has been developed (press temperature exceeds 700°C in some recent optical glass of low dispersion and high refractive index). The glass can be cut and polished for use as pre-form material for forming; however spherical and aspheric pre-form material with flawless surfaces can also be produced directly from molten glass.

High-precision molded aspheric lenses are produced by preventing reactions between softened glass and die surfaces.

11.1.3 Future Prospects

Objective lenses with a numerical aperture (NA) of 0.85, intended for next-generation large-capacity optical disks require extremely high accuracy. They feature a P–V (peak-to-valley) value of 0.05 μm max., a lens thickness accuracy of ±1 μm max. and a displacement of 1 μm max. between axes of aspheric surfaces on both sides. Aspheric lenses of higher accuracy are required for use in a variety of optical devices and will continue to be developed further.

Gradient Index Lens (1978) 11.2

Gradient index lenses (Fig. 11.6) have a refractive index distribution that collects and focuses light rays. The refractive index distribution is created by an ion exchange treatment on the surface of glass bars. Lenses are normally convexed or concaved. However, refractive index lenses are rod shaped with flat ends, which mean they can be produced easily in large amounts. In addition, erected images can be created easily with these lenses. Gradient index lenses are used for the coupling of optical fibers or lasers. Arrays of multiple gradient index lenses are used as imaging lenses of copy machines and scanners.

11.2.1 Background of Development

Since around 1966, full-scale research has been performed for communication systems incorporating optical fibers, where a core with a high refractive index is enclosed by a cladding with a low refractive index. For higher communication capacity of optical fibers, the "refractive index distribution type," in which refractive index changes continuously, is more effective than the "step index type," in which refractive index changes step by step. Against this backdrop, the Nippon Sheet Glass Company, Ltd. and Nippon Electric Company, Ltd. developed in 1968, a "SELFOC," a gradient index distribution fiber incorporating the ion exchange method.

Following shifting manufacturing trends in optical fibers for long-distance communication to "single-mode fibers" with higher capacity, SELFOC lenses have been developed for use as gradient index lenses. The products are used as coupling lenses in optical communication systems and in the imaging systems of copy machines.

11.2.2 Characteristics

11.2.2.1 Characteristics and Specifications

Light rays travel in a straight line in homogeneous substances, while they are refracted at the boundary of substances with different refractive indexes (on the surface of lenses and prisms, etc.). Refractive indexes in gradient index lenses are distributed in a parabolic shape, as shown in Fig. 11.7. Internal light rays travel in a sinusoidal waveform with a wavelength of P, forming an image at every P/2 node. The inverted image (reversed image of the original object) and the erected image (facing the same direction as the original object) are formed alternately.

11.2.2.1.1 Selfoc Micro Lens

SML (Selfoc Micro Lens) is a cylindrical lens with a diameter of 0.25–4 mm. Light rays are bent by convexed or concaved surfaces in normal lenses. However, the SLM is flat on both ends, making it very easy to grind the surface. This contributes to better mass productivity. Lenses with various imaging characteristics can be created from the same material by changing the length of lenses (Fig. 11.8).

The SML is used widely as coupling lenses of optical fibers and light sources (such as LD) in the optical communication field, etc. (Fig. 11.9). The refractive power may be further enhanced by processing the cut ends of the SML into spherical shapes (Table 11.1).

11.2.2.1.2 Selfoc Lens Array

SLA (Selfoc Lens Array) is an array of SMLs that form erected images of the same magnification (Note 11.3)

Note 11.3 Image that is the same as the original object in terms of orientation and size. If lenses that form erected images of the same magnification are arrayed, images formed by adjacent lenses can be superimposed. Images formed by normal lenses are inverted and cannot be superimposed.

(Fig. 11.10). They are used in readout systems of copy machines and scanners because they are capable of projecting the line images in manuscripts as they are.

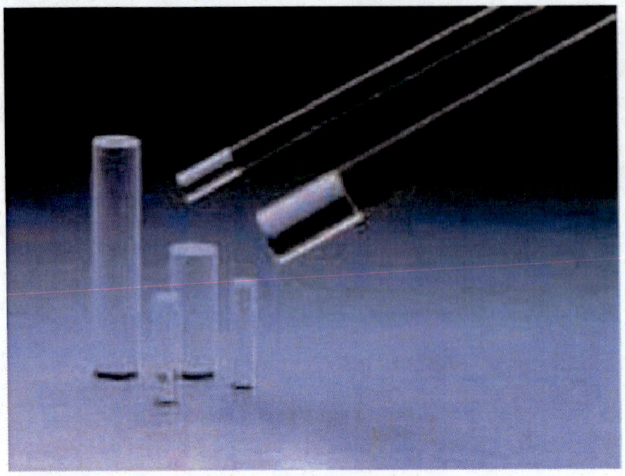

Fig. 11.6 Gradient index lens. Refractive index varies inside the glass. Therefore, it functions as a convex lens (although it is a flat-ended rod)

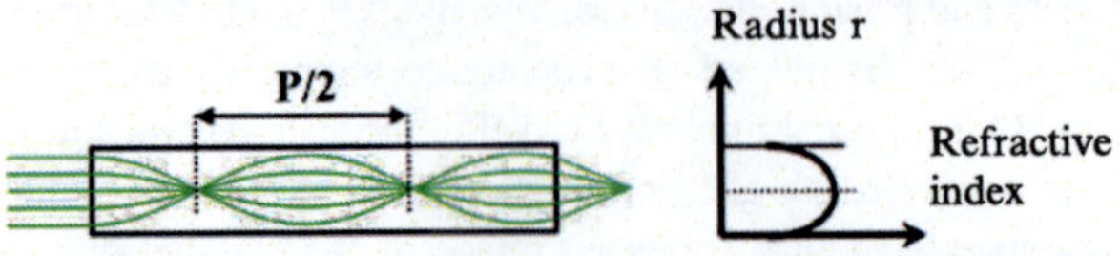

Fig. 11.7 Refractive index distribution in the gradient index lens. Refractive indexes are the highest on the optical axis (*center line*) and lower in the peripheral area, exhibiting a parabolic shape. Because of this, internal light rays travel in a sinusoidal waveform with a wavelength of P, forming an image at every P/2 node

Optical systems incorporating the SLA are much smaller than those incorporating conventional lenses. Scanning time is also shorter as images are brighter. Furthermore, the cost is low because the manufacturing process is simple.

11.2.2.2 Manufacturing Method

The rod, base material of SML, is made by cutting a rod of several tens of mm in diameter out of a glass block and thermal stretching the rod. The base materials of SLA are produced in large amount through a continuous process where molten glass is directly spun (Fig. 11.11) (Note 11.4).

SELFOC glass composition is characterized by a large amount of univalent elements (M_2O, M indicates metal). Univalent elements are fixed in the form of univalent elements M^+ in the glass network (irregular network structure of SiO_2, etc.) under normal temperatures. M^+ moves freely in the network when the temperature becomes high.

Refractive index distribution is realized with an "ion exchange" treatment, in which the base rod is immersed in a high-temperature dissolved salt. On the surface of the base rod, M^+ in glass and univalent ions (Na^+, K^+, etc.) dissolve and ions from the salt are exchanged (Fig. 11.12). Following the ion exchange, M^+ concentration on the rod surface becomes low, inducing concentration gradient. M from the salt diffuses inward radially due to the concentration gradient, inducing a concentration distribution. If the rod is pulled out of the dissolved salt at the point where the concentration distribution is in a parabolic shape and annealed, the refractive index distribution corresponding to the concentration distribution is fixed.

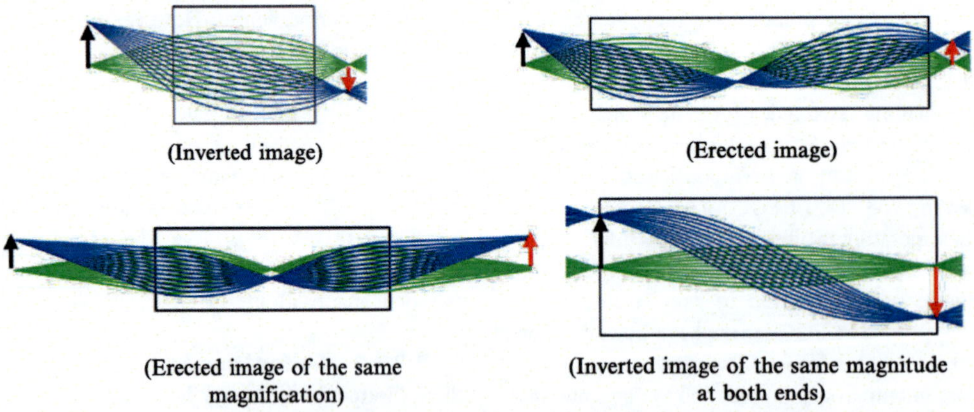

Fig. 11.8 Image formation by gradient index lens. The lens focal length can be changed by adjusting its length. It is easy to form erected images and form images on cut ends

Note 11.4 Glass does not have a distinctive melting point where the solid phase changes to the liquid phase. However, glass has a glass-transition point where substantial changes in expansion rates are observed. In normal oxide glass, the glass-transition point is approximately 400–600°C.

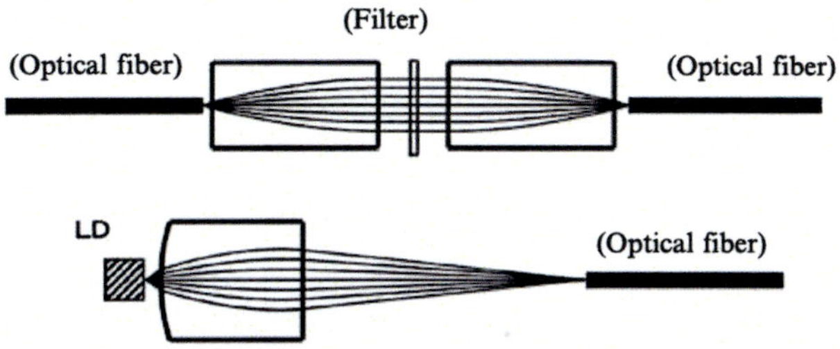

Fig. 11.9 Application of SML. (*Top*) Two optical fibers are coupled via filters. (*Bottom*) Ray of light generated from LD (diode laser) are transmitted to an optical fiber

Table 11.1 Characteristics of SML (wavelength 1,550 nm)

Type of lens	SLW10	SLW18	SLW20	SLH18
Diameter of lens	1.0 mm	1.8 mm	2.0 mm	1.8 mm
Refractive index at the center		1.59		1.634
Length of lens (0.25P)	2.63 mm	4.81 mm	5.34 mm	3.76 mm

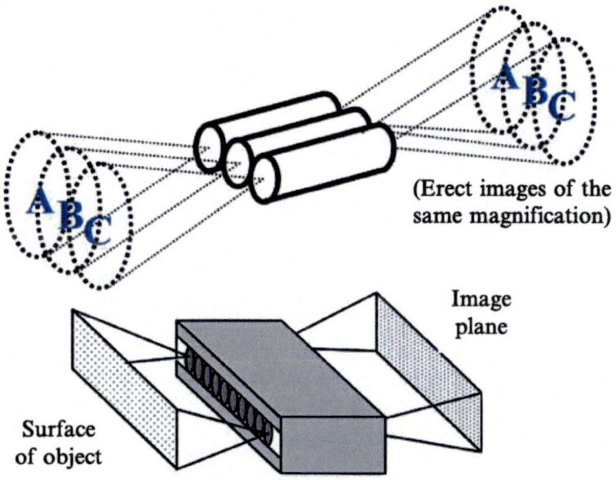

Fig. 11.10 SLA. If small lenses that form erected images of the same magnification are arrayed, linear images are formed by using a compact optical system

After the ion exchange, the rod is cut into the length appropriate for the application and both cut ends are ground. For production of SLA, the rods are fixed in a frame in one or two rows so as to cut them into a width corresponding to the specified lens length. The cut ends are then ground.

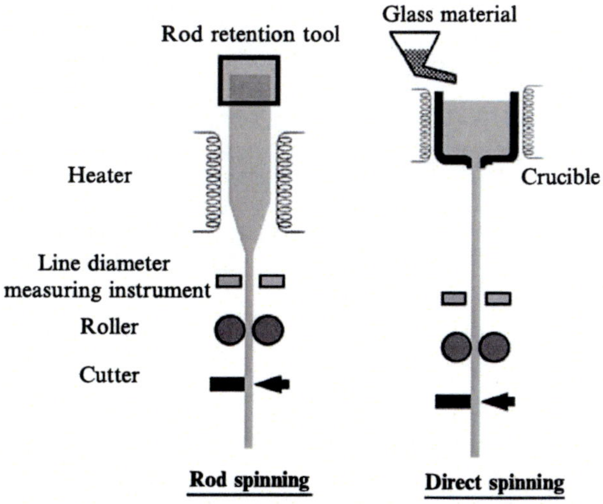

Fig. 11.11 Manufacturing method of base glass for gradient index lenses. In rod spinning, a rod of several tens of mm in diameter is cut out of a glass block and is thermal-stretched. In the continuous process, molten glass is spun directly

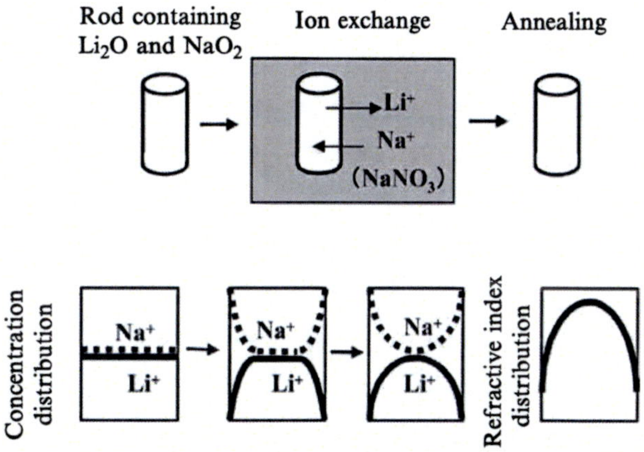

Fig. 11.12 Ion exchange. Ion exchange observed when a rod containing Li_2O, a high-refractive index component, is immersed in $NaNO_3$ dissolved salt

11.2.3 Future Prospects

SMLs are key components in optical communication systems. Miniaturization of SLA contributes to improvements in resolution. The external dimension of the parts that were first commercialized was 1.1 mm, followed by 0.9 mm and 0.6 mm products. Currently, 0.3 mm products are being developed.

Optical Fiber (1980) 11.3

Silica glass-based optical fibers are made of transparent glass. They are composed of a central core that has a high refractive index (through which light travels) and cladding, which covers the central axis and confines light within the core. The concentric ring structure is shown in Fig. 11.13. The refractive index of the core is maintained higher than that of the clad to confine the light inside the core for transmission. This is shown in Fig. 11.14. Optical fibers coated with a plastic sheath are called "optical fiber wires" or "optical fiber core wires," while optical fiber core wires coated with a protective layer are called "optical fiber cords." A multiple optical fiber core wires that are strengthened by a protective coating called a sheath or reinforcement are called "optical cables," which are mounted on telephone poles. Signal attenuation is smaller in optical fibers than in metal cables, which use electrical signals for communication. In addition, optical fibers feature broadband, allowing long-distance transmission of a large amount of data at high speeds. Transmission media are essential for the Internet communication of today and are used in optical transmission systems and optical devices.

11.3.1 Background of Development

For telecommunications before the 1980s, transmission systems utilizing metal cables were used. Telephones were the mainstream in terms of telecommunications. Optical fiber communication technologies improved dramatically following the introduction of practical optical fibers in the early 1980s and the application of a reliable semiconductor laser as the light source. Communication networks throughout the entire country have been completed in Japan. FTTH (Note 11.5) that connects the networks to households via optical fibers on telephone poles has begun to expand as shown in Fig. 11.13. Current data communications including the Internet service are supported by optical fibers, which began to be supplied to the market following the advance in mass production technologies such as the VAD method. This enabled the production of low-priced optical fibers to the market that are (1) low in loss, (2) broadband and (3) free from electromagnetic induction.

11.3.2 Characteristics

11.3.2.1 Characteristics and Specifications

In the 1970s, when it was proved that optical fibers were low-loss transmission media, Multi Mode Fiber (MMF) in 0.85 μm wavelength bands was used for optical communication. Soon after that, the single mode fiber transmission was developed for high-speed long-distance transmission, which led to the commercialization of the Single Mode Fiber (SMF). Optical fibers have wavelength dispersion (signal speed variation due to wavelength), which is attributable to the material, silica glass, and the basic structure. The transmission system was designed for communication, using the zero-dispersion wavelength, which becomes 1.3 μm in combination with the conditions of single mode transmission. The system that performs as an optical relay in backbone

Note 11.5 Data communication service via optical fiber intended known as fiber-to-the-home (FTTH). The general term for the communication service in which optical fibers are connected to households to provide the integrated service that includes telephone, internet service, broadcasting, etc. The transmission speed is higher than conventional metal transmission and currently services run at 100–1,000 Mbps. NTT (Nippon Telegraph and Telephone Company) in Japan started "B-Flets," a constant connection service, in 2001. This service has more than two million subscribers at the present of 2009.

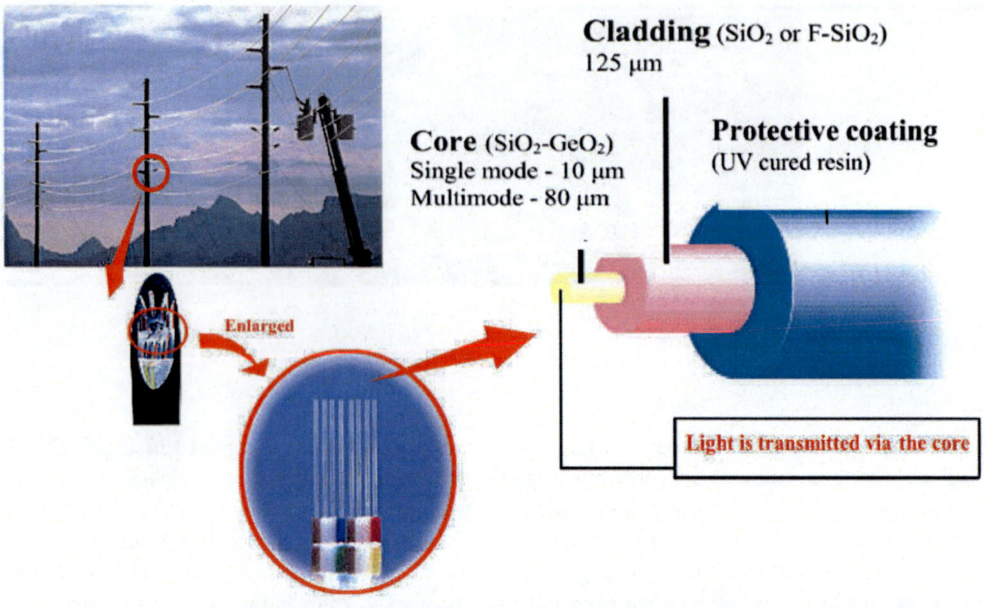

Fig. 11.13 Structures of optical cables and optical fibers in use. Optical cables connect the user to the internet provider are supported by telephone poles (*aerial cables*), a cross-sectional view of an optical cable and a picture of an optical fiber core wire under the coating. The illustration is a pattern diagram showing enlarged optical fiber core wires

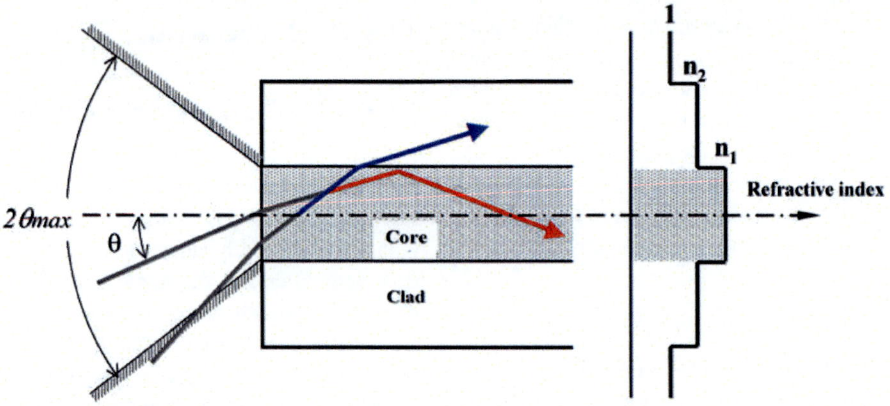

Fig. 11.14 Transmission of light by optical fiber. The light that enters the optical fiber is reflected entirely at the boundary of the core with high refractive index and cladding with low refractive index and travels in the axial direction. Note that the incident angle is limited

networks by using a transmission wavelength of 1.55 μm was developed following the release of an optical fiber amplifier (amplified band 1.55 μm) that can amplify the light without modification in the second half of the 1980s, because the loss of silica glass fiber is the lowest at the wavelength of 1.55 μm. Responding to this development, Dispersion Shifted Fiber (DSF) made by shifting zero-dispersion wavelength to 1.55 μm was developed. In the 1990s, telecommunications needs expanded tremendously following the expansion of the Internet, increasing the needs for large transmission capacities. Responding to these needs, wavelength division multiplexing (WDM) technology, which enables transmission in high speeds and density through existing optical fibers, was developed. Wavelength division multiplexing has 8 or 16 channels, for example, at an interval of 0.8 mm (equivalent to 100 GHz) in a wavelength of 1.55 μm that is the amplifier bandwidth of optical fiber amplifiers. Consequently, WDM transmit information via virtually 8 or 16

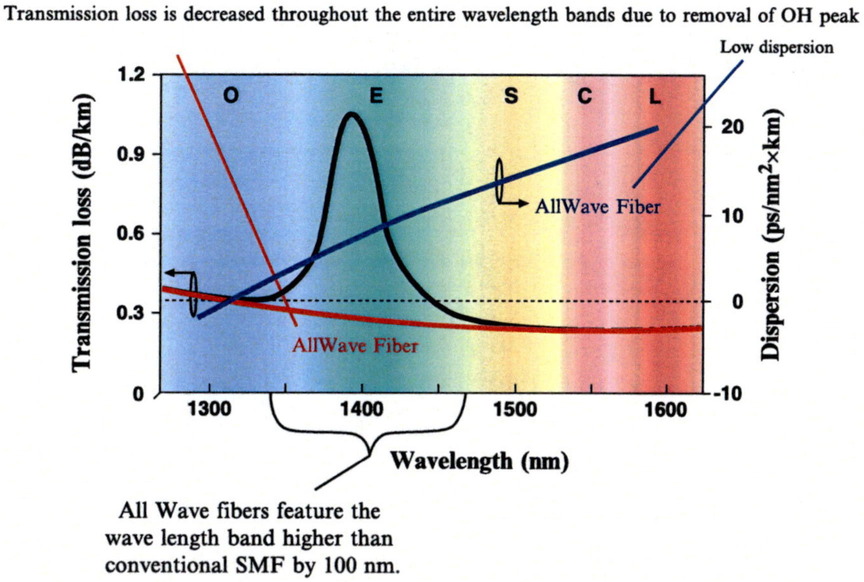

Fig. 11.15 Example of an optical fiber made by removing OH radicals for reduction of optical loss: ALLWave Fiber. The figure shows characteristics of ALLWave Fiber (which is made by removing moisture) during the process of optical fiber production so as to reduce loss due to the absorption of OH radicals around 1,400 nm. The *black line* indicates the transmission loss in conventional SMF, while the *red line* indicates the loss in ALLWave Fiber. The *blue line* indicates wavelength dispersion of ALLWAVE Fiber, which is smaller than the values of conventional SMF

independent fibers (although a single fiber is actually used). In single channel transmission, high-speed transmission without influence on signal waveforms was possible if the transmission is performed in the zero-dispersion wavelength. However, when multiple channels are transmitted in the case of high density by narrowing the wavelength interval, interference due to overlapping of adjacent channels occurs at the zero-dispersion wavelength. The interference occurs because the signals are modulated for data transmission and each of the signals has spectrum width. In order to solve this problem, Non Zero Dispersion Shifted Fiber (NZDSF) with zero-dispersion wavelength slightly deviated from 1.55 μm and Ultra Low Slope Dispersion Shifted Fiber (ULS-DSF) with low dispersion slope have been developed.

Optical fibers are regulated by the International Telecommunication Union (ITU) for global application. MMF, SMF, DSF, CSF (submarine cables with core diameters slightly larger than normal), NZDSF and ULS-DSF are, respectively, specified by ITU-G651, ITU-G652, G653, G654, G655 and G656. Mixing of OH radicals in optical fibers is unavoidable because of the manufacturing process. Absorption by OH radicals occurs at a wavelength around 1.4 μm, hence, this bandwidth was not previously used as a communication wavelength band. However, following the recent advance in technologies for removal of OH radicals intended for the reduction of loss, the SMF capable of optical transmission at wavelength bands from 1.3 μm through 1.55 μm has been developed (refer to Fig. 11.15 and Table 11.2 for examples). The SMF is regulated by ITU-G652.C, D.

Optical fiber core wires are made by coating the surface of silica glass with resin. Optical fibers that are further protected by reinforcing materials for use in severe environment are called optical cables. Optical fiber core wires are used for interior wiring, connection of devices and components inside devices. They have sizes of 0.25 or 0.9 mm (Fig. 11.16) (Note 11.6). The tape core wires (Fig. 11.17) are made by arranging 0.25 mm optical fibers in a parallel manner. Optical cables contain multiple optical fibers and have rigid structures for outdoor use. The cables have a steel wire at the center called the tension member, which protects optical fibers from strong external forces such as bending force and tensile force applied during installation work. They are wrapped in a sheath layer so as to protect them from external forces such as heavy loads and lateral pressure applied to underground optical fibers (Fig. 11.18). Tape slot cables, which are made by mounting tape core wires on a slot, are used as all-purpose multicore cables (Fig. 11.19).

Optical cables of various structures and types are in the market.

Note 11.6 0.25 mm products are sometimes called optical fiber wires.

Table 11.2 Characteristics of an optical fiber made by removing OH radicals for reduction of optical transmission loss: ALLWAVE Fiber

Optical properties			
Transmission loss (before processing into cables)			
Wavelength (nm)	Transmission loss (dB/km)		
1,310	≤0.34 (typical value 0.32)		
1,383	≤0.31 (typical value 0.28)		
1,490	≤0.24 (typical value 0.21)		
1,550	≤0.21 (typical value 0.19)		
1,625	≤0.24 (typical value 0.20)		
1,285–1,330	≤ Transmission loss at 1,310 nm + 0.03		
1,360–1,480	≤ Transmission loss at 1,385 nm + 0.04		
1,525–1,575	≤ Transmission loss at 1,550 nm + 0.02		
1,460–1,625	≤ Transmission loss at 1,550 nm + 0.04		
OH peak loss after hydrogen aging test[a]			
1,838 ± 3	≤0.31 (typical value 0.28)		
Bending property			
Bending diameter (mm)	Number of winding(s)	Wavelength (nm)	Increase in loss (dB)
32	1	1,550	≤0.05
50	100	1,310	≤0.05
	100	1,550	≤0.05
60	100	1,550	≤0.05
	100	1,625	≤0.05
Discontinuous property of transmission loss	≤ 0.05 dB at 1,310 nm, 1,550 nm		
Mode field diameter	9.2 ± 0.4 μm at 1,310 nm 10.2 ± 0.5 μm at 1,550 nm		
Cutoff wavelength	Cable cutoff wavelength (λoc): ≤ 1,260 nm		
Dispersion property	Zero-dispersion wavelength: 1,302 ~ 1,322 nm Zero-dispersion slope: ≤ 0.090 ps/nm² km (typical value 0.087 ps/nm² km)		
Dimensions			
Clad diameter	125.0 ± 0.7 μm		
Clad non-circularity	≤1.0%		
Coating eccentricity measurement	≤12 μm		
Core/clad eccentricity measurement	≤0.5 μm (typical value <0.2 μm)		
Diameter of coating	245 ± 5 μm		
Length	≤50.4 km		
Environmental property			
Temperature dependence of transmission loss	≤0.05 dB/km (at 1,310 nm, 1,550 nm, 1,625 nm, –60°C ~ increase of loss at +85°C)		
Temperature/humidity cycle test	≤0.05 dB/km (at 1,310 nm, 1,550 nm, 1,625 nm, –10°C ~ increase in loss at +85°C, 95% RH)		
Water immersion test at 23°C	≤0.05 dB/km (at 1,310 nm, 1,550 nm, 1,625 nm, +23°C ± increase of loss at 2°C)		
Accelerated deterioration (temperature) test at 85°C	≤0.05 dB/km (at 1,310 nm, 1,550 nm, 162 5 nm, +85°C ± increase of loss at 2°C)		
Mechanical property			
Proof test	0.7 GPa (100 kpsi)		
Tensile strength	≥3.8 GPa (550 kpsi) (length of specimen 0.5 m, medium value prior to aging)		
Coating removal force	≥1.3 N, <8.9 N		
Other characteristics (typical values)			
Group refractive index	1.467 at 1,310 nm 1.468 at 1,383 nm 1.468 at 1,550 nm		
Dynamic fatigue factor (n_d)	20		
Loss in fusion splicing	<0.02 dB (bonding of ALLWAVE ZWP fibers)		

Optical and mechanical properties, dimensions and supplemental data of ALLWAVE Fiber. The characteristics conform to ITU-G652.D, the international standard

[a]A test performed by simulating an environment where installed cables are exposed to hydrogen for a long period

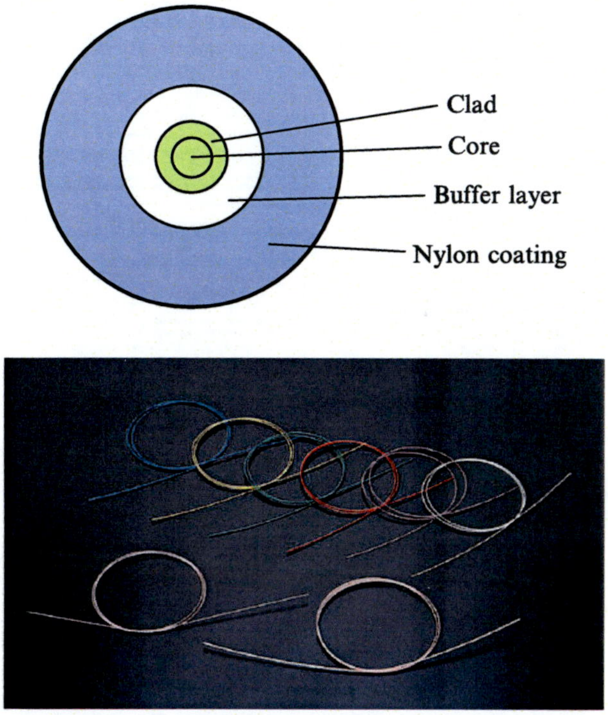

Fig. 11.16 Cross-section of optical fiber core and appearance of finished products. Made by coating fibers immediately after fiber drawing with nylon. Optical fiber core wires of 0.9 mm in diameter are used most widely as core wires for optical cables in Japan

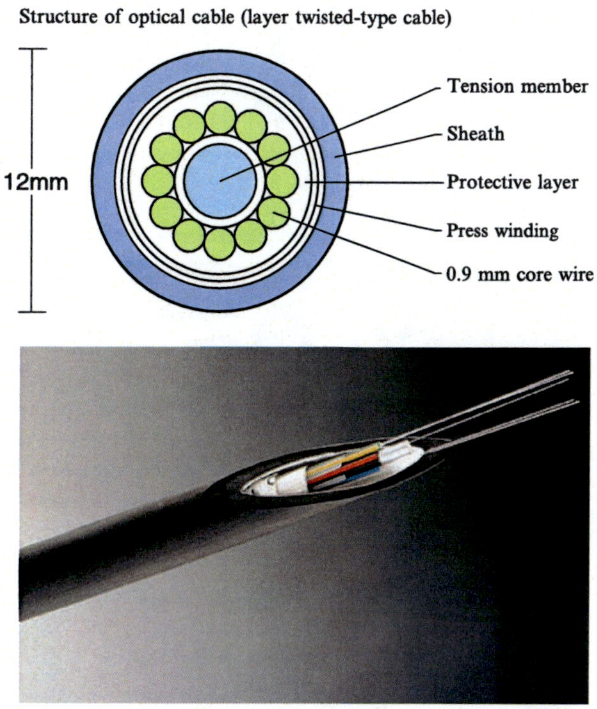

Fig. 11.18 Cross-sectional view and picture of a layer twisted LAP sheath cable. LAP sheath (Laminated Aluminum Polyethylen Sheath) is used for coating. The cable has a wire called the tension member in the *center* and optical fiber core wires that surround the tension member

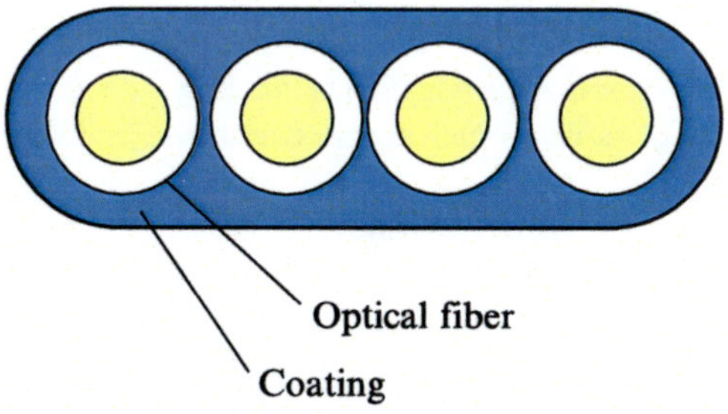

Fig. 11.17 Cross-sectional view of the tape core wire. Optical fiber core wires are arranged in *parallel* and are coated all together. There are 2-core, 4-core and 8-core types

11.3.2.2 Manufacturing Method

Optical fibers are manufactured by the MCVD (Modified Chemical Vapor Deposition) method, the OVD (Outside Vapor Phase Deposition) method and the VAD (Vapor Phase Axial Deposition) method. The VAD method is incorporated in Japan. In the MCVD method and the OVD method, there is a void in the central area. The VAD method (in which everything is deposited on the tip of the seed rod) attains high purity and is suitable for mass production. The VAD method is explained below. The VAD method is the quartz-based optical fiber manufacturing method developed in Japan. It was released by Nippon Telegraph and Telephone Public Corporation (currently NTT) in 1977. In the VAD method, cylindrical glass preform (called soot) is synthesized continuously in an axial direction. The method is suitable for mass production.

$SiCl_4$ and $GeCl_4$, the raw materials, are continuously fed into an oxyhydrogen burner to synthesize the soot with a flame hydrolysis reaction (Fig. 11.20). The soot is deposited in the axial direction, at semi-sintered state, on the tip of a rotating rod. Consequently, the soot is deposited on a free surface in an open space, making it possible to synthesize large-sized cylinder. The soot is inserted in a furnace, dehydrated in the chlorine atmosphere and is made transparent (Fig. 11.21). This method is characterized by a structure without a hole in the central area. Because there is no discontinuous area in the center, the intrusion of impurities during the manufacturing process is avoided. Furthermore, this

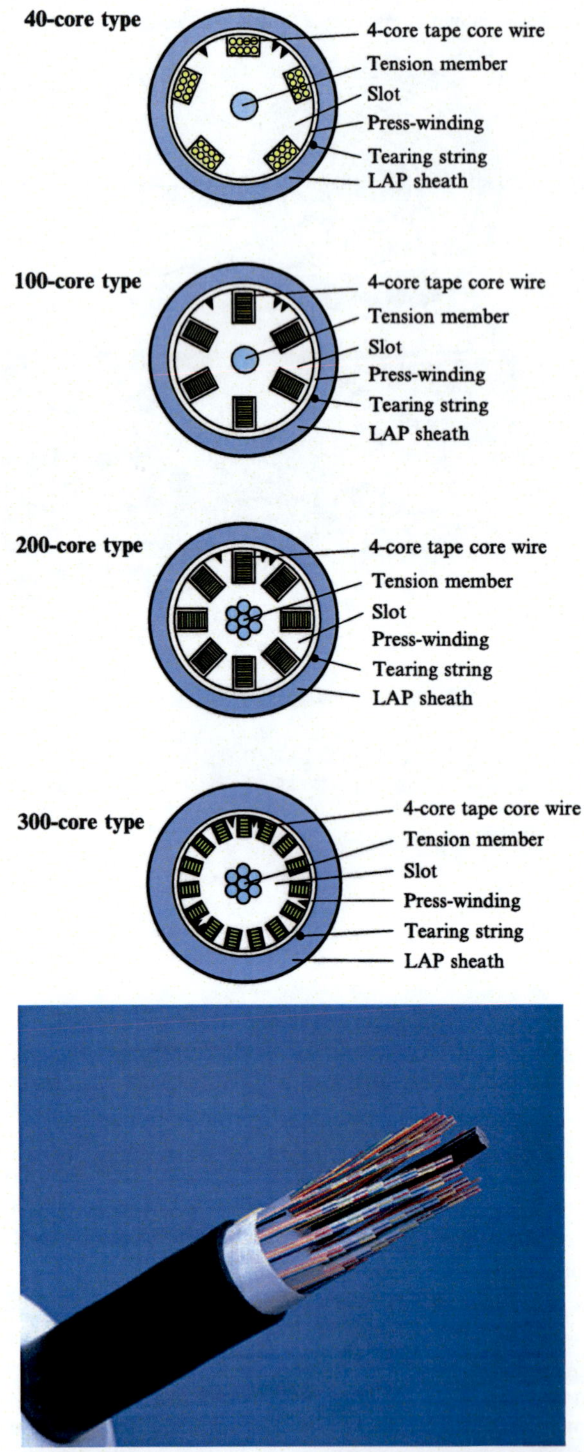

Fig. 11.19 Cross-sectional views and picture of a multiple core tape slot type cable. The cables are composed of 4-core tape core wires that are slipped into the slots (spacers), and each type consists of multiple numbers of the 4-core tapes. The picture shows an all-purpose optical cable

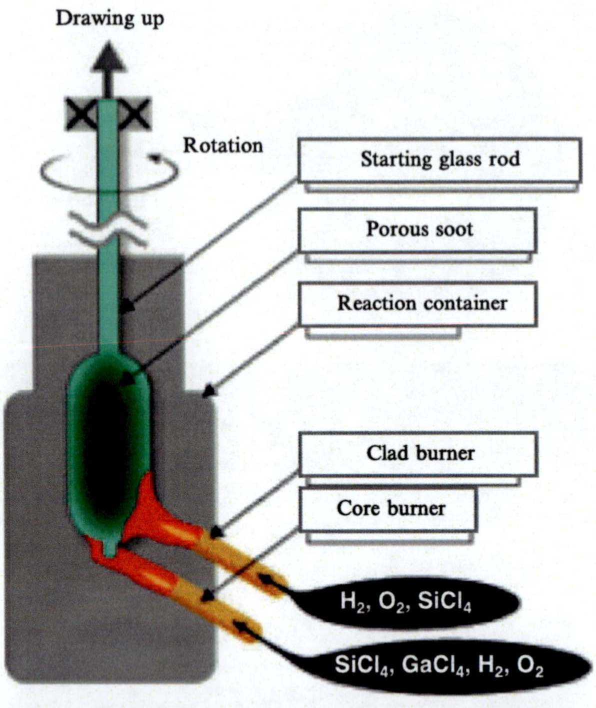

Fig. 11.20 Illustration of the VAD method. Raw materials are reacted in the oxyhydrogen flame from the burner for the generation of glass particles. Glass particles are deposited on the start glass rod, and a porous body called the soot is formed by gradually pulling up the rod

method does not require the collapse process (Note 11.7). It is a superior mass production technology.

Silica glass-based optical fibers are made by melting the preform in a furnace and then pulling down the preform to narrow it to the standard diameter of 125 μm. Preform made by the MCVD method and the VAD method is placed vertically in a furnace. This is shown in Fig. 11.22. The preform, heated at about 2,000°C in a furnace, melts and has low viscosity. It begins to melt from the tip and drops down due to its own weight. The molten glass is wound around a bobbin (Note 11.8) at a high speed and the outer diameter is controlled by adjusting the winding speed.

Optical fibers feature wavelength dispersion determined by both the wavelength dependency of refraction observed in the silica glass itself and the structure dispersion attributable to the structure designed for single mode transmission. Dispersion due to materials cannot be controlled, but the

Note 11.7 The process where a cylindrical void is collapsed by thermal treatment to create cylindrical glass.

Note 11.8 A cylindrical jig on which an optical fiber is wound.

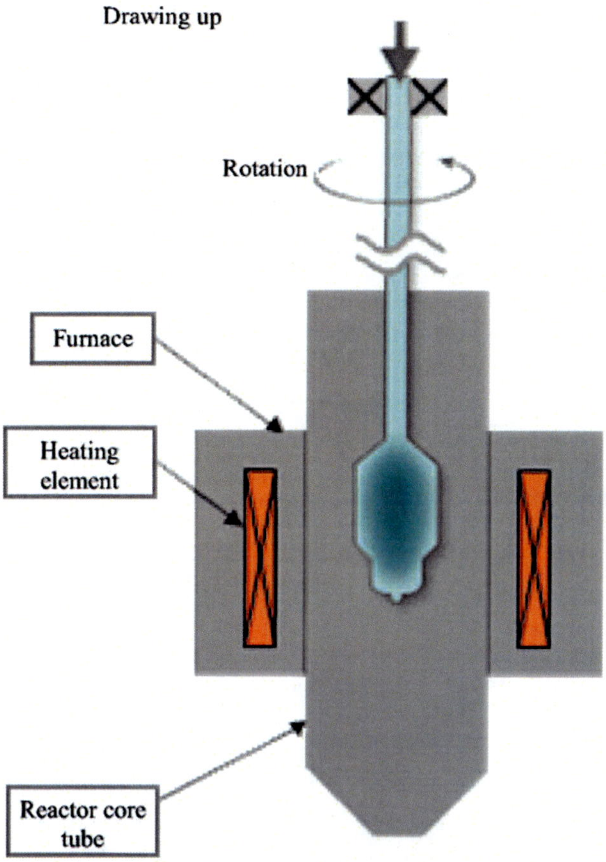

Fig. 11.21 Preform creation by the vitrification process. The porous body called the soot is processed at high-temperature and dehydrated ambient, followed by heating process and densification for creation of transparent glass called the optical fiber preform

structure dispersion can be controlled to some extent by adjusting refraction profiles of the core and the clad while maintaining the single mode condition. The refractive index varies depending on the dopant. It increases when Ge or P is used as the dopant, while it decreases when F is used as the dopant, for example. With respect to profiles, the ratio of contribution to structure dispersion increases when the core system is slimmed. Figure 11.23 shows wavelength dispersion of optical fibers incorporating various dopants and profiles. Optical fibers are used not only for communication but also in the output portion of laser processing devices and medical devices. They have been developed to address the needs of various fields. Standardization for each application has been discussed.

11.3.3 Future Prospects

The room temperature semiconductor laser and the low-loss optical fiber realized in 1970 marked the start of optical communications. The advent of the erbium doped fiber amplifier (EDFA) (Note 11.9) in the 1980s and the subsequent developments of various optical fibers and optical parts led to the innovation of optical communication technologies that have been supporting the current information technology based society. The optical fiber network has also been supporting the backbone of mobile and smart phones that have been expanding rapidly. Following the evolution of Internet communication, optical fibers are now connecting private houses as well. The photonic crystal fiber has been studied. It is a new optical fiber based on a new principle and featuring refraction indexes of multidimensional periodic structure. Fibers resistant to bending force will soon be commercialized (although they are not introduced in this section). It is important to establish an information-based society where information is transmitted to every corner of the earth by an internet protocol (IP) network created by the fusion of broadcasting and communication. Optical fibers will continue to develop for expansion of FTTH and for use as the basic technology that supports the development of high-speed, high-capacity optical transmission technologies and optical signal processing technologies.

Note 11.9 The 1.55 μm signal light can be amplified by adding Er, rare earth, to the core of a quartz-based fiber and light to the fiber using a semiconductor laser. In long-distance optical communication prior to the introduction of EDFA, light was transformed to electricity for amplification and electricity was transformed to light for transmission via optical fibers. Following the introduction of EDFA, signal light can now be relayed and transmitted without transformation to electricity.

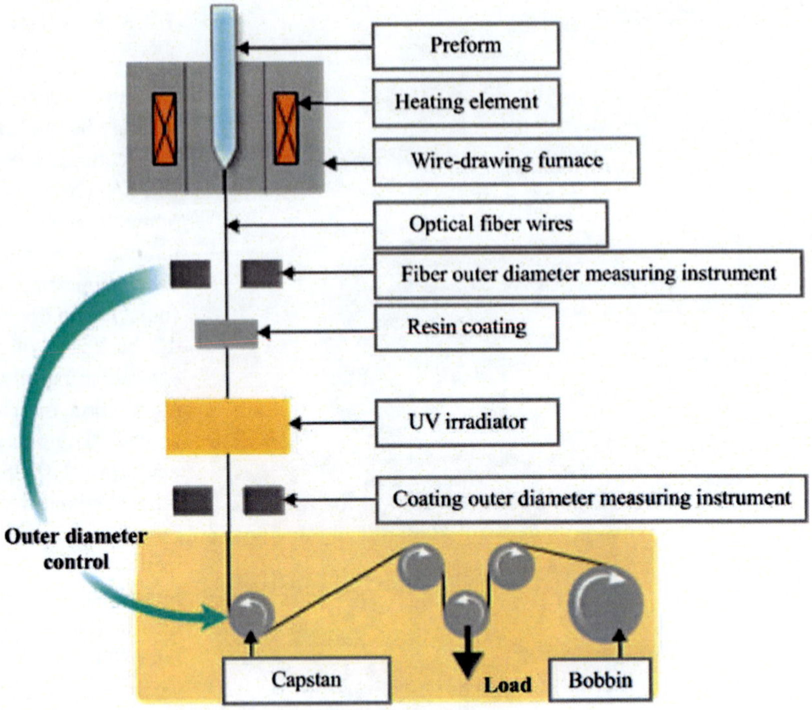

Fig. 11.22 Wire drawing process. The tip of the preform is heated to 2,000°C or higher and melt in a wire-drawing furnace, and the molten glass is continuously wound to produce glass fibers. The optical fibers are coated with resin to protect the sensitive glass surface

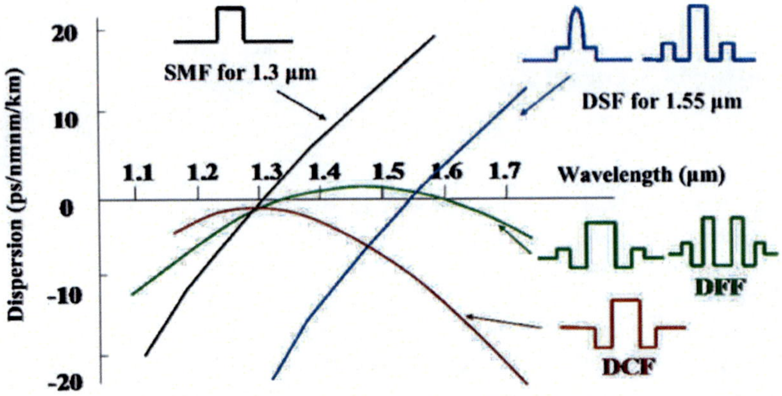

Fig. 11.23 Relationship between profiles and dispersion of optical fibers. Wavelength dispersion of optical fibers is controlled by adjusting the type and concentration of the dopant as well as the structure of the core and the clad. The graph shows the profile and wavelength dispersion of each optical fiber

Optical Fiber Amplifier (1995) 11.4

In optical fiber communication, an optical signal or light pulse is transmitted through a fiber. The decrease in the intensity of the signal over a length of transmission fiber is known as loss or also called attenuation. This loss in the signal intensity is attributable to branching, adsorption and molecular level irregularities in the fiber which cause scattering. Optical amplification is required to compensate for the loss. Optical fiber amplifiers, whose amplification media are optical fibers that have their cores doped with rare-earth ions, are used for optical amplification. Optical fiber amplifiers can amplify the incident light with high efficiency based upon a very simple principle. In addition, it matches up so well with the optical fiber in the transmission route because the amplification media itself is an optical fiber. Presently, the erbium-doped optical fiber amplifier (EDFA) in particular, plays a very important role in the large capacity optical communication as the wavelength band of amplifiable light corresponds with the low-loss wavelength band of the transmission fiber.

11.4.1 Background of Development

In line with the explosive growth of the information and communication industries triggered by the global expansion of the Internet, requirements for high capacity and high speed optical communication systems continue to grow in both the trunk transmission systems (long-distance communication) and the branch transmission systems [CATV (Note 11.10), FTTH (Note 11.11)]. The WDM (wavelength division multiplexing) (Note 11.12) communication is indispensable for building high-capacity, high-speed transmission systems that satisfy these requirements. In the WDM transmission system, signal lights of different wavelengths have to be amplified all together. The optical fiber amplifier (Fig. 11.24) is the most appropriate for this purpose among many types of optical amplifiers. EDFA, which amplifies light in the wavelength band where the transmission loss of the optical fiber is minimum (around 1.55 μm), plays a crucial role in both trunk transmission systems and branch transmission systems, as shown in Fig. 11.25. The current high-speed, high-capacity optical communication system via WDM communication would have been impossible without the EDFA.

11.4.2 Characteristics

11.4.2.1 *Basic Structure of Optical Fiber Amplifier*

Figure 11.26 shows the basic structure of the optical fiber amplifier. It is composed of an optical fiber whose amplification media incorporates a rare earth ion-doped core, an excitation light source that excites rare earth ions, a WDM coupler that multiplexes excited light and signal light for their entry into the rare earth-doped optical fiber, an isolator that prevents laser vibration, an output monitor coupler that controls the output of the optical fiber amplifier and a light

Note 11.10 CATV stands for "Cable Tele-Vision", meaning a cable broadcasting service that distributes television broadcasting via a network connected by electrical wires, such as coaxial cables. The system was originally developed so that TV broadcasting can be viewed in areas where the reception of ground-based TV broadcasting is poor. However multi-channel, telephone and high-speed Internet services have been offered through the system in recent years.

Note 11.11 FTTH stands for "Fiber to the Home", which is a high-speed communication service that utilizes optical fiber and is intended for families. Telephone, Internet and TV services can be offered as one package by connecting homes using optical fibers.

Note 11.12 "Wavelength Division Multiplexing" allows various information to be transmitted simultaneously via a single optical fiber by putting together signals that vary slightly in wavelength (multiplexing). This makes it possible to increase, without much difficulty, the transmission capacity.

detecting device. The type of rare earth ions applied to rare earth-doped optical fibers is selected according to the wavelength band of the signals used in the optical communication system. In actual production, wavelength bands causing a small loss in transmission fibers are used, and the amplifier employs fibers doped with Er (erbium), Pr (praseodymium) or Tm (thulium), which is selected according to the wavelength band, as shown in Fig. 11.27. In respect to WDM communication, in which the intensity of signals of each wavelength needs to be the same, a gain equalizer is used to planarize the wavelength dependency of gains in the rare earth-doped optical fiber.

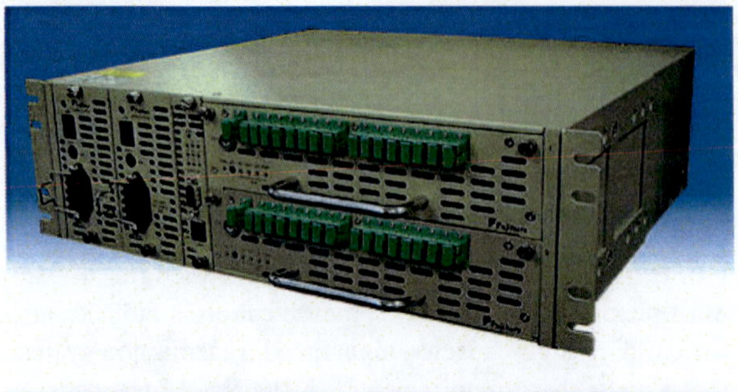

Fig. 11.24 Erbium-doped optical fiber amplifier (EDFA). In this picture, a EDFA and the light distributor are integrated and mounted on the distribution and transmission system. The input signal light is amplified and distributed into multiple signals for output

11.4.2.2 Principle of Optical Fiber Amplifier

The principle of optical fiber amplification is explained using EDFA as an example. Figure 11.28 shows the energy level of the Er^{3+} ion. The absorption band of the Er^{3+} ion is around 0.98 and 1.48 μm. When excitation light of these wavelengths enters the erbium-doped optical fiber (EDF), the Er^{3+} ions absorb the light, and the electrons are raised from the ground state to the excited state. After the elapse of a certain period of time the exited Er^{3+} ions emit light with a wavelength (1.5–1.6 μm in erbium) equivalent to the difference between the excited state and the ground state and

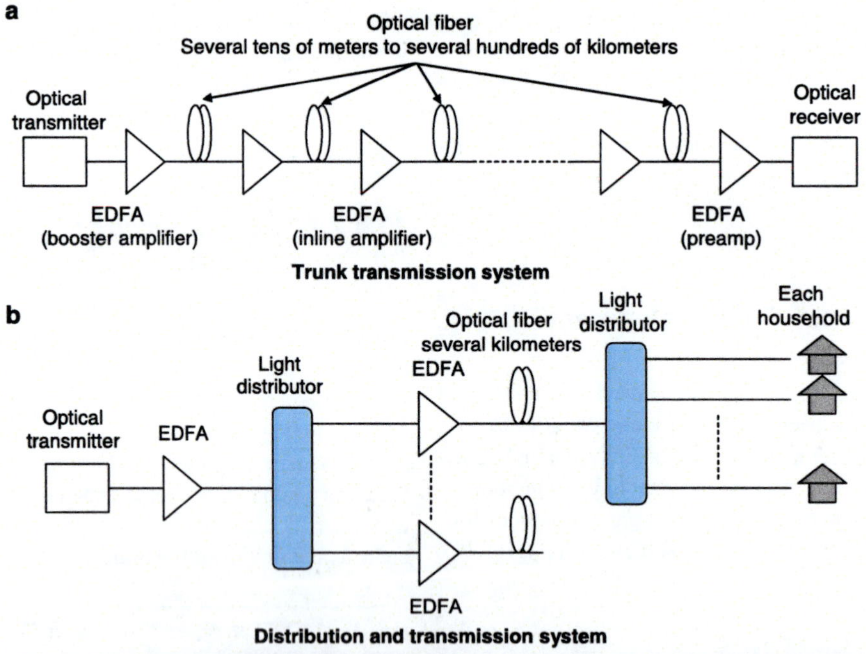

Fig. 11.25 Example of optical fiber amplifier application. (**a**) In trunk transmission systems, amplifiers are installed at an interval of several tens of to several hundreds of kilometers so as to amplify the signals attenuated in the transmission fiber. (**b**) In distribution and transmission systems, amplifiers are installed to compensate for the loss during transmission to households as well as the loss due to distribution. The loss increases in proportion to the number of branches, requiring high-output amplifiers

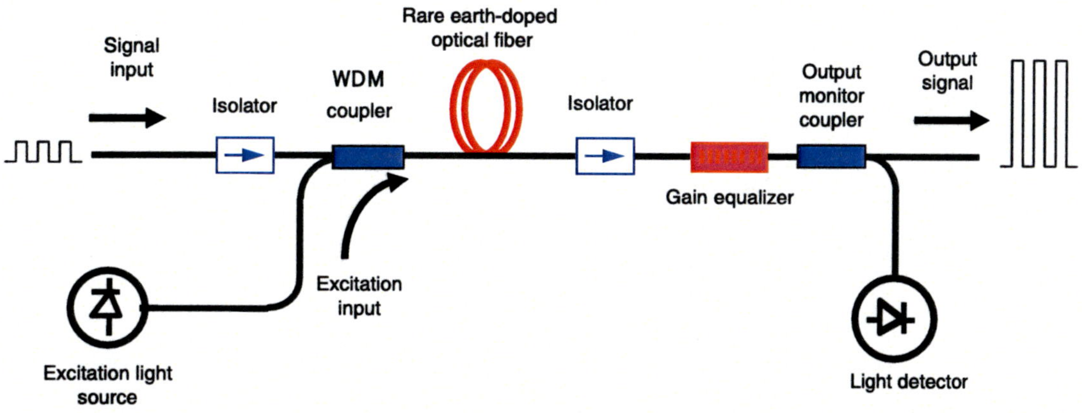

Fig. 11.26 Basic structure of optical fiber amplifier. The rare earth-doped optical fiber is used as the amplification media. Excited light is applied to the rare earth-doped optical fiber and then signal light is input in order to amplify the signal light

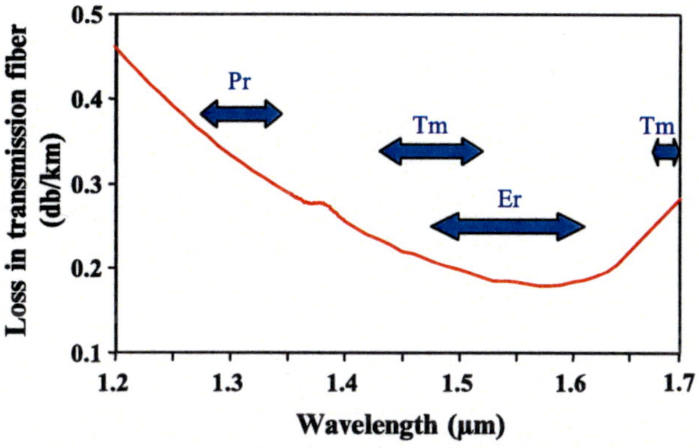

Fig. 11.27 Dependency of transmission fiber loss on the wavelength and amplification band of optical fiber amplifiers. Rare earth added to the core is selected according to the wavelength band. Er (erbium) is used for 1.5–1.6 μm, Pr (praseodymium) for 1.3 μm and Tm (thulium) for 1.45–1.5 μm and 1.67 μm

returns to the ground state. If the intensity of the excited light is strong enough at this point, a population inversion state is achieved where the number of Er^{3+} ions under the excited state exceeds that at the ground state. If the signal light enters during this state, induced emission (Note 11.13) is observed in the Er^{3+} ions under the excited state, amplifying the intensity of the signal light.

Note 11.13 When electrons in the excited state are exposed to external light, having an energy equivalent to the energy difference between the ground level and the excitation level, the electrons release energy in the form of light and then relax to a low-energy level. This phenomenon is called induced emission.

11.4.2.3 Characteristics of Optical Fiber Amplifiers

As explained above, rare earth-doped optical fiber amplifiers operate under a very simple principle and the signal light is amplified by merely inputting excited light. The gains range from 20 dB (× 100) to 40 dB (× 10,000) or more. The signal light is amplified without the process of converting the signal light to electricity, and therefore, the device structure is very simple when compared with conventional amplifiers that require light/electricity conversion. The amplification efficiency against the excited light is extremely high at 80%, achieving high-efficiency amplification. In addition, they can be connected with optical fibers used in existing transmission circuits easily and with low-loss, thus achieving high compatibility as the amplification media itself is an optical fiber and the surrounding parts also incorporate optical fibers.

11.4.3 Future Prospects

Optical fiber communication system growth is expected to continue into the future focusing on higher capacity and speed. Under such circumstances, the enhancement of optical fiber amplifiers for amplification of broader bands is

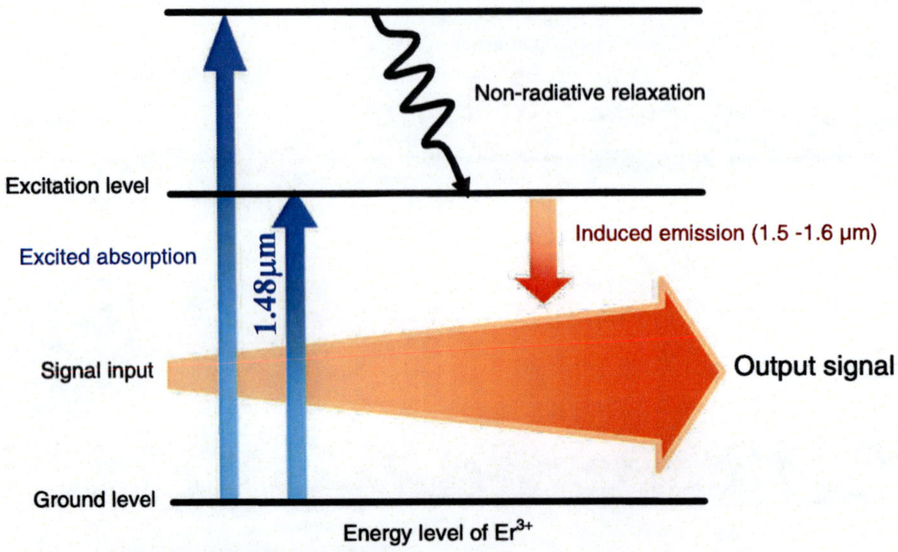

Fig. 11.28 Energy level of Er^{3+} ion. When signal light is applied to Er^{3+} ions that have absorbed excited light and are in a population inversion state, the signal light is amplified by induced emission

a critical issue for the utilization of as many WDM signals as possible. As for the technologies that will be made with a larger bandwidth, a variety of methods are being discussed. Methods include changing of the host glass material, the utilization of multiple amplifiers with different amplification bands, arranging them in cascade connection or parallel connection and the fiber Raman amplifier using transmission fiber nonlinearity. Optical fiber communication systems incorporating these amplifiers are expected to be developed.

Zirconia Ferrule Used in Connectors for Optical Communication Devices (1978)

11.5

11.5.1 Optical Connectors: Zirconia Ferrule

Zirconia ferrules (Fig. 11.29) are used in optical connectors that allow for simple connection and disconnection of optical cables. They are small but important parts, indispensable in "Fiber to the Home" (FTTH) products. A large number of them are used in connecting the users' devices as well as devices in business establishments. The ferrule is required to have a function that maintains the precise connection of fibers to an accuracy of 1 μm maximum variability, high process reproducibility in preparing ferrules with the same dimensions, and connection/disconnection performance as well as a high reliability during installation. The high dimensional accuracy and high reliability allows repeated disconnection/connection cycling that is made possible by using zirconia ceramics.

11.5.2 Background of Development

In response to the rapid increase in distributed information due to the expansion of the internet in recent years, discussions on the advancement of network technology have been proceeding at a rapid rate. Optical fiber connectors are the interface parts that play important roles in achieving reliable and flexible optical communication networks. For economical and flexible optical communication networks of the future, optical fiber connectors of high performance, high reliability and low cost will be more important than ever.

In the earliest days of optical communication connectors during the period from the late 1970s to early 1980s, they were primarily used in the main trunk lines of optical communication systems and technology development was focused on the improvement of performance and reliability. Since the beginning of the 1990s, when optical communication lines began to be utilized not only for the trunk lines but also by subscribers and buildings, the need for lower prices equivalent to the prices of electric connectors have been added to these requirements. The DWDM method (Note 11.14) began to be widely adopted in the twenty-first century and the demands for compact and high density packaging type connectors have been increasing. The method is currently applied to optical communication connectors and devices.

11.5.3 Characteristics

11.5.3.1 Characteristics and Specifications

Major specifications of the zirconia ferrule are shown in Fig. 11.30. The single mode (SM) ferrule is applicable to the single mode optical fiber cables, which features a core (which allows light transmission) size that is as small as 10 μm in diameter. When light that tends to go straight, such as with laser beams, is input into the fiber, signals can be transmitted with nearly no dispersion, allowing long-distance and ultra-high speed transmission. They are therefore used in trunk communication lines and networks that connect telephone stations, etc. Meanwhile, the multi-mode (MM) ferrule is used in multi-mode optical fiber cables, which features a core (that allows light transmission) size as thick as 50 or 62.5 μm. The light is reflected inside the cable during transmission, and the signals are dispersed and blurred. They are therefore not suitable for long-distance or ultra-high speed transmission. However, they are inexpensive and applied to Ethernet and LAN, which do not require long-distance transmission.

Y_2O_3-partially stabilized zirconia is used because it is suitable for high convex spherical PC (Note 11.15) due to low Young's modulus, and also high resistance to brittle fracture and high reliability due to high fracture toughness.

Note 11.14 DWDM stands for Dense Wavelength Division Multiplexing. It is one type of the communication technology incorporating optical fibers.

Note 11.15 P C stands for physical contact. It is a technology for stacking optical fibers together to prevent Fresnel reflection.

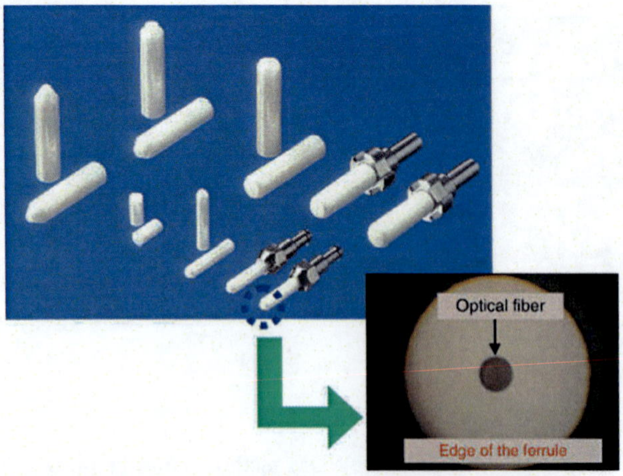

Fig. 11.29 Zirconia ferrule. They are used in optical communication connectors and devices. The optical fiber is inserted into the hole on the optical ferrule and is bonded. The contact face is polished and inspected before they are assembled with connector parts and delivered as a corded assembly

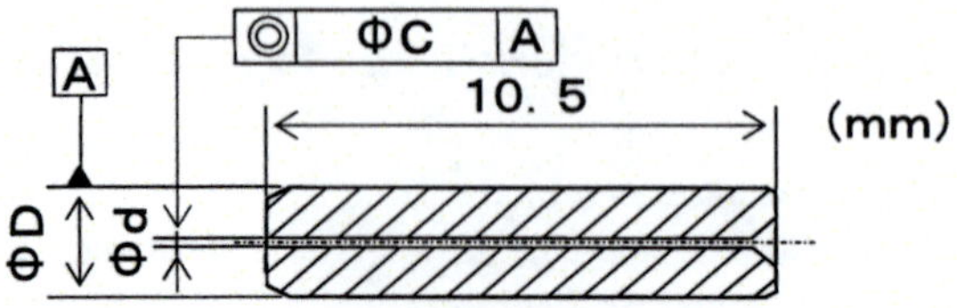

	Single mode	Multi mode
Concentricity Φ C	≦ 1 μm	≦ 6 μm
Insidediameter Φ d	$0.1255^{+0.001}_{0}$	$0.127^{+0.005}_{0}$
Outside diameter ΦD	$2.499^{\pm 0.0005}$	$2.498^{\pm 0.002}$

Fig. 11.30 Comparison of accuracy between SM ferrule and MM ferrule

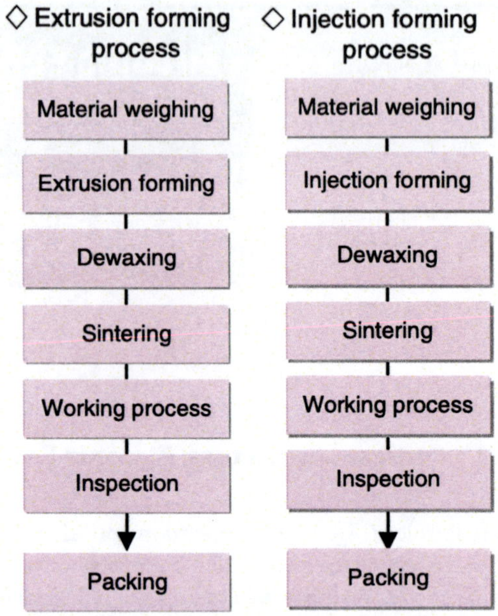

Fig. 11.31 Ferrule manufacturing process. The zirconia material and the binder are weighed and formed in each process. The green body becomes a product after going through dewaxing, sintering and working processes

11.5.3.2 Manufacturing Method

Figure 11.31 shows examples of the manufacturing methods of zirconia ferrules. Currently, two main methods have been adopted in order to manufacture ferrules. One is the extrusion molding method and the other is the injection molding method.

In the extrusion molding method, zirconia powder is weighed and agitated with a binder that consolidate the zirconia powder into a required shape, and then the mixed powder is put into an extrusion molding machine to form a bar-shaped green compact. During this process, pilot holes are created on the bar-shaped green compact by the pins on the extrusion molding machine. Production of the ferrule material is completed by sintering the green compact in a high-temperature furnace in order to create a dense body. High dimensionally accurate ferrule products are manufactured by performing additional processes on the material, responding to customer needs and requirements.

In the injection molding method, zirconia powder is weighed with a thermoplastic resin binder, and the pellets, divided into a certain amount and kept at room temperature, are melted into a clayey material by heating in the injection molding machine. The clayey zirconia fluid is filled into the injection molding machine by compression. The machine is also equipped with core pins for creating pilot holes. After the injection, the temperature is lowered in order to harden the material and remove the green compact. When removing the green compact, a part (called the gate, on the injection port) used in the manufacturing process is removed together with the green compact. This part is cut off before sintering. Production of the ferrule material is completed by sintering the green body in a high-temperature furnace. High dimensionally accurate ferrule products are manufactured by performing additional processes on the material, according to the end product requirements.

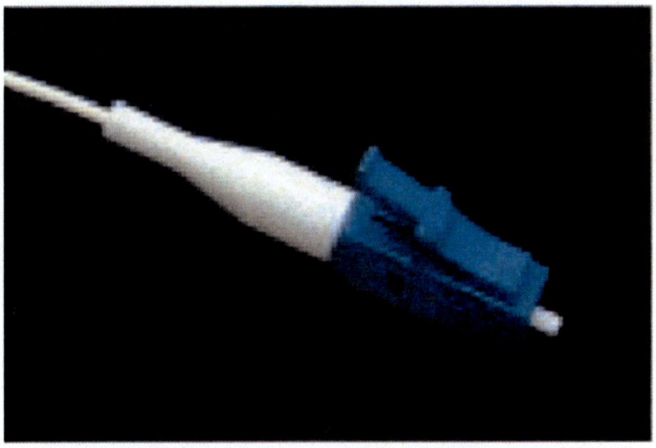

Fig. 11.32 LC-type optical connector

11.5.4 Future Prospects

In respect to optical communication systems, optical connectors were mainly applied to cable terminals and the input/output of optical devices. Responding to the expansion of the DWDM system, the MU type (Note 11.16) and the LC type (Fig. 11.32) (Note 11.17), which are small and of low cost, have been rapidly expanding in this field. In light of the sophisticated systems of the future and increasing need for high-density packaging, it is expected that ϕ1.25 mm zirconia ferrules will be widely used in the future.

Literature

1. Angle MA, Blair GE, Maier CC (1974) US Patent 3,833,347 **(11.1)**
2. Murata H, Koizumi T, Niizeki Y (2003) History of optical fiber, memoire of developers. Kogyo Tsushin Co., Ltd., Tokyo (in Japanese) **(11.2)**
3. Okubo K (1994) Optical fiber technology in ISDN era. Rikogakusha, Tokyo (in Japanese) **(11.3)**
4. Furukawa Electric Co., Ltd. Homepage: http://www.furukawa.co.jp/fiber/jp/index.htm **(11.3)**
5. Suematsu Y (ed) (1986) Optical fiber applications technologies. Nikkei Gijutu Tosho (in Japanese) **(11.3)**
6. ITU-T (International Telecommunication Union–Telecommunication Standardization Sector). Homepage: http://www.itu.int/ **(11.3)**
7. Miki T, Sudo S (2002) Optical communication technology handbook. Optronics, pp 248–259 (in Japanese) **(11.5)**

Note 11.16 It is an optical connector, developed by NTT (Nippon Telegraph and Telephone Corporation), incorporating a ϕ1.25 mm zirconia ferrule.

Note 11.17 It is an optical connector, developed by Lucent Technologies, Inc., incorporating a ϕ1.25 mm zirconia ferrule.

Part III

Transportation and Aerospace

Automobile Engines and Car Electronics

12

Alumina based products that take advantage of its electrical insulation characteristics such as those used in automobile parts, has been brought about by improvements in mechanical strength through grains refinement as well as by advances in ceramic processing such as forming and sintering. As for spark plugs (Sect. 12.2), ceramics are required to provide high insulation properties under explosive combustion caused by high-voltage application. High-purity alumina has also been used as a structural material in cutting tools, mechanical seals and so on, in addition to the application to spark plug insulators.

Full-fledged application of structural ceramic parts to aircraft started during wartime in Germany in the middle of the twentieth century. Companies tried to use ceramic materials in fighter jet engines. After the war, this attempt was taken over by England and the U.S. companies and research and development of structural ceramics, intended for use in aircraft jet engines, automotive gas turbine engines and compact gas turbine generators, became an active field.

In Japan, usage of ceramics in automobile engines began to be highly anticipated in the 1980s, and a driving test of a diesel engine vehicle mounted with a non-cooled all-ceramic engine was introduced on television. This event received a great response, accelerating the development of ceramic engines and leading to the commercialization of the ceramic glow plug (Sect. 12.1) and the hot plug. The ceramic glow plug, in which the heater element is insulated by ceramics, is used to assist in the starting of a diesel engine (Sect. 12.3). Earlier ceramic glow plugs were integrated by hot pressing silicon nitride around a metallic heating element, while current ceramic glow plugs have evolved into all-ceramic elements made by using ceramics to manufacture the internal resistor and the lead. Silicon nitride was also used to produce abrasion resistant parts that contact the cam, such as rocker arms and tappets. In respect to the ceramic turbo charger (Sect. 12.4), a dynamic component, its usability as a structural component was confirmed due to its mechanical strength, while previous ceramic components were characterized by thermal resistance and abrasion resistance. From the commercialization of ceramic turbo chargers, ceramics have reached a new stage as structural materials, overcoming brittleness.

Meanwhile, the widespread use of automobiles has led to problems such as air pollution. Photochemical smog, which was first observed in Los Angeles California, in the 1940s, began to cause problems frequently in Tokyo Japan, and other areas in the first half of the 1970s. The photochemical smog, which is caused by NO_x generated through oxidizing nitrogen at high temperatures, is partly attributable to the improvement of thermal engine efficiency. NO_x is removed by reduction and the majority of exhaust gas denitration systems used in stationary equipment such as boilers employ the ammonia reduction method. The reduction method that utilizes gasoline was incorporated into automobiles. In the cordierite-based honeycomb for exhaust gas purification (Sect. 12.5), the engine exhaust gas is led to catalyst layers containing platinum, etc. and is reduced through reaction with unburnt gas for purification. Ceramics used to support the catalysts are extrusion-formed bodies having multiple holes and are shaped like honeycombs. The partition walls need to be thinner for high purification performance.

Responding to the progress in electronics applications, the range of ceramic products applied to automobiles has been expanded and diversified in terms of types. Details of ceramic components (sensors, control units, motors, antennas, etc.) used to control automobiles are also explained here.

The thermistor type temperature sensor, which uses stabilized zirconia as a heat sensitive element, was the first exhaust gas temperature sensor (Sect. 12.6) mounted in automobiles. The perovskite oxide thermistor element, which was developed later using $YCrO_3$ as the base composition, incorporates a structure that minimizes heat capacity so as to realize high-temperature responsiveness. The element is capable of measuring a wide range of temperatures and is currently applied widely to prevent the overheating of catalysts and turbo chargers, as well as controls various exhaust gas purification systems and for other purposes.

Y. Imanaka et al. (eds.), *The Ceramic Society of Japan, Advanced Ceramic Technologies & Products*,
DOI 10.1007/978-4-431-54108-0_12,

The exhaust gas oxygen sensor (Sect. 12.7) is used in exhaust gas purification systems and is mounted on the upstream-side and the downstream-side of exhaust gas. Zirconia, a solid electrolyte, is used. The oxygen partial pressure is detected via the electromotive force generated by oxygen ion movement. High reliability including a long service life, in addition to functions as sensors, is necessary for exhaust gas oxygen sensors. Furthermore, they are designed with considerations for thermal resistance, mechanical shock resistance, thermal shock resistance, water resistance and chemical resistance to ensure performance in severe environment.

The knock sensor (Sect. 12.8) is a sensor for detecting knocking caused by the shock wave generated by an abnormal pressure increase in the engine. Engine block vibration generated by knocking is detected by a piezoelectric element. This knock sensor system, which suppresses the generation of knocking by using the signals from the sensor, is mounted on nearly all automobiles, from light-duty vehicles to luxury vehicles.

The ferrite magnet (Sect. 12.9) features a comparatively high magnetic force, low cost and chemical stability and is widely used in compact motors. Following the higher performance of magnets through the development of technologies for the entire process, LaCo-based Sr ferrite, developed in 1997, substantially improved the saturated magnetization and anisotropy field. Industrial applications started in 1999, realizing further miniaturization, weight reduction and higher performance motors. Further evolution is expected in the future through development of various technologies including designing. The Low Temperature Co-fired Ceramics (LTCC) substrate (Sect. 12.10) is a multilayer ceramic substrate that incorporates internal wirings applied with low-resistivity metallics such as Ag and Cu. It was first commercialized as a circuit substrate for supercomputers and then began to be actively developed for applications in automobiles as it features high thermal resistance and is small and light. In 1995, the mass production of Electronic Control Unit (ECU) substrates for Antilock Brake System (ABS) began, followed by the commercialization of a variety of electronics substrates. Under the current progress in electronic control technologies, its application range is expected to expand further. Automobile navigation systems employ microstrip patch antennas that incorporate dielectric bodies made of barium titanate or magnesium titanate as GPS antennas (Sect. 12.11). For the miniaturization of components, the dielectric constant of ceramics is an important factor in material selection. Antennas with a dielectric constant of 60–95, have in recent years become mainstream. Improvement in dielectric constants, as well as miniaturization, has resulted from these improvements.

Ceramic Glow Plug (1985)* 12.1

In diesel engines, stable ignition is not easily ensured and the starting performance is low at low temperatures. This is because ignition is performed by injecting fuel into the chamber with air which is raised to a high temperature through adiabatic compression. Due to this reason, the engine must be started after preheating the air inside the cylinder using a glow plug designed for cold starts. Ceramic glow plugs are made of sintered ceramics and feature rapid-heating performance, heat resistance and durability. Low compression engines (Note 12.1) have been increasing in recent years as a means of dealing with the exhaust gas pollutants produced by diesel engines. Cold start performance, however, is lower in low compression engines, requiring a strong ignition source. This increases the demand for ceramic glow plugs.

12.1.1 Background of Development

In diesel engines, starting is difficult when the temperature is low and the engine is cold as ignition is performed by injecting fuel into the combustion chamber with air raised to a high temperature by adiabatic compression. The glow plug, used as an auxiliary plug for the starting of diesel engines, is mounted on the engine head, with the tip of the heater element projected toward the sub-combustion chamber or the chamber, as shown in Fig. 12.1. It is used as an ignition source for starting and in order to stabilize combustion. Metal glow plugs were for many years predominated. They are composed of a metallic heating coil enclosed in a metallic tube and a heater element that is filled with magnesia powder and wrapped around the coil. Ceramic glow plugs, first used in the mid 1980s to quickly increase temperature through the improvement of thermal conductivity in the area around the heating element, which were made by embedding a metallic heating coil in an insulating sintered ceramic. In the second half of the 1990s, the second-generation ceramic glow plugs, which incorporate ceramic resistors for high-temperature heating, were developed, proving the superiority of ceramic glow plugs over metal glow plugs in terms of heating temperature.

Diesel engines, which are characterized by low CO_2 emissions, have been reexamined in recent years as internal combustion engines for passenger cars. Meanwhile, stricter regulations requiring the imperative reduction of emissions have been established in Japan, the U.S. and Europe, as diesel engines emit large amounts of NO_x and PM (Note 12.2). NO_x reduction through the lowering of compression ratios is often adopted as one of the measures toward improving engine combustion. However, when the compression ratio of diesel engines has been lowered, temperature rise through adiabatic compression becomes less prominent, lowering the starting performance and increasing the need for glow plugs with higher rapid-heating performance. Since the release of the third-generation ceramic glow plugs (Fig. 12.2) in 2005, the demand for ceramic glow plugs has been gradually rising.

12.1.2 Characteristics

12.1.2.1 *Characteristics of Each Generation*

The structure and conduction performance of ceramic glow plugs from each generation are shown in Fig. 12.3. A ceramic glow plug is, roughly speaking, composed of a ceramic heater element, a metal fitting mounted on the engine head

*The number in parenthesis indicates the year that the product was first commercialized.

Note 12.1 Ratio of volume with the piston at the bottom dead center to volume with the piston at the top dead center. An index indicating the ratio of compressing the inhaled air.

Note 12.2 Particulate matter exhausted after combustion in engines consisting mainly of black smoke (soot), unburned organic components and sulfur compounds.

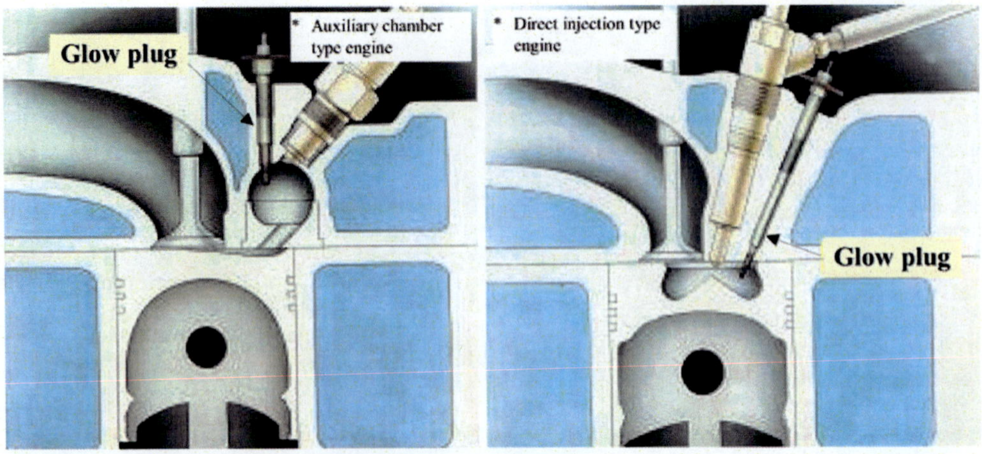

Fig. 12.1 Glow plug mounting position on diesel engines. The glow plug is mounted on the engine head and the heater element on the tip is projected toward the combustion chamber

Fig. 12.2 Third-generation ceramic glow plug. An auxiliary plug for the starting of diesel engines. The heater element on the tip is made of ceramic and heats up to 1,350°C

and the part that carries current from the terminal to the element.

The first-generation ceramic heater element was made by placing a metallic heating coil inside a densely sintered ceramic so as to achieve a temperature increase faster than that of metal glow plugs. The temperature reached 800°C in 3 s after a voltage application of 11 V. This time was much shorter than the previous shortest time (5 s) it took for metal glow plugs to reach the same temperature. The first generation ceramic heater elements, however, were mounted with a controlling resistance to prevent excessive temperature rise inside the metal fitting. This is because if the battery voltage was directly applied, the saturation temperature exceeds the limits of heat resistance of the insulating ceramic material and the metallic heating coil, resulting in premature disconnection. Standard-grade silicon nitride was used as the material of the insulating ceramic.

The second-generation ceramic heater element used highly heat-resistant silicon nitride instead of standard silicon nitride as the material of the insulating ceramic. The material of the heating element of the internal resistor was also changed to a ceramic. Although the heating performance was slightly lower than the first-generation, the heat resistance had improved substantially to 1,350°C from 1,200°C with the first-generation, allowing for the removal of the controlling resistance, which was needed in the first-generation.

The third-generation achieved a rapid temperature increase; 1,000°C in 2 s with an 11 V source by lowering the resistance value while maintaining the heat resistance from the second-generation. Metallic tungsten, which was used in the second-generation as the material for the lead of the internal resister, was replaced by ceramic in the third-generation, realizing an all-ceramic element and improving the reliability. A composite of tungsten carbide and silicon nitride was used as the material for the ceramics of the internal resistors in the second and third-generations, and the resistance value could be controlled by adjusting the content of the conductive substance, tungsten carbide.

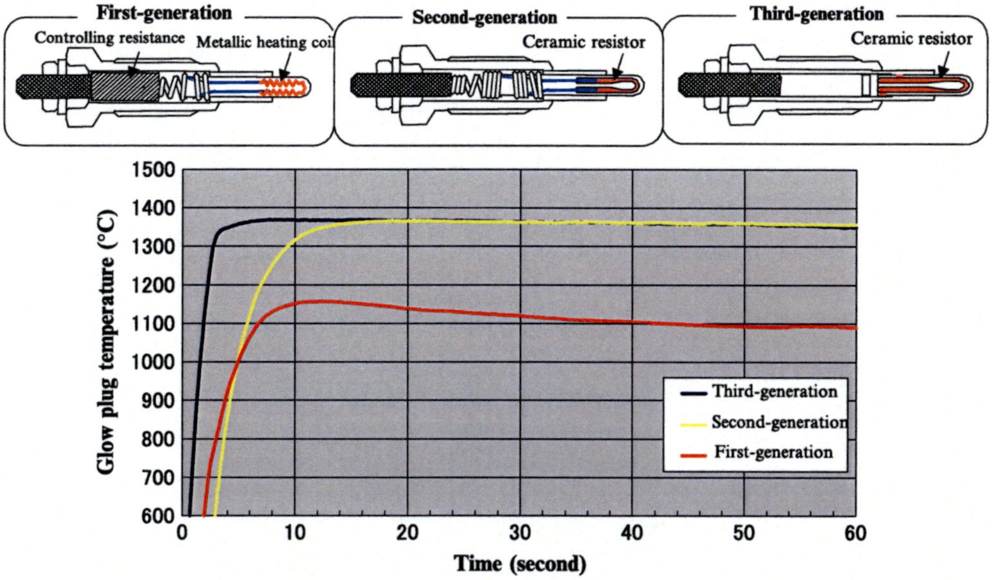

Fig. 12.3 Thermal performance of glow plug. Ceramic glow plugs are characterized by their high-temperature heating performance. The demand has been increasing for rapid temperature rise type plugs that enhance engine start performance

12.1.2.2 Manufacturing Method

The process flow of a ceramic heater element is shown in Fig. 12.4. Insulating ceramic material is made by weighing silicon nitride and sintering additives, and pulverizing and mixing them for some period of time in a trommel (Note 12.3). In this process, an organic binder is also added to the slurry, which is then spray dried and the powder is press formed by dies into the respective shape. In the first-generation ceramic heater element, the metallic heating coil and the insulating ceramic material are pressed in a mold so that the coil is enclosed in the ceramic, and then de-waxed and sintered by hot pressing. After sintering, a cylindrical ceramic heating element is shaped by polishing.

In respect to the second and third-generation ceramic heating elements, materials for internal resistors are weighed, pulverized, mixed and spray dried, in the same manner as the insulating ceramic material in order to produce dry powder. The powder is added to a thermoplastic resin and kneaded to create a highly viscous mixture, which is formed into resistors by the injection molding that features high form accuracy. Then, the formed internal resistor and the insulating ceramic material are pressed in a die so that the resistor is enclosed within the ceramic material in the same manner as the first-generation. It is then de-waxed and sintered by hot pressing and polished to produce the ceramic heating element.

Note 12.3 A kind of rotary sieve, a rotary screen.

12.1.3 Future Prospects

The demand for ceramic glow plugs that feature high-temperature heating is estimated to expand in response to the trend toward diesel engines of lower compression ratios. In respect to the elevation of glow heating temperature intended for the improvement of engine start performance, further enhancement may be difficult from the point of view of thermodynamics as the grain boundary phase of silicon nitride is the glass phase. A breakthrough is needed so that the advantages of ceramics may be utilized to a greater extent.

In addition, it is believed that the detailed control of diesel engines is required in order to clear stricter emission control standards. Specifically, a method of controlling fuel injection timing based on combustion pressure data is currently being discussed. If this method is realized, glow plugs equipped with a pressure detecting function will be required.

In Japan, diesel automobile vehicles have not been very popular due to their engine noise and smell from their exhaust emissions. However, they feature high levels of low-speed torque and are suitable for city driving. They also will certainly be appraised in Japan if the means for clearing the emission control standards are established.

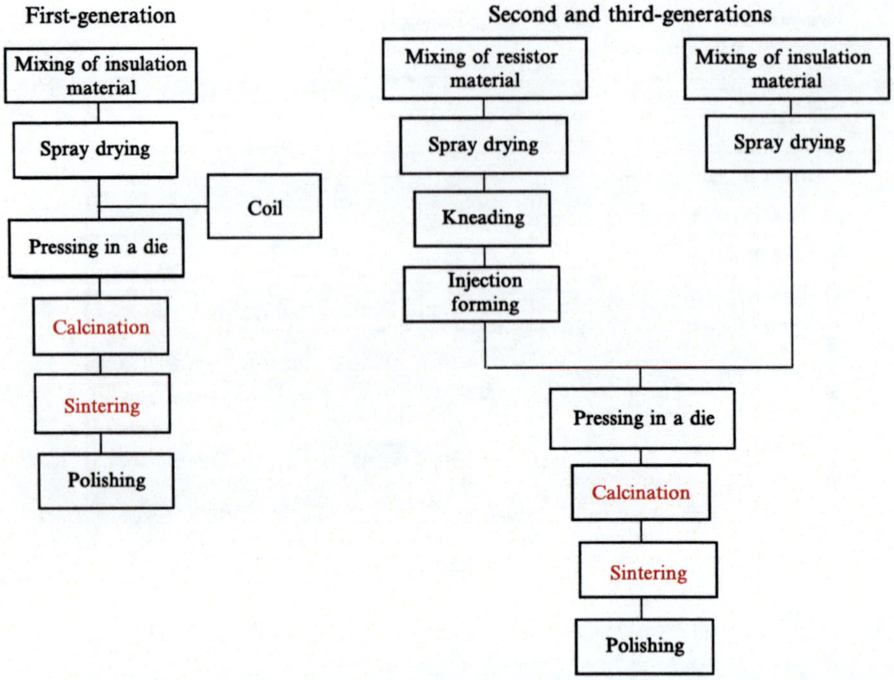

Fig. 12.4 Process flow of ceramic heater element. The metallic heating coil or the injection formed internal resistor is press-formed in a die to create an isolated section

Spark Plug Equipped with Resistor (1971) 12.2

The spark plug is mounted on an engine and extends into the combustion chamber. It is the functional part that ignites gas (fuel/air) mixtures and is required to release sparks steadily under every imaginable condition. Once the engine is started, it is exposed to high temperatures and pressures caused by the explosive combustion. Then, in an instant, it is rapidly cooled by introduction of a gas mixture. In addition, high voltages that sometimes exceed 30 kV are applied to it at the time of ignition. These actions are repeated in high-speed cycles (Fig. 12.5). Spark plugs are called the "heart" of any engine because of this reason. Insulating ceramics that are used in spark plugs will be introduced below, with the focus placed on the manufacturing method.

12.2.1 Background of Development

Ceramics have been incorporated in engines (internal combustion engines) for many years. It is believed that Jean-Joseph Étienne Lenoir (France) was the first to use ceramics in the ignition device (spark plug) of a gas engine in 1860 (Note 12.4).

Since that time, spark plugs have been indispensable parts of engines, and as a result spark plugs have evolved in line with engines. During the past years, internal combustion engines, represented by gasoline engines, have evolved remarkably with the aim of high output, high fuel economy and low pollution. The temperature and pressure inside the combustion chamber has increased, while the voltage applied to produce sparks has also increased following this evolution, leading to the usage of spark plugs under even more severe environments and requiring improvements in material properties such as thermal conductivity, mechanical strength and insulation property.

Note 12.4 Electrical noise is caused due to sparks generated at the spark gap, causing negative effects or audible noise on radios and control devices. Resistance is applied, in general, to the circuits subjected to high voltage in order to reduce the noise. In some cases, resistors are also mounted on the plug cord or the cap.

12.2.2 Characteristics

12.2.2.1 *Characteristics of Products*

The structure of the spark plug is shown in Fig. 12.6. The inside of the alumina ceramic, an insulator, is sealed under a high temperature with a resistor material consisting mainly of glass, while the outside is sealed by plastically-deformed main metal part, which is made of low-carbon steel. A variety of characteristics (strength) are required of the alumina ceramic, which is the key material, so as to ensure performance during the use as well as that in the manufacturing process which involves bonding.

Figure 12.7 shows the characteristics required of the spark plug. Characteristics that reflect the characteristics of the insulator are indicated in blue letters. The spark plug insulator is required to possess the properties of alumina ceramic, such as mechanical strength, the thermal property, electrical strength, chemical stability, etc., and needs to be stable as well as in good proportion.

Spark plugs, which are classified into as many as 2,000 part numbers or types, are selected depending on applications and service conditions. The type that incorporates noble metal alloy welded on the discharge portion of electrodes for improvement of ignition performance and lengthening of the service life has been used more widely in recent years (Fig. 12.8).

Spark plugs equipped with a resistor, in which the resistor can be positioned at the location nearest the area where sparks are discharged, have been widely used because they can prevent electrical discharge noise (Note 12.4) more effectively and at a lower cost in comparison with spark plug

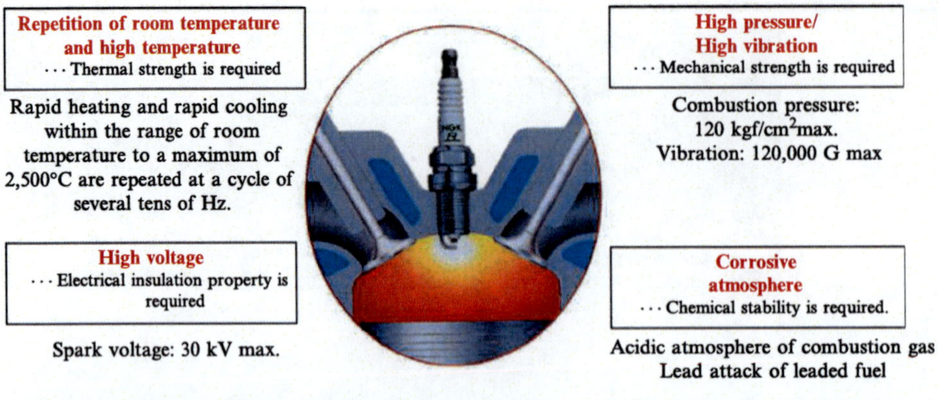

Fig. 12.5 Spark plug service environments

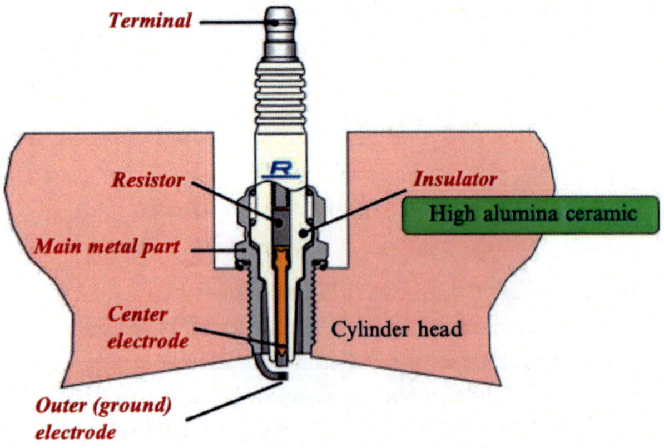

Fig. 12.6 Structure of the spark plug. Spark plugs are affixed to engine cylinder heads via the threads formed on the main metal part. The spark gap on the tip is projected into the combustion chamber and ignites the gas mixture

cords and caps. The monolithic type spark plug, which is currently in the mainstream, is made by filling composite material made of glass powder, metal powder and carbon powder between the center electrode and the terminal. The softening of the material at a high temperature of approximately 900°C and subsequent sealing then takes place. Although it is difficult to evaluate the discharge-noise-prevention performance of plugs equipped with a resistor as this varies depending on environments and conditions, an example of the its effect is shown in Fig. 12.9.

12.2.2.2 Manufacturing Method

The alumina is made of low soda alumina with a Na_2O content of less than 0.05%. It is preferable that Na_2O content is low because Na lowers the insulation property and deteriorates the furnace and the setter.

Although each manufacturer decides its own composition of the sintering additives with consideration towards voltage resistance performance, mechanical strength, thermal conductivity, etc., it is generally made from a mixture of mineral materials such as SiO_2/CaO/MgO, setting the content of Al_2O_3 at approximately 95% (Note 12.5).

The slurry, made by wet-mixing sintering additives into alumina at a specified ratio, is added with organic binders to enhance dispersion performance and ensure the strength of the green body. The slurry is granulated by spray drying and then is pressed in a rubber mold to create the green body. The outer circumference of the green body is cut by a grinder for preciseness before sintering. There are two types of furnaces; the tunnel furnace, in which setters are piled on a carriage in layers for continuous sintering, and the roller hearth kiln, in which setters are placed in a single layer on a roller conveyor. Sintered insulators are assigned their part no., etc. by inorganic pigment ink and coated with glaze (glass), which is melted under a temperature of approximately 900°C. The glaze prevents contamination of the insulator, which is extruded from the main metal part and prevents leakage of sparks by strengthening the adhesion with the plug cap. Then the insulator from the above is sealed within the inner hole of the insulator.

Meanwhile, the main metal part is made by rough forming the low-carbon steel through cold forging. The outer circumference is cut to create an accurate form. Then, a ground electrode is mounted by welding, screw threadings are formed by rolling and the plating is applied. The finished metal part is assembled with the insulator (which has already

Note 12.5 In general, spark plugs that are installed on the cylinder head by using screws sized less than M 12 (ISO metric thread) in diameter are called the small diameter type. The length of screws is 19 mm in general, and the ones that use longer screw are called the long reach type.

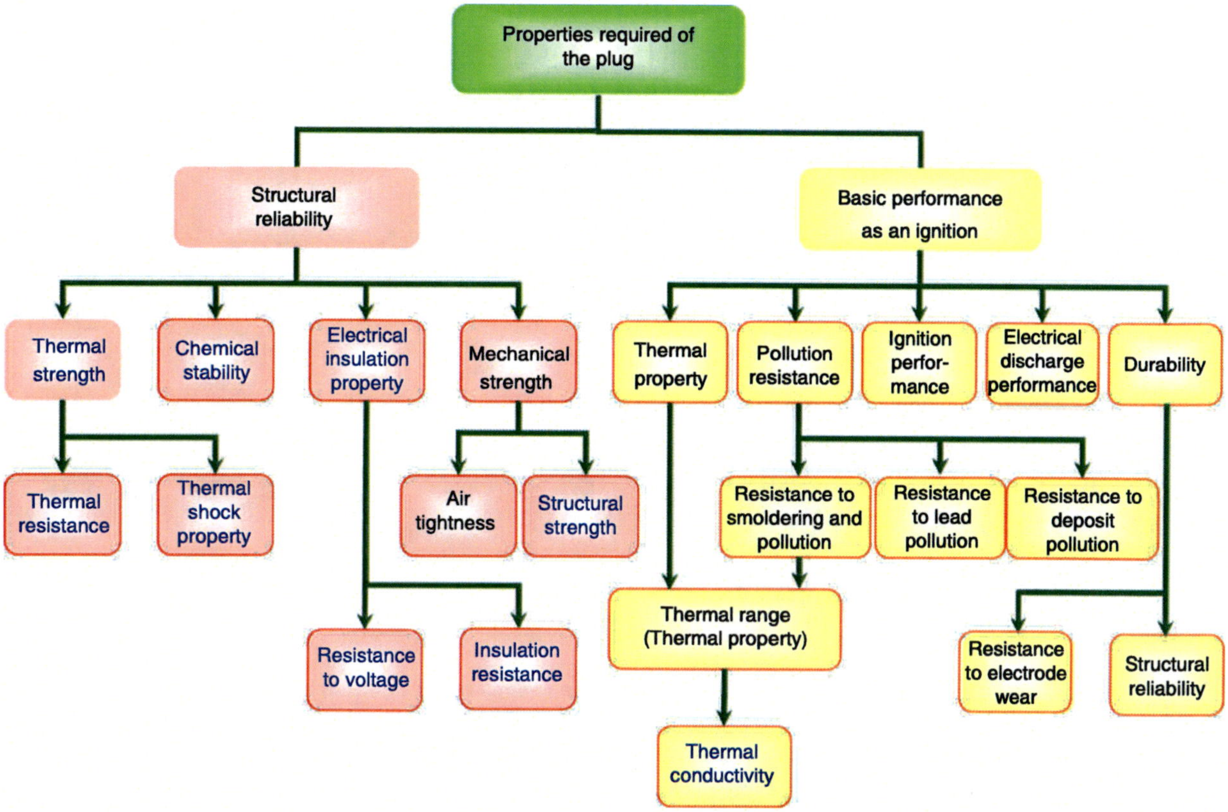

Fig. 12.7 Characteristics required of spark plugs. There are many characteristics required for the proper functioning of spark plugs. These characteristics need to be well-balanced. Many of the characteristics are affected by the characteristics of the insulator ceramic

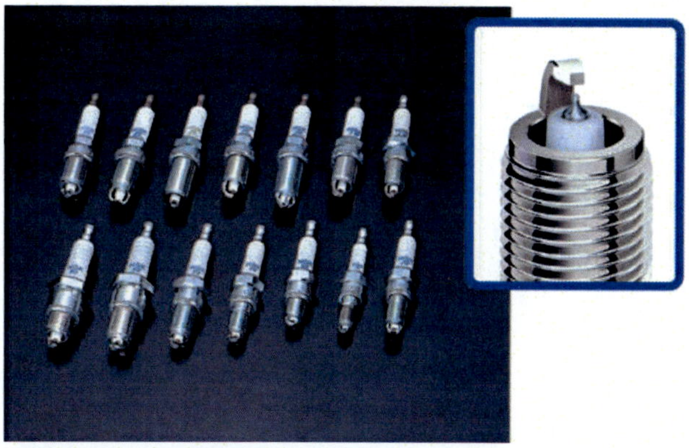

Fig. 12.8 Spark plugs and Ir electrodes. Spark plugs, which are classified into as many as 2,000 part numbers, are selected depending on applications and service environments. The ones that use precious metals such as the Ir alloy on the electrode tip have recently been widely used

been bonded) and is mechanically caulked. Then, the ground electrode is bent to create a spark gap and the gasket is caulked for completion.

12.2.3 Future Prospects

The following spark plug technologies will be required in line with the development of engines:

1. Reduction in diameter, increase in reach (Note 12.5)
2. Improvement of ignition performance
3. Extension of service lifespan
4. Addition of sensing function (Note 12.6)

Note 12.6 A function that measures combustion pressure waveforms and feeds them back to the control of engine, etc. is being discussed.

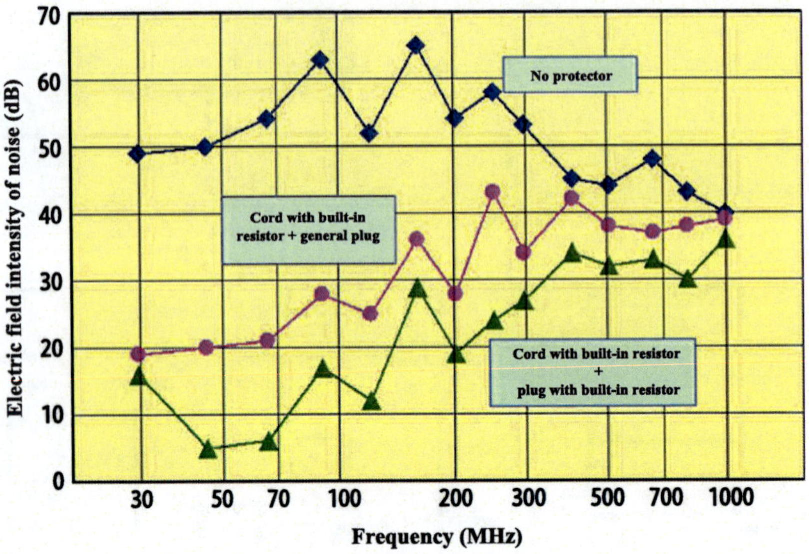

Fig. 12.9 Example of passenger car electrical noise level measurement. The noise level drops remarkably when a cord and a spark plug, both with a built-in resistor, are used. This is in comparison with that without any noise preventing mechanism

Among these requirements, the reduction of diameter and the increase of reach largely rely on the improvement of the material properties of insulator ceramics as well as the improvement of dimensional accuracy. Therefore, developments of ceramic technologies hold the key to the improvement of engine characteristics.

Thermally Insulated Diesel Engine (Not Yet Commercialized) 12.3

During the cold war in the 1970s, the fact that the majority of heat resistant metals such as nickel, chrome, tungsten and titanium were produced in communist countries presented a serious problem to non-communist countries. Meanwhile, the development of new materials in place of these heat resistant metals was important. These new materials (i.e. ceramics) included silicon, nitrogen and carbon materials, all of which can be found in abundance. Development of insulated diesel engines was promoted actively by companies including Cummins Inc. in the U.S. Insulated diesel engines were developed to improve durability, enhance power output ratio and reduce the weight of engines by removing water cooling device (radiator), which is the jugular of the engine, by incorporating ceramic materials (Note 12.7).

In Japan, awareness of the general energy crisis arose as a result of the second oil shock. In order to enhance engine efficiency and respond to the diversification of fuels, the development of high-efficiency insulated diesel engines that incorporate ceramics was promoted under the supervision of the Ministry of International Trade and Industry [MITI, in 2001 changed name to the Ministry of Economy, Trade and Industry (METI)] and led by New Energy Development Organization (NEDO). Kyocera developed a thermally insulated diesel engine that incorporated a newly developed silicon nitride (Note 12.8) ceramic and succeeded in using it in a car engine to drive along the Kinkowan bay in Kagoshima Prefecture, Japan. The scene was broadcast by NHK TV during its New Year's program in 1982 (Note 12.9), and many research institutes thus began to focus on the development of ceramics following this event.

12.3.1 Background of Development

Ceramic engines were first developed in the U.S. in the 1970s, as shown in Table 12.1. Efforts were made to use ceramics as engine components were promoted while targeting the gas turbine engine and the diesel engine. Cummins Engine Co., Inc. (a major diesel engine manufacturer in the U.S.) signed a contract with the U.S. government in regard to the insulated diesel engine development program. Meanwhile, in Japan, development of components was initiated by Kyocera, in collaboration with Isuzu Motors Ltd. (which had the largest share in small diesel engines in Japan during that time).

Kyocera decided that in-house testing using actual engines should be performed, instead of relying only on user comments for the design and evaluation of components. This was decided in order to lead the above program to a successful conclusion. Therefore, the company received a subsidy for the development and research of important technologies (1978–1980) from the MITI jointly with Isuzu Motors, and designed and prototyped ceramic diesel engines. The first model was designed after the model of the Cummins Engine, having a silicon nitride liner inserted inside a water-cooling type cylinder block made of cast iron and the piston and the head plate made of silicon nitride. The engine was assembled using these components, and durability tests and performance tests were conducted. Later, the company designed a more advanced non-water cooling type of engine made entirely of ceramics. In the late autumn of 1981, it was contacted by NHK television about the possibility of introducing ceramic engines on its New Year's TV program, and it decided to manufacture all-ceramic engines in response to that request.

Note 12.7 About 25% of the energy from fuel is used effectively as power. More than half of the energy is wasted as a result of water cooling and lost in the exhaust gas. The radiator was removed to improve the efficiency of the engine by reducing the energy wasted in cooling.

Note 12.8 A compound consisting of nitrogen and silicon, indicated as Si_3N_4. Alumina and rare earth oxide are added for liquid-phase sintering. The material features high insulation property and high thermal shock resistance due to low thermal conductivity and low thermal expansion coefficient.

Note 12.9 Japan Broadcasting Corporation, NHK TV: Japan's national public broadcasting organization.

Table 12.1 History of ceramic diesel engine development

Year	Events
Around 1970	Development of ceramic engine began in the U.S.
1977	Kyocera began development of glow plug
1978–1980	Isuzu Motor and Kyocera jointly designed and developed prototypes (subsidy for development and research of important technologies, the Ministry of International Trade and Industry)
1981	Glow plug employed by Isuzu Motors
Autumn of 1981 to New year of 1982	Prototype of all-ceramic engine was produced Mounted on a passenger car for demonstration run
1982	Ceramic hot plug used in Aska of Isuzu Motors
1983	Glow plug employed by Mitsubishi Motors and Mazda
1984	Visited by Japan's Showa Emperor (1.2 L 4-cylinder flat engine was prototyped)

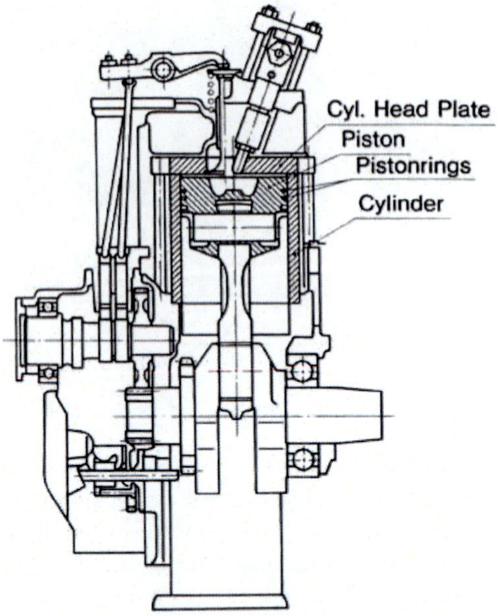

Fig. 12.10 Cross-sectional view of ceramic engine KC201. A non-water cooled engine type incorporating silicon nitride SN220, in which the majority of the engine components (piston, cylinder, head plate and ceramic turbo) are made of ceramics

12.3.2 Characteristics

12.3.2.1 Characteristics and Structure

After manufacturing the components and performing durability tests, a 2.8 L 3-cylinder engine, having the cylinder, piston head and head plate made of silicon nitride, was assembled and the ceramic engine was mounted in a passenger car. This engine was based on the air-cooled 3-cylinder diesel engine (manufactured by Mitsui DEUTZ Diesel Engine Co., Ltd., Model F3L913, 2,800 cm^3 displacement, 58 horse power). The air-cooling fins were removed from the model and the silicon nitride cylinder had a smooth surface. The appearance and the structure of the engine are shown in Figs. 12.10 and 12.11, respectively. Kyocera succeeded in a demonstration run conducted at the end of 1981 and the test vehicle drove on a volcanic rock road in Sakurajima, Japan during the New Year's feature program (Fig. 12.12). The impact of the world's first driving of a vehicle with a "ceramic engine" was substantial, attracting attention to fine ceramics and enhancing their appraisal.

Later, Kyocera modified the components and created a prototype for a 1.2 L 4-cylinder flat engine (KC401) and conducted not only a demonstration run but also a 100-h endurance test run at full power as well as other performance tests. The appearance and the structure of this engine are shown, respectively, in Figs. 12.13 and 12.14. In 1984, Japan's Showa Emperor visited the Kokubu Plant at Kyocera and observed the modified ceramic engine. Specifications of these engines are indicated in Table 12.2.

Fig. 12.11 All-ceramic non-cooling type engine. This engine was based on the air-cooled 3-cylinder diesel engine (manufactured by Mitsui DEUTZ Diesel Engine Co., Ltd., Model F3L913, 2,800 cm^3 displacement, 58 horse power). The air-cooling fins were removed from the model and the silicon nitride cylinder had a smooth surface

12.3.2.2 Development of Materials and Components

Silicon nitride, featuring higher heat resistance and lower thermal expansion coefficient compared with other ceramics, has high thermal shock resistance. The development of

Fig. 12.12 Vehicle mounted with the all-ceramic engine speeding along a lava road in Sakurajima. An all-ceramic engine car, Isuzu Aska mounted with a 3-cylinder all ceramic engine KC201 shown in Figs. 12.10 and 12.11, speeds through a volcanic rock road along the Kinko Bay. Seen in the back ground is the Sakurajima Island of Kagoshima Prefecture. The scene was broadcast during the 1982 New Year program by NHK TV

Fig. 12.13 KC401 Engine, a modification of the air-cooled 4-cylinder flat engine. The *black area* at the *bottom* of the picture is the ceramic cylinder

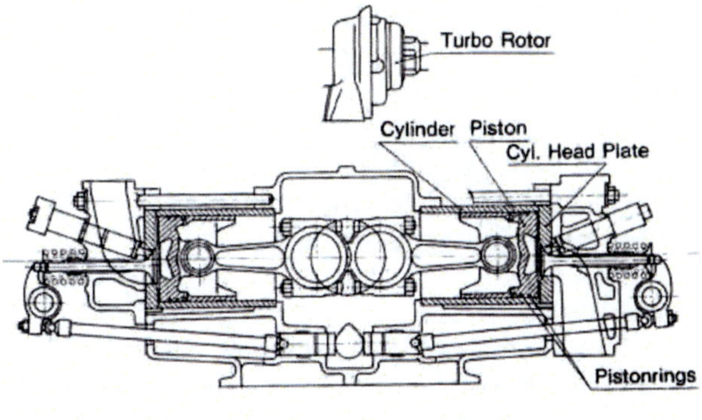

Fig. 12.14 Structural drawing of the KC401 engine. The *hatched areas* in the drawing indicate the ceramic parts

Table 12.2 Specifications of ceramic engines

	KC201	KC401
Engine type	4-Cycle diesel engine	
System	Direct injection	
Inner diameter	100 mm (diameter) ×120 mm (stroke)	80 mm (diameter) ×60 mm (stroke)
Cylinder layout	In-line 3-cylinder	4-Cylinder flat
Capacity (cm^3)	2,837	1,236
Max. output (kW/rpm)	–	33.1/4,500
Max. torque (N m/rpm)	–	84.3/2800
Compression ratio	17	17.5

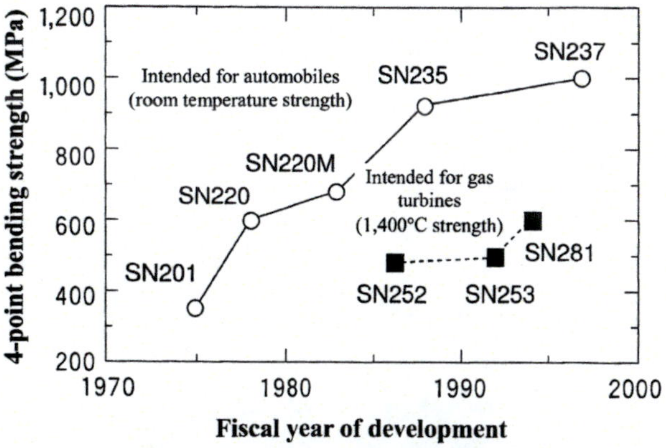

Fig. 12.15 Development process of silicon nitride materials. Compositions and processes with priorities focusing on cost and room temperature strength were employed for materials intended for automobiles, while compositions featuring enhanced high-temperature strength and oxidation resistance were developed for use in gas turbines. These parts were used by AGT and ATTAP in the U.S. as well as in the CCT project in Japan

materials was carried out by using a rare-earth based sintering additives instead of the magnesia alumina-based sintering additives that was used previously. Later, silicon nitride materials were classified into two types. These types were developed parallel to each other and classified as materials intended for automobiles, with a focus on cost, whereas materials intended for gas turbines, focused on performance. Figure 12.15 shows the development process. In respect to materials for automobiles, materials and processes were selected with a priority on cost. In addition, emphasis was placed on achieving high strength. As for materials for gas turbines, the priority was placed on performance and development was performed aiming to satisfy high-temperature resistance and oxidation resistance (Note 12.9).

Kyocera selected the hot plug and the glow plug shown in Fig. 12.16, which are the main components in the swirl chamber and reflect the characteristics of ceramics more than any other component, as the target of development in the early stage. Development was carried out in cooperation with Isuzu Motors, whom showed an interest in developing ceramic components for diesel engines.

Development of the glow plug began in 1977. Originally, a heating circuit was created by printing tungsten paste on a formed alumina sheet, and the heater was produced by sintering the sheet wrapped around a round alumina bar. It turned out that the resistance value of this prototype increased after repeated temperature rise cycles and it could not stand the thermal shock caused by fuel injection.

For the next prototype, silicon nitride featuring high thermal shock resistance was used, as shown in Fig. 12.17. The heating element was embedded in the silicon nitride. The hot press method was employed after a series of discussions. A tungsten coil was used to reduce thermal stress between the silicon nitride and the heating element. The main frame of the manufacturing method and materials was decided, but a number of problems were detected. These problems included low durability confirmed by an overload durability test and the deterioration of characteristics with age. However, through the optimization of materials and structures, the glow plug, which had a 2 s temperature rise, passed a 50,000 temperature rise cycle test and a 100 h durability test. This has been used by Isuzu Motors and Mitsubishi Motors since 1981 and 1983, respectively, and later by Mazda Motor Corporation.

The second version of the ceramic component was the ceramic hot plug shown in Fig. 12.18. It was first used in the

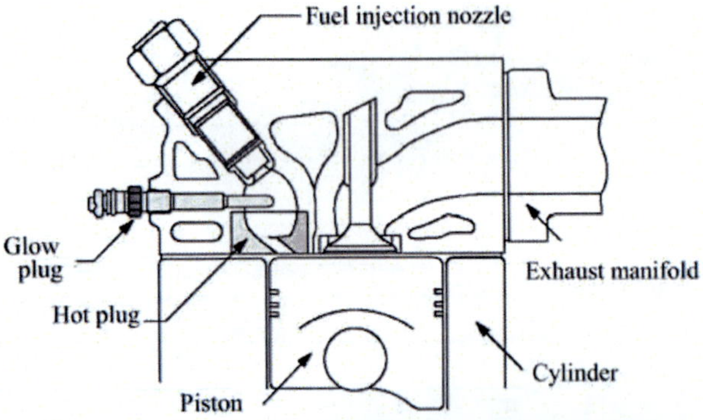

Fig. 12.16 Structure of diesel engine cylinder head. In order to improve engine start performance of small diesel engines, a silicon nitride glow plug capable of quick temperature increase and a silicon nitride hot plug, featuring low heat capacity, low thermal conductivity and high temperature increase performance, were used

Fig. 12.18 Silicon nitride hot plug. Temperature drop of combustion gas, which occurs at the time of cold start of small engines, was reduced and the starting performance was improved because of its insulation performance and low heat capacity

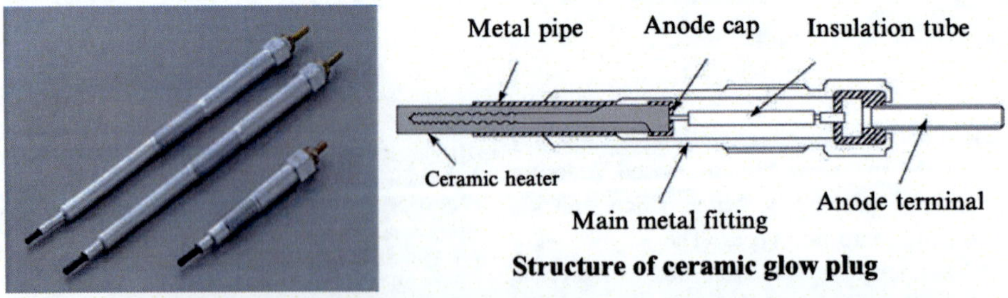

Fig. 12.17 Glow plug made of silicon nitride (*left*) and its internal structure (*right*). The diesel engine glow plug was manufactured by the hot press method. The heater is made of tungsten, a metal featuring thermal expansion properties as low as silicon nitride. It increases the temperature more quickly in comparison with conventional metal glow plugs, improving the engine start performance dramatically

Aska of Isuzu Motors in 1982. Following the introduction of the component, idling noise was reduced and cold start performance was improved as temperature in the swirl chamber was increased due to the insulation property. Mazda, which incorporated the product later, reported a reduction in diesel particulate emission, while Toyota Motors, which used self-developed products, reported a 10% increase in engine output.

12.3.3 Future Prospects

By utilizing silicon nitride featuring high thermal resistance and thermal shock resistance, the materials are expected to be effective for thermal engine components that require high thermal efficiency as well as components that support utilization of diversified fuels other than fossil fuels. This will offer a means for coping with energy and environmental problems.

12.4 Ceramic Turbo Charger (1985)

Expectations for new material development increased beyond one's expectation in the 1980s, when "ceramic fever" griped society which resulted from the development of the ceramic engine. This fever pushed forward the developments of various ceramic components for automotive engines. The ceramic turbo charger was one of the most impressive accomplishments among the products developed during this period. The development not only enhanced the responsiveness of the turbocharger but also proved the reliability of ceramics, which are conventionally thought of as brittle materials, to the marketplace by the commercialization of ceramics for use in dynamic structural components.

12.4.1 Background of Development

A compact and high output engine is created if the pressure of the mixed gas injected into the combustion chamber is increased by supercharging in an internal combustion engine. Supercharging the intake gases is done by a pump called a supercharger, which directly utilizes the engine's spinning crackshaft to compress the intake gas. This contrasts with a turbocharger, which uses the flow of exhaust gas to compress the intake gas. The turbocharger is characterized by its simple structure and compact size.

Turbochargers were developed based on the patent obtained by Alfred J. Büchi in 1905. They were mounted on Boeing B17 bombers in 1938, and began to be widely used in the diesel engines of trucks in the mid 1960s. Vehicles mounted with turbochargers won a series of car races in the 1970s, and then they began to be used in commercially available gasoline-powered vehicles.

The turbocharger is characterized by its ability to generate a large power output from a compact engine. It however suffers from what is called turbo lag, which is a slight delay before the reaching of a high engine speed after stepping on the acceleration pedal. Ceramics were used to reduce the weight of the rotor and reduce the delay or turbo lag.

12.4.2 Characteristics

12.4.2.1 Materials

The turbocharger is made by fitting two impellers on a shaft. One of the impellers (a turbine wheel) is rotated by the exhaust gas, and the other (a compressor impeller) increases the pressure of air fed to the engine (Fig. 12.19).

Nissan Motor Co., Ltd. released the first mass production vehicle equipped with a turbocharger in Japan in 1979. The turbine rotor of the vehicles was made of Ni-based heat resistant alloy (GMR235). Later in 1985, the world's first ceramic turbocharger that incorporated silicon nitride was commercialized, following studies on the utilization of ceramics (Fig. 12.20). Impellers are normally made of aluminum, but some of them are made of light-weight CFRP (carbon fiber reinforced plastics).

Ceramics were used mainly because of the lightweight property, as well as their high thermal resistance, which was important in order to resist degradation in the high-temperature exhaust gas. Titanium-based intermetallic compounds (TiAl), which also possess lightweight and thermal resistance, were also considered as a suitable material for this application.

In general, ceramics are brittle materials, and breakage during driving may result in leakage of engine oil and result in an engine fire. This risk was avoided by mounting the ceramic rotor on the tip of the metallic shaft so that oil leakage is prevented by the metallic shaft in the case of breakage of ceramic components. In regard to the breakage of ceramics, careful inspections for defects were carried out during the manufacturing process. Over-speed tests were performed on all finished products and guarantee tests were performed to remove and scrap low-strength products. As explained above, the ceramic turbos were developed on the basis of a number of technologies, including high-temperature bonding technology, technology of the forming rotors of complicated shapes and the reliability guarantee technology that involves fracture mechanics.

Fig. 12.19 Operation mechanism of the turbocharger. The exhaust gas from the engine flows across an impeller which is connected to turbine that compresses the gas fed into the engine resulting in supercharging

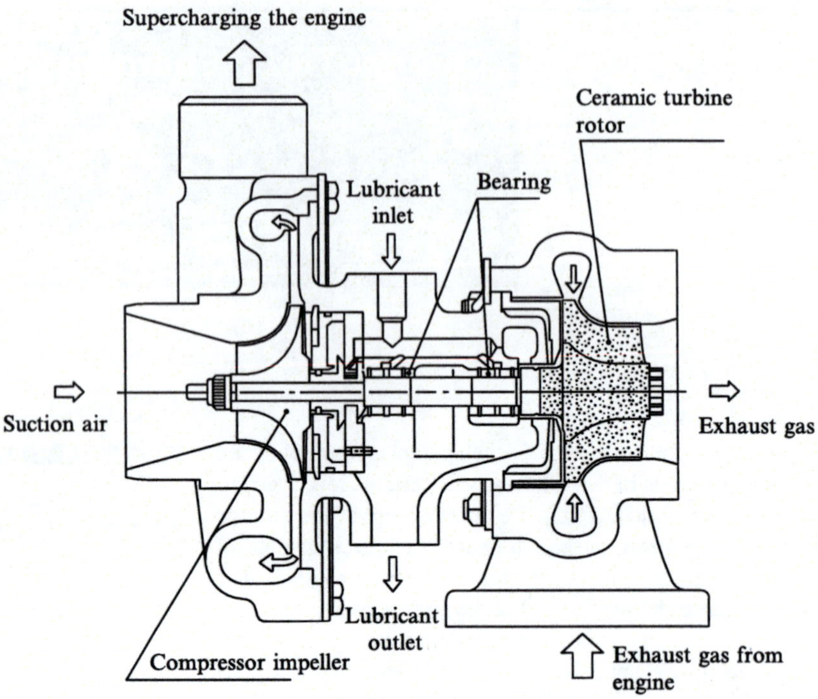

Fig. 12.20 A ceramic turbo charger assembled in the housing. The ceramic turbine on the *right side* is bonded on the metallic shaft and the metallic shaft fitted with screws is inserted into the aluminum impeller on the *left side*, fixing the end by nuts

12.4.2.2 Manufacturing Method

An example of the ceramic rotor manufacturing method is shown in Fig. 12.21. The ceramic rotor is generally made by bonding a ceramic wheel on a metallic shaft and fitting an aluminum impeller on the other side of the shaft with screws.

The ceramic wheel is manufactured by injection molding or slip casting. Basically, silicon nitride powder is mixed with sintering additives such as rare earth oxide (Note 12.10), and then a molding agent (Note 12.11) is added to the mixture to form the wheel. Cracks and strains tend to be generated during de-waxing in the injection molding process and during drying in the slip casting process. Manufacturers have their own know-how with respect to these processes and the preceding forming process.

It then goes through sintering and machining and is bonded on the metallic shaft before the rotor is completed. The machining is performed mainly to ensure dimensional accuracy of the shaft in preparation for bonding. If necessary, parts of the blades are removed by grinding for balanced rotation.

Note 12.10 Because silicon nitride features low sinterability, additives are blended in silicon nitride for sintering. The additives remaining in the sintered body cause the deterioration of the properties of silicon nitride, and therefore, rare earth oxides are normally used in consideration of their sinterability and material properties.

Note 12.11 As non-clay ceramics have low plasticity, their formability is compensated by adding organic substances. The organic substances are removed by heating. However, the strength of the green compact decreases, resulting in cracks and deformation, when the temperature is increased to the range of 300–500°C. This is the range in which the organic evaporates and thermal decomposition occurs. Forming agents are selected based on the know-how of each manufacturer.

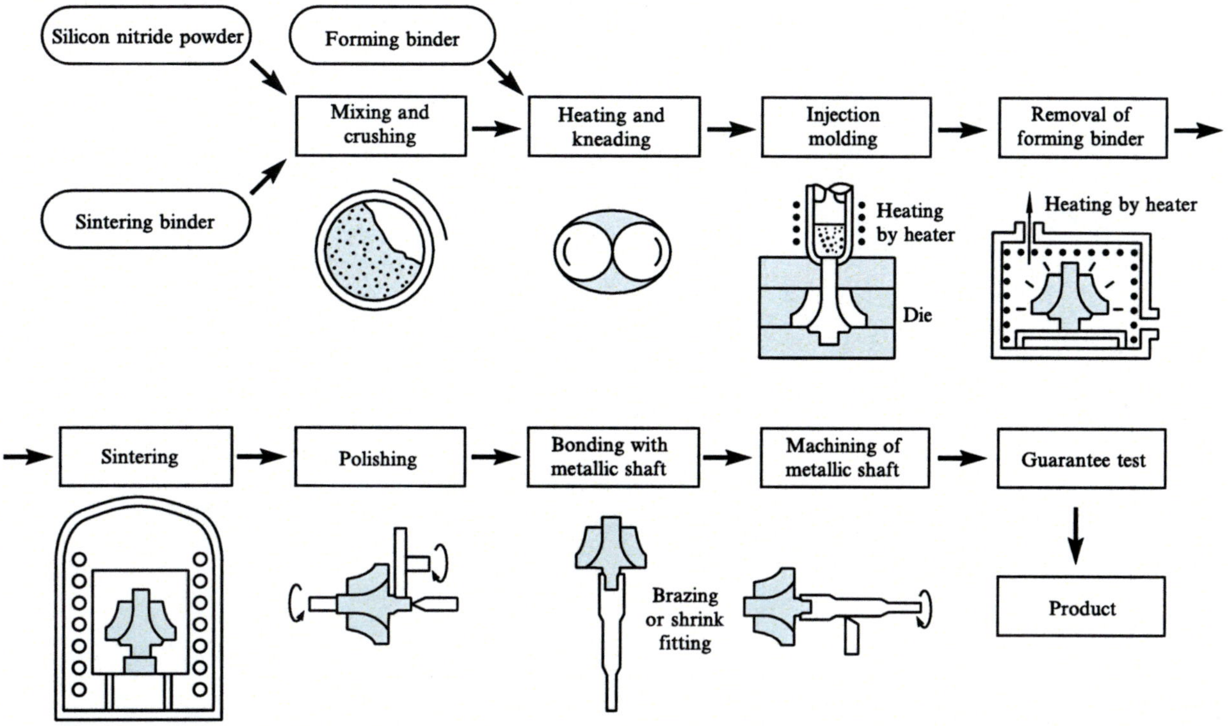

Fig. 12.21 Example of ceramic rotor manufacturing method. The wheel is produced by injection molding, using resin as the molding agent, and is bonded with the metallic shaft

12.4.2.3 Performance of Product

Acceleration characteristics of ceramic turbochargers are compared with that of metallic rotors in Fig. 12.22. The size of the ceramic rotor is the same as the size of the conventional metallic rotor. However, the density of the silicon nitride ceramic is approximately 3.2 Mg/m^3, while the density of the conventionally-used heat resistant metallic alloy is approximately 8.2 Mg/m^3. Acceleration performance was improved by approximately 36% due to this weight reduction.

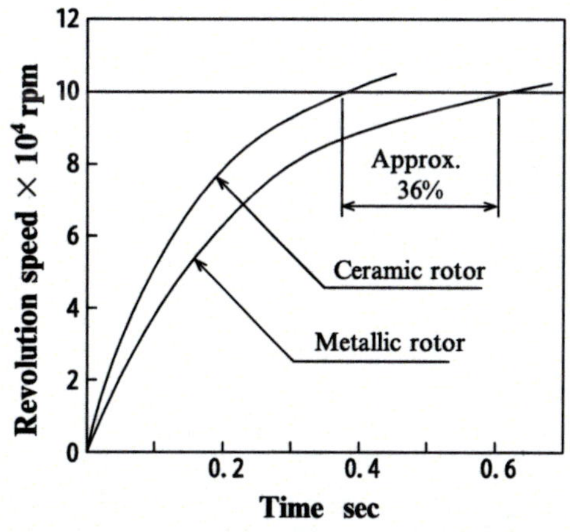

Fig. 12.22 Acceleration characteristics of the ceramic rotor, in comparison with the metallic rotor. Rotor revolution speed increases at a higher rate when ceramics are used

12.4.3 Future Prospects

This ceramic turbocharger is not currently used in new automobiles, and the production is limited to the quantities needed for replacement. This is perhaps because of a shift in the marketplace from high performance-oriented products to products focusing on gas conservation due to environmental and safety concerns.

Cordierite Honeycomb for Automotive Catalytic Converters Used for Exhaust Gas Purification (1976) 12.5

Honeycomb ceramics (Fig. 12.23) are extrusion-formed bodies and are structured like honeycombs, having many holes like a beehive enclosed by thin partitions. There has been a trend toward strict regulations on HC (hydrocarbon) (Note 12.12), CO and NO_x, harmful components in exhaust gas from gasoline-powered vehicles in Japan, Europe and the U.S. since the 1970s. In response to this trend, exhaust gas purification systems that incorporated catalysts were established called catalytic converters. Honeycomb ceramics that are made of cordierite having excellent thermal resistance and thermal shock resistance due to its low thermal expansion properties, are widely used as catalyst supports.

12.5.1 Background of Development

In response to the global concerns about air pollution, many countries including Japan, the U.S. and Europe, introduced emission control regulations on gasoline-powered automobiles, starting from the Muskie Act (Note 12.13) established in 1970 in the U.S. Since then, regulations are becoming stricter year after year. Developments and improvements of vehicle emission purification devices have been promoted in line with the constant improvements of automobile engines.

Vehicle emission purification devices became dramatically more advanced by the incorporation of catalytic converters. In particular, the emergence and improvement of the three-way catalyst, which is capable of removing three harmful components, HC, CO and NO_x, simultaneously, are contributing largely to the technologies that address the strict regulations of today.

Note 12.12 Generic name of compounds that consist of carbon atoms (C) and hydrogen atoms (H). Also refers to a specific set of atoms plus other atoms.

Note 12.13 The "Clean Air Act Amendment of 1970" proposed by Edmund S. Muskie, Senator of the U.S. Democratic Party in 1970.

Honeycomb ceramics are used widely and in large amounts as the catalyst support in catalytic converters. Figure 12.24 shows the location of a catalytic converter in an automobile.

12.5.2 Characteristics

12.5.2.1 *Characteristics and Specifications of Honeycomb Ceramics*

Honeycomb ceramics is a generic name that indicates ceramic products that are structured like honeycombs, having a number of uniformly distributed square cells. In catalytic converters for automobiles, γ-alumina containing precious metal is normally coated on the inner surface of the cells of the honeycomb ceramics for three-way catalyst function. Contact efficiency with emissions to be purified is improved and airflow resistance is reduced, reducing the pressure drop in engine output as much as possible. This advantage results from the structural characteristics of honeycomb ceramics.

In terms of actual application, thermal resistance to high emission temperatures, thermal shock resistance to quick heating and cooling that are frequently repeated, mechanical strength during embedding into the metallic casing as well as against vibrations from vehicles and the adhesion performance with the catalyst are required of catalyst carriers. Porous cordierite, that has a low coefficient of the thermal expansion, is used as the support of catalyst for vehicle emission purification. Characteristics of the honeycomb support are indicated in Table 12.3. Honeycomb structures are generally classified by the thickness of the cell partition and the number of cells, and the partition thickness is indicated by mil units (1 mil = 1/1,000 in.) and the number of cells is indicated by the number in 1 $in.^2$ (cpsi). Historically, the 12 mil 200 cpsi honeycomb was first employed, and the 6 mil 400 cpsi products were in the main stream for many years after 1980. Products with thin partitions of 4 or 3 mil were recently commercialized.

Fig. 12.23 Cordierite honeycomb and catalytic converter for purification of vehicle emission. The honeycomb supports a catalyst for the purification of vehicle emissions is manufactured by extrusion forming cordierite featuring high thermal shock resistance

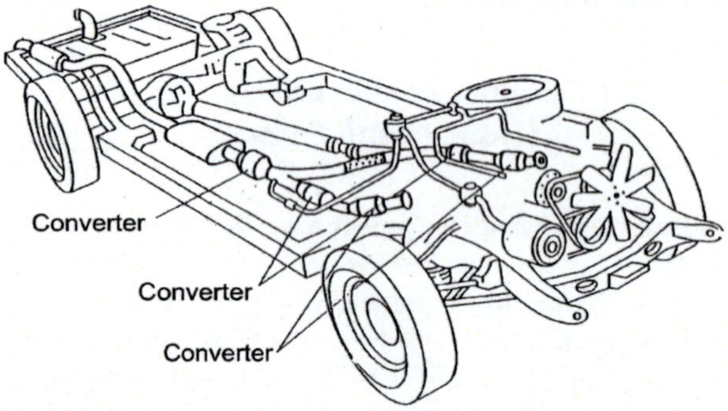

Fig. 12.24 Location of the catalytic converter in an automobile. The exhaust gas from the automobile engine passes through the catalytic converter and is purified while it is hot before being released in the air via the muffler

Table 12.3 Characteristics of honeycomb carriers (example)

	12 mil/ 200 cpsi	12 mil/ 300 cpsi	6 mil/ 400 cpsi
Aperture ratio (%)	68.9	62.7	75.0
Pressure loss ratio (%)	73	115	100
Geometric surface area (cm^2/cm^3)	18.5	21.6	27.3
Bulk density (g/cm^3)	0.51	0.62	0.43

12.5.2.2 *Manufacturing Method*

In 1975, when honeycomb ceramics were first used as the catalyst support for vehicle emissions, they were manufactured by the extrusion method, corrugation method, emboss method, etc. Since 1977, when products with more cells, 12 mil/300 cpsi, were needed for the improvement of purification performance, only the extrusion method has

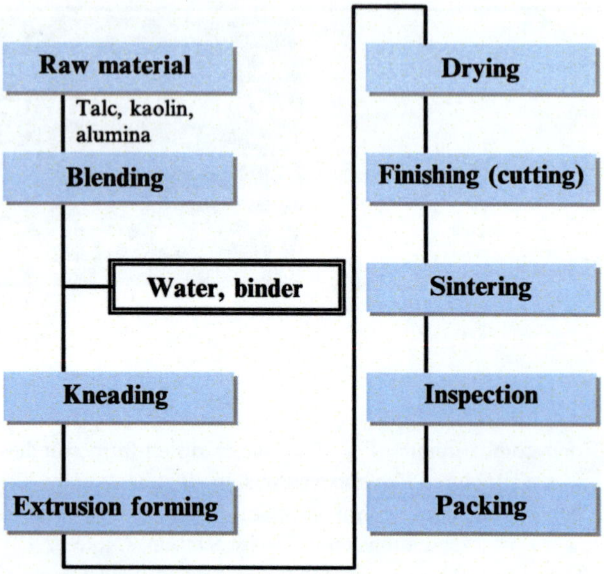

Fig. 12.25 Manufacturing process of honeycomb ceramics. The raw material, made by blending talc, kaolin and alumina, is mixed with water and binder and kneaded into a clay-like state. It is formed by an extrusion machine, dried and sintered in order to manufacture the product

been employed. The extrusion method is suitable for creating homogeneous and high-strength honeycombs and features high productivity. The outline of the process is shown in Fig. 12.25. The raw material is ground to an appropriate size and blended to achieve a specified chemical composition, and then added with water and a binder, kneaded, put into an extrusion machine and extruded out of a die. The extrusion formed honeycomb is dried and sintered to complete the honeycomb ceramics. The raw materials are talc ($3MgO{\cdot}4SiO_2{\cdot}H_2O$), kaolin ($Al_2O_3{\cdot}2SiO_2{\cdot}2H_2O$) and alumina ($Al_2O_3$). These raw materials are blended to create cordierite ($2MgO{\cdot}2Al_2O_3{\cdot}5SiO_2$).

The honeycomb extrusion process is characterized by the extrusion die. The structure is shown in Fig. 12.26. The clay-like mixture (kneaded substance) enters the die through the feeding hole on the back of the die, and then enters honeycomb-structured slits on the outlet side, expanding in crosswise directions. The mixture is compressed, with each piece sticking to each other and forming a honeycomb structure.

12.5.3 Future Prospects

Emission control standards are being strengthened substantially in Japan, the U.S. and in Europe. Honeycombs with thin-partitions and numerous cells are effective for the further improvement of emission purification performance. Reduction in partition thickness leads to catalyst weight

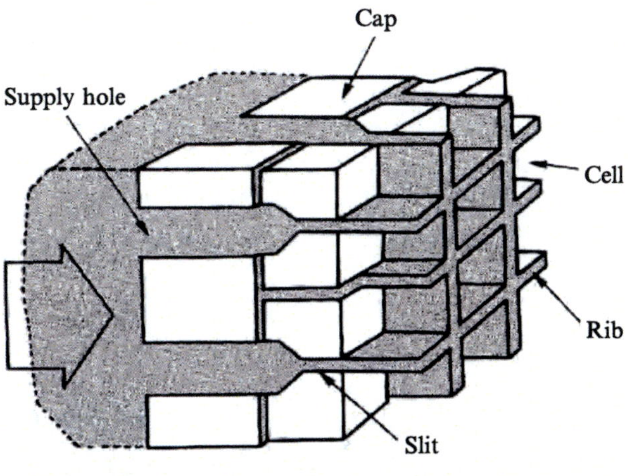

Fig. 12.26 Extrusion die for honeycomb ceramics. The clay (kneaded substance) enters the die through the feeding hole on the back of the die and then enters honeycomb-structured slits on the outlet side, expanding in cross-wise directions. The clay is compressed with each piece sticking to each other and forming a honeycomb structure

reduction, which means reduction in heat capacity, making it possible to heat the honeycombs in a shorter time by means of hot exhaust gas to the optimal temperature needed for the efficient catalysis. In addition, the pressure drop across the catalytic converter of the exhaust gas can also be reduced. The partition thickness has also become thinner. It has become as thin as 6–4 mil, to 3 mil, and then to 2 mil, which is as thin as a piece of tissue. They have also been commercialized (Fig. 12.27). When the number of cells increases, the geometric surface area increases, resulting in an increase in the area of the catalytic conversion of the exhaust gas and improving the purification performance. The exhaust gas processing performance which is improved by thinning the partitions and increasing the number of cells is also utilized to reduce the amount of expensive metal catalysts and to downsize the catalytic converters.

In response to active developments of green diesel vehicles of today, utilization of honeycomb ceramics are expected to expand remarkably in their use as carriers of DOC (diesel oxidation catalyst) and SCR [selective catalyst reduction method (Note 12.14)] in diesel engine emission purification devices.

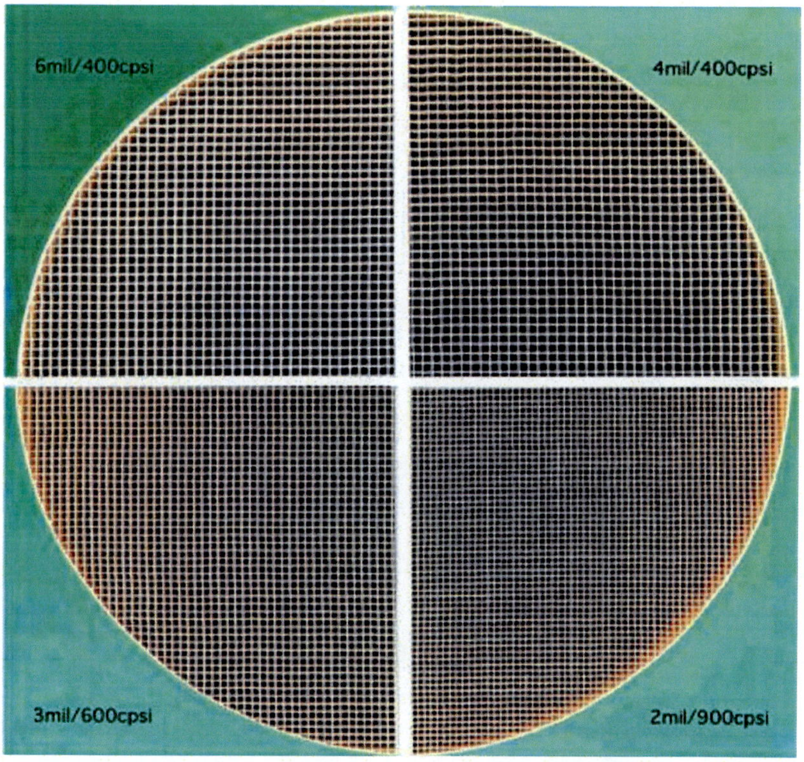

Fig. 12.27 Various cell structures. Structure of honeycombs with square-shaped cells. The geometric surface area increases as the partitions become thinner. They become as thin as 6–4 mil, 3 or 2 mil, and the number of cells increases

Note 12.14 A catalytic method in which NO_x is reduced to nitrogen and water. The reduction needs to be performed in an oxidizing atmosphere, and therefore, reducing agents such as ammonia and urea need to be added.

Exhaust Gas Temperature Sensors (1975) 12.6

Thermistor-type temperature sensors (Note 12.15) for controlling the automobile exhaust gas using a heat sensitive element which is made of stabilized zirconia were developed and commercialized (Fig. 12.28) because mounting an overheat alarm device onto the catalyst converter was a necessity with the implementation of emission control regulations.

Later, a thermistor element (heat sensitive element) that incorporated $YCrO_3$-based perovskite oxide, which is capable of measuring a wider range of temperature, was developed. Thermistor elements that can cover a much wider range of temperatures from –40°C to 900°C have been developed in recent years. Exhaust gas temperature sensors are currently used to protect turbo chargers and catalysts as well as to control a variety of exhaust gas purification systems, contributing to the reduction of fuel consumption and purification of exhaust gas.

12.6.1 Background of Development

Exhaust gas temperature sensors were first mounted on vehicles to monitor the temperature of catalytic converters, which were introduced following the implementation of automotive emission control regulations. In the case of unburnt gaseous fuel mixture were to flow into the catalytic converter due to a misfire, a rapid burning reaction occurs on the catalyst, resulting in an abnormally high temperature in the catalytic converter. Mounting an overheat alarm is required by those regulations in order for the temperature of the catalyst to be monitored constantly by a temperature sensor. The alarm on the front panel is activated when the catalyst temperature exceeds a preset temperature value, alerting the driver to abnormalities and thus allowing for stopping of the vehicle while the temperature is still in a safe range. The thermistor-type temperature sensors, using stabilized zirconia as the heat sensitive element, were developed and commercialized for this purpose. Later, the demand for exhaust gas temperature sensors declined for a while as the mounting of the catalyst overheat alarm, which was required only in Japan, became unnecessary. Exhaust gas temperature sensors are currently widely used in order to protect turbos and catalysts as well as to control a variety of exhaust gas purification systems that incorporate DPF (Note 12.16), NO_x catalyst, SCR (Note 12.17) catalysts, etc., contributing to the reduction of fuel consumption and purification of exhaust gas (Fig. 12.29).

12.6.2 Characteristics

12.6.2.1 *Characteristics and Specifications of Products*

The temperature sensors for overheat alarms of earlier years were only required to detect high temperatures for the activation of the alarms. However, the temperature sensors intended for protection and control are required to measure a wide range of temperatures accurately and continuously. A variety of metal oxides such as the fluorite-type, the corundum-type and the spinel-type have been used materials for thermistors for years. However, the range of measurable temperatures is limited with thermistors made from these materials. Thermistors that incorporate perovskite oxide based on $YCrO_3$, a p-type semiconductor, and that cover a wide range of temperatures have been developed. Some of the products

Note 12.15 NTC is type of a thermally sensitive resistor. The NTC (Negative Temperature Coefficient) thermistor, in which electrical resistance drops as the temperature increases, is used in exhaust gas temperature sensors.

Note 12.16 Stands for Diesel Particulate Filter.

Note 12.17 Stands for Selective Catalytic Reduction. The method of reducing nitrogen oxide (NO_x) in exhaust gas to nitrogen by using ammonia, etc. as the reducing agent.

that have been developed recently cover an extremely wide range of temperatures, –40°C to 900°C, in response to the needs of the OBD (Note 12.18) employed in the U.S. Figure 12.30 shows characteristics of thermistors covering a wide temperature range, as examples. They are NTC thermistors (Note 12.15), and the B constant (Note 12.19) has been adjusted to 2,000–8,000 K. The resistance value drops exponentially from 100 to 0.1 kΩ as the temperature increases.

The temperature coefficient α (Note 12.19) of the thermistor changes several percent/°C, and its absolute value is one order of magnitude larger than that of the platinum resistance temperature detector. The elements can also be downsized, and therefore, they feature high sensitivity and responsiveness. Exhaust gas temperature sensors for automobiles are required to be resistant to high temperatures of 1,000°C and need to be designed to resist severe conditions such as gravel from road surfaces, the splashing of water, vibration and impact. Meanwhile, temperature sensors used in temperature controls are required to have quick responsiveness. The sensor structure with the least possible heat capacity is preferable in this respect. Because of these reasons, the heat-sensing portion of the sensor is structured by inserting the thermistor element into a thermal resistant metallic tube that is thinner at the end and is enclosed in a thermal resistant oxide, realizing both high toughness and responsiveness (Fig. 12.31).

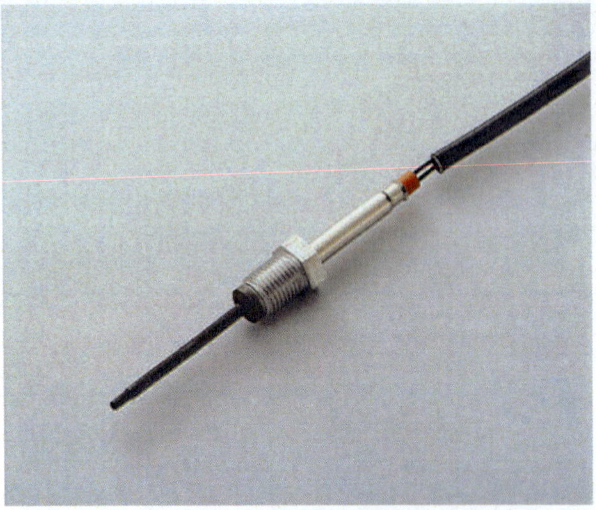

Fig. 12.28 Exhaust gas temperature sensors. Exhaust gas temperature sensors are used to protect turbo chargers and catalysts as well as to control a variety of exhaust gas purification systems, contributing to the reduction of fuel consumption and purification of exhaust gas

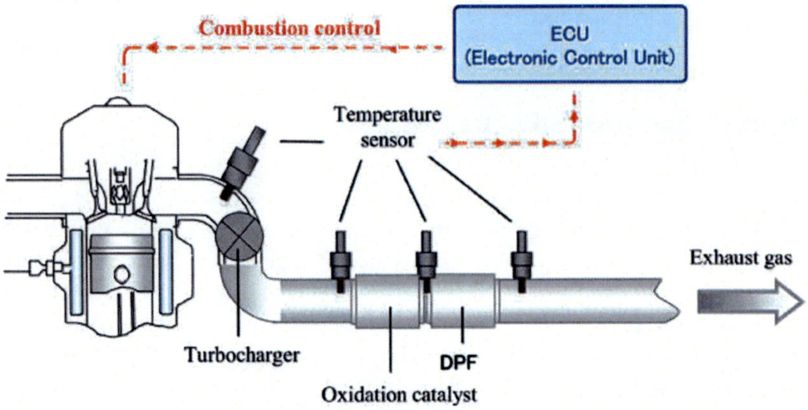

Fig. 12.29 Exhaust gas temperature sensors mounting method: example. Exhaust gas temperature sensors are mounted in front of and in the back of the turbo and the catalyst and continuously monitor the exhaust gas temperature. Signals from the temperature sensors are transmitted to the ECU and are used to control the combustion and the exhaust gas purification systems

Note 12.18 On-board Diagnosis. On-board self-diagnosis for detecting abnormality in the system.

Note 12.19 A constant that indicates a thermistor's sensitivity to temperature. The relationship between temperature and resistance value is approximately obtained by the formula below.

$$RR = R_0 \exp\left\{B\left(1/T - 1/T_0\right)\right\}$$

R: thermistor resistance value [Ω] at temperature T[K]
R_0: thermistor resistance value [Ω] at reference temperature To[K]
B: thermistor constant [K]
Relation between temperature coefficient α and B constant is expressed as below.
$\alpha = -B/T^2$

12.6.2.2 Manufacturing Method

Thermistors are produced in a manufacturing process generally used by ceramics, which includes mixture of materials, temporary calcination, crushing, granulation, forming and sintering. Characteristics of the thermo-couple and the platinum resistance temperature detector depend on the properties of the substances to be used, while the characteristics of thermistors are not decided simply by the properties of the materials to be used and change substantially depending on the ceramic grain diameter, grain boundary, vacancy concentration, segregation, etc. Therefore, strict process control is required in order to avoid impurities and adjustment of the segregation, as well as for the accurate control of conditions for temporary calcination and sintering in thermistor production.

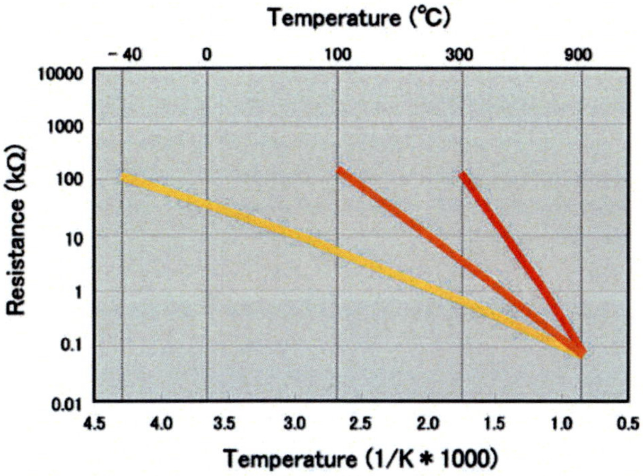

Fig. 12.30 Characteristics of thermistors covering a wide temperature range: examples. Thermistor elements are selected with consideration for the temperature range that needs to be measured. The slope of the graph indicates B constant, a constant indicating the characteristics of the thermistor

12.6.3 Future Prospects

Temperature measurement is the most basic type of measurement in automobiles. Exhaust gas temperature sensors were explained in this section, however various temperature sensors made for various purposes, such as the temperature sensors for intake air, water, oil, EGR (Note 12.20), etc., are used in automobiles. A majority of automotive temperature sensors, unlike temperature sensors that serve other purposes, are used under severe conditions and require high-level technology. At the same time, the cost must be reasonable.

Improvements in accuracy, applicable range and high-temperature resistance of exhaust gas temperature sensors are expected to continue. For the purpose of the further reduction of fuel consumption and purification of exhaust gas, exhaust gas temperature sensors will continue to play important roles as sensors for combustion control and exhaust gas purification systems.

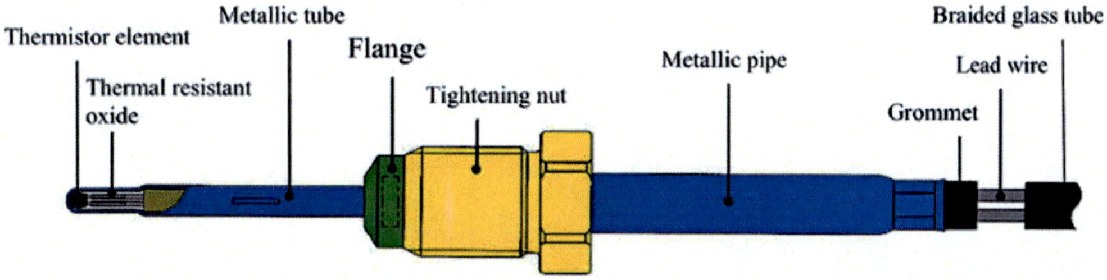

Fig. 12.31 Structure of exhaust gas temperature sensors. The thermistor element is inserted in a thermal resistant metallic tube that is thinner at the end and is enclosed in a thermal resistant oxide, realizing both high toughness and responsiveness

Note 12.20 Stands for Exhaust Gas Recirculation. Recirculation of exhaust gas (for reducing NO_x).

Exhaust Gas Oxygen Sensor 12.7

Automotive oxygen sensors used in exhaust gas purification systems that use a three-way catalyst (Note 12.21) are important functional components that are essential for the combustion control of internal combustion engines. The zirconia oxygen sensor is mounted on the upstream side as a combustion control sensor that detects exhaust gas concentrations, while a zirconia oxygen sensor for OBD (On-Board-Diagnostics)-II is mounted after the catalyst in the downstream. Zirconia oxygen sensors are mounted in zones where the signals change quickly and purification ratios of three types of gases, HC, CO and NO_x, are the highest. If the signals from the oxygen sensors move back and forth within a range from the richest side of the stoichiometric ratio (fuel rich) to the leanest side (fuel lean), highly purified exhaust gas is emitted continuously.

Responding to strengthening of governmental automobile regulations, oxygen sensors are required to be of high performance, and also need to be designed with considerations for high reliability, including thermal resistance, mechanical shock resistance, thermal shock resistance, water resistance, and chemical resistance.

12.7.1 Background of Development

Automotive oxygen sensors monitor exhaust gas composition to effectively utilize a three-way catalyst after fuel combustion in engines (Fig. 12.32) to reduce emissions. The zirconia oxygen sensor is mounted in the upstream of exhaust gas as a combustion control sensor that detects exhaust gas composition, while an oxygen sensor for OBD (On-Board-Diagnostics)-II is mounted behind the catalyst in the downstream (Fig. 12.33).

This sensor was released for the first time in the world by Bosch in 1976 and was mounted on Volvo vehicles. The sensors were later employed by GM and Ford in the U.S., followed by Toyota and Nissan in Japan. Currently, oxygen sensors are mounted on combustion control systems in cars located in nearly all developed countries.

12.7.2 Characteristics

12.7.2.1 Principle of Oxygen Sensor

Oxygen sensors are made by using zirconia as a constituent material, a solid electrolyte that is capable of detecting oxygen gas concentration differences. The solid electrolyte is a substance in which ions such as O^{2-} can move freely. Using the solid electrolyte as a diffusion barrier, the space is divided into chamber A and chamber B having a platinum electrode on each side. When there is a difference in oxygen concentration between the two chambers, oxygen ions diffuse from the side with higher oxygen concentration toward the low side, reducing the difference in oxygen concentrations. The shift of O^{2-} involves a shift of electric charge (e^-) similar to a battery and develops an electromotive force (Fig. 12.34) that is measured. The electromotive force developed here is expressed by the following formula, which is called the Nernst equation (Note 12.22).

$$E = (RT/4F) \cdot \ln(P_a / P_b)$$

Note 12.21 A catalyst that converts three types of gases, hydrogen carbon (HC), carbon monoxide (CO) and nitrogen oxide (NO_x), which are contained in the exhaust gas and causes pollution, to water vapor (H_2O), carbon dioxide (CO_2), nitrogen (N_2) and Oxygen (O_2).

Note 12.22 $E = (RT/4F) \cdot \ln(Pa/Pb)$ R: gas constant, T: absolute temperature, 4: oxygen gas molecule constant. The constant is 4 because O_2 is displaced by two O^{2-} ions, which are displaced by four electrons. F: Faraday constant, *Pa*: oxygen partial pressure of chamber A, which is higher in oxygen concentration, and uses normal atmospheric concentration. *Pb*: oxygen partial pressure of chamber B, which is lower in oxygen concentration. This is the oxygen concentration within the exhaust gas.

The electromotive force is obtained by placing actual air excess ratio (λ) and oxygen partial pressure in the Nernst equation. The oxygen partial pressure changes suddenly at around $\lambda = 1$, causing a rapid change in electromotive force. This rapid change in electromotive force is as high as 1 V, which can be used as the ON/OFF signal, without any special electrical circuit, making it possible to detect the air excess ratio of the exhaust gas at $\lambda = 1$.

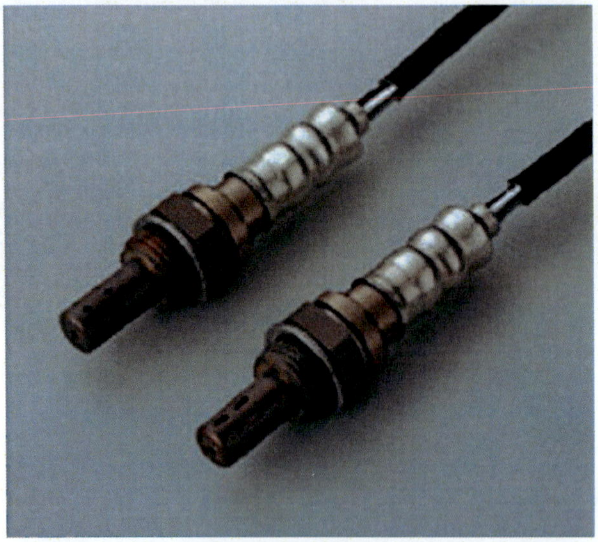

Fig. 12.32 Oxygen sensors. Automotive oxygen sensors used in exhaust gas purification systems that use a three-way catalyst, are important functional components that are essential for combustion control of internal combustion engines

12.7.2.2 Air–Fuel Control by Oxygen Sensor

Figure 12.35 shows the purification ratios of three types of gases, hydrocarbon (HC), CO and NO_x, using a three-way catalyst, and the signals from the oxygen sensor for fuel injection control, which is a key component of the system.

The figure indicates that there is a zone where the purification efficiency of the three types of gases are maximized within the range where the signals from the oxygen sensor change dramatically. This zone is called the window. If the signals from the oxygen sensors move back and forth within this range, between the richest side of the stoichiometric ratio (fuel rich) and the leanest side (fuel lean), the exhaust gas will remain in the highest purification efficiency range.

12.7.2.3 Structure of the Oxygen Sensor

Oxygen sensors are important functional components and are required to have both high performance and reliability. This includes having long-term warrantees of 15-years or 150,000 miles (240,000 km).

As shown in Fig. 12.36, the sensing element of the zirconia oxygen sensor is enclosed in a main metal fitting made of thermal resistant stainless steel. The element, which is exposed to exhaust gas, is covered by a protector made of stainless steel, which has high thermal resistance and oxidation resistance.

Automobile oxygen sensors are subject to severe environments such as an engine vibration, vibration impact from road surfaces, gravel and the splashing of water. Therefore, they

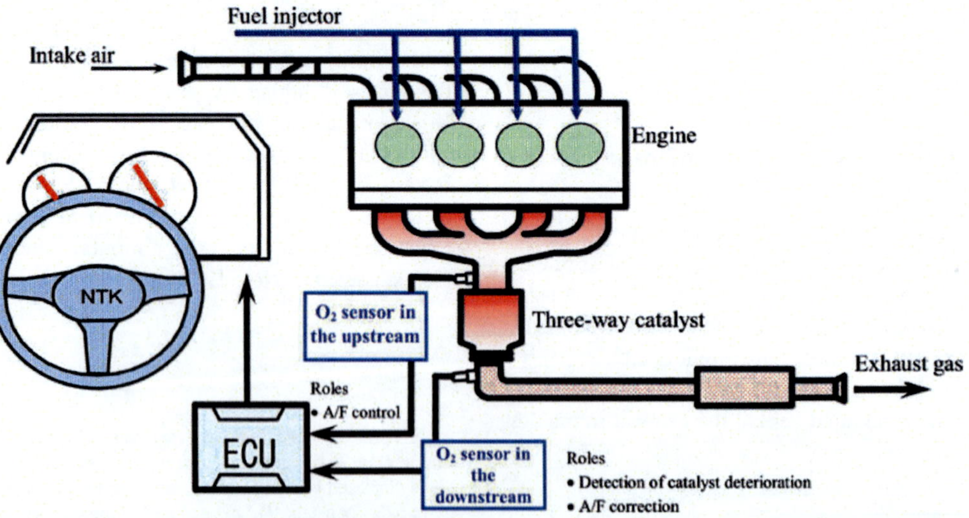

Fig. 12.33 A/F control system incorporating an oxygen sensor. The sensor for combustion control is mounted in the upstream, while an oxygen sensor for OBD-II is mounted behind the catalyst in the downstream. The zirconia oxygen sensor is used in combination with the three-way catalyst, which is important for combustion control in response to the current regulations

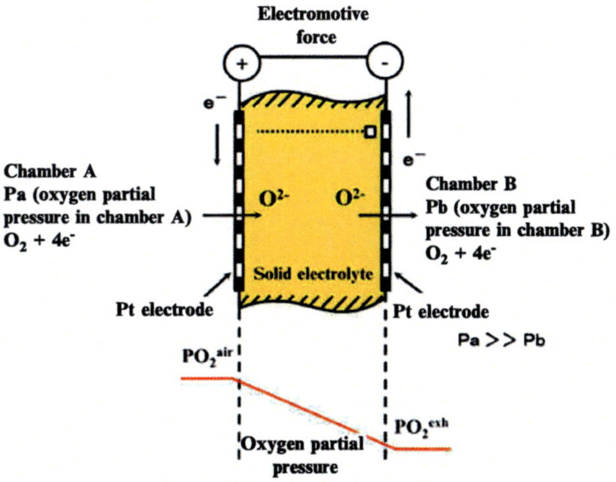

Fig. 12.34 Principle of oxygen sensor. When there is a difference in the oxygen concentration (oxygen partial pressure) between chamber A and chamber B, oxygen ions move from the side with the highest oxygen concentration toward the other side, reducing the difference in oxygen concentrations

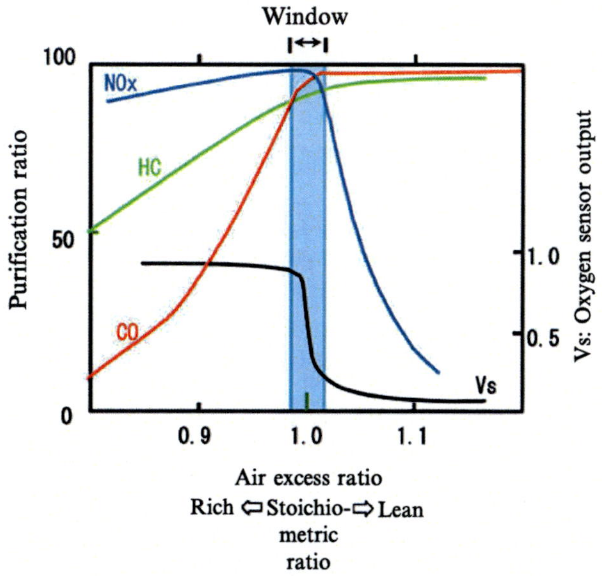

Fig. 12.35 Basic characteristics of the oxygen sensor and purification ratios of a three-way catalyst. This shows the purification ratios of three types of gases, HC, CO and NO_x, using a three-way catalyst. $\lambda = 1$ indicates theoretical air fuel ratio, an air–fuel mixture ratio ideal for perfect combustion. The highest purification efficiency is achieved in this area

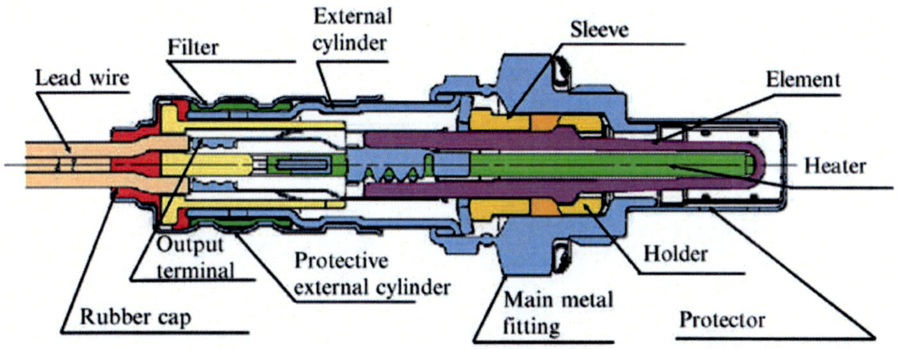

Fig. 12.36 Cross-sectional drawing of the oxygen sensor. Following the strengthening of exhaust gas regulations, these components need to be designed in consideration for high performance and reliability

need to be designed to be thermal resistance, mechanical shock resistance, thermal shock resistance, water resistance, chemical resistance, etc.

12.7.3 Future Prospects

In response to the strengthening of governmental automobile exhaust gas regulations, the introduction of electric vehicles and fuel cell vehicles may be realized earlier than expected in order to reduce automobile emissions. However, issues such as the building of infrastructure and production costs remain unsolved. Therefore, improvements in existing gasoline engines to lean-burn types and hybrid types will be promoted. Oxygen sensors are also used in the exhaust gas systems of these engines. In addition, emission control regulations will be established in countries such as China and India in the future, following the U.S., Europe and Japan. Because of these reasons, the global demand for oxygen sensors is expected to expand further in the future.

12.8 Knock Sensor (1983)

Engine combustion normally converts an air and fuel mixture into gaseous reaction products within the combustion chamber by means of flame propagation (Note 12.23) and the chamber pressure rises gradually. However, if self-ignition (Note 12.24) occurs in the unburnt portion of the fuel mixture (Note 12.25) during the compression step, shock waves are generated due to the sudden pressure increase, causing a knocking-like noise. This phenomenon is called knocking. Intense knocking may damage the engine due to melting of the piston.

The knock sensor is directly mounted on the engine block. When the engine block vibrates due to knocking, a large force is applied to the piezoelectric element inside the knock sensor. Due to this large force, the piezoelectric element generates a larger than normal electrical charge. The Electronic Control Unit (ECU) (Note 12.26) installed on the vehicle detects the charge and delays the ignition timing (Note 12.27) of the engine in order to reduce the knocking.

12.8.1 Background of Development

Responding to the social needs for energy conservation, various improvements are being made in the automobile industry so as to meet the demands for engines that feature low fuel consumption and high output. However, these improvements resulted in more frequent engine knocking. In order to solve this problem, the knock control system has been developed and employed. This system detects knocking by using a knock sensor. The ECU mounted on the vehicle receives signals from the sensor and delays engine ignition timing in order to reduce knocking.

Meanwhile, when no knocking is detected, the ignition timing fires accordingly to its pre-determined optimized value near the knocking limit.

12.8.2 Characteristics

12.8.2.1 *Principle and Types*

The knock sensor is directly mounted on the engine block. The mounting position and connection of the knock sensor is shown in Fig. 12.37. When the engine block vibrates due to knocking, a large force is applied to the piezoelectric element within the knock sensor. Due to this large force, the piezoelectric element generates a larger than normal electrical charge.

Knock sensors are classified into two types, the resonance type and the non-resonance type. Figure 12.38 shows the appearance and structure of each type of knock sensor.

12.8.2.1.1 Resonance Type

The resonance type of knock sensor has a vibrator inside the sensor. The vibrator is made by bonding a piezoelectric element on a diaphragm (a thin piezoelectric ceramic plate is bonded on a metallic diaphragm). The electric charge is generated due to the flexing of the vibrator.

The frequency in which knocking occurs, varies depending on the engine type, operating conditions and fuel. Therefore, different types of vibrators are used depending on the engine. Engines require sensors prepared exclusively for each engine model.

Note 12.23 The phenomenon in which combustion of fuel mixture spreads gradually from the point of ignition.

Note 12.24 The phenomenon where a fuel mixture of high temperature and pressure inside the engine combustion chamber are ignited naturally, without ignition by the spark plug.

Note 12.25 Mixture of fuel and air that is used to drive the engine.

Note 12.26 Stands for Electronic Control Unit. An electronic control device mounted on vehicles to control the engine, which responds to signals from various sensors.

Note 12.27 The timing of igniting the compressed fuel mixture by the spark plug (a part mounted on gasoline engines to ignite the gas mixture by sparks generated by application of high voltage).

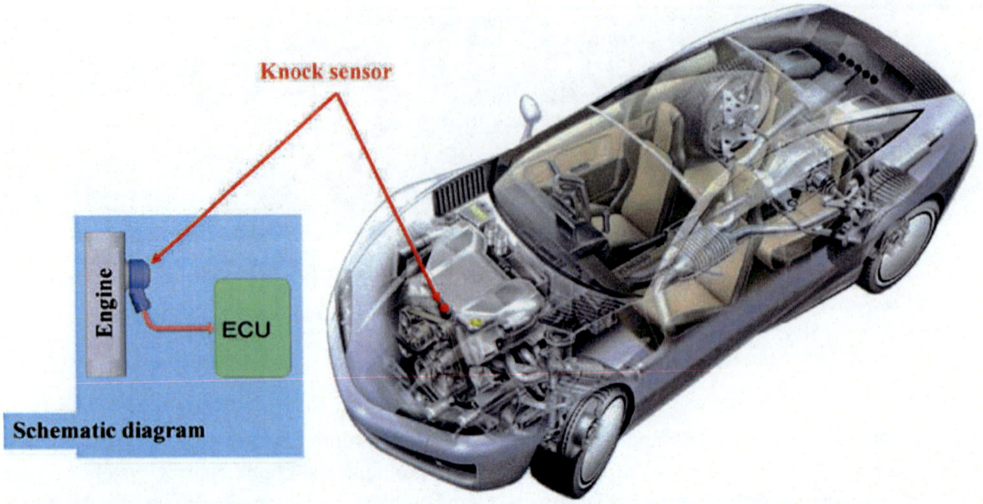

Fig. 12.37 Mounting position and connection of the knock sensor in automobiles. The knock sensor is mounted directly on the surface of an engine to detect vibration of the engine. The sensor is connected to the ECU and the signals are processed by the ECU

Fig. 12.38 Appearance and structure of knock sensor. In the resonance type, the electrical charge is generated due to flexing of the vibrator (a thin ceramic plate bonded on a metallic diaphragm). In the non-resonance type, the electrical charge is generated through compression of the donut-shaped ceramic element by the "weight"

	Photograph of item	Structural drawing
Resonance type		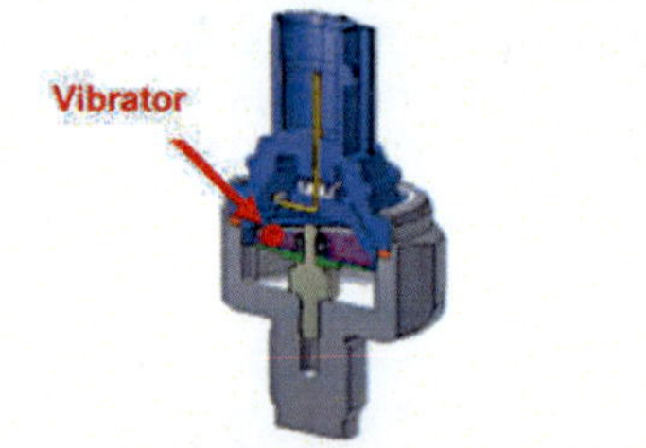
Non-resonance type		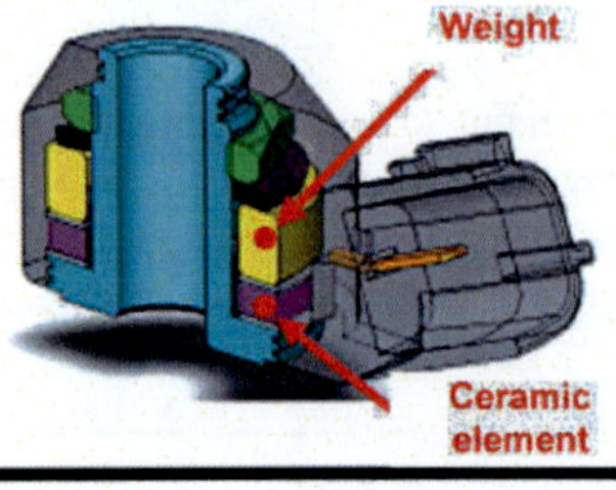

12.8.2.1.2 Non-resonance Type

The non-resonance type of knock sensor is a kind of an acceleration sensor and its electrical signal is nearly flat throughout a wide range of engine vibration frequencies. Therefore, they do not have to be prepared exclusively for different types of engines. The same knock sensor can be applied to a wide variety of engines. In the non-resonance type knock sensor, the electrical charge is generated through compression by a weight on a donut-shaped piezoelectric element.

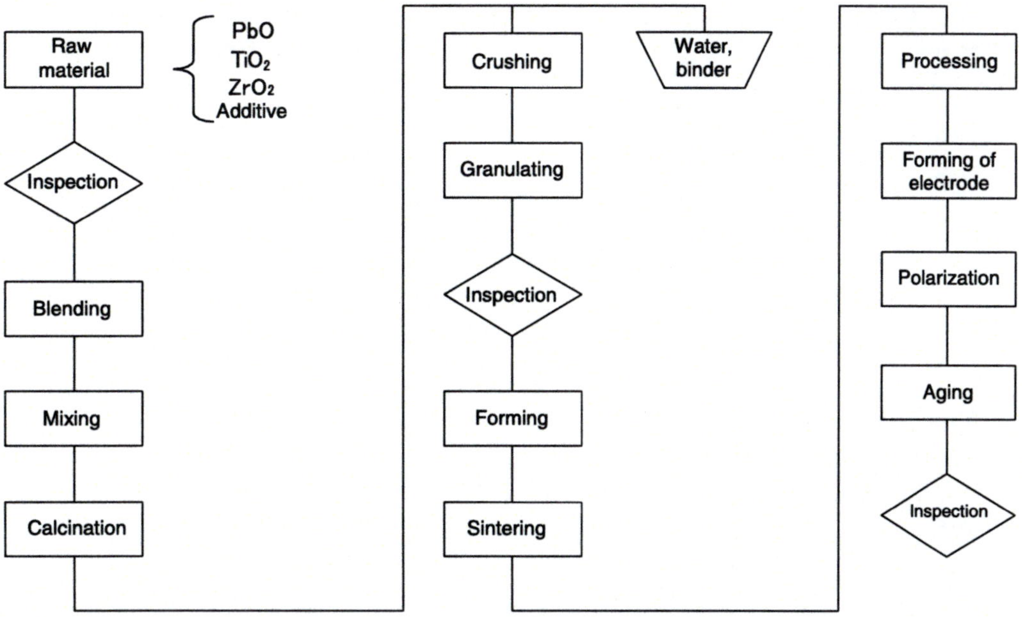

Fig. 12.39 Manufacturing process of knock sensor element. Raw materials (PbO, TiO_2, ZrO_2, additives) are mixed, calcined and mixed with water and binders, which is crushed and granulated. The powder is formed, sintered, and processed to an intended shape and then electrodes are formed by using silver paste, followed by polarization and aging

12.8.2.2 Manufacturing Method of Knock Sensor Piezoelectric Element

In consideration for low cost and high productivity, piezoelectric ceramics are generally made with a solid-state reaction method. The production process of piezoelectric ceramics with the solid reaction method is shown in Fig. 12.39. First, the materials (PbO, TiO_2, ZrO_2, additives) are blended and mixed. Then, the material is temporarily calcined to create a solid-solution through solid reaction. The solid solution is mixed with water, binders, etc. It is then crushed for granulation. The ceramic powder is then formed to a shape that is similar to the sensor shape. The formed body is then sintered. In this process, the formed body is magnetized and densified by completing chemical reactions of elements that have not yet reacted. Then the formed body is processed to the intended shape by grinding and then the electrodes are formed using silver paste. The isotropic microcrystal axes of the ceramics called domains are directed in the same direction by applying a high electrical field at elevated temperature to pole the domains to provide uniform piezoelectric characteristics. Finally, aging is performed so as to stabilize the characteristics of the ceramics.

12.8.2.3 Manufacturing Method of Knock Sensor

Figure 12.40 shows the manufacturing method of the nonresonance type knock sensor that is currently being widely used. The insulator, electrode, element, weight and washer are tightened on a metal fitting with a nut. Then the unified

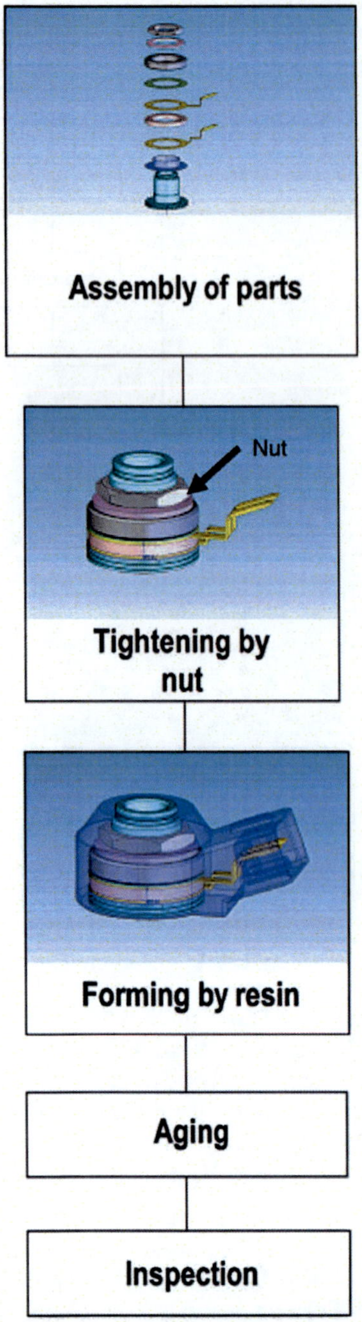

Fig. 12.40 Manufacturing process of non-resonance knock sensor. The insulator, electrode, element, weight and washer are tightened on a metal fitting by a nut. Then the unified body is processed to the shape of a sensor by means of injection molding using nylon resin. This is followed by aging for the stabilization of the sensor characteristics

body is processed to the shape of a sensor by injection molding using nylon resin, followed by aging for stabilization of the sensor characteristics.

12.8.3 Future Prospects

Today, the knock control system is used in nearly all vehicles mounted with gasoline engines, ranging from light-duty vehicles to luxury vehicles. Meanwhile, the knock sensor is employed by nearly none of diesel engine vehicles. However, following the expansion of the diesel common rail system, the knock sensor began to be used to detect the vibration in diesel engines. In Europe, where the percentage of diesel vehicles is as high as 50%, vehicles mounted with a knock sensor are expected to increase.

12.9 Ferrite Magnets for Compact Motors (Around 1960)

Ferrite magnets are made using complex oxides that consist of Fe and Sr or Ba. These magnets generate lower magnetic force compared with rare earth metal magnets that are represented by NdFeB magnets. However, they feature high cost performance and chemical stability. Ferrite magnets are widely used in compact motors (Note 12.28). Compact DC motors that incorporate ferrite magnets are widely used in the automobile industry in particular due to low cost and high reliability of ferrite magnets. Improvement in the performance of ferrite magnets was believed to be limited for some time in the past, due to the limitation in theoretical value of the characteristics. However, this limitation was overcome by the development of the LaCo-based strontium ferrite magnet in 1997. Since this development, the performance of ferrite magnets has been improving along with motor design technology. This has resulted in contributions to the downsizing and the improvement in performance of compact motors.

12.9.1 Background of Development

Motors are used to convert electrical energy to mechanical energy using magnetic attractive force and repulsive force. Magnetic fields can also be generated by electrical magnets. However, the efficiency of motors is improved by using permanent magnets, which continue to generate magnetic fields without an energy supply. Figure 12.41a shows ferrite magnets for compact motors. Figure 12.41b shows the appearance of a DC motor fitted with a brush. The magnet is assembled inside the motor as shown in Fig. 12.41c.

The performance of a magnet is indicated by residual magnetic flux density (B_r), which expresses the intensity of the magnetic force, and coercive force (H_{cJ}), which expresses stability against external magnetic fields. Magnets are subjected to a backward magnetic field generated by the coil during motor operation, and demagnetization and deterioration of magnetic force occur when the H_{cJ} value is low. The tendency of demagnetization also depends on the shape of the magnet, and the phenomenon tends to occur more frequently in thinner magnets. If the magnet features a high H_{cJ} value, it is less likely to be demagnetized and the magnet can be thinned for further downsizing of the motor diameter. Meanwhile, higher B_r value means a larger amount of magnetic flux, which contributes to improving the torque of the motor, allowing downsizing of the motor length while achieving the same torque. Improvement of magnet performance leads to performance improvement, size reduction and the weight reduction of motors.

Note 12.28 In general, compact motors are sometimes classified as motors with an output less than 70 W. However, in a broad sense, they are small motors that can be carried in one hand. Compact motors indicated here refer to the latter.

12.9.2 Characteristics

12.9.2.1 Performance and Applications

Ferrite magnets that are widely used are made of strontium ferrite ($SrFe_{12}O_{19}$) or barium ferrite ($BaFe_{12}O_{19}$), which has a hexagonal crystal structure. Barium ferrite was developed in 1952, and strontium ferrite, which features higher performance, was released in 1963. Ferrite magnets, that are made by using iron oxide as the major material, are inexpensive.

Figure 12.42 shows characteristics of permanent magnets that are widely used today. Compact motors vary in applications, performance and sizes. Magnets are selected based on the intended use, placing priority on magnetic properties, cost or reliability. For applications where the priority is placed on high performance and downsizing rather than on cost, NdFeB sintered magnets and NdFeB bond magnets are selected. Meanwhile, ferrite magnets are selected for applications where the priority is placed on low cost.

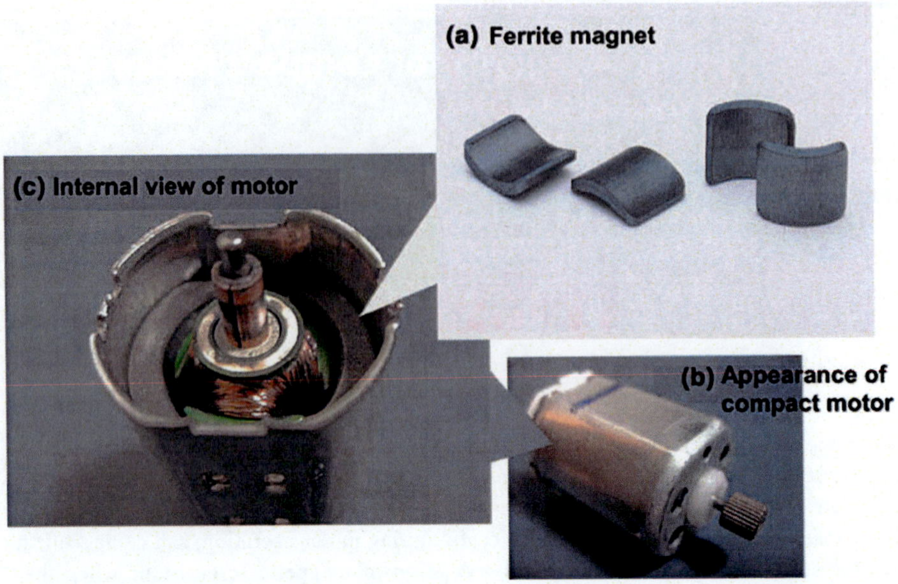

Fig. 12.41 Example of ferrite magnets intended for compact motors. (**a**) Ferrite magnets for DC motors fitted with a brush. The completed ferrite magnets are not magnetized. They require an applied external magnetic field for magnetization in the intended direction of use. (**b**) A DC motor fitted with a brush, which was assembled in a drier. (**c**) Ferrite magnets assembled in the motor

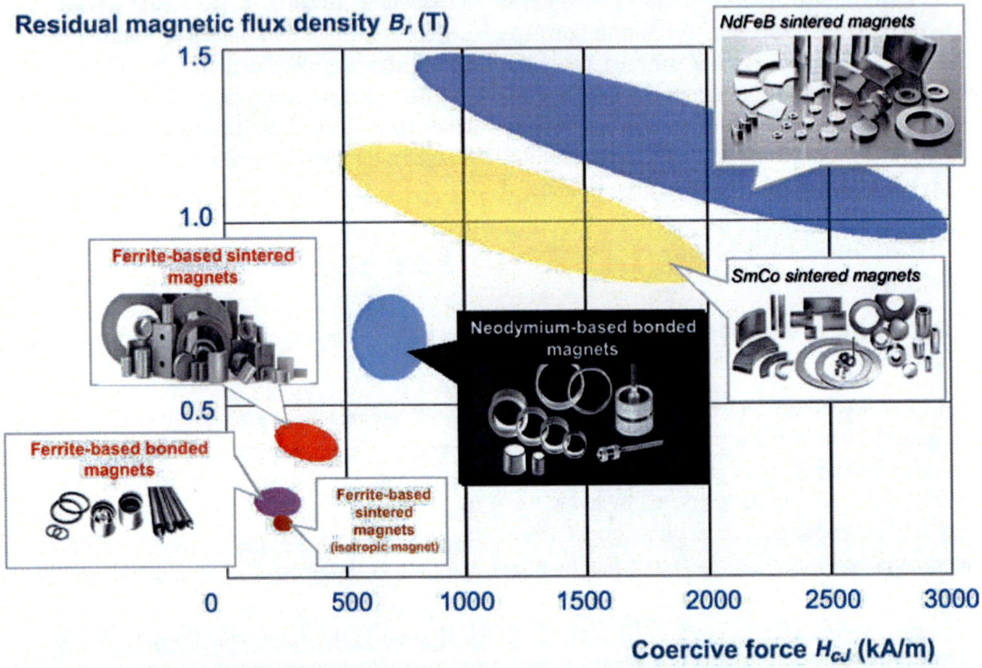

Fig. 12.42 Performance of typical magnet materials. The strongest magnet at present is the NdFeB sintered magnet. The SmCo sintered magnet is extremely expensive but features high thermal resistance. It is easy to form the neodymium bonded magnet into the intended shapes. Ferrite magnets are inferior to the magnets mentioned above in terms of characteristics, but they are superior in terms of cost-performance

Ferrite magnets are oxides and are also characterized by their chemical stability. Compact ferrite magnet motors fitted with a brush; satisfy the needs of automotive motors, which are required to be of low cost and high reliability. Ferrite magnets are used in large amounts in compact motors for automobiles.

Characteristics of strontium ferrite magnets are shown in Table 12.4. Required B_r, H_{cJ} values vary depending on the application and design of the motor, and therefore, magnets belonging to the same class usually show a variation in their properties. As indicated in the table, B_r and H_{cJ} are typically opposites of each other.

Table 12.4 Magnetic property of ferrite magnets

Al amount	B_r (mT)	H_{cJ} (KA/m)	(*BH*) max. (KJ/m³)
Small	440	263	36.7
Medium	420	318	33.4
Large	400	358	30.3

An example showing the magnetic properties of an anisotropic strontium ferrite magnet made with the wet method. (*BH*)max (Note 12.29) is the maximum energy product. The figures show the values of materials of the same grade or class, and magnets featuring an optimal property balance are selected depending on the motor to be designed. The property balance is normally adjusted by adding Al_2O_3, etc.

12.9.2.2 Manufacturing Method

The manufacturing process of ferrite magnets is shown in Fig. 12.43. Iron oxide is mixed with strontium carbonate or barium carbonate, which is dispersed and temporarily calcined to create the ferrite. The calcined particles are processed into sub-micron dimensions during the crushing process.

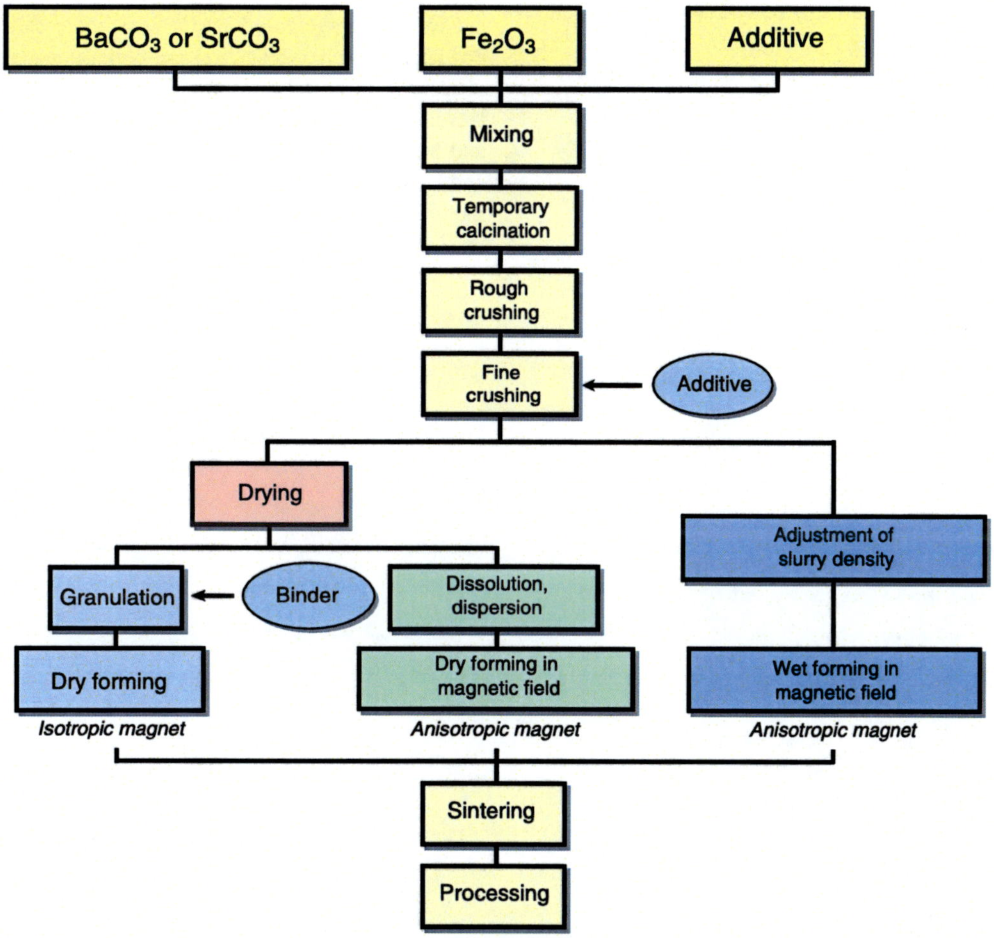

Fig. 12.43 Manufacturing method of ferrite magnets. The manufacturing process flow of three types of magnets are shown; namely the dry isotropic, dry anisotropic and wet anisotropic magnets. Anisotropy of crystal orientation is achieved by means of forming within a magnetic field

Note 12.29 The maximum value of the product of magnetic flux density (*B*) on the demagnetization curve of the magnet and the corresponding magnetic field (*H*). The maximum energy that can be achieved from a magnet is proportional to the maximum energy product of the magnet, and therefore, the maximum energy product is used to express the magnet performance. In order to enhance the value, residual magnetic flux density B_r and coercive force H_{cJ} need to be high.

Then the material is compressed and formed to an intended shape. This process is characterized by forming in a magnetic field. Ferrite magnets exhibit uniaxial anisotropy (Note 12.30) and tend to be magnetized in the c-axis direction of the hexagonal crystal. B_r is improved when the crystal orientation is unified by forming in a magnetic field. The magnet in which there is unified crystal orientation is called the anisotropic magnet, while the magnet in which there are randomly oriented crystals is called the isotropic magnet. The forming methods include the dry forming method which features lower cost and the wet forming method which has higher performance. In the dry forming method, magnetic powders are formed easily. Meanwhile, in the wet forming method, the slurry containing magnetic powders is compressed and formed while being dehydrated. In applications where low cost is especially important, anisotropic ferrite magnets made with the dry method are often used. On the other hand, in applications in which high performance is required, anisotropic ferrite magnets made with the wet method are used. The ferrite magnet is completed by sintering the formed body and processing the sintered body into the intended shape.

12.9.3 Future Prospects

In response to the market needs, the performance was improved by fully utilizing composition control technologies, crushing control technologies, fine structure control technologies. The industrial ferrite magnets' characteristics reached a steady-state in the mid 1990s. This is because improvements in the ferrite magnets previously meant to push the characteristics of ferrite magnets toward the theoretical value of strontium ferrite was nearly reached. However, a breakthrough was observed in 1997. It was confirmed that saturated magnetization and anisotropy field (theoretical limit value of B_r and H_{cJ}, respectively) could be improved by adding both La and Co into strontium ferrite. The LaCo-based strontium ferrite magnet was commercialized in 1999, which resulted in motor downsizing, weight reduction and performance improvement coupled with an advancement in motor design technology. Developments of LaCo-based ferrite magnets have continued and higher performance magnets that feature B_r of 470 mT and H_{cJ} of 380 kA/m have been recently introduced. Compact motors that incorporate ferrite magnets are expected to be smaller, lighter and higher in performance.

Note 12.30 Characteristics of ferromagnetic bodies in which magnetization tends to occur in a certain direction (axis of easy magnetization) and in which magnetization is difficult in the perpendicular direction (axis of difficult magnetization). Both the ferrite and NdFeB magnets are characterized by uniaxial anisotropy.

LTCC Substrate (1995) 12.10

LTCC is an abbreviation for Low Temperature Co-fired Ceramics and is a type of ceramic multi-layer circuit substrate, which uses silver (Ag) or copper (Cu) internal wiring that has low electrical resistivity for high electrical conductivity. In order to improve the transmission speed of electrical signals, Ag and Cu were used instead of tungsten (W) and molybdenum (Mo), which are used in conventional alumina ceramic substrates and have both a higher melting point and higher electrical resistance. The name "low-temperature" ceramics originated from the low sintering (or firing) temperature of approximately 900°C, which is lower than the typical sintering temperature (approximately 1,600°C)of Al_2O_3 multilayer ceramic substrates that use Mo or W wiring. The LTCC was first commercialized for use as the circuit board in supercomputers, in which the priority is placed on high-speed signal transmission, taking advantage of the low electrical resistance of the Ag and Cu and low dielectric constant of ceramics. Later, the material began to be used in automobiles, taking advantage of thermal resistance of ceramics, and communication devices that are require compactness and light weight. The application range of LTCC substrates is expected to expand, responding to growing needs for electronic control devices for automobile safety and enhancement of environmental concerns.

12.10.1 Background of Development

The LTCC was first commercialized as the MCM (Multi Chip Module) used in supercomputers, whose development began in the 1970s. In order to mount ICs directly and in high density on the substrate for high electrical signal transmission speed, the LTCC substrate incorporating Cu embedded wiring was used instead of printed circuit boards and/or HTCC (High Temperature Co-fired Ceramics) substrates that were being used at the time.

In respect to the applications of ceramic substrates in automotive wiring boards, until the 1980s, single-layer HIC (Hybrid IC) substrates were made by forming conductor and resistor circuits on Al_2O_3 ceramics by the thick film method, because they required higher heat-resistance due to their usage inside automobile engine compartments; the single-layer HICs were used as substrates for ignition, etc. with comparatively simple circuits.

Later, advanced electronic control systems with larger circuits became essential in solving environmental issues including pollution restrictions, and the conventional HIC substrates were not large enough to accommodate the additional circuitry. Responding to this trend, multilayer ceramic substrates such as the HIC multilayer substrate, made by screen-printing the circuits on an alumina substrate, and the HTCC multilayer substrate, made by forming Cu wiring on an alumina substrate, which has multilayer W conductor wirings, and forming a LaB_6-based cermet resistor on the surface, began to be used. However, both of them failed to enter the main stream because they both lacked long-term reliability and the ability to be mass produced. Therefore, the LTCC substrate, a compact high-integration ceramic substrate featuring high accuracy and high reliability, was introduced. The substrate featuring high assembly performance was created by using the LW (length–width) zero shrinkage sintering technology, which was revolutionary at the time, and by employing conductor and resistor technologies used to manufacture conventional Ag-based HIC. The structure is shown in Fig. 12.44.

In respect to automotive application, the substrate was first mass produced for use in ECU (Electro Control Unit) of ABS (Antilock Brake System) in 1995 (Fig. 12.45), followed by application to engine control ECU substrates (Fig. 12.46) and ECU substrates of TCU (Transmission Control Unit).

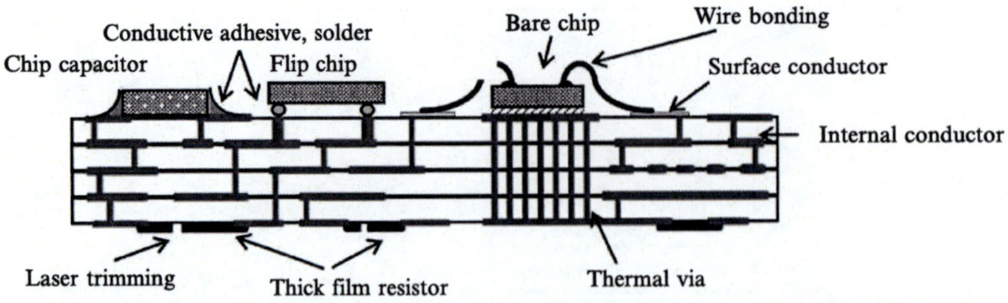

Fig. 12.44 Appearance of ABS module incorporating LTCC in the ECU substrate. The IC on the surface was mounted and wire bonded with Au wire. Then, heat from the IC was released from the back side through the thermal via and the resistor was printed on the back to achieve a high-density structure

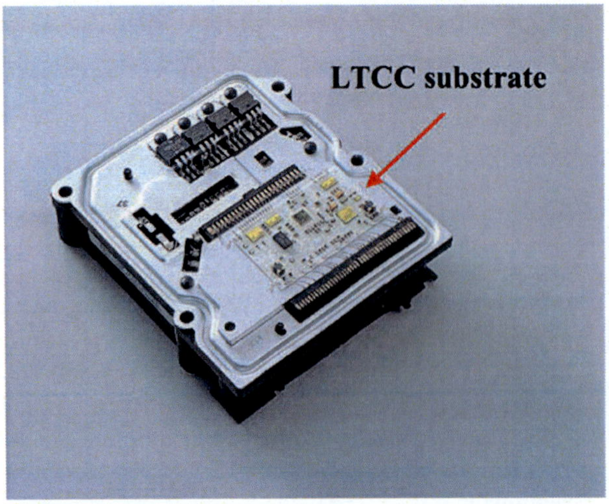

Fig. 12.45 Appearance of TCU module incorporating LTCC in the ECU substrate

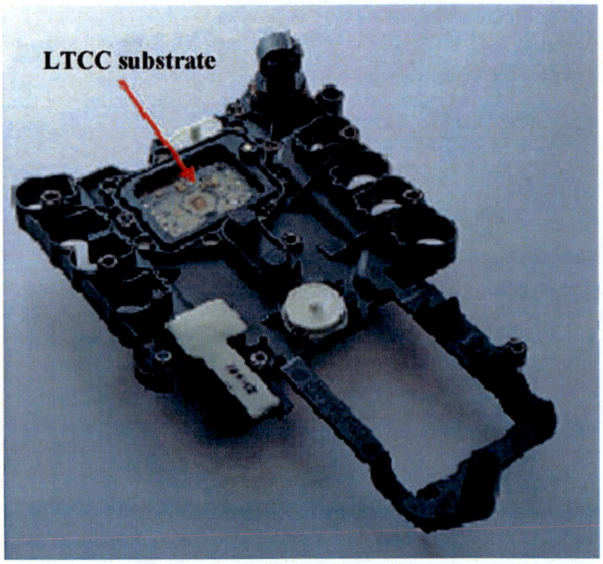

Fig. 12.46 Structure of automotive LTCC substrate

12.10.2 Characteristics

The sintering temperature of ceramics used as the LTCC substrate is approximately 900°C as Ag and Cu are used as internal wiring materials. It also needs to function as the circuit substrate, and the surface conductor needs to be connected to passive components by soldering, conductive adhesives, etc. and wire-bonded with ICs using Au wire so that it can be mounted with semiconductors and passive components such as capacitors. Therefore, a plated film that ensures the high adhesion properties of Ni, Au, etc. on the surface conductor and has a clean surface needs to be formed, while the resistor needs to be formed directly on the substrate for downsizing and density enhancement. In addition, materials free of environmental pollutants such as Pb must be used, responding to the trend toward environmental load reduction in recent years.

The LTCC substrate which satisfied all of the requirements was commercialized, using Ag as the conductor, a cermet resistor made of RuO_2 and glass as the resistor and a compound made of Al_2O_3 filler and CaO–Al_2O_3–SiO_2–B_2O_3-based glass as the insulating ceramic.

In addition to the characteristics mentioned above, the LTCC substrate is characterized by the "zero shrinkage sintering," in which shrinkage occurs in only the Z-axis direction and with no shrinkage occurring in the X, Y-axis directions, realizing extremely accurate post-sintering dimensions.

12.10.2.1 Characteristics of Ceramics, Conductors and Resistors

Main characteristics of the LTCC ceramics are indicated in Table 12.5, and characteristics and applications of the conductor and the resistor are indicated in Table 12.6.

12.10.2.2 Flatness of Substrate and Dimensional Accuracy

The most distinctive characteristics of the LTCC substrate are the superior flatness and dimensional accuracy, which were achieved at the same time by adopting the zero shrinkage sintering method, in which the laminated body does not shrink in the plane and shrinks only in the thickness direction. This was realized during sintering by utilizing non-sinterable layers on top and bottom of the sintered substrate to limit shrinkage. One of the examples is shown in Table 12.7.

Table 12.5 Characteristics of ceramic materials used in LTCC

Items	Characteristics
Chemical composition	$CaO–Al_2O_3–SiO_2–B_2O_3–Al_2O_3$
Sintering temperature	890°C
High density	2.9 g/cm^3
Bending strength	270 MPa
Dielectric constant	7.7
Thermal expansion coefficient	5.5×10^{-6}/°C

Table 12.6 Characteristics of paste materials' electrical conductivity

Types	Material	Major applications
Metallic conductor	Ag	Wiring conductor
	Au	Conductor for wire bonding
Resistor	RuO_2	Sheet resistance : 10–300 kΩ/□

12.10.2.3 Manufacturing Method

The green sheet lamination method, which is normally employed to manufacture multilayer ceramic substrates, is used in principle, while various methods are employed for sintering. The zero shrinkage sintering, the main feature, is performed using three different methods, namely the pressurizing method, the non-pressurizing method and the non-binding method. Figure 12.47 shows the manufacturing method of typical pressurized zero shrinkage sintering method.

12.10.3 Future Prospects

Further developments of the LTCC substrates are expected in the field of automotive multilayer ceramic ECU substrates. By developing unique package technologies, which cannot be realized by other packages, substrates, downsizing, high-density mounting with high-reliability will be realized in order to meet future more restrictive requirements such as vehicle weight reduction and advanced electronic control.

Table 12.7 Characteristics of LTCC substrate

Characteristics	Characteristic value
Dimensional accuracy of substrate	±0.05%
Flatness of substrate	5 μm/4 mm □
Max. substrate size	203.2 mm × 203.2 mm
Plating type	Ni/Au, Ni/Pd/Au
Accuracy of resistor printed on surface	Resistance value: ±1%
	Temperature coefficient of resistance: ±100 ppm

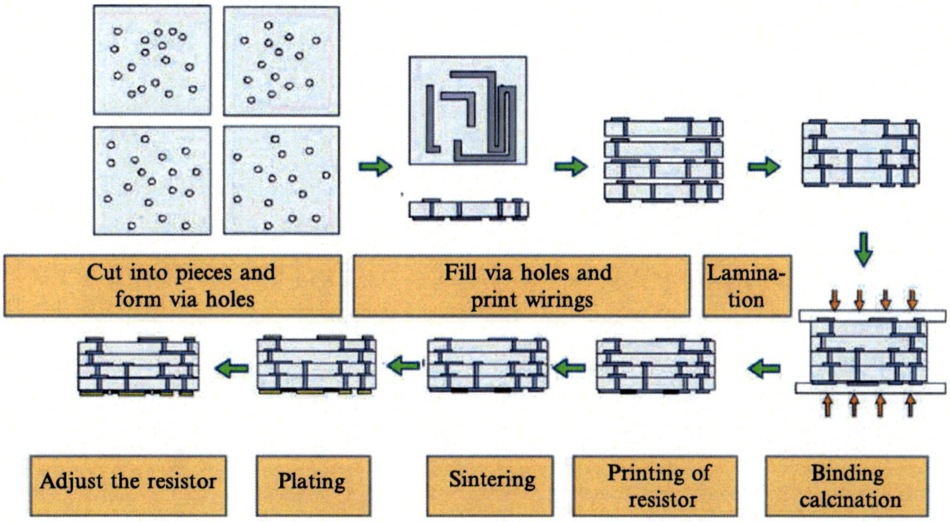

Fig. 12.47 Manufacturing method of a LTCC substrate

Dielectric Patch Antenna for GPS (1996) 12.11

Positional information devices incorporating a Global Positioning System (GPS) (Note 12.31) have been expanding rapidly. Antennas used in GPS positioning systems represented by car navigation systems are made from ceramics. They are also required to be compact and of low loss, as well as to have sufficient sensitivity for the reception of GPS signals. Antennas for the GPS systems are composed of microstrip antennas called the "patch antennas (Note 12.32)." The patch antenna has an electrode formed on each side of a square ceramic of 4–6 mm in thickness, with a power supply probe attached to one of the electrodes. By using ceramics as the structure, on which electrodes are formed, the wavelength of electric waves for transmitting GPS signals are shortened due to the dielectric constant of ceramics, making it possible to downsize. Demands for patch antennas have been increasing and they are incorporated in small and simple navigation systems called Personal Navigation Devices (PND) (Note 12.33) and in mobile phones. The patch antennas currently used are smaller and thinner as well as have higher reliability compared with previous ones.

Note 12.31 Global Positioning System: this system was developed by the U.S. Military for the U.S. armed forces to provide positional, velocity and time information and was released for consumer use. The accuracy is lowered intentionally for public usage.

Note 12.32 An antenna with limited band and width directionality. It is made by etching the metal pasted on an insulated structure. They are of a simple 2D structure, are of low cost and are applicable to various polarized waves.

Note 12.33 Stands for Personal Navigation Device and indicates a simplified portable car navigation system. Generally, the display is 4-in. or smaller (Familiar type is sized 7–8-in.). It is characterized by its low price compared to the general car navigation system. The majority of sensors for determining positions are for GPS only.

12.11.1 Background of Development

Responding to the rapid expansion of positional information devices incorporating the GPS positioning system (Fig. 12.48) (represented by car navigation systems), antennas in the systems are made from ceramics, requiring compactness and low loss, as well as have sufficient sensitivity for the reception of GPS signals. Previously, antennas used resins or air as dielectrics in the electrical circuit. They were very large and were not very reliable. These issues were resolved by means of the introduction of antennas that use ceramics as the structure. Demands for patch antennas have been increasing and they are incorporated in small and simple navigation systems called the PND and are also incorporated in mobile phones. The patch antennas currently used are smaller and lower in height and have higher reliability compared with previous ones.

12.11.2 Characteristics

12.11.2.1 Characteristics of Products and Ceramics

The Dielectric constant (ε_r) of the ceramic selected for use depends on the antenna size, where ε_r of conventional antennas are 20–40, or higher in the case of barium titanate, or magnesium titanate. Recent compact and thin antennas have ε_r of 60–95, and barium titanate (which satisfies requirements) is used as the material.

Antennas for GPS systems utilize a microstrip antenna called a "patch antenna" (Fig. 12.49). The conventional patch antenna has an electrode formed on each side of a 25 mm × 25 mm ceramic of 4–6 mm in thickness, with a power supply probe attached to one of the electrodes. By using ceramics as the base material (on which electrodes are

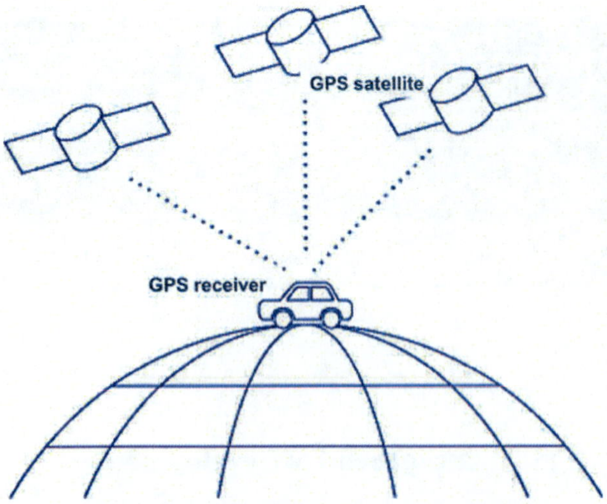

Fig. 12.48 Principle of positioning by GPS. This system was developed by the U.S. armed forces to provide positional, velocity and time information and was released for consumer use. The time delay of electric waves transmitted from the satellite is measured to determine the current position based on the distance from the orbit

Fig. 12.49 Patch antennas for GPS. Dielectric ceramics made from barium titanate, magnesium titanate, etc. are used in the antenna. An electrode is formed on each side of the structure, with a power supply pin on one side

formed), the wavelength of electric waves for transmitting GPS signals are shortened due to the high ε_r of ceramics, making it possible to downsize.

12.11.2.2 Manufacturing Method

Ceramic sintered bodies are manufactured using the dry pressing method. The main material such as barium titanate is blended and mixed with additives and the mixture is ground. Then the binder that helps to form intended solid shapes from the mixture of powders is added and mixed. The spray method is employed for granulation to create a powder with high fluidity and uniformity. It is important that ceramics used as antennas have uniform sintered density and composition distribution. The powder is formed into intended shapes by a dry press method and sintered in a zone with stable and uniform temperature distribution during the sintering process.

An example of the antenna manufacturing method is explained below. An electrode that functions as the antenna GND is printed on one side of the ceramic structure, using silver paste by the screen printing method. On the other side of the ceramic structure, an emitting electrode that determines antenna frequency and polarization of received signals is formed by a similar method. The printed electrodes are dried to volatilize the solvent in the paste and heat treated. GPS antennas receive high frequency signals of 1.54542 GHz, and current loss occurs in electrodes due to the influence of electric conductivity and skin effect in the electrodes. The loss needs to be reduced as much as possible, and therefore, the electrodes are formed as thick as possible using paste of high silver content (if a silver paste is used). A pin with a length that is greater than the thickness of the ceramic is inserted from the emitting electrode side of a through a hole, which is located slightly off the center of the ceramic. The pin and the emitting electrode are electrically connected by soldering. The pin extruding from the GND side is not connected with the GND. The completed antenna is mounted on printed substrates and the substrate pattern and the antenna pin are electrically connected by soldering. The manufacturing flow is shown in Fig. 12.50.

12.11.2.3 Performance and Specifications of Products

Explained below is an example of ceramics used in GPS antennas. A ceramic substrate with $\varepsilon_r = 20$ and external dimensions of 18 mm × 18 mm × 4 mm is used. The antenna frequency closely depends on the dielectric constant of the ceramic and the size of the electrodes. In an antenna of this size, a gain of −2 to 0 dB is needed in the area above the antenna (zenith direction), where the gain of the antenna is the highest. Some other examples are shown in Table 12.8.

Fig. 12.50 Manufacturing flow of GPS antennas. After blending and granulation, the powder is formed by the dry pressing method and calcined. An electrode is formed on each side of the ceramic and a power supply pin is attached to complete the product

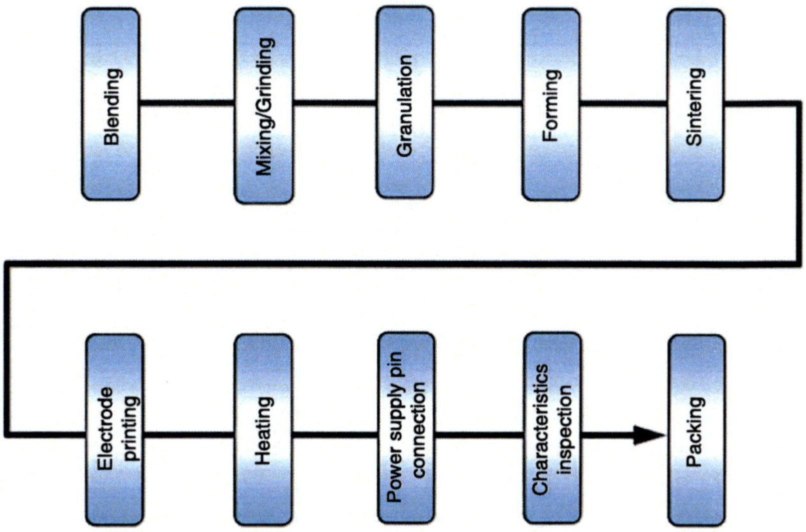

Table 12.8 Dimensions and characteristics of GPS antenna electrodes (example)

Antenna size (mm)	Gain[a] (dBic)
25×25×5	3.0
18×18×4	1.0
14×14×4	0.0
12×12×3	–2.0

[a]Values are based on a 50 mm×50 mm ground plane

12.11.3 Future Prospects

The GPS systems are also used in ships, motor bikes and bicycles as well as in vehicles. They are also used to provide positional information to walkers. The terminal devices are becoming smaller and easier to carry. Demands are high for the miniaturization of patch antennas mounted on the ceramics. High dielectric constants with low loss are competing material characteristics required of the ceramic material used in antennas. In conventional car navigation systems, antennas for the GPS system, the vehicle information and communication system (VICS) (Note 12.34), Electronic Toll Collection (ETC) system (Note 12.35), the Dedicated Short Range Communications (DSRC) system and wireless LAN are all independent. An antenna that covers all ranges of frequencies and polarization is required. In order to realize such antennas, antenna design technologies as well as material technologies that enable distribution of more than two different dielectric constants in one ceramic, for example, are needed.

Note 12.34 Stands for the vehicle information and communication system. The system displays traffic information received from FM multiplex broadcasting and roadside transmitters by marks and letters.

Note 12.35 Stands for the Electronic Toll Collection system, a non-stop automatic toll collection system that allows users of toll roads to pass through toll gates without stopping the vehicle.

Literature

1. Sugimoto M et al (1996) Ceramics 31:320–323 (in Japanese) (12.1)
2. Kagawa J et al (2007) LEMA 486:24–34 (in Japanese) (12.1)
3. Nishio N (1984) Knowledge and characteristics of spark plugs. Sankaido, p 233 (in Japanese) (12.2) (12.3)
4. Kato (1984) Insulator materials and inorganic materials used in spark plugs. Industry Research Institute, pp 64–67 (in Japanese) (12.2) (12.3)
5. Arai H (1995) History of automobile developments (Part 1). Sankaido (in Japanese) (12.3)
6. Goko M (2000) Turbocharger. Sankaido, Tokyo (in Japanese) (12.4)
7. The Editing Committee of Ceramic Component Application to Engines (ed) (1990) Ceramic component application to engines. Uchida Rokakuho Publishing, Tokyo (in Japanese) (12.4)
8. Kamigaito O (1987) Ceramic engine. Maruzen, Tokyo (in Japanese) (12.4)
9. Itoh T (1990) Ceramic turbocharger. Tojusha, Tokyo (in Japanese) (12.4)
10. Gas Turbine Society of Japan. http://wwwsoc.nii.ac.jp/gtsj/2000/g_103.html (12.4)
11. Kitagawa J (1990) Ind Mater 38(9):49–53 (in Japanese) (12.5)
12. Umehara K (1998) Ceramics 33:530–533 (in Japanese) (12.5)
13. Abe F (2001) Future Mater 1(7):26–33 (in Japanese) (12.5)
14. 75-Year history of NGK insulators (Nihon Gaishi), pp 437–438 (1995) (in Japanese) (12.5)
15. Nishio K (1999) Sensors for engine control. Sankaido, Tokyo, pp 83–100 (12.7)
16. Nishio K (1999) Engine control sensor. Sankaido, Tokyo, pp 59–65 (12.8)
17. Masuzawa K (1998) Electron Month 35–39 (12.9)
18. Kenjyou T, Sadotomo S, Kimura G (2006) All about latest compact motors in illustrations and diagrams. Gijutsu-Hyohron Co., Ltd., Tokyo (12.9)
19. Nishigaki S, Yano S, Fukuta J, Fukaya M, Fuwa T (1985) ISHM 85 225–234 (12.10)
20. Nishigaki S, Fukuta J, Yano S, Kawabe H, Noda K, Fukaya M (1986) ISHM 86 429–407 (12.10)

Aviation, Aerospace, and Transportation

13

Transportation and aerospace cover a wide range of fields including ground transportation vehicles, aircrafts and space satellites. Ceramic products are used in a variety of applications within these fields. The characteristics required for these component vary depending on the application. Roughly speaking, the components are required to have high thermal resistance, light weight and high mechanical properties. Resin, which has low thermal resistance, cannot be used independently. Ceramics are superior to metals in terms of their light weight. Therefore, the market for ceramics has been expanding in these fields. Four ceramic components, among various components employed by transportation and aerospace, are explained in this section.

Carbon/carbon (C/C) composite was used in the rocket nozzle of the orbit transfer apogee motor of a space rocket (Sect. 13.1). In order to create this composite, rayon fibers with low specific weight were used as the carbon fibers, and were Rosetta-laminated by using phenol resin preimpregnated into the fibers called "prepreg" and then formed. They were then carbonized and graphitized. Pitch impregnation was performed in the densification process and subsequently carbon deposition was applied by the C-CVD method to improve air tightness. Ceramic matrix composite (CMC) (Sect. 13.2), featuring light weight and high fracture energy, is a promising high-temperature material for use in aerospace. Long-fiber-reinforced SiC/SiC composite has been developed for application in a variety of structural elements within aerospace engines. In this section, application to jet engine afterburner flaps, jet engine tail cones, rocket engine thrust chambers, etc. were introduced as examples. Further applications to commercial aircraft are anticipated.

The C/C composite has been used in the brake disk portion of aircraft brakes for more than 40 years (Sect. 13.3). Brakes incorporating the C/C composite are lighter by 40–50% than conventional brakes and the service life is two to three times longer in comparison with steel brakes. The C/C composite is produced by reinforcing the carbon matrix by carbon fibers, featuring a density of approximately 1/4 of iron, high thermal conductivity, high thermal resistance, etc. It is applied to a number of components within the aerospace field.

Earth observation satellites use huge telescopes for real-time observation of the earth's surface. The most important component in a telescope is the primary mirror that initially gathers light. Low-expansion glass primary mirrors, which have been conventionally used are typically several meters in diameter and weigh several hundred kilograms. There was much demand for downsizing and weight reduction of these primary mirrors. SiC, featuring a low thermal expansion coefficient and thermal conductivity that is as high as 100 times that of glass, reduces strains caused by the temperature differences within the material and remarkably improves the mirror accuracy. SiC also has high rigidity and stiffness in comparison to its own weight. SiC, manufactured by the CVD method has been commercialized thus far. High-strength reaction-sintered SiC (Sect. 13.4) is currently being developed with an aim to commercialize large caliber/diameter reflecting mirrors.

Y. Imanaka et al. (eds.), *The Ceramic Society of Japan, Advanced Ceramic Technologies & Products*,
DOI 10.1007/978-4-431-54108-0_13, © Springer Japan 2012

Carbon Fiber Reinforced Carbon Composite for Space Rockets (1987)* 13.1

Development of carbon fiber reinforced carbon composite [C/C composite (Note 13.1)] began soon after carbon fiber was commercialized for industrial use. It was first used as a material for the brakes of the supersonic airplane, Concorde. The material is used in the brakes of large passenger planes throughout the world (B767, B777, etc.) as well as F1 racing cars, and approximately 80% of the C/C composites are used for the production of aircraft brakes. The composite is also used as the heat-resistant material in the space shuttle and is applied to the nose cap and the leading edge of the space shuttle. The composite is also employed within hot press molds, jigs for semiconductor production equipment and trays in industrial furnaces. The demand for the composite is on the rise as an industrial material. Explained below is the C/C composite adopted as the C/C rocket nozzle of the H-I rocket apogee motor for spacecraft. The apogee motor for the orbital maneuvering system was assembled in the engineering test satellite "Kiku-5" (approximately 550 kg) mounted on the H-I rocket, which was launched by the National Space Development Agency of Japan in August 1987. The apogee motor was installed in the center of the satellite and was ignited at the apogee after the satellite was positioned on the extended elliptical orbit, playing an important role of putting the satellite into a stationary (circular) orbit. The Aerospace Division of Nissan Motor Co., Ltd. (currently IHI Aerospace) developed the apogee motor for the H-I rocket and succeeded in its production for the first time in Japan.

13.1.1 Background of Development

The C/C composite is light and retains high strength at high-temperatures in non-oxidizing atmospheres of up to 2,000°C, which is not observed in other materials. The application of the C/C composite rocket nozzle, manufactured by making full use of these properties, on the apogee boost motor (ABM, Fig. 13.1) and the three-stage rocket motor of the H-I rocket was examined. In multistage rockets, weight reduction of motors for the top stage contributes directly to the weight of the satellite, and weight reduction of the motors is essential. For example, requirements for motors of the three-stage rocket are expressed in the form of speed multiplication per stage. The thrust of rocket motor, rocket propellant weight (Note 13.2) and rocket motor structure are determined by the weight of the third stage in which a satellite is launched into orbit around the earth by the engines mounted in the first and the second stages. A series of trade-offs were performed to realize the design that satisfies these requirements. The background on adoption of the C/C composite is explained here. For many years, abrasion properties were used to evaluate the materials for performance in order to protect the structure against combustion gas from the solid rocket fuel. Surface temperature rise of the structural material was prevented by latent heat of vaporization for a certain period of time by using materials with low thermal conductivity. Silica glass/phenol composite (Note 13.3) and carbon/phenol composite (Note 13.4), insulation composites have low sublimation rates and function as a structural body until the temperature rises, were used for the nozzle. In this case, the initial thickness of the nozzle wall was designed with these considerations for their loss, strength reduction due to temperature rises and the required structural thickness.

The silica glass/phenol composite has a thinner char layer and larger loss amount, while carbon/phenol composite has a

*The number in parenthesis indicates the year that the product was first commercialized.

Note 13.1 Composite material made of carbon fiber and carbon matrix.

Note 13.2 Weight of solid fuel used by solid-fueled rockets. Propellant of the rocket engine is the substance injected from the rocket engine and the substance that generates energy supplied to the injected substance. In chemical rockets, both fuel and oxidant are called the propellant. Propellant is normally made by kneading ammonium perchlorate and metal powder, as oxidants, into polybutadiene-based synthetic rubber.

Note 13.3 Composite made of silica fiber and phenol resin.

Note 13.4 Composite made of carbon fiber and phenol resin.

Fig. 13.1 Apogee motor for launching 550 kg-class stationary satellite. *Nozzle—C/C composite*: when a conventional CFRP material was used, the thickness was 10 mm in order to resist the heat at 3,000°C. In this nozzle, the C/C composite made by hardening the carbon fiber with thermal-resistant carbon matrix is used, reducing the thickness to 3–4 mm and substantially enhancing the thrust force accuracy

smaller loss amount and a thicker char layer when the two are compared. The required thickness is nearly the same between the two. The cost of the latter is higher but it is light as its specific gravity is 1.7 g/cm^3.

Meanwhile, graphite-based materials have been used to manufacture nozzle throats because the material loss by sublimation due to hot gas flow is minimal. If the entire nozzle can be produced from this material, the thickness added balances out the loss and carbonization will be unnecessary and substantial weight reduction will be realized. However, the adoption of the material was not realized because it is difficult to produce for large and thin parts, and a defect-free surface cannot be produced. In addition, the material is not resistant to the environmental loads caused by vibration and impact as well as the load applied during operation.

The problems mentioned above can be solved by adopting the C/C composites. They are several times stronger than graphite materials and can be formed and processed into a thin bell shape (Note 13.5) without much difficulty. Moreover, the size of defects on the surface is smaller than the gap in the weave texture and erosion due to hot gas flow is limited because C/C composite consists of graphite. Strength is enhanced as the temperature rises, allowing for the reduction of the structural required thickness as well as the minimum thickness determined by mechanical workability. The adoption of the C/C composites was decided by the above mentioned reason.

13.1.2 Characteristics

The Rosetta laminate (Note 13.6) was formed from phenol resin prepreg consisting of rayon-based carbon fiber, featuring low specific gravity of 1.4 g/cm^3. Then the carbon fiber reinforced polymer (CFRP) material was carbonized and graphitized, and pitch impregnation was performed during the densification process. Furthermore, carbon was deposited by the C-CVD method (Note 13.7) to ensure airtightness of the completed nozzle. Detailed manufacturing flow of the C/C rocket nozzle is shown in Fig. 13.2 and characteristics of C/C composites are listed in Table 13.1.

13.1.3 Future Prospects

The C/C composite was used in the cone at the nozzle exit of the H-I rocket, as indicated above. However, the material was not used in the next H-II rocket.

The C/C composite is estimated to be applicable to the nozzle throat and the exit cone (e.g. ABM). In the H-IIA rocket that followed the H-II rocket, a three-dimensional (3D) C/C nozzle throat has been employed, starting from the initial model, so as to improve reliability of the graphite that

Note 13.5 A bell shaped body with a thickness of 3–10 mm and made of C/C composite.

Note 13.6 To stack carbon fiber fabrics like cards layered by shifting each card slightly from the one below.

Note 13.7 Chemical vapor deposition (CVD) method applied to carbon (normally, methane gas is pyrolyzed).

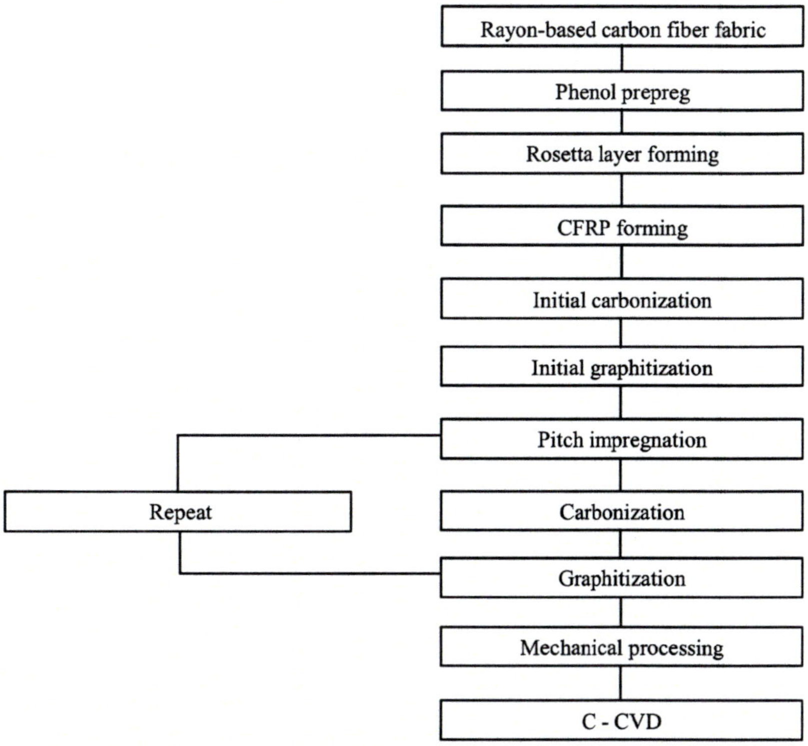

Fig. 13.2 Flow of manufacturing C/C-ABM rocket nozzle. *Key points in selection of materials*: carbon fiber: the rayon-based carbon fiber was selected because the density is lower compared with PAN-based carbon fiber. Consideration was also given to the influence of accidents such as the one in the U.S., in which a nozzle made of the PAN-based carbon fiber broke during combustion. *Key points in the process*: (1) developments of the methods for preventing interlayer peeling and deformation of CFRP formed nozzles during the initial carbonization and graphitization were the key points. (2) The C-CVD process was employed to improve gas permeability of the nozzle. It was important to make the carbon deposited on the surface as thin as possible to prevent increase in the nozzle weight, and therefore, a two-stage processing was adopted

Table 13.1 Characteristics of C/C composite for space rockets

Characteristics	Values
Bulk density (g/cm^3)	1.53–1.60
Apparent density (g/cm^3)	1.72–1.75
Apparent porosity (%)	9.7–11.2
Tensile strength (MPa)	39.2–68.6
Tensile elastic modulus (GPa)	14.7–18.6
Tensile elongation (%)	0.46–0.50
Compressive strength (MPa)	34.3–49.0
Compressive elastic modulus (GPa)	13.7–17.6
Interlaminar shear strength (MPa)	8.0–11.5

was used in the throat. The three-dimensional C/C composite has also been used in the throat in all stages of the M-V rocket (Note 13.8), starting from the fifth model (Fig. 13.3). The technology has been established for the stable production of high-density C/C composite at 1.95 g/cm^3.

Fig. 13.3 Material for nozzle throats of large rockets

Application of the material to the exit cone has not been realized since its adoption on the AMB. However, the C/C composite features characteristics required for performance improvements that cannot be realized by other materials, including light weight and high strength at ultrahigh temperatures. Judging from the fact that the C/C composite is used as the nozzle exit cone of liquid-fueled rockets in Europe and the U.S., exit cones made of C/C composite will be used in both solid-fueled and liquid-fueled rockets in the near future.

Note 13.8 The M-V rocket (μ-V) is a three-stage rocket developed jointly by the Institute of Space and Astronautical Science (ISAS) of the Ministry of Education, Culture, Sports (MEXT), Science and Technology and the Aerospace Division of Nissan Motor Co., Ltd (currently IHI Aerospace Co., Ltd.) and operated by the Institute of Space and Astronautical Science (ISAS) of the Japan Aerospace Exploration Agency (JAXA) as the successor of the ISAS of MEXT. The rocket was used to launch satellites and planetary exploration spacecrafts from the launching facility located at the Uchinoura Space Center. The rocket retired after the launch of SOLAR-B at 6:36 p.m. on September 23, 2006. It was the last and most advanced rocket in the μ series, solid-fueled rockets made in Japan. Its design originated from the "pencil" rockets that were developed by Dr. Itokawa.

Ceramic Composite Materials for Aerospace High-Temperature Parts (1994) 13.2

Ceramic matrix composites (CMCs) are lighter than heat-resistant metallic materials and feature higher fracture energy than monolithic ceramic materials. Long-fiber-reinforced SiC/SiC composites have been developed for use as aerospace engine structural components. A variety of engine parts have been prototyped using these materials, demonstrating their practicality. Typical prototype parts, including the afterburner flaps for jet engines, tail cones and rocket combustors, as well as the CMC manufacturing process are introduced below.

13.2.1 Background of Development

Non-oxide ceramic materials, featuring higher strength at elevated temperature than metallic materials, were expected to be applicable to high-temperature structural components of turbine engines, etc. However, normal monolithic sintered ceramics tend to be broken in brittle fractures (easy crack propagation) and are not suitable for use as structural components. As a means of solving the issue of brittleness, the long-fiber-reinforced CMC (hereinafter called CMC), which incorporates silicon carbide in both the fibers and in the matrix, has been developed. CMC exhibits pseudoplastic deformation behavior (Note 13.9), which is similar to the behavior of metallic materials. It is superior in specific strength and thermal resistance, and research has been promoted for the application of the material to high-temperature parts of aerospace engines. The material is applicable to high-temperature parts of jet engines and rocket engines.

Note 13.9 The pseudoplastic deformation behavior of CMC is realized by the fibers that are pulled out of the matrix that bridging cracks in the matrix. A large amount of fracture energy is absorbed when the fibers are being pulled out. Microscopic fractures are generated in the matrix by deformation and plasticity cannot be maintained.

13.2.2 Characteristics

13.2.2.1 For Afterburner Flaps of Jet Engines

The afterburner is a mechanism used to inject additional fuel to the exhaust portion of a jet engine for combustion, which is needed to temporarily increase engine thrust greatly. For activation of the afterburner, the shape of the throttle of the engine exhaust nozzle needs to be changed. The moving part used for this purpose is called the afterburner flap. If CMC is used, the weight is reduced to 1/3 compared with a part made of nickel alloy. The part has been commercialized in Europe and the U.S. and its application to engines is expected also in Japan. Figure 13.4 shows an engine mounted with a portion of it made of CMC, which was used for the static firing test.

13.2.2.2 Tail Cones for Jet Engines

The combined cycle engine developed by the HYPR project (Note 13.10) incorporates a tail cone made of CMC in the core engine. The part was subjected to 1,700°C combustion gas (Fig. 13.5) without cooling in a static firing test, proving the soundness of the part.

13.2.2.3 Rocket Engine Thrust Chamber

Parts are being developed for rocket engine thrust chambers of propulsion systems, altitude control systems for space

Note 13.10 Through "Research and Development of Propulsion System for Supersonic Transport Aircraft" (1989–1998) by the Agency of Industrial Science under the Ministry of International Trade and Industry, the research was also commissioned to private companies through NEDO. The hybrid cycle engine operates in combination with a turbo jet core engine at low speeds of up to Mach 3 and in combination with a ramjet engines at high speeds of up to Mach 5.

Fig. 13.4 A jet engine mounted with afterburner flaps made of CMC. This was the first domestic engine that was mounted with CMC parts. The combustion test on ground was successful, proving that CMC was applicable to actual products. The four parts indicated by numbers 1–4 were made of CMC

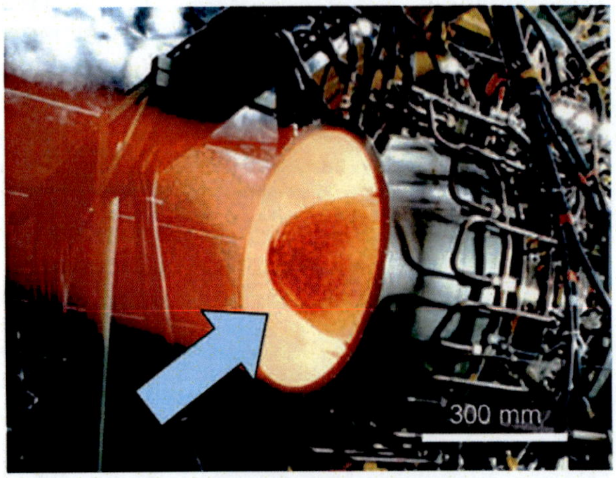

Fig. 13.5 The CMC tail cone for HYPR core engines. A 15-min combustion test was performed setting the turbine inlet temperature at 1,700°C. The soundness of the material was examined after the test

transportation, etc. If the niobium-based metallic material is replaced by CMC, the weight of parts will be reduced to 1/3 or less. The parts must resist extremely high temperatures and pressures during combustion even for a short period of several minutes. The material must also have high density so as to prevent leakage of the combustion gas. Combustion tests were performed under 0.7 MPa nitrogen pressure and 4.5 MPa water pressure to confirm that the developed parts are free from leakage (Fig. 13.6), and it was confirmed that

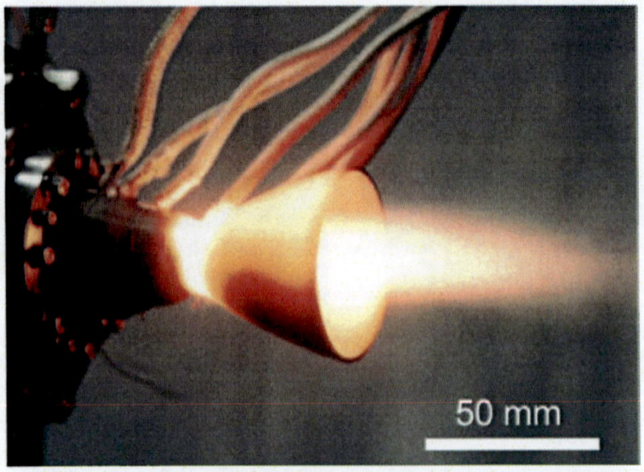

Fig. 13.6 Combustion test of the rocket engine combustion nozzle. A dense matrix is formed by repetition of the PIP process. A combustion test was performed by using the fuel intended for rockets (hydrazine). Leakage of combustion gas or substantial damage to parts was not observed

they are applicable to actual products. Research has been continued for test flights that are expected to be performed in the near future.

13.2.2.4 Other Parts

In addition to the above-mentioned parts, developments and evaluation tests of turbine shrouds, exhaust guide vanes, noise absorption panels, turbine-disk blade integrated parts (blisks), etc. (Fig. 13.7) intended mainly for use as high-temperature parts in jet engines have been performed in the ESPR project (Note 13.11) and others.

13.2.2.5 Manufacturing Method and Characteristics

CMC is made by (1) creating a fabric formed body (woven fabric) first and then (2) forming the matrix. Mechanical properties of CMC are substantially affected by the texture of fabrics, and consequently, the design of the fabric formed

Note 13.11 The ESPR (Research and Development of Environmentally Compatible Propulsion System for Next-Generation Supersonic Transport) project (1999–2004) by the Ministry of Economy, Trade and Industry included three technology subjects to establish the basic technology of engines for next generation supersonic transports: Low noise, NO_x reduction, and CO_2 reduction.

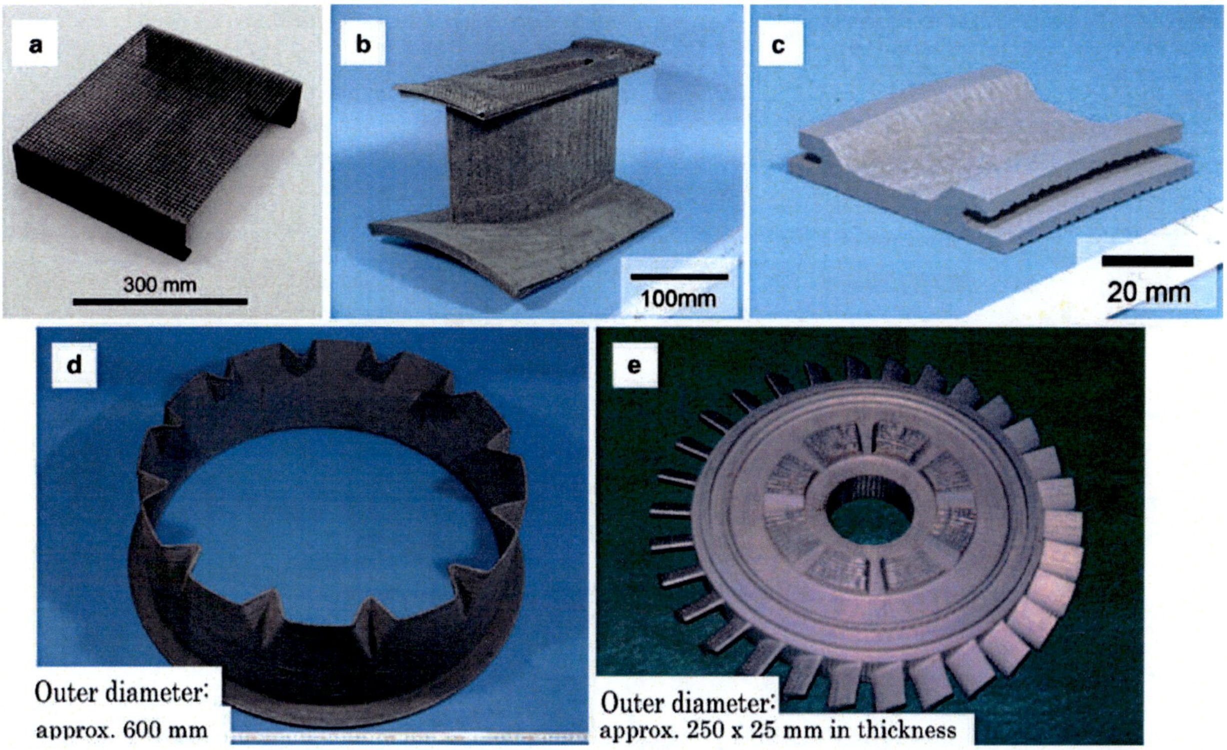

Fig. 13.7 Prototypes of engine parts made of CMC. (**a**) Sound absorption panel for HYPR engine, (**b**) exhaust guide vane for ESPR engine, (**c**) turbine shroud, (**d**) exhaust air duct and (**e**) turbine blisk

body is important. CMC is a difficult workable material, like sintered ceramics. Therefore, the compacted body should be prepared by net-shape forming if at all possible. An interfacial layer is formed on the surface of fibers that constitute the formed body for the proper weak bonding between the fibers and the matrix.

The matrix forming processes include (2a) the chemical vapor infiltration (CVI) process, which uses the vapor phase as the raw material and (2b) the polymer impregnation and pyrolysis (PIP) process, which uses the liquid phase as the raw material. By proper combination of these matrix forming processes, materials that are comparatively high in initial elastic modulus, high in breaking strain and favorable in the design of machines can be prepared (Fig. 13.8). Figure 13.9 shows an example of the process combination flow and an optical micrograph of the texture of the CMC produced by the processes.

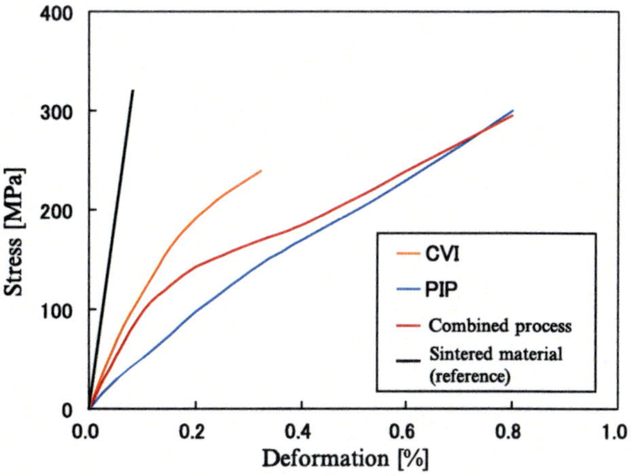

Fig. 13.8 Diagram showing stress—deformation of CMC. By combination of the CVI and PIP, materials featuring a high initial elastic modulus and a large breaking strain can be produced

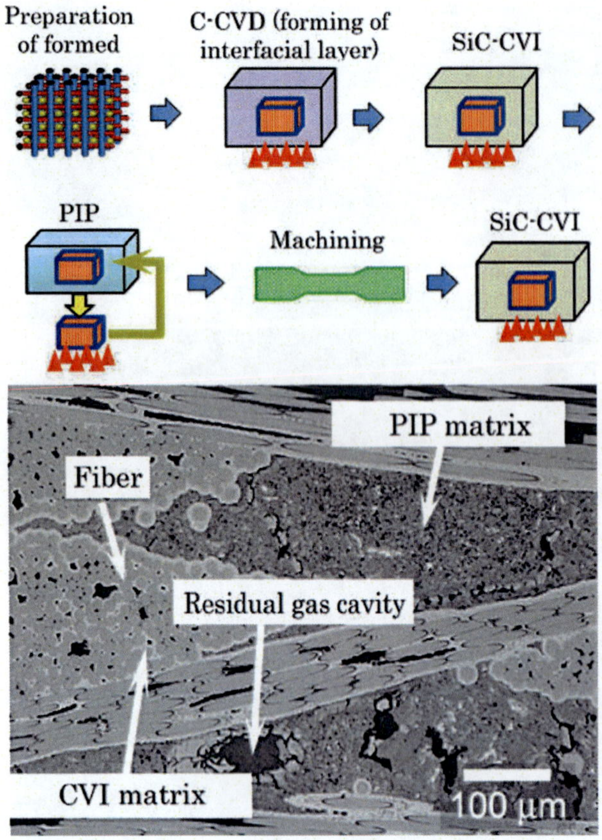

Fig. 13.9 The CMC manufacturing process and macro–micro texture. The voids in the fabric formed body are classified roughly into (1) small voids inside fiber bundles and (2) large voids between fiber bundles. The matrix can be created efficiently by densifying small voids by CVI (gas-phase raw material) and large voids by PIP (raw material: molten polymer)

13.2.3 Future Prospects

The CMC produced by the processes explained above has superior properties but it requires numerous production processes and is expensive to produce. Recently, the wet-type excitation and impregnation process (which requires a short time for processing and much less material cost) has been developed to replace the PIP process.

An increase of engine combustion temperature by means of the adoption of materials with high heat resistance and the reduction of engine weight by means of the adoption of light materials are both effective for the efficient generation of thrust using limited fossil fuels. Application of CMC to aerospace engines will continue.

C/C Composite Brake for Aircraft (1971) 13.3

Carbon/Carbon composite (C/C composite) is a carbon matrix material strengthened by carbon fibers. The material is light, with a density of approximately 1/4th of iron and has excellent thermal properties such as high specific heat, thermal conductivity and thermal resistance.

The C/C composite brake was first used as brake disks on fighter jets, which require light brakes with high performance, in 1971. It was commercialized as a light and high-performance aircraft brake capable of absorbing a high amount of kinetic energy in a short time.

The weight can be reduced by 40–50% compared with steel brakes and the service life is two to three times longer than that of steel brakes. It is widely used in fighters, which need light brakes with high performance, as well as commercial aircrafts, aiming to reduce fuel consumption through weight reduction and to reduce maintenance cost through the adoption of brakes with a longer service life.

13.3.1 Background of Development

In the aerospace industry, which continues to seek weight reduction, composite materials such as CFRP (Note 13.12) have advanced remarkably. The carbon matrix composite reinforced by carbon fibers was developed around 1960. In 1971, the material was used in brakes for high-performance fighters and was commercialized as the C/C composite brake .

Later, the material began to be used in commercial and military aircraft including the Concorde, the supersonic jetliner. It has been used for more than 40 years, and a lot of the material has been used for aircraft brakes (Table 13.2).

Note 13.12 Carbon fiber reinforced plastics: a carbon fiber composite that uses carbon fibers as reinforcement fibers and epoxy resin as the matrix material. It is widely used in the sports industry in golf club shafts, tennis rackets, etc., and is also used as the primary structural material of aircrafts.

13.3.2 Characteristics

13.3.2.1 Characteristics

Aircraft C/C composite brakes are multiple-disk hydraulic brakes and the friction material is made of C/C composite (Fig. 13.10).

A weight reduction of 40–50%, compared with steel brakes, can be achieved, realizing weight reduction of approximately 900 kg in the B747-400, a large aircraft (Table 13.3).

The friction coefficient needed for braking is 0.15–0.30 in C/C composite, which is slightly lower than that in steel brakes. The wear volume is 1/2–1/3 compared with steel brakes and the service life is two or three times longer.

Characteristics of typical brake friction materials are shown in Table 13.4. The temperature rise of the C/C composite at braking is set at 600–800°C, which is higher compared with friction materials used in automobiles and railway cars, so as to exhibit excellent thermal properties with light weight. However, the strength deteriorates gradually due to oxidization during repeated use at temperatures over 500°C. Therefore, oxidation resistant coating of ceramics such as SiC and phosphorus compounds is applied on the side surface of internal and outer circumferences.

In addition, heat insulation countermeasures are performed on parts of wheels and brakes to prevent loss of strength in light and high-strength aluminum alloy due to the influence of heat transfer and heat radiation arising from the temperature rise during braking.

C/C composite brakes are priced higher than steel brakes by 1.8–3.8 times, but the maintenance cost for replacement is lower because of the longer service life and fuel consumption is reduced due to the light weight. Therefore, the life cycle cost (Note 13.13) of C/C composite brakes is estimated as comparable to that of steel brakes.

Note 13.13 Life cycle cost: the total cost is obtained by adding the costs arising in all stages from the start of development to the end of the service life, including designing, trial production, production and operation. It is referred to as LCC.

Table 13.2 Major aircraft that incorporates C/C composite brakes

Commercial airplanes			Military aircraft	
CONCORDE	MD-11	CL604	F-14	JSF-35
A319	MD-90	Saab 340	F-15	C-17
A320	B747-400	Saab2000	F-16	B-1B
A321	B757	Fokker100	F/A-18	Rafail
A330	B767-300	Gulfstream IV	B-2	Rafail D&M
A340	B777	Gulfstream V	F-22	Mirrage 2000
A340-5/600	B787	EMB 170/175	F-117	Saab JAS-39
A380	CL600	EMB 190/195	X-37	

Adoption of C/C composite brakes reduces weight, enhances energy absorption capacity and extends the service life of brakes. They have been used in many aircraft including high-performance fighter jets and commercial airplanes. They are also used in other vehicles such as TGV and formula 1

Table 13.3 Weight reduction by application of C/C composite brake

Model name	Number of brakes/aircraft	Weight reduction/plane (kg)
B747-400	16	820
A310	8	590
B767	8	400
A320	4	300
B757	8	160
F-15	2	55

Weight of aircraft can be reduced by using C/C composite brakes, and the weight will be reduced by about 900 kg at maximum if they are used in large aircraft mounted with many brakes

13.3.2.2 Manufacturing Method

The general manufacturing method of C/C composite disks is shown in Fig. 13.11. Sheets of carbon fiber cloth impregnated with resin are laminated and short fibers are mixed with resin to create a formed body by press forming. The formed body is placed in an inert gas and the resin on the matrix is carbonized at a temperature of approximately 1,000°C. The formed body has numerous voids and its strength is low at this stage.

Table 13.4 Typical brake friction materials and characteristics of C/C composite for aircraft

Characteristics		C/C composite	Steel	Copper
Density (g/cm^3)		1.6–1.8	7.85	8.96
Tensile strength (MPa)		70–100	670	210
Bending strength (MPa)		100–150	–	–
Izod impact strength (J/m)		300	–	–
Thermal expansion coefficient (×10^{-6}/°C)		1–8	14	16
Specific heat (J/(g°C)		1.4	0.47	0.38
Thermal conductivity (J/(m s °C))	Thickness direction	10–15	43	385
	Radial direction	20–30		
Melting temperature (°C)		3,500	1,535	1,083

Physical, chemical and mechanical properties of C/C composite are compared with those of other friction materials. C/C composite features low density and excellent thermal properties such as high specific heat, thermal conductivity and melting temperature

Fig. 13.10 C/C composite brakes installed on main axles of aircraft. C/C composite, featuring light weight, high heat capacity and long service life, is used in brake disks (as friction material)

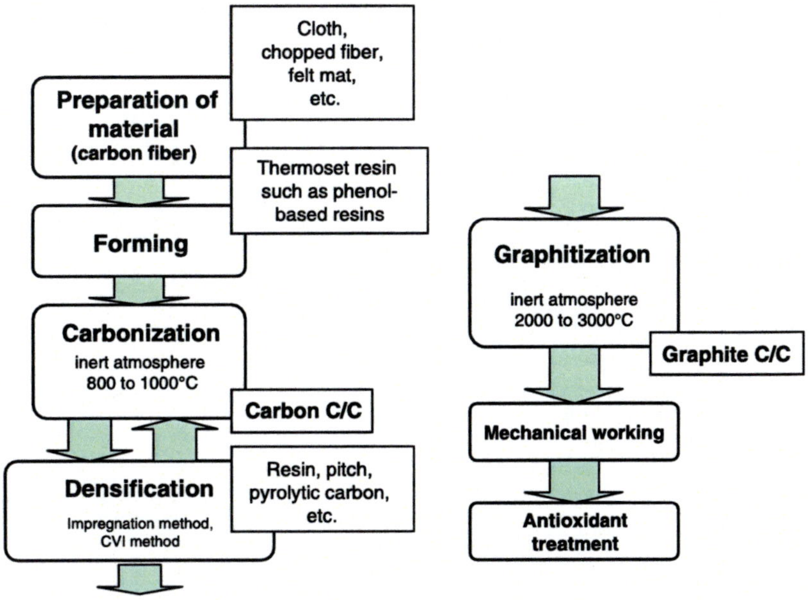

Fig. 13.11 General manufacturing method of C/C composite disks. Sheets of carbon fiber cloth impregnated with resin are laminated and formed. After heat treatment and carbonization, impregnation of pitch and carbonization are repeated for densification. The product is completed after graphitization, mechanical working and antioxidant treatment

Resin and pitch impregnation, carbonization and CVI (Note 13.14) are repeated several times for densification. Through these procedures, the voids are reduced and the density is enhanced, attaining higher strength. Finally, the formed body is graphitized at a high temperature of above 2,000°C, improving the friction properties and physical properties of the friction material. Several months are normally needed in order to complete the manufacturing process.

13.3.3 Future Prospects

C/C composite brakes are widely used in aircraft because of their light weight, high performance and long service life. Other than aircrafts, the application will expand to automobiles, etc. in the future if the quality is stabilized and the cost is reduced further through improvements in the manufacturing method.

Note 13.14 Chemical vapor infiltration: hydrocarbon gas such as methane and propane is infiltrated into a preheated carbon fiber formed body to deposit the carbon generated by thermal decomposition for an improvement in density with this method.

SiC Mirror for Earth Observation Stationary Satellites (2003) 13.4

Stationary satellites orbit above the earth and allow real time earth observation. However, they orbit at an altitude of 36,000 km and a large telescope with a diameter of one to several meters in diameter is required in order to realize a ground resolution of several tens of meters. Primary mirrors (Note 13.15) used in this kind of telescope were conventionally made of glass, but primary mirrors made of glass weighed several hundred kg, increasing the mass and development cost of satellites. For example, the primary mirror of the Hubble Space Telescope is 2.4 m in diameter and weighs 830 kg. High-strength reaction-sintered SiC features high thermal conductivity with a Young's modulus and thermal stability that is higher than that of glass. The bending strength and fracture toughness are also high, and therefore, the weight of a reflecting mirror will be around 1/5th when compared with a glass mirror. SiC is free of porosity and can be polished in the same manner as glass. High-strength reaction-sintered SiC is a promising material for large space telescopes.

13.4.1 Background of Development

Since the well-known Landsat No. 1 satellite was launched in 1972 for global observation, various optical cameras have been developed for actual use as well as for research. Currently, images of clouds transmitted from the weather camera on the weather satellite Himawari No. 6 are shown in weather forecasts many times a day in Japan. The camera mounted on the weather satellite is set at a ground resolution (size of 1 pixel) of 1 km, which allows for the identification of clouds, and images of the entire globe seen from space are sent to earth every hour. When a typhoon approaches Japan, the images are shot and transmitted at a speed of 1 image every 30 min. Meanwhile, more refined images with a ground resolution of approximately 1–20 m are distributed by Spot Image Corporation and Space Imaging, Inc. and anyone can purchase these images. When, for example, a large disaster such as the massive tidal wave (Tsunami) that struck Indonesia occurs, damage is estimated from the images of the site shot before and after the tidal wave. These satellites orbit around the earth approximately 15 times a day in a lower orbit at an altitude of approximately 1,000 km above the earth. The satellites move along a certain orbit, while the earth rotates. Therefore, the distance of the orbits of the 15 flight paths are approximately the same and the distance of the orbits above the equator is approximately 2,700 km. The minimum line-of-sight distance from the height of the satellites to ground is about 500 km, and the orbits are selected so that the satellites return to the original position in 5–20 days. The earth cannot be observed when the ground is covered by clouds, and it sometimes takes several weeks to months to observe a certain point on the earth. If the images can be shot from a stationary orbit at a space resolution of 10–50 m, as with weather satellites, real-time disaster observation will be possible. If locations of ships can be monitored by this method, suspicious ships and pirate ships can be watched for safe navigation. What is preventing the realization of cameras that can be used for these purposes? It is the diffraction phenomena attributable to the wave nature of light, which limits the angles that can be distinguished by the diffraction phenomenon. If the opening size of a telescope is expressed as D and wavelength of light as λ, the limit angle that can be distinguished is expressed as 2.44 λ/D (rad). The height (h) of a stationary satellite is 36,000 km, and in order to distinguish an object with a diameter (d) of 20 m, the angle expressed as $d/h = 0.6$ μrad (0.1 arcsecond) needs to be resolved. Light has a wavelength of approximately 0.38–0.75 μm, and therefore, a telescope with an opening size (D) of approximately 2.4 m is needed. Before the Subaru Telescope was built in Hawaii, a telescope with an aperture of 1.88 m in the Okayama Astronomical Observatory, the largest astronomical telescope in Japan, was used for a long time. Judging from these circumstances, launching a satellite

Note 13.15 The mirror used to collect light initially in a reflecting telescope. It is usually the largest mirror in a telescope.

mounted with a high resolution telescope is a challenge. Furthermore, in order to create a proper image by collecting the light, the shape error of the reflecting mirror has to be kept below 1/10 (several tens of nanometers) of the wavelength and a surface roughness of below 1 nm RMS (Note 13.16). When the mirror is enlarged to the size of the globe, the allowable error is about the size of a curbstone, and the surface roughness has to be below the level of a gravel road, which requires an extremely high level of technology. However, a telescope with an aperture of 2.4 m has already been launched into space and is in actual use, e.g. Hubble Space Telescope (Fig. 13.12). The Hubble Space Telescope was carried to space in 1990 by the Space Shuttle and has been transmitting excellent observational data for the past 17 years, during which the telescope underwent repair work in orbit. ULE, low thermal expansion glass (Note 13.17), is used as the reflecting mirror of the telescope mounted on the Hubble Space Telescope. The reflecting mirror features a donut-shaped honeycomb structure, as shown in Fig. 13.13. It is assembled under the force of gravity, mounted on a

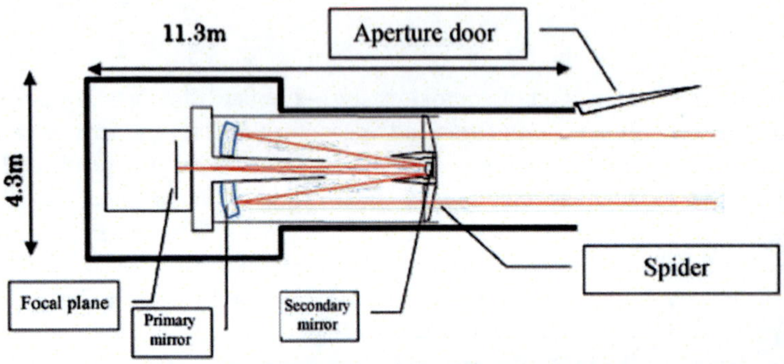

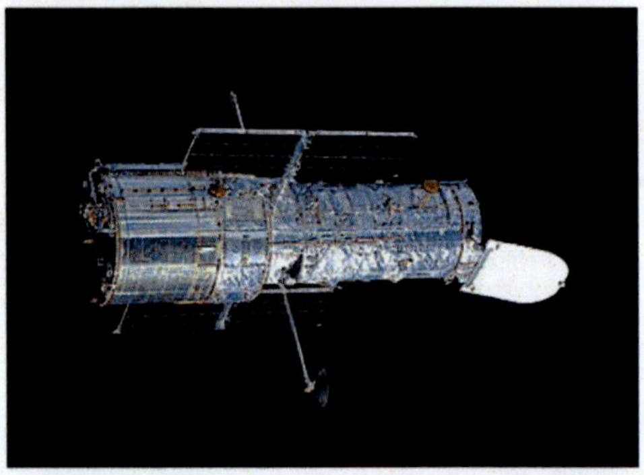

Fig. 13.12 The Hubble Space Telescope and the optical system. The Hubble Space Telescope in orbit. The Hubble Space Telescope is a huge telescope itself, and the major component of the telescope is the primary mirror, which is 2.4 m in diameter. The light collected by the primary mirror is reflected by the secondary mirror and the image is focused on the focal plane (picture released by NASA)

Fig. 13.13 Primary reflecting mirror in the Hubble Space Telescope. The primary mirror of the Hubble Space Telescope is made to minimize its weight by joining thin glass plates using a special method on *top* and *bottom* of donut-shaped honeycomb glass plates, but the mass is still 830 kg (picture released by NASA)

Note 13.16 Stands for root mean square and is one of the indexes used to express the level of error.

Note 13.17 Glass materials featuring a thermal expansion coefficient of 10^{-6} or lower are usually called low expansion glass. Glass materials of special blending and glass ceramics in which part of the glass is recrystallized have been commercialized.

satellite and is carried to the zero-gravity environment of space. Therefore, deformation under its own weight cannot be neglected, with respect to the entire telescope optics as well as to the mirror. The hole in the center was made to avoid the influence of gravity, to reduce the weight of the entire mirror and to avoid influence on the structure. Because of these measures, the mass of the primary mirror mounted on the Hubble Space Telescope was reduced to 830 kg. However, it is still as heavy as a light automobile. The satellite weighs 11 tons because of the measures taken to support the mirror stably and accurately so that it can withstand vibration and shock at the time of blastoff.

SiC is a material that is expected to further reduce the weight of space telescopes. Ultra-low expansion (ULE) glass has a linear expansion coefficient of approximately 3×10^{-8} K^{-1}, which is approximately 1/100th compared with 2.5×10^{-6} K^{-1} of SiC, and it is thermally stable. However, in the actual environment where telescopes are used, the telescope temperature changes dramatically depending upon the satellites position towards the sun. Under this environment, temperature variation in the SiC mirror is 1/100 or less compared with that of glass mirror, because thermal conductivity of SiC is higher than glass by more than 100 times, which means thermal strain of a SiC mirror is smaller than that of a mirror made of low expansion glass. In addition, the specific rigidity (Young's modulus/density) parameter is several times higher in SiC, and a reflecting mirror made of SiC is lighter when the structure is the same. Because of these characteristics, SiC is the optimal material for large reflecting mirrors. If SiC is used, it is estimated that the weight of a reflecting mirror with a diameter of 1 m will be less than 80 kg and the one with a diameter of 2 m will be less than 200 kg.

13.4.2 Characteristics

The laser altimeter of the asteroid probe "Hayabusa" launched in 2003 was the first space optical system that employed a SiC mirror in Japan. The reflecting mirror was small and 13 cm in diameter, but it was essential for landing on the asteroid Itokawa. A 70 cm reflecting mirror was used in the Infrared Rays Astronomical Satellite "Akari," which was launched in 2006. These mirrors achieved considerable success as SiC mirrors, but they were made by the CVD process (Note 13.18), which requires much time and is expensive, and it is believed that further scaling up will be difficult. The manufacturing method of reflecting mirrors by using high-strength reaction-sintered SiC, which is currently under development, is shown in Fig. 13.14. A mixture of carbon

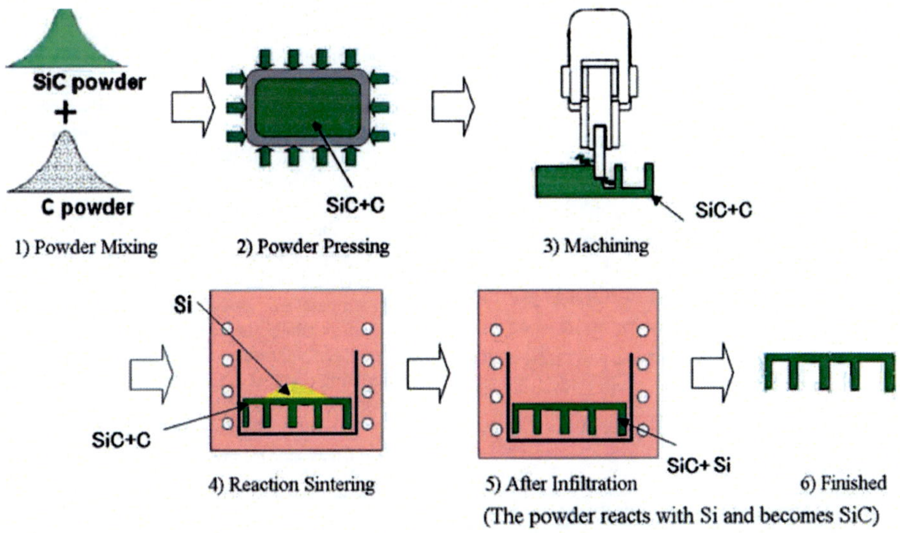

Fig. 13.14 Manufacturing method of preform for high-strength reflecting mirrors. Manufacturing process of reaction-sintered SiC

Note 13.18 Stands for chemical vapor deposition. Gas containing the component of the thin film to be produced is supplied on a base substance heated in the reaction furnace in order to deposit the material through chemical reaction on the base surface or in the vapor phase. It is also called chemical deposition.

powder and SiC powder is formed by the cold isostatic pressing method. The formed body is soft and machined to create a shape close to that of the finished product. Then, silicon is infiltrated into the furnace to react with unbound carbon to create SiC and to fill in the porosity with the excess silicon at the same time. The surface of the sintered product is ground. The created material is dense and free of pores and can be polished into the shape and surface roughness required of the mirror surface.

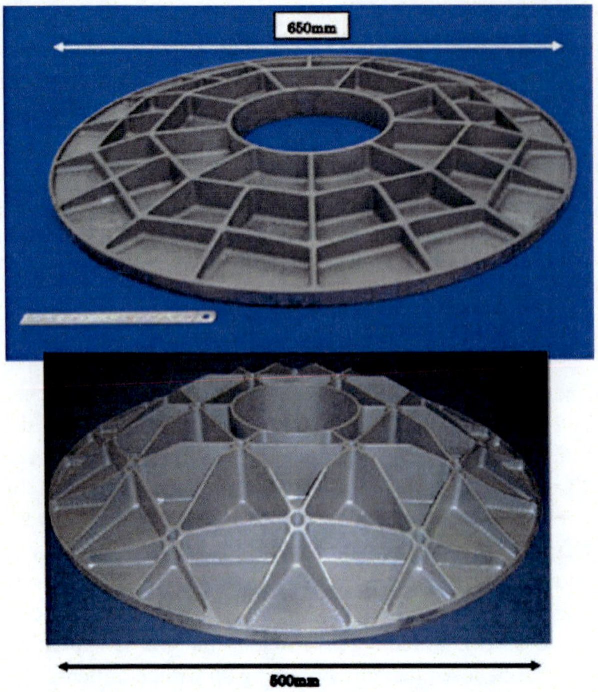

Fig. 13.15 SiC reflecting mirror. Prototype of a base material for the mirror surface made of high-strength reaction-sintered SiC (Note 13.19). Upper side is a base material of a plane mirror with a diameter of 650 mm, and lower side is a base material for a spherical mirror with a diameter of 500 mm

13.4.3 Future Prospects

Figure 13.15 shows prototypes of a spherical mirror with a diameter of 500 mm and a plane mirror with a diameter of 650 mm. The base material of large and light reflecting mirrors made of high-strength reaction-sintered SiC are currently limited to 3 mm in mirror surface thickness, 3 mm in back rib thickness and 100 mm in rib height. The various shapes of new products are on verge of becoming a reality with the constraint of the production equipment being 1,200 mm. The technology has reached a level capable of producing a stationary earth observation sensor with a ground resolution of approximately 40 m.

With an aim of creating a reflecting mirror of the same dimensions as the Hubble Space Telescope, research has been performed on preparing large reflecting mirrors by tiling together 1 m mirrors, to reduce the mirror thickness needed for further weight reduction.

Literature

1. Yasuda E, Tanabe Y, Akatsu T (1999) Ceramics 34:266–269 (in Japanese) **(13.1)**
2. Hosomura T, Sobue T (1986) J Jpn Soc Compos Mater 12(3):103–108 (in Japanese) **(13.1)**
3. Corporate brochure, Aerospace Division of Nissan Motor Co., Ltd. (in Japanese) **(13.1)**
4. Yamauchi H, Hayakawa H, Suzuki S (2007) IHI Eng Rev 40(1): 27–30 (in Japanese) **(13.1)**
5. Ishizaki M, Shioda M, Miyahara K, Sasa T, Araki T, Masaki S, Imamura R, Onabe H (1994) Production technology and application of fiber reinforced ceramic composite materials. Tech Rep IHI Corp 34(6):418–422 (in Japanese) **(13.2)**
6. Nishikiori T, Masaki S (2000) Development of TiAl alloy and CMC composite materials for the improvement of aerospace engine performance and parts manufacturing technology. J Jpn Inst Met 64(11):992–998 (in Japanese) **(13.2)**
7. Ishizaki M (2004) Application of SiC/SiC composite material to high-temperature parts of aerospace engines. In: Symposium on ultra-high-temperature materials XVIII, pp 42–51 (in Japanese) **(13.2)**
8. Murata H, Nakamura T, Tanaka Y (2006) Development of new production method of CMC for aircraft engines. Tech Rep IHI Corp 46(3):101–108 (in Japanese) **(13.2)**
9. Tsuji K (1995) Abstract book of the 33rd aircraft symposium, pp 31–34 (in Japanese) **(13.3)**

Note 13.19 Reflecting mirrors are made by polishing the surface of the base material for the mirror surface. The material prior to the polishing process is called the base material.

Part IV

Ceramics in the Energy Sector

14 Energy

Ceramic components used in "generation and transmission of power" in the electricity field are explained in this section. Figure 14.1 shows the flow of electricity generated at power plants in which electricity travels from substations to consumers (factories and houses). In Japan, 99% of the electricity is generated at large-scale power plants, of which 59% of this electricity is produced at thermal power plants, 34% at nuclear power plants and 6% at water power plants in 2005. Meanwhile, a wider range of power source systems are being developed rapidly following partial deregulation of the electric power market. The generation of electricity using renewable energy such as solar battery, wind power generation and geothermal power generation, as well as new energy such as fuel cells and thermoelectric generation are attracting greater attention. Power storage technologies for sodium–sulfur batteries and superconductive power storage are also in the process of commercialization. Electricity generated at these power plants are boosted for effective long-distance transmission and then the voltage is lowered at substations along the route to 100 and 200 V for general use, reaching the destinations via the power transmission networks covering every corner of the Japan.

The six themes explained in this section cover ceramics, which are positioned as key components among the materials used in the energy field related to the "generation and transmission of electricity." High reliability is required of energy related components because they generally affect society on a large scale. Therefore, the key components made of ceramics and used in energy devices are commercialized after strict, long-term reliability tests.

"Insulators for transmission lines and substations (Sect. 14.1)" are ceramics that are used to support transmission lines and to store and support conductive wires at substations. Larger products are required responding to the increase in demand for electricity, and they are among the largest ceramic structures. Needs for electricity have been increasing rapidly in China, India and Southeast Asia, where the economic development is expanding. Under these circumstances, Japanese technologies are being utilized in many countries. "Varistors for electricity (Sect. 14.2)" are made of zinc oxide ceramics, which are used in arresters that absorb voltage spikes caused by lightening, etc. in substations and power distribution facilities. The resistance value of varistors declines when high voltage due to lightening, etc. is applied. This results in high electric currents and usage of protection devices such as transformers and electrical power breakers to enhance the safety of the power transmission network.

"Uranium dioxide fuel (Sect. 14.3)" is used in nuclear power plants and pellets of the fuel, approximately 1 cm in size, are packed in fuel rods that are about 4 m in length. One pellet produces energy equivalent to the energy produced from two barrels of oil or more and is capable of supplying electricity equivalent to that used by one household over 6 months. Nuclear power plants, which do not emit CO_2, do not contribute to global warming. However, security and conservation should be considered adequately in the future.

"B_4C control material for nuclear power generation (Sect. 14.4)" is used in nuclear reactors to control nuclear fission reactions and to shield against excessive neutron generation. In order to control the output of reactors, the number of neutrons inside reactors needs to be adjusted in order to control electricity generation.

"Sodium–sulfur batteries (Sect. 14.5)" are secondary batteries and use β alumina as the solid electrolyte that separates the positive electrodes (+) and negative electrodes (–). It is used to store electricity as well as for load leveling of electricity for daytime supply and nighttime supply at large-scale power plants.

"High-temperature superconducting cables (Sect. 14.6)" are superconducting wires capable of transmitting large amounts of electricity with low loss and are classified by the material composition into Bi-based and Y-based cables. High-temperature superconducting wires are housed inside insulated pipes in which liquid nitrogen at a temperature of approximately –200°C is circulated. The full-scale introduction of this technology is expected in 2015.

Y. Imanaka et al. (eds.), *The Ceramic Society of Japan, Advanced Ceramic Technologies & Products*,
DOI 10.1007/978-4-431-54108-0_14, © Springer Japan 2012

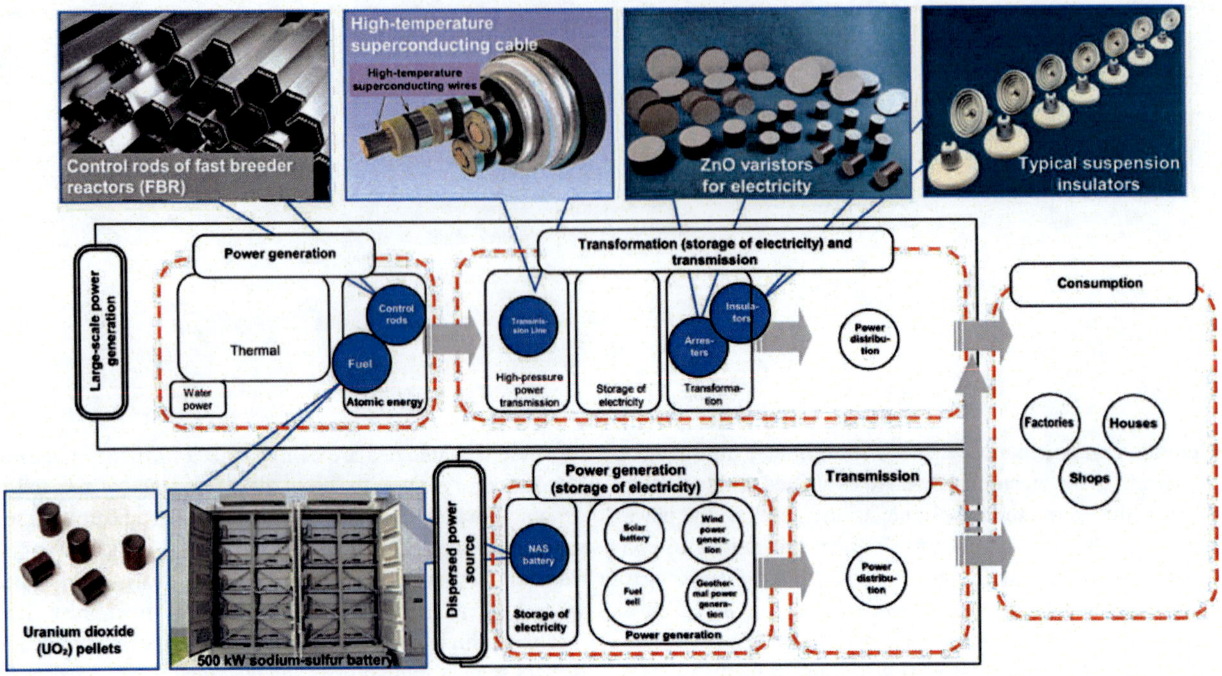

Fig. 14.1 Schematic of electricity generation and distribution highlighting ceramic component usage

Insulator for Transmission Line and Substation (1899)* 14.1

Conductors and insulators are indispensible in the utilization of electricity. Insulators (Note 14.1) are used to isolate electric power transmission lines from supporting pillars and also to mechanically support the transmission lines. In Japan, an 11 kV high-voltage transmission line was established for the first time in 1899. Demand for electricity increased dramatically in response to the upturn in social life and economic activity. Under these circumstances, the transmission voltage was increased and long-distance transmission lines were constructed. Insulators are required to be larger and stronger in response to the increase in the capacity of the transmitted power. Various shapes of insulators are designed so as to meet different environments. The majority of insulators used in Japan are made of porcelain and manufactured by sintering a mixture of china stone, feldspar, clay and alumina at high temperatures of approximately 1,300°C. Among the currently used suspension insulators, the strongest one is capable of supporting a load of 530 kN. In view of the 1,000 kV transmission in the future, verification tests are conducted at a substation facility incorporating hollow porcelain bushing (Note 14.1) of 11.5 m in height.

14.1.1 Background of Development

In Japan, following the adoption of insulators by the communication and electric light fields, Koriyama Silk Yarn Spinning and Hiroshima Hydroelectric started the 11 kV overhead power line business in 1899 (32nd year of the Meiji era). Insulators made in the U.S. were used to support the power lines. Since 1907, a series of power lines of 55, 66 and 77 kV were constructed and the era of intermediate power lines started. The insulators were imported from the U.S. and Germany at the beginning, but high-voltage pin-type insulators made in Japan began to be used soon after that. Responding to the increase in voltage of transmitted power, suspension insulators that could be connected for use became popular. Suspension insulators are mainly used in transmission lines, while support insulators and hollow porcelain bushings are mainly used in substations. The materials, shapes and manufacturing methods of insulators were improved in response to needs. High quality insulators have contributed to a stable power supply, which is a necessity for society and economic activities.

14.1.2 Characteristics

14.1.2.1 Insulators for Transmission Lines

Suspension insulators for transmission lines are used by connecting the needed number of insulators, depending on the voltage of transmitted power and environmental conditions (Note 14.2). Suspension insulators are designed so that a compressive stress is applied on the porcelain area where a tensile load is applied to the cap and the pin. The structure of a suspension insulator is shown in Fig. 14.2. The bottom surface of the suspension insulator is corrugated (watershed fins) or wavy so as to ensure sufficient insulation along the porcelain surface between the cap and the pin. In order to maintain electrical properties in a harsh environment, deeply corrugated insulators are used in areas where the surface contamination of insulators is estimated to be severe. A suspension insulator applicable to a tensile load of 820 kN has been

*The number in parenthesis indicates the year that the product was first commercialized.

Note 14.1 Insulators are electrical isolators and hollow isolators with open ends are called porcelain bushings. The structures, materials, shapes and sizes vary depending on the application.

Note 14.2 Electrical insulation properties of outdoor insulators is lowered by condensed moisture containing electrolytes on their surface. This is called the "contamination of insulators". Insulators are contaminated by various substances, including sea salt particles which are carried by strong wind from the sea.

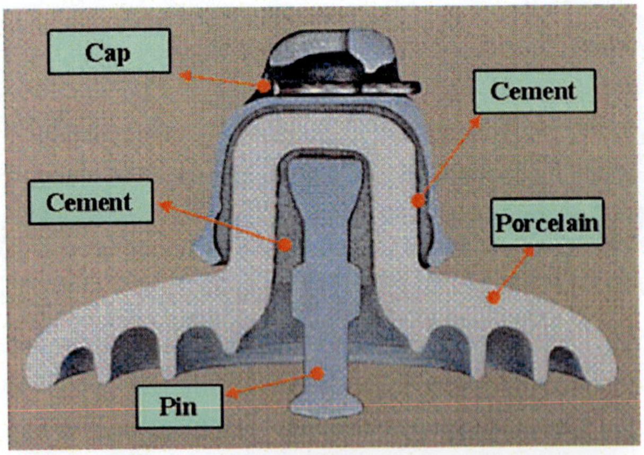

Fig. 14.2 Structure of a suspension insulator. High voltage stability and mechanical strength are required for suspension insulators. Therefore, each component is designed in consideration to meet these requirements. Standard type suspension insulators have a tensile strength of 165 kN and a dielectric breakdown strength of 140 kV or higher. This indicates the electric field under which the internal porcelain part can maintain its electrical isolation

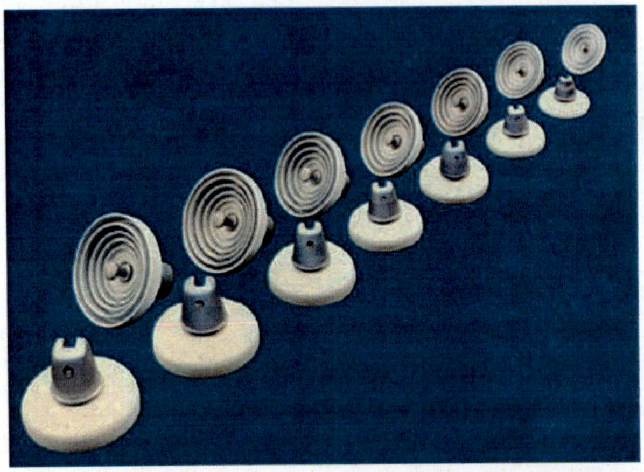

Fig. 14.3 Typical suspension insulators. Suspension insulators featuring mechanical strength of 165, 210, 330, 420, 530, 680, 820 kN, from *right* to *left* (254, 280, 320, 340, 380, 440, 460 mm in diameter). Insulator supporting loads of up to 530 kN have already been in put into practical use

developed, as indicated in Fig. 14.3. In actual transmission lines, suspension insulators support loads of up to 530 kN.

Fig. 14.4 1,000 kV class bushing. A porcelain bushing filled with gas, 11.5 m in height. It is currently being tested in an experimental study on a 1,000 kV power transmission system in Japan

14.1.2.2 Insulators for Substations

At substations, hollow porcelain bushings and bar-shaped support insulators are used. Porcelain bushings are used to house substation equipment and connect components as well as to connect to high voltage terminals on oil tanks and gas tanks. Bar-shaped support insulators are used to support metallic parts at the contact points of bus bars and disconnecting switches. Based on the developments of porcelain materials and advanced design technologies, compact porcelain bushings and insulators with high earthquake resistance, have been realized. Porcelain bushings for substation equipment that is used up to the 1,000 kV have an integral design. Figure 14.4 shows a porcelain bushing (Note 14.3) used with 1,000 kV class equipment. The porcelain bushing is 11.5 m in height and is the largest porcelain product in the world.

14.1.2.3 Manufacturing Method

Insulators generally consist of the following three components:

- Porcelain (major component that provides electrical insulating performance and mechanical strength)
- Metallic part (connects insulator or porcelain bushings to structure)
- Cement (bonds porcelain to the metallic part)

The porcelain is generally made from china stone, feldspar, clay and alumina. The raw materials are ground, blended and dehydrated to reduce the amount of water. This is then kneaded in vacuum and formed into the shapes of insulators. Suspension insulators for transmission lines are made by the pull-down method, in which the raw material is placed into a plaster mold, which forms the top surface of the insulator, and the material is pressed by a metal mold, which forms the pleated bottom surface. Insulators and porcelain bushings for substations are formed by cutting bar-shaped or cylinder-shaped materials. After forming, they go through a drying process and then are sintered at about 1,300°C to create a porcelain part. The porcelain part is bonded with a metallic part by using high-quality cement and completed. Strict inspections and quality control are performed at each process so as to produce insulators and porcelain bushings that feature high electrical insulation properties and mechanical strength. Porcelain for insulators is generally called alumina-containing porcelain and contains double digit percentages of quartz, mullite and corundum crystal. The major characteristics of the material are shown in Table 14.1.

Table 14.1 Characteristics of porcelain (for power transmission) used for insulators

Item	Value
Specific gravity	2.5–2.6
Bending strength	120–150 MPa (unglazed) 160–190 MPa (glazed)
Dielectric breakdown strength	38–46 kV/mm

Test method: IEC 60672-2 (1997)

14.1.3 Future Prospects

In Japan, a 1,000 kV transmission line was constructed in the area covered by Tokyo Electric Power Co., Inc. and is currently being operated at 500 kV. An experimental study on the substation equipment is currently in preparation for the voltage increase in the future. In China and India, where the economic growth is remarkable, construction projects utilizing ultra-high voltage power transmission systems (AC 1,000 kV class, DC ±800 kV class) are in progress. The possibility of concentrating technologies from various countries to create an international standard in this field has been studied. Research for the practical use of ultra-high power transmission has advanced in Japan and Japanese technologies are expected to contribute to the realization of global standards. Japanese technologies in insulators and porcelain bushings will certainly play important roles in the realization of projects in these countries.

Note 14.3 Used to connect high voltage terminals on substation equipment, insulators such as oil and insulating paper and gas fill the central area, which also consists of the conductor that transmits electricity. This is in addition to the outer area, which has a porcelain bushing.

Varistor for Power System (1977) 14.2

In modern society, a stable supply of electric power is essential and a reliable power distribution is required. Arresters play an important role by protecting power distribution equipment such as transformers and switchboxs from transient abnormal voltage (surge) caused by lightning in the power transmission system. ZnO varistors for power systems (Note 14.4) (Fig. 14.5) are functional ceramics in which zinc oxide is the major component and which shows nonlinear resistance characteristics (Note 14.5). They are applied to arrestors of various voltage classes. The resistance value of ZnO varistors drops remarkably in response to voltage increase. Therefore, they conduct minimal electric currents under normal operating voltage, but conduct large currents when the resistance value drops significantly due to high voltage caused by surges. ZnO varistors have proven long-term reliability and are not degraded by continuous voltage application. Furthermore, they are capable of absorbing high-energy surges (energy tolerance) without failure. For these reasons, ZnO varistors are widely used.

14.2.1 Background of Development

Compact ZnO varistors for low voltage systems were developed around 1970 with additives such as bismuth oxide (Bi_2O_3) and antimony oxide (Sb_2O_3). Large ZnO varistors for power systems were developed later and they began to be used in arrestors for power systems in the 1970s. Arrestors were previously made by assembling a silicon carbide (SiC) element or a conventional nonlinear resistor, and a series gap. However, following the development of ZnO varistors which exhibit superior nonlinear resistance characteristics, gapless arrestors were realized, contributing to the miniaturization of arrestors and the enhancement in their protection performance.

Figure 14.6 shows functions of arrestors at substations as an example. The arrestors protect the equipment by absorbing lightning based voltage surges within the power systems. Varistors shown in Fig. 14.5 are laminated in multiple layers, the number of which depends on the voltage class. They are incorporated in arresters.

14.2.2 Characteristics

14.2.2.1 Characteristics of Varistors

ZnO varistors are characterized by unique nonlinear resistance characteristics, which are attributable to the potential barriers (Note 14.6) formed on the grain boundary of ZnO crystals in the sintered body. The following three properties are required of varistors for power systems:

1. Current–voltage nonlinear resistance characteristics
Nonlinear current–voltage characteristics where the resistance value changes substantially responding to minor changes in voltage that permits the flow of very low leakage currents under normal operating voltage and the flow of large currents, so as to absorb surge energy, responding to high voltage.

2. Energy tolerance
There is an energy tolerance so as to absorb excess surge energy without electrical or mechanical break down when subjected to excess surge energy that is generated within the power transmission system.

3. Service life properties

Note 14.4 The word varistor is a combination of variable resistor, a resistor with variable resistance.

Note 14.5 The characteristics where the resistance value is high at low voltages and drops quickly when the voltage exceeds a specified value.

Note 14.6 A kind of wall, generated by discontinuous crystals or a constituent bias. It blocks flow of electrons.

ZnO varistors have a service life of more than several tens of years and does not deteriorate (increase in leakage current) even if the operating voltage of the system is applied continuously.

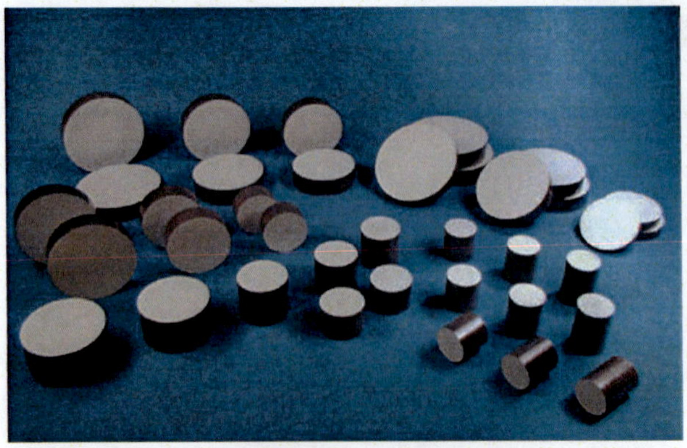

Fig. 14.5 ZnO varistors for power system. Disk-shaped nonlinear resistance ceramics, Φ 30–100 mm, are applied with an insulating coating on the circumference and have electrodes formed on both faces. They are used in arrestors of various voltage classes

14.2.2.2 Manufacturing Method

The manufacturing method of ZnO varistors is shown in Fig. 14.7. ZnO varistors are manufactured by powder processing, thermal treatment and then part fabrication. In the powder treatment process, additive components such as bismuth oxide and antimony oxide as well as binders are added to the major component, ZnO. They are mixed homogeneously, granulated and press formed into the intended disk shapes. The green bodies are densified through the binder removal process, in which the binder component is removed at a temperature of 400–500°C, and by the sintering process, which is performed at a temperature of 1,000–1,200°C. The exterior surface of the sintered body is applied with an electrical insulating coating and aluminum electrodes are formed on both faces by the thermal spray process before completion of the ZnO varistor device.

ZnO varistors have grain boundary layers with a different chemical composition based upon the additives that has a higher resistivity than does the polycrystalline ZnO grains (see electron microscopic pictures in Fig. 14.8). This is the main microstructural feature that allows for non-linear

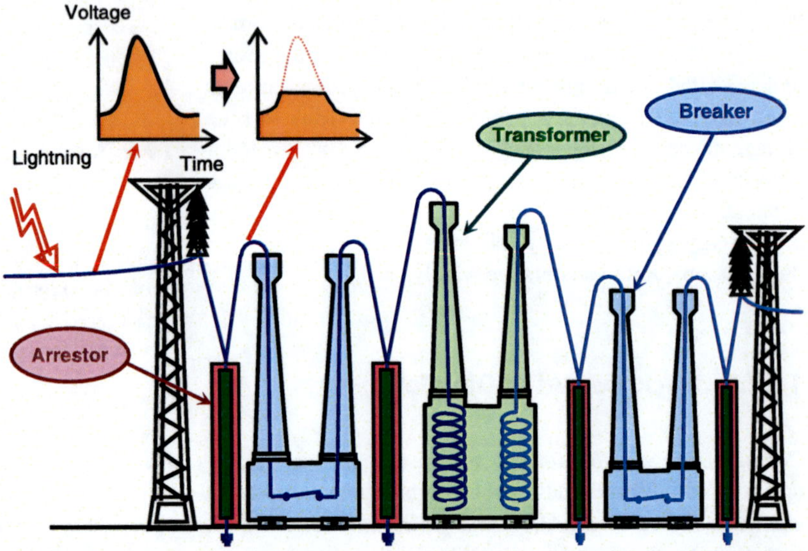

Fig. 14.6 Configuration of substations and functions of arrestors. Arrestors absorb abnormal voltage (surge) generated in the power system by lightning, to protect equipment such as transformers and breakers

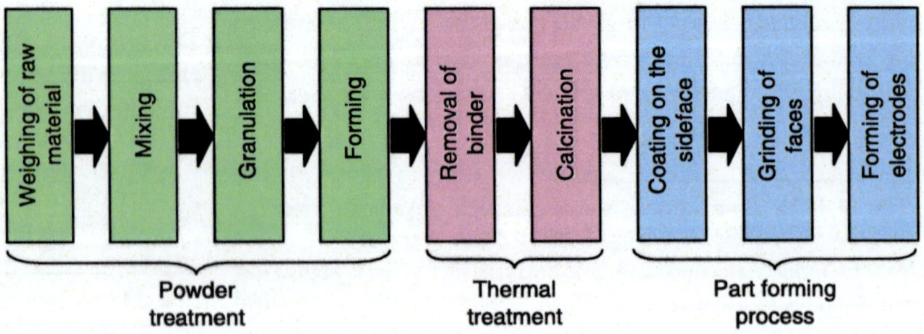

Fig. 14.7 Manufacturing process of ZnO varistors. ZnO varistors are manufactured by three different processes: powder treatment, thermal treatment and part formation

resistance characteristics. For this reason, the diameters of ZnO polycrystalline grains need to be controlled strictly by adjusting the amount of additives and sintering temperature. This ensures accurate control of their operating voltage (voltage at 1 mA current).

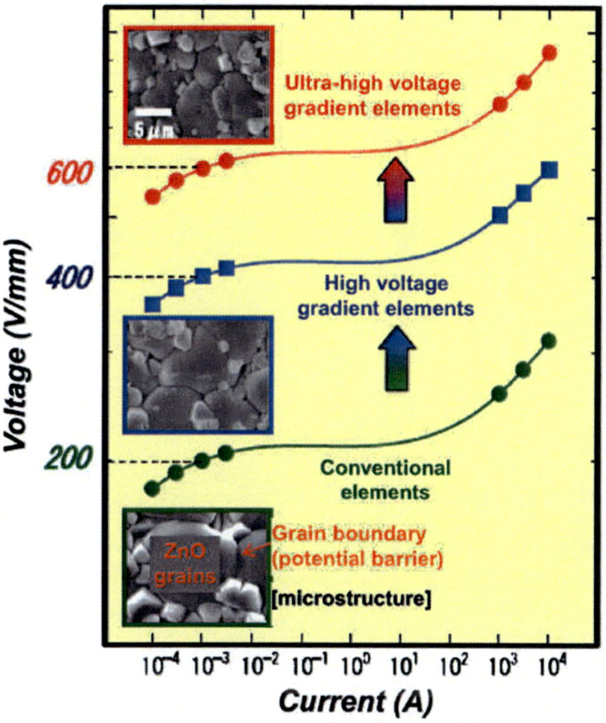

Fig. 14.8 Enhancement of voltage gradient in varistors and their microstructures. ZnO varistors with a high voltage tolerance enables the downsizing of arresters, have been developed by reducing the diameters of the polycrystalline grains

14.2.3 Future Prospects

High-voltage type ZnO varistors have been developed in recent years as shown in Fig. 14.8. High-voltage type means higher varistor operating voltage, which makes it possible to reduce the number of varistor elements laminated in an arrester and directly leads to the size reduction of the varistors.

Since the early stages of development of ZnO varistors for power systems, conventional devices with an operating voltage of 200 V/mm class were used. In the late 1990s, high-voltage type elements of the 400 V/mm class were commercialized, and ultra-high-voltage type elements of the 600 V/mm class began to be applied recently.

Figure 14.8 shows pictures of microstructures of varistor elements. The potential barriers between ZnO grains with nonlinear resistance characteristics are increased in the current path by reducing diameters of ZnO crystal grains through the adjustment of additive components, thus enhancing the operating voltage and realizing varistor devices at a high voltage.

The operating voltage of the ultra-high-voltage type devices is three times that of conventional elements. Therefore, the number of varistors laminated in arresters can be reduced to 1/3. This realizes a substantial size reduction in the arresters.

However, when the voltage on a ZnO varistor increases, the volume of the element that actually responds to the surge decreases, requiring enhanced surge energy tolerance. Figure 14.9 shows the results of a temperature distribution in varistors during absorption of surge energy conducted using an infrared camera. These observations were performed as part of efforts to improve energy tolerance. If heat distribution during the absorption of surge energy by the varistors is strongly heterogeneous, thermal stress is generated due to concentration of current in high-temperature portions and

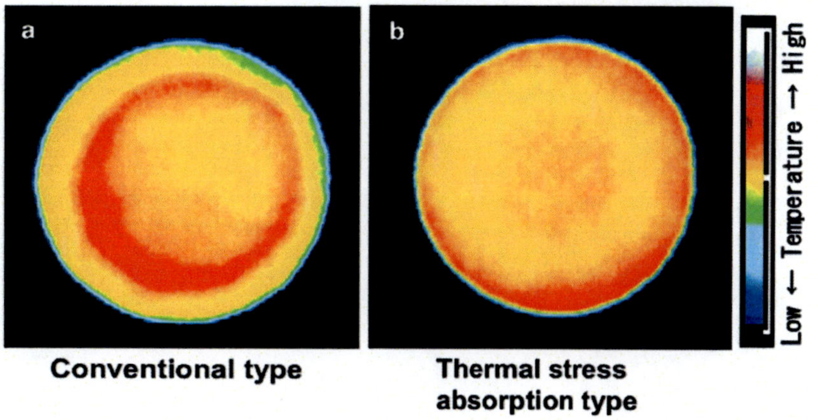

Fig. 14.9 Temperature distribution in ZnO varistor. Heat generated during energy absorption is homogenized and thermal stress is reduced to improve the energy tolerance of ZnO varistors: (**a**) conventional type (**b**) thermal stress absorption type

temperature difference between devices. This results in a drop in the devices' energy tolerance. Heat distribution is homogenized by improving thermal treatment conditions, during the varistor manufacturing process in order to enhance the energy tolerance.

Voltage and energy tolerance of ZnO varistors for power systems will continue to be enhanced, contributing to the further downsizing of arresters and substation equipment as well as to the stable supply of electricity.

Uranium Dioxide Fuel (1957) 14.3

The nuclear reactor core, which is the center of reactors used in nuclear power generation, consists of multiple fuel assemblies. Uranium dioxide fuel in the form of uranium dioxide (UO_2) pellets is inserted into a fuel rod, which is approximately 1 cm in diameter and 4 m in length and comprises part of the fuel assembly. The uranium dioxide fuel is used under severe conditions (high temperature, for long periods of time) and as such, the shape and quality of the fuel changes during use. Even under such severe conditions, fission products are required to remain within the UO_2 pellets so as to prevent release of radioactive isotopes into the external environment. Therefore, the uranium dioxide fuel is processed and inserted into a fuel rod in adherence to strict quality control standards in order to attain high density, precise dimensions, and strict moisture composition. A single UO_2 pellet has the capacity to generate electricity equivalent to the amount consumed by a single household for a period of 6–8.5 months.

14.3.1 Background of Development

In nuclear power generation, power is generated by utilizing energy that is generated by the triggering of atomic fission (Note 14.7) from fissile isotopes such as ^{235}U of uranium (U) inside the reactor. Uranium dioxide (UO_2) fuel is inserted into a fuel rod consisting of a mixture of uranium isotopes in the pellets, which comprises part of the fuel assembly used as the fuel of the reactor. Uranium dioxide fuel is being used by 55 reactors in Japan (as of the end of December 2005) and 434 reactors around the world (as of the end of December 2004). Reactors are classified into three types based on the type of coolant; the light-water reactor, the heavy-water reactor and the gas reactor. The light-water reactors [BWR (Note 14.8) and PWR (Note 14.9)] are in operation in Japan. Concepts on the BWR fuel assembly and the PWR fuel assembly as well as the pellet filling layouts are shown in Fig. 14.10.

14.3.2 Characteristics

14.3.2.1 Characteristics of Uranium Dioxide Ceramics

Uranium dioxide is a face-centered (fluorite type) cubic crystal structure with a lattice constant of 0.547 nm. The melting point is approximately 2,800°C and it is free from phase transformation until the temperature reaches the melting point. The theoretical density of a single crystal is 10.96 g/cm³, but the density of the uranium dioxide fuel, a polycrystalline sintered bodies with a grain diameter of 5–10 μm, is typically 95–97% of the theoretical density.

14.3.2.2 Manufacturing Method

Uranium dioxide powder is compressed by a pressing machine to create a cylinder-shape green pellet. The green pellet is sintered in a reducing atmosphere at 1,700–1,800°C in order to create a sintered pellet. The manufacturing process is shown in Fig. 14.11.

Note 14.7 The phenomenon where ^{235}U, ^{239}Pu or ^{241}Pu in the fuel absorbs neutrons and splits into two fission fragments, releasing two to three neutrons and energy.

Note 14.8 BWR is an abbreviation of boiling water reactor. The steam generated by boiling the coolant in a reactor is supplied directly to the turbine to generate power.

Note 14.9 PWR is an abbreviation of pressurized water reactor. Atmospheric pressure of approximately 160 atm. is applied to the primary cooling system to create pressurized water, which is supplied to a steam generator to turn the water in the secondary cooling system into steam, which is then supplied to the turbine for power generation.

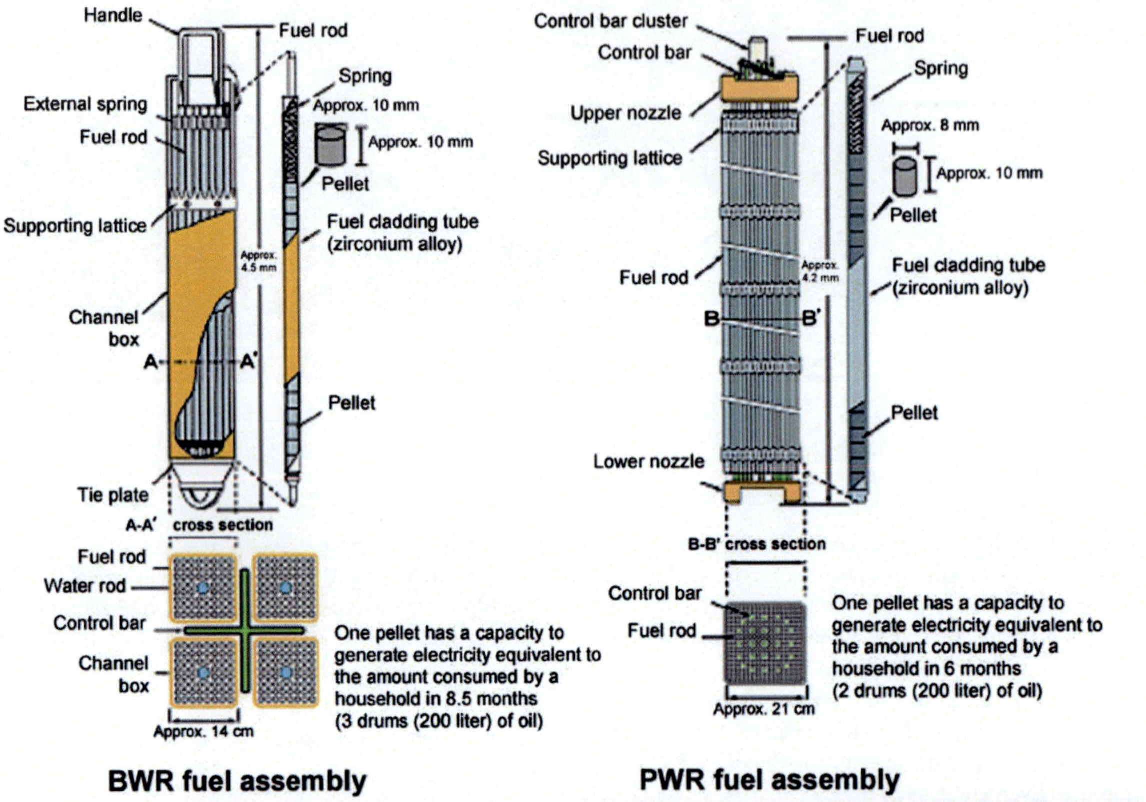

Fig. 14.10 BWR fuel assembly and PWR fuel assembly. The BWR fuel assembly contains 60–74 fuel rods, while the PWR fuel assembly contains 179–264 fuel rods. The uranium dioxide fuel is inserted into the fuel rod, which is approximately 1 cm in diameter and approximately 4 m in length, consisting of uranium dioxide (UO_2) pellets (*sources*: Agency for Natural Resources and Energy "Atomic Energy 2005," "Atomic Energy/Energy" drawings 2007 published by the Japan Atomic Energy Relations Organization)

Fig. 14.11 Manufacturing process of uranium dioxide (UO_2) pellets. Uranium dioxide pellets go through a series of processing steps such as powder mixing, where uranium dioxide powder is mixed with binders and formed using a rotary press, sintering takes place in a reducing atmosphere, grinding using a centerless grinder, and then visual inspection

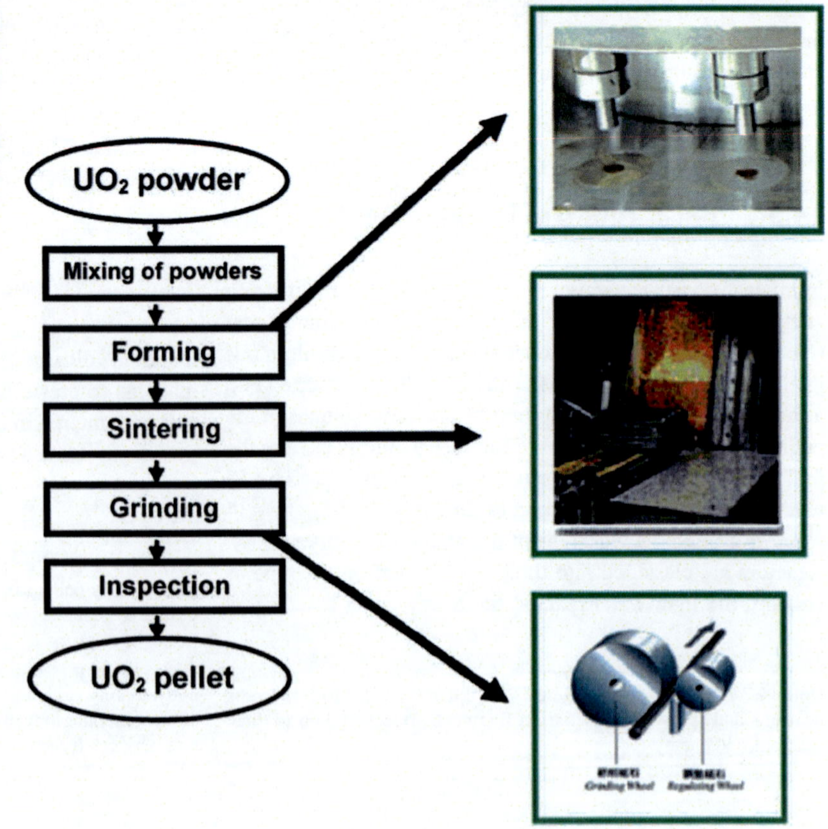

In the forming process a rotary press is used. It is similar to the tableting machine used to manufacture medication. The uranium dioxide powder is required to be of high fluidity in order to improve density, to reduce deformation during sintering and to enhance productivity. In order to improve fluidity of the powder, during the granulation process binders are added to enhance mixing. The green pellet is put in a container made of molybdenum alloy and is placed in a sintering furnace, which is filled with a reducing atmosphere of mixed gas containing hydrogen and nitrogen and set at 1,700–1,800°C. The sintering process takes from 4 to 8 h. A sintered pellet with a theoretical density of 95–97% with a O/U ratio of 2.00 is created by this sintering process. The outer circumference of the sintered pellet is ground by a centerless grinder in order to attain a precise outer diameter. All UO_2 pellets go through a visual inspection and are inserted into the fuel rod in adherence with strict quality control standards that includes specifications on moisture.

14.3.2.3 Product Performance and Specifications

The appearance of UO_2 pellets used in BWR and PWR fuel assemblies are shown in Fig. 14.12. They are cylindrical and both the height and the diameter are approximately 10 mm (PWR pellets are slightly smaller). BWR pellets are chamfered, while PWR pellets have a dish (indentation) on the end and are chamfered. The outer circumference of UO_2 pellets is ground as specified in units of micrometers. This is because the gap needs to be accurate in order to enhance thermal conductivity between the UO_2 pellets and the internal surface of the fuel rod cladding tube and in order to reduce the contact force between the UO_2 pellets and the cladding tube.

The uranium dioxide fuel undergoes atomic fission in the reactor and the generated heat is transmitted to the coolant via the fuel rod cladding tube. The amount of heat generated [linear fuel output (Note 14.10)] reaches a maximum of 440 W/cm. Therefore, the center temperature of the uranium dioxide fuel reaches a temperature as high as 1,800°C. Due to the temperature rise and output variation during use, crystal growth, crack and the rearrangement of pellet fragments are observed in the UO_2 pellets, as shown in Fig. 14.13.

Fuel assemblies are used for 4–7 years in a reactor. The energy generated from the uranium dioxide fuel during the period is expressed as burnup (Note 14.11). Electric power

Fig. 14.12 Uranium dioxide (UO_2) pellet. Uranium oxide powder is formed, sintered and ground to create cylindrical pellets. BWR pellets are chamfered, while PWR pellets have a dish (dent) on the face and are chamfered

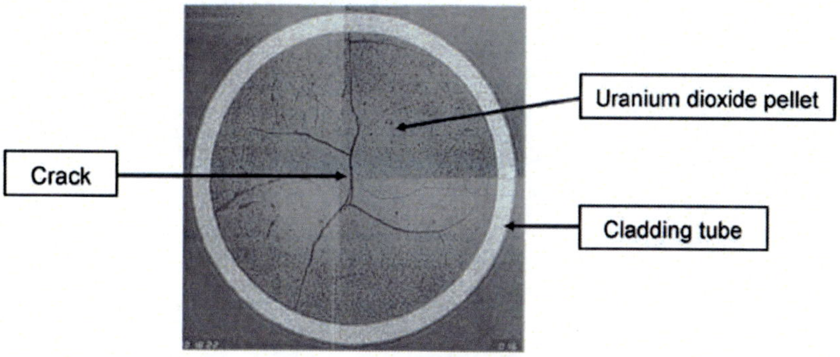

Fig. 14.13 Cross-sectional metallographic picture of a spent fuel rod. The picture was taken by cutting a fuel rod after use in a reactor and grinding the cut end. The uranium dioxide pellet is cracked and is in contact with the inner surface of the cladding tube

Note 14.10 Amount of heat generated per unit length is transmitted to the coolant via the fuel rod cladding tube. The unit is W/cm or kW/m.

Note 14.11 The energy generated from 1 ton of uranium contained in fuel. The unit is MWd/t or GWd/t.

companies try to enhance the burnup to improve energy efficiency. However, the burnup is determined by national regulations. As for BWR and PWR fuel assemblies used in Japan, the uranium dioxide fuel is allowed to be used up to 55,000 MWd/t.

The uranium dioxide fuel crystal structure is damaged by atomic fission and affected by radiation damage caused by fission fragments. The majority of fission fragments remain in the UO_2 material. The gas component (fission gas: Xe, Kr) in fission fragments remain inside the UO_2 pellet when the temperature and burnup are low. When the temperature and burnup are high, however, the gas diffuses within the pellet and is segregates at the grain boundaries. This forms cracks along the grain boundaries and the fragments of the UO_2 pellets fill the voids in the fuel rod, as shown in Fig. 14.14. The increase in the internal pressure of the fuel rod caused by the build up of fission gas products may result in damage to the cladding tube. Therefore, fission gas products must be kept inside the UO_2 pellets.

14.3.3 Future Prospects

The fuel for light-water reactors often consists of UO_2–Gd_2O_3, a mixture of UO_2 and gadolinium oxide (Gd_2O_3) powder. Gadolinium oxide, has a high neutron-absorbing capacity, is mixed to a maximum concentration of 10% in order to control reactivity variation during fission. The manufacturing method and characteristics are nearly the same as for those of uranium dioxide fuel.

In France, Germany and Switzerland, MOX (Mixed OXide) fuel, made by mixing PuO_2 with UO_2, is also used in power generating reactors. The fuel containing plutonium (Pu) at a maximum concentration of 10% is produced by reprocessing spent fuel and is used in light-water reactors. In Japan, electric power companies have been working on the plutonium-thermal project.

The fast breeder reactor (FBR), which is being developed as the next-stage reactor after the light-water reactor, uses the MOX fuel that contains 20–30% Pu. It is a high-performance fuel and the burnup is targeted at 100,000 MWd/t.

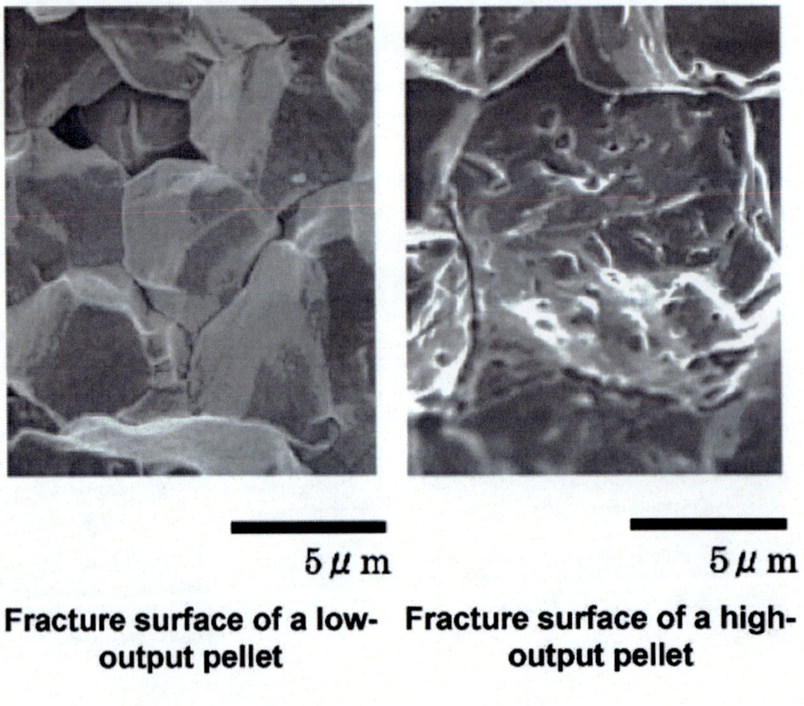

Fig. 14.14 Detailed observation of grains on the fracture surface of used uranium dioxide fuel pellets. As for pellets used under low power output, the surface of crystal grains remains the same as that after production. As for the pellet used under high power output, pores are seen on the surface of crystal grains

B_4C Control Material for Atomic Energy (1970) 14.4

B_4C (born carbide) is a comparatively light (2.5 g/cm^3) and an extremely hard (H_V=4.1×10^4 MPa) ceramic. B_4C also has a high neutron-absorbing capacity because it contains a large amount of boron. ^{10}B (approximately 20% content in natural boron), an isotope of boron, absorbs neutrons via the (n, α) reaction. The ^{10}B content in boron can be enhanced to approximately 90% by enrichment, and therefore, B_4C is used as the control material for nuclear fission reactions as well as the neutron shielding material in various reactors. Natural B_4C powder (^{10}B: approximately 20%) is used as the control material in boiling water reactors (Note 14.12) (BWR), which account for approximately 60% of commercial electrical power generating reactors. A B_4C sintered body, in which the ^{10}B content is adjusted, is used as the control material for fast breeder reactors (Note 14.13) (FBR) that have been developed to establish technology for a nuclear fuel cycle. B_4C has also been used in experimental reactors as the control material and shielding material (including composites).

14.4.1 Background of Development

Atomic energy, characterized by a stable electrical power output, high economic potential and low environmental loads, is an essential energy source when considering the reduction of global warming gases. In 2006, 55 commercial nuclear power stations were in operation in Japan and more than 30% of total energy production was generated by atomic energy. In Japan, which is poor in natural resources, atomic energy measures have been promoted in order to ensure a stable electrical power supply and the effective use of resources with the focus being on the nuclear fuel cycle and safety. With a view to further promote development and utilization of atomic energy, consumption of plutonium generated as a by-product in commercial light-water reactors (the plutonium-thermal project) and the development of fast breeder reactors has been promoted.

14.4.2 Characteristics

14.4.2.1 Characteristics of B_4C

The B_4C crystal is rhombohederal and regular icosahedral boron is formed on the apex of each rhombohederal unit cell that has three carbon atoms on a diagonal line, as shown in Fig. 14.15. It is known that a portion the central carbon chain on the diagonal line tends to be replaced by boron (composition changed to $B_{13}C_2$), while the icosahedral boron (black circle in particular) can be partially replaced by carbon. As explained above, in boron–carbon compounds and solid solutions, the crystal framework remains the same and the elements on lattice points only are interchanged. Responding to the B–C substitution, the distance between faces changes slightly, but the number of peaks and the intensity ratio remain nearly the same according to the X-ray diffraction pattern of the powder. This makes research on the phase diagram of the B–C system challenging, resulting in differences in phase diagrams made by researchers.

B_4C is comparatively light (density=2.5 g/cm^3) and extremely hard (Vickers hardness=4.1×10^4 MPa). It is a super hard substance similar to diamond and c-BN.

B_4C also has a high neutron-absorbing capacity. This characteristic is attributable to the high boron content of 78.3%

Note 14.12 A type of light-water reactor, which releases the heat generated by atomic fusion by using H_2O (light-water that slows down neutrons). In the BWR reactor, the steam is generated by directly boiling the coolant is supplied to the turbine for power generation.

Note 14.13 Uranium and plutonium, which are obtained by reprocessing spent nuclear fuel, are used in this reactor. The plutonium is bred using high-speed (high-energy) neutrons. Uranium is effectively utilized in this reactor.

Fig. 14.15 Structure of B_4C crystal. The B_4C crystal is rhombohederal and regular icosahedral boron is formed on the apex of each rhombohederal unit cell that has three carbon atoms on a *diagonal line*. The central portion of the carbon chain on the *diagonal line* tends to be displaced by boron (composition changed to $B_{13}C_2$)

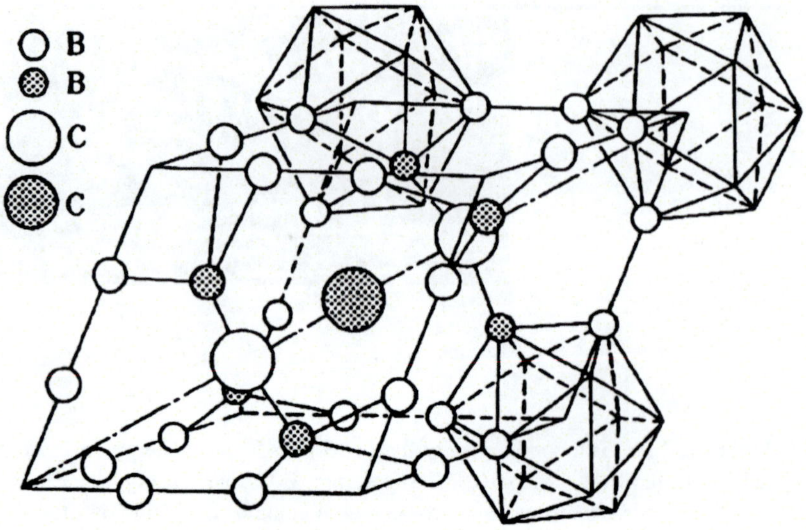

Table 14.2 General B_4C manufacturing methods

Reaction process	Remarks
1. Direct synthesis method $4B+C\rightarrow B_4C$	• Experimental manufacturing method
2. Reduction method of boron oxide (a) Carbon reduction method of boron oxide $2B_2O_3+7\,C\rightarrow B_4C+6CO$ (b) Mg reduction method of boron oxide in the presence of carbon $2B_2O_3+6\,Mg+C\rightarrow B_4C+6MgO$	• Industrial manufacturing method. The B/C ratio tends to deviate from 4.0 • Suitable for production of B_4C powder with concentrated ^{10}B
3. Reaction of boron nitride with carbon $4BN+C\rightarrow B_4C+2\,N_2$	
4. Solid phase carbonization of boron chloride $4BCl_3+6\,H_2+C\rightarrow B_4C+12HCl$	
5. Gas-phase carbonization of boron chloride (a) $4BCl_3+CH_4+H_2\rightarrow B_4C+12HCl$ (b) $4BCl_3+CO+7\,H_2\rightarrow B_4C+12HCl+H_2O$	• Manufacturing method of boron carbide film or high-purity fine powder

(natural composition). ^{10}B (content of approximately 20% in natural boron), an isotope of boron, absorbs neutrons via (n, α) reaction.

$$^{10}_{5}B + ^{1}_{0}n \rightarrow ^{4}_{2}He + ^{7}_{3}Li$$

Many of the neutron absorbing materials absorb neutrons and release harmful γ-rays, while ^{10}B does not release γ-rays while absorbing neutrons. The ^{10}B content in boron (approximately 20% in natural boron) can be enhanced to approximately 90% by means of enrichment. The concentration can be adjusted depending on the application.

B_4C, like other carbides and borides, has low sinterability. The hot-press method is generally employed to manufacture high-purity formed bodies.

14.4.2.2 Manufacturing Method

Table 14.2 shows the general manufacturing method of B_4C. The "carbon reduction method of boron oxide" is most frequently employed for industrial use. The ingot is created by applying heat to the raw material in an electric furnace after it is ground, refined and granulated to produce B_4C powder. A large amount of CO gas is generated in the process of creating the ingot, while B_2O_3 has a low melting boiling point. Therefore, volatilization loss during the heating process is substantial, causing deviation of the B/C ratio from 4.0. Stainless steel pipes filled with B_4C powder consisting of natural boron, which is produced with this method, are used as control rods in boiling water reactors (BWR) (refer to Fig. 14.16). In the Mg reduction method for boron

Fig. 14.16 Control rod for boiling water reactor (BWR). The neutron absorption rods, stainless steel pipes filled with B_4C (boron carbide) powder (grain), are housed within the blades and move up and down in the spaces between fuel assemblies (illustrations from Hitachi, Ltd. Website/"control rod structure")

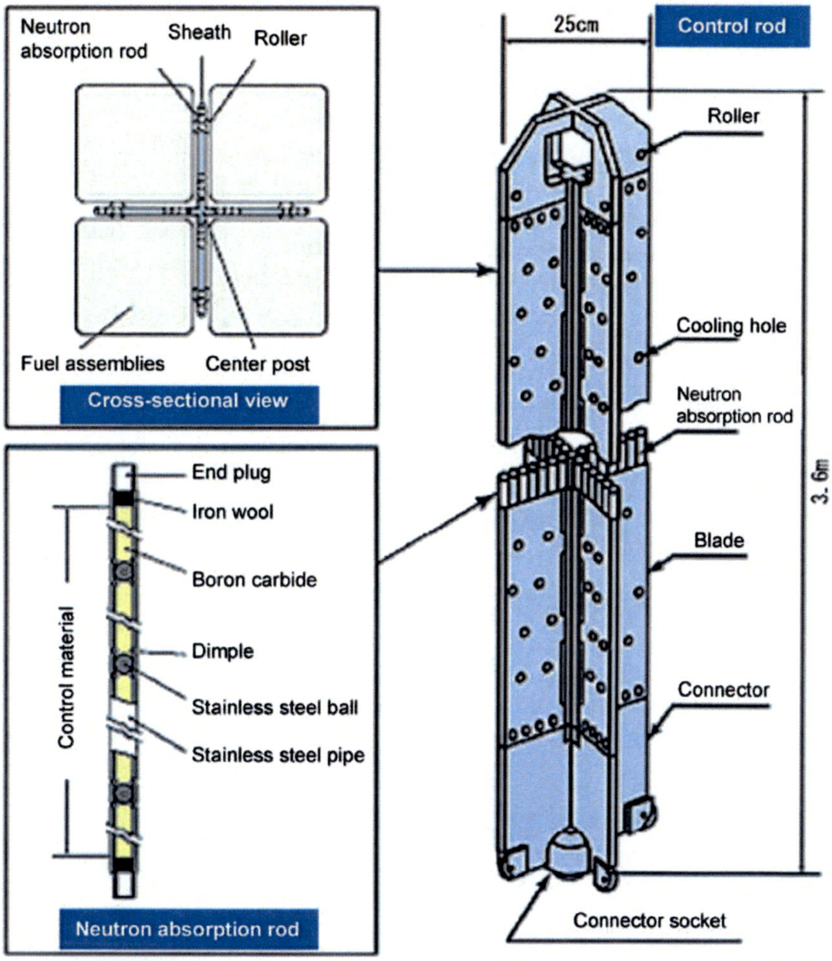

Fig. 14.17 Control rods for fast breeder reactors (FBR). An exhibition model showing the cross-sectional area of the reactor core of the experimental fast breeder reactor called "Joyo." The reactor core is composed of hexagonal fuel assemblies and control rod assemblies. The control rod assembly (*central area*), a bundle of cladding tubes filled with cylindrical B_4C sintered bodies, is surrounded by the fuel assemblies

oxide, B_4C is created at a comparatively low temperature. The Mg reduction of B_2O_3 involves the so-called "thermite" reaction, and the sharp reaction needs to be controlled for the stable production of B_4C. The cost is high with this method. However, if the reaction is properly controlled, the volatilization loss during the reaction can be reduced and high-purity fine powder of B_4C can be produced using comparatively small-scale equipment. Control materials used in fast breeder reactors (FBR) are currently under development in Japan. They are the cylindrical pellets produced by sintering, using the hot-press method and the ^{10}B content-adjusted B_4C fine powder, which is produced by this manufacturing method. The pellets are inserted into stainless steel cladding tubes, which are bundled together to create the control rod assembly (refer to Fig. 14.17).

14.4.3 Future Prospects

The neutron absorption cross-sectional area of B_4C is extremely large, while neutron absorption involves the (n, α) reaction, which causes problems such as swelling attributable to the retention of generated He in B_4C crystals. With respect to BWR control rods, adoption of hafnium, which allows for a longer-period of usage in reactors when compared with B_4C, has been studied. However, B_4C is more stable than other boron compounds at high temperatures and there is a rich amount of irradiation data on B_4C. For these reasons, B_4C, designed with the consideration for the adjustment of B/C composition ratios and grain diameters, is used in the aforementioned reactors. In addition to it, B_4C has been commercialized as a control and shielding material while the commercialization has been studied in experimental reactors (including composites containing graphite, metals, resins, etc.).

14.5 Sodium–Sulfur Battery (2002)

A sodium–sulfur battery incorporates a β alumina tube, which allows for Na ion diffusion, and has high-energy density and high efficiency. The full-fledged development began in 1980 in Japan, and the product was commercialized through developments of material technologies related to the high-performance β alumina tube. Commercialization of the product was realized by a number of technologies, including technologies relating to carbon cathode materials and corrosion resistance metallic materials, metal/ceramic joining technology, technologies relating to safety structure electric cell materials and production technology as well as modularization technology and the battery and system control technologies. Sodium–sulfur batteries, capable of storing a large amount of electricity equaling several tens of MWh, are currently used by many households. Application has expanded for the effective use of renewable energy.

14.5.1 Background of Development

The basic principle of the sodium–sulfur battery originated in a patent applied for in 1967 by Ford Motor Company of the U.S. In Japan, development of a new battery for storage of electricity was started in 1980 as the Moonlight Project of NEDO, following the establishment of the new battery development project for electric vehicles in 1969 by the Agency of Industrial Science and Technology at the Ministry of International Trade and Industry. In the 1980s, when development was started, the maximum demand for electricity in the daytime during the mid summer continued to grow at a rate of 5% per year. And storage of large amounts of electricity in suburban areas was considered to be effective in avoiding the construction of new power plants and power distribution facilities. It was estimated that the system for the storage of electricity needed to be 75% or higher in charge–discharge efficiency, 15 years or more in durability and lower than the construction cost of a pumped storage hydroelectric power plant.

The sodium–sulfur battery, which companies and researchers continued to develop for several decades, was commercialized following successful developments of the material and mass production technologies of the solid electrolyte (β alumina tube), the heart of the battery. Large, high-efficiency batteries were realized by means of reducing the resistance of the β alumina tube, while the cost reduction of the industrial product was realized by means of the development of battery manufacturing technology. Figure 14.18 shows the sodium–sulfur battery system (Note 14.14) and the module battery (Note 14.15), the basic unit.

14.5.2 Characteristics

14.5.2.1 Characteristics of the Sodium–Sulfur Battery

The sodium–sulfur battery is a high-temperature-operated secondary battery made by using Na and sulfur as active materials for the anode and cathode with beta-alumina solid electrolyte ceramics as the separator between the active materials. The operating temperature is set at around 300°C, with consideration for the ion conductivity of beta-alumina and the melting point of active materials. The durability is 15 years and the charge–discharge efficiency of the entire system including the power conditioning system (PCS) is 75% or

Note 14.14 An electricity storage system composed of a modular battery, a power conditioning system and a transformer. It converts the AC power in the system to DC power for storage in a battery. A function to protect electrical power users against an instantaneous voltage drop in the electrical distribution system can be added by means of combing it with a high-speed switch.

Note 14.15 A type of sodium–sulfur battery. A number of electric cells (electromotive force: approximately 2.1 V) are series-parallel arranged in an insulated container. The series-parallel configuration is changed due to the voltage and current specifications. The battery is operated at high-temperature. A heater is used to increase the temperature.

Fig. 14.18 The 50 kW modular battery and the 500 kW system. The rated output of the modular battery is AC 50 kW (360 kWh). This can be optionally set at a higher output for a short-period of time. Multiple electric cells are connected in series/parallel and stored in an insulated container to maintain the operating temperature. How the batteries are configured depends on the system output and electric energy, and the power conditioning system (PCS)

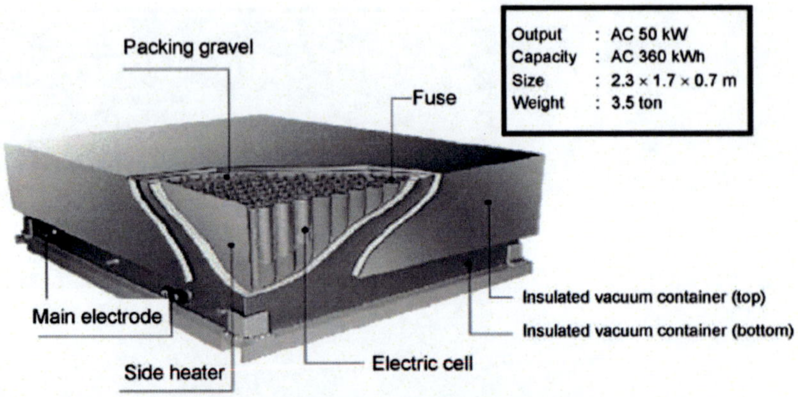

50 kWh modular battery

500 kWh sodium-sulfur battery system

higher. The energy density of the modular battery is 100 kWh/m^3 or higher. The battery is characterized by high-energy density and high-efficiency. Figure 14.19 shows the internal structure of the electric cell and its elements. The unit cell is composed of the anode consisting of Na and storage parts, the cathode consisting of sulfur and carbon felt that functions as the structural material of the electrode, solid electrolyte (beta-alumina tube) that separates the anode from the cathode, and the cathode container that also functions as the collecting electrode for the cathode. The electric cell is fully airtight and a bottomed aluminum tube called the "safety pipe" is put inside the beta-alumina tube. The safety pipe controls the gap from the beta-alumina tube so as to ensure safety by minimizing direct reaction between sodium and sulfur, with consideration for the possibility of damage to the solid electrolyte.

The high-efficiency, high-energy density electric cell was realized not only by the development of the β-alumina tube but also by breakthroughs in various technologies including the technology relating to the carbon felt electrode structural material, metal/ceramic air-tight joining technology and thin film forming technology as well as the development of active materials with high-temperature corrosion resistance.

Beta alumina is classified into the β type ($Na_2O{\cdot}11Al_2O_3$) and the β'' type($Na_2O{\cdot}5$–$7Al_2O_3$) depending on the crystal structure. A difference in Na concentration in the non-stoichiometric compound changes the amounts of Mg and Li added to stabilize the crystals. The β'' type contains more Na and its resistance for ionic conductivity is higher in comparison with the β type. Historically, development began with the β type, which was doped with Li, and the composition was optimized based on the β'' type doped with Mg, which is easier to handle during production. This was followed by the reduction of resistance and enhancement in strength.

The beta alumina is made of hexagonal plate crystal and has anisotropy in Na ion conduction. The material technology for controlling the shapes of crystals and homogenization of crystals was developed so as to reduce the resistance of β alumina tubes, polycrystalline sintered bodies, as a structure.

Fig. 14.19 Internal structure of an electric cell. Composed of a solid electrolyte tube (β-alumina tube), the anode containing sodium and parts and the cathode containing sulfur and carbon felt. It is completely sealed to ensure air tightness

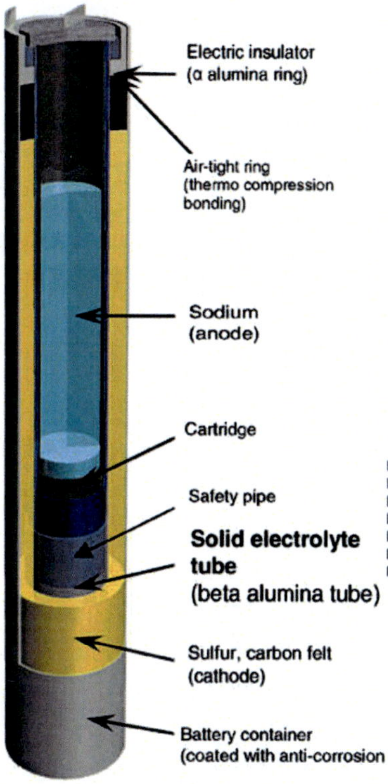

Table 14.3 Performance of the β alumina tube

Size [mm]	Φ59×470
Resistance for Na ion conductivity [Ω cm at 623 K]	2.2
Strength of material (radial crushing strength) [MPa]	>300
Bulk density [g/cm³]	3.24

In order to develop beta alumina tubes for industrial use, the following technological breakthroughs were made:

1. Good balance between high Na ion conductivity and high strength (crystalline orientation control, homogenization of crystals)
2. Materials suitable for mass production and industrial use
3. Development of a mass production process for large and thin β alumina tubes

Table 14.3 shows the performance of the beta alumina tube and Fig. 14.20 shows the changes in the reduction of beta alumina resistance.

14.5.2.2 Manufacturing Method

The well-known zeta process is an organic solvent-based process requiring repeated crystal synthesis and grinding and was not suitable for the manufacturing of industrial products. Therefore, the direct synthesis process was developed. This is a process in which a formed body is created by using homogeneously mixed granular materials through the water solution process and in which crystal synthesis and densification are completed by means of a single sintering process. During this process, alumina, Mg and Na are homogeneously mixed and crushed. Na source dissolves in water solution and the Na moves with water during the drying operation of the granulation process, depositing again on the surface of granules. This results in the reduction of strength caused by the abnormal growth of crystals due to segregation of the Na component during the sintering process. This problem was solved using a binder that can be used in high-alkali water solutions and by adopting pH control which utilizes the buffering action of Na in order to homogeneously disperse the Na in the granules.

Following the development of the dry forming technology such as the drybag method and application of the sintering conditions that can be controlled in large sintering furnaces, economic mass production of high-performance β alumina tubes was realized for the first time in the world. Figure 14.21 shows the zeta process and the direct synthesis process.

14.5.3 Future Prospects

The sodium–sulfur battery is used for electricity storage that can stably store a large amount of electricity equaling several tens of MWh. The surrounding environment has changed since the 1980s when full-fledged development started.

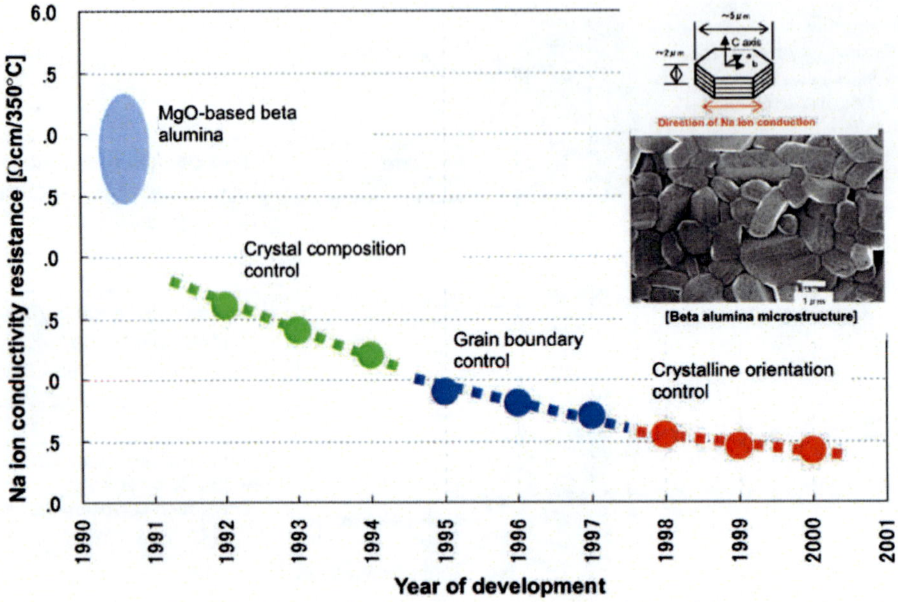

Fig. 14.20 Change in Na ion conductivity resistance. The beta alumina features anisotropy crystal structure and ion conductivity in the direction vertical to the *c*-axis. The resistance of beta-alumina tubes was reduced by adjusting the crystal composition (chemical composition), reducing the resistance in crystal grain boundary and controlling the crystalline orientation

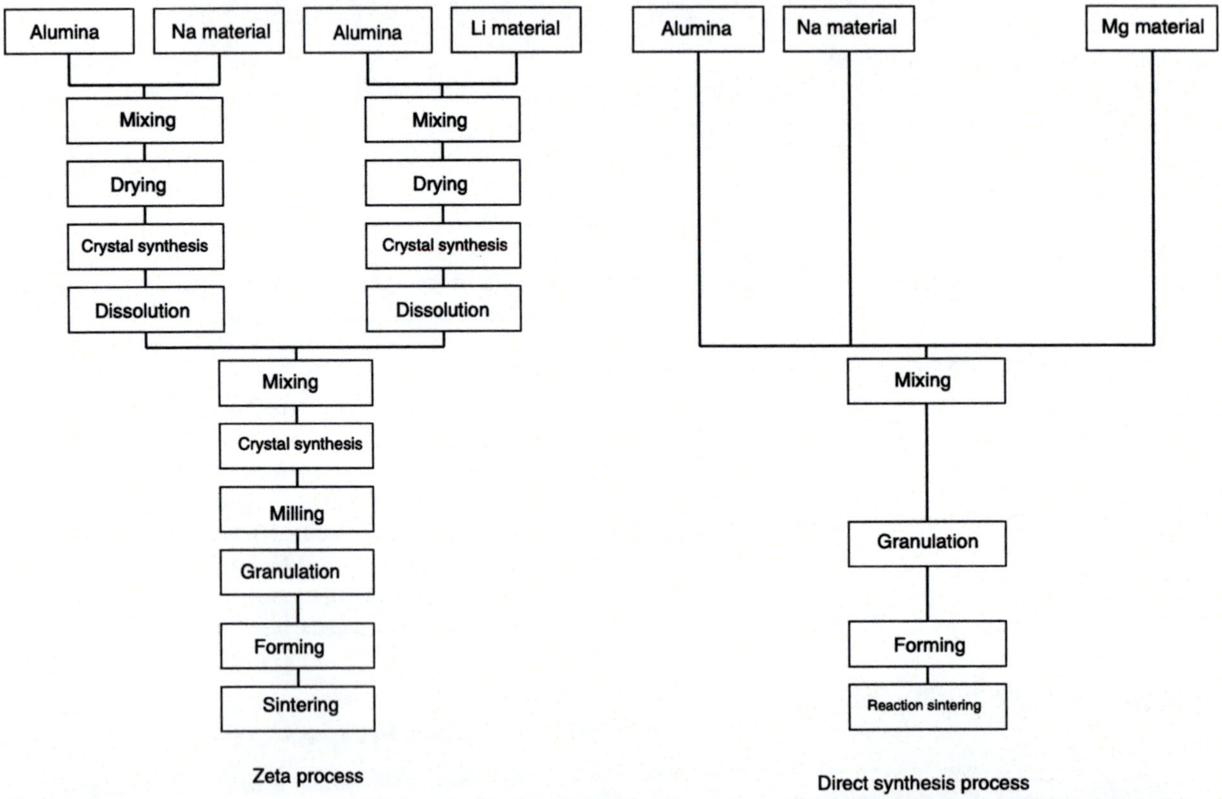

Fig. 14.21 Comparison of beta alumina manufacturing method. The conventional process (zeta process), shown on the left side, is a complicated process that requires repeated crystal synthesis and powder milling. The newly developed direct synthesis process is shown on the right side. The material is homogeneously mixed, and beta alumina crystal synthesis and dense sintering are completed with a single reaction sintering process

Recently, it has been used in enhancing the quality of electrical power through the prevention of instantaneous voltage drops, in addition to initial purposes that include load leveling and offsetting reductions in the electrical supply.

This application is expected to expand to various fields and the battery will be used, for example, for the stabilization of system power responding to the increase in wind power generation. The battery, incorporating functional ceramics and having high-efficiency and high-energy density, will contribute to the effective utilization of limited resources and renewable energy.

14.6 High-Temperature Superconducting Cable (2000)

The high-temperature superconducting (Note 14.16) power cable utilizes a high-temperature superconducting wire as the conductor, whose resistivity decreases dramatically to a superconducting state at liquid nitrogen temperatures. The cable is capable of transmitting higher currents when compared with existing cables. It has low loss and is expected to contribute to CO_2 reduction. Two types of high-temperature superconducting wires have been commercialized for use in high-temperature superconducting cables. One is the Bi-based superconducting wire, which is made by mounting a filament, an oxide consisting of Bi, Sr, Ca, Cu and O in a silver alloy tape. The other is the Y-based superconducting wire, which is made by forming a thin film containing Y, Ba, Cu and O on a metallic substrate tape. Verification tests of high-temperature superconducting cables have been conducted in various countries including Japan. The technological feasibility has nearly been established for commercialization of the products.

14.6.1 Background of Development

Responding to global warming, Japan is shifting from large-scale energy consumption to more efficient use of energy. Meanwhile, home appliances are increasing in size, air conditioning equipment is spreading and computers and communication devices are increasing within our presently advanced information based society. Under such circumstances, the percentage of electricity of total energy consumption has been increasing, and it is believed that the reduction of loss in power generation and power transmission is essential for CO_2 reduction. The high-temperature superconductor, discovered in 1989, changes to a superconducting state (zero resistance) at the liquid nitrogen temperature (77 K), while the conventional metallic superconductor changes to a superconducting state at the liquid helium temperature (4.2 K). Therefore, the high-temperature superconductor is effective in the reduction of cooling costs. It is believed that economical power systems will be realized by using the high-temperature superconductor in power generation and transmission devices.

The high-temperature superconducting cable, which is one of the superconducting power devices, is used in a conductor where the current flows through a high-temperature superconducting wire and is capable of having a high current flow via a small cross-sectional area with low loss. The power transmission loss is reduced by half when compared with conventional power transmission cables that incorporate copper as the conductor. Furthermore, light and compact large-capacity transmission lines are now possible. Conventional copper power cables in urban areas pass through a cable tunnel with a diameter of approximately 2 m. The high-temperature superconducting cables can run through a cable conduit with a diameter of 150 mm. This is installed underground and substantially reduces the cable construction cost.

14.6.2 Characteristics

14.6.2.1 Characteristics of High-Temperature Superconducting Cables

The structure of a high-temperature superconducting cable incorporating high-temperature superconducting wire is shown in Fig. 14.22. The superconducting cable consists of cable cores and the insulation tube that encloses the cable cores. The cable core is made by wrapping multiple high-temperature superconducting tapes around a twisted cylindrical copper wire called a former, which is enclosed by an electrical insulation layer, a superconducting shield layer and a protective overcoat layer. Liquid nitrogen flows through the

Note 14.16 The superconducting state is maintained at the liquid helium temperature (boiling point: 4.2 K) in metallic superconductors, while the superconducting state is maintained at the liquid nitrogen temperature (boiling point: 77.3 K) in the oxide superconductor.

insulation tube and cools down the cable core to the temperature of liquid nitrogen. High thermal insulation properties are required of the insulation tube so as to allow the flow of liquid nitrogen at approximately −200°C, without evaporation. It is made of a waved stainless steel double tube and an insulation material is filled between the tube walls, and the space between the walls is evacuated.

Furukawa Electric Co., Ltd. installed a 500 m Bi-based superconducting cable at the Central Research Institute of Electric Power Industry and conducted a field test (Fig. 14.23). Sumitomo Electric Industries, Ltd. also installed a cable at a substation in Albany in the U.S., and has been conducting power transmission tests. Field tests are also being performed in the U.S., Korea and China.

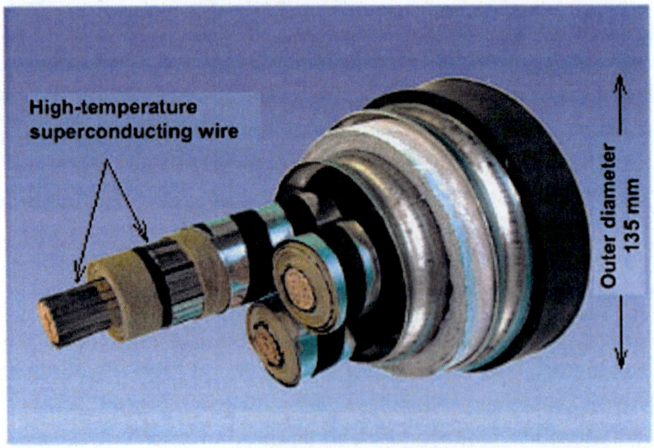

Fig. 14.22 High-temperature superconducting cable. By using high-temperature superconducting wires in conductors that transmit current, high currents can be flown with low loss. In order to maintain the superconducting state, the cable is enclosed in an insulated flexible pipe in which liquid nitrogen flows

14.6.2.2 Manufacturing Method

Two types of high-temperature superconducting wires have been developed for use in high-temperature superconducting cables (Fig. 14.24). One is the Bi-based superconducting wire, which is made by inserting oxide powder consisting of Bi, Sr, Ca, Cu and O into silver alloy pipes, which are drawn to reduce cross-sectional areas and then rolled to the shape of tape. The tape is heat-treated at a temperature higher than 800°C for an extended period so as to sinter the oxide material and form multiple superconducting filaments inside the silver alloy base metal in the longitudinal direction to create the products.

The other is the Y-based superconducting wire, which is currently being developed for commercialization. It is made by forming an intermediate layer, a superconducting layer and a protective overcoat layer on a metallic substrate such as a hastelloy substrate, a nickel-tungsten substrate. The intermediate layer is the base on which a superconducting layer is epitaxially grown (Note 14.17), and an oxide film with controlled crystalline orientation is formed on the intermediate layer. Oxide crystals consisting of Y, Ba, Cu and O (YBCO) are in-plane aligned on the intermediate layer and form the superconducting layer. The PLD (pulse laser deposition) method, the CVD (chemical vapor deposition) method and the MOD (metal organic deposition) method are employed to form the superconducting layer. The protective

Fig. 14.23 Verification test of a 500 m high-temperature superconducting cable. With the aim of validating technological feasibility of the high-temperature superconducting cable, a field test using the world's longest Bi-based superconducting cable was performed in 2005 in a NEDO project

Note 14.17 A method in which a crystalline film having a preferential orientation is grown on the substrate crystals by means of deposition.

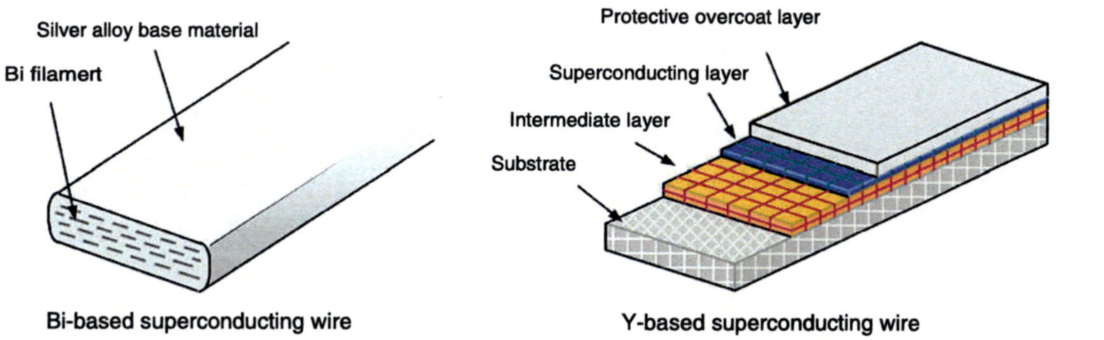

Fig. 14.24 High-temperature superconducting wire. Two types of superconducting wires, the Bi-based superconducting wire and the Y-based superconducting wire, are currently commercialized as high-temperature superconducting wires that become superconducting when cooled to liquid nitrogen temperatures

overcoat layer, which prevents intrusion of water and release of oxygen out of the oxide layer, is made of silver by sputtering or vacuum vapor deposition.

14.6.3 Future Prospects

The Bi-based superconducting cables are used in actual power systems for power transmission tests. In order to commercialize the cables, however, the AC loss (Note 14.18) that occurs when an alternating electrical current is flown through the cable needs to be reduced. In the Y-based cable, which is currently being developed, the AC loss can be reduced to 1/6 of the loss in the Bi-based superconducting cable. The low AC loss is attributable to the thin superconducting layer (YBCO) of the Y-based superconducting wire, which is only 1 μm thick, and also due to the width which was cut from 10 to 2 mm using lasers. Commercialization of the Y-based superconducting cable is expected to reduce the loss in transmission lines, which is essential for further reductions in CO_2.

These superconducting cables will be introduced in full scale beginning around 2015, when currently used power cables reach the end of their service life.

Note 14.18 When fluctuating magnetic fields or currents are applied to a superconductor that changes to a zero-resistance superconducting state, under a certain temperature, energy is consumed due to the internal flux change. This is a phenomenon called "AC loss."

Literature

1. Supervised by Ryo Kawasaki, Written and edited by Yoshiyasu Ito (2005) Personal dispersion energy system. Yokendo, Tokyo (in Japanese) **(14)**
2. Shibagaki K, Ito M, Hasegawa S, Tachibana T (1995) 75 year's history of NGK Insulators, Ltd., Nagoya (in Japanese) **(14.1)**
3. Institute of Electrical Engineers of Japan (1983) Gaishi (in Japanese) **(14.1)**
4. Ando H et al (2006) Toshiba Rev 61(10):66–69 (in Japanese) **(14.2)**
5. Ando H et al (1999) Ceramics 34(3):156 (in Japanese) **(14.2)**
6. Tanabe T et al (2004) J Atom Energ Soc Jpn 46:566–572 (in Japanese) **(14.3)**
7. Une K et al (2004) J Atom Energ Soc Jpn 46:728–735 (in Japanese) **(14.3)**
8. Mori K (2004) J Atom Energ Soc Jpn 46:838–844 (in Japanese) **(14.3)**
9. Nuclear Power Engineering Corporation (2002) 2001 Report on Safety Tests of Fuels including High Burnup Fuels (Comprehensive evaluation of BWR high burnup fuel), March 2002 (in Japanese) **(14.3)**
10. Samsonov GV et al (1956) Fiz Met Metallovad Acad Nauk 3:309 **(14.4)**
11. Elliott RP (1961) Armor Research Report. USA **(14.4)**
12. Tamaki S (1989) Ceramics 24:533–537 (in Japanese) **(14.4)**
13. New Energy and Industrial Technology Development Organization (2005) 2004 Report on Results of Commissioned Operation "Research and development of basic technologies for AC superconducting power devices" (in Japanese) **(14.6)**
14. Mukoyama S (2007) Yttrium-based high-temperature superconducting power cable. Monthly OHM, February 2007 issue (vol. 94–2) (in Japanese) **(14.6)**
15. Weber CS et al (2005) Overview of the underground 34.5 kV HTS Power Cable Program in Albany, NY. IEEE Trans Appl Supercond 15(2):1793–1797 **(14.6)**

Part V

Ceramic Products for Manufacturing Industries

Production of Raw Materials 15

The processing of materials including crushing as well as thermal processing such as melting, annealing and sintering are indispensable in the production of metallic and ceramic materials. The performance of milling devices and thermal treatment furnaces are continuously enhanced for improved control of these processes. Ceramic products have been contributing largely to the improvement of industrial devices for material production, taking advantage of thermal resistance and abrasion resistance. Ceramic products that play important roles in supporting the material manufacturing industry are introduced in this section.

Crucibles made of mullite-based ceramic have been used for laboratory analysis for many years. The material varies depending on the application, and alumina-based ceramics are used for high temperature processing. Crucibles made of high purity alumina, high purity magnesia or high purity zirconia, featuring high corrosion resistance and thermal resistance, have been used in recent years as thermal reaction containers for various advanced materials (high-performance glass, positive terminal materials of lithium-ion batteries, phosphors, etc.) (Sect. 15.1). Setters are used to support parts during thermal treatment at high temperatures where refractories have been used conventionally. The material used in the setters for electronic components are selected with considerations for their mechanical properties, corrosion resistance, reaction performance, thermal shock resistance, etc. In the case of calcination of lead-based piezoelectric materials, setters made of magnesia or zirconia are used.

Crushing media (Sect. 15.2) are used to crush and disperse ceramic powder in ball mills. Conventionally, alumina media are used mainly for ceramics. In recent years, high-speed mills have been used to process electronic ceramic powders intended for multilayer ceramic condensers, because precise composition control, fine grain diameters and narrow particle size distribution are required for electronic based ceramic powders. Y-TZP crushing media featuring high abrasion resistance have been introduced in response to the growing needs in this area.

In resistance heating elements (Sect. 15.3), Joule heat generated from electricity passing through the element is utilized for heating. They are used in industrial furnaces for the thermal treatment of metals and production of electronic components. The maximum service temperature of resistance heating elements varies depending on the used material [graphite (C), molybdenum disilicide ($MoSi_2$), silicon carbide (SiC), Ni–Cr]. Heating elements are selected with considerations for their service life, temperature, ambient temperature, and volatiles from processed substance, price, etc.

Ceramic products used in firing furnaces (Sect. 15.4) include shelves for supporting products and containers for holding products during thermal treatment in the calcination furnace. They are used for processing of ceramic products such as tableware, tiles and sanitary goods as well as electronic components such as ferrite and ceramic condensers, and also in the thermal treatment of semiconductors. Taking advantage of high strength and high thermal resistance of silicon carbide-based refractories, the thickness of furnace shelves has been reduced and the material has been used as beams, contributing both to energy saving and the reduction of calcination times. Since silicon carbide gradually changes to silicon dioxide due to oxidization during use in an oxidizing ambient at high temperature, various countermeasures have been taken.

Y. Imanaka et al. (eds.), *The Ceramic Society of Japan, Advanced Ceramic Technologies & Products*,
DOI 10.1007/978-4-431-54108-0_15, © Springer Japan 2012

Crucibles and Setters (1913)* 15.1

Crucibles have been used for many years as laboratory porcelains (Note 15.1) in which specimens for chemical analysis are heat treated and are normally made of a mullite-based material (Fig. 15.1). In recent years, crucibles have been used as containers for the thermal treatment of advanced materials, including high-performance glasses, cathode materials of lithium-ion batteries, and phosphors (Fig. 15.2). Such applications are increasing. The materials of the containers intended for thermal treatment are required to be high in corrosion resistance and thermal resistance and at the same time to be of high purity. Alumina- , magnesia- and zirconia-based materials are selected depending on the application.

Setters, used for sintering of electronic ceramics and thermal treatment of metallic parts, are required to be high in corrosion resistance and mechanical properties (Fig. 15.3).

15.1.1 Background of Development

Crucibles for laboratory use have been manufactured in Japan since 1913 and are used for the treatment of specimens for chemical analysis, etc. Mullite-based materials are generally used to fabricate the crucibles, but alumina-based materials are employed for the crucibles used at high temperatures. In recent years, an increasing number of crucibles are used for the thermal treatment of advanced materials featuring high-performance (Note 15.2), in addition to conventional applications such as the pretreatment of specimens for chemical analysis and thermal treatment. The key points in the thermal treatment of advanced high-performance materials are (1) accurate compositional control, (2) homogenization of microstructure and (3) uniformity of material properties. In order to satisfy these requirements, crucibles and setters need to have the following characteristics: (1) no discoloration or mixing with thermally-treated objects or sintered objects, (2) high air-tightness and creep resistance and (3) high thermal uniformity. Because of these reasons, crucibles and setters are now required to feature high thermal resistance, high corrosion resistance and high purity. Mullite-based materials cannot fulfill these requirements, and therefore alumina- , magnesia- and zirconia-based materials are currently used, depending on the application.

15.1.2 Characteristics and Manufacturing Method

15.1.2.1 Characteristics

Mullite-based materials are used to manufacture crucibles that are used for the treatment of specimens for general chemical analysis and thermal treatment at relatively low temperatures. Meanwhile, alumina- , magnesia- and zirconia-based materials are used to manufacture crucibles that are used for the high-temperature treatment of high-performance glasses and metals as well as for the thermal treatment of high-purity inorganic compounds, because mullite-based materials do not provide sufficient thermal resistance or corrosion resistance. The major characteristics of these materials are indicated in Table 15.1. For thermal treatment in a variety of atmospheres, an alumina-based material, featuring high air-tightness and chemical stability, is usually used. The magnesia-based material is used for melting active metals and the thermal treatment of alkali oxide and alkaline-earth oxide. Recently, the majority of the materials are used to manufacture large-size thermal treatment containers that are employed to produce advanced materials including cathode materials of lithium-ion batteries and phosphor materials, rather than to manufacture crucibles.

*The number in parenthesis indicates the year that the product was first commercialized.

Note 15.1 Evaporating dishes, crucibles, mortars, pestles, boats, Tammann tubes, combustion tubes, which are used for chemical analysis, tests and measurements, are called laboratory ceramics.

Note 15.2 Electronic devices such as piezoelectric components, cathode materials for lithium-ion batteries, phosphors, and biomaterials.

Fig. 15.1 Crucibles: laboratory ceramics (mullite-based). They have been manufactured in Japan since 1913 and are mainly used for chemical analysis

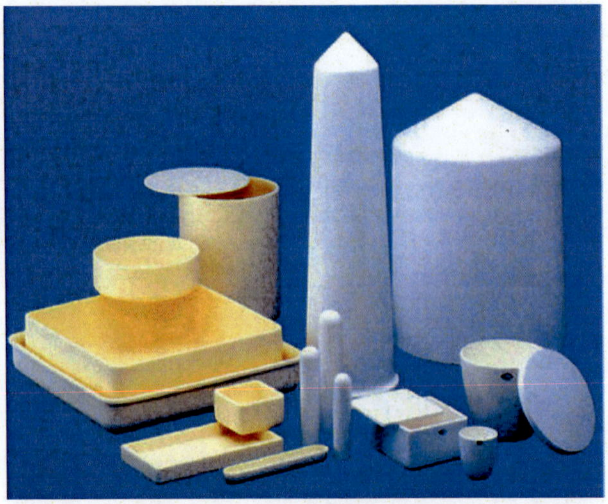

Fig. 15.2 Containers for thermal treatment (alumina-based). They are being used for thermal treatment of industrial materials, and various shapes and sizes are required depending on the application

Fig. 15.3 Setters (alumina-based, zirconia-based). They are free from deterioration and deformation caused by reaction with calcined objects, allowing repeated use under severe conditions. Depending on the sintered object, the surface is ground and the flatness is improved

Refractories were mainly used to manufacture setters, but these setters lack corrosion resistance which makes them unusable in various applications, following the improvement in performance of thermally-treated materials. Therefore, setters made of dense materials are currently used for the sintering of electronic ceramics. Setters are generally required to have high mechanical properties and corrosion resistance. Materials for setters are selected with considerations for reactions between the thermal-treated material and the setter material as well as the stability of mechanical

Table 15.1 Characteristics of alumina-, magnesia- and zirconia-based materials (representing values)

Materials	Alumina (Al_2O_3)	Magnesia (MgO)	Zirconia ($ZrO_2+Y_2O_3$)
Chemical composition (mass %)	99.5	99.6	99.8
Bulk density (g/cm^3)	3.9	3.4	5.6
Bending strength (MPa)	250	240	210
Coefficient of thermal expansion ($\times10^{-6}$/K)	8.1	13.5	10.2
Thermal conductivity (W/(mK))	36	42	3

properties such as thermal shock resistance under various service conditions. The alumina-based setters have broad utility and are suitable for a variety of materials, but magnesia- and zirconia-based setters are used for sintering of piezoelectric materials.

Examples of applications of crucibles and setters made of each material are listed in Table 15.2.

15.1.2.2 Manufacturing Method

The outline of the manufacturing process is shown in Fig. 15.4. Raw materials are blended to create a specified composition, which is wet-crushed to a certain particle size and used for formation. As for conventional mullite-based products, clay of kaolin is used as the raw material. However, raw material powders having high-purity and sinterability are used to manufacture high-performance alumina-based

Table 15.2 Applications of each materials and types of crucible

Crucible applications/types	Mullite-based	Alumina-based	Magnesia-based	Zirconia-based
For chemical analysis and thermal analysis	○	○		
For glass dissolution		○		
Thermal treatment of materials for electronic parts such as piezoelectric bodies and dielectric bodies		○	◎	◎
Thermal treatment of materials for electrodes of lithium-ion batteries		◎		
Crucible for single-crystal growth		○		
Thermal treatment of phosphor materials		◎		
Dissolution of basic slags and metals			◎	
Thermal treatment of ceramic materials		○		○
Thermal treatment of metallic materials		○	○	○
Thermal treatment of NaS battery materials			◎	

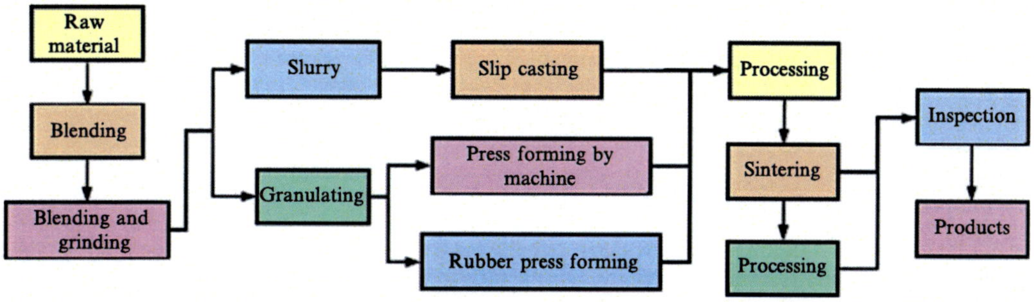

Fig. 15.4 Manufacturing process of crucibles and setters. Raw materials are blended to create a specified composition, wet-milled and granulated, and then are slip casted and press formed. The formed bodies are sintered and additional processing is performed, if necessary, depending on the application

crucibles and setters, so as to achieve high corrosion resistance and mechanical properties. The raw materials are mixed, crushed and dispersed in water to create the slurry of a specified particle size which is used for slip casting. Meanwhile, the slurry is mixed with a binder and is spray-dried to create dry granular powder that is used for press forming. Crucibles are usually made by slip casting, but rubber press forming is sometimes employed in response to the diversification of shapes. Setters are press-formed by using dies. The products are completed by sintering the formed bodies and grinding them depending on the application. Careful attention is paid to the flatness and surface roughness of setters as deformation and sticking due to reactions needs to be avoided.

15.1.3 Future Prospects

Performance of advanced materials including electronic materials will continue to be improved and diversified, requiring accurate control of material compositions. In order to minimize variations in compositions caused by thermal treatment, high corrosion resistance as well as mechanical properties including high thermal shock resistance is required of crucibles and setters. In response to the expected improvements in the performance of advanced materials, materials for crucibles and setters will be switched from refractories to dense materials, while crucibles and setters made of new materials that are applicable for advanced materials, will be developed.

15.2 Grinding Media

Grinding or milling media are used to grind and disperse ceramic powders typically using alumina or zirconia media depending upon the material to be ground. Maintaining chemical purity with a fine, sharp particle size distribution for the milled powder is needed in order to manufacture e.g. advanced materials represented by multilayer ceramic condensers. Media stirrer mills (Note 15.3) featuring high grinding/dispersion efficiency are used to meet these requirements. The mills are used to stir the media at high speed and are required to possess good mechanical and abrasion-resistance properties. The Y-TZP (yttria-stabilized zirconia) media fully satisfy the requirements and are used to treat raw material powders when producing multilayer ceramic condensers.

15.2.1 Background of Development

Traditionally, raw materials of ceramics are milled using ball mills containing grinding media. Large-sized alumina media are mainly used as the grinding media in the ball mills, regardless of dry milling or wet milling. Various types of materials in recent years have been ground and media is chosen depending on the material to be ground and its application.

Meanwhile, advanced high-performance materials represented by multilayer ceramic condensers are being miniaturized and improved at a rapid pace. Raw material powders used to manufacture them are required to be of high purity and have both fineness and a sharp particle size distribution. Conventional ball mills are incapable of processing raw material powders featuring these characteristics, and media stirrer mills featuring high grinding/dispersion efficiency began to be used. Mills of this type are capable of stirring small-size media at a high speed, realizing high milling/dispersion efficiency, but the media and the internal surface of the mills are subjected to high loads. If the shock resistance of the media is low, they will be cracked or chipped during the crushing/dispersion process, resulting in trouble. If the abrasion resistance of the media is low, powders will be generated due to abrasion from the media, resulting in the mixing of undesirable impurities into the processed powder. Due to these reasons, media featuring high shock resistance and high abrasion resistance need to be used to process the powder in manufacturing advanced high-performance materials.

Note 15.3 The mill is filled with milling media such as balls, pebbles and beads is stirred by using a stirring bar or a rotary disk to give motion to the media for grinding and dispersion.

15.2.2 Characteristics

15.2.2.1 Characteristics of products

Media made of various materials are used, but the particle sizes of milled powders and abrasion properties of media vary largely depending on the material of the media when milling is performed under the same conditions. Figure 15.5 shows the specific surface area of the powder and abrasion rate of the media, which was obtained by ball-milling $BaTiO_3$ powder using various kinds of media under the same condition. The grinding efficiency (specific surface area of powder after milling) is proportional to the density of the media, while the abrasion rate varies depending on the material of the media. Because high milling efficiency and abrasion resistance are required for the media, the Y-TZP media are the most suitable for processing powders used to manufacture advanced high-performance materials. The Y-TZP media are positioned as the de facto standard with respect to the processing of $BaTiO_3$ powder (Fig. 15.6). The milling efficiency of the Y-TZP media changes when there is a difference in the size of the media.

Figure 15.7 shows the relationship between the average particle diameter and milling time, which was obtained by treating $BaTiO_3$ powder using the Y-TZP media of various

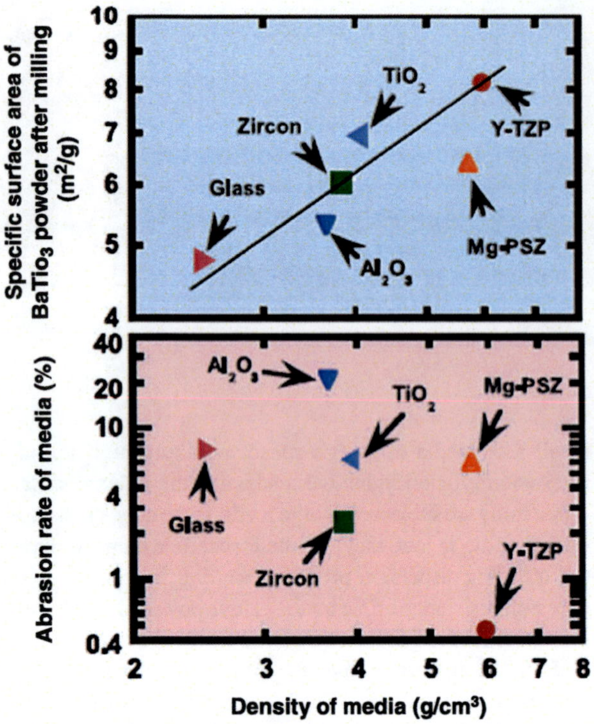

Fig. 15.5 Specific surface area of $BaTiO_3$ powder and abrasion rate of each media, obtained by milling under the same condition. Milling efficiency improves when the density of media increases, while the abrasion rate of media varies depending on the material of the media and is not affected by the density of the media

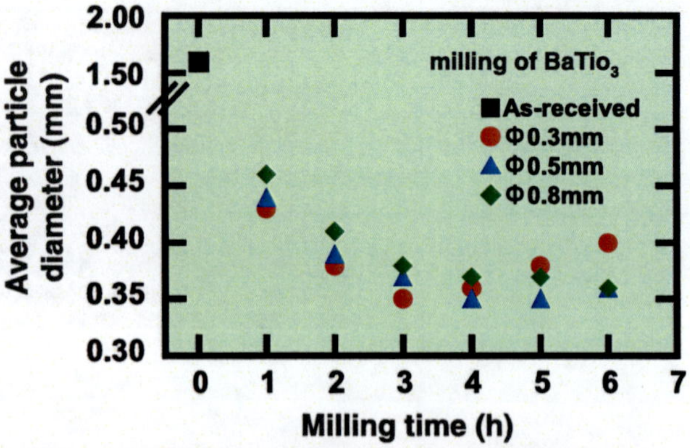

Fig. 15.7 Relationship between the average particle diameter of $BaTiO_3$ and grinding time or media sizes. Particle sizes become small in a shorter period of time when small-sized media are used. Powders generated from media due to abrasion are minimized and the amount of impurities in the processed powder may be reduced

Fig. 15.6 Picture of media made of T-TZP (yttria-stabilized zirconia). Media with sizes less than ϕ 0.1 mm have been developed

sizes and with the same condition. The particle sizes decreased in a shorter period of time when small-sized media was used. Therefore, high-efficiency processing will be possible if the media size has been optimized, depending on the particle size of raw material powder and the particle size to be attained. In addition, the amount of powder generated due to the abrasion of media and that mixed into the material will be minimized as the processing time is shortened. The media used to process advanced high-performance materials are required to have the following characteristics, in addition to good mechanical properties: (1) sharp media size distribution, (2) spherical uniformity, (3) smooth surface and (4) to be free from superficial and internal defects. Accurate control of the particle sizes of the milled power will be possible if these requirements are satisfied.

The Y-TZP media are not applicable to all of the materials and the media are selected based on the material and its application. Currently, alumina media with various degrees of purity as well as silicon nitride media are used to process non-oxide powders (Tables 15.3 and 15.4).

15.2.2.2 Manufacturing Method

The outline of the manufacturing process is shown in Fig. 15.8. The raw materials are blended to create a specified composition, which is wet-milled, granulated to a specified size and formed. Y-TZP, the raw material, is processed into powder by employing liquid phase methods such as the hydrolysis method and the neutralization method, so as to refine the powder and uniformly disperse Y_2O_3 in ZrO_2. The

Table 15.3 Characteristics of various types of media

Material	Alumina			Zirconia	Silicon nitride
Composition	93% Al_2O_3	93% Al_2O_3	93% Al_2O_3	Y_2O_3–ZrO_2	92% Si_3N_4
Bulk density (g/cm^3)	3.7	3.8	3.9	6.0	3.2
HV10[a]	1,200	1,500	1,800	1,250	1,400
Fracture toughness (MPa m$^{1/2}$)	3.5	4.0	3.5	6.0	6.0
Abrasion resistance characteristics: Grinding of SiO_2	A	B	S	S	A
Abrasion resistance characteristics: Grinding of Al_2O_3	S	A	S	A	A
Abrasion resistance characteristics: Grinding of $BaTiO_3$	B	B	S	S	S

S excellent, *A* very good, *B* good
[a]Vickers hardness measured with a load of 10 kgf

Table 15.4 Application of various media and examples

	Alumina	Zirconia	Silicon nitride
Porcelains, tiles, sanitarywares, etc.	◎		
Electronic part material	○	◎	○
Glass, glaze, pigment	◎		
Materials of phosphors and battery electrodes		◎	
Pigment, paint	○	◎	
Medicine, agriculture, food		◎	
Fine powder for polishing, abrasive grain for CMP		◎	
Raw material of fine ceramics		◎	○
Non-oxide raw materials			◎
Other high-performance materials		○	○

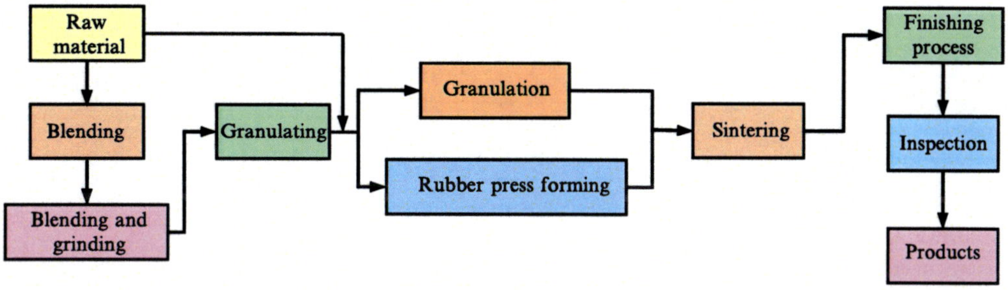

Fig. 15.8 Manufacturing process of grinding media. The raw materials are blended to create a specified composition, which is wet-milled to a specified size and granulated or formed by the rubber press, depending on the size of the media. The formed bodies are sintered and go through the finishing process, in which the initial abrasion is removed before completion

particles are formed to specified spherical shapes by the rubber pressing method or the tumbling granulation method (Note 15.4) and are sintered. The sintering conditions including the temperature are adjusted depending on the usage of the media, even if the material is the same. Generally speaking, when it is used for wet milling, the grain size is designed to be small so as to improve abrasion resistance. On the other hand, when they are used for dry milling, the grain size is designed to be larger than the one for wet milling so as to reduce chipping due to the mutual impact of grains in the media. The sintered media go through the finishing process, which removes initial abrasion before completion of the product.

Note 15.4 A forming method in which spherical grains are formed by spraying water on the powder, so as to utilize the aggregation phenomenon of powder, while rolling or vibrating the powder.

15.2.3 Future Prospects

Usage of powders consisting of nano-size initial particles that are manufactured through chemical processes such as the liquid phase method is increasing. Following this trend, the major role of media will shift from "grinding to create fine powder" to "dispersion by dissolution of powders in which nano-size initial particles are agglomerated." Because of this reason, sizes of the media are becoming smaller and smaller responding to particle sizes of powders. Microscopic media having sizes less than ϕ0.1 mm have been developed, and the need for microscopic media is expected to increase in the future.

Resistance Heating Elements (1923) 15.3

A number of different types of heat treatment methods such as resistance heating, arc heating, induction heating, microwave heating, gas heating and plasma heating are generally used for heat treatment of materials. Resistance heating elements are heated by the Joule heat (Note 15.5) generated by using electricity as the energy source and have been used for many years. Heat treatment is used primarily in the industrial field, although lamp heaters and parts of far-infrared heaters are also used by consumers. Resistance heating elements for industrial use are introduced below.

15.3.1 Background of Development

Electrical heating elements in industrial furnaces are used in a variety of industries such as heat treatment of metals, electronic components, ceramics and chemicals. For examples, they are used as heating elements in the process of manufacturing semiconductors, multilayer nickel condensers, liquid crystals, which are used in products that are essential for modern life, such as PCs, mobile phones, liquid crystal display (LCD) televisions and photovoltaic power systems.

15.3.2 Characteristics

15.3.2.1 Types of Resistance Heating Elements

There are many types of resistance heating elements, but they are roughly classified into two categories, the metallic heating elements and the non-metallic heating elements. The metallic heating elements are represented by Ni–Cr-based, Fe–Cr–Al-based, molybdenum-based, tungsten-based, platinum-based and $MoSi_2$-based elements. Meanwhile, silicon carbide, graphite, zirconia, and lanthanum chromite are used in the non-metallic heating elements. Four major heating elements are introduced below (Table 15.5).

15.3.2.2 Major Heating Elements and Their Characteristics

15.3.2.2.1 Ni–Cr-Based Heating Elements

The Ni–Cr-based heating elements are typical metallic heating elements and are the most popular because they are inexpensive and easy to manufacture. Two types, one is composed of Ni 80% and Cr 20%, whereas the other is composed of Ni 60% and Cr 17% (plus Fe), are most widely used. The oxide film (Cr_2O_3) formed on the surface of the heating element functions as a protective film.

15.3.2.2.2 $MoSi_2$-Based Heating Element

The $MoSi_2$-based heating elements are made by forming a SiO_2 (glass) film on the surface of $MoSi_2$, a metallic material. The SiO_2 film functions as a protective film, which enables usage of the element at high temperatures. Therefore, this metallic heating element can be used at high temperatures of up to 1,900°C under normal atmospheric conditions. However, the service temperature is lower under reducing atmospheres and high-vacuum conditions, in which the SiO_2 film is degraded. The disadvantages of this element include brittleness at room temperature, low thermal shock and softening/deformation at high temperature ranges. They are normally shaped like the letters U or W.

15.3.2.2.3 Silicon Carbide-Based Heating Element

Non-metallic heating elements are represented by the silicon carbide-based heating elements (Fig. 15.9), which feature high thermal resistance and corrosion resistance. As for the

Note 15.5 Amount of heat generated when current I (A) is supplied to an object with resistance R (Ω). The value can be obtained by "$Q=I^2Rt$," which is called Joule's law.

Table 15.5 Major heating elements and their characteristics

Heating elements	Max. service temperature (ambient) (°C)	Specific resistance (Ω cm)	Density (g/cm³)
Ni–Cr	1,000–1,200°C (atmosphere)	108×10^{-6}–112×10^{-6}	8.25–8.4
SiC	1,650°C (atmosphere)	0.02–0.1 (1,000°C)	3.2
	1,300°C (hydrogen)		
	1,450°C (nitrogen)		
C	2,600°C (non-oxidizing atmosphere)	8×10^{-6} (20°C)	2.2
	2,200°C (vacuum)		
$MoSi_2$	1,900°C (atmosphere)	0.3×10^{-6} (20°C)	5.5
	1,350°C (hydrogen)		
	1,600°C (nitrogen)		

Heating elements are selected according to their characteristics. The maximum temperature, specific resistance and density are indicated herewith

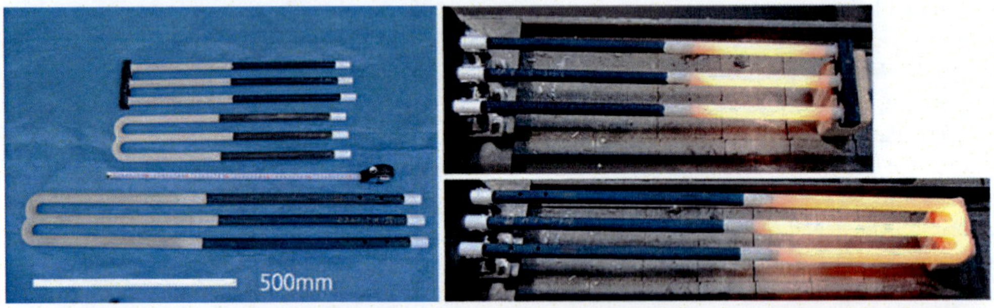

Fig. 15.9 Shapes of silicon carbide-based heating elements and the heated elements. The heating elements shown in the figure, named three-phase heating elements, are the most popular among the SiC-based heating elements. They are used as heaters of float bath furnaces, which are used to manufacture building materials, automotive glass and LCD glass. The production amount has been increasing in recent years

electrical properties, they have negative characteristics up to approximately 600°C and positive characteristics at higher temperatures. The heating elements, with the maximum service temperature of 1,600°C, can be used over a versatile temperature range. The silicon carbide-based heating elements also have a SiO_2 film that functions as a protective film generated on the surface, as in the case of $MoSi_2$-based heating elements. However, the resistance increases over time, approaching the end of service life. They are often bar-shaped, U/W-shaped, or spiral-shaped.

15.3.2.2.4 Graphite Heating Element

This heating element has the maximum service temperature of approximately 2,600°C in a non-oxidizing atmosphere such as in an inert gas. However, the elements cannot be used in an oxidizing atmosphere because they react with oxygen and water vapor at temperatures above 400°C. They can be easily processed into the shape of a bar, U, W, spiral, etc. The purity can be improved, allowing for the application to semiconductors. Furthermore, graphite-based heating elements are used not only for resistance heating but also for arc heating, and the demand for them as electrodes in steel manufacturing is high (Fig. 15.10).

15.3.2.3 Notes of Caution in Selecting Heating Elements

15.3.2.3.1 Selection of Material

In selecting heating elements, considerations should be given to the following points: (1) service temperature, (2) atmosphere, (3) volatile gas emission from the heat treated object, (4) price of the heating element, (5) batch-type or continuous-type of treatment, (6) amount to be processed, (7) heat capacity of the object to be processed, (8) weight and heat capacity of the jig that carries the object to be processed and (9) temperature ramping rate. Generally, heating elements are primarily selected according to the conditions 1–4, and the amount and dimensions of the heating elements are decided according to the conditions 5–9.

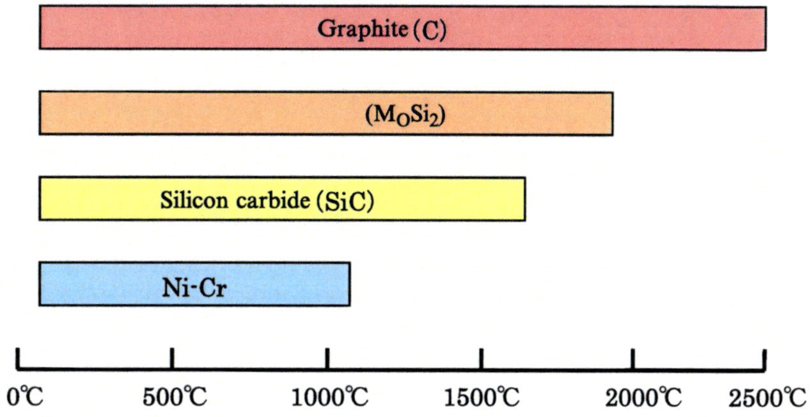

Fig. 15.10 Service temperatures and applications of heating elements. Service temperatures, important factors in selecting heating elements, are as follows: Ni–Cr < silicon carbide < $MoSi_2$ < graphite

15.3.2.3.2 Harmful Substances

Caution is required for substances that degrade the heating elements. Typical harmful substances are explained below:

1. Sulfur: caution is required because sulfur reacts with nickel, producing a low-melting point compound.
2. Halogen (Note 15.6): halogen such as chlorine and fluorine reacts with the protective film, preventing the generation of a film, and shortens service life of heating elements. Careful attention should be paid to halogen, which is very active and harmful to heating elements.
3. Alkali metals (Note 15.7) with low-melting point metals: alkali metals such as sodium and potassium as well as low-melting point metals such as zinc, lead and tin react with the heating element protective film and lower the melting point.
4. Acid, alkali: SiC and C, ceramic heating elements, are resistant to many substances, but they react with metals, and therefore, caution is required when heating metals.
5. Water vapor: water vapor should be discharged from the furnace at low temperatures as water vapor in the atmosphere accelerates oxidation degradation.

Note 15.6 Generic name of elements belonging to the 17th column in the periodic table. Fluorine (F), chlorine (Cl), bromine (Br) belong to the family.

Note 15.7 Alkali metals, including Na and K, belong to the first family and have common chemical properties.

15.3.3 Future Prospects

The heating elements explained above have been used traditionally as raw materials. Responding to the technological innovations in recent years and developments of various new products, a wider range of performance is required of heating elements. Their industrial applications will further expand by appropriate matching of their required performances with the heat treated materials.

Products Used in Firing Furnaces Consisting of Silicon Carbide-Based Refractory (1940)

15.4

Various items are used in firing furnaces that consist of refractories such as the shelves on which the products are placed (Fig. 15.11) and containers in which the products are placed within during calcination. They are used for calcining ceramic products such as dishes, tiles and sanitary wares and electronic parts such as ferrite ceramic condensers, as well as for the thermal treatment of electronic material powders and semiconductors. By utilizing thin plates of silicon carbide-based refractories that have high strength and high thermal resistance such as beams and saggers; the conservation of energy and shorter sintering time have been realized within a wide variety of industries. As a result, the types of silicon carbide-based refractories have also increased with improvements in their properties.

15.4.1 Background of Development

Firing or heat treatment is a major production process for a number of companies including ceramic manufacturers. Developments in the application of silicon carbide to tools used in the furnace started in the 1940s, and utilization of shelves made of oxide bonded silicon carbide, made by mixing silicon carbide grains with clay, for firing of heavy products such as insulators and sanitary wares began in the 1950s. In those days, there were problems such as the deformation of shelves due to load, but they continued to be used in furnaces for some time. In the late 1960s, a new material made by bonding silicon carbide grains with silicon nitride was commercialized, improving the creep resistance. The oxide bonded and silicon nitride bonded materials have both been used in a variety of fields. However, there was demand for silicon carbide-based refractories with improved properties, hence improvements have continued. Oxide bonded silicon carbide with a higher creep resistance, which is being widely used even today, expanded rapidly in early 1980s as shelves used in firing ceramic products such as dishes and tiles. Responding to an increasing demand for materials from ceramic manufacturers in neighboring countries, shelves made of oxide bonded silicon carbide based on the Japanese technology, began to be manufactured in foreign countries in the 1990s. The products, featuring properties equivalent to those made in Japan, began to enter Japan. Meanwhile, refractory manufacturers in Europe, America and Japan promoted the research and development of silicon carbide-based refractories featuring better properties, resulting in the commercialization of a variety of refractories based on recrystallized silicon carbide (Note 15.8), silicon impregnated silicon carbide (Note 15.9), high-strength silicon nitride bonded silicon carbide and pressure-less sintered silicon carbide, etc. These refractories are five times stronger than conventional ones made of oxide bonded silicon carbide, realizing reduction in thickness and weight of the furnace tools. This was done in response to the demand for energy conservation and efficient firing.

15.4.2 Characteristics

15.4.2.1 Characteristics of Products

Silicon carbide based refractories that feature properties such as high strength and high thermal resistance. The properties and microstructures are shown in Table 15.6. Long beam-shaped products and large products (Fig. 15.12) are made while taking advantage of their high strength. A variety of tools for furnaces have been commercialized. In regard to the phenomenon where silicon carbide is oxidized and converted

---

Note 15.8 Silicon carbide-based refractory are made by sintering silicon carbide grains at a temperature of 2,000°C or higher.

Note 15.9 Silicon carbide-based refractory with porosity of nearly 0%, are made by impregnating the pores between silicon carbide grains with silicon at a temperature of 1,414°C or higher, which is the melting temperature of silicon.

Fig. 15.11 The inside of a furnace with shelves made of silicon carbide. Silicon carbide shelves separate dishes to be calcined in numerous layers

to silicon dioxide in an oxidizing atmosphere during actual use, residual pores in the product are blocked by silicon to reduce the porosity to zero to prevent the diffusion of oxygen inside the refractories. Oxide films are formed on the surface in order to prevent diffusion of oxygen.

15.4.2.2 Manufacturing Method

Manufacturing methods of silicon carbide-based refractories are classified by the forming method, as shown in Fig. 15.13. The major forming methods include press forming, extrusion forming and slip casting. Panels and block-shaped refractories are made by press forming, elongated objects such as pipes and beams are made by extrusion forming, and saggers and complicated shapes are made by slip casting. Several different types of silicon carbide powders are used. For example, the primary material consists of various grain sizes that are mixed and kneaded with auxiliary material to prepare the bonding phase for forming. Then the formed body is fired to bond silicon carbide grains and to add the strength as the

Table 15.6 Properties of silicon carbide-based refractories

Types	Oxide bonding	Nitride bonding	Recrystallization	Nitride bonding (high strength products)	Silicon impregnation
Bending strength (MPa)	50	60	100	200	250
Bulk density	2.8	2.7	2.7	2.8	3.0
Porosity (%)	10	11	17	5	0
Thermal expansion coefficient $(\times 10^{-6}/K)$	4	4	4	4	4
Thermal conductivity (W/(mK))	15	15	50	35	200
Microstructure	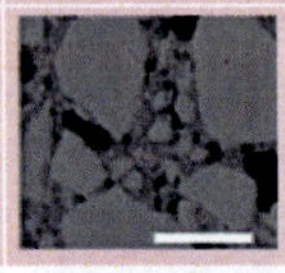	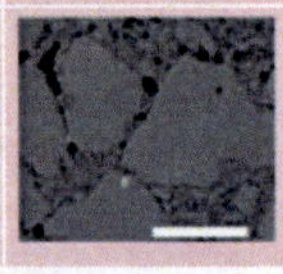	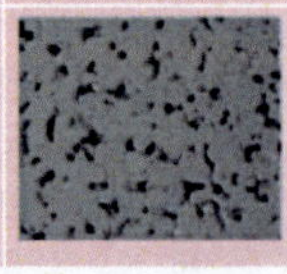		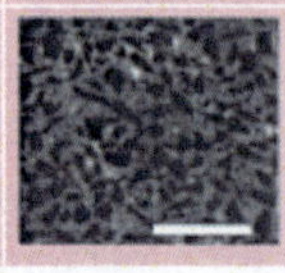

Silicon carbide-based refractories feature high strength and the thermal conductivity is higher compared with other refractories

Fig. 15.12 Examples of furnace products made of silicon carbide. Furnace related products used during firing with various shapes have been commercialized. Large panels and multilevel racks of complicated shapes are also manufactured

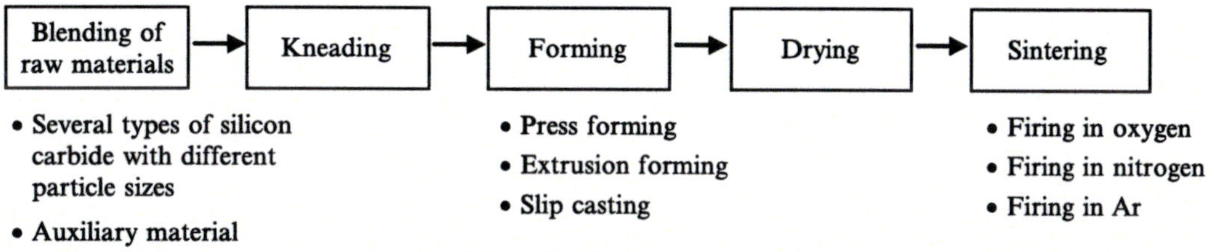

Fig. 15.13 Manufacturing methods of silicon carbide-based refractories. Silicon carbide materials of various particle sizes are mixed and kneaded, which is formed by a method appropriate for the shape of the intended product

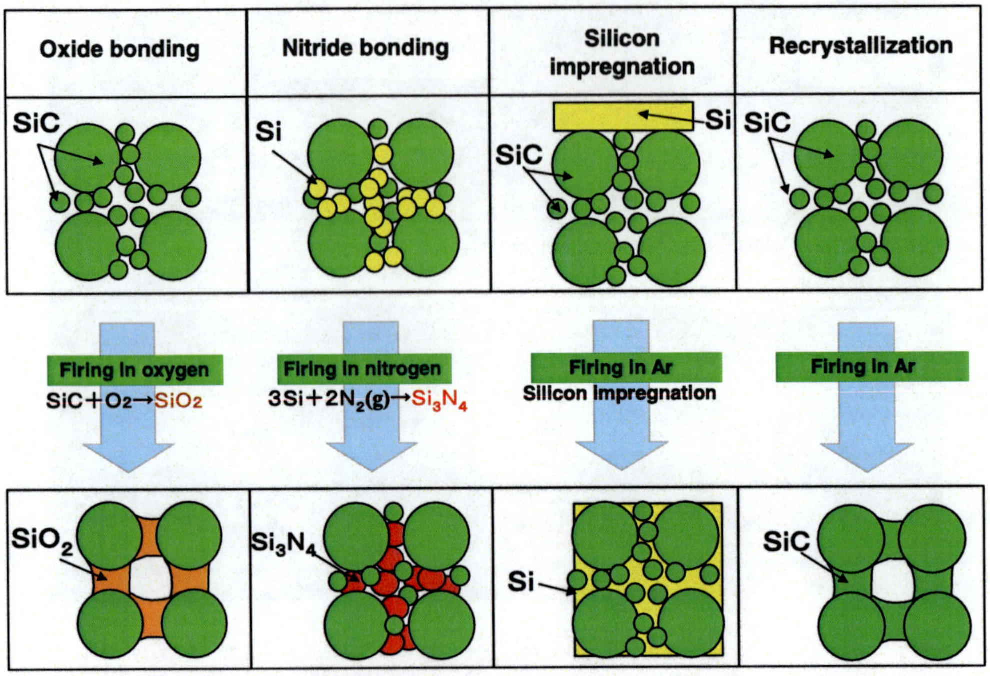

Fig. 15.14 The firing processes of silicon carbide-based refractories. Silicon carbide grains sinter together during the firing process

refractory. Illustrations of the phenomena observed during the firing process including oxide bonding, nitride bonding and silicon impregnation are shown in Fig. 15.14.

15.4.3 Future Prospects

Responding to the surging fuel prices and needs for CO_2 reduction, energy saving is becoming more important than ever. Silicon carbide-based refractories, featuring high strength, allows the production of thin and large furnace tools. Furthermore, they are expected to be utilized as tools for minimizing energy usage during the heating and firing processes, because the thermal conductivity is higher compared with that of other refractories.

Literature

1. Japan Electro-Heat Association, Electro-Heat 2003, No. 128, Kazuhisa Sakamoto (Resistance Heating Elements) (15.3)

Processing of Mechanical Components and Raw Materials

16

Ceramics are used for various purposes in the manufacturing industry because of their superior material properties (high hardness, thermal resistance, abrasion resistance). Bearings, cutting tools, grinders and abrasive materials are among the various ceramic products widely used in mechanical components and for material processing that are introduced in this section.

Hybrid bearings (Sect. 16.1) that use silicon nitride ceramics as the rolling elements (balls or rollers) are lighter and feature higher abrasion resistance, seizure resistance and thermal resistance when compared with conventional steel bearings. They are also employed as spindles on machinery, turbo pumps on spacecraft and in a diverse range of equipment. Demands for ceramic bearings will continue to increase responding to the needs for bearings with longer service life and higher reliability. There is a large need for lower costs in addition to higher performance.

Ceramic cutting tools (Sect. 16.2) are normally made of alumina-based ceramics or silicon nitride-based ceramics. They were commercialized in 1938, 10 years after the introduction of carbide tools. There were many problems at the beginning of commercialization. This included breakage during cutting work as the toughness and strength of the material was low. Following active research and development, technologies for the purification and sintering of materials advanced and mechanical properties were enhanced. As a result, these tools are now currently used in a wide range of fields. Light and durable new materials used by the automobile and aircraft industries tend to be difficult-to-cut. Ceramic tools featuring high thermal resistance and abrasion resistance are also expected to play important roles within these fields. Applications will expand to high-speed and high-efficiency cutting as well as dry cutting and mist cutting, which are of importance due to environmental concerns.

Diamond tools (Sect. 16.3) are widely used for cutting nonferrous metals such as aluminum alloy and copper alloy as well as nonmetallic materials such as lumber and ceramics. Natural diamond is expensive and varies in its mechanical properties. Inexpensive and high quality polycrystalline diamond, which was developed in 1975 by General Electric, is widely used for cutting. Following the enhanced efficiency and accuracy of cutting processes in recent years, materials to be cut are increasing in variety and are becoming harder making them more difficult-to-cut. Responding to this situation, new polycrystalline diamond products with higher strength, abrasion resistance and thermal resistance are being developed through the refining of microstructure and removal of bonding agents.

The grinding process is one of the precision processing methods in which the processed item is cut by a grinder that rotates at a high speed. This cutting process is applied to components of products in all industrial fields, such as automobiles, aircrafts and vessels as well as familiar products such as air conditioners and refrigerators. For production of semiconductor elements, silicon carbide, alumina and silica abrasives are used, respectively, in the slicing, lapping and polishing processes in order to create a flat silicon wafer (Sect. 16.4). In the polishing process, silica grains of 100 nm or less are normally used to remove the affected layer from the surface of wafers and grinding pads made of resin are used for grinding. The integration technology of ICs continues to advance and the quality standard for the polishing products is very strict. Abrasives are required to be free from heavy metals and contain few metallic impurities. In consideration for the environment, research on the recycling of abrasives is in progress.

Y. Imanaka et al. (eds.), *The Ceramic Society of Japan, Advanced Ceramic Technologies & Products*,
DOI 10.1007/978-4-431-54108-0_16, © Springer Japan 2012

Ceramic Bearing (1984)* 16.1

Hybrid bearings (Note 16.1) made of silicon nitride ceramics are lighter, harder and have higher abrasion resistance compared with conventional steel bearings. They also have excellent properties in terms of seizure and the binding of rolling elements (balls or rollers that are made of silicon nitride ceramics) compared with their counterparts (metals) under high-speed turning. They can be used at high temperatures and are widely used as bearings for high-performance, high-reliability machine tools.

Ceramic bearings excellent properties and high reliability have been highly evaluated. They are used in various applications, including expensive equipment such as turbo pumps (Note 16.2) in the NASA Space Shuttle as well as in general-purpose equipment such as in motors and turbo chargers.

16.1.1 Background of Development

Silicon nitride features high mechanical strength, abrasion resistance and thermal resistance and is a popular material that represents engineering ceramics. Densification of silicon nitride under pressure-less sintering, using yttria as the sintering agent became possible in 1974. Toshiba Materials Co., Ltd. commercialized silicon nitride ceramics intended for bearings through a joint development with JTEKT Corporation (formerly Koyo Seiko Co., Ltd.), which was started in 1984.

Figure 16.1 shows silicon nitride ceramic products intended for bearings usage. Hybrid bearings made of silicon nitride ceramics feature lighter weight as well as higher rigidity, abrasion resistance, seizure resistance and thermal resistance (refer to Table 16.1) compared with conventional steel bearings, and are widely used as bearings for high-performance high-reliability machine tools. A spindle motor of a machine tool is shown in Fig. 16.2. The excellent properties and high reliability have been highly evaluated. They are used for various purposes, including in the application to high-end equipment such as turbo pumps in the NASA Space Shuttle (refer to Fig. 16.3) as well as in application to general-purpose equipment such as in motors and turbo chargers.

16.1.2 Characteristics

16.1.2.1 Characteristics of Products

The load bearing property under repeated stress is higher in silicon nitride ceramics than in other ceramic materials (refer to Table 16.2). Ceramic bearings, used under cyclic fatigue and abrasion are required to be free from defects. Therefore, process design as well as the material composition are important factors that affect performance. Silicon nitride ceramics are composed of needle-like grains based upon cross-section observation of their grain boundary phase. It is important to determine and control the optimal microstructure (refer to Fig. 16.4) for bearing materials.

Silicon nitride ceramics used for bearings are classified as ceramic materials for bearings by the standards established by ASTM International (formerly the American Society for Testing and Materials). Classification is made based upon major characteristics, including bending strength, hardness, fracture toughness and pore size. These classifications are used when users select a material class that is appropriate for their particular application. The standard material produced by Toshiba Materials Co., Ltd., for example, satisfies the requirements of Class I (refer to Table 16.3). Silicon nitride materials with high thermal conductivity have been prepared by utilizing grain boundary control technology that

*The number in parenthesis indicates the year that the product was first commercialized.

Note 16.1 Bearings that are composed of different materials, where rolling elements (balls or rollers) are made of ceramics and the inner and outer rings are made of bearing steel (SUJ2).

Note 16.2 Bearings used in a pump that supplies liquid oxygen in the form of injected fuel. They have excellent functionality even at ultra-cold temperatures of −196 °C.

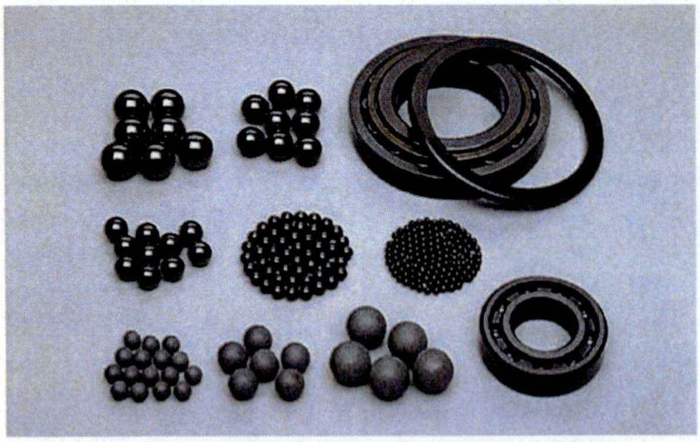

Fig. 16.1 Silicon nitride ceramic products used in bearings. The bearing materials and balls were made of ceramics and are finished products

also possess high abrasion resistance. Through process innovation, materials with properties that exceed the standards have also been developed and commercialized as bearing materials.

16.1.3 Future Prospects

Greater demands for improvements in bearing performance, including extending their service life, will become more important in the future. Reflecting this trend, expectations for wider usage of silicon nitride ceramics have become higher. However, greater utilization of this material is sometimes forsaken due to its higher cost compared to steel. The higher price is attributed to the small quantity of

Table 16.1 Comparison of silicon nitride ceramics versus bearing steel (SUJ2) characteristics

Items	Units	Silicon nitride	Bearing steel	Characteristics of ceramic bearings
Thermal resistance	°C	800	180	Retains high strength at high temperatures
Density	Mg/m³	3.24	7.8	Reduces centrifugal force of rolling elements (balls or rollers)→improves service life, prevents temperature rise
Thermal expansion coefficient	K^{-1}	3.2×10^{-6}	12.5×10^{-6}	Change in internal gaps due to temperature rise is limited→prevents vibration, small change in pre-compression amount
Hardness	HV	1500	750	Limited deformation in contact areas of rollers→high rigidity
Young's modulus	GPa	320	208	
Poisson's ratio		0.29	0.3	
Abrasion resistance		High	Low	Can be used in special environments such as in acid solutions and alkali solutions
Magnetic property		Nonmagnetic material	Ferromagnetic material	Variation in rotation due to magnetization in intense magnetic field is limited
Conductivity		Insulator	Conductor	Prevents electrical corrosion
Bonding condition of materials		Covalent bonding	Metal binding	Adhesion of roller contact areas due to oil film breakage is limited

Silicon nitride ceramics are superior to bearing steel in thermal resistance, wear resistance and abrasion resistance. They are light, nonmagnetic and insulating

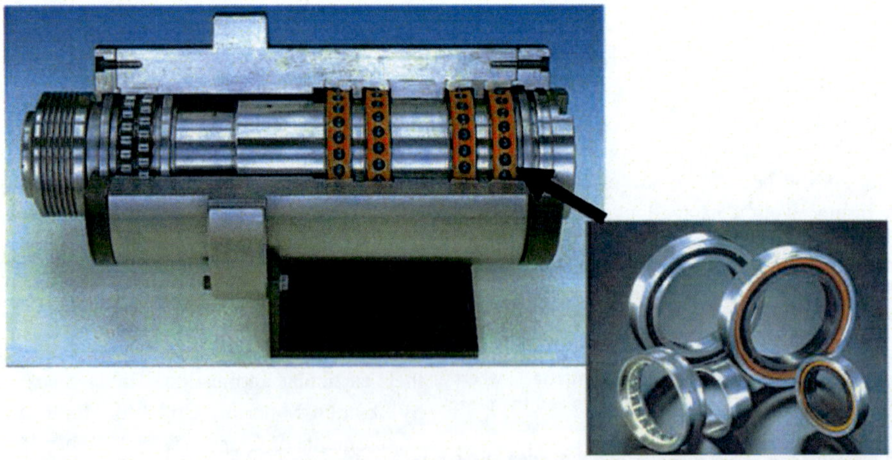

Fig. 16.2 Spindle motor. Hybrid bearings and the structure inside a spindle motor

Fig. 16.3 Space shuttle. The space shuttle launching

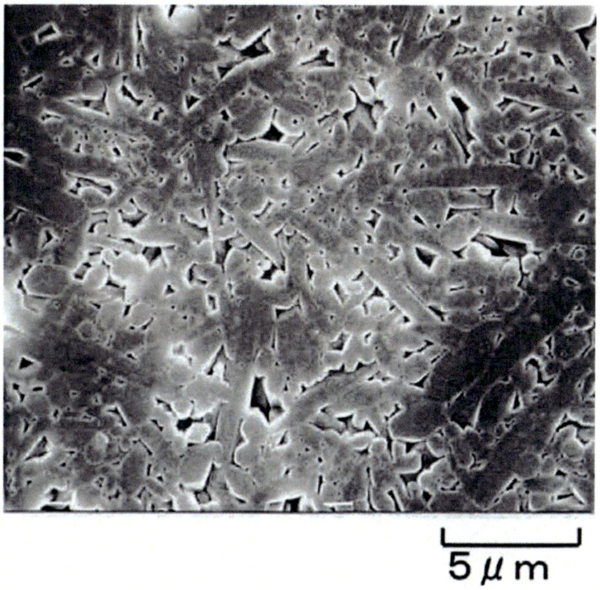

Fig. 16.4 SEM photograph showing microstructure of silicon nitride ceramic. The SEM observation was performed by etching the grain boundary after mirror-like polishing of the cross-sectional surface. Silicon nitride grows in needle-like shapes, but the grain boundaries are evenly distributed and have dimensions several microns

Table 16.2 Results of load bearing property tests for various ceramic materials

Test conditions		
	Oil lubrication	Water lubrication
Lubrication	Spindle oil	Running water
Counterpart material	3/8 in. bearing ball (SUJ2)	3/8 in. silicon nitride ball (Si_3N_4)
Load	Number of repeated stress × step up every 10^7 cycles	
Speed	1,200 rpm	

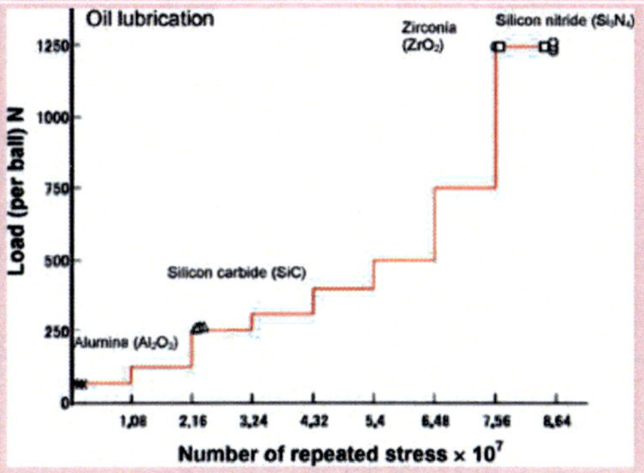

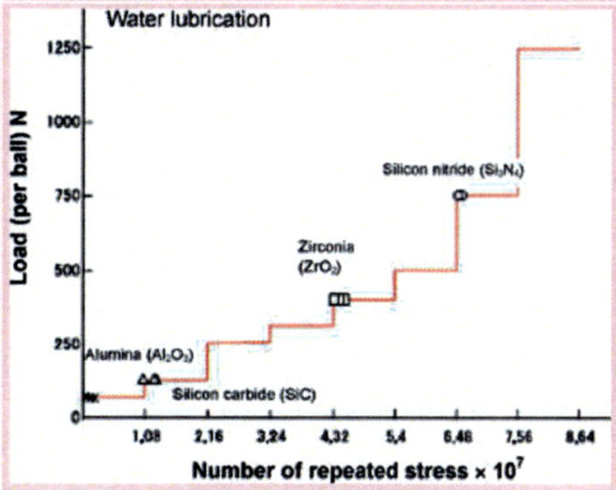

Silicon nitride ceramics features the highest load bearing properties among various ceramic materials

nitride ceramics produced and the manufacturing process, which requires special manufacturing equipment unlike oxide ceramics. For further expansion of applications, a breakthrough in manufacturing cost as well as new applications to high-end equipment with greater reliability is also required.

Table 16.3 Material classifications by ASTM and comparison Toshiba Materials' standard material

Characteristics	Methods	Units	Material classes			Standard class (treated with TSN-03 HIP)
			I	II	III	
Bending strength	3-Point bending (average)	MPa	≥900	≥800	≥600	1,100
	Weibull modulus		≥12	≥9	≥7	15
Hardness	Vickers HV 20 kg		≥1,460	≥1,360	≥1,300	1,500
Fracture toughness	IFR method	MPa $m^{1/2}$	≥6.0	≥5.0	≥5.0	7.0
Pore size (cross sectional)	SEM observation	μm	≥10	≥10	≥25	<1

WK5634 Standard Specification for Silicon Nitride Bearing Balls (ASTM F2094)

Ceramic Cutting Tools (1953) 16.2

Cutting tools are used to cut metals and are made of cemented carbide, cermet and ceramics. They are used in extremely severe environments because they are in contact with hard materials (mainly metals) during cutting work. The tools are exposed to high stress, high temperature and are subject to severe friction (including chemical reaction). For these reasons and also due to their superior properties such as high hardness, thermal resistance and abrasion resistance ceramic materials have been drawing attention for a long period of time, e.g. alumina ceramics were used for the first time in the 1950s as materials for cutting tools. Typically, cutting tools are made of alumina-based and silicon nitride-based ceramics. The quality of the products depends upon the characteristics of the material, and composite technology and special production methods such as the pressurized sintering are utilized. Cutting tools are typical ceramic products that have been developed in line with the progress of materials (Fig. 16.5).

16.2.1 Background of Development

Cutting tools are used in severe environments characterized by high temperatures and high stress. The following characteristics are material requirements:

1. High hardness, especially at high temperature
2. Chemical stability that prevents reaction with the work material
3. High mechanical strength, especially resistant to defect formation and thermal shock

From the view point of the above requirements, ceramics are believed to be more appropriate than other materials for usage in cutting tools. Cutting tool development began in the 1930s over 80 years ago and ceramic cutting tools were first commercialized in 1938, about 10 years after the commercialization of carbide tools, according to sources. Ceramic tools in those days were low in toughness and the bending strength was as low as 300 MPa, resulting in unexpected breakage during cutting work. Applications were also limited. However, the mechanical properties were improved substantially due to the development of technologies related to the creation of high-purity materials as well as sintering technologies, which led to utilization of alumina ceramics and silicon nitride ceramics. Research and development was actively carried out especially in Japan. New sintering technologies such as hot-pressing (Note 16.3), gas pressure sintering and hot-isostatic pressing (HIP method) (Note 16.4) were introduced, and studies on composition and developments of microstructure control technologies were carried out. This contributed to the improvement of the penetration rate of ceramic tools in the marketplace. Meanwhile, machine tools featuring high rigidity were developed and the cutting speed was increased. In line with these improvements, ceramic tools began to be applied within a wider range of industries including the automobile industry, which uses cast iron, and the aircraft industry, which uses heat resistant alloys. Application of ceramic cutting tools are expected to expand further to dry cutting and mist cutting, which have been attracting attention because of their environmental conversation appeal.

Note 16.3 Hot pressing is one type of the technology used for sintering ceramics. Sintering is carried out while applying mechanical pressure in a single axial direction by jigs made of carbon, etc. The direction of pressure application is anisotropic, limiting the shapes that can be processed. Products are made by sectional cutting and are thus expensive.

Note 16.4 Another type of the sintering technology adopted for the densification of ceramics. Sintering is promoted by applying high pressure gas which exceeds 1,000 times that of atmospheric pressure. The products are isotropic, and therefore, there is no limit in respect to shapes. In gas pressure sintering (GPS), gas pressure of several hundred times that of atmospheric pressure is applied. With this sintering method, the effect of controlling decomposition is utilized, rather than the pressure.

Fig. 16.5 Ceramic tools and holders. Representative examples of ceramic tools (*left*). Materials and shapes are selected depending on cutting conditions, etc. Ceramic tools are mounted in special holders when they are used for cutting (*right*)

Table 16.4 Characteristics of ceramic tools (examples)

Materials	Alumina	Alumina-carbide compound	Silicon nitride	Sialon
Hardness (GPa)	18–20	20–22	16–17	17–20
Strength (MPa)	700–1,000	800–1,200	1,000–1,200	800–1,200
Fracture toughness (MPa m$^{1/2}$)	3–4	4–6	5–7	4–6
Thermal expansion coefficient (×10^{-6} K^{-1})	8	7–8	3–4	3–4

Table 16.5 Applications of ceramic tools to cutting materials

			Ceramics			
Work materials	Diamond	cBN	Alumina	Alumina-carbide compound-based	Silicon nitride	Sialon
Cast iron		o	o	o	o	o
Nonferrous metal, Aluminum alloy	o					
Heat resistant alloy			o	o	o	o
Super-hard material (quenching steel, chilled cast iron)		o	o	o		

16.2.2 Characteristics

16.2.2.1 Classification of Products and Their Characteristics

Ceramic tools currently in the market are classified as follows by material: (1) alumina-based materials, (2) alumina-carbide compound materials, (3) silicon nitride materials and (4) sialon materials.

There are also tools made of cermet (intermediate between ceramics and metals), cubic boron nitride (cBN) and diamond. They are not explained here because they are manufactured using different methods.

Typical material properties of the products mentioned above are indicated in Table 16.4 and the types of work materials are indicated in Table 16.5. Alumina-based materials, which are sintered by adding a small amount of additives such as magnesia and zirconia, feature high chemical stability and superior abrasion resistance. Some of them are pressure-less sintered, but many of them are manufactured by HIP sintering and are white in color. These are called white ceramics and are comparatively inexpensive. Alumina-carbide materials, made by the combined addition of carbide, are inferior to alumina-based material in terms of sintering properties. They become expensive when hot pressing and HIP sintering are applied. They are called black ceramics due to their color. Although titanium carbide is normally used, molybdenum carbide, titanium nitride, tantalum carbide, titanium carbon nitride and silicon carbide are also sometimes used. The hardness and strength of materials were improved by utilization of compound materials and the tools can now be used to cut difficult-to-cut materials. Both silicon nitride and SiAlON ceramics, characterized by their needle-shaped crystal grains, have low thermal expansion properties and are used to manufacture tools that need to have high thermal shock resistance. Silicon, the major component of silicon nitride, tends to react with steel, resulting in substantial wear under certain cutting conditions. As a countermeasure, products coated with alumina have also been commercialized for improvement in performance. Sialon, made by solid solution of alumina component on silicon nitride grains, is used to cut thermal resistant alloys because it features both high thermal resistance and chemical wear resistance.

16.2.2.2 Manufacturing Method

Cutting tools have developed in line with improvements in materials formulation and manufacturing methods of the materials. High-purity fine ceramic powders must be used and excessive additives must be excluded during sintering. Intended properties are attained by devising sintering methods and process conditions. Measures must be taken to sinter perfect and dense bodies without defects such as pores (porosity) and structural variations. Figure 16.6 shows the current manufacturing flow of ceramic tools. They are normally sintered at high temperatures of 1,500 °C or higher. In early stages of development, pressure-less sintering was adopted for production because mass production and shaping are easy with this method. Currently, hot pressing, gas pressure sintering and HIP sintering methods are adopted so as to maximize the properties. Hot pressing is a superior sintering method, where mechanical pressure is applied using jigs made of carbon, etc. directly to the material to facilitate sintering. Caution is required because pressure is applied in a single axial direction, sometimes generating anisotropy. The gas pressure sintering method, where sintering is facilitated by utilizing gas pressure, several hundred times that of atmospheric pressure, is mainly applied to silicon nitride. The HIP sintering method is similar to the gas pressure sintering method, but much higher pressure (several thousands of atmospheric pressure) is applied during this method in order to achieve complete densification. This method enables the production of fine and homogeneous sintered bodies and is widely used to produce ceramic tools. Products made by both of these methods are isotropic and free from the problems associated by anisotropy.

16.2.3 Future Prospects

Ceramic tools were conventionally used in high-efficiency processing in order to minimize energy consumption, as well as high-speed dry processing in particular. The environmental load will remain a serious issue for industry, and new materials featuring light weight and toughness are needed by the automobile and aircraft industries, etc. These materials tend to be difficult-to-cut, and tools capable of cutting these new materials both at high speed and efficiently will be required. Ceramic tools, featuring superior thermal resistance and abrasion resistance, show potential utility for use as tools for cutting difficult-to-cut materials and are expected to be used more frequently for dry cutting, mist cutting, etc.

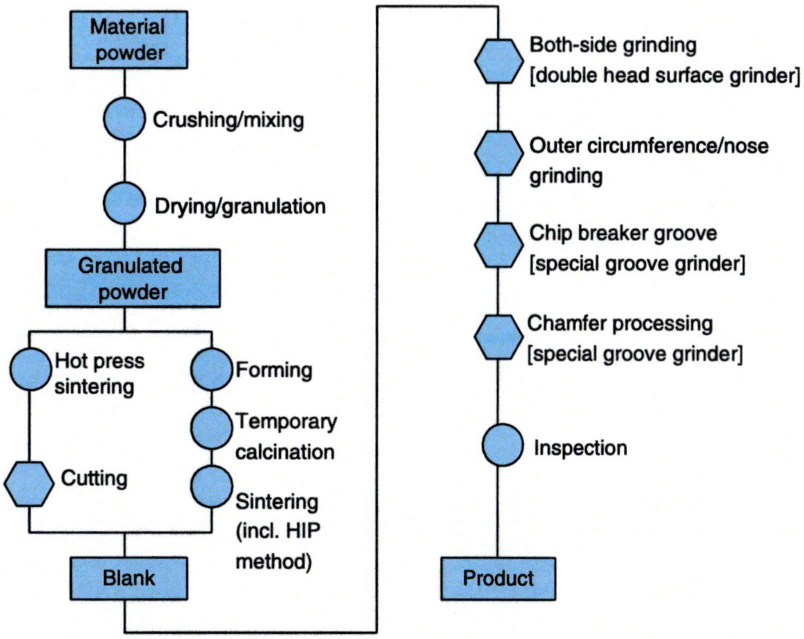

Fig. 16.6 Manufacturing process of ceramic tools. Raw material powders are blended, spray-dried and granulated. The granules are processed to their intended shapes by press forming (as-blended powders are used in hot pressing method) and are then sintered. Nearly complete densification results from pressure application during sintering which is important for tool life

Diamond Cutting Tools (1975) 16.3

Diamond cutting tools are widely used to cut nonferrous metals such as aluminum alloy and copper alloy and nonmetallic materials such as plastic, timber and ceramic green bodies. In particular, polycrystalline diamond is made by sintering diamond powder at ultrahigh pressure and temperature. It possesses both high hardness and toughness, and it is used as a cutting tool and surface abrasion tool. Materials are becoming diversified and harder to cut following improvements in efficiency and accuracy of machine processing, and needs for diamond cutting tools have been increasing.

16.3.1 Background of Development

Cutting operations using diamond cutting tools have been increasing responding to improvements in efficiency and accuracy of cutting in machine, electrical and electronics industry. Also, for example, as a result of an increase in the use of aluminum alloy for the reduction of vehicle body weight and the need for the efficient processing to enhance productivity, the need for diamond cutting tools has been increasing within the automobile industry.

16.3.2 Characteristics and Manufacturing Method

16.3.2.1 Characteristics

Diamond has the highest level of hardness and thermal conductivity among all substances. Because of these characteristics, natural diamonds have been used as tools for decades. However, natural diamond is expensive and does not provide constant mechanical properties. In 1955, General Electric (GE) succeeded in synthesizing diamond for the first time by using equipment that generates ultra-high pressure and temperature. Powder-type synthetic diamond began to be used as the raw material of grinders. In approximately 1975, GE developed a sintering method for synthetic diamond powder using metal binding materials such as cobalt. Later, varieties of polycrystalline diamond were developed and began to be widely used for cutting operations due to their low price and high quality (Fig. 16.7). The total production of diamond cutting tools in Japan amounted to sales of 11 billion yen in 2007. The production amount includes tools that incorporate single crystal diamonds used for ultrahigh precision processing. In this section, tools incorporating polycrystalline diamonds, which share the majority of the production amount, are introduced.

Figure 16.8 shows the typical microstructure of a polycrystalline diamond compact (PDC). Diamond particles with a diameter of several μm to several tens of μm are bound together and the gaps between the diamond particles are filled with a metal binding material such as cobalt. In contrast, single crystal diamond tends to break easily in certain directions and is low in toughness due to cleavage (Note 16.5). Meanwhile, polycrystalline diamond is an assembly of crystals having various directions and is less affected by cleavage, which means higher toughness and strength when compared with single crystal diamonds. Because of these characteristics, PDC is used in turning tools (Note 16.6), millers (Note 16.7) and drills for cutting materials in various fields.

Note 16.5 A phenomenon where a crack proceeds due to bond cleavage in a certain plain direction of a single crystal. It occurs on the plane where the binding force between atoms is weak within the crystal structure.

Note 16.6 Tools attached to turning machines and surface grinders and used for cutting.

Note 16.7 Cutting tools in which the tools are rotated for cutting. Cutting blades are equipped on the outer circumference or at the ends.

Fig. 16.7 Polycrystalline diamond tools. (*Left*) Cutting chips used in blade-replaceable turning tools. (*Upper right*) A blade-replaceable miller and chips. (*Bottom right*) Drill bits where polycrystalline diamond (*black area*) is pasted on the tip of tools only

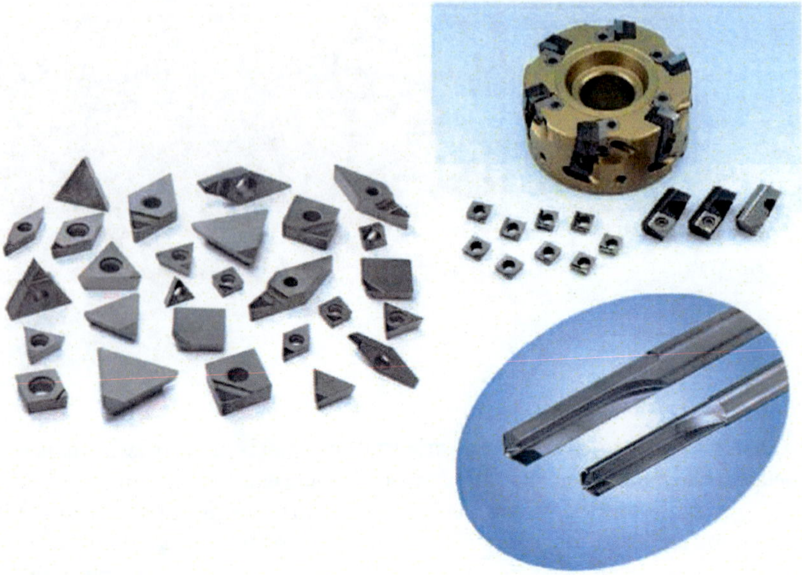

Fig. 16.8 Microstructure and illustration of microstructure of a polycrystalline diamond compact (PDC). The *gray area* indicates the diamond particle and the *white gray area* indicates the binding material. Diamond particles are bonded strongly by the binding material

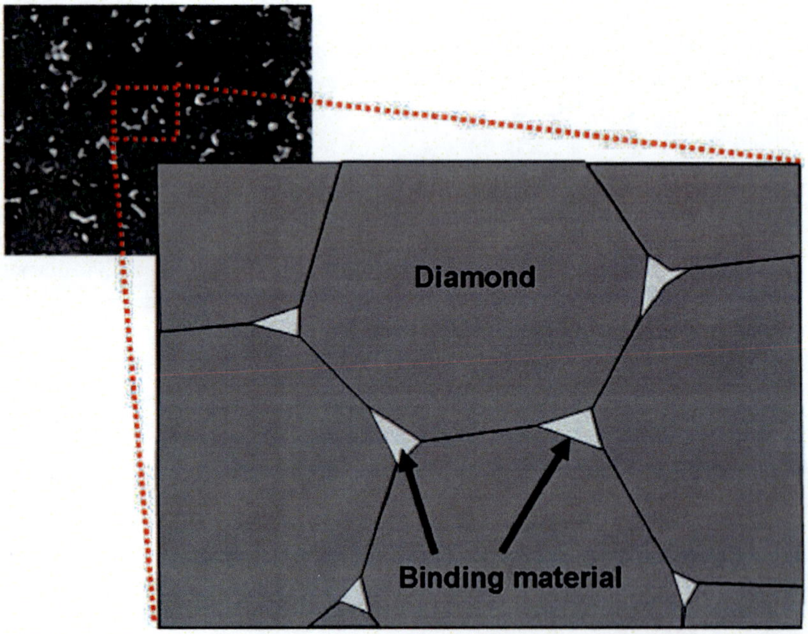

16.3.2.2 Manufacturing Method

Figure 16.9 shows the manufacturing process of polycrystalline diamond tools. Diamond powder and metal binding material such as cobalt are mixed. The mixture is pressurized to about 50,000 atm. and heated to around 1,500 °C using equipment that generates ultrahigh pressures and temperatures. In this process, the metallic bonding material melts down and fills the gaps between diamond particles. Diamond particles melt slightly within the liquid state binding material and are deposited between particles, resulting in bonding between diamond particles. Through repetition of this process, diamond particles are grown and bonded strongly to become a dense sintered body. The equipment that generates ultrahigh pressures and temperatures, which is used in the sintering process, is indispensible for the production of polycrystalline diamond and is a major mass production technology. Refer to specialized books for details of the equipment.

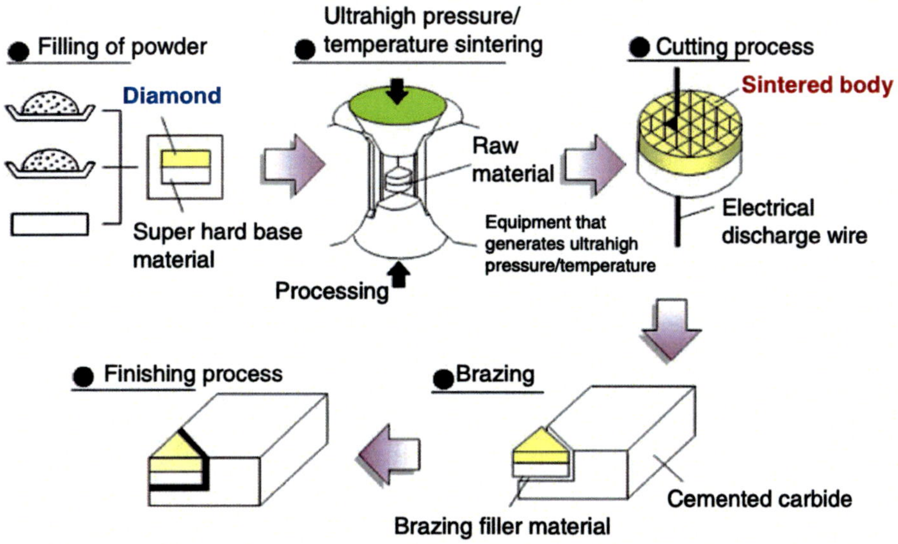

Fig. 16.9 Manufacturing process of polycrystalline diamond tools. Polycrystalline diamond is sintered by using equipment that generates ultrahigh pressures and temperatures. The sintered body is expensive. It is cut into small pieces and used on the tip of tools only

The sintered body is cut into small pieces by wire discharge processing (Note 16.8), etc. The small pieces cut from the sintered body are brazed on the main bodies made of cemented carbide, etc. and the blade tips are processed using diamond grinders. This completes the tools.

16.3.3 Future Prospects

In recent years, materials are becoming diversified and harder to cut following improvements in efficiency and accuracy responding to activities concerning environmental issues. For example, the ratio of difficult-to-process materials used in aircraft components have been increasing rapidly responding to the needs for an improvement in the strength and weight reduction of aircrafts. The requirement placed on diamond tools has also risen. Manufacturers have been developing new products, aiming to realize higher strength and abrasion resistance, by refining the composition, etc. (Fig. 16.10). The conventional polycrystalline diamond contained metal binding material and the thermal resistance was low at around 700 °C, limiting the service condition.

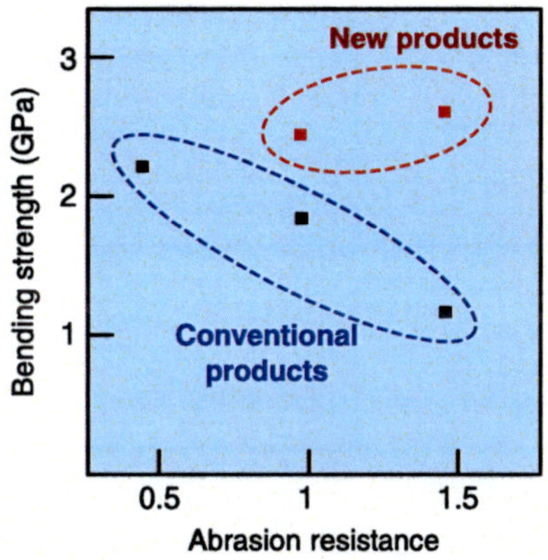

Fig. 16.10 Characteristics needed for new diamond tools. New products are being developed, aiming to realize higher strength and abrasion resistance, by refining the microstructure, etc.

Developments of high-strength polycrystalline diamond, made by using nonmetallic binding material or made without using binding agents (direct graphite conversion method) are in progress.

Through these activities, polycrystalline diamond will continue to serve the needs of various cutting works and contribute to industrial development.

Note 16.8 Products are processed by ark discharge, using metallic wires made of brass, etc. as electrodes.

Abrasives Used to Manufacture Semiconductor Devices (1950) 16.4

Semiconductor devices are used in home appliances, electronics and within the infrastructure that supports our information-based society. For the production of silicon wafer based semiconductor devices, various kinds of abrasives are essential for the planarization of silicon wafers, a semiconductor material. Development of abrasive manufacturing technologies, as well as semiconductor devices, is essential for the future of today's information-based society.

Meanwhile, a huge amount of energy is consumed in order to produce abrasives, which may affect the environment. For the realization of a ubiquitous society and the sustainable development of society as a whole, research for a technology that enables the recycling of abrasive resources, etc. needs to be promoted.

16.4.1 Background of Development

Semiconductor devices are used in home appliances such as personal computers (PCs), mobile phones and digital cameras as well as in the infrastructure that supports an information-based society. They play extremely important role in enhancing the quality of our living environment.

Integration technology for integrated circuits (IC) used in semiconductor device fabrication, has been developed remarkably year after year, enabling downsizing and improvement in the speed of home appliances, etc. through high density integration technology.

The IC is created by transferring the circuit design onto a silicon wafer (thin disks shown in Fig. 16.11). For accurate transfer of the circuit, the wafer needs to be extremely flat and free from stain. For efficient production of semiconductor devices, efficient polishing or flattening of silicon wafers was required, which led to the development of processing methods and abrasives.

16.4.2 Characteristics and Manufacturing Methods

16.4.2.1 Characteristics of Products

In the general silicon wafer manufacturing process (Fig. 16.12), abrasives are used in the slicing, lapping and polishing processes. Processing of wafers by abrasives is explained step-by-step below.

A single crystal Si ingot is grown from high-purity silicon melt by means of the CZ method (Czochralski method) (Note 16.9). The outer circumference of the ingot is ground to create a cylindrical ingot, which is sliced into wafers. Slicing of the wafers is performed by the wire saw, a mechanical device that cuts the ingot using a piano wire while pouring viscous abrasive liquid continuously onto the wire.

High-speed cutting is required for slicing so as to improve productivity, and silicon carbide abrasive, featuring extremely high hardness and chemical stability, is used in this process. Silicon carbide, a ceramic consisting of carbon (C) and silicon (Si), is mainly synthesized outside Japan. In the process of slicing, minor cracks are generated on the surface of wafers. The cracks are called the damaged surface layer, which blocks normal operation of semiconductor devices and needs to be removed in the post process. The sliced wafers go through the beveling process, chamfering to avoid chipping and cracking of the end face, before going to the next process.

Sliced wafers vary in thickness depending on their diameter. In order to equalize the thickness, the wafers are ground

Note 16.9 The Czochralski method. Silicon is melted in a crucible made of quartz. The single crystal ingot is grown from molten silicon. This method is mainly applied to manufacture single crystal ingots of large diameter.

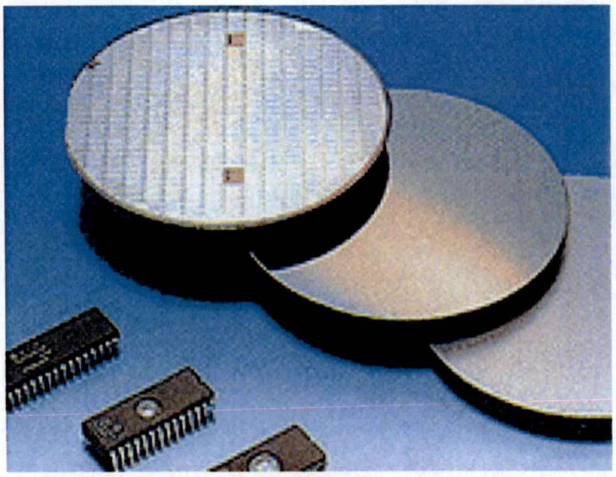

Fig. 16.11 Wafers and integrated circuits (ICs). Multiple layers of circuits are formed on thin and flat disk-shaped silicon wafers that make up a device

Fig. 16.13 Alumina-based abrasives (SEM). Microstructure of abrasives composed mainly of alumina, which is softer than silicon carbide

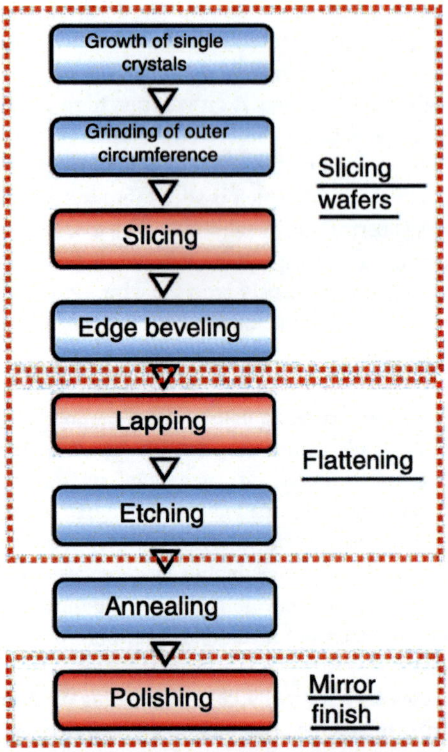

Fig. 16.12 An example of silicon wafer manufacturing process flow. Ceramic-based abrasives are used in the slicing, lapping and polishing processes

under a grinding pad while applying mechanical pressure with abrasive liquid. This grinding process is generally called lapping. In the lapping process, alumina abrasive (Fig. 16.13) is used. Alumina is softer than silicon carbide. The alumina abrasive has the grinding performance required in lapping and reduces the generation of damaged surface layers at the same time. Wafer thickness is equalized by lapping, but damage and minor roughness still remain on the surface layer of wafers.

In the etching process, the outer surface of the wafer is dissolved in chemical solution and no mechanical pressure is applied, removing the damaged surface without creating a new damaged surface layer.

During the finishing process, minor roughness and deformations on the outer surface of wafers as well as the damaged surface layer are removed. In this process, silica particles of 100 nm or smaller are used mainly as the abrasive (Fig. 16.14). A soft grinding pad made of resin is used to prevent the generation of a damaged surface layer due to mechanical stress. This process is called the polishing process.

Abrasives used for cutting and grinding in the slicing and lapping processes, where the wafers are cut and ground by mechanical force, need to be sharp to some degree and hard enough to resist the force applied from outside.

Silicon carbide and alumina are harder than silicon and are used to process products with mechanical force application. Meanwhile, silica particles are used in the polishing process, where both mechanical force and chemical reaction are utilized.

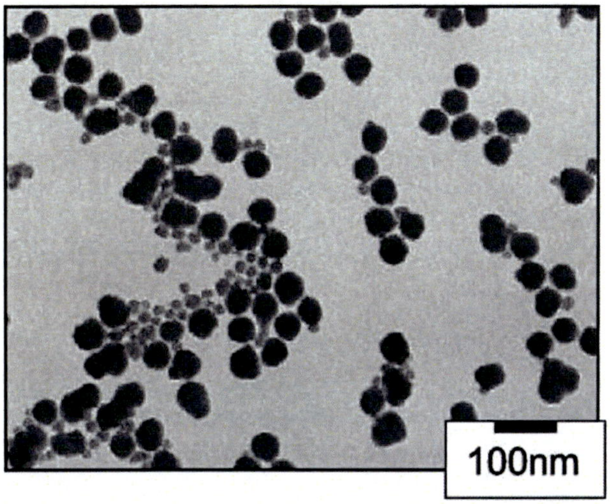

Fig. 16.14 Silica (silicon dioxide)-based abrasives (colloidal silica). In the polishing process, polishing liquid consists of fine silica particles and chemical solution to utilize mechanical force and chemical reaction

16.4.2.2 Manufacturing Method

Silicon carbide is generally synthesized from silica stone and coke, utilizing the Acheson Method (Note 16.10). Alumina is produced by melting raw material, synthesized from bauxite, at a high temperature. A large amount of electrical energy is required for the synthesis of both silicon carbide and alumina.

Synthesized silicon carbide and alumina are produced in the form of large blocks. Abrasives with appropriate particle size distribution are produced by crushing the blocks and sorting the particles by size in classifiers, etc.

16.4.3 Future Prospects

Abrasives are essential for the production of semiconductor devices and are a key element in today's information-based society. Meanwhile, a large amount of energy is consumed to produce abrasives, which raises concern about the influence on the environment.

The diameter of silicon carbide particles applied to the wire saw for cutting an ingot into wafers was originally #600 (approximately 20 μm), but it has been reduced to #1500 (approximately 8 μm). The number of wafers created from an ingot increases as the diameter of the piano wire used for cutting along with the diameter of abrasive particles decrease. Research on further reduction in diameters of the wires and the particles is expected to come about for more effective utilization of resources.

Technology development on the integration of ICs has been advancing and the quality standard of polished wafers is strictly controlled. Abrasives used in the polishing process, in particular, are required to be of extra high purity. The demand is very high for the reduction of metallic impurities that affect the characteristics of devices.

For the realization of a ubiquitous society, semiconductor production technologies need to be developed further. Therefore, quality improvement of wafers, the basis of semiconductor devices, is essential. At the same time, the development of next-generation abrasives is expected for the production of high-quality wafers. Research on the recycling of abrasives also needs to be promoted in consideration of the environment for the sake of lasting development that does not waste resources.

Note 16.10 The Acheson method. The mixture of carbon (C) and silica (SiO_2) materials are made to react in high temperature by means of direct energizing, etc. so as to manufacture silicon carbide.

Literature

1. Fine Ceramics Utilization Technology (1982) Science Forum Inc., 315 pp (in Japanese) **(16.2)**
2. Itano H (1995) New ceramics. Powersha Inc., 129 pp (in Japanese) **(16.2)**
3. Urashima K (2007) Powder and powder metallurgy technology strategy map. Japan Society of Powder and Powder Metallurgy (in Japanese) **(16.2)**
4. The Industrial Diamond Association of Japan (ed) (2007) Comprehensive diamond technologies. NGT, pp 12–20 **(16.3)**
5. Okita et al (2008) SEI Tech Rev 151:96–99 (in Japanese) **(16.3)**
6. Kakutani H, Irifune T (2004) SEI Tech Rev 165:68–74 (in Japanese) **(16.3)**
7. Homepage of Fujimi Incorporated **(16.4)**
8. Kobayashi A, Kasai T, Moriwaki T et al (1994) Practical technologies of super precision production technology system. Fuji Techno System, Tokyo **(16.4)**
9. Takata S et al (2000) Semiconductor silicon industry of the 21st century. Kogyo Chosakai Publishing Co., Ltd., Tokyo **(16.4)**

17 Semiconductor Production

Semiconductor devices are used in electronics products such as PCs, mobile phones and game machines. Applications have been expanding further to automobiles, home appliances, etc. in line with the progress in IT technologies. The integration degree of semiconductor devices has become denser year after year responding to advances in integration capability, process design rules, and productivity has been improved by means of the increase in the diameter of silicon wafers. The process design rule of 45 nm between metallization lines and a wafer diameter of 300 mm are currently in the main stream, and semiconductor manufacturing equipment is required to produce larger wafers of 450 mm in the future with higher accuracy. The microfabrication technologies and processing technologies developed for semiconductor devices are incorporated in flat panel displays and their production technologies, contributing to the recent rapid progress of the display industry.

Optical materials for the production of semiconductor exposure apparatuses (Sect. 17.1) are used to produce lenses for steppers, which are essential for the production of high-integration IC chips. Optical systems that incorporate stepper lenses are made of quartz glass and fluorite featuring extremely high transmittance and require large lenses.

In the 1970s, quartz was mainly used for equipment such as furnace tubes for thermal treatment of semiconductors. Following the development of dense silicon carbide-based materials, these materials, featuring high thermal resistance and thermal conductivity, have been used as furnace quartz reactor tubes. Furthermore, following the improvement in purity of the materials and the development of high-purity SiC film coating technology, and because the thermal expansion properties of silicon carbide are similar to that of silicon, silicon carbide began to be used to produce the ports where wafers are loaded for thermal treatment (Sect. 17.2). The reaction sintering process is employed to produce these silicon carbide based materials. Dense silicon carbide/silicon composite materials with high strength are produced in the process of impregnating silicon carbide containing carbon components with molten silicon.

The basic microfabrication process includes film formation, lithography and etching of various materials. The process is repeated to manufacture intended circuits. Halogen gas plasma, which is generally extremely corrosive, is used in the microfabrication process, and therefore, ceramics featuring high corrosion resistance, thermal resistance, abrasion resistance and chemical stability are used as the components. In particular, aluminum nitride and alumina are used as the ceramic heater (Sect. 17.3) and the electrostatic chuck (Sect. 17.4) on which wafers are directly placed. They are the core components of the semiconductor manufacturing equipment. Metallic electrodes for heating and wafer chucking are buried in the ceramic heater to form the electrostatic chuck. The electrodes are co-fired with the ceramics. Patterns and positional accuracy of the buried electrodes directly affect the soaking properties of heaters and wafer chucking properties of electrostatic chucks. Therefore, these design requirements are as important as those involving the control of the thermal and electrical properties of ceramic materials.

A wide variety of wafer handling stages for semiconductor exposure apparatuses of today is equipped with an air slide (a static pressure gas bearing) (Sect. 17.5). The air slide enables precise and high-speed movement of stages because it is free from the deterioration of accuracy and the generation of particles attributable to abrasion or heating due to friction, which frequently occurs in rolling bearings. Ceramics, mainly alumina, are used as components due to their high rigidity against metallic materials, low thermal expansion properties and light weight.

Liquid crystal displays, which currently represent flat panel displays, were mainly used in laptop PCs in the 1990s. At the time, mother glass substrates were approximately 300 mm×400 mm in size. Following numerous technological innovations, the liquid crystal displays began to be used in large TVs and the size of mother glass substrates of today far

Y. Imanaka et al. (eds.), *The Ceramic Society of Japan, Advanced Ceramic Technologies & Products*,
DOI 10.1007/978-4-431-54108-0_17, © Springer Japan 2012

exceeds 2,000 mm×2,000 mm. Aluminum/silicon carbide composite materials (MMC) (Sect. 17.6) are used as active components of large glass substrates. The material is made by impregnating a silicon carbide preform with molten aluminum, making it possible to manufacture large components by the near net shape method. The material is as light as aluminum and its rigidity is higher than that of cast iron, realizing the substantial weight reduction of components.

Optical Materials for Apparatus for Semiconductor Lithography (1980s)* 17.1

The use of semiconductor integrated circuits (ICs) has become an essential part of modern life. The evolution of apparatus for semiconductor wafer lithography is indispensible for the development and production of further advanced ICs. To be more precise, the exposure wavelength is being reduced to print ICs more accurately with higher resolution. The ArF excimer laser (emission wavelength: 193 nm) (Note 17.1) is currently employed in advanced systems. The optical glass used in precision optical apparatuses cannot be used because it does not fully transmit the light in this wavelength range. As optical materials for lenses in advanced apparatus for semiconductor lithography; silica glass and fluorite, featuring extremely high transmittance in short wavelength and high durability, are used. These optical materials are essential for the production technology of semiconductor ICs that utilize light.

17.1.1 Background of Development

Semiconductor ICs are used in nearly all of the communication devices and home appliances around us. It is not an exaggeration to say that ICs are indispensable for our daily lives. Responding to the advancement in information communication and the increase in the amount of information to be handled, ICs are required to achieve higher levels of integration.

In order to improve the integration levels of ICs, the resolution of optical lithography for printing IC patterns needs to be enhanced. One of the most effective solutions is the shortening of the wavelength of light used for printing (exposure wavelength). The internal structure of the stepper (Note 17.2), a widely used semiconductor exposure apparatus, is shown in Fig. 17.1. The projector lens, which comprises the heart of the apparatus, has a total length of more than 1 m and consists of more than 20 different lenses. Figure 17.2 shows changes in the exposure wavelength of apparatus for semiconductor lithography and transmittance of optical materials used in the manufacture of projector lenses. In the second half of the 1970s, the exposure wavelength that was most used was the UV g-line (436 nm), and projector lenses were made from optical glass. However, optical glass was no longer applicable since the second half of the 1980s, when the wavelength started to be shortened and the KrF excimer laser (248 nm) began to be used. This is because optical glass does not fully transmit the laser light, as shown in Fig. 17.2. In order to produce advanced ICs, silica glass and fluorite (featuring extremely high transmittance in the excimer laser wavelength range) are essential.

17.1.2 Characteristics and Manufacturing Method

17.1.2.1 Characteristics and Specifications of Products

The most important property required of materials for projector lenses is the high transmittance of light at the exposure wavelength. The transmittance per 1 cm is 99.8% or higher in both silica glass and fluorite. In order to realize this performance, the concentration of impurity needs to be controlled to below ppb (10^{-9}) order. In addition, the lenses are required to resist radiation of strong excimer laser beams for

*The number in parenthesis indicates the year that the product was first commercialized.

Note 17.1 Stands for *Ex*cited D*imer* Laser. and it provides ultraviolet laser with intense energy.

Note 17.2 A sequential shift exposure system. The IC is printed on a large wafer by repeated exposure of a small area and repositioning the wafer until all the surface has been exposed.

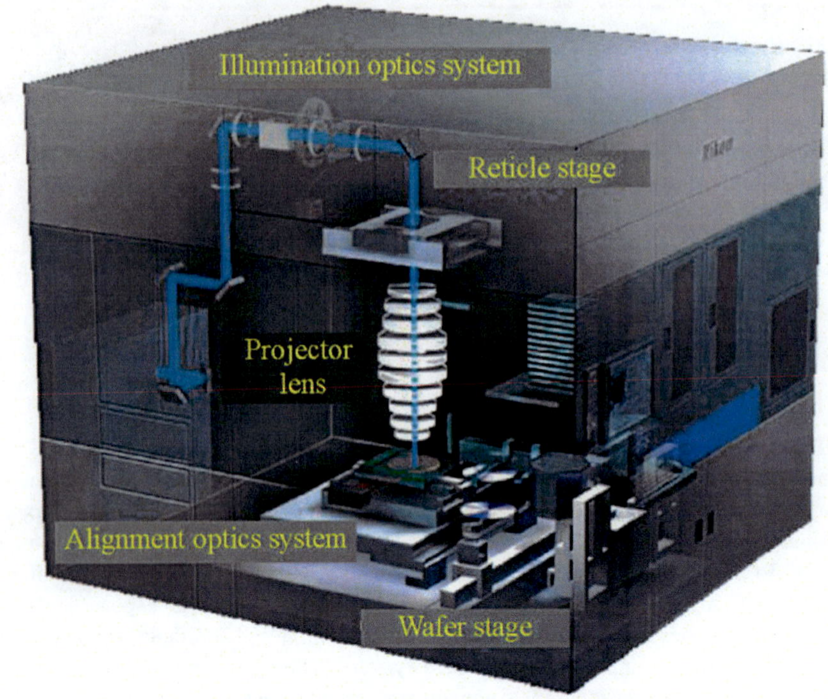

Fig. 17.1 Internal structure of the stepper (rough sketch). The projector lens in the center is the heart of the stepper. It is difficult to comprehend from the sketch, but the total lens length is more than 1 m and the number of lenses exceeds 20. The resolution of the first mass-produced g-line machines were 1 μm, while the latest mass-produced ArF machines are 65 nm (0.065 μm)

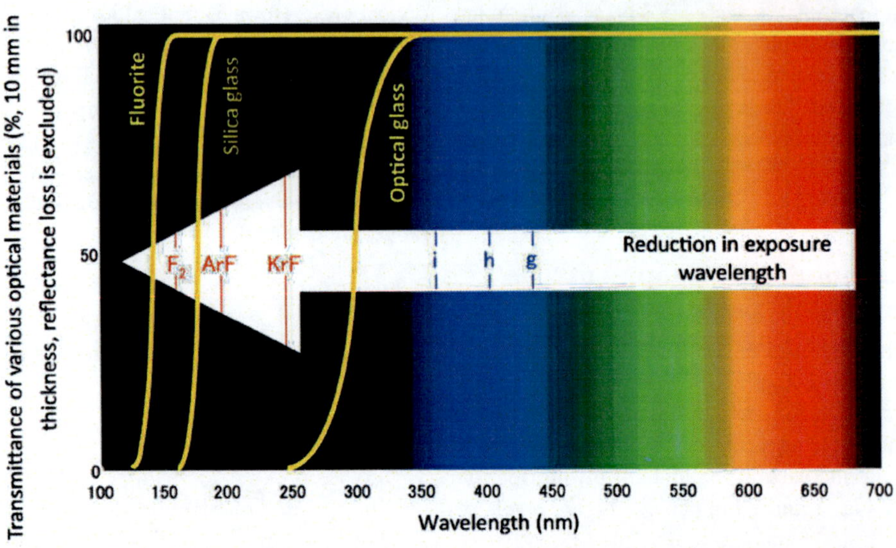

Fig. 17.2 Changes in exposure wavelength. Exposure wavelength of the stepper starts from the g-line, and optical glass cannot be used when the wavelength is shorter than ArF, KrF (g-line: 436 nm, i-line: 365 nm, KrF: 248 nm, ArF: 193 nm, F_2: 157 nm)

a long period. Figure 17.3 shows silica glass used to manufacture projector lenses. As the picture indicates, the maximum outer diameter of lenses exceeds 300 mm. Such large components have to be manufactured with extremely high homogeneity (refractive index, transmittance).

17.1.2.2 Manufacturing Method

The "oxygen–hydrogen flame hydrolysis method" is one of the typical manufacturing methods of silica glass for projector lenses. It is manufactured by using the equipment shown in

Fig. 17.4, using a high-purity silicon compound ($SiCl_4$, etc.) as the raw material. In this method, contamination from the environment is limited because the compound surface is protected by flame and the impurity content is extremely low because the glass does not contact any container. In addition, the method produces highly homogeneous glass because the glass is produced while it is rocked back and forth and side to side. During the final process step slow cooling is performed to remove thermal stress and enhance the quality of the lens material.

Fig. 17.3 Silica glass for projector lenses. The maximum outer diameter of silica glass intended for projector lenses exceeds 300 mm. High quality material of this size has to be produced

Manufacturing methods of fluorite for projector lenses are produced by the "Bridgeman–Stockbarger method." This method employs a crystal growing furnace having a high-temperature chamber and a low-temperature chamber for production (Fig. 17.5). Calcium fluoride, the raw material in powder form, is placed inside the crucible and is turned into melt (liquid) in the high-temperature chamber (at a temperature higher than the melting point of calcium fluoride). Then the crucible is moved slowly toward the low-temperature chamber (at a temperature in which the melt is solidified) to grow crystals from the bottom. Crystals with extremely high purity are grown because of a phenomenon that excludes impurity (segregation). Then slow cooling is performed for the same purpose as for silica glass.

Silica glass and fluorite manufactured in the methods explained above go through precise physical property measurements and strict inspections required for optical materials before they can become projector lenses.

17.1.3 Future Prospects

High levels of manufacturing technologies have been established for the silica glass and fluorite intended for projector lenses used in apparatus for semiconductor lithography.

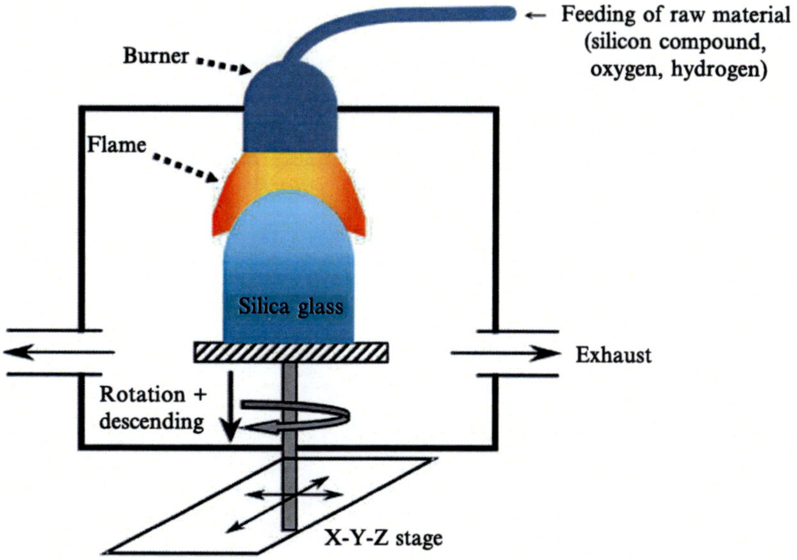

Fig. 17.4 Silica glass manufacturing method. The oxygen–hydrogen flame hydrolysis method. The raw material is fed into the flame of a burner to produce silica glass by hydrolysis reaction and deposition at high-temperature

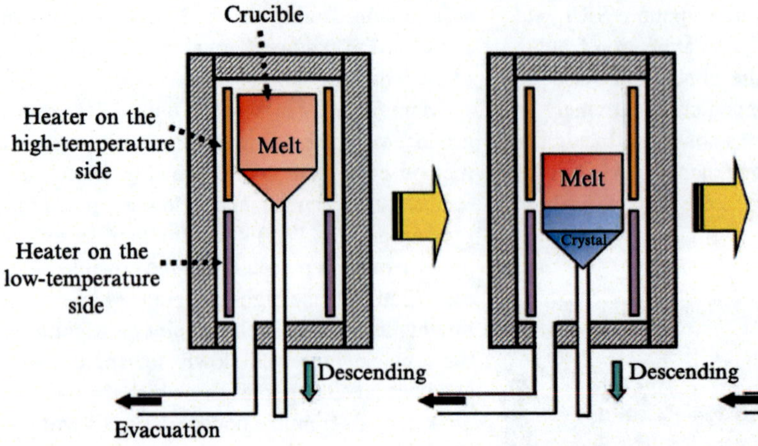

Fig. 17.5 Fluorite manufacturing method. The Bridgeman–Stockbarger method. The crucible descends from the high-temperature chamber to the low-temperature chamber to grow crystals

However, optical designers of projector lenses are demanding higher quality and properties, and further efforts for developments and improvements will be required as long as the optical lithography exists. Cost reduction by improvement of manufacturing efficiency is also a big issue which needs to be resolved.

High-Purity SiC Ceramic Components Used in Semiconductor Thermal Treatment Equipment (1977) 17.2

SiC ceramics have been used in the thermal treatment equipment of semiconductors for more than 30 years. The material was originally used in tube furnaces to replace quartz process tubes but the applications expanded largely following developments of technologies for creating high-purity base materials and high-purity SiC-CVD films. These developments were made possible due to its stability at high temperatures up to 1,200°C, the high purity equivalent to quartz products and the thermal expansion coefficient close to that of Si wafers. These characteristics matched the conditions used in semiconductor LP-CVD processing equipment.

Further development of SiC components, responding to the increase in wafer diameter (ϕ450 mm), with added value (such as high-precision processing) will be required in the future.

17.2.1 Background of Development

In the 1970s, quartz components were the main material used in the thermal treatment equipment of semiconductors. SiC components were first used in tube furnaces to replace quartz reaction tubes. It was because SiC, featuring high thermal resistance and thermal conductivity, was suitable for the purpose. Since the 1980s, when materials of high purity were developed, the material has been used in boats that carry wafers (Fig. 17.6).

Following the increase in size of silicon wafers to 6–8 in., conventional horizontal type thermal treatment furnaces were replaced by vertical type thermal treatment furnaces. The vertical type was superior in terms of uniformity of heat inside the diffusion furnace and was expected to reduce thermal slip of silicon wafers (crystal defects). The ratio of SiC components is higher in the vertical type thermal treatment furnace. This is partly because thermal deformation at a high temperature of 1,200°C is smaller in the SiC components when compared with quartz components. When they are free from deformation, silicon wafers can be loaded automatically, leading to an improvement in productivity.

Following the development of a technology to coat the base material with a high-purity CVD-SiC film (Note 17.3), the replacement of quartz components with SiC ceramic components was promoted. Until that time, SiC components were mainly used in high-temperature thermal treatment, but the components began to be used in the low-temperature low pressure CVD (LP-CVD) processes (Note 17.4) because of the abrasion resistance against hydrofluoric acid, thermal expansion coefficient, etc. of the SiC components. This is because they were expected to be more effective than quartz components in the reduction of particles (Note 17.6).

SiC components are widely used in the thermal treatment equipment for semiconductors largely because of two reasons: (1) high purity and (2) characteristics of SiC components were more suitable for the series of thermal treatment processes than those of quartz components (refer to Fig. 17.7 and Table 17.1).

Note 17.3 A high-purity SiC film formed on the surface of base material by the Chemical Vapor Deposition Method. The film prevents the diffusion of metallic impurities from the base material.

Note 17.4 A process in LSI production that employs bonding a SiN or poly-Si film (Note 17.5) on a wafer by the CVD method. It is called the LP-CVD method because the CVD process is generally performed under a low pressure of 10–1,000 Pa (0.1–10 Torr).

Note 17.5 Polycrystal Si film (for information, wafers are single crystal Si). Used as gate electrodes in LSI production. The film is formed at 500–600°C, using SiH_4 (monosilane) gas.

Note 17.6 In LSI production, particles are largely the cause for device failures that in reduce yield. In the LP-CVD process for bonding SiN or poly-Si films on wafers, SiC, which features thermal expansion coefficient closer to wafers compared with quartz, is expected to reduce particles that result from the peeling of films.

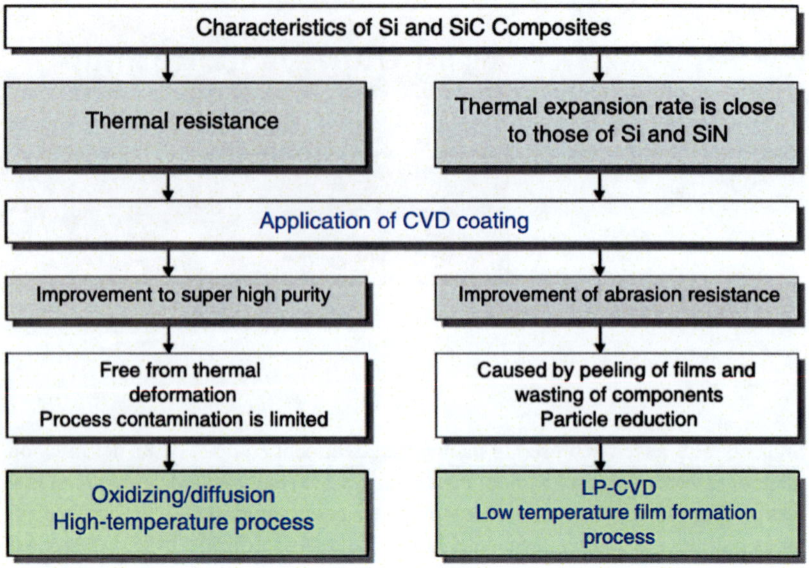

Fig. 17.6 Characteristics of high-purity SiC and its application to semiconductor processes. SiC components are widely used in semiconductor thermal treatment equipment largely because they feature high purity and their characteristics are more suitable for the series of thermal treatment processes when compared with quartz components

Fig. 17.7 SiC products used for semiconductor thermal treatment equipment. On the *front side* is a horizontal type boat for wafer heat treatment, and behind it are components for vertical types. From *right* to *left*, they are a furnace tube, a vertical boat and an example of a mounting method

Table 17.1 Mechanical and thermal properties of SiC compared to quartz

Item[a]	Unit	Reaction bonded SiC	Quartz
Bulk density	g/cm^3	3.02	2.2
Specific heat	J/(g K)	0.69	0.70
Bending strength	MPa	280	105
Thermal expansion coefficient[b]	10^{-6}/K	4.3	0.5
Thermal conductivity	W/(m K)	175	<2
Electrical resistance	Ω cm	10^{-1}	10^{18}
Thermal shock resistance, ΔT	°C	400	900

They were originally used as furnace tubes because of high thermal conductivity. However, the application was expanded to semiconductor thermal treatment because the thermal expansion coefficient is larger than quartz by one digit and is close to that of SiN and Si

[a]Values other than the thermal expansion coefficient were obtained at room temperature

[b]Average values obtained in temperatures from room temperature to 1,000°C

17.2.2 Characteristics and Manufacturing Method

17.2.2.1 Characteristics of Products

The SiC material used in the semiconductors thermal treatment equipment is reaction bonded SiC, a composite made of metallic Si and SiC (refer to Fig. 17.8). There are other types of sintered SiC, but they are made by adding sintering additives and are not easily improved in terms of purity. It is easy to attain high purity in the reaction bonded SiC, which is manufactured by impregnating the purified base material with high-purity metallic Si. In addition, the reaction bonded SiC, in which the strength of the bonded areas is enhanced to the level of the strength of the base material in the process of Si impregnation, allows production of complicated components. The reaction bonded SiC was selected for use in the thermal treatment of semiconductors because (1) it can be purified to a high level and (2) it is comparatively easy to process into various shapes.

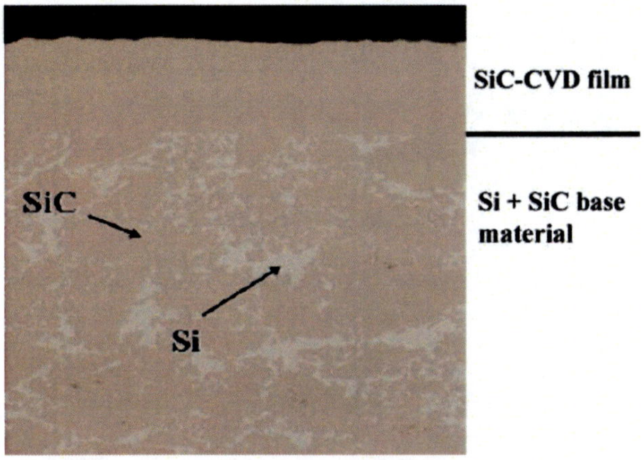

Fig. 17.8 Microstructure of reaction bonded SiC (with CVD film on surface). The reaction bonded SiC base material is a composite material containing 5–20% of Si. The *colored area* indicates SiC and the *white area* indicates Si

17.2.2.2 Manufacturing Method

Figure 17.9 shows the manufacturing method of a general reaction bonded SiC material. The raw material, α-SiC powder of comparatively high-purity, is mixed with a binder. The mixing method of the raw material and the binder is selected depending on the forming method. The forming methods include hydrostatic molding, cold isostatic pressing (CIP forming), casting and extrusion forming. The formed body is calcined in inert atmosphere. The calcined bodies are processed into components, which are bonded to form various shapes such as boats. The assembled components are then purified at a high temperature and then are impregnated with

Fig. 17.9 Manufacturing process of high-purity reaction bonded SiC

high-purity Si (reaction sintering). The material is called reaction bonded SiC because the carbon (C) in the binder is made to react with Si.

The base material impregnated with Si becomes dense and strong because of the secondary SiC generated by the reaction and accumulation of Si within the voids. Then, grooves for mounting silicon wafers are processed using diamond tools. The products are delivered after going through washing and other processes. Some of them are coated with a high-purity SiC film by the CVD method, depending on the application. At the final stage, dimensions, etc. are inspected before the products are packed and delivered.

Fig. 17.10 Ends of grooves chamfered by machining. It is expected that particles generated by friction are reduced by chamfering the ends of grooves upon which silicon wafers are mounted

17.2.3 Future Prospects

Silicon wafers with a diameter of 12 in. (ϕ 300 mm) are currently the main stream. Therefore, components used in thermal treatment are becoming larger, and they are required to be higher in purity and accuracy responding to scaling-down of integrated circuits.

Figure 17.10 shows the edges of grooves for mounting silicon wafers. The edges are chamfered by mechanical machining. Recently there has been an increase in the need for value-added products, such as the ones explained above. The edges were chamfered to reduce particles generated by friction. This is because recent silicon wafers are larger in size and are therefore heavier in terms of weight. In the future, when the size has increased to 18 in. (ϕ 450 mm), even more accurate processing will be required.

Utilization of conventional SiC components used for semiconductor thermal treatment was expanded due to the advantage (characteristics) of the material itself. Further expansion, however, cannot rely on the advantage of the material itself. The level of added value will be the key point in expanding applications. The precision processing explained earlier is essential in this respect. It may be no exaggeration to say that future progress depends on precision processing.

Ceramic Heaters for Semiconductor Manufacturing Equipment (1997)

17.3

Semiconductor devices are produced by using semiconductor manufacturing equipment through the repetition of film formation and etching processes of a semiconducting wafer. In the process of producing devices, corrosive halogen plasma gases are used. The durability of the parts is therefore important. In addition, temperature of the heater on which wafers are directly placed needs to be uniform across the wafer to achieve films of uniform thickness. Aluminum nitride, featuring high resistance to corrosive gas and high thermal conductivity, is a suitable material for the heater on which Si wafers are directly placed for heating within the chemical vapor deposition (CVD) environment (Note 17.7).

17.3.1 Background of Development

Semiconductor devices, which are used in PCs and mobile phones, have also been used in flash memory, etc. in recent years and such applications are on the increase. Responding to the needs of new products, technologies for highly-integrated and high-speed semiconductors continue to advance. Currently, devices incorporating Cu interconnection featuring the interconnection design rule or half-pitch metalized line spacing of 45 nm [which is an index for LSI (large scale integration) miniaturization] are being mass-produced.

Although semiconductors are advancing to achieve higher integration and miniaturization, semiconductor devices are basically made by repeating microfabrication processes, including thin film formation on the Si wafers by the lithography method (Note 17.8), the dry etching method (Note 17.9), and the CVD method, etc. The technology to control the Si wafer temperature is believed to be the key technology that affects the success of microfabrication. Accurate temperature control of the wafer at high-temperatures, which were not previously required, are needed in the thin film formation process. In view of maintenance frequency and COC (cost of consumable parts) reduction, resistance that prevents temporal change under the corrosive gas environment is required. Responding to these circumstances, improvements including the change of materials have been made to the semiconductor manufacturing equipment (particularly to the parts inside the chamber).

Note 17.7 A method for depositing a solid material on a substrate from a gas reactant by a chemical reaction. The gas reaction is induced by means of thermal energy that can be enhanced by plasma.

Note 17.8 A photosensitive substance that is partially exposed then dissolved with a solution to create patterns. The technology is utilized to create layers for printing on semiconductors and integrated circuits.

Note 17.9 The material is etched by reactive gas, ions and radicals in this method. Normally etching is done by a chemical reaction of a solid with a reactant to form a gaseous product.

17.3.2 Characteristics and Manufacturing Method

17.3.2.1 *Characteristics and Specifications of Products*

The ceramic heater developed by NGK Insulators, Ltd., Japan uses aluminum nitride (AlN) as the ceramic material. Table 17.2 shows the characteristics of typical AlN materials. AlN features high corrosion resistance to halogen gases that are normally used in the semiconductor manufacturing process. In the CVD process, corrosive NF_3, CF_4, ClF_3, etc. are generally used as the cleaning gas for the removal of by-products generated during film formation. Figure 17.11 shows the results of a corrosion test that were performed by NGK using NF_3 plasma. The figure indicates that AlN is more resistant to halogen gases such as NF_3, compared with other materials such as silicon, aluminum, quartz and SiC. In addition, AlN features high thermal conductivity and is suitable as a heater material.

The cylindrical part (shaft), which is made of AlN and supports the AlN heater plate, forming a gas tight joint. Figure 17.12 shows the appearance of a ceramic heater

Table 17.2 Material properties of aluminum nitride

Material		Aluminum nitride	
Characteristics		High thermal conductivity type	High-purity type
Purity	%	95	99.9
Bulk density	g/cm^3	3.33	3.26
Open porosity	%	0	0
Thermal conductivity	(W/(mK))	170	90
Specific heat	J/(kg K)	750	750
Coefficient of linear expansion	$\times10^{-6}$/°C	5.7 (40–1,000°C)	5.7 (40–1,000°C)
Bending strength @ RT	MPa	400	360
Young's modulus	GPa	300	300
Fracture toughness, K_{IC}	MN/m$^{3/2}$	3.4	2.3
Thermal shock temperature, ΔT	°C	500	350
Melting point	°C	>2,000	>2,000

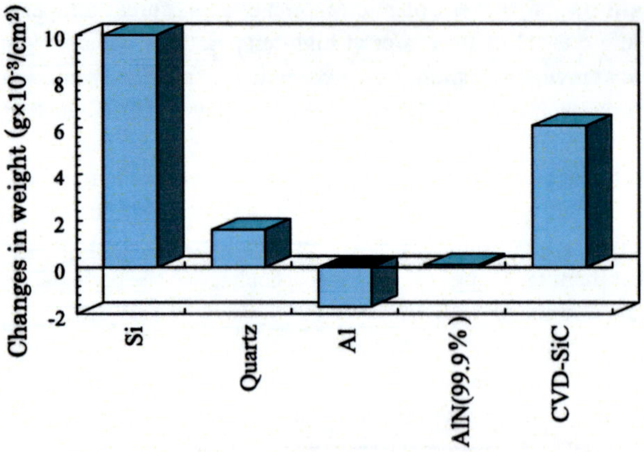

Fig. 17.11 Corrosion resistance to fluorine plasma gas (gas: NF_3 plasma output: ICP800W test temperature: 200°C). Aluminum nitride features high corrosion resistance to fluorine plasma gas as noted by its negligible mass change in fluorine gas

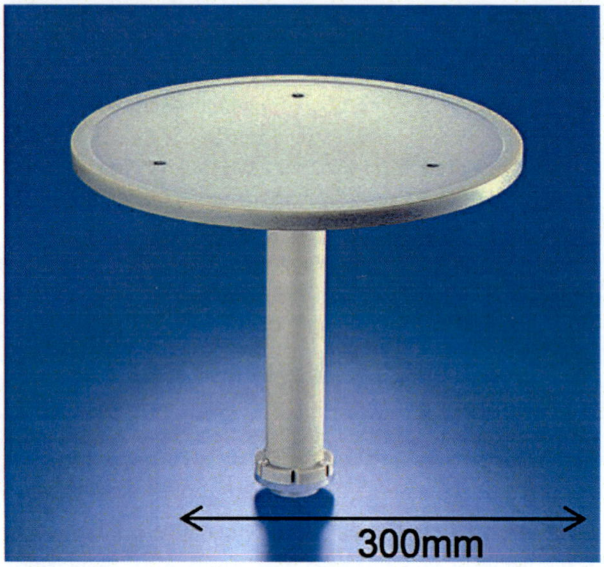

Fig. 17.12 Appearance of a ceramic heater

mounted with a shaft, and Fig. 17.13 shows the outline of the structure of plasma CVD equipment. The metal terminal that supplies current and voltage to the heating element and high-frequency electrodes can be positioned inside the ceramic shaft under this system. By incorporating the structure in which the ceramic heater plate and the ceramic shaft are integrated, exposure of metal parts in the chamber was avoided, protecting the metal parts perfectly from corrosive atmosphere. Figure 17.14 shows temperature distribution on the heater. The temperature is evenly distributed, which indicates the heater is suitable for the film formation process incorporating the advanced CVD method. The heater can handle film formation at temperatures up to 800°C.

17.3.2.2 Manufacturing Method

The AlN heater that NGK Insulators, Ltd., Japan has developed consists of a metallic heating element inside an AlN plate and co-fired. The resistive heating element can be buried inside the ceramic plate accurately and flawlessly by using the hot press method to sinter the ceramic plate. In addition,

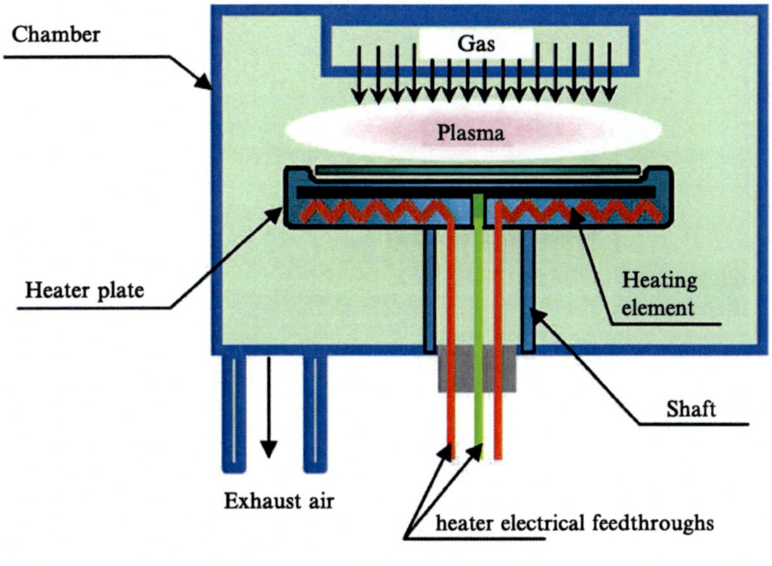

Fig. 17.13 Outline of the structure of plasma CVD equipment. The Si wafer is placed on the heater and plasma is formed directly above the heater to deposit material on the wafer

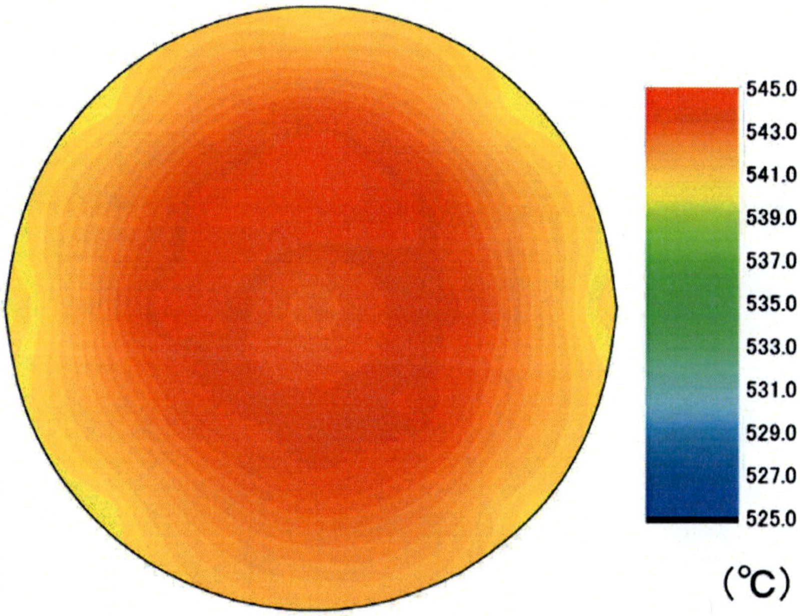

Fig. 17.14 Temperature distribution on the ceramic heater. The temperature range at 540°C is approximately 2°C

plate-shaped high-frequency electrodes and electrostatic chuck electrodes as well as multiple heating elements and electrodes can also be buried by employing the newly-developed forming and sintering technologies. Because of the technologies, a piece of ceramic plate is equipped with various functions such as the high-frequency electrode function for the generation of plasma, the electrostatic chuck function and the heating function capable of adjusting temperature distribution in multiple areas in addition to the heating function, realizing a very compact and high-performance heater.

17.3.3 Future Prospects

Following the adoption of a ceramic heater, durability, reliability and operability of the heater mounted on the CVD device have been improved remarkably. They are currently widely used for heating the wafers in the semiconductor manufacturing process. The production amount of semiconductors is estimated to increase in the future and the heaters are expected to be used as parts that support the production.

Aerostatic Bearings (Air Slides) (1980) 17.4

The aerostatic bearing used to guide precision measuring instruments and ultraprecise processing machines is a type of sliding bearing that uses compressed air as the lubricant. An aerostatic bearing is composed of a fixed guide axle and a slider that moves along the axle. A pressurized gas fills the gap between the guide axle and the slider by compressed air supplied from the holes on the slider, lifting the slider from the guide axle. The aerostatic bearing is free from the deterioration of accuracy due to wear and features stable positioning accuracy because it is a non-contact bearing. It can be operated by a small driving force and is used widely in devices that need to be of high positional accuracy and reliability.

17.4.1 Background of Development

Sliding bearings utilizing lubricant oil are generally used as guide elements of machine tools. Meanwhile, the aerostatic bearings used as guide elements in precision measuring instruments and ultraprecise processing machines are also sliding bearings but they use compressed air instead of oil as the lubricant.

Higher accuracy and speed are required of the mechanical systems such as the reticle (Note 17.10) scan stage and the wafer scan stage used in the semiconductor exposure equipment. Therefore, aerostatic bearings are widely incorporated in conventional optical steppers (Note 17.11) used in atmosphere.

The extreme ultraviolet (EUV) (Note 17.12) and electron beam lithography (EBL) (Note 17.13), candidates for the next-generation light sources, are used in the vacuum environment, and therefore, ball bearings are incorporated in the X–Y stage. However, the guide elements in ball bearings tend to wear and deteriorate resulting in problems with positioning accuracy and also cause trouble that directly affected the exposure accuracy, such as dust and heat due to wear. The stages used in instruments require high speed for increased throughput and the stage device incorporating ball bearings can no longer respond to these needs. Because of these reasons, application of hydrostatic bearings has been expanding to semiconductor exposure apparatuses.

17.4.2 Characteristics and Manufacturing Method

17.4.2.1 Characteristics and Specifications of Products

Stainless steel, aluminum, stone, ceramics are used as the structural material of hydrostatic bearings. Ceramics are used mainly as the material of the guide axle.

Note 17.10 A photo mask used to expose circuit patterns on wafers in the semiconductor device manufacturing process. The reticle is basically a negative (original plate) used to print circuit patterns on wafers. Chromium patterns are created on substrates made of clear silica glass. Electron beam lithography, etc. is sometimes used for production.

Note 17.11 A reduced projection exposure device, a kind of semiconductor manufacturing equipment. The name comes from the projection exposure process, in which reticle patterns are reduced and project-exposed on wafers with projector lenses while shifting (stepping) the patterns over the wafers (which are coated with a photosensitive resist) in the process of printing circuit patterns on silicon wafers. An aligner moves the reticles and wafers into the proper position, and a scanner is used for exposure of the wafers with a reticle.

Note 17.12 Ultraviolet rays featuring short wavelength (approximately 10 nm) are used in the semiconductor (IC, LSI) exposure processes, responding to the recent miniaturization and densification. In the early days, g-line (436 nm) was used as the light source. Following the miniaturization, the wavelength was shortened from g-line to i-line (365 nm), KrF excimer laser (248 nm), ArF excimer laser (193 nm) and finally to F_2 excimer laser (157 nm). For shorter wavelengths, immersion exposure is also performed.

Note 17.13 A lithography exposure process that employs electron beams for semiconductor production (IC, LSI) is called the e-beam lithography (EBL). Accelerated electron beams are also used in familiar products such as picture tubes and discharge display devices of plasma TVs, and is used in electronic microscopes, etc. within the research field. It is used to manufacture semiconductor masks and systems, which are used to create fine patterns by direct drawing on wafers.

Characteristics of ceramics and aerostatic bearings are listed below, with respect to the ceramic air slide (Fig. 17.15) and the ceramic X–Y stage (Fig. 17.16).

17.4.2.1.1 Characteristics of Ceramics

1. Thermal expansion coefficient is 1/2 of metals and the stability against temperature changes is high.
2. The stiffness is typically 1.5 times higher than metals (such as stainless steel), indicating high resistance to deformation.
3. Hard and difficult to machine, but suitable for high-accuracy processing.
4. Free from rust, which contributes to easy maintenance of devices.
5. Even if the material is dented, the dent is usually concaved, which ensures position accuracy to some extent.
6. The specific weight is 1/2 or less of metals, contributing to the weight reduction of devices.
7. Nearly free from secular change.

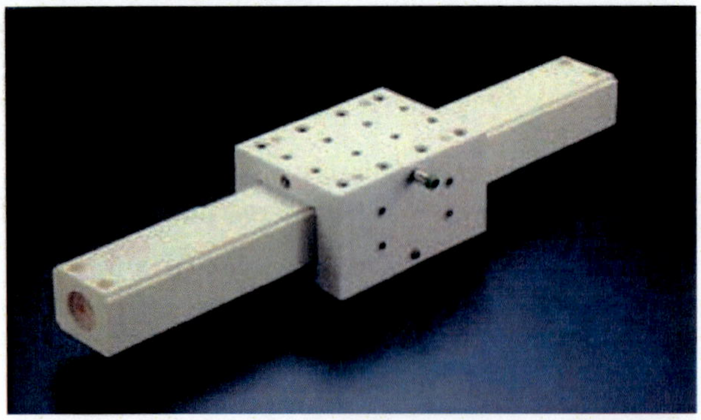

Fig. 17.15 Ceramic air slide. A single-axle air slide incorporating alumina ceramics is the basic component of 3D measuring machines, etc.

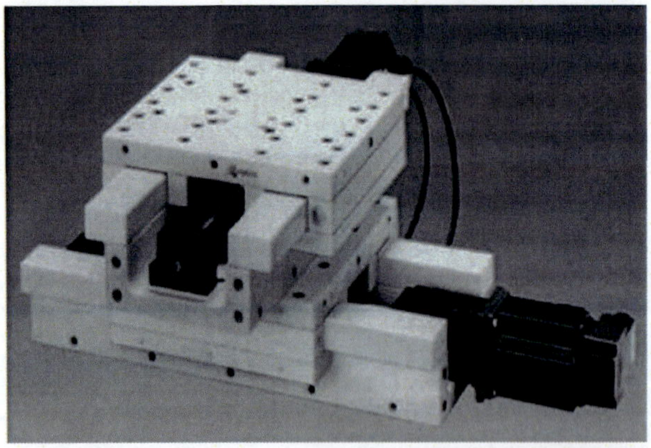

Fig. 17.16 Y–Y stage made of ceramics. An example of a double-axle X–Y stage incorporating alumina ceramics is used in semiconductor manufacturing equipment, etc.

17.4.2.1.2 Characteristics of Aerostatic Bearings

1. Compressed air is supplied, not requiring lubricants such as oil.
2. Does not require maintenance work.
3. Free from stick-slip.
4. Non-contact and free from accuracy deterioration due to wear.
5. Free from friction and can be operated by a small driving force.
6. The air is blown toward the sliding portion, which means no sucking of dust and stable accuracy.

As explained above, the aerostatic bearings are superior to others in terms of straightness, positioning and low maintenance, as well as are suitable for use in devices that need to be high in accuracy and reliability.

17.4.2.2 Manufacturing Method and Structure

The majority of aerostatic bearings are the restrictive type, which is shown in Fig. 17.17. It is composed of a fixed guide axle and a slider that moves along the axle. A compressed gas lubricating layer is created in the gap between the guide axle and the slider by compressed air supplied from the holes on the slider, lifting the slider from the guide axle. The air is blown out through small holes, grooves or pores, which are generally called "restrictors." The grains on the bearing surface are classified into the "inherent restrictors," the "orifice restrictors," the "porous restrictors," and the "surface restrictors," etc. The characteristics vary depending on the type, and an appropriate type is selected for each purpose. Numerical calculations by the finite element method, etc. is performed for designing the restrictors, aiming to enhance stiffness.

Because of the structure of aerostatic bearings, all of the supplied compressed air is discharged via the gap along the bearing surface, as explained above. Therefore, in order to use the aerostatic bearings in a vacuum environment, they need to incorporate air evacuation capability into the bearing. A new concept hydrostatic bearing, which incorporates a differential evacuation mechanism to address the issue, is introduced below (Fig. 17.18). Figure 17.18 shows a common air pad with multiple restrictors. Outlet pores (orifices) are created in the central area of the air pad and surface restrictors are created in a reticular pattern. The compressed

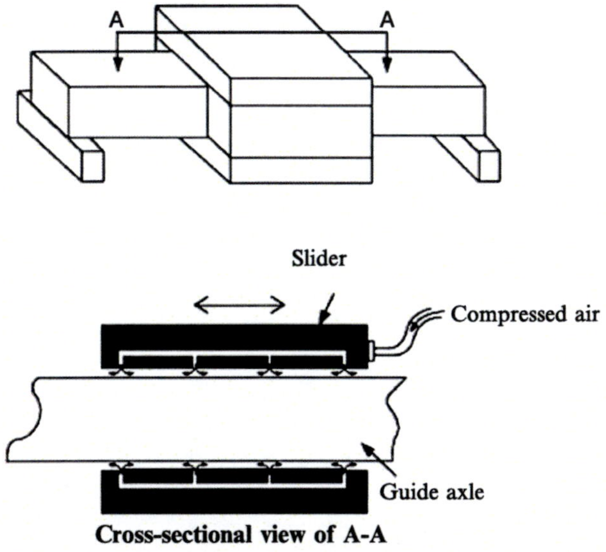

Fig. 17.17 Aerostatic bearings. Appearance and cross-sectional view of a standard type aerostatic bearings show a guide axle and a slider that moves along the guide axle. The slider is lifted away from the guide axle by the compressed air supplied from holes in the slider

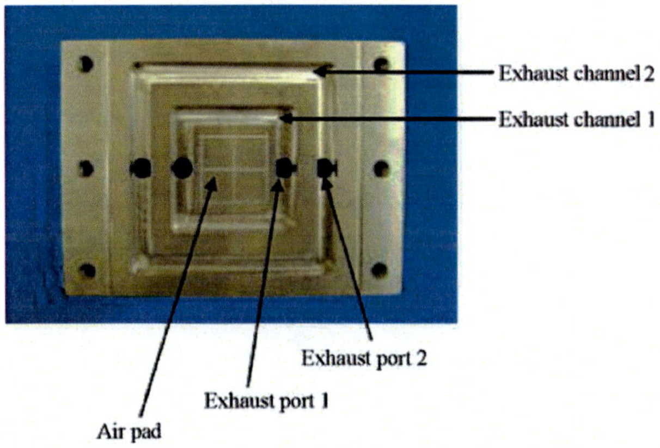

Fig. 17.18 An air pad applicable for use in a vacuum environment. The air pad applicable to the vacuum environment has double exhaust channels surrounding the air pad. The supplied compressed air flows into the exhaust channels and is exhausted out of the exhaust port by an external vacuum pump

air in the reticular pattern area generates a lifting force against the fixed axle. The supplied compressed air flows into exhaust channel 1 and is discharged out of exhaust port 1 by an external vacuum pump. The gas that flows into exhaust channel 2 on the second stage is discharged out of exhaust port 2. The double exhaust channels that surround the air pad are called guard rings.

Fig. 17.19 Hydrostatic bearing applicable to the vacuum environment. It has sliders composed of four panels; a top panel, a bottom panel and side panels, which are positioned on the air pad applicable to the vacuum environment. The exhaust port for the discharging of the supplied air via the exhaust pipe in the fixed axle is mounted on the edge of the fixed axle

17.4.3 Future Prospects

Responding to the improvements in performance and functions of large scale integration circuits (LSIs), the enhancement in memory capacity and the increase in functions of logistic devices, the LSI pattern has been scaled down at a rate of 70% every 3 years. The optical lithography technology allows resolution of patterns up to the size of approximately 1/2 of the light source wavelength, and light sources with shorter wavelengths are used more frequently. For example, the KrF excimer laser and the ArF excimer laser feature wavelengths of 248 and 193 nm, respectively, which allows microfabrication up to a 100 nm pattern. For patterns smaller than 100 nm, the F2 excimer laser, EUV, EBL, etc. are used currently.

Meanwhile, the hydrostatic bearing applicable to the vacuum environment (Fig. 17.19), which satisfies the specifications required of the mechanical systems (high accuracy, high speed, dust-free, low thermal generation), has sliders composed of four panels, top panel, bottom panel and side panels, positioned on the air pad applicable to the vacuum environment (Fig. 17.18), and the exhaust port for discharging the supplied air via the exhaust pipe in the fixed axle is mounted on the edge of the fixed axle. It is mainly used as a transfer arm in the vacuum environment. Double-axle type X–Y stages applicable to the vacuum environment have also been developed and they will continue to be used in wider fields as the basic components in the next generation of semiconductor exposure apparatuses.

Electrostatic Chuck (1990) 17.5

Electrostatic chucks are the core component of semiconductor manufacturing equipment and are used in the large-scale integration (LSI) manufacturing process to produce integrated circuits by immobilizing Si wafers on wafer stages in a vacuum (Fig. 17.20). Practical use of ceramic electrostatic chucks began in the 1980s. The superior thermal resistance, abrasion resistance and chemical stability of ceramics were essential for electrostatic chucks, which are used repeatedly under the severe processing environment. Ceramic electrostatic chucks were commercialized in the 1990s for use as stages in dry etching equipment such as high density plasma–chemical vapor deposition (HDP–CVD) (Note 17.14) equipment and physical vapor deposition (PVD) (Note 17.15) equipment. The types of equipment that require ceramic electrostatic chucks increased following the advance in LSIs and technological innovation in semiconductor manufacturing equipment. Ceramic electrostatic chucks are currently used in various semiconductor manufacturing equipment that involve processing in a vacuum, such as ion implantation, ashing, and wafer inspection.

17.5.1 Background of Development

The most important roles of electrostatic chucks in semiconductor manufacturing equipment are the temperature control of Si wafers and particle (Note 17.16) prevention during the manufacturing process. The Si wafer temperature during process varies depending on the application and ranges from room temperature to 600°C. Dry etching is explained here as an example. A plasma is generated with several thousands of watts of power during the dry etching process above the wafer to remove material from the wafer surface. The wafer temperature needs to be controlled at about 80°C by transmitting the plasma heat from the Si wafer to the refrigerant effectively and evenly. This is needed because the quality of etching depends on the temperature during the etching process (Fig. 17.21).

In older etching equipment, the Si wafer was fixed on the stage by physically pressing the wafer edge against the stage. However in this method, only the edge region in contact with the stage transfers heats properly and the temperature was lower in the outer circumference of the Si wafer. More precise control of the Si wafer temperature with greater uniformity was required following the miniaturization of LSI wiring patterns, which led to utilization of electrostatic chucks.

If an electrostatic chuck is used, the Si wafer can be fixed by electrically attracting it from the back side. This method has the following advantages compared with mechanical attachment:

1. The entire surface of the Si wafer is in physical contact with the stage and Si wafer surface heating from the plasma process is evenly transferred to the refrigerant. Therefore, the temperature of the Si wafer surface can be controlled more evenly.
2. Cooling efficiency is enhanced because the plasma heat is transferred from the wafer surface to its back side due to uniform contact between the chuck and wafer. Therefore, plasmas with higher power intensities can be generated, increasing the throughput (Note 17.17) by shortening the etching time.
3. The Si wafer is attracted to the chuck from its back side, without touching the working surface (front side). Therefore, particles falling on the working surface of the Si wafer can be prevented, increasing the yield rate of semiconductor devices resulting from fewer failed devices.

Note 17.14 A type of processing equipment used to deposit insulation films of SiO_2, etc. between LSI circuit wirings.

Note 17.15 A type of processing equipment used to deposit films for metal wiring in LSIs.

Note 17.16 Fine dust or particles that may cause circuit disconnection. If fine dust falls on a circuit during the semiconductor manufacturing process, a device failure may occur leading to the drop in yield rate.

Note 17.17 Processing capacity per unit time.

17.5.2 Characteristics and Manufacturing Method

17.5.2.1 Characteristics and Specifications of Products

The electrostatic chuck is composed of a ceramic part with an outer diameter nearly equivalent to that of the Si wafer and a disk-shaped metallic plate used for cooling. An internal electrode for attracting the Si wafer is formed inside the ceramic part during manufacturing. When a Si wafer is placed on the electrostatic chuck and voltage is applied to the internal electrode, the Si wafer is electrically attracted and immobilized within vacuum.

Fig. 17.20 Electrostatic chuck. Ceramic electrostatic chucks feature high thermal resistance, abrasion resistance and chemical stability and are used in various semiconductor manufacturing equipment. The ones on top are the products assembled with a metal plate for cooling

Ceramic electrostatic chucks are largely classified into two types depending on the Si wafer attraction method.

The Coulomb type electrostatic chuck has an electrical resistivity of the dielectric layer of 1×10^{16} Ω cm or higher. Coulomb type electrostatic chucks generally incorporate high-purity ceramics such as alumina and aluminum nitride, and therefore, the metal contamination (Note 17.18) is limited. In addition, the time constant for attraction and release of the Si wafer is short, increasing the throughput. Meanwhile, a voltage of 2.0 kV to 3.0 kV needs to be applied to the dielectric layer for chucking.

The second type is the Johnson–Rahbek (J–R) type electrostatic chuck (Note 17.19), in which the electrical resistivity of the dielectric layer is $1\times10^{9-12}$ Ω cm. In the J–R type electrostatic chuck, additives such as metal oxides are added to ceramic materials such as alumina and aluminum nitride to reduce the electrical resistivity. Therefore, particles from the chuck area are a concern for metallic contamination of the wafer. In addition, it needs to be designed by setting the electrical resistivity at an appropriate level, because the time constant for attraction and release of the Si wafer depends on the chuck's electrical resistivity. Consequently, a voltage of 1.0 kV or less is sufficient for chucking.

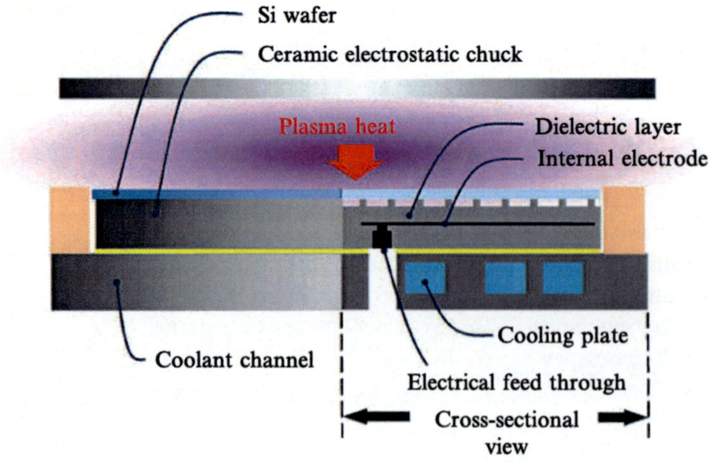

Fig. 17.21 Operation of ceramic electrostatic chucks. Electrostatic chucks are used to fix Si wafers in the vacuum chamber of semiconductor manufacturing equipment. The heat generated by plasma is transferred to the refrigerant that flows inside the cooling plate, maintaining the Si wafer at a constant temperature

Note 17.18 Contamination that is caused by metallic particles which may cause the device to fail which then reduces the yield rate of semiconductor devices.

Note 17.19 Johnson–Rahbek type electrostatic chuck. Electrical resistivity of the dielectric layer is 10^{9-12} Ω cm. The Si wafer is attracted to the chuck when a high voltage is applied to it.

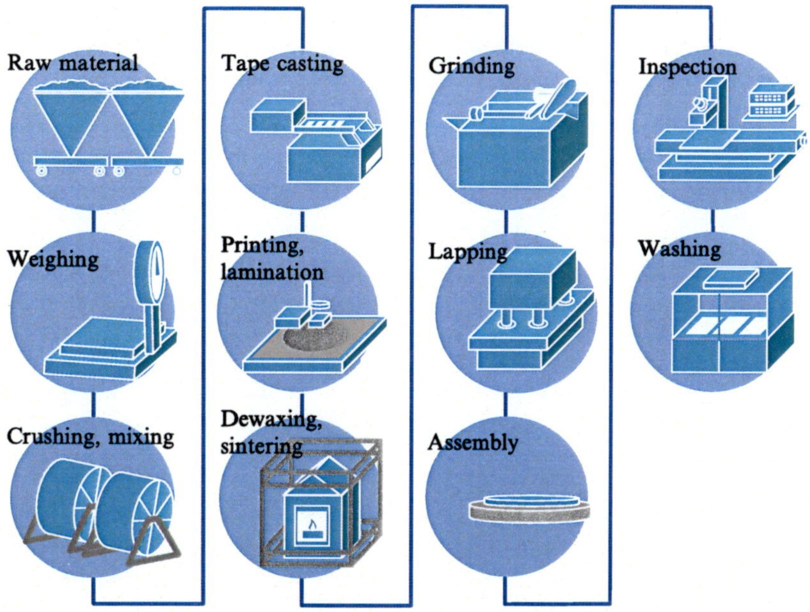

Fig. 17.22 Manufacturing process of ceramic electrostatic chucks. The main material, alumina, is crushed, mixed and tape casted to create a green sheet. Internal electrodes are printed on green sheets, which are laminated, dewaxed and sintered to create a sintered body. After forming, they are assembled with cooling plates before completion of the products

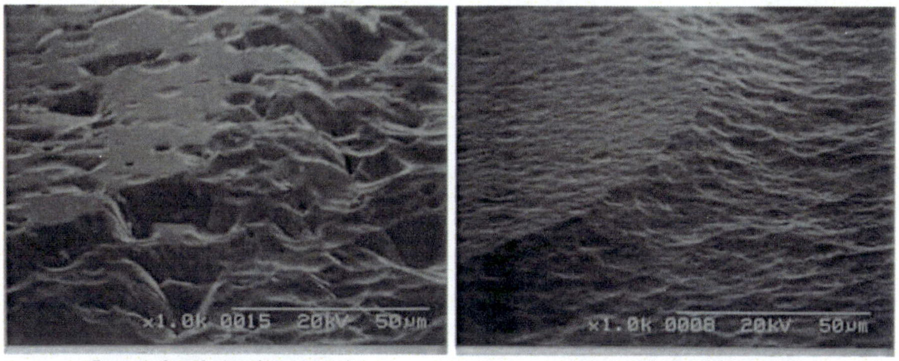

Fig. 17.23 Peripheral plasma etching of ceramics for electrostatic chucks. Conventional ceramics are etched where exposed to plasma generated from advanced dry etching equipment (*left*). Etching is minimal in the advanced materials (*right*)

17.5.2.2 Manufacturing Method

Figure 17.22 shows the manufacturing process of a J–R type electrostatic chuck consisting mainly of alumina. For production of J–R type electrostatic chucks, titanium oxide, etc. are added to the alumina material in order to control the its electrical resistivity to a specified value. The mixture is crushed and mixed to create a green sheet. Multiple green sheets, printed with internal electrodes, are laminated, formed and sintered. A metal electrode is fixed on the electrical feedthrough by brazing. It is assembled with a metal plate for cooling and then goes through precision cleaning intended for semiconductors.

17.5.3 Future Prospects

Electrostatic chucks continue to evolve responding to the advances of semiconductor device technologies. For example, plasma etching resistance was not important for conventional dry etching equipment, but conventional materials could not be used in advanced equipment because of etching by an intense high energy plasma. To solve this issue, ceramic materials featuring higher plasma etching resistance were developed recently for use in electrostatic chucks (Fig. 17.23). However, the resistance of these materials may not be sufficient enough to satisfy the needs of next-generation equipment.

Furthermore, the specification regarding reduction of particles that fall on Si wafers has been increasing. To solve this issue, new surface processing technologies and cleaning technologies for the reduction of particles to the utmost limit will be required.

MMC for LCD Manufacturing Equipment (1990) 17.6

Metal matrix composites (MMC) (Note 17.20) are composite materials made by homogeneously dispersing ceramic reinforcement materials such as silicon carbide in a metal matrix such as aluminum. They are light weight equivalent to aluminum alloy and high rigidity equivalent to iron. They are used widely in Liquid Crystal Display (LCD) manufacturing equipment, semiconductor manufacturing equipment, etc. because the thermal expansion coefficient can be reduced by controlling the types and amounts of the matrix and the reinforcement material.

MMC can be processed into large products exceeding 2 m and are used as movable parts such as the stages of LCD manufacturing equipment responding to the increase in panel sizes.

17.6.1 Background of Development

The market for large liquid crystal panels grew substantially in the 1990s for use in laptop PCs, etc. In the late 1990s, the market of desktop PC monitors grew rapidly following the shift from Cathode-Ray Tubes (CRT). The majority of them are currently intended for TVs, and TV panels have been growing in size more rapidly than PC panels.

For manufacturing of liquid crystal panels, high-efficiency production through increases in the size of mother glass substrates (Note 17.21) and speed of the manufacturing process has been promoted so as to realize increases in panel size and reduction in price at the same time (Fig. 17.24).

Note 17.20 Stands for metal matrix composite.

Note 17.21 Liquid crystal panels are made from a mother glass substrate. Multiple panels are formed on a large mother glass substrate, which is cut into pieces. The cost per panel is reduced by producing multiple panels at one time. Because of this reason, sizes of mother glass substrates continue to increase. At present, the shift from the 9th generation (2,450 mm × 3,050 mm) to the 10th generation (2,880 mm × 3,280 mm) is under way.

17.6.2 Characteristics and Manufacturing Method

17.6.2.1 Characteristics of Products

To realize high-efficiency production of liquid crystal panels, components that constitute LCD manufacturing equipment, moving parts such as stages and tables in particular, are required to be larger than 2 m in diameter and feature lightweight and high rigidity so as to realize high speed and high accuracy.

MMC are composite materials made by homogeneously dispersing ceramic reinforcement materials such as silicon carbide in a metal matrix such as aluminum. They feature a light weight equivalent to aluminum and high rigidity equivalent to cast iron. Fabrication of MMC does not involve change of form such as the sintering shrinkage observed in the production of ceramics, making it possible to create large products exceeding 2 m in diameter. Thermal expansion coefficient can be controlled by adjusting the types and ratio of the matrix to the reinforcement material.

17.6.2.1.1 Characteristics of MMC (Table 17.3)

1. Large size, integration: large sizes exceeding 2 m are possible. The sizes can be increased as units, realizing high processing accuracy (Fig. 17.25).
2. Lightweight, high rigidity: weight of cast iron parts can be reduced substantially because the material features a light weight equivalent to aluminum and rigidity equivalent to cast iron. (specific weight is 1.1 times of aluminum, 0.4 times of cast iron, elasticity is 3.7 times of aluminum, 2.3 times of cast iron, in the case of non-pressure infiltration material).
3. Low thermal expansion, high thermal conductivity: change in accuracy of components due to temperature change can be reduced substantially (thermal expansion coefficient is 0.2 times that of aluminum, 0.6 times of cast iron, thermal conductivity is 1.2 times of aluminum, 3.7

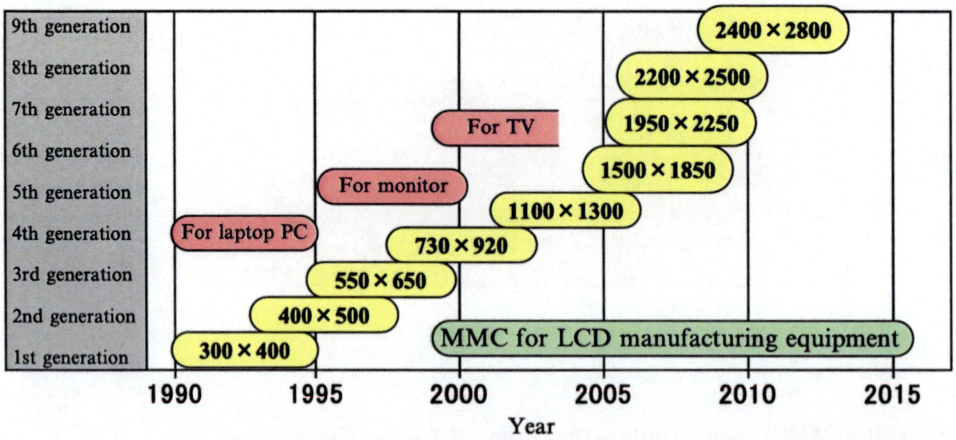

Fig. 17.24 Transition of mother glass substrates for liquid crystal panels. Responding to the rapid increase in size and drop in price of liquid crystal panels, mother glass substrates are becoming larger during the past few years

Table 17.3 Characteristics of typical MMC

Item	Unit	MMC			Comparison	
		Casting	Pressure metal cementation method	Non-pressure metal infiltration method	Aluminum (A5052)	Cast iron (FC250)
Metal	vol.%	Al (70)	Al(35)	Al(30)		
Ceramics	vol.%	SiC (30)	SiC (65)	SiC (70)		
Density	10^3 kg/m^3	2.8	3.0	3.0	2.7	7.3
Elasticity	GPa	125	230	265	71	114
Thermal expansion coefficient	10^{-6}/°C (RT –200°C)	14	8	6	23	10
Thermal conductivity	W/(m K)	150	160	172	138	47

Fig. 17.25 MMC stage for LCD manufacturing equipment. Responding to the increase in size of liquid crystal panels, MMC featuring large size, lightweight and high rigidity is used to manufacture moving components, such as stages, of LCD manufacturing equipment (picture: 1 m × 1 m × 0.05 m, large size of more than 2 m is possible)

times of cast iron, in the case of non-pressure infiltration material).

4. High vibration damping: high vibration damping is realized due to the internal friction attributable to composition of the metal and ceramic reinforcement materials. Therefore, the settling times of components that operate in high-speed can be reduced.

17.6.2.2 Manufacturing Method

MMC is manufactured by casting, metal cementation, etc. (Fig. 17.26).

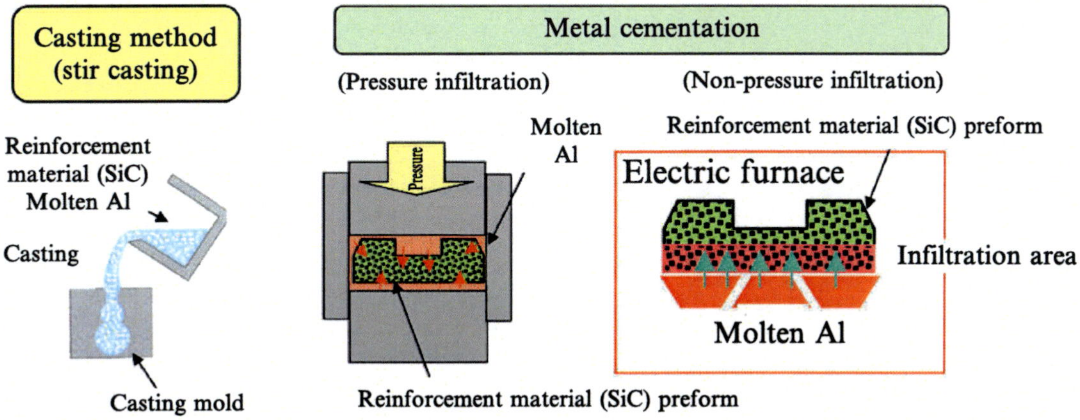

Fig. 17.26 Manufacturing method of typical MMC. MMC is manufactured by casting or metal cementation (pressure infiltration, non-pressure infiltration). In the casting method, reinforcement materials are dispersed in the molten metal, which is filled into cast iron for solidification. Molten metal, the matrix, is infiltrated in a porous preform made of reinforcement materials, etc. In the metal cementation method, the molten metal is infiltrated into a porous preform. There are two methods, the pressure infiltration method where pressure is applied for infiltration and the non-pressure infiltration method where the molten metal is infiltrated spontaneously without pressure application

17.6.2.2.1 Casting (Stir Casting)

Molten metal is stirred while adding and dispersing reinforcement materials in it. It is filled in a cast iron and solidified. It can be made into large and complex shapes. Content of reinforcement materials is limited to 30% or less because fluidity of the molten metal decreases if the percentage increases.

17.6.2.2.2 Metal Cementation

Molten metal, the matrix, is infiltrated in a porous preform made of reinforcement materials, etc. There are two methods, the pressure metal cementation method where the molten metal is infiltrated by pressure application and the non-pressure metal infiltration method where the molten metal is infiltrated spontaneously without pressure application. The non-pressure filtration method was developed by Lanxide of the U.S. Wetting properties of the molten metal and reinforcement materials are improved by adding penetration enhancer to infiltrate molten metal in a preform without pressure application. In both methods, the concentration of reinforcement materials can be increased to 70% in volume, realizing composition into near-net-shapes, depending on the shape of the preform.

17.6.3 Future Prospects

Sizes of mother glass substrates for liquid crystal panels continue to increase. The 9th generation is sized 2,450 mm×3,050 mm, while the 10th generation is sized 2,880 mm×3,280 mm. It is expected that MMC, featuring lightweight and high rigidity, will continue to contribute to the advance in the LCD industry.

Literature

1. Jinbo H (1999) N Glass 14(3):41–46 (17.1)
2. Mizugaki T (2004) Mater Integr 17(11):55–59 (17.1)
3. Horiuchi Y (2007) Ceramics 42:421–425 (17.2)
4. Maeda K (2000) Introduction to semiconductor process. Kogyo Chosakai Publishing Co., Ltd., Tokyo (17.3)
5. Shimura S, Ohashi T, Watanabe K (1999) Fluorination of AlN ceramics, silicon and silica in an inductively coupled NH_3 plasma. Finish Adv Ceram Glasses Ceram Trans 102:341–349 (17.3)
6. Mizutani N, Mizuno T, Ushigoe R et al (2000) Development and commercialization of ceramic parts for semiconductor manufacturing equipment. Fine Ceramic Report, Japan Fine Ceramic Association, Tokyo 18(7) (17.3)
7. Kawasaki K (2004) Application of AlN to parts for semiconductor manufacturing equipment. Ceramics 39(9):688–691 (17.3)
8. Nihon Ceratec Co. Ltd., HP. http://www.ceratech.co.jp/ (17.6)

Part VI

Ceramic Products for Residential Construction Sector

Building Materials and Glass

18

Many different types of construction materials are used in buildings and houses including ceramic-based materials, glasses, ceramics and metals. With respect to materials used in exterior walls, architectural design places a high priority on appearance and durability. Interior materials are also required to be fire resistant, thermal insulators and resistant to high humidity in order to realize a safe and comfortable living environment. Construction materials such as glass fiber products and glass construction materials used in buildings and houses are introduced in this section. Electrical noise absorbing materials used as exterior materials and walls of test chambers are also introduced. They are essential for an IT-based society, where various kinds of electronic devices are used.

ALC (autoclaved lightweight concrete) is produced in a vessel called an autoclave that is used to create high-temperature and high pressure conditions for materials such as cement raw materials. The ALC is a typical exterior wall material in Japan (Sect. 18.1). It features a number of important functional properties including light weight, fire resistance, seismic resistance, thermal insulation and soundproof properties. It is also characterized by high processability and workability. Many of these functions are attributable to tobermorite, which is prepared using foams and formed at high temperature and high pressure. This material was invented in Sweden. It has been improved to match Japanese architecture and will continue to be used as a construction material.

Plaster board (also known as drywall or sheetrock) is used as an interior material. It is environmentally-friendly because industrial by-products such as flue-gas desulfurized gypsum and by-product gypsum are used as raw materials (instead of natural gypsum). Plaster boards also feature functions that realize safe and comfortable living interiors due to its fire resistance, soundproofing and thermal insulatory properties (Sect. 18.2). Plaster boards also prevent the spreading of fires.

Glass wool has been used widely in place of carcinogenic asbestos. The material is used as an insulation material in automobiles and ships, in air conditioning ducts of buildings, as thermal insulations for hot water pipes, as an insulation material in houses and as noise absorbing material in movie theaters (Sect. 18.3). Eighty-five percent of the raw material consists of recycled glass produced from collected bottles. Glass wool is nonflammable as well as features high thermal insulatory properties and is effective for improving cooling/heating efficiency. Thermal insulating ceramic products contribute to CO_2 reduction.

Functional construction materials are comparatively new materials that have been introduced responding to the needs from society. When thermal insulation and air tightness are enhanced in buildings or homes, problems such as high humidity or excessive dryness occur. The sick house syndrome has become a social problem. Interior materials were developed quickly to address these issues. Humidity control properties are explained below in detail. The key point in terms of technology is the microstructure of the material. Taking advantage of the difference in moisture absorption/release from the surface due to the difference in pore diameters, products capable of maintaining humidity at a comfortable level have been produced by controlling pore diameters. Humidity control construction materials unlike air conditioners, control humidity without using energy. They thereby contribute to measures against global warming (Sect. 18.4).

Insulated glass is made by adding heat-absorbing materials and light reflective materials to sheet glass so as to reduce heat generated by sunlight. The double-glazed glass contains a layer of air or a vacuum between layers of glass so as to improve thermal insulation properties (Sect. 18.5). Insulated glass transmits visible light, while the transfer of heat between indoor and outdoor is controlled. Loss of energy in buildings is caused mainly by thermal loss via openings (windows). Insulated glass features high heat shielding/insulation properties and is attracting attention as a construction material that will contribute to energy saving and CO_2 reduction.

Safety glass includes reinforced glass, which is made by heating sheet glass and quenching the surface of sheet glass to form a compressed layer so as to enhance fracture strength,

Y. Imanaka et al. (eds.), *The Ceramic Society of Japan, Advanced Ceramic Technologies & Products*,
DOI 10.1007/978-4-431-54108-0_18, © Springer Japan 2012

and laminated glass, which has a resin layer sandwiched between two glass panels so as to prevent the scattering of glass chips and penetration at the time of fracture (Sect. 18.6). Laminated glass is represented by automobile windshields. Performance has been enhanced by improvements of manufacturing methods and structures. Safety glass was originally developed to ensure safety of people at the time of fracture. The focus of safety glass is now on the prevention of fires and crime.

Glass ceramics used as construction materials are superior to natural marble stone and granite in terms of mechanical and chemical properties and feature high whiteness and diffuse reflectance (Sect. 18.7). They are made by carefully reheating a vitrified material for partial crystallization. They have diffuse reflectance that is twice that of natural stone and thus are capable of reflecting light in all directions. They can be processed to be bendable and are used widely as tough materials that decorate walls and pillars of buildings, offices, subway stations and underground shopping malls.

The ferrite electromagnetic wave absorber absorbs electrical waves or noise effectively (Sect. 18.8). Similar to other ceramic products, it is made by forming and sintering raw material powders that have been mixed completely. The products are processed into tiles. Electronic and electrical devices are common place in modern life. This material is indispensible for use in electromagnetic anechoic chambers that evaluate the electrical noise characteristics of these devices. The material has also been commercialized for use as a measure against broadcast wave reflex disturbance due to high-rise buildings.

Autoclaved Lightweight Concrete (1963)* 18.1

ALC (autoclaved lightweight concrete) is a lightweight foamed concrete, which is made by steam curing at high temperatures and at high pressures. Outside of Japan, it is also known as AAC (autoclaved aerated concrete), gas concrete or cellular concrete.

The ALC technology was introduced to Japan from Europe in 1962 and further developed responding to the needs of Japanese buildings. ALC is a precast construction material (Note 18.1) characterized by lightweight, high fire resistance, earthquake resistance, high thermal insulation characteristic, sound adsorbing and high workability. The material is widely used in exterior walls, partitions, floors and roofs of houses, offices, shops, plants, warehouses and skyscrapers. The high performance characteristics of ALC are attributable to its porous structure composed of a countless air bubbles with a diameter of approximately 1 mm and calcium silicate hydrate (tobermorite) matrix generated by means of autoclave curing. The safety of buildings is enhanced due to the performance of this material as well as due to its construction.

18.1.1 Background of Development

ALC was developed in Sweden in the 1920s (patented by J.A. Eriksson, 1923–1925). The ALC production technology, enhanced through improvements in raw materials and the production process, expanded across Europe in the 1940s. The technology was introduced to Japan in 1962. The characteristics of ALC matched the needs of the times. This was when buildings began to be positioned as industrial products, while the construction of high-rise buildings was promoted as an effective use of land and demands for an improvement in the quality of buildings increased. The technology imported into Japan was improved, utilizing technologies exclusive to Japan. This was done in order to meet the requirements needed for building construction in Japan. For example, a majority of ALC products in Europe are brick-shaped and are similar to stone and brick construction materials. This is because European buildings are built from stones, bricks and mortar. Meanwhile, in Japan, the ALC material was developed to contain steel reinforced prefabricated panels.

18.1.2 Characteristics

18.1.2.1 Manufacturing Method

An example of an ALC manufacturing process is shown in Fig. 18.1. The major raw materials of ALC are silicate and calcite. Silica stone and silica sand are used as the silicate-based materials, while calcite and Portland cement are used as the calcite-based materials. Aluminum metallic powder, a foaming agent, and a surfactant are also used (Fig. 18.1-1A). Plaster is sometimes added. Raw material types and blending ratios vary depending on the manufacturing method.

Raw materials are processed into a slurry (Fig. 18.1-2), which is poured into a mold (Fig. 18.1-3) in which steel reinforcement bars (Fig. 18.1-1B) are arranged. The mixing of materials as well as the viscosity and temperature of the slurry are strictly controlled because they substantially affect the foaming, hardening and cutting step during the post processing. The volume of slurry in the mold swells by twofold due to hydrogen gas generated by the chemical reaction of aluminum metallic powder mixed within the slurry. At the same time, the slurry is hardened due to the hydration of the calcite-based material. The slurry is then removed from the mold when it has partially hardened and can be handled

*The number in parenthesis indicates the year that the product was first commercialized.

Note 18.1 Construction materials that are manufactured at plants and transported to the construction site for the purpose of installation. In contrast, the method in which concrete is poured into fabricated forms built on the construction site is called the cast-in-place method, which is applied during the building of general concrete structures.

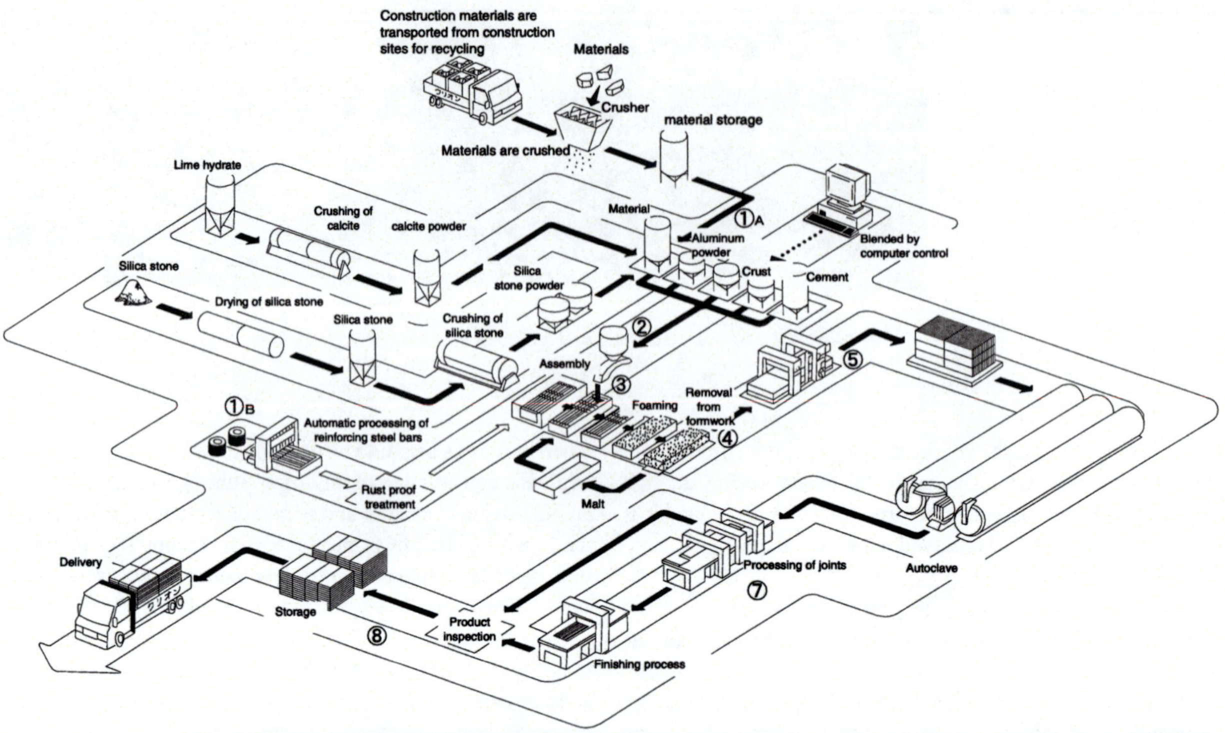

Fig. 18.1 ALC production process. Production of ALC is characterized by the process of creating a porous material by foaming of the raw material slurry and the process of generating stable reactant by means of curing within the autoclave. The production process varies slightly depending on the manufacturer and the plant operation

(Fig. 18.1-4), cut into panels (Fig. 18.1-5) by using a wire cutter and placed inside an autoclave for curing in saturated steam at 180°C for several hours (Fig. 18.1-6). By means of curing in an autoclave, calcium silicate hydrate (tobermorite), the major component of the solid phase, is formed.

Tobermorite is usually expressed by the chemical formula suggested by Megaw and Kelsey, $Ca_5(Si_6O_{18}H_2)\cdot4H_2O$. In actual tobermorite, the Ca/Si ratio varies from 1.0 to 0.7, according to other sources. The Ca/Si ratio of α-C_2S hydrate (Note 18.2), which is generated at the beginning of the reaction, and C–S–H with high Ca/Si ratio (Note 18.3) gradually decreases due to the hydrothermal reaction caused by the dissolution of silicate within the autoclave, changing to tobermorite. Therefore, the reaction is regulated by dissolution of silicate. The crystal growth of tobermorite substantially affects the properties of ALC, and caution is taken in the selection of raw materials to be used (silicate-based material in particular) and in the adjustment of the powders.

Metal joints for mounting are attached (Fig. 18.1-7) after curing and cutting processes are performed but before delivery.

Note 18.2 $Ca_2(SiO_4H)(OH)$.

Note 18.3 Amorphous calcium silicate hydrate. It is a gel-like substance in which the composition ratio of CaO, SiO_2, H_2O varies. It is generated by the hydration of cement in ambient temperature and hydrothermal reaction of CaO and SiO_2.

18.1.2.2 Characteristics of Materials

ALC consists of approximately 20 vol.% solid and 80 vol.% pores. The pores consist of air bubbles of 0.1 mm or larger created by hydrogen gas generation via chemical reaction of metallic aluminum (Fig. 18.2) and voids of 0.1 mm or smaller in the solid phase. The volume ratios of air bubbles and voids are 50% and 30%, respectively. As for the solid phase, 13–16 vol.% is tobermorite (Fig. 18.3) while 4–7 vol.% is unreacted silicate.

The physical properties of ALC depend largely on the bubbles and tobermorite. The bulk density of ALC is approximately 500 kg/m^3 (450–550 kg/m^3 according to JIS), which is approximately 1/4 of concrete. The thermal conductivity is approximately 1/10 of concrete.

Tobermorite shows lower drying shrinkage and higher thermal stability compared with C–S–H in typical concrete.

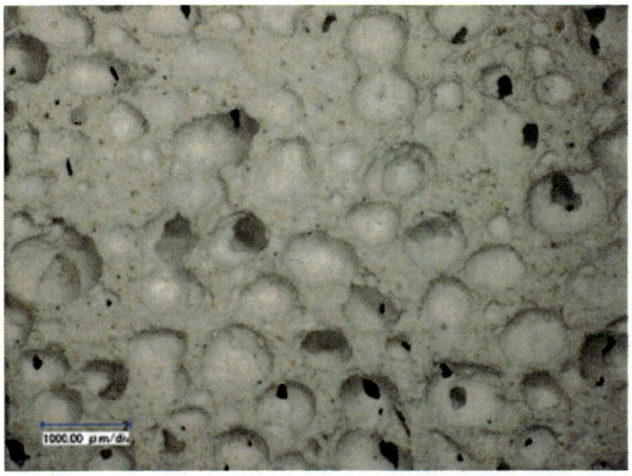

Fig. 18.2 Foams in ALC. Gas foam generated by the foaming agent (aluminum metallic powder) is homogeneously distributed in the ALC panel

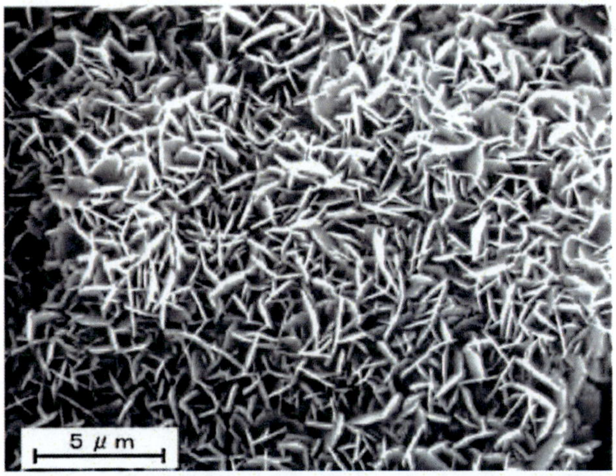

Fig. 18.3 Tobermorite observed on the surface of ALC foams. Tobermorites created by curing within the autoclave are plate-shaped or bamboo leaf-shaped crystals and they combine to form a stacked playing card like structure

In addition, ALC has high fire resistance and workability because of the interaction between the solid phase and the gas cavity.

18.1.2.3 Characteristics of ALC as Construction Material

As explained above, ALC is used in Japan as panels that have reinforced steel bars arranged inside (Fig. 18.4). Standard

Fig. 18.4 Example of a reinforced steel bar arrangement inside an ALC panel. Reinforcing steel bars inside the ALC panel ensure strength of the panel. The thick steel bars and the pipe nuts in the picture are used as an anchor for installation on building members

panels are 600 mm in width, and the thickness varies depending on the application and the area in which the panel is used. Panels that are 35–75 mm in width are called thin panels and are mainly used as exterior walls and refractory investment materials in wooden buildings and low-rise steel framed structures (Note 18.4). Thick panels with a thickness of 75–180 mm are widely used in exterior walls, partitions, floors, roofs, etc. of medium-sized buildings to high-rise buildings [steel framed and reinforced concrete structures (Note 18.5)] (Fig. 18.5).

ALC is light-weight and has been widely used as a general construction material because it reduces the load on building members (Note 18.6) and the construction period. The material also has high insulation properties, fire resistance and earthquake resistance. Fire resistance and earthquake resistance of ALC became widely recognized by the general public following the Great Fire of Sakata City in Yamagata

Note 18.4 Steel framed buildings. Large structures are built with steel frames. Members such as pillars and beams are made of steel frames, on which the wall, floor, ceiling and roof panels are connected.

Note 18.5 Reinforced concrete building. Large structures such as pillars, beams, slabs and walls are all made of reinforced steel bars and concrete to form an integrated structure.

Note 18.6 Sometimes simply called members. They are large structures that support buildings, i.e. foundations, pillars, beams, walls and floors, however, the contents vary depending on the building structure. In wooden buildings and steel framed buildings, the members are the foundation, pillars and beams, while concrete walls (supporting walls) and floors (slabs) are included as the members of reinforced concrete buildings.

Fig. 18.5 Examples of buildings that use ALC panels on exterior walls

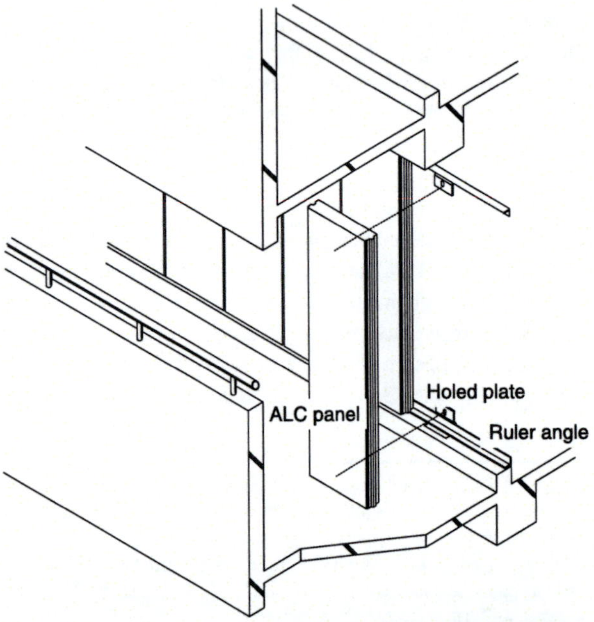

Fig. 18.6 Example of attaching a vertical wall panel using the locking method. Built-in anchors at the *top* and *bottom* of the panel are connected with building members using special metallic parts. The figure shows the locking method applied to skyscrapers

Prefecture in 1976 and the Great Hanshin–Awaji Earthquake in 1995 both in Japan, when ALC buildings were undamaged by the fires and the earthquake.

ALC has high fire resistance mainly because it is the inorganic material and because thermal shift is restricted due to insulation properties that are attributable to its high porosity and also because water vapor generated upon heating is released slowly, preventing explosion.

The earthquake resistance of exterior walls was enhanced following the application of the dry panel installation method. In particular, the vertical wall locking method, in which built-in anchors at the top and bottom of the panel are connected with building members using special metallic parts that allows for high in-plane deformation during an earthquake without damaging the wall (Figs. 18.6 and 18.7).

18.1.3 Future Prospects

Specifically designed panels are made by cutting the surface and L-shaped corner panels for outer/inner corners responding to the needs of diversified construction styles. The locking method was developed and established as the standard method to improve the safety of buildings.

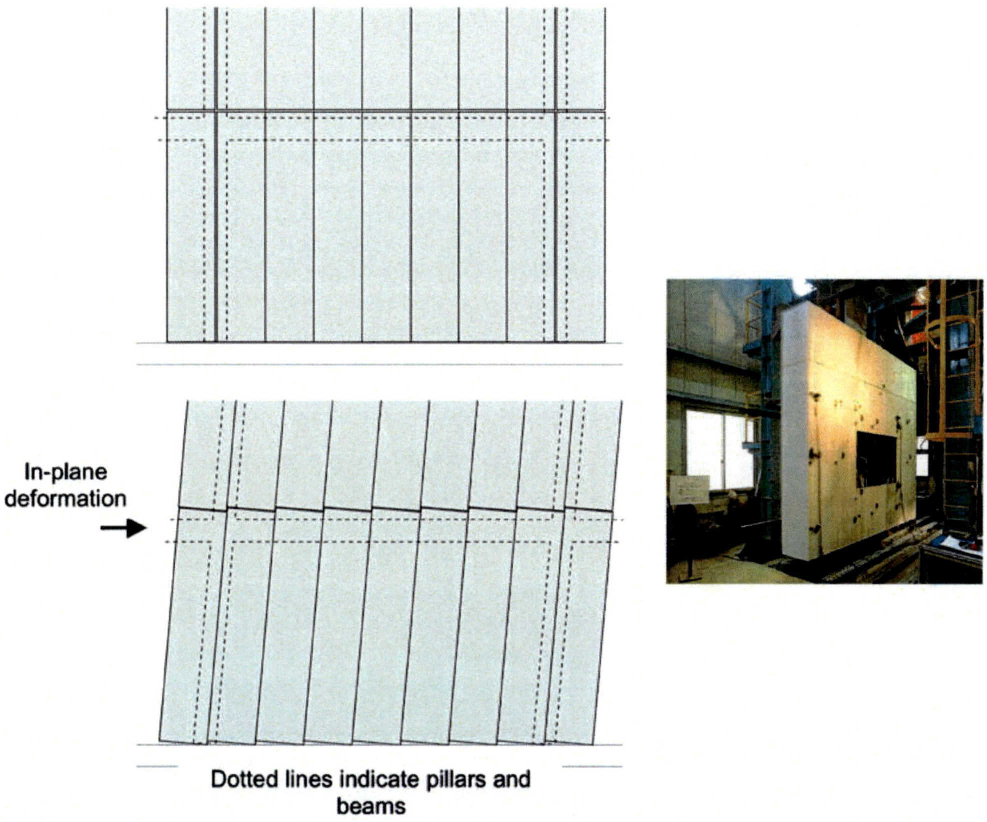

Fig. 18.7 Movement of wall panels installed by the vertical locking method. When the locking method is applied, panels swing independently responding to the in-plane deformation force, avoiding excessive load on panels and realizing high integrity. The picture shows the scene of an in-plane deformation test

"Safer," "more comfortable" and "environmentally-friendly" will continue to be the important characteristics for buildings. Technologies based on ALC materials and the installation of ALC panels are expected to be further enhanced responding to the building needs for higher toughness, durability, insulation properties and material recyclability.

Plaster Board (1921) 18.2

Human beings have used plaster since the beginning of civilization such as in ancient Egypt and Mesopotamia. It is known that plaster was used to fill the joints of stones used to build the pyramids. Figure 18.8 shows a stone of natural plaster (Note 18.7) ore and Fig. 18.9 shows carvings made of plaster in ancient Egypt. Plasterboard also know as sheetrock or drywall are made by sandwiching plaster between sheets of thick paper. It was invented in the U.S. in 1902 and production in Japan began in 1921. Figures 18.10 and 18.11 show some examples of plasterboards that are currently used. In foreign countries, the majority of plaster is made from natural plaster ore, while plaster produced in Japan is mainly made from by-products such as flue-gas desulfurized gypsum (Note18.8).

18.2.1 Background of Development

Three major characteristics of the plasterboard are it has fire resistance, soundproofing property and high thermal insulation ability. Dry wall materials are nearly free from thermal expansion or shrinkage and are easy to cut. Because of these characteristics, plasterboard has been used widely as fire resistant interior building material instead of plywood since the 1950s. The JIS (Japanese Industrial Standard) was established for plasterboard in 1951, which further promoted its popularization. Figure 18.12 shows the relationship between the plasterboard delivery amount per m^2 of newly constructed buildings and the area damaged by fire.

In 2005, the production amount of plasterboard in Japan was approximately 5.04 million tons or 570 million m^2, while 3.69 billion m^2 of plasterboard, 6.5 times of the amount produced in Japan, was consumed in North America. Approximately 7.5 billion m^2 of plasterboard are used in about 50 countries throughout the world. They are rarely imported or exported due to their heavy weight.

Note 18.7 Japan imports plaster ore mainly from Thailand, Mexico and Australia.

Note 18.8 Not only flue-gas desulfurization gypsum but also phosphate gypsum, titanium gypsum and hydrofluoric gypsum are also by-products used for the production of gypsum.

18.2.2 Characteristics

18.2.2.1 *Nonflammable Properties*

Plasterboard with a thickness of 12.5 mm is classified as a nonflammable material, while the 9.5 mm thick board is classified as quasi nonflammable material by the Building Standards Act in Japan. This is because gypsum, the core material of plasterboard, contains water in a crystalline form and its weight is approximately 20% of the total weight of the calcium sulfate dihydrate which is gypsum. The release of water upon heating from the dihydrate is expressed as follows:

$$CaSO_4 \cdot 2H_2O \rightarrow CaSO_4 \cdot 1/2H_2O$$
$$+3/2H_2O \rightarrow CaSO_4 + 1/2H_2O$$

18.2.2.2 *Manufacturing Method*

Figure 18.13 shows the plasterboard manufacturing process and Fig. 18.14 shows the plasterboard forming line. The outline of the manufacturing process is as follows:

1. Gypsum is heated in a heating furnace and processed to modeling plaster.
2. It is then crushed within a grinding machine to an appropriate particle size.
3. It is then stored in a modeling plaster tank.

Fig. 18.8 Natural plaster ore

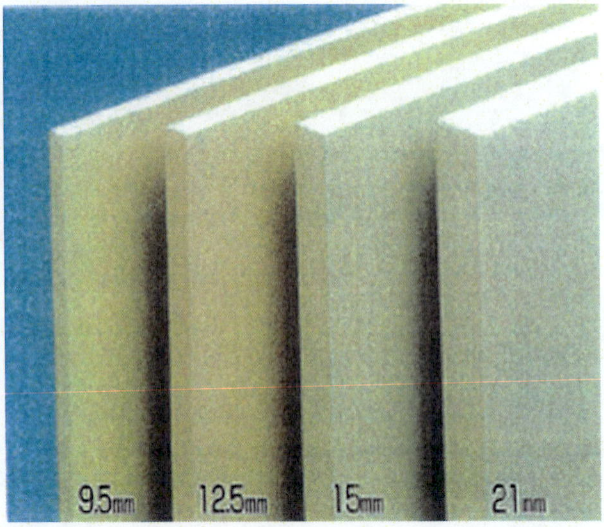

Fig. 18.10 Plasterboards

Fig. 18.9 Plaster carvings in ancient Egypt

Fig. 18.11 Decorated plasterboard

4. Modeling plaster, water, admixture and additive are kneaded in a mixer and the mixture is flown between two horizontal gypsum liner boards.
5. It is then formed in to a specified thickness using a forming roll.
6. The modeling plaster goes through hydration and is hardened on the conveyer belt.
7. The quasi hardened board is cut roughly at the end of the belt.
8. It goes through a dryer and excess moisture is removed.
9. It is finally cut into specified sizes and the products are then completed.

Fig. 18.12 Relationship between the plasterboard delivery amount per m² of newly constructed buildings and the area damaged by fire in Japan

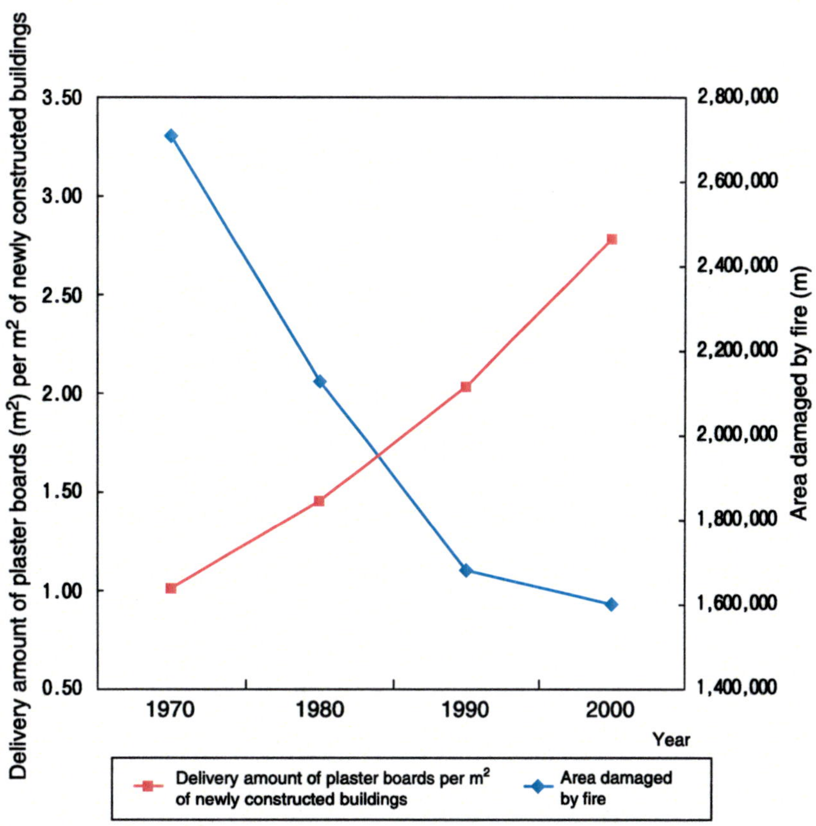

18.2.3 Future Prospects

Plasterboard will likely become thicker and harder responding to the needs for greater fire resistance, higher soundproofing property, etc. In addition, systematic methods involving fire resistance structures and sound shielding structures will be developed responding to the needs of the future.

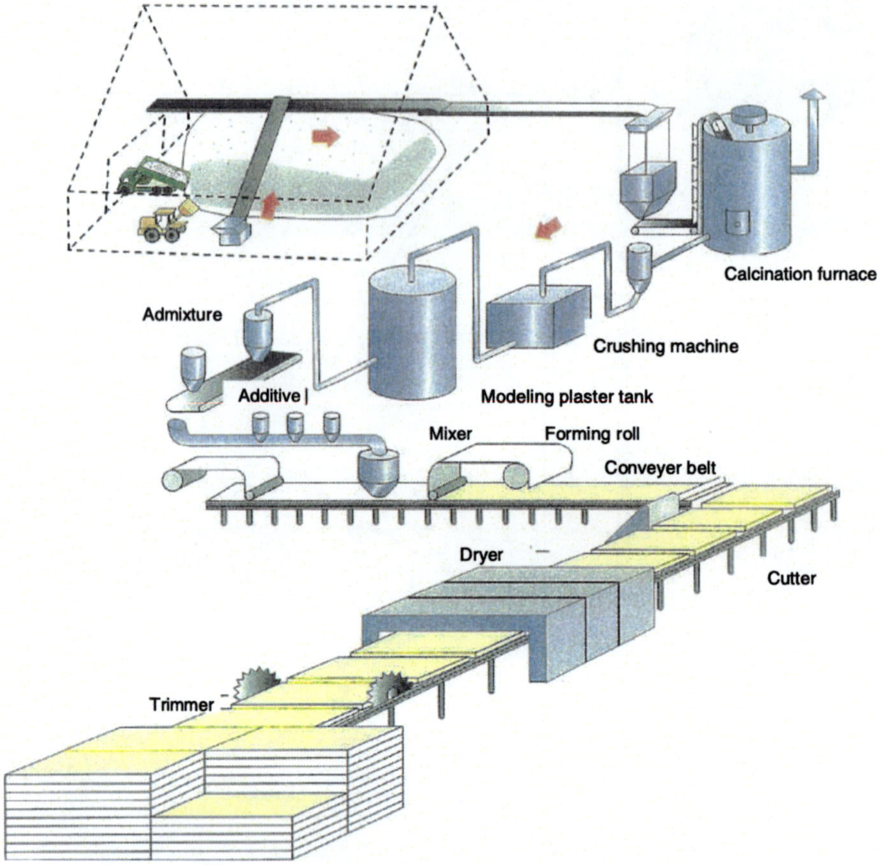

Fig. 18.13 Plasterboard manufacturing process

Fig. 18.14 Plasterboard forming line

18.3 Glass Wool for Thermal Insulation (1948)

Glass wool (consisting of thin glass fibers) is widely used as thermal insulation material in ships, automobiles, air conditioning ducts of buildings, water supply pipes, cold storage warehouses and houses as well as for noise absorbing material in studios, concert halls and movie theaters. Figures 18.15 and 18.16 show examples of glass wool applications. Approximately 85% of all glass wool is made of recycled cullet (waste glass) produced from glass bottles. Glass wool is normally produced by the centrifugation method. In this method, molten glass is fed into a high-speed rotating spinner with several tens of thousands of small holes. The glass fibers are extruded out of the holes by the centrifugal force with a diameter of several micrometers by applying pressure using gas jets. High-performance glass wool with smaller fiber diameters is also produced, and production amounts of high-grade products (HG product) have been increasing in their use as insulation materials in houses.

18.3.1 Background of Development

The industrial production of glass wool is said to have begun in 1916 in Germany. Following the trade suspension of asbestos thermal insulator from Canada during World War I, glass wool was mass produced and used as insulation materials for railway engines and battleships.

In Japan, production of glass wool started in around 1937 responding to the increase in the demand for military-grade glass wool. It was used as a substitute for asbestos. However, production stopped after the end of World War II. Production was recommenced in 1948 and glass wool was used as thermal insulation material in refrigerators. The materials also began to be used in ships, rice cookers, molds for the production of automobile, air conditioning ducts of buildings, water supply pipes, thermal insulation materials for cold storage warehouses and so on. Production amounts of glass wool increased substantially following its use as thermal insulation in houses. Glass wool is also effective for sound absorption and is used in the soundproofing of the walls of the Shinkansen (bullet train) and on expressways.

18.3.2 Characteristics

18.3.2.1 Characteristics of the Product

Glass wool is made of short glass fibers and is nonflammable. Approximately 85% of the raw material is recycled cullet produced from glass panels, bottles, picture tubes (panel portions), and fluorescent tubes (treated) collected from drop-off sites. The material obtained approval under the "Wide Area Certification System" established by the Ministry of the Environment (Note 18.9) as one of the measures to reduce construction waste in Japan.

While asbestos is a carcinogen, glass wool is not carcinogenic to humans. According to the evaluation of carcinogenicity of man-made mineral fibers conducted in October 2001 by "IARC (Note 18.10)," glass wool is classified as Group 3 (non-carcinogenic to humans).

Note 18.9 Manufacturers obtain certification from Japan's Minister of the Environment to collect wastes from a wide range of areas and recycle the wastes as raw material or dispose of it properly under this system (Wastes Disposal and Public Cleansing Act, revised edition: implemented on December 1, 2003). If a manufacturer obtains the certification in combination with delivery companies handling its products, they may handle the wastes without obtaining the licenses for collection, transportation and disposal of industrial wastes.

Note 18.10 International Agency for Research on Cancer, an outside organization of WHO (World Health Organization), is the most prestigious carcinogen evaluation organization in the world.

Fig. 18.15 Air conditioning ducts are wrapped with glass wool bonded with an aluminum film

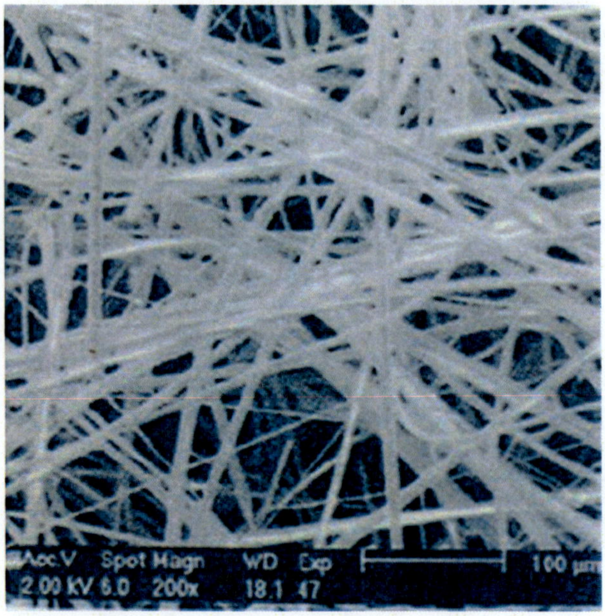

Fig. 18.17 Glass fibers of conventional glass wool

Fig. 18.16 Walls of a wooden house are fitted with HG glass wool

18.3.2.2 Manufacturing Method

Thin glass fibers have been manufactured by various methods such as the disk method, the spray method, the squirt method and the blaze method. Currently, glass wool for thermal insulation and noise absorption is normally produced using the centrifugation method, with the exception of glass fibers intended for special applications. The current centrifugation technology was developed by Owens Corning Inc. of the U.S. and separately developed by Saint-Gobain Corp. of France. In both technologies, molten glass is fed into a high-speed rotating spinner having several tens of thousands of small holes. The primary glass fibers extruded out of the holes by centrifugal force are extended to fibers with a diameter of several microns by applying pressure using gas jets, etc.

18.3.2.3 Performances and Specifications of the Product

Thermal performance of glass wool improves as the average diameter of the fiber decreases. Fibers are, therefore, becoming smaller in diameter. In conventional glass wool, the average glass fiber diameter is about 7 µm. At the present, high-performance glass wool (high grade = HG) with an average fiber diameter of about 4 µm is produced in large amounts for use in home insulation.

Pictures of fibers in normal products and HG products are shown in Figs. 18.17 and 18.18. The thermal insulation performance of HG glass wool (density: 16 kg/m^3) is equivalent to that of conventional glass wool (density: 24 kg/m^3) (Table 18.1).

Fig. 18.18 Glass fibers of HG glass wool. The fibers are smaller than those of conventional glass wool

Table 18.1 Comparison of glass wool performance

Name of material	Density (kg/m^3)	Thermal resistance [m^2 K/W (25°C)]
Glass wool thermal insulation material for houses (conventional product)	16	2.2 or higher
	24	2.6 or higher
HG glass wool thermal insulation material (HG: high grade)	16	2.6 or higher
	24	2.8 or higher

The thermal insulation performance of conventional glass wool (density: 24 kg/m^3) is equivalent to that of HG glass wool (density: 16 kg/m^3) with thickness of 100 mm

18.3.3 Future Prospects

Energy saving is urgently needed now due to abnormal climate change resulting from greenhouse gases. Glass wool has a high CO_2 reduction effect, high performance to cost ratio and nonflammable properties. This superior thermal insulation material will continue to contribute to the reduction of global warming. Glass wool provides very high performance when used as the core material for vacuum type thermal insulation in refrigerators, vending machines and other items.

Functional Construction Materials (Humidity Control) (1998)

18.4

In response to improvements in individual homes' air tightness and higher thermal insulation needs for energy conversation, functional building materials for humidity control have been attracting attention to regulate interior relative humidity (RH) passively, that is, by dampening swings in RH. To effectively respond to interior air humidity changes, interior building materials need to have high moisture absorption/release capacity with high responsiveness. Control of the interior building material pore structure is the key point in achieving the required performance. In addition, these materials need to be of sufficient quality (in terms of strength and earthquake resistance) and of appropriate design that satisfies the needs for finishing. In addition, materials with few additives such as formaldehyde are needed since formaldehyde is believed to be the cause of the sick house syndrome (an allergic reaction to one's living environment). Furthermore, building materials need to be objectively assessed and certified to ensure they are suitable for interior living environments.

18.4.1 Background of Development

In respect to energy consumption in Japan, the amount of energy consumed by industry has leveled off, while the consumption by households has been increasing. The increase in energy consumption by the civilian sector is attributable to individual use of electricity for maintaining a comfortable living environment.

Meanwhile, air tightness and thermal insulation performance of houses have been improved (as energy-saving measures) responding to the oil crisis and higher prices for energy. However, these types of houses with high air-tightness tend to have wide swings in their relative humidity such as the air being too humid or too dry due to poor ventilation. Under these conditions, problems such as condensation, mold and tick infestation are a nuisance. At the same time, volatile organic compounds [VOC (Note 18.11)] in indoor air has been identified as a cause of the "sick house" syndrome.

At present, mechanical air handling equipment such as air conditioners are mainly used to control air temperature and recently for air purification. However, these machines consume a large amount of energy and their operation runs counter to global environmental preservation measures. Functional building materials including humidity-control materials (Note 18.12) are in focus as passive materials for solving humidity-related issues without energy consumption.

18.4.2 Characteristics

18.4.2.1 Functions and Characteristics of Products

Humidity-control building materials have a function that naturally adjusts indoor humidity levels. They naturally absorb moisture to lower the humidity when the indoor humidity is high and release moisture to increase humidity when the indoor humidity is low.

Indoor humidity incessantly changes due to daily activities such as cooking and bathing as well as due to changes in the natural environment due to weather and seasonal changes. For effective humidity control responding to various humidity changes, materials need to have both high moisture absorption/release capacity and high responsiveness.

Note 18.11 Stands for Volatile Organic Compounds, representing several hundred volatile organic compounds. WHO defines VOC as organic compounds in the atmosphere with a boiling point ranging from 50°C to 260°C. The Japanese Ministry of Health, Labour and Welfare presented guideline values for permissible indoor air concentrations of 13 substances including formaldehyde.

Note 18.12 Construction materials that are used mainly to reduce changes in relative humidity in building interior living or working space.

These two points are important factors that need to be considered in designing the humidity control materials.

Adjustments of the pore structure in indoor materials are important for the control of moisture absorption/release properties (Note 18.13) (capacity, speed), which affects the humidity control performance. It was confirmed through past research that management of the moisture absorption/release capacity is important in achieving superior humidity control performance. Management of the moisture absorption/release capacity involves the control of fine pores dimensions of several to 10 nm where capillary condensation occurs and control of macro range pore size of 30 nm or larger which regulate the diffusion of water vapor.

In general, the strength of the material decreases if the pore size increases. If the material is sintered at high temperatures so as to improve the strength, microscopic pores of the raw material decreases, lowering the moisture absorption/release properties. Therefore, raw material selection and sintering temperatures are the key points in realizing a construction material satisfying both appropriate strength and adequate moisture absorption/release properties.

In order to satisfy these two requirements, natural clay mineral allophane, etc. with fine pores was used as raw materials. The raw materials were sintered at a temperature lower than the temperature normally used for normal tiles (approximately 1,100°C or higher) to realize compatibility between basic properties including strength and superior moisture absorption/release properties. This functional construction material has been commercialized for use as an interior wall finishing materials (Figs. 18.19 and 18.20).

18.4.2.2 Performance of the Product

Test methods for the evaluation of moisture absorption/release properties (speed, capacity), which affects relative humidity control performance, is specified in "Moisture absorption/release property test method for humidity-control construction materials—Part 1: Temperature response method. The moisture absorption/release test is based on standards set for temperature change (JISA1470-1)" and "Equilibrium moisture content measuring method for construction materials (JISA1475)" in accordance with Japanese Industrial Standards (JIS).

Figures 18.21 and 18.22 shows test results for the comparison of humidity control construction materials and other materials in moisture absorption/release amount and in humidity variation control performance, respectively (test conditions not conforming to JIS). Figure 18.21 indicates that moisture absorption/release amount of the humidity-control construction materials is higher than that of other materials. Figure 18.22 indicates that humidity-control construction

Fig. 18.19 Examples of functional construction materials (humidity control construction materials). Above are examples of functional construction materials made by high-temperature sintering of natural clay mineral allophane containing fine pores. The sizes of fine pores in the materials are designed to absorb and release moisture effectively

Fig. 18.20 Example of functional construction materials (humidity control construction materials) used in a living room. This living room incorporates functional construction materials made by the high-temperature sintering of natural clay mineral allophone containing fine pores (used as interior walls). The material satisfies requirements for construction materials (strength and earthquake resistance) and the design requirements needed for finished materials

Note 18.13 The property of materials attributed to the absorption and releases of moisture.

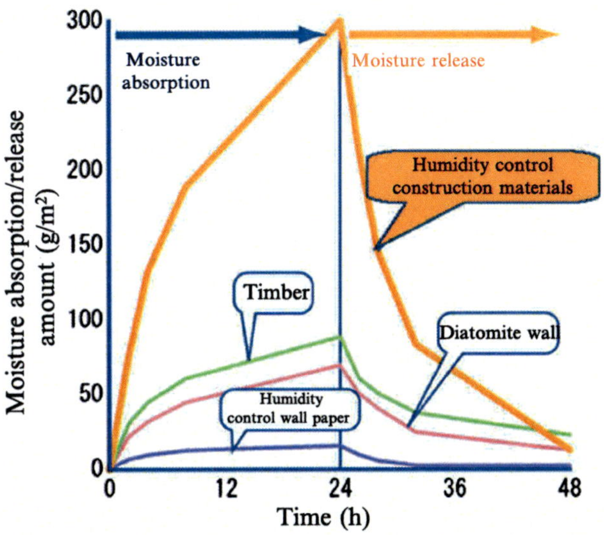

Fig. 18.21 Moisture absorption/released amount for various materials including humidity control construction materials. Test method: equalize the specimens in a constant temperature and humidity chamber set at 25°C to 50% RH. The specimens are then placed in a constant temperature and humidity chamber set at 25°C to 90% RH and the absorbed moisture amounts are measured for 24 h. Finally, specimens are placed in a constant temperature and humidity chamber set at 25°C to 50% RH to measure the amount of moisture released

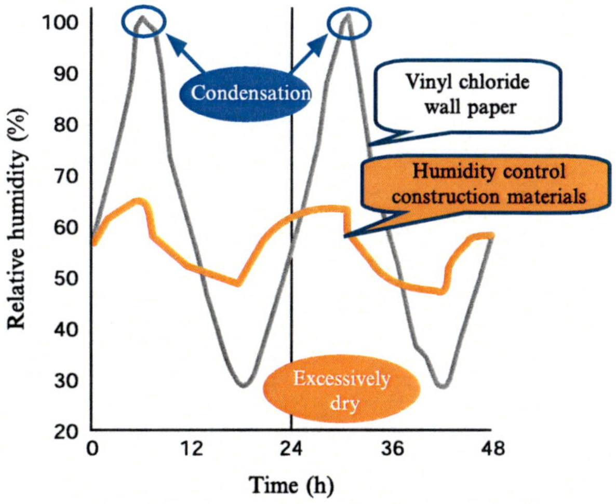

Fig. 18.22 Relative humidity change comparison for different materials. Test method: specimens are equalized initially at 15°C to 60% RH and placed in an airtight container, where temperature and humidity chamber are controlled during a 24-h cycle within 5–25°C. The temperature inside the airtight container is measured using a thermo-hygrometer

materials show higher performance than the other materials in reducing humidity changes. It was confirmed that condensation and excessively dry air are reduced and the humidity remains constant at comfortable conditions (40–70% RH) under the set conditions.

18.4.3 Future Prospects

Functional construction materials are used not only to control humidity but also to remove harmful chemical substances. The Building Center of Japan operates a certification system for construction materials that reduce the pollution of indoor air by volatile organic compounds. According to the certification system, the standard is set and implemented that building materials should not release volatile organic compounds into air regardless of changes in the indoor environment conditions.

In terms of indoor environments, the materials should be benign. The material should not cause an allergic reaction triggered by chemical substances nor permit growth of microorganisms so as to provide a healthy and comfortable living environment. Meanwhile, reduction of environmental loads is also required. Further development will continue for new functional construction materials that contribute to the reduction of environmental loads and to the creation of healthy and comfortable living environments.

Thermal Insulation Glass (1970) 18.5

Today, when the reduction of CO_2 emissions to prevent global warming is a worldwide concern, insulation of entryways and windows in houses and buildings is very important for the reduction of cooling and heating loads. Thermally insulated glass is expected to play an important role in buildings. These products include heat absorbing glass and heat reflecting glass, which have high heat shielding performance, as well as double glazed glass and evacuated glass, which enclose a thin space or a vacuum layer between two glass sheets to reduce heat transfer between outdoors and indoors. Types of glass and the substance between the two sheets can be selected depending on the application so as to enhance the performance and functions. Therefore these products can be used in a variety of buildings, contributing substantially to reduced energy usage.

18.5.1 Background of Development

Worldwide effects are on-going against global warming, and energy-saving measures are promoted also in buildings including houses, aiming to reduce CO_2 emissions. However, people seek more comfortable living environments and the amount of energy consumed by air conditioners continues to increase. Therefore, the reduction of energy consumption by air conditioners is important for reduction of CO_2 emissions.

A significant amount of air conditioning energy in buildings is mainly attributable to heat loss from these enclosed openings (windows) or doors. With respect to houses, about 40% of total heat loss is attributable to these openings. Responding to these circumstances, sheet glass that absorbs or reflects sunlight (light rays) as well as double glazed glass with dry air enclosed between the two sheets has been expanding rapidly for the purpose of the reduction in cooling and heating loads.

18.5.2 Characteristics

In a broad sense, heat absorbing glass and heat reflecting glass denote insulated glass. However, glass with high thermal resistance and heat insulation performance (Note 18.14) of 4.0 W/m^2 or lower is defined as insulated glass in the sheet glass industry. The recently the revised Energy Saving Act of Japan classifies window glass into three energy-saving grades. The Flat Glass Association of Japan recommends first grade glass (heat transmission coefficient of 2.7 W/m^2 K or lower), named "eco glass®," for use in entryways or windows as insulated glass.

18.5.2.1 Heat Absorbing Glass, Heat Reflecting Glass

The heat absorbing glass is transparent sheet glass manufactured by adding a small amount of iron, cobalt, selenium or nickel to sheet glass composition to produce green, blue, gray or bronze colors, as shown as examples in Fig. 18.23. The glass absorbs an appropriate amount of visible light and sunlight in the infrared range to reduce the cooling load. It also features anti-glare and decorative performance due to its color. The ultraviolet absorption rate is comparatively high, preventing discoloration and deterioration.

Note 18.14 Evaluation of thermal insulation performance is based on heat transmission coefficient or heat transmission resistance (inverse of heat transmission coefficient). Heat transmission coefficients indicate the quantity of heat that passes through a glass window when there is a difference between the indoor temperature and the outdoor temperature. Smaller heat transmission coefficients and larger heat transmission resistance indicate higher heat insulation properties. Such is expressed by the quantity of transmitted heat (Watt) per 1 m^2 to produce a difference of 1° (Kelvin).

Fig. 18.23 Application of heat absorbing glass and heat reflecting glass. Glass with heat shield performance, including the heat absorbing glass (*left picture*) made by adding a small amount of metallic component to a sheet glass composition so as to create color, and reflecting glass made by coating metal oxides films on the surface of sheet glass (*right picture*), reduces heat and by blocking sunlight, leading to reduction of cooling loads during summer

Heat reflecting glass is manufactured using two methods. In one method, the glass is made by coating metal oxide layers on the surface to reflect sunlight and reduce the cooling load. Then the glass is coated with metal oxides with heat absorbing characteristics thus combining heat reflecting and heat absorbing properties which provides a superior energy-saving glass. This coated glass has a half-mirror effect and it reflects visible light. This is indicated in Fig. 18.23.

In the other method, the glass is made by coating float plate glass (incorporating the sputtering method) with a homogeneous and durable metallic film that features high sunlight reflection performance. The glass features appropriate sunlight shielding performance (Note 18.15) and saves energy substantially by reducing cooling and heating loads. As the manufacturing method allows application of multiple films, a variety of reflected colors are offered on the exterior side.

18.5.2.2 Low-Emissivity Coating Glass (Low-E Glass)

The glass is normally used as a component of double glazed glass (explained below) rather than as a single glass sheet so as to take advantage of its low emissivity. Low-E glass is made by coating transparent sheet glass with a low emissive film that has high sunlight reflecting properties. The glass is manufactured by (1) the CVD (chemical vapor deposition) method, where a metal oxide film of tin, etc. is coated on the glass, and (2) the sputtering method, where multiple films such as a metallic film of silver with high sunlight reflecting properties, a metal oxide film and a nitride film on the glass.

18.5.2.3 Double Glazed Glass

A schematic view of the double glazed glass is shown in Fig. 18.24. A metallic part called the spacer is placed between glass sheets to create a hollow space for air with low thermal conductivity in order to improve the heat insulation performance. Double glazed sheets are normally 12–18 mm in total thickness. The glass also maintains the temperature on the inside sheet, preventing condensation and cold drafts (Note 18.16) around the window when the outside temperature is low.

The manufacturing process is shown in Fig. 18.25. Glass sheets of Low-E glass are cut into specified sizes and go through a cleaning process. A pre-processed spacer containing desiccant is bonded on one of the sheets via the primary seal, which is usually made of butyl rubber and has low gas permeability. The other sheet is bonded on the spacer and pressed and the edges of glass sheets are applied with a secondary seal.

Note 18.15 Heat shield performance is evaluated based on solar radiation heat acquisition rates or shielding coefficients. The solar radiation heat acquisition rate indicates the rate of solar radiation heat that flows into room among the solar radiation heat falling on the glass window. It is also called the solar radiation intrusion rate. The shielding coefficient is the relative value obtained by setting the solar radiation heat acquisition rate of a 3 mm float plate glass (transparency) at 1.

Note 18.16 During cold winter nights, window glass surface temperature on the indoor-side decreases due to the heat loss through the window. The indoor air temperature near the window glass decreases and it has higher density, generating a downward current of air. This type of air current is called a cold draft.

Fig. 18.24 Schematic view of general double glazed glass. Two glass sheets are laminated via a spacer containing desiccant. Edges of the glass sheets are sealed with high-quality dual sealant

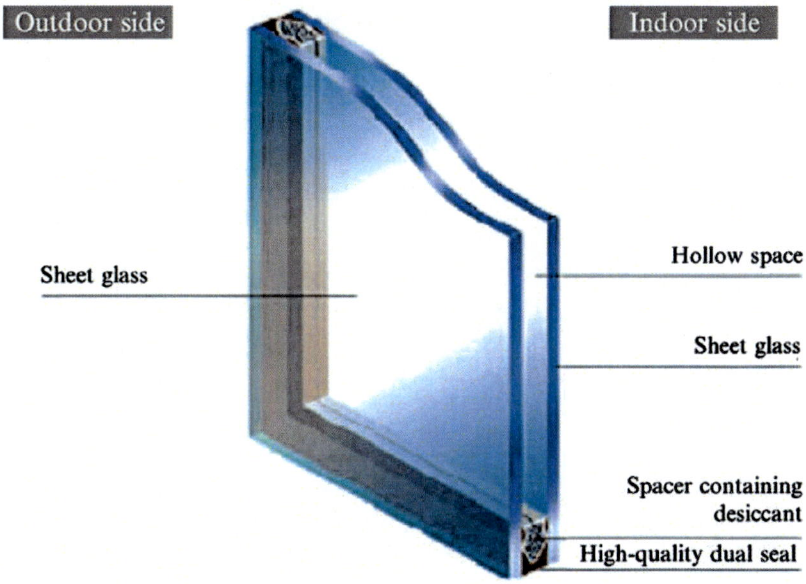

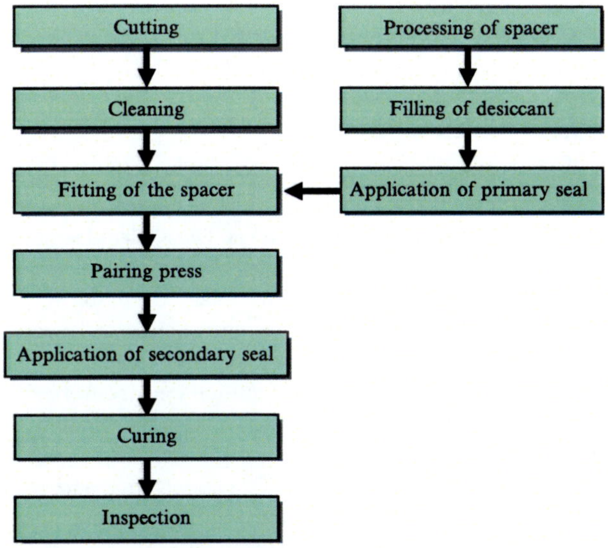

Fig. 18.25 Manufacturing process of double glazed glass. Glass sheets cut into specified sizes are cleaned and a spacer containing desiccant is fitted on the glass and bonded by the primary seal. The other glass sheet is bonded on the other side of the spacer to form a double layer and the edges of the glass are filled by the secondary seal. The product is completed after curing and inspections take place

This is typically made of silicone and has high adhesion properties as a sealing material. The product is completed after curing and is inspected.

In normal double glazed glass, 2/3 of the heat transmitted in the air space is generated by radiation. Therefore, thermal insulation performance is enhanced using the Low-E glass explained above. If the Low-E glass is used on the indoor side, the heat escaping from the indoors is substantially reduced in winter. If the Low-E film is coated on the exterior side, sunlight is reflected substantially, improving the cooling effect during the summer. High performance double glazed glass is applicable for various purposes. Furthermore, heat absorbing glass, tempered glass, or laminated glass are sometimes used in combination with each other.

If the hollow space is filled with a rare gas such as argon and krypton, which have lower thermal conductivity than air, higher thermal insulation performance will be achieved.

18.5.2.4 Evacuated Glass

Heat is transmitted by conduction, convection and radiation. Higher thermal insulation performance is realized by reducing these components. The evacuated region is made by evacuating a 0.2 mm layer (instead of a space containing air) between two glass sheets explained above. This is shown in Fig. 18.26. This vacuum layer reduces the conduction and convection of heat, and the special metallic film of high heat resistance (Low-E film) reduces radiation. This produces thermal insulation performance that is extremely high.

This thermally insulated glass saves energy by reduction of air conditioning loads and reduces condensation, which is realized by maintaining room temperature on the inside glass sheet, and has soundproofing performance that cannot be achieved by the double glazed glass. The thickness is only 6.2 mm. Therefore it can easily replace single sheet glass, which is already used in existing houses.

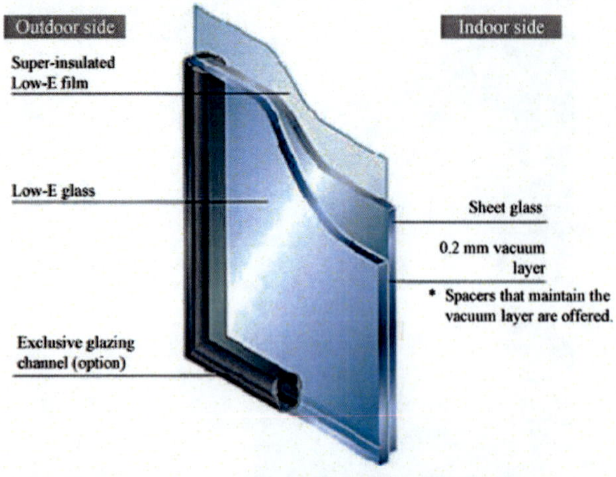

Fig. 18.26 Schematic view of evacuated glass. The double glazed glass has a 0.2 mm vacuum layer maintained by a micro spacer. The edges are sealed by low-melting point glass

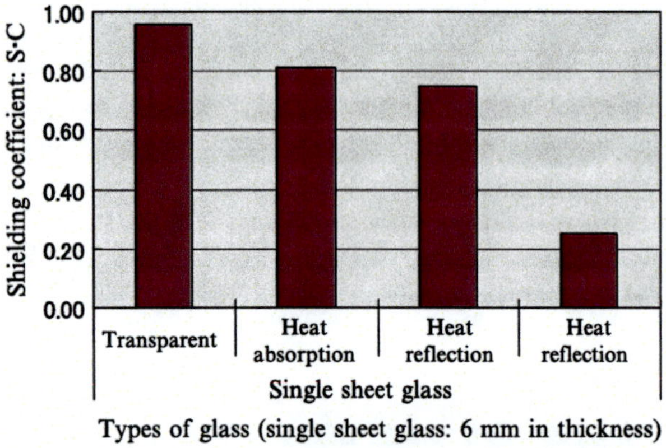

Fig. 18.27 Comparison of solar heat shield performance of glass. Heat absorbing glass and heat reflecting glass reduce the amount of heat (*light*) that passes through the glass by absorbing and reflecting solar radiation heat, reducing cooling loads and providing heat insulation during summer time

18.5.2.5 Performances and Specifications of Products

Heat absorbing glass and heat reflecting glass (with heat shield performance) are capable of blocking 20–70% of sunlight, resulting in the reduction of cooling loads in the summer. This is shown in Fig. 18.27. These types of glass reflect approximately 30–40% of visible light, providing a half-mirror effect.

Meanwhile, double glazed glass and evacuated glass reduce sunlight related heat by 20–50% depending upon the type of glass to be used, as shown in Fig. 18.28. Thermal insulation performance of general double glazed glass is two times higher than that of transparent single sheet glass while that of evacuated glass is four times higher than that of transparent single sheet glass. Due to its higher insulation performance, condensation on the inside glass occurs when the outdoor temperatures are below the freezing point under the specified indoor environment (indoor temperature: 20°C, indoor relative humidity: 60%, indoor natural convection, outdoor wind speed: 3.5 m/s), and therefore, the possibility of condensation is limited in daily life.

18.5.3 Future Prospects

Insulated glass is used widely in general houses and buildings. According to a survey by the Flat Glass Manufacturing Association of Japan, the shipping amounts of double glazed glass intended for architectural use was 16 million m^2 in 2006, which was three times the amount delivered 10 years earlier. The association estimates that CO_2 emissions per year will be reduced by about 17 million tons (converted to 34 million 200-L drums of kerosene) if all of the windows in Japanese houses are replaced by eco glass®. This amount is equivalent to 10% of the total CO_2 emitted from Japanese houses in 2003 and 23% of the total greenhouse gases required to be reduced by the Kyoto Protocol. Meanwhile, the research and development of films with thermal insulation characteristics have been actively investigated. Various types of insulated glass are expected to be created so as to easily provide insulation for every building window.

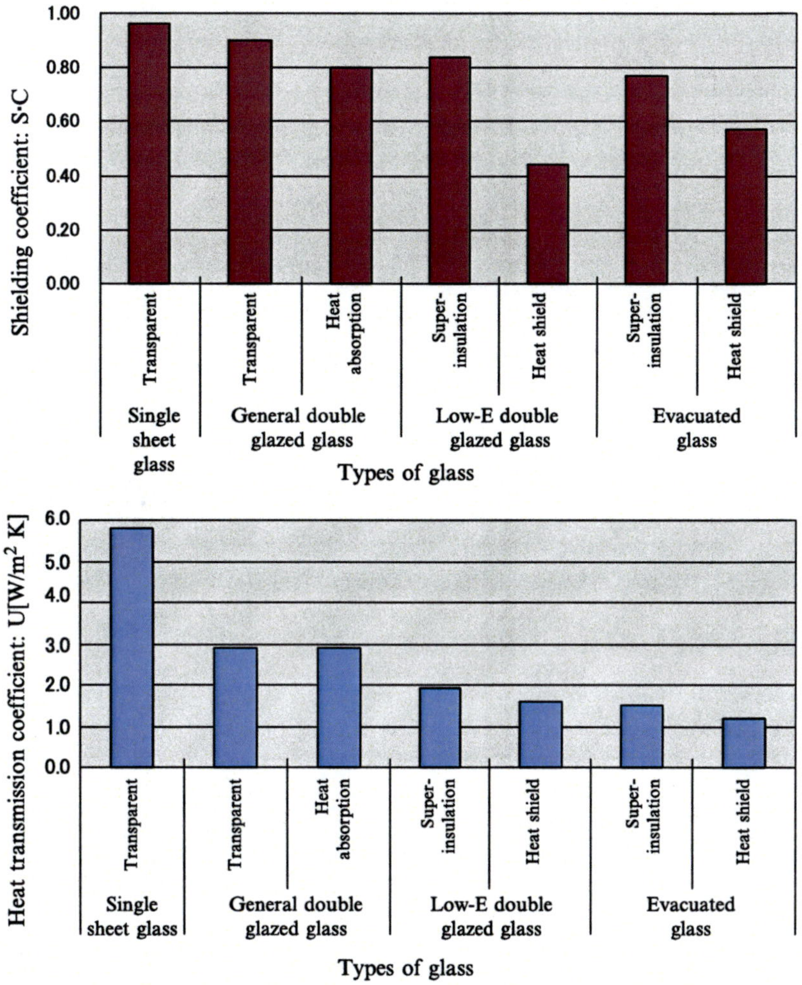

Fig. 18.28 Comparison of shielding performance and heat insulation performance of double glazed glass and evacuated glass. The shielding performance of double glazed glass and evacuated glass is affected by the type of sheet glass. Shielding performance is enhanced by using sheet glass applied with a special metal oxide film with high sunlight reflectance. Heat insulation performance is substantially improved by the use of double glazed glass. The heat insulation performance of transparent double glazed glass is two times that of transparent single sheet glass while that of evacuated glass is four times that of transparent single sheet glass

Safety Glass (1930) 18.6

There are two types of safety glass: tempered or toughened glass and laminated glass. Tempered glass is shatter-resistant and when it breaks, it breaks into small pieces, preventing serious injury. Laminated glass, having an intermediate film between two glass panels, prevents objects from passing through the broken glass and also prevents glass fragments from scattering (Fig. 18.29). The role of the safety glass has been changing with the times. It was originally used to protect human bodies in the case of automobile collisions and prevent shattering when impacted by balls and stones. Today, safety glass is also used for protection against theft, prevention of breakage due to disasters such as typhoons and earthquakes as well as prevention of hazards and injuries caused by glass fragments.

18.6.1 Background of Development

18.6.1.1 Building Windows and Entryways

The architectural field can be roughly classified into two categories, that is, large public buildings including offices, public halls and museums (called buildings hereinafter) and houses. As public buildings are used by the general public, it is recommended that safety glass (laminated glass, reinforced glass) be used in doorways so as to prevent injury due to the collision of people against the glass (Note 18.17).

With respect to houses, crimes including theft (breaking and entering a house) have been increasing, while the ratio of arrests has been declining since the end of the 1990s. Burglars intrude into houses by breaking window glass in the majority of cases. The laminated glass, one type of such a safety glass, began to be used so as to prevent such crimes as crime-prevention glass (Note 18.18).

18.6.1.2 Automobile Windows

Glass plays an important role in the safety of automobiles. Safety includes preventive safety, which involves undistorted vision, and collision safety, which involves the protection of occupants at the time of collision. The windshield glass is the most important aspect regarding collision safety in automobiles. It was stipulated in 1959 that safety glass should be used in windshields. The use of partially-tempered glass was legislated in 1969 to ensure transparency even at the time of breakage. The use of laminated glass was legislated in 1985 to ensure occupant safety during collisions.

18.6.2 Characteristics

18.6.2.1 Characteristics and Specifications of Products

The meanings and levels of "safety" have been changing with the times. The safety level of windshields was enhanced substantially following the application of laminated glass instead of tempered glass.

As for safety glass used in buildings and houses, the meanings (or range) of "safety" changed from the protection of human bodies at the time of collision and reduction of injury due to fragments of glass broken by impact with

Note 18.17 "Safety Design Guideline on Glazed Openings," stipulated by the Building Guidance Division, Housing Bureau, Construction Ministry, Noticed on May 31, 1986 (Revision noticed on April 4, 1991).

Note 18.18 "List of construction parts with high crime prevention performance" (disclosed in 2004) disclosed by the "Government-Public Joint Conference on Construction Parts with High Crime Prevention Performance" (established in 2002).

Float plate glass

This glass is widely used, but the fragments have sharp edges and are very dangerous if broken. The impactor scatters glass fragments and passes through the glass.

Laminated glass

A special film is sandwiched between two glass panels, and therefore, the scattering of glass fragments and the passing of the impactor through the glass are rarely observed. If the thickness of the intermediate film is increased, the penetration resistance also increases.

Tempered glass

The strength is about 3 times that of float plate glass. The glass is broken into small grains, preventing serious injury. The impactor penetrates the glass.

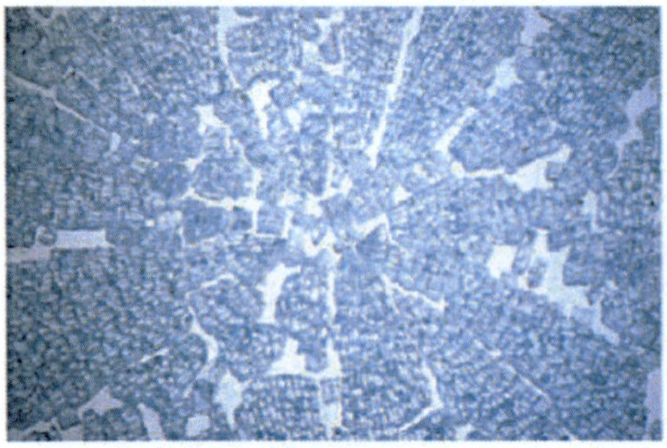

Fig. 18.29 Comparison of broken states of sheet glass

airborne balls and stones to disaster resistant glass for the reduction of damage caused by typhoons and earthquakes and crime-prevention glass for the prevention of theft. In line with the changes in meanings of "safety," tempered glass is being replaced by laminated glass.

Tempered glass is required to pass several safety tests such as (1) the falling ball impact test, (2) the broken condition test and (3) the shot bag impact test specified by JIS R 3206 (Figs. 18.30 and 18.31). The test conditions and acceptance criteria directly related to the safety of 5 mm tempered float glass are explained below. The tests are performed using glass sheets with a thickness of 5.0±0.3 mm. In the falling ball impact test (1), glass sheets dimensioned 610 mm × 610 mm are fixed in to a specified frame with a specified method, and a steel ball of 1,040 ± 10 g is dropped from a height of 100 cm above the specimens without applying force. The number of broken specimens should be no more than one out of six specimens. In the fragment condition test (2), glass sheets dimensioned equivalent to the product are impacted by a specified hammer or punch at specified points. The number of fragments (counted by the specified method) in the 50 mm × 50 mm area in the range where the fragments are largest (excluding the specified area) should be 40 or more. In the shot bag impact test (3), glass sheets dimensioned 1,930 mm × 864 mm are fixed in specified frames by a specified method, and a specified impactor (45 ± 0.1 kg, generally called a shot bag) is freely dropped using the pendulum method from a height of 300 mm above the specimens to apply an impact on the central areas of the specimens. If the specimens do not break, the height of the impactor is increased to 75 cm and it impacts the central area of the same specimens. If the specimens do not break, further increase of

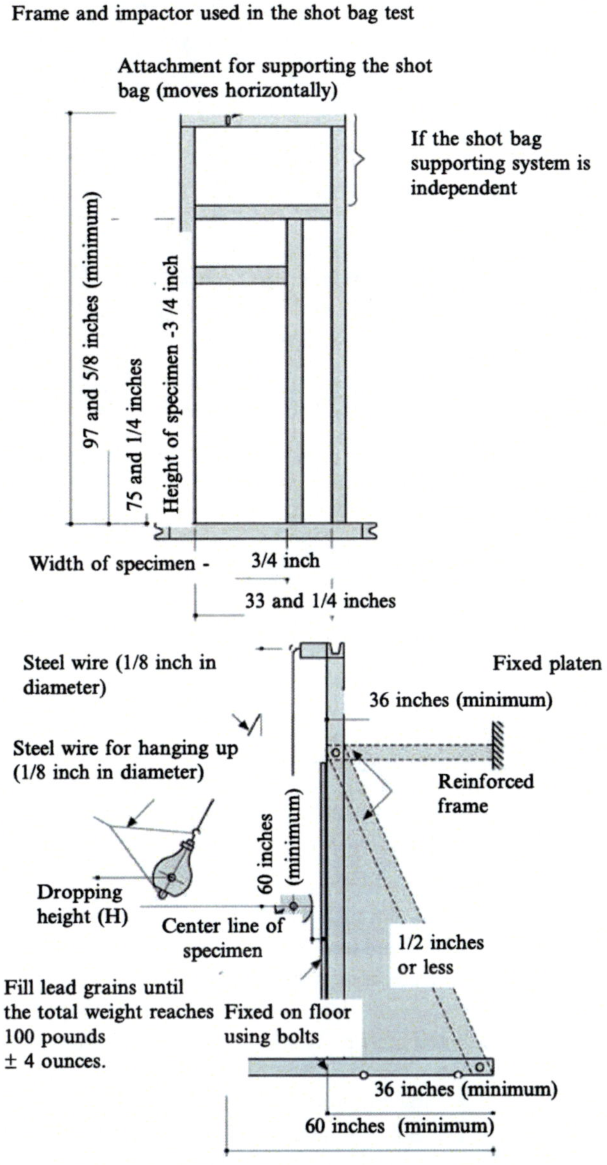

Fig. 18.30 Shot bag test equipment

the height to 120 cm and repeat the test. All of the four specimens are required to satisfy the specified standards (not broken or, if broken, the total weight of the ten largest broken fragments should be less than the specified weight).

Laminated glass is required to pass (1) the thermal resistance and light resistance tests, and (2) the shot bag impact test specified by JIS R 3205. Laminated glass has an intermediate film (polymeric material) sandwiched between two glass sheets, as shown in Fig. 18.32. It is tested for heat resistance and light resistance (1) to ensure long-term optical performance and safety of the building components. In the shot bag impact test (2) used to confirm safety, glass sheets dimensioned 1,930 mm × 864 mm are fixed in specified frames using a specified method, and a specified impactor (45 ± 0.1 kg) is freely dropped using the pendulum method (from a height of 120 mm for the II-1 specimen, 75 cm for the II-2 specimen and 30 cm for the III specimen) so as to apply an impact on the central areas of the specimens. The specimens are required to satisfy the specified standards (not broken or, if broken, the broken area is not an opening that allows the free passage of a ball 75 cm in diameter).

The standard heights for the shot bag tests of tempered glass and laminated glass are specified in the Safety Design Guidelines on Glazed Openings based on building types and areas. For example, the 120 cm glass is applicable to the areas around the entrance to the lobby of buildings (such as a public hall) that require safe designs. With respect to crime-prevention performance, tests are performed according to overseas standards, since JIS does not evaluate glass for crime-prevention.

18.6.2.2 Manufacturing Method

18.6.2.2.1 Tempered Glass

A glass sheet is placed inside a convection furnace and heated to 650–700°C, which is close to glass softening temperature. Immediately after heating, air is blown uniformly on both sides of the glass sheet for quenching. The glass surface becomes colder than the inner area, forming stable compressive stress layers on the surface (Note 18.19). Through this type of processing, the strength of the glass is improved and the wind pressure resistance is increased to three times that of float plate glass (Fig. 18.33).

18.6.2.2.2 Laminated Glass

In order to produce laminated glass, an intermediate film made of transparent and adhesive resin is sandwiched between two glass sheets, which is kept for about 1 h in an autoclave (pressurized furnace) for pressure bonding at 120–130°C and 1.47 MPa (approximately 15 kgf/cm^2) (Fig. 18.34).

Note 18.19 When a glass sheet is broken by a bending force, the pressed side is concaved, generating compressive stress. The other side is convexed, generating tensile stress. Glass is resistant to compression force and sensitive to tensile force. Therefore, compressive layers are created on the surface of tempered glass to enhance its strength in order to reduce the tensile force on the surface.

Float plate glass (5 mm)
Large glass fragments fall down and a large opening is created.

Laminated glass (3 mm + 3 mm)
Scattering of glass fragments is limited and the impactor seldom penetrates through the glass.

Reinforced glass (5 mm)
Fragments of the broken glass scatter and the impactor penetrates through the glass.

Fig. 18.31 Comparison of shot bag test results

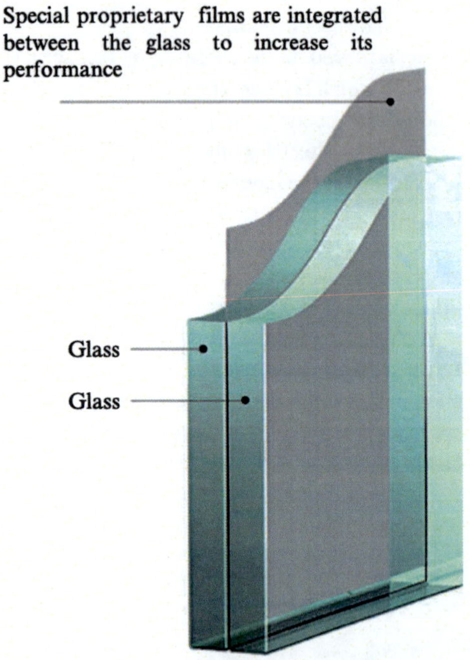

Fig. 18.32 Structure of laminated glass

18.6.3 Future Prospects

18.6.3.1 Tempered Glass

The tempered glass represents safety glass and has been used widely since the second half of the twentieth century. High-strength glass will continue to be used widely in the architectural field. Following improvements in manufacturing methods, tempered glass more than three times stronger than float glass of the same thickness, has been produced for use as safety glass with fireproof performance. Further technological improvements and new applications are expected in the future.

18.6.3.2 Laminated Glass

Safety performances required of laminated glass have changing with the times, from collision safety and scattering prevention, to disaster prevention and crime prevention (Fig. 18.35). It is anticipated that even higher performance will be required in the future. In the past, Japanese users,

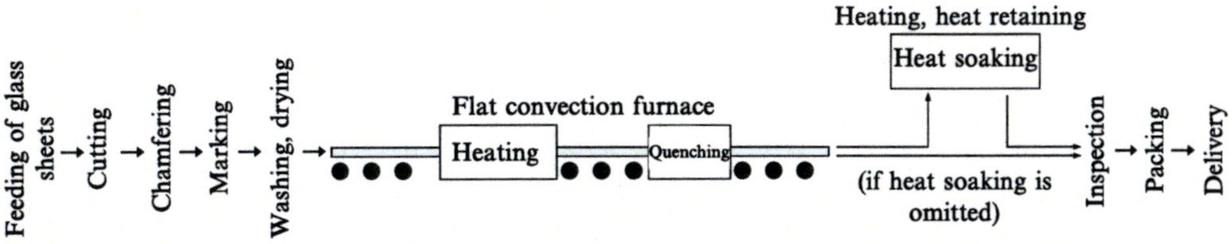

Fig. 18.33 Manufacturing process of tempered glass for architectural use

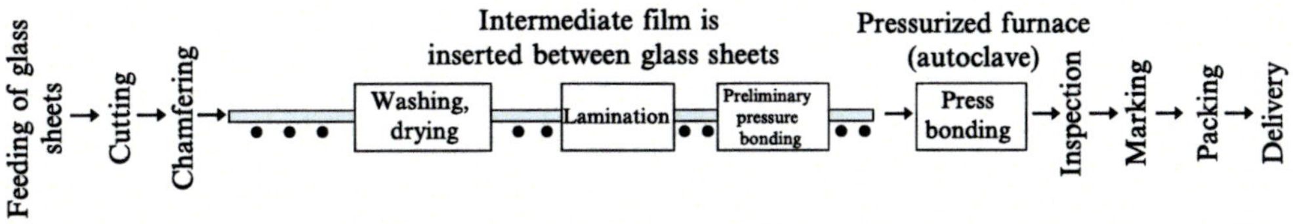

Grade	SECU-RE™ level I SECU-RE Coolverre™ level I	SECU-RE™ level II SECU-RE Coolverre™ level II	SECU-RE™ level III SECU-RE coolverre™ level III	SECU-RE™ level IV
Thickness of special film mil is the measure of special film thickness. 1mil = 1/1000 inch The illustrations show images of glass cross sections based on the 15 mil film, and the thickness is actually controlled based on the total thickness of the film.	0.76 mm [30mil] (Nominal thickness 0.8 mm) 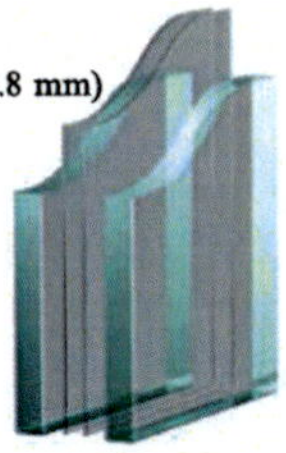	1.52 mm [60mil] (Nominal thickness 1.5 mm) 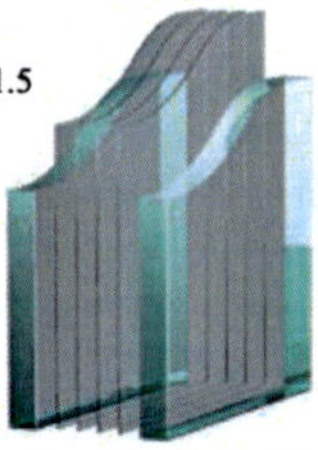	2.28 mm [90mil] (Nominal thickness 2.3 mm) 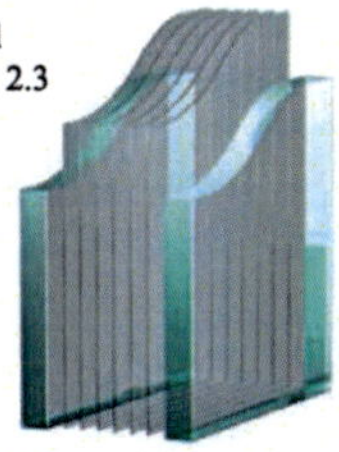	Polycarbonate <lexan> sheet 3 mm 
Overseas regulations on crime prevention (*)	Confirms to EN356 P2A	Conforms to EN356 P4A	Conforms to EN356 P5A	Conforms to EN356 P6B
Standards on performances and specifications stipulated by the Flat Glass Association of Japan (*)	Conforms to P2A/P2K	Conforms to P4A/P3K	Conforms to P5A/P3K	——

Fig. 18.35 Comparison of crime prevention performance of laminated glass. "SECU-RE™" and "SECU-RE Coolverre™" are trademarks of Asahi Glass Co. Ltd., which correspond to "crime prevention glass" and "crime prevention & UV-IR cut glass," respectively

unlike users overseas, were satisfied with the glass even if the crime prevention performance and the performance to prevent the intrusion of flying objects during typhoons were relatively low. However, in the future it is likely that the users will demand higher glass performance responding to crime prevention e.g. involving handguns and damage resistant from natural disasters such as typhoons and earthquakes. Therefore, demands for the research and development of new materials and new manufacturing methods are expected to increase.

Crystallized Glass Ceramics Used as Construction Materials (1974)

18.7

Crystallized glass ceramics are made by the heat treatment of glass. Crystals are homogeneously precipitated inside the glass. They have unique properties due to the precipitated crystalline phase and the remaining glass phase. Crystallized glass ceramics used as construction and building materials (Fig. 18.36) have higher mechanical and thermal impact resistance than natural stones (such as marble stone and granite). They also have high whiteness and diffuse reflectance as well as superior chemical resistance against acid and alkali. Therefore, crystallized glass ceramics have been used widely to cover the walls and pillars of buildings, offices, subway stations, underground malls, etc. since the 1970s.

18.7.1 Background of Development

Natural stone such as marble stone and granite have a unique characteristic where the growth of cracks caused by mechanical shocks is blocked by grain boundaries between the crystalline material. However, the cracks remain internal because the structure is heterogeneous. Marble stone is an assembly of calcite and its strength, hardness and chemical resistance are relatively low. Granite is composed of various minerals such as quartz, feldspar and mica. Mica with low thermal shock tends to peel off easily. Meanwhile, if glass is used as the exterior wall material, growth of cracks caused by mechanical and thermal shock cannot be blocked as glass does not have grain boundaries unlike natural stone, although no intrinsic cracks exist in glass as the structure is homogeneous. These disadvantages were resolved by creating crystallized glass ceramics, which have a structure typical of glass with precipitated crystals inside the glass.

Crystallized glass ceramics used as construction materials show unique properties, which were realized by a combination of the crystalline phase created homogeneously inside the glass through thermal treatment in combination with the remaining glass phase. Following the development of white crystallized glass ceramics with the patterns of high-quality white marble stone, crystallized glass ceramics of various colors can be created now by coloring the remaining glass phase by the intended color.

18.7.2 Characteristics

18.7.2.1 Characteristics of Products

Crystallized glass ceramics used as construction materials have a higher level of mechanical and thermal impact resistance than natural stones (such as marble stone and granite). They also have high whiteness and diffuse reflectance as well as superior chemical resistance against acid and alkali.

There are two types of crystallized glass ceramics for use as construction materials; (1) the surface crystallized glass ceramics (Note 18.20) made by fusing glass grains and growing crystals from the grain boundaries toward the inside of the glass grains utilizing the crystal seeds generated on the fused boundaries and (2) the volume crystallized glass ceramics (Note 18.21) made by growing multiple crystal cores homogeneously inside sheet glass using nucleating agents such as titanium oxide and zirconium oxide (TiO_2, ZrO_2) to grow crystals on these crystalline seeds by increasing the temperature.

Note 18.20 Crystallized glass ceramics in which crystals are grown by utilizing crystal seeds generated on fused boundaries of glass grains.

Note 18.21 Crystallized glass ceramics are made by growing multiple crystalline seeds homogeneously inside sheet glass using nucleating agents such as titanium oxide and zirconium oxide that initiate crystalline growth at higher temperature.

Fig. 18.36 Crystallized glass ceramics used on pillars of a subway station. The material is made of surface crystallized glass ceramics with high whiteness and diffuse reflectance

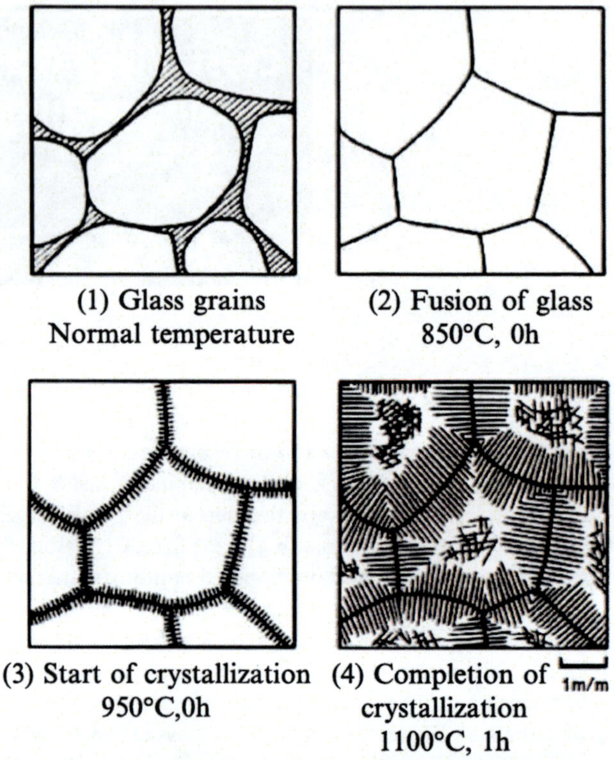

Fig. 18.37 Crystallization process by the accumulation technique. Glass grains with a diameter of 1–6 mm are homogeneously accumulated inside the refractory formwork and heated in the furnace to complete crystallization

18.7.2.2 Manufacturing Method

18.7.2.2.1 Surface Crystallized Glass Ceramics

Na_2O–CaO–Al_2O_3–SiO_2-based glass made by increasing CaO content of soda-lime glass for easy nucleation of crystals is melted, and the melt is poured into water to create glass grains with a diameter of 1–6 mm. Incorporating the crystallization process called the accumulation technique shown in Fig. 18.37, (1) Glass grains are homogeneously accumulated inside the refractory formwork and heated in the furnace, and then (2) the glass grains are softened and deformed at around 850°C and fuse together. (3) At about 950°C, needle crystals of β-wollastonite ($CaO{\cdot}SiO_2$) start growing from the fused boundary of the glass grains toward the inside of the grains, and (4) crystallization is completed in 1 h at about 1,100°C, creating a flat sheet composed of 30% crystal phase and 70% glass phase (in terms of weight). If the flat sheet is reheated on a fireproof curved die, the residual glass is softened and the sheet is transformed into a curved sheet intended for use as pillar covers. The surface of flat and curved sheets can be polished to a mirror finish. The largest product size is 900 mm × 1,800 mm × 15 mm. Crystallized glass ceramics with various colors and patterns are created using glass melts by adding pigments or by performing crystallization thermal treatment by mixing glass grains and inorganic pigments.

18.7.2.2.2 Volume Crystallized Glass Ceramics

Volume crystallized glass ceramics are made by adding nucleating agents, e.g. TiO_2, ZrO_2, to the MgO–ZnO–Al_2O_3–SiO_2-based glass composition. Glass melt, which is melted at around 1,600°C, is formed into a glass sheet with a thickness of 7 mm via the roll forming method. The glass sheet is placed on a refractory plate and the crystallization heat treatment is performed on the glass sheet. At about 750°C, countless numbers of crystalline seeds nucleate inside the glass, and forsterite ($2MgO{\cdot}SiO_2$), gahnite ($ZnO{\cdot}Al_2O_3$) and rutile (TiO_2) precipitate after 1 h at 1,050°C, creating a crystallized glass ceramics sheet with a 15% crystalline phase and 85% glass phase composition (in terms of weight). Because of the high content of the glass phase, the material can be molded easily. The surface of the material after crystallization is smooth like a mirror because of the surface flow during the crystallization heat treatment. Polishing is also not required. The maximum product size is 900 mm × 1,800 mm × 7 mm.

18.7.2.3 Characteristics

Characteristics of crystallized glass ceramics and natural stone for construction materials are compared in Table 18.2.

Table 18.2 Comparison of characteristics of crystallized glass ceramics and natural stone for construction materials

	Surface crystallized glass ceramics	Volume crystallized glass ceramics	Marble stone	Granite
Whiteness (L value)	89	93	59	66
Diffuse reflectance (%)	80	89	42	44
Thermal expansion coefficient (10^{-7}/K)	62	73	80–260	50–150
Density (g/cm^3)	2.7	2.5	2.7	2.7
Bending strength (MPa)	50	59	17	15
Vickers hardness (Hv0.2)	530	580	150	70–720
Acid resistance (1% H_2SO_4) (%)	0.08	–	10.2	1.0
Alkali resistance (1% NaOH) (%)	0.05	–	0.30	0.10
Water absorption rate (%)	0.00	0.00	0.30	0.35

18.7.2.3.1 Surface Crystallized Glass Ceramics

Surface crystallized glass ceramics show high bending strength, which is about three times higher than that of natural stone, and high Vickers hardness, making them resistant to scratches. In addition, the light passing through the glass phase is reflected by numerous β-wollastonite crystals precipitated inside, realizing high whiteness of 89% and diffuse reflectance of 80%, which is approximately twice that of natural stone. Wall coverings made of crystallized glass ceramics reflect 80% of the incident light to various directions. If the materials are used on the walls and pillars of subway stations, bright illuminated spaces without shadows will be realized.

18.7.2.3.2 Volume Crystallized Glass Ceramics

Volume crystallized glass ceramics have homogeneously-distributed small crystal grains sized 0.5–1.5 μm and the crystallinity is low. However, the whiteness and diffuse reflectance are high at 93% and 89%, respectively, because there is a big difference between the refractive index of precipitated crystals and that of the glass phase. The materials are also used to realize well illuminated spaces. They are mainly used as interior materials because of their light weight.

18.7.3 Future Prospects

Various types of colored crystallized glass ceramics for construction materials with optical functions are manufactured by this crystallization process, so called accumulation technique. If the transmitted light is scattered on the boundaries where the glass is crystallized, the material will look as if it is a luminescent material. If the boundaries of fused glass grains are prepared to contain light storage crystals, which absorbs and stores the energy of light and then spontaneously releases the energy in the form of light, fantastic blue–green light will be emitted.

Ferrite Electromagnetic Wave Absorber (1969) 18.8

The ferrite electromagnetic wave absorber (Note 18.22) (Fig. 18.38) is a solid body made by sintering a mixture of iron oxide (Fe_2O_3) and bivalent metal oxides (such as NiO and ZnO) at high temperatures of one thousand several hundred degrees Celsius. It is processed into the form of tile for use in various applications. Ferrite electromagnetic wave absorbers, developed in the 1960s, have been used in the communication field for TV broadcasting, wireless communications and various radars as well as the EMC (Note 18.23) field for the prevention and evaluation of noise from electric and electronic devices. The material is essential for electromagnetic anechoic chambers used for the evaluation of noise from electric and electronic devices.

18.8.1 Background of Development

It has been reported that electromagnetic wave absorbers were already commercialized in the early 1940s. The initial research and development were performed with a focus on resistive wave absorbers and dielectric wave absorbers made by using carbon, graphite, etc. The new electromagnetic wave absorber containing ferrites was developed in the second half of the 1960s by professors Kunihiro Suetake, Yoshiyuki Naito and Yasutaka Shimizu at the Tokyo Institute of Technology. They focused their attention on the magnetic relaxation (the loss caused by the delay in magnetization in relation to the rapid change of magnetic fields) of ferrite in the high-frequency range to develop electromagnetic absorbing materials. Ferrite electromagnetic wave absorbers have been used in the communications field for TV broadcasting, wireless communications and various radars as well as the electromagnetic compatibility (EMC) field for the prevention and evaluation of noise from electric and electronic devices.

Composite wave absorbers made by combining a ferrite electromagnetic wave absorber and a dielectric loss body have been commercialized for use in electromagnetic anechoic chambers intended for EMC tests, which is shown in Fig. 18.39. Ferrite electromagnetic wave absorbers are essential for electromagnetic anechoic chambers used for EMC tests. Ferrite electromagnetic wave absorbers have also been used to prevent false echo images in air traffic control radar caused by the ventilation towers of the Tokyo Bay Aqua-Line (Fig. 18.40) and to prevent the reflex disturbance of TV airwaves (TV ghost) attributable to skyscrapers such as the Tokyo Metropolitan Government Buildings (Fig. 18.41).

18.8.2 Characteristics

18.8.2.1 Characteristics of Products

The ferrite electromagnetic wave absorber is a solid body made by sintering a mixture of iron oxide (Fe_2O_3) and bivalent metal oxides (such as NiO and ZnO) at high temperatures of one thousand several hundred degree Celsius. It is expressed with the chemical formula $M^{II}O{\cdot}Fe_2O_3$ and processed for use into the forms of tiles. Ferrite electromagnetic wave absorbers have physical properties similar to that of porcelain tiles used as exterior building materials. Therefore, its chemical stability is equivalent to that of porcelain tiles and secular change of its complex permeability is limited, featuring high durability.

As for complex permeability of ferrite ($\mu_r = \mu'_r - j\mu''_r$), the value of μ''_r, the loss term, becomes high in the VHF band (30–300 MHz), and frequency characteristics μ'_r and μ''_r

Note 18.22 A substance that absorbs and attenuates incident electromagnetic energy and converts the energy to thermal energy.

Note 18.23 (Electro-magnetic compatibility)

The balance between not having devices emit electromagnetic waves that may affect other devices (emission issues) and the normal operation of devices without them being affected by external electromagnetic waves (immunity issues).

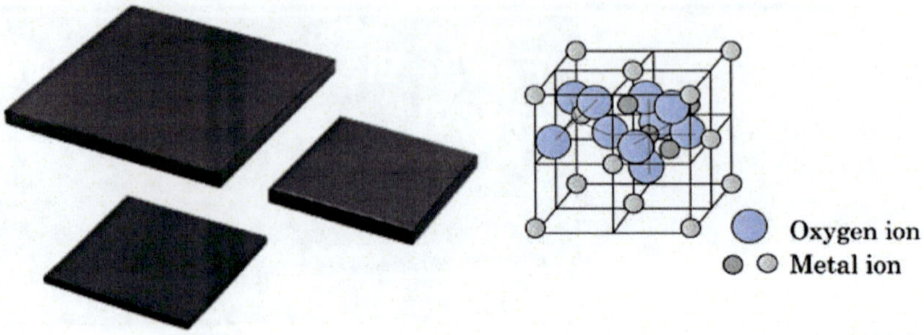

Fig. 18.38 Ferrite electromagnetic wave absorber. A ferrite electromagnetic wave absorber is a solid body made by sintering a mixture of iron oxide (Fe_2O_3) and bivalent metal oxides (such as NiO and ZnO) at high temperatures of one thousand several hundred degree Celsius. It is processed for use into the forms of tiles, etc.

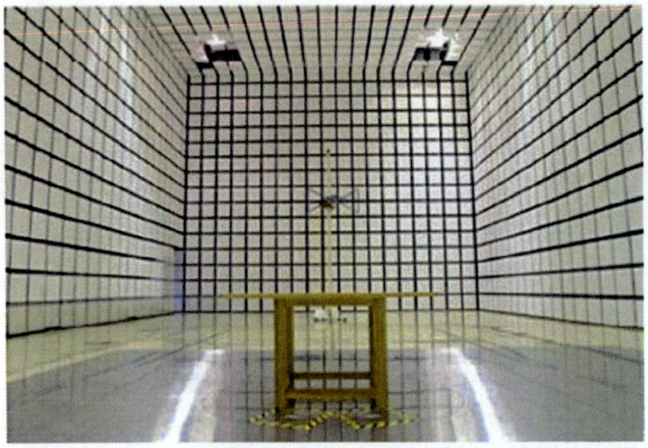

Fig. 18.39 Electromagnetic anechoic chambers used for EMC tests. The floors of electromagnetic anechoic chambers for EMC tests are required to have wave reflective properties equivalent to that of metals or the ground of the earth itself. Composite wave absorbers (containing ferrite electromagnetic wave absorbers and a dielectric loss body) are mounted on the five walls

Fig. 18.41 Ferrite electromagnetic wave absorber for prevention of TV ghost. Ferrite electromagnetic wave absorbers are used to prevent TV ghost images caused by skyscrapers taller than 200 m, such as the Tokyo Metropolitan Government Buildings and Shinjuku Park Tower Hotel

Fig. 18.40 Ferrite electromagnetic wave absorber for the prevention of false echo images in air traffic control radar. A ferrite electromagnetic wave absorber is used to prevent false echo images in air traffic radar at Haneda Airport caused by the ventilation towers of the Tokyo Bay Aqua-Line

satisfy the conditions required of wave absorbers. Therefore, the material provides superior wave absorption performance in a wide range of frequencies. Ferrite electromagnetic wave absorbers are normally manufactured by blending Ni–Cu–Zn, Mn–Zn or Ni–Zn-based bivalent metal oxides.

Figure 18.42 shows wave absorption characteristics of Ni–Cu–Zn ferrite electromagnetic wave absorbers. Ni–Cu–Zn ferrite electromagnetic wave absorbers are mainly used in electromagnetic anechoic chambers for EMC tests. This is because the material can increase the return loss in low frequency ranges of the VHF band and the thickness can be reduced to about 6 mm.

18.8.2.2 Manufacturing Method

Ferrite is manufactured by various methods, which are classified into three types—the dry method, the wet method and the co-precipitation method. In Japan, ferrite for industrial use is normally made via the dry method. The manufacturing method of ferrite electromagnetic wave absorbers is shown in Fig. 18.43, taking the Ni–Cu–Zn ferrite electromagnetic wave absorber as an example.

The main material of ferrite is Fe_2O_3 and the Ni–Cu–Zn ferrite electromagnetic wave absorber contains Fe_2O_3 at about 70 wt%. Responding to improvements in the performance of ferrite electromagnetic wave absorbers, higher purity Fe_2O_3 is required. With respect to other materials that are used, impurities vary depending on the type of ferrite and the intended use. Therefore, adoption of these materials is decided in accordance with the intended use.

The selected raw materials are weighed and mixed by the wet method or the dry method. During this process, the raw materials are controlled with consideration for compositional changes caused by calcination (1st sintering), grinding, and sintering (2nd sintering). The chemical reaction of ferrite is promoted during calcination so as to stabilize the shrinkage rate of formed bodies during sintering. Grinding after calcination is performed to promote sintering reactions and so as to adjust the particle size distribution in order to improve the density of the formed body. Then the ground powder is granulated and formed into tile shapes by applying a specified pressure. The formed body is heated at a high temperature of one thousand several hundred degree Celsius for the sintering so as to complete the chemical reaction of the ferrite and enhance its density and mechanical strength. Microstructure of the ferrite is formed, meaning determination of the magnetic properties (complex permeability), during sintering. Therefore, the heat treatment conditions and the atmosphere during the sintering process are important factors. Finally, the edges of the tile-shaped ferrite electromagnetic wave absorber are polished to prevent the deterioration of magnetic properties.

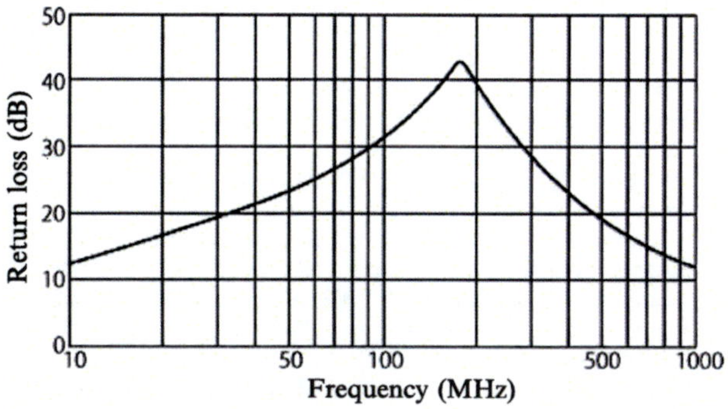

Fig. 18.42 Wave absorption characteristics of Ni–Cu–Zn ferrite electromagnetic wave absorbers. Ni–Cu–Zn ferrite electromagnetic wave absorbers are mainly used in electromagnetic anechoic chambers for EMC tests because the material can increase the return loss in low frequency range of the VHF band (30–300 MHz). The thickness can be reduced to about 6 mm

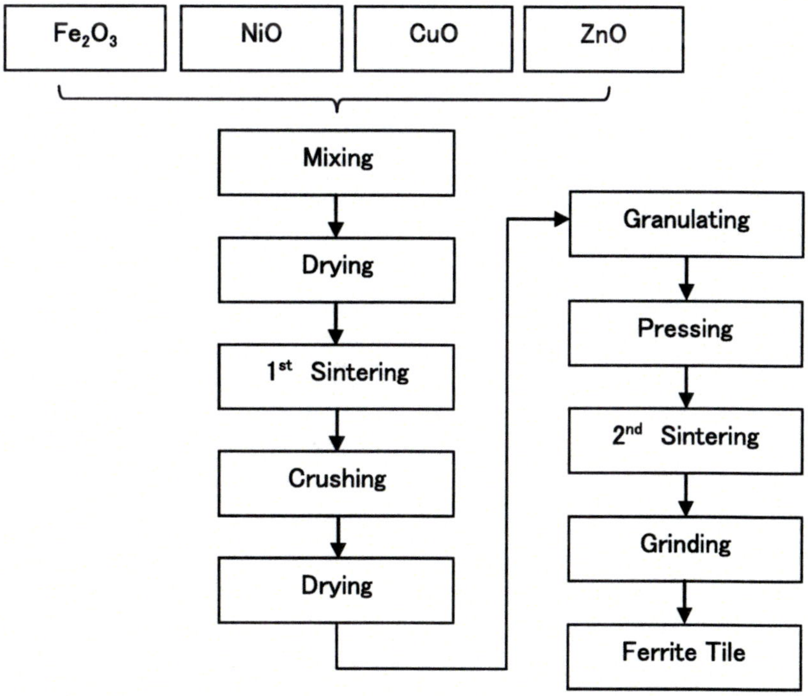

Fig. 18.43 Outline of ferrite electromagnetic wave absorber manufacturing method (example of Ni–Cu–Zn ferrite electromagnetic wave absorber). Amounts of metal oxides (Fe_2O_3, NiO, CuO and ZnO) that comprise the ferrite are controlled. The ferrite electromagnetic wave absorber is manufactured following the manufacturing method

18.8.2.3 Performances of Products

The most important performance required of ferrite electromagnetic wave absorbers is the input impedance characteristic. The frequencies for the evaluation of radiation noise from electric and electronic devices range from 30 MHz to 40 GHz, which is extremely wide. Therefore, it is rare that the ferrite electromagnetic wave absorber is used independently in electromagnetic anechoic chambers for EMC tests. Due to this reason, a composite wave absorber is used. This is made by combining a dielectric loss body (foam mixed or impregnated with a dielectric material such as carbon) with a ferrite electromagnetic wave absorber. Impedance matching of each material is important for composite wave absorbers. Therefore, they are designed so that the ferrite electromagnetic wave absorber effectively absorbs waves in low frequencies ranging from 30 to 300 MHz and so that the dielectric loss body effectively absorbs waves in high frequencies above 300 MHz. As explained above, the input impedance characteristic is very important for ferrite electromagnetic wave absorbers.

18.8.3 Future Prospects

Following implementation of the EMC directive in 1996 and strengthening of regulations on large devices [ISM (Industry, Science, Medical)], demands for electromagnetic anechoic chambers capable of evaluating large/heavy equipment have been increasing. Demands for electromagnetic anechoic chambers (used to perform measurements for EMC measures and evaluations) are expected to increase in the future. The performance of ferrite electromagnetic wave absorbers will continue to be improve responding to increasing demands for higher-performance electromagnetic anechoic chambers. Applications will expand to communication devices as measures to prevent communications failures. We anticipate that ferrite electromagnetic wave absorbers will continue to be used and develop further.

Literature

1. Glass Fiber Association of Japan (1986) History and future of glass fibers. Editorial Committee of 25th Anniversary Magazine, pp 67–68 (in Japanese) **(18.3)**
2. Fukunaga H (2002) Ceramics 37:6–9 (in Japanese) **(18.4)**
3. Fukunaga H, Yokoyama S, Kitamura K (2005) Environ Res Eng 52:128–135 (in Japanese) **(18.4)**
4. Watamura S, Shibazaki Y (2000) Ind Mater 48(7):18–22 (in Japanese) **(18.4)**
5. Fukunaga H, Yokoyama S (2005) Environ Res Eng 52:59–64 (in Japanese) **(18.4)**
6. Suzuki A, Kawai S, Tanabe S (2002) Abstract of the convention of the Architectural Institute of Japan (Hokuriku), D-2, pp 995–996 (in Japanese) **(18.4)**
7. Consumer Goods Industries Bureau, Ministry of International Trade and Industry (1996) For comfortable houses that provide affluence and abundance, Tokyo (in Japanese) **(18.5)**
8. Wada M, Ninomiya M (1995) Ceramics 30:846–850 (in Japanese) **(18.7)**
9. Yamamoto R (2006) Eco material handbook. Maruzen, pp 134–136 (in Japanese) **(18.7)**
10. Yamane M (1999) Glass engineering handbook. Asakura Publishing, pp 587–589 (in Japanese) **(18.7)**
11. Suetake, Naito, Takeda. Magnetic-type resistance film wave absorptive wall. Shingakukai Microwave Research Document (1967.1) (in Japanese) **(18.8)**
12. Shimizu, Sugiura, Ishino, Inui. Advanced method for shielding and absorption of electromagnetic wave. Nippon Gijutsu Tosho (in Japanese) **(18.8)**
13. Hiraga, Okutani, Ojima (1986) Ferrite. Electronic Material Series, Maruzen (in Japanese) **(18.8)**

Housing Products

19

Ceramic products that are used in houses and closely related to our daily lives are introduced in this section.

Sanitary wares that consist of water handling products have the following characteristics, such as, high cleanability, corrosion resistance and wear resistance and are used as toilet basins, sinks, etc. (Sect. 19.1). Needs for improvements in cleanability have been increasing and efforts are being made to develop new shapes and materials. One of the technologies related to quality is surface smoothing technology. A homogeneous glass layer is created on the glaze layer so as to maintain properties for a long period of time.

Ceramic tiles are used not only in bathrooms and restrooms but also as exterior wall materials that are essential for buildings (Sect. 19.2). In order to enhance the stain-proof performance of exterior wall tiles, a superhydrophilic stain-proof layer is coated on the surface and a water containing film remains on the surface after getting wet. Hydrophobic components attributable to exhaust gas from automobiles and plants adhere to this water layer. However, the adhesion force is weak and the components are easily washed away by rainwater, etc.

People are paying more attention to drinking water in recent years. Tap water purifiers are widely used in many households. Water purifiers have built-in filters (Sect. 19.3) to remove chlorine and trihalomethane, etc. in tap water by using a double-layered activated carbon. Individual bacteria, bacteria coliform, iron, etc. are removed completely using a ceramic filter. Ceramic filters have honeycomb-like structures as they need to be compact and to have a large surface area. They are composed of a film portion that removes impurities and a honeycomb-shaped base material portion.

Single-lever (one handle) mixer faucets are widely used because they allow for the easy adjustment of water flow rate and temperatures and feature high operability (Sect. 19.4). The cartridge inside the faucet contains multiple porous ceramic disks. The faucet water flow rate is adjusted by changing the position of the disks via the lever. Ceramic disks are generally made of alumina as high abrasion resistance and corrosion resistance are required of the disks. Silicon carbide is sometimes used as surface coating that improves sliding performance.

Y. Imanaka et al. (eds.), *The Ceramic Society of Japan, Advanced Ceramic Technologies & Products*,
DOI 10.1007/978-4-431-54108-0_19, © Springer Japan 2012

Sanitary Ware (1914)*

19.1

As the name indicates, sanitary wares remain sanitized for a long period of time. They have been used widely for households equipment such as toilet bowls and wash basins since the after WWII. These products belong to ceramic ware called old ceramics, which are made of natural materials such as clay, china stone and feldspar and are coated with glaze. Since they are complex in shape, large and are subjected to water, optimization of the materials as well as improvement of the manufacturing technologies has been promoted.

In recent years, the environmental needs for "water saving" and sanitary needs for "high cleaning performance" have been increasing for household equipment, and research and development have been produced new functions.

19.1.1 Background of Development

Sanitary wares are easy to clean because of the smooth surface. They are also free from corrosion and abrasion. They have been used in houses and public facilities for many years as sanitary equipment such as toilet basins, urinals, wash basins and hand wash basins. This is shown in Fig. 19.1. In Japan, research began on manufacturing methods, in 1912 at the earthenware laboratory of Nippon Toki Ltd. with a focus on the English style toilet bowl (Note 19.1). The flush urinal was released to the Osaka market in 1914. The applications expanded rapidly after the World War II in line with the efforts to solve the issues of insufficient supply of houses, to improve the living standard and to upgrade environmental sanitation. They are currently used in most living spaces. Approximately 5.9 million sanitary wares were produced in Japan in 2007.

*The number in parenthesis indicates the year that the product was first commercialized.

Note 19.1 Currently Noritake Company Ltd.

19.1.2 Characteristics and Manufacturing Method

19.1.2.1 *Characteristics*

Sanitary wares are much heavier than dishes, and if they are made of porcelain that is high in glass content, they deform under their own weight during the calcination process, making production difficult. If they are made of china that is low in glass and high in aggregate, the areas not coated with glaze are water absorbing, resulting in the absorption of polluted water or fracture due to freezing. Therefore, a green body called a melt green body, which has the properties between that of china and porcelain, featuring small deformation during calcination and low water absorbing properties, is used to produce sanitary wares. Currently in Japan, sanitary wares are normally made of a melt green body, which is specified in the JIS A5207 standard (Note 19.2).

Typical properties of sanitary wares are listed in Table 19.1.

19.1.2.2 *Manufacturing Method*

Sanitary wares are generally made by the slip casting method (refer to Fig. 19.2). In slip casting, the raw material containing clay, china stone, feldspar, silica stone, etc. is blended with water and crushed to create slip, which is fed into a mold made of plaster, etc., where the water is absorbed to produce a formed body. In slip casting, solid casting (Note 19.3) and drain casting

Note 19.2 Established in 1953. It specifies that "penetration of ink to be 3 mm max." with regard to "quality of earthen ware."(revised in 2005).

Note 19.3 In this forming method, a mold is used to create a shape of a certain thickness and then the slip is added from both sides to create a formed body (Fig. 19.2).

Note 19.4 In this forming method, slip is fed into a mold, which is designed to have a space larger than the thickness. When a certain amount of slip is formed on the mold, the remaining slip is drained (Fig. 6.2.1-2).

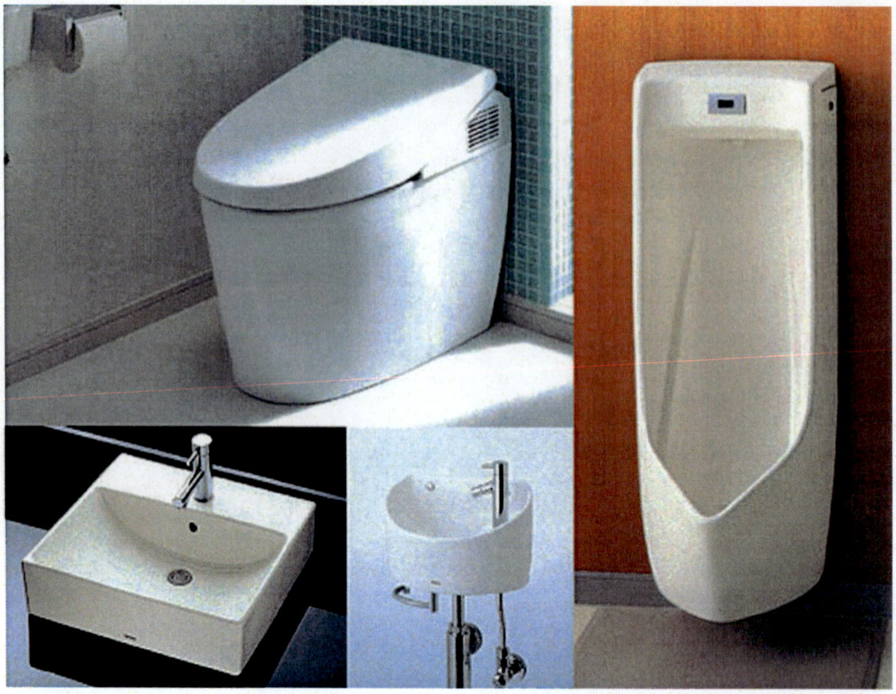

Fig. 19.1 Sanitary wares used in restrooms

Table 19.1 Typical properties of sanitary wares

Specific weight	g/cm³	2.5
Hardness	Mohs	6.5
Water absorption rate	%	0.03
Bending strength	MPa	80
Thermal expansion coefficient	10^{-7}/°C	68

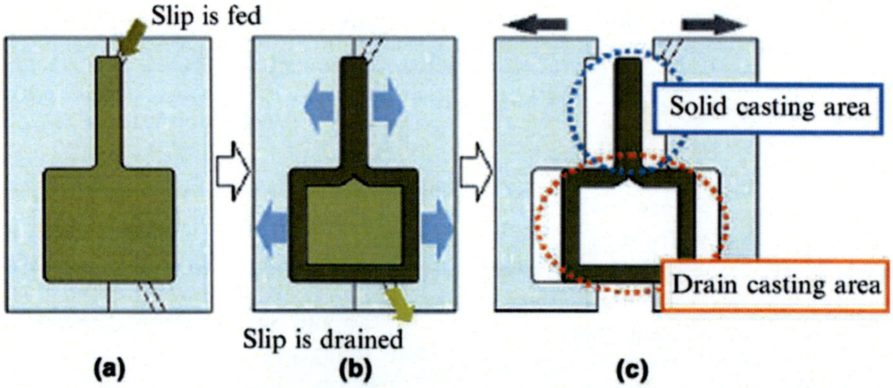

Fig. 19.2 Manufacturing process incorporating the slip casting method (example). (**a**) Slip is fed into the mold, (**b**) it is formed for a specified period and then the remaining slip is drained, (**c**) the slip is compacted and then is removed from the mold to create a formed body

(Note 19.4) can be performed at the same time, making it possible to manufacture sanitary wares such as toilet bowls featuring complex shapes with a water supply channel, a bowl and a drain channel within a limited configuration of a mold.

Figure 19.3 shows the general production flow of sanitary wares. The formed body made using slip casting is dried and coated with glaze of slip with the spray method. The formed

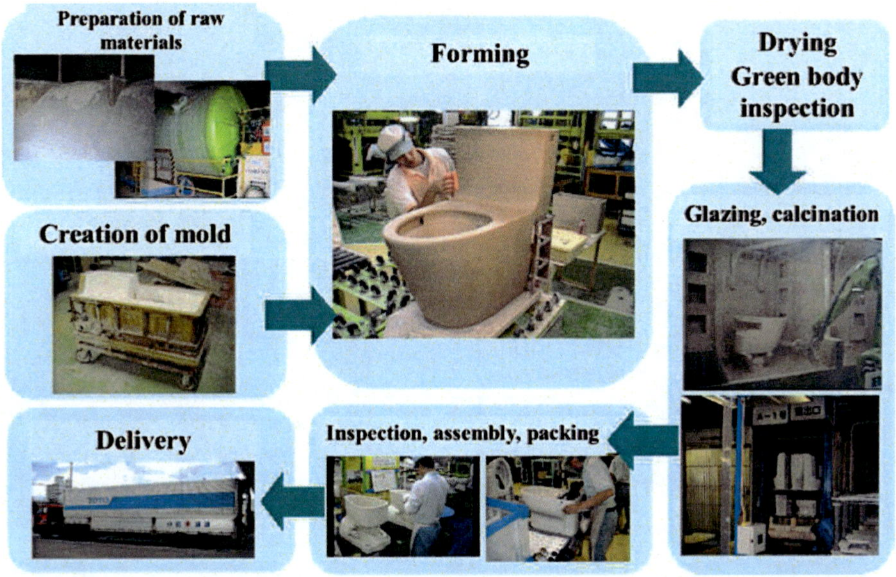

Fig. 19.3 Production flow of sanitary ware (toilet bowl)

body applied with glaze is calcined at about 1,200°C for 0.5–1 day to create the product.

The pressure forming method, where porous resin mold is used for pressurized formation and dehydration, has been employed recently for the improvement of productivity. Following the introduction of the resin mold, production amount per day and service life of the mold have been improved remarkably when compared with the plaster mold.

19.1.3 Future Prospects

Sanitary wares of today are required to feature high "water saving" and "cleaning" performance. With respect to the "improvement of cleaning performance," multiple approaches are being made in regard to the shapes and the materials.

As for the shapes, the areas including the edge of toilet bowls are being improved to easy-to-clean shapes (Fig. 19.4).

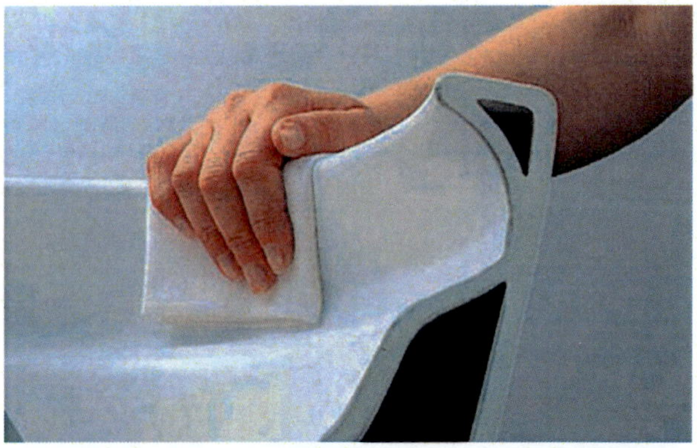

Fig. 19.4 Toilet bowl without curved edge. The shape of the edge (which gets dirty easily and causes a bad smell) was improved. Following the improvement, water outlets on the back of the edge were discontinued. A swirling jet of water is discharged horizontally from a water outlet

As for materials, antifouling technologies have been studied to keep the surface of sanitary wares free from soiling, urinary calculus, water stain, virus, etc., and some of the technologies have been commercialized. The technologies include the antimicrobial technology where antibacterial agents such as silver are mixed in the glaze layer, the water

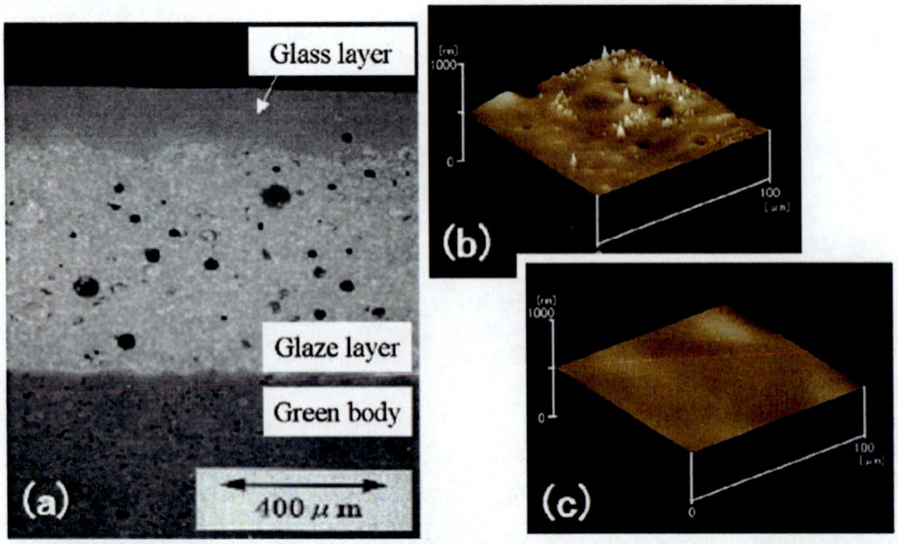

Fig. 19.5 Antifouling technologies for sanitary wares. (**a**) Cross-sectional SEM image of a product incorporating the surface smoothing technology, where a homogeneous glass layer is coated over a conventional glaze layer, (**b**) AFM image of the surface of a conventional glaze layer, (**c**) AFM image of the surface of a product incorporating the surface smoothing technology

repellent technology where the surface of a sanitary ware is coated with water repellent treatment and the surface smoothing technology (Fig. 19.5) where a homogeneous glass layer is added on the conventional glaze layer for longer maintenance of smoothness.

Efforts will be continued for the development of new technologies that realize further reduction of environmental loads (which includes water saving and cleaning agent saving by improvement of cleaning performance) and to make possible a rich and comfortable life.

19.2 Superhydrophilic Stain-Resistant Ceramic Tiles (1999)

Ceramic tiles have 4,000 years of history and are widely used as finishing materials in buildings throughout the world. They are used as interior and exterior materials in Japan. Ceramic tiles feature higher durability and stain-resistance performance compared with other materials, but black streaks or stain are sometimes observed on them after long term atmospheric exposure. Superhydrophilic stain-resistant ceramic tiles were made by improving the stain-resistant performance of conventional ceramic tiles for exterior use. The tiles are coated with silica particles of several tens of nanometers to prevent adhesion of hydrophobic fine particles such as particles or soot contained in exhaust gas from automobiles and plants, which are believed to cause stains on buildings in urban areas. Even if they are stained, the stains are easily washed away by rain, retaining the buildings and landscapes original appearance.

19.2.1 Background of Development

Ceramic tiles have a history of over 4,000 years and historic buildings made of the tiles still remain beautiful due to the superior durability of the tiles. Ceramic tiles began to be used in Japan in line with the introduction of western architecture into Japan after the Meiji Restoration. They are major construction materials and used in exterior walls, bathrooms and restrooms due to their superior durability (Fig. 19.6). Ceramic tiles are more stain-resistant than other exterior materials due to the following properties: (1) they feature appropriate hydrophilic properties, (2) they hardly absorb water, absorb few stains, (3) they are resistant to deterioration because they are sintered at 1,200°C or higher and (4) they are resistant to acid rain due to chemical resistance.

However, stains are observed on the surface of exterior ceramic tiles around projections such as parapets (Note 19.5), head boards (Note 19.6), exhaust vents and signboards as well as around openings such as sash windows of modern buildings in urban areas. Stains usually contain hydrophobic components (such as oil) and are not easily washed away by rain.

In order to prevent these stains, sealing materials resistant to stain are used or considerations are given to the structure of window ledges under sash windows to prevent water from flowing on the exterior wall. Moreover, responding to the needs for exterior ceramic tiles with higher stain-resistant properties, ceramic tiles featuring higher hydrophilic properties and stain-resistant properties than conventional ones have been developed.

19.2.2 Characteristics and Manufacturing Method

19.2.2.1 Characteristics of Ceramics

Superhydrophilic stain-resistant ceramic tiles are made by coating a film with a thickness of 100–300 nm, consisting of hydrophilic silica particles that are 10–30 nm in diameter, to create a fractal structure (Note 19.7) (Fig. 19.7). Hydroxyls that adhere to silica particles absorb moisture in the air, forming a water molecule film on the surface of ceramic tiles. The surface constantly remains hydrophilic. The water molecule film prevents generation of static electricity on the surface, reducing the possibility of stain adhesion on the surface. At the same time, stains are easily washed away by rain because the surface is hydrophilic (Fig. 19.8). Therefore, high stain-resistant performance is maintained all day and all night, regardless of the season. Running costs such as the cleaning cost is reduced, leading to a reduction of the environmental load due to stain-resistant properties.

Note 19.5 Lower walls on the outer circumference of building roofs. Intended for prevention of falling and water intrusion.

Note 19.6 Parapets and construction materials mounted on the upper ends of verandah walls. Installed to ensure waterproof property and improve design quality.

Note 19.7 The structure that becomes similar to the structure of the entire body when it is enlarged (self-similarity).

Fig. 19.6 Example of construction using ceramic tiles. Ceramics are widely used as finishing materials of buildings

19.2.2.2 Manufacturing Method

They are manufactured by forming a superhydrophilic stain-resistant layer on normal exterior tiles. Aqueous suspension, made by mixing inorganic binder in colloidal silica fine particles of 10–30 nm, is coated and sintered at about 600°C on the tiles to form a thin film.

Fig. 19.7 Scanning electron micrograph of superhydrophilic surface. The surface is covered by fine particles of several tens nanometers in diameter, featuring fine texture

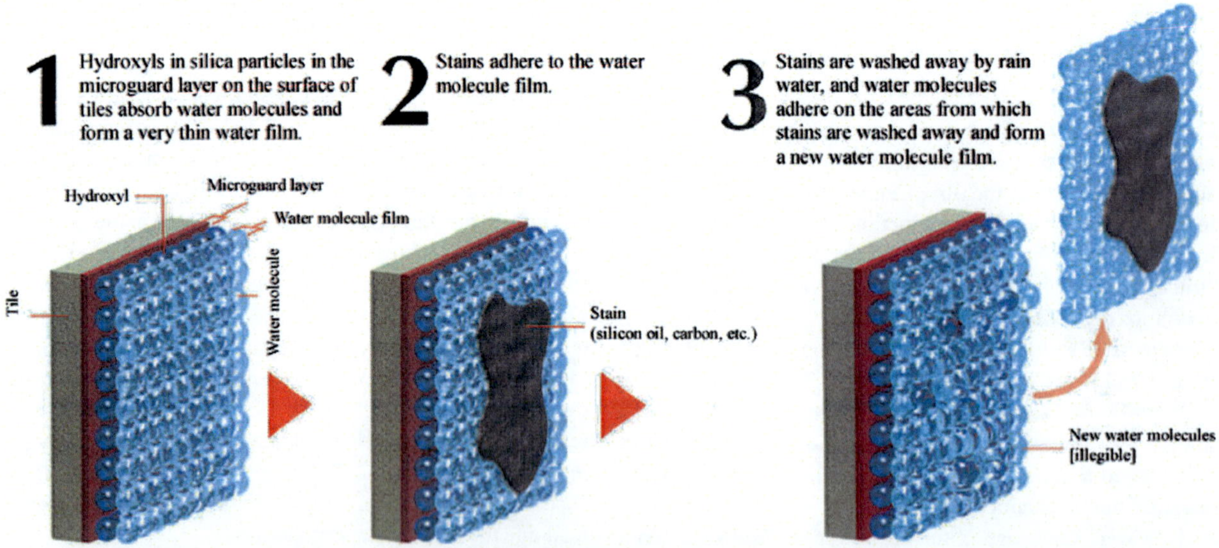

Fig. 19.8 Stain-resistant mechanism of superhydrophilic surface. Water vapor in the air adheres to the superhydrophilic surface, forming a water film. Stains adhere to the water film, and therefore, can be easily washed away by rain water, etc.

19.2.2.3 Performance and Specifications of Products

Stain removal performance was evaluated using silicon oil, as stains in the urban areas consist mainly of hydrophobic substances. It is difficult to completely remove stains on normal tiles using water. However, stains on superhydrophilic stain-resistant ceramic tiles can be almost completely removed using water (Fig. 19.9).

Results of outdoor exposure tests, which were performed to evaluate the products under actual use conditions, are shown in Fig. 19.10. Normal ceramic tiles and superhydrophilic stain-resistant ceramic tiles were bonded on flexible boards, filling the joints with silicon-based sealing material, which were mounted on exposure benches with stainless roofs angled at 10°. Silicon-based sealing material was used because it tends to be stained most severely according to the report by Kishima et al. [2]. Progress of stain formation was observed and the water contact angle against the surface was measured in respect to the two types of ceramic tiles. In both types, hydrophilic property was comparatively high with a water contact angle of 10° or less in the early stage of the test. However, the water contact angle of normal ceramic tiles exceeded 70° after 1 month and reached 120° after 6 months, indicating the progress of stains in line with the increase of the water contact angle. Meanwhile, the water contact angle of superhydrophilic stain-resistant ceramic

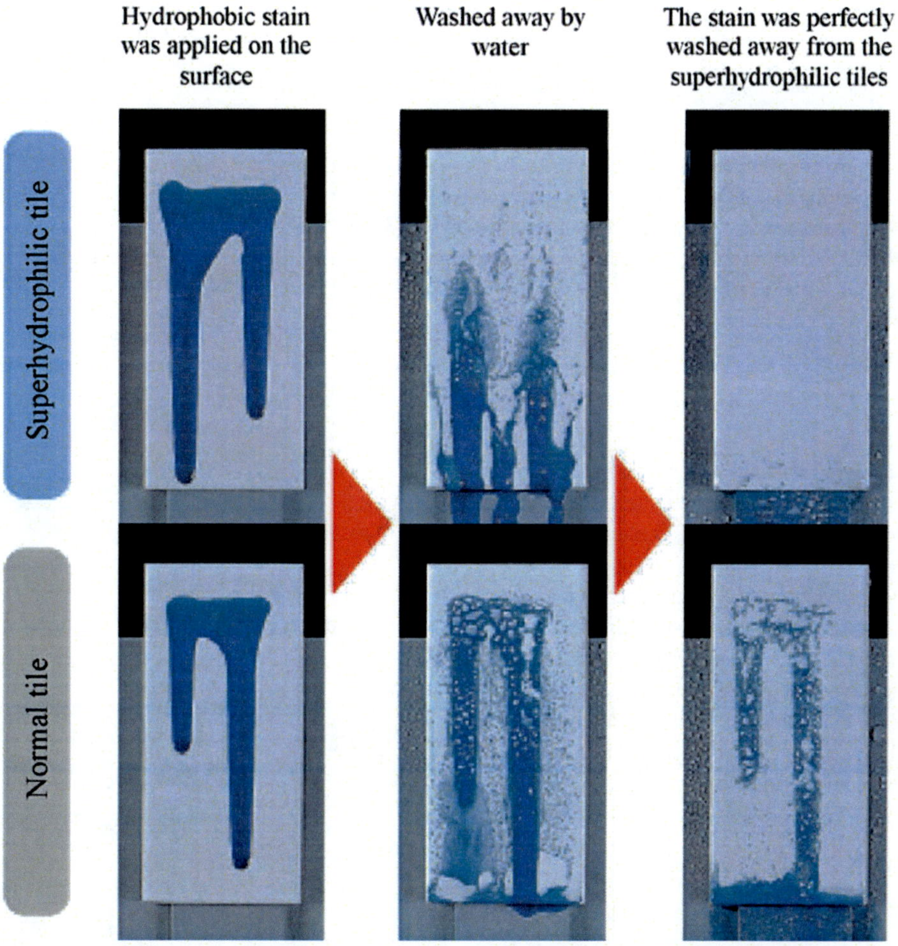

Fig. 19.9 Comparison of tiles in the performance of hydrophobic stain removal. Tiles are applied with silicon oil, hydrophobic stain, and was washed with water. The stain on the superhydrophilic tile was removed easily

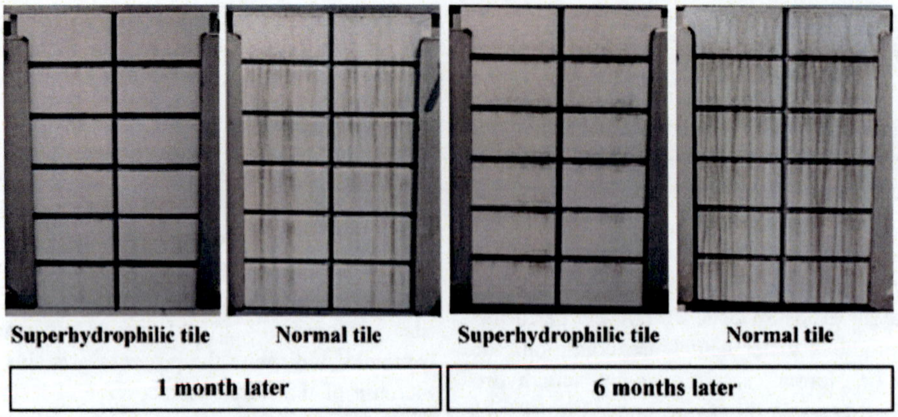

Fig. 19.10 Stains on tile surfaces during accelerated exposure test. Difference in stain formation was observed between the two types of tiles in the accelerated exposure test as well. Stain adhesion is clearly less in the superhydrophilic tile

tiles increased over time. However, it was about 40° after 6 months, showing a minimal sign of stains. It was proved that the progress of stains is extremely slow in superhydrophilic stain-resistant ceramic tiles, at about 1/10 compared with normal ceramic tiles (Fig. 19.11).

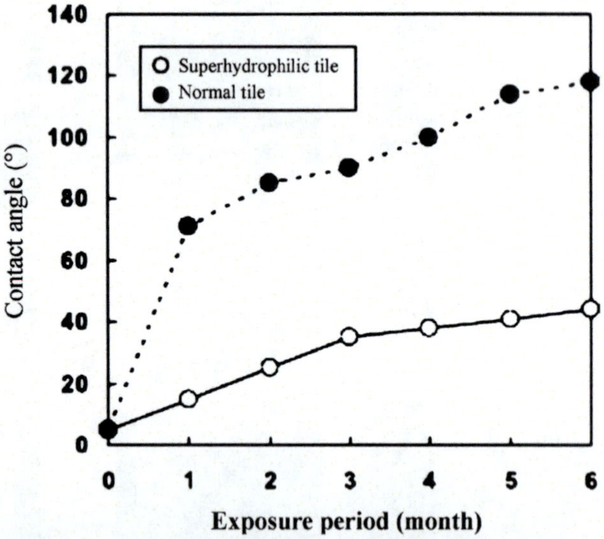

Fig. 19.11 Change of contact angles against tile surface during accelerated exposure test. In the accelerated exposure test, the water contact angle against the tile surface increases responding to the hydrophobic components seeping out of the sealing material. The rate of increase in the water contact angle is lower in superhydrophilic tiles than normal tiles

19.2.3 Future Prospects

Superhydrophilic stain-resistant tiles have been used for nearly 10 years and the interest in stain resistant exterior materials has been steadily increasing. The interest has been expanding to ceramic siding materials (Note 19.8), ceramic paint and ceramic window glass. Comprehensive anti-stain measures, including improvement of building structures for the proper draining of rain water and selection of appropriate sealing materials, have also been promoted.

It is expected that construction materials including stain-resistant tiles will be further advanced as well as technologies for eliminating sources of pollution.

Note 19.8 Made by forming, curing and hardening the materials consisting of cement and fibers. The materials can be formed into various shapes and 70% of the exterior finishing materials used today are ceramic sidings.

Water Purifier Filter (2001) 19.3

Water purifiers have been used widely by consumers who want to ensure tap water safety and good taste. Filters are used in water purifiers and ceramic filters that can be recycled after usage are attracting attention. Ceramic filters for water purifiers need to be compact and have a large membrane area, and therefore, the honeycomb structure is adopted. They are made from alumina, which is sintered to achieve a high density in order to enhance its strength. Therefore, the ability of the filter to remove impurities for months regardless of the water temperature, the chlorine concentration or the water pressure is important. They are manufactured using a typical ceramic processing method. However, strict quality control inspections are performed because they are used for the treatment of water that consumers drink.

19.3.1 Background of Development

People are more and more concerned about drinking water quality with keywords being purity and taste. "Tap water quality in Japan should be number one in the world based upon its taste." Following this idea, water purifiers for home use have been used widely in order to thoroughly remove impurities from tap water. Water purifiers for home use generally consist of activated carbon that removes chlorine and filters that remove impurities. Filters are classified into two types; organic polymer filters and inorganic ceramic filters. Figure 19.12 shows the structure of a water purifier that incorporates a ceramic filter. Figure 19.13 shows a schematic diagram of the filtration structure.

19.3.2 Characteristics and Manufacturing Method

19.3.2.1 Characteristics and Specifications of Products

Ceramic water purifying filters are required to remove bacteria such as cryptosporidium (Note 19.9) and iron released from water pipes completely. Ceramic filters are generally sintered at high temperatures enhancing their high pressure resistance and thermal resistance. They also need to maintain their impurity removal over a long lifetime.

Water purifiers are required to treat a large volume of water continuously over a specified period. They must be compact and have a large membrane area. A honeycomb structure is adopted to realize a large membrane area. In order to reduce permeation resistance during the passage of water, water purifiers are divided into two parts; the membrane area for removal of impurities and the base material area featuring a honeycomb structure. The filter has dimensions of 90 mm diameter × 100 mm in length in consideration for use within domestic kitchens. The filter material used is alumina. Its durability is not affected by factors that vary depending on the region, such as changes in water temperature or differences in the chlorine concentration.

19.3.2.2 Manufacturing Method

Ceramic filters for water purifiers are made using the extrusion molding method, with alumina being the main raw material.

Note 19.9 A single-cell parasite belonging to the cryptosporidium family, eucoccidiida, sporozoea. *Cryptosporidium parvum*, which enters the body and inhabits the intestines of humans and cattle, can be found across the globe. Recently, waterborne transmission via drinking water is an issue of concern. Infection rates are high in infants and during the summertime.

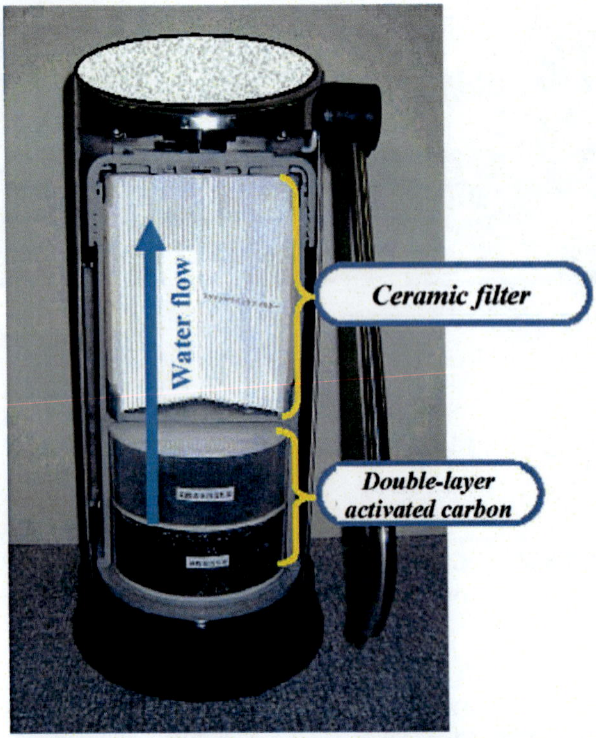

Fig. 19.12 Configuration of a water purifier. The supplied water passes through two layers of activated carbon; the activated carbon layer for removal of chloride and the activated carbon layer for removal of trihalomethane, etc., and goes into the ceramic filter that removes bacteria and iron

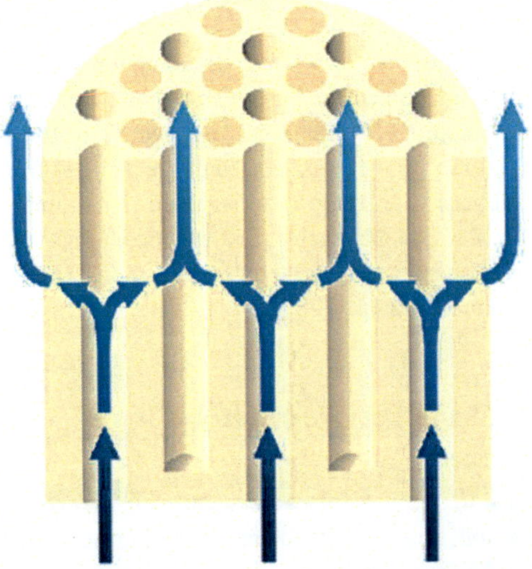

Fig. 19.13 Configuration of the filter. Tap water flows into the water channel (where impurities such as bacteria and iron are captured on the surface of the membrane) and diffuses into the adjacent transmission channel

The extrusion molding method is appropriate for the mass production of homogeneous and high-strength honeycomb structures. The outline of the manufacturing method is shown in Fig. 19.14.

The raw material is crushed to a specified grain size and is blended to create a specified chemical composition. It is then mixed and kneaded with water and binder, then formed into a honeycomb structure using an extrusion molding machine. The formed body is dried and calcined to complete the single-layer base material. Membranes with fine pores that completely remove impurities are coated on the internal surface of honeycomb cells, which is followed by the sintering process. Fine pores have dimensions of 10 μm in the base material and 0.1 μm in the membrane, for example. Figure 19.15 shows the appearance of a ceramic filter for water purifiers and the cross-sectional view of the multilayer structure. After formation of the membrane, it is cut into a specified size and the cut end is sealed with glaze.

Calcination is performed three times during the manufacturing process in consideration of the safety of consumers, as the tap water processed by the filter is drunk by consumers. The raw materials are completely sintered during each process so as to eliminate elution or dissolution of raw materials in the permeating water. Fine pores in the membrane have dimensions of 0.1 μm because the filter is required to thoroughly remove bacteria coliform or general bacteria that may be present in tap water. During the final inspection of the completed filters, a pinhole inspection is preformed to confirm the eradication characteristic and water-flow rate is checked, in addition to an appearance inspection and the measurement of the dimensions. These are conducted on all products.

Furthermore, periodic eradication performance certification tests are conducted by an external inspection agency. This test is called the bacteria challenge test (Note 19.10), in which indicator bacteria are added to water and filtrated under a specified condition to verify the eradication performance of filters.

Ceramic filters for water purifiers are manufactured with the production method of general ceramic products, but they are checked more strictly for performance in comparison with filters intended for industrial use. This is because they are intended for use at home by general consumers.

Note 19.10 A test performed to decide the diameter of fine pores depending upon the leakage rate of bacteria, which is confirmed by filtrating bacteria with a known size. Indicator bacteria (*Pseudomonas diminuta*) are used to determine the nominal pore diameter at 0.22 μm. Indicator bacteria (*Serratia marcescens*) are used to determine the nominal pore diameter at 0.45 μm (both of them are short bacilli with length-to-diameter ratio of 1.5:1).

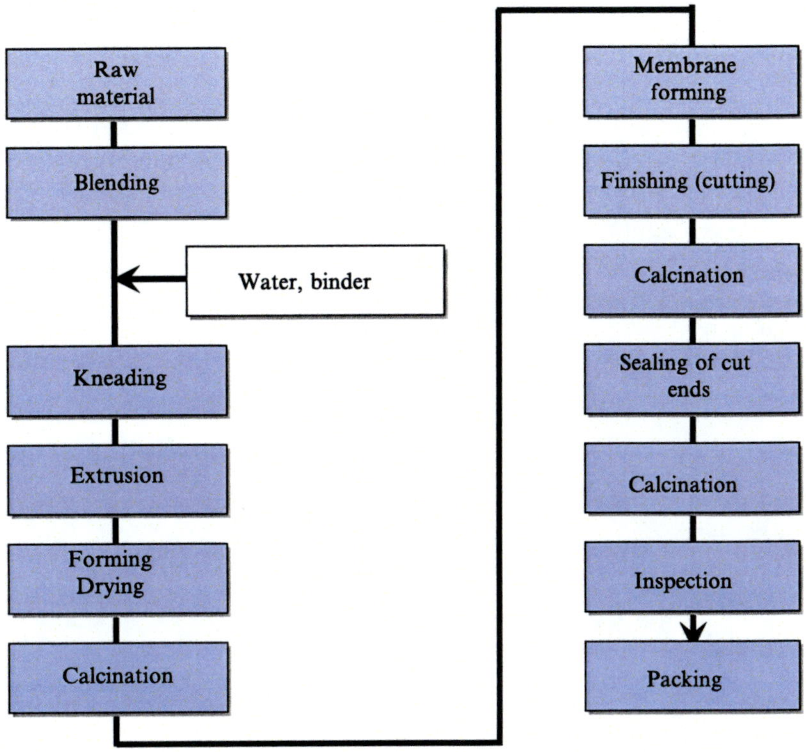

Fig. 19.14 Ceramic filter manufacturing process. Alumina, the raw material, is blended and kneaded with water and binder to create a clayey mixture. The mixture is formed by an extrusion machine, dried and calcined. Then the membrane is formed and calcined, and the glaze is sintered to seal the channels at the ends

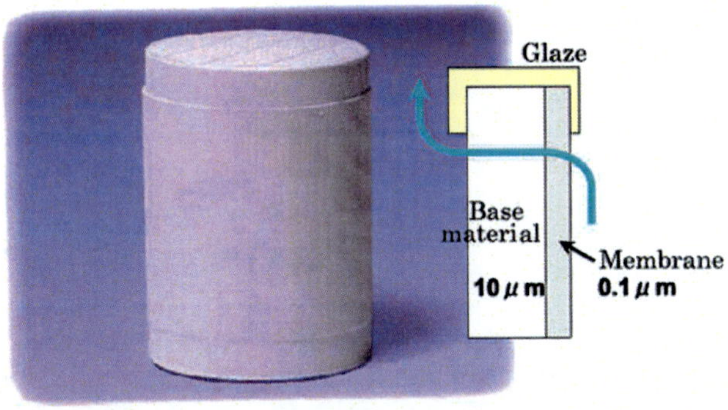

Fig. 19.15 Ceramic filter for water purifier. The filter is cylindrical and its cross section features a honeycomb structure with cylindrical cells. The internal surfaces of cells are coated with membranes that capture impurities and are double layered. The double-layer structure is incorporated to enhance water flow rate

19.3.3 Future Prospects

Filters of water purifiers are classified roughly into two types in terms of materials; organic filters and ceramic filters. Filters of water purifiers are generally replaced periodically to be used continuously as water purifiers because the amount of water flow decreases due to the clogging by impurities. The majority of water purifiers are used at home. Used filters are disposed of as industrial non-combustible waste. Therefore, environmental consideration about their disposal needs to be considered in the future. Ceramic filters can be collected, crushed and recycled into ceramic ware, bricks, etc. and, therefore, are eco-friendly.

The recycling of ceramic filters will be expanded in the future to include industrial filters, not only focusing on water purifiers for home use.

Faucets (1981) 19.4

Single-lever tap water mixer faucets are widely used as faucets because water discharge volume and water temperature can be adjusted by simply moving the lever vertically and horizontally. Hot water and cold water are mixed in a cartridge built into the faucet, which is operated to discharge and control water flow. The cartridge contains two to three pieces of flat ceramic disks, whose positions are changed for water flow and temperature adjustment. The disks take advantage of the high hardness, abrasion resistance and corrosion resistance as well as the high smoothness (surface roughness and flatness) of ceramics. The ceramic disks are made from alumina or silicon carbide ceramics, which are mass-produced by press forming raw material based upon powders. The disks are smoothed by the lapping process after sintering. The ceramic disks are required to feature higher sliding performance and accuracy responding to the recent improvements in the operating performance and downsizing of the handles.

19.4.1 Background of Development

Single-lever mixer faucets (Fig. 19.16) are designed so that water discharge volume and temperature can be controlled by operating the lever with one hand. Therefore, the valve needs to feature (1) sealing performance (Note 19.11) for water shutoff, (2) sliding performance for reduction of operating force, (3) durability that enables repeated operation and (4) corrosion resistance that allows for use in water. Ceramics are superior to metallic materials in terms of hardness, corrosion resistance and abrasion resistance. Ceramics are more suitable than other materials for satisfying these requirements because they can be processed into products with high-accuracy and smoothness (Note 19.12). Valves are normally made of alumina, but some of them are coated to improve sliding performance or are made of silicon carbide, which features high sliding performance.

Note 19.11 Ceramic discs in use move in a straight line and also rotate. The coefficient of static friction as well as the coefficient of dynamic friction in both directions of motion is required to be low.

Note 19.12 Smoothness in this case means the accuracy of both the surface roughness (center line average roughness) and overall flatness.

19.4.2 Characteristics and Manufacturing Method

19.4.2.1 Characteristics of Products

Cartridges, which are shown in Fig. 19.17, are mounted at the bottom of single-lever mixer faucets. The internal structure of the cartridge is shown in Fig. 19.18. The cartridge contains layers consisting of two to three ceramic disks. This is shown in Fig. 19.19. The diameter ranges from 20 to 40 mm and the thickness is between 3 and 10 mm. Water discharge amount and temperature can be controlled by operating the lever to move the locations of holes in the three disks vertically and horizontally.

19.4.2.2 Manufacturing Method

Figure 19.20 shows the flow of the manufacturing process. The ceramic material is crushed and blended with binders. The mixture is spray-dried by using a spray drier to create spherical powders. Dies configured to the shapes of ceramic disks are used to compress the powder by press forming machines to create formed bodies. In press forming, variations in density of powders affect the dimensional accuracy of sintered bodies. Therefore, measures including the division within the die are used to create formed bodies with homogeneous density. The dry press forming method allows for continuous forming and is suitable for mass production. Formed bodies are heated in a calcination furnace at a high temperature to create sintered bodies. The sintered bodies are ground on both sides. Both sides of the disks are normally machined simultaneously by using double-ended grinders. Then, lapping (grinding) is performed because the faces of ceramics that slide against each other need to be smooth. In the lapping process, ceramic disks are pressed against fixed platens (made of copper or tin) of a lapping machine, where the ceramic disks are rotated while being sprayed with diamond abrasive grains. The surface roughness of Ra 0.2 μm or less and flatness of 0.3 μm or less are achieved in this process. After the process, the disks go through ultrasonic cleaning for the removal of extraneous matter. Dimensions and appearance of the products are checked before completion of the ceramic disks.

19.4.3 Future Prospects

It is expected that single-lever mixer faucets will be more widely used in the future. Improvement in the sliding performance of ceramic disks is required for higher operability of the handles. For that purpose, coating is applied on the surface of ceramics, or materials such as silicon carbide featuring a friction coefficient that is lower than that of alumina have been used recently. In addition, new materials with higher performance and lower cost are being investigated. The cartridges have also been downsized for flexibility in design. Responding to these improvements,

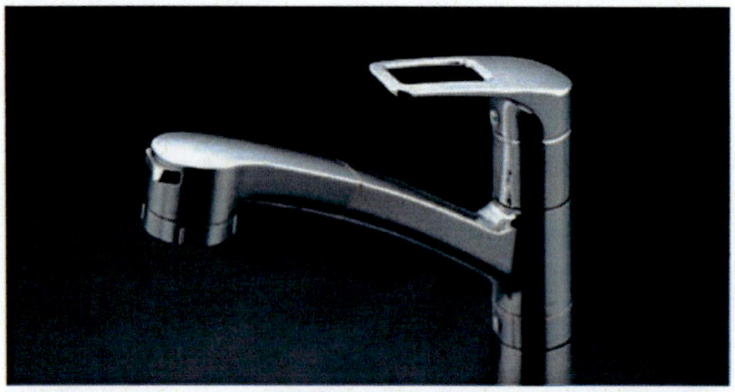

Fig. 19.16 Single-lever mixer faucets. Water temperature and discharge volume can be controlled by moving the lever *vertically* and *horizontally* with one hand

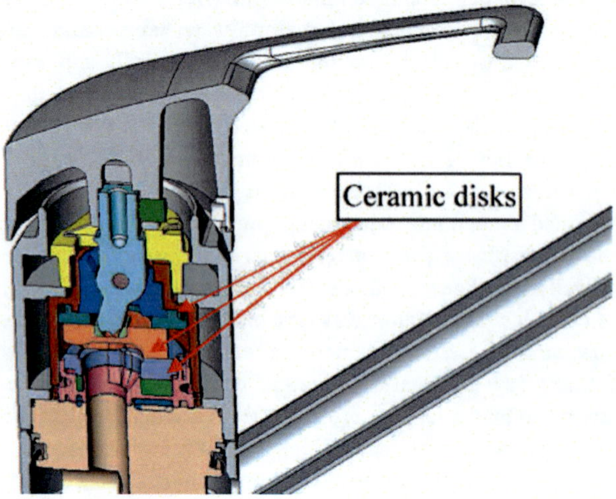

Fig. 19.18 Internal structure of a cartridge. Three ceramic disks are moved by the lever to control water temperature and flow rate

Fig. 19.17 Cartridges for single-lever mixer faucets. Examples of cartridges in single-lever mixer faucets. They have been downsized by improvements in design

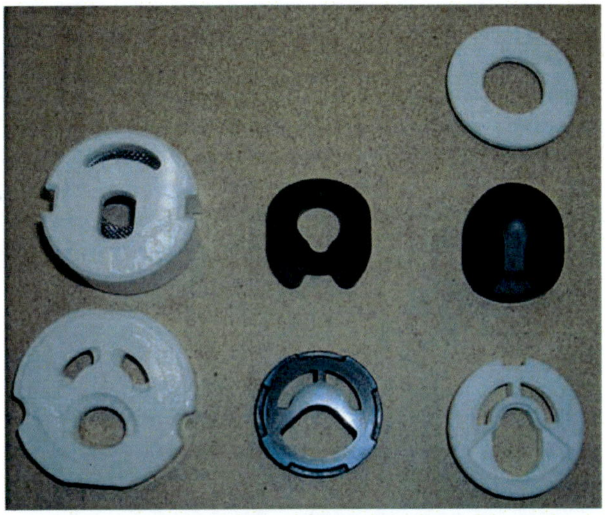

Fig. 19.19 Ceramic disks. Examples of ceramic disks. Hot water and cold water, respectively, flow through the holes on both sides of the parts with three holes, which are shown at the *bottom* of the picture, are mixed and then flows out through the hole in the *center*. Coated alumina or silicon carbide disks are used for improvement of the sliding performance, in addition to alumina

(1) Raw material
↓
(2) Blending
↓
(3) Spray drying (spray drier)
↓
(4) Press forming
↓
(5) Calcination
↓
(6) Grinding
↓
(7) Lapping
↓
(8) Inspection

Fig. 19.20 Manufacturing process of ceramic disks. Raw material powder is press formed and sintered, which goes through the grinding and lapping processes before completion of the product

ceramic disks are required to be smaller in size, more complex in shape with accurate dimensions. Production technologies need to be further improved to address these requirements.

Literature

1. Kitsutaka Y (1989) J Struct Construct Eng Architect Inst Jpn 404:15–24 **(19.2)**
2. Kishima T et al (1999) Summaries of technical papers of annual meeting, The Architectural Institute of Japan, pp 1063–1064 **(19.2)**

Part VII

Ceramics for Medical and Sanitation Sector

Bio-Medical Related Products 20

Metals, resins and ceramics are used as bio-medical materials that enhance bodily functions due to bone and joint related illness are an important market for ceramics. The characteristics of ceramics vary widely depending on the area of usage. Bio-medical related ceramic products that are used in bones and joints are introduced in this section.

Calcium phosphate-based ceramics, hydroxyapatite and β-tricalcium phosphate, have been commercialized as artificial bones because they are non-toxic, feature high affinity with bones and integrate easily with bones.

Hydroxyapatite is the major inorganic component in bones and has been commercialized as an artificial bone prosthetic material that directly integrates with living bones (Sect. 20.1). The strength (porosity) and shapes are adjusted to supplement various bone defects and reconstruct bone tissues. Dense bodies are used in areas where strength is required, while madreporic bodies with high porosity are used in areas where they required integration with living bone tissues. They are also used to supplement bone replacement throughout the body and are processed into various shapes and sizes.

β-Tricalcium phosphate is an artificial bone prosthetic material. It is absorbed by bones and replaces autologous bones, unlike hydroxyapatite, which is rarely absorbed in living bodies (Sect. 20.2). It is an amadreporic body featuring high porosity and is easy to process during surgical operations. The products are offered in forms of granules or blocks. It is used in orthopedic surgeries to supplement large bone defects caused by injuries or the removal of bone tumors.

Bioactive bone paste is made by mixing calcium phosphate powder, which is also called calcium phosphate-based cement, and a water-based hardening solution. It is a self-curable bone prosthetic material that transforms into hydroxyapatite in living bodies when it is applied on defective bones or broken bones (Sect. 20.3). The bone paste can be easily injected by syringes, making it possible to supplement complex bone defects. The products are widely used for orthopedic surgeries, plastic surgeries and cranial nerve surgeries.

In the case of artificial tooth roots, pure titanium featuring high affinity has been developed for clinical use. Ceramic materials that have been developed are superior to pure titanium in their affinity with living bodies (Sect. 20.4). With respect to ceramic tooth implants, products made of single crystal, polycrystalline alumina and hydroxyapatite were developed in Japan. They have been approved by the appropriate ministry in Japan and have been used successfully as implants. Artificial tooth roots need to be designed with sufficient consideration for service environments as are other bio-medical materials when used in the mouth.

Artificial joints are widely used to treat a variety of joints in which functionality has been lost due to accidents or diseases. The service life is limited because of overloading or repetitive motion. Issues concerning extension of service life of materials remain unresolved. For the extension of service life, it is effective to reduce abrasion of joints (bearings) made of polyethylene resin. Artificial joints that use alumina ceramics, which feature higher wear resistance than metallic materials, in femoral heads of artificial hip joints have been commercialized (Sect. 20.5).

Artificial teeth made of ceramics are characterized by high chemical stability and hardness. Color tones similar to those of natural teeth can be reproduced and they are free from substantial discoloration and staining (Sect. 20.6). Ceramic teeth (porcelain teeth) as well as resin containing teeth made of acrylic resin are widely used.

Y. Imanaka et al. (eds.), *The Ceramic Society of Japan, Advanced Ceramic Technologies & Products*,
DOI 10.1007/978-4-431-54108-0_20, © Springer Japan 2012

Hydroxyapatite a Prosthetic Material for Artificial Bone (1985)* 20.1

Basic research on hydroxyapatite, a major inorganic component of bones, started in the 1970s as an artificial bone or prosthesis. Commercial production began in 1985 with the approval of the Ministry of Health, Labor and Welfare in Japan. The material's biocompatibility with bones is not observed with metallic materials or polymers. It is currently used mainly in bone transplants performed in orthopedic operations. Advanced artificial bones must not only restore bone defects but also regenerate bone tissue. Artificial bone materials, which have higher porosity has been achieved by controlling the pore structures, have been commercialized that facilitate regeneration and reconstruction of bone tissue (Figs. 20.5 and 20.6). This implanted material is also integrated with the surrounding bone tissue at an early stage.

The QOL (quality of life: a scale that indicates a sense of fulfillment and satisfaction as a human being) of patients will be enhanced by application of artificial bones.

20.1.1 Background of Development

For the reconstruction of defective bones brought about as a result of fractures, bone tumors or the removal of bones during operations on organs, metallic materials, polymers and ceramic materials are used. Among the ceramic materials, hydroxyapatite [$Ca_{10}(PO_4)_6(OH)_2$, indicated as HAp hereinafter] is the major inorganic component of bones. It was proved that artificially synthesized HAp bonds directly with bones by Dr. Jarcho of the U.S. and Dr. Hideki Aoki of Japan in the early 1970s. Following this demonstration that proved the superior biocompatibility (Note 20.1) of the HAp ceramics with living bodies, developments were started for the commercialization of these materials as a prosthetic artificial bone. HAp-based artificial bones are approved by the Ministry of Health, Labor and Welfare in Japan as medical devices. Artificial bones shaped and structured to match bone defects in various areas of the body are used clinically in orthopedic, cranial nerve and dental surgeries. This is shown in Figs. 20.1 and 20.2.

20.1.2 Characteristics

20.1.2.1 Characteristics and Speci cations

Artificial bones currently used are roughly classified into two structures—a dense body with a relative density of 99% or higher and a porous body with cavities of several hundred μm. Both types are directly bonded with bones [bone healing (Note 20.2)]. The dense body is a high-strength sintered body with a compressive strength of about 730 MPa and is used to reconstruct bones that need to be strong, such as spinal bones, limb bones, and small bones such as ossicles. The porous body has cavities that allow for easy penetration, allowing the integration of transplanted material with bone tissues. As shown in Fig. 20.3, the distribution of cavities and strength can be controlled by changing the porosity, which allows for the selection of materials depending on the area of usage and purpose.

20.1.2.2 Manufacturing Method

HAp is synthesized by means of several processes: wet method, dry method and hydrothermal method, and so on. Details of each synthesis method are not explained here, but the obtained HAp powder is press formed to create green

*The number in parenthesis indicates the year that the product was first commercialized.

Note 20.1 It is free of toxicity and does not cause xenobiotic reactions such as loculation, tissue inflammation and destruction.

Note 20.2 The phenomenon where the transplanted material bonds directly with bone tissue without the intervention of soft tissue.

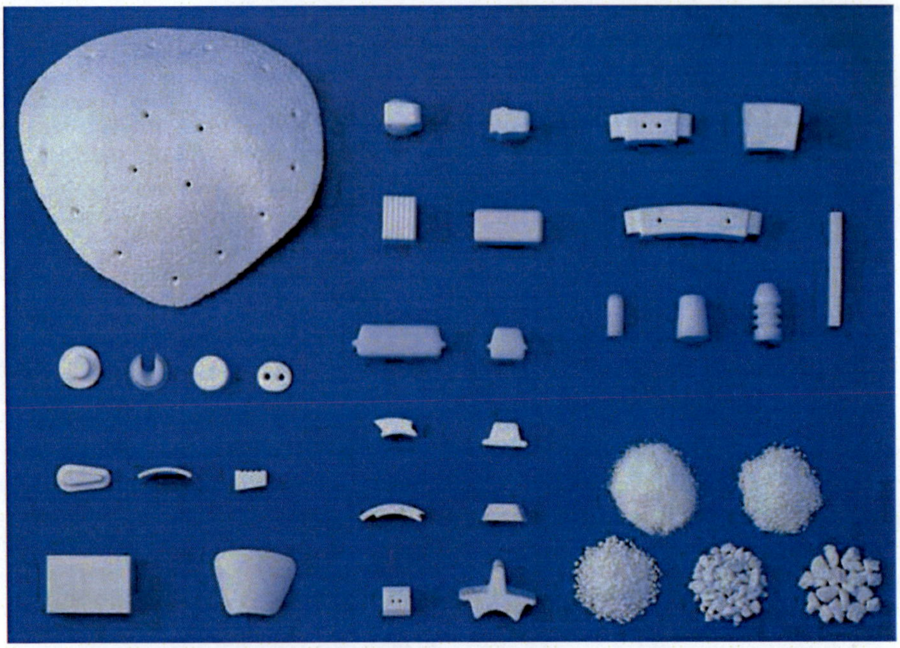

Fig. 20.1 Examples of HAp artificial bone products. Shapes suitable for various applications and portions are offered so as to cover all possible bone defects throughout the body

pellets, which are sintered at 1,000°C or higher to produce a dense body. Meanwhile, the HAp powder is mixed with a forming agent and an organic substance, and then the mixture is sintered to remove the organic substance that creates a porous body with a specified porosity and pore structure. During biomaterials synthesis caution is needed in order to avoid incorporation of impurities (heavy metals are toxic to living bodies in particular) in the raw materials and during each production process. Unlike general industrial products, the materials have to be produced in a clean room and go through a sterilization process. The final products also have to be guaranteed to be biologically clean. The manufacturing processes of artificial bones and inspections to be performed in each process are shown in Fig. 20.4.

20.1.3 Future Prospects

The intended goal of artificial bones, which have been studied and developed since the 1970s and used clinically since the second half of the 1980s, has changed from the "restoration of bone defects" to the "regeneration of bone functions." In order to attain this goal, material structures that function as a foothold for more active bone regeneration are needed. The porous body shown in Fig. 20.5 is an HAp artificial bone with 85% porosity and triple pore structure that allows easy intrusion of bone cells and tissues. In 4 weeks after transplanting the bone, bone tissue is actively regenerated inside the pores and the material is integrated with the surrounding tissue (Fig. 20.6). The material has high porosity and is not very strong at the beginning. It has been confirmed that the material integrates with bone tissue at an early stage and that the strength of the transplanted region recovers to the level of strength as that seen in the original bone. QOL of patients will improve if clinical applications of the artificial bones with this porous structure increase.

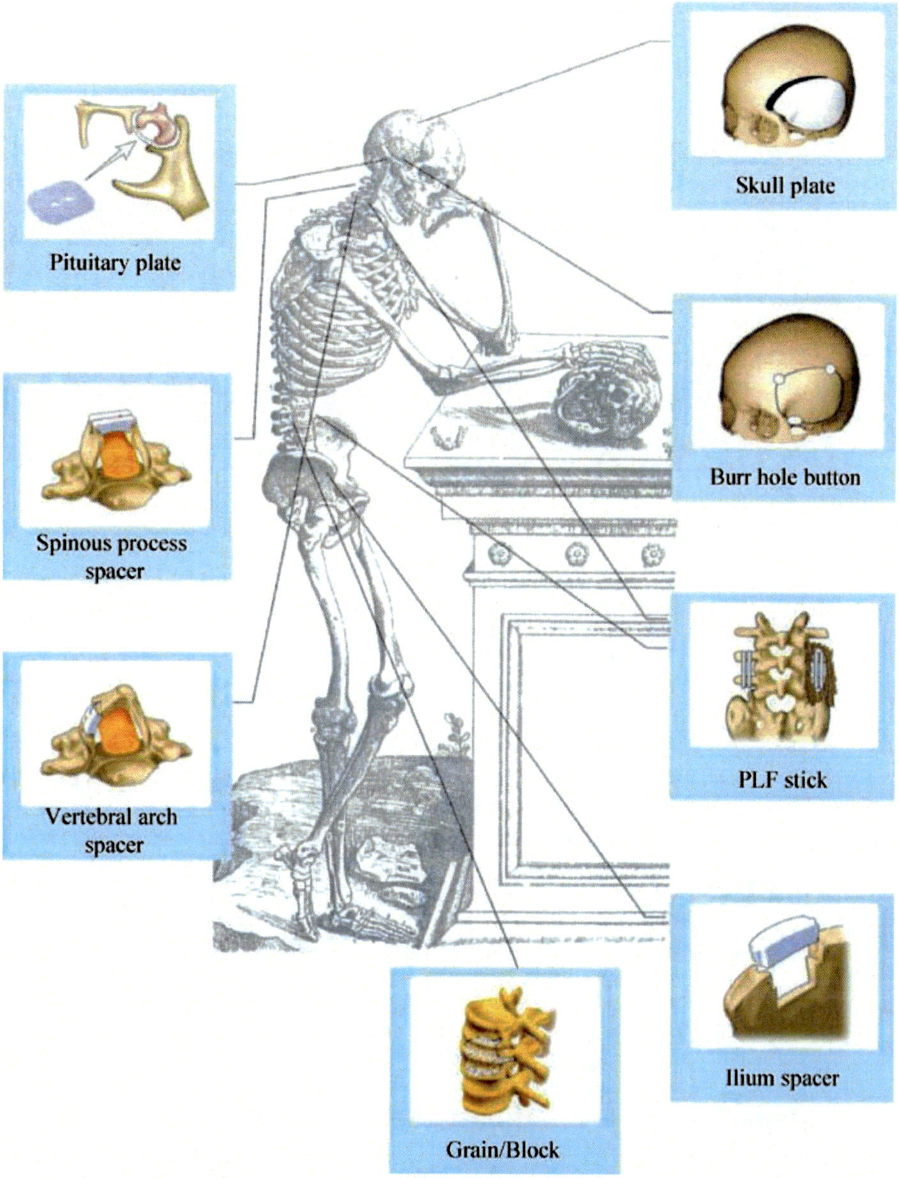

Fig. 20.2 Examples of HAp artificial bone application. Used in various fields such as orthopedic, cranial nerve and dental surgeries

Porosity (%)	0	5	15	30	40	50	60
Compressive strength (MPa)	730	540	240	175	65	30	15

Surface structure

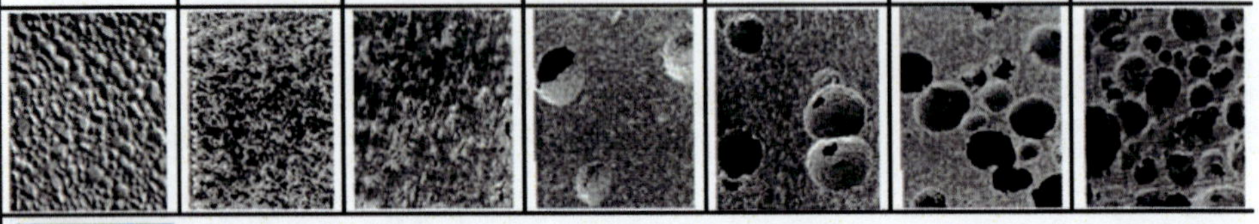

Purpose of use, example

Fig. 20.3 Porosity, strength and structure of pores of HAp artificial bones. If the porosity is increased, pores in the material increase, resulting in a decrease in the strength. Materials with optimum porosity can be selected depending on the portion and purpose of application

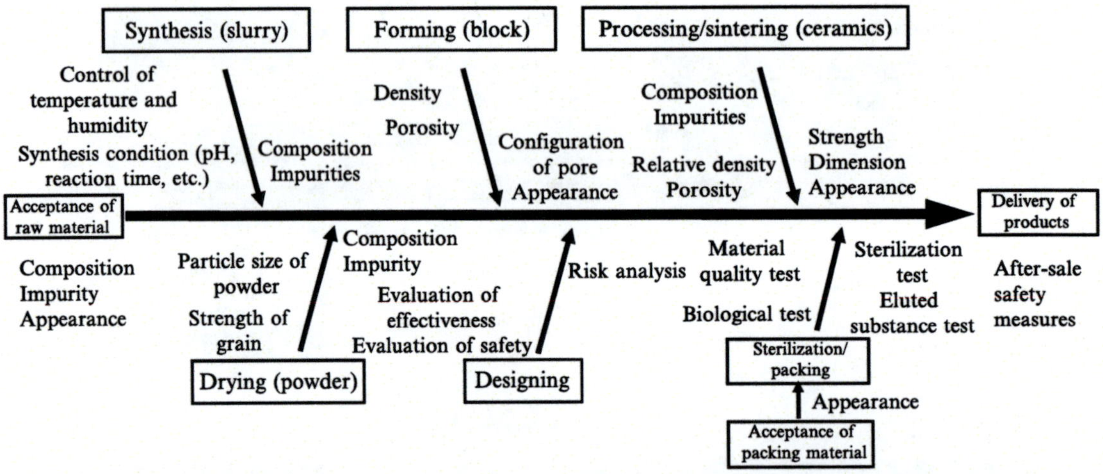

Fig. 20.4 Example of manufacturing processes of artificial bones and steps to be performed during each process. Care is needed to avoid the mixing of contaminants within the raw materials. This is also true during each production process of artificial bones used as medical devices. The final products need to go through a sterilization process so as to guarantee that they are biologically clean

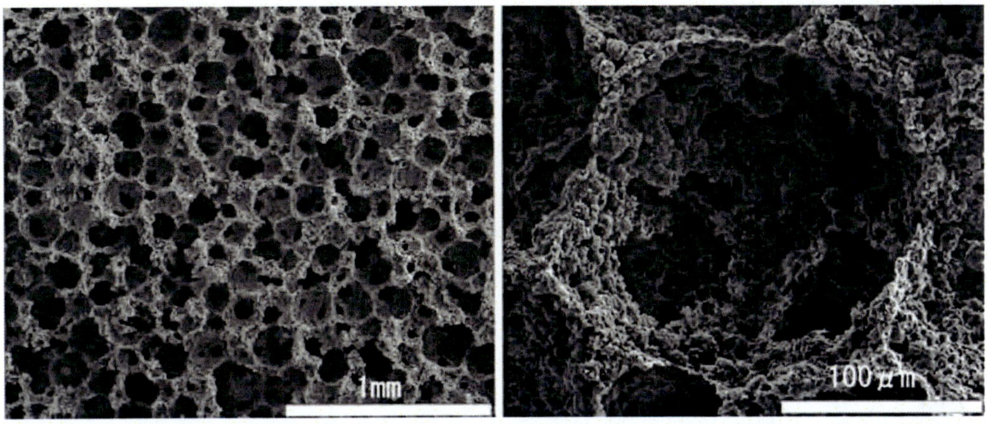

Fig. 20.5 Structure of HAp artificial bone featuring 85% porosity and triplex pore structure. It contains homogeneous macro pores with an average diameter of 250 μm, connecting holes that connect pores and micro pores sized several μm with high cellular adhesiveness. The average compressive strength is about 2 MPa, which is sufficient for normal operation

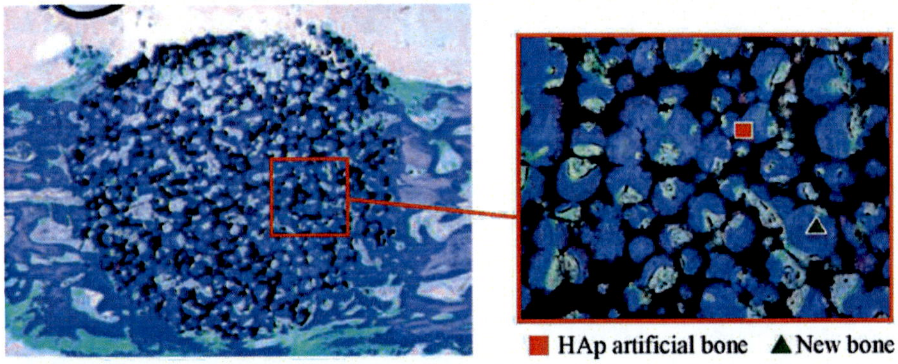

Fig. 20.6 Bone formation of an HAp artificial bone with 85% porosity. Four weeks after transplanting, new bone indicated by the *bluish purple* stain is formed within nearly every pore of the material. The material is integrated with bone tissues at an early stage and the strength of the transplanted region recovers to the strength of the original bone

β-Tricalcium Phosphate a Prosthetic Material for Artificial Bone (1999) 20.2

Tricalcium phosphate is used as a prosthetic material for artificial bone. It is made by synthesizing high-purity β-tricalcium phosphate using the mechanochemical method (Note 20.3) and then by adding foam stabilizer and surfactant to the synthesized powder, which is sintered to create porous artificial bones in the form of blocks or granules. The produced porous body contains macro-sized pores of 100–400 μm that are interconnected as well as micro-sized pores. The fine pores are believed to provide a good foothold for the growth of cells. The material is gradually replaced by autologous bone after implantation, and therefore, it is capable of responding to the growth change in infant patients. The pores are larger than those of autologous bones, allowing for sufficient blood flow. Therefore, it can be used on large bone defects (where the blood flow is normally poor) and is used as artificial bones in orthopedic surgery.

20.2.1 Background of Development

In the field of the orthopedic surgery, when a large bone defect is generated due to external injuries or the removal of bone tumors, fresh autologous bone is generally transplanted to fill in the missing bone. However, surgical extraction has to be performed on a healthy portion in order to obtain autologous bone for implanting. Furthermore, the amount of bone that can be obtained in this manner is limited. Due to these reasons, artificial bone prosthetic materials that replace autologous bones have been developed.

Biocompatible hydroxyapatite has been generally used as an artificial bone prosthetic material. Hydroxyapatite is not absorbed by the body nor replaced by autologous bones and thus remains permanently in the body, sometimes affecting the growth of bones in infant patients. Therefore, efforts have been made to develop high-purity ceramic artificial bones, which do not interfere with the growth of bones and that are capable of being eventually replaced by autologous bones.

Note 20.3 Chemical reactions caused by physical energy. The reactants are mixed and react in a rotary ball mill. This technique utilizes the energy generated from collisions of ceramic balls against one another for chemical reaction.

20.2.2 Characteristics

20.2.2.1 Characteristics and Speci cations

An artificial bone prosthetic material is a β-tricalcium phosphate porous body consisting of macro-sized pores of 100–400 μm, that are interconnected with one another, and micro-sized pores. The pores are spherical or oval shaped, and the material is characterized by the complex shapes of pores. The β-tricalcium phosphate porous body has a porosity of approximately 75% and a compressive strength of 2–3 MPa. The strength is adjusted in consideration for easy processing at the operation site for surgical implant (Fig. 20.7).

The crystalline phase of this porous body was confirmed as β-tricalcium phosphate by X-ray diffraction analysis. And the β-tricalcium phosphate was confirmed to be of high purity by the impurity analysis of the product. The β-tricalcium phosphate porous body is ideal for use as an artificial bone prosthetic material because it is eventually replaced by autologous bones.

The usability of this β-tricalcium phosphate porous body as an artificial bone prosthetic material was evaluated. An experiment was performed by implanting β-tricalcium phosphate porous bodies in the tibiae of beagle dogs. The β-tricalcium phosphate porous bodies were in the process of being replaced by autologous bones and trabecular bones were being reconstructed when the dogs were observed 2 and 24 weeks after surgery (Fig. 20.8). In another experiment, the β-tricalcium phosphate porous bodies were imbedded in the tibiae of rabbits for histological observation (Note 20.4). Osteoclast-like cells (Note 20.5) were observed on the surface of β-tricalcium phosphate, indicating the replacement of β-tricalcium phosphate. Osteoblastic cells (Note 20.6) lining on the surface of new bones were also observed, indicating active new bone formation. Through these experiments, it was confirmed that β-tricalcium phosphate is replaced by autologous bones and eventually by neonatal bones (Fig. 20.9).

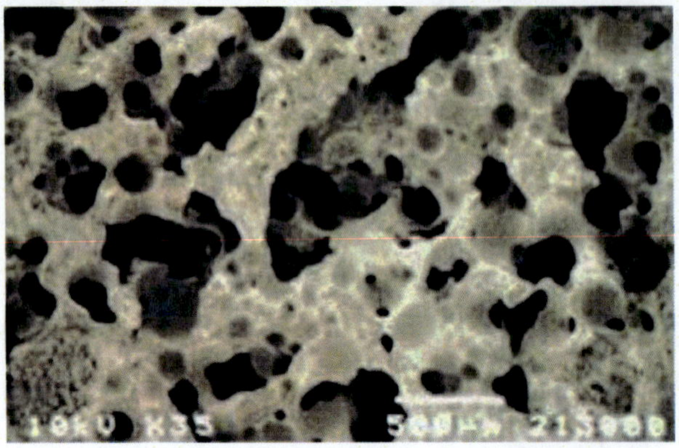

Fig. 20.7 Pore structure in the β-tricalcium phosphate porous body

Figure 20.10 shows X-ray images of a finger on which a bone tumor operation was performed. The pictures were taken before the surgery, just after the surgery and 4 months after the surgery. After curettage of enchondroma, a 1:1 mixture of autologous bone and β-tricalcium phosphate porous granules was used to fill in the defects. The β-tricalcium phosphate were gradually replaced by autologous bone and normal bone formation was in progress when it was observed 4 months after the surgery.

Figure 20.12 shows X-ray images of a femoral fibrous dysplasia case. After curettage of enchondroma, a 4:1 mixture of β-tricalcium phosphate granules (75 cm^3) and autologous bone was used to fill in the defects. Active bone resorption was observed before the surgery, but a clear image indicating calcification was observed 6 months after the surgery.

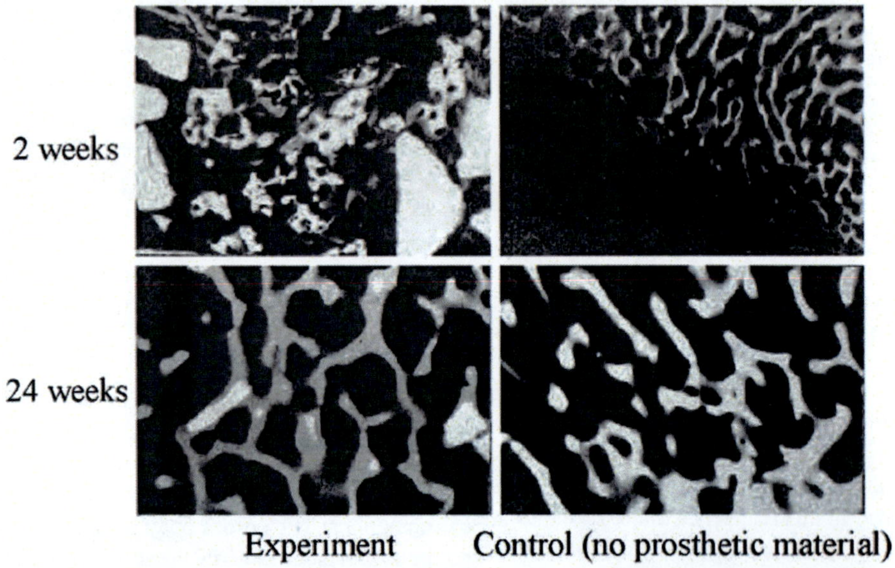

Fig. 20.8 CMR image of the tibiae of a beagle dog (courtesy of Jikei University)

Note 20.4 Observation by microscopes. Lesions, etc. can be observed at the cell level or the tissue level.

Note 20.5 Multinucleated giant cells containing several to several tens of nucleus. The cells absorb old bone in bone tissues and release proteins that promote bone formation.

Note 20.6 Cells that form bones in bone tissues.

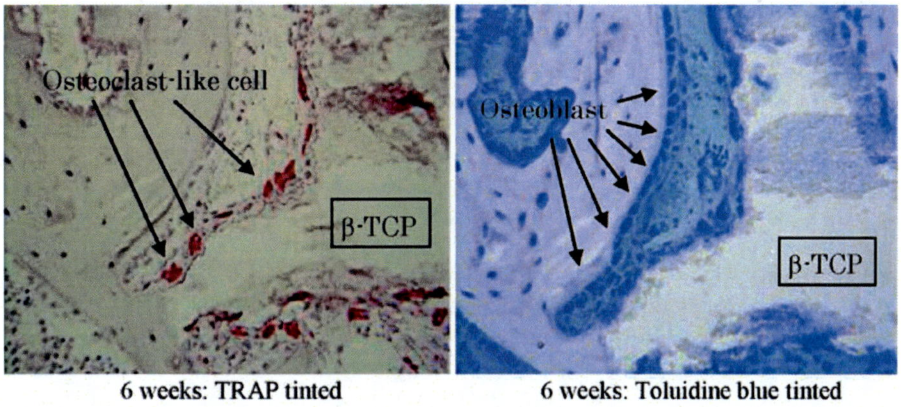

Fig. 20.9 Results of histological observation of a rabbit shinbone

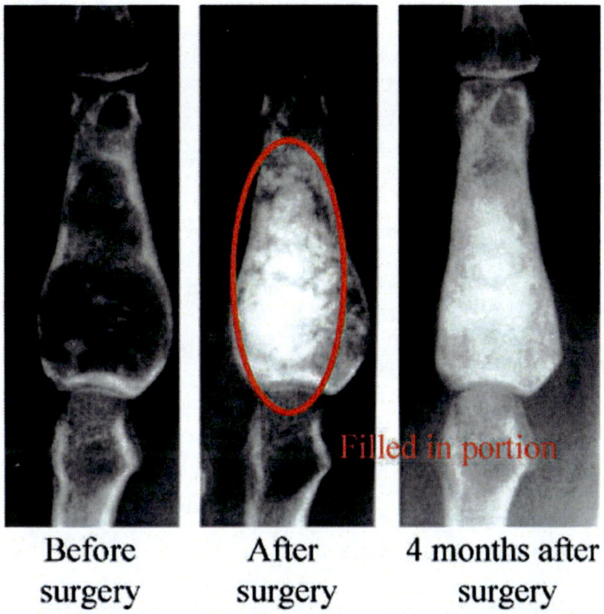

Fig. 20.10 Enchondroma in proximal phalanx of the left middle finger (50-year-old female): X-ray pictures showing the conditions before the surgery, just after the surgery and 4 months after the surgery (courtesy of Jikei University)

Fig. 20.11 The granular product and the block-shape product

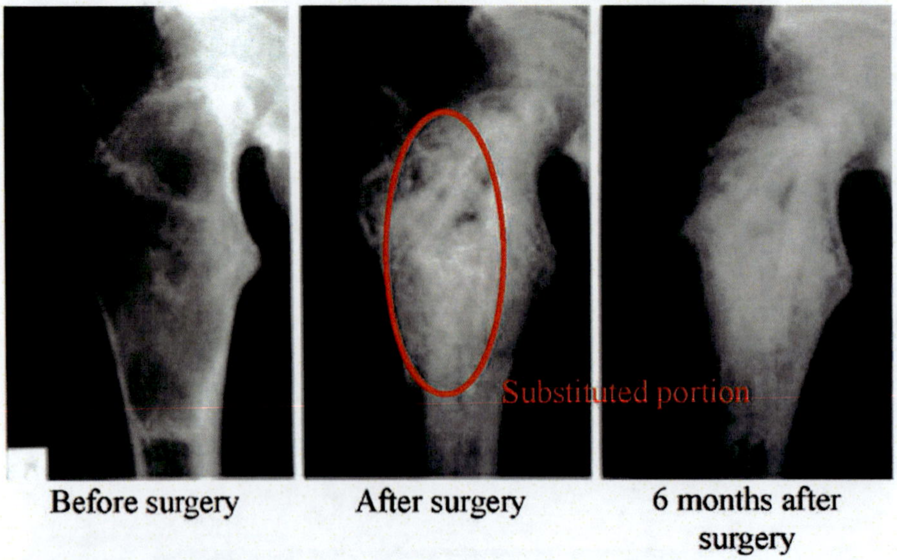

Fig. 20.12 A femoral fibrous dysplasia case (54-year-old male): X-ray pictures showing the conditions before the surgery, immediately after the surgery and 6 months after the surgery (courtesy of Jikei University)

20.2.2.2 Manufacturing Method

Pure water is added to high-purity calcium carbonate and calcium hydrogen phosphate then ball milled with ceramic milling media for 24 h, and the milled slurry is calcined at 750°C to create β-tricalcium phosphate. Surfactant, foam stabilizer and pure water are added to the β-tricalcium phosphate to produce foam body. This foam is then sintered at 1,050°C to produce a β-tricalcium phosphate porous body. The material is processed into blocks and granules for use (Fig. 20.11).

20.2.3 Future Prospects

Artificial bone materials (β-tricalcium phosphate porous bodies) for prosthesis have been commercialized are also attracting attention as a scaffold for culturing bones in the field of regenerative medicine.

Differentiated (or undifferentiated) stem cells are disseminated on the β-tricalcium phosphate porous body to obtain bone cells from which bones are formed. According to the results of animal experiments, the material is superior in bone quality, is applicable to large defects and enables stronger bonding with autologous bones when compared with conventional artificial bone prosthetic materials.

In the near future, these new achievements will contribute to the vitality of an ageing society.

Bioactive Bone Paste (1990) 20.3

Bioactive bone paste is made by kneading calcium phosphate-based powder and a water-based sclerosing solution. The paste is a self-curing bone prosthetic material and is used to fill in bone defects and fractures. The material has high moldability and is applicable to bone defects of complicated shapes. The kneaded material hardens due to hydration hardening and changes to bone-like hydroxyapatite with low crystallinity and is biocompatible, bonding directly with bones. The material was introduced on the market in 2000 and has been used widely in orthopedic, plastic, cranial nerve surgeries, and so on. The material is expected to be employed as a carrier of DDS (Note 20.7) and as a cell scaffold (Note 20.8) in regenerative medicine.

20.3.1 Background of Development

In recent years, sintered bone prosthetic materials such as HAp [hydroxyapatite: $Ca_{10}(PO_4)_6(OH)_2$], β-TCP [tricalcium phosphate: $Ca_3(PO_4)_2$] are generally used to repair bone fractures and defects caused by diseases and accidents. Sintered bone prosthetic materials, which are implanted into bones and are comparatively strong, are highly evaluated as restorative materials. However, they need to be designed based on CT images and trimmed during surgical operations when they are used to substitute for complicated bone defects and fractures. Doctors typically seek flexible materials that fit complex-shape with limited preparation time. The bioactive bone paste is in the form of a moldable paste and responds to these needs.

The bioactive bone paste (also called calcium phosphate cement or apatite cement: indicated as bone cement hereinafter) has been studied since 1976, when Kadoma et al. discovered a hydration hardening method for α-TCP [tricalcium phosphate: $Ca_3(PO_4)_2$]. In the beginning, high temperatures of 80°C or higher and several hours were required for hardening. The temperature and time for hardening were substantially reduced by means of using an organic acid as a sclerosing solution. (Bone paste is explained in this section although products intended for dental uses have also been released.) Research and development continued to create a bone paste made by adding components other than acid to the powder and sclerosing solution as the acid stimulates the living body. Component systems of current bone pastes are represented by the system made by adding TeCP [tetracalcium phosphate: $Ca_4(PO_4)_2O$] to α-TCP and the system made by adding MCPM [calcium dihydrogen phosphate monohydrate: $Ca(H_2PO_4)_2{\cdot}H_2O$] to α-TCP. In addition, a component system that uses TeCP/DCPA (calcium hydrogen phosphate: $CaHPO_4$) or DCPD (calcium hydrogen phosphate dihydrate: $CaHPO_4{\cdot}2H_2O$) as the major component was reported in 1985 by Brown, Chow et al. Several types of bone paste were commercialized based on these materials and have been used clinically in orthopedic, plastic, cerebral surgeries (two types of products have been released in Japan).

20.3.2 Characteristics

20.3.2.1 Characteristics and Speci cations

The bone paste is made by kneading calcium phosphate-based powder and water-based sclerosing solution. The material is processed to the form a moldable paste by adjusting the powder–liquid ratio. It fills in complex-shaped bone defects and fractures perfectly without gaps, which is the major characteristic of the product. The moldable-type material can be formed into intended shapes easily. The paste-type material can be filled inside the affected areas using syringes and needles. Figure 20.13 shows examples of these products and Fig. 20.14 shows the outline of the application method.

Note 20.7 Drug delivery system. The technology used to deliver drugs to the target area effectively and intensively.

Note 20.8 Scaffold used to promote the multiplication of cells.

The bone paste hardens after kneading and application, then it gradually transforms into hydroxyapatite. Examples of reactions of the α-TCP/TeCP component system and the TeCP/DCPA (DCPD) component system are shown below:

- α-TCP/TeCP: $2Ca_3(PO_4)_2 + Ca_4(PO_4)_2O + H_2O \rightarrow Ca_{10}(PO_4)_6(OH)_2$
- TeCP/DCPA (DCPD): $2Ca_4(PO_4)_2O + 2(CaHPO_4 \cdot 2H_2O) \rightarrow Ca_{10}(PO_4)_6(OH)_2\ (+4H_2O)$

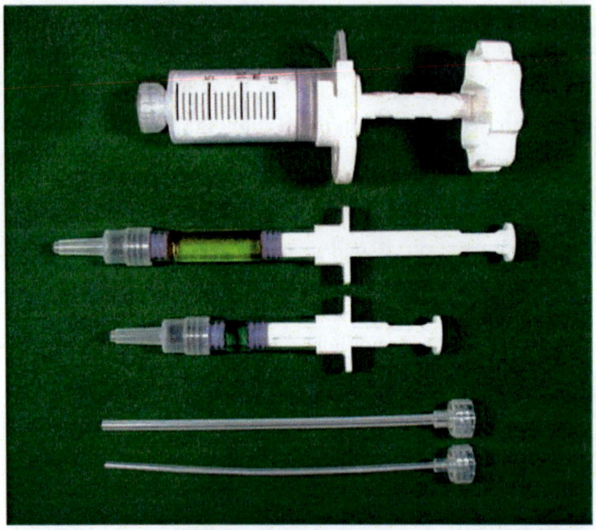

Fig. 20.13 Products of bioactive bone paste. From top, "kneading and injection device containing Ca–P powder," "sclerosing solution," "viscosity adjusting solution," "injection needle (large)" and "injection needle (small)"

Hardening time and strength (compressive strength) are the basic characteristics of the bioactive bone paste. The hardening time is an important factor that affects the filling operation performed by doctors during surgeries. Hardening time (conforming to JIS T-6602) after the commencement of kneading, varies depending on the product and temperature. However, it is generally takes 5–20 min at body temperature (37°C). The reaction is generally slow at low temperatures and fast at higher temperatures. Figure 20.15 shows an example of the relation between the temperature of TeCP/DCPA bone paste and hardening time.

Transformation into hydroxyapatite continues after hardening. The strength increases over time, reaching its highest strength within several hours to several days. Figure 20.16 shows the time dependence of strength in TeCP/DCPA bone paste. During actual use, it is not as strong as the cortical bone (Note 20.9) (80–200 MPa). However, it is strong enough to replace spongy bone (Note 20.10) (several MPa).

20.3.2.2 Manufacturing Method

Manufacturing method of the raw material varies depending on the product. The raw material powder is basically made by mixing two or more types of calcium phosphate compounds. The sclerosing solution is water-based and additives such as salt and polysaccharide are added to promote hardening and enhance operability. Properties of the powder substantially affect the reaction and properties of the hardened material. For example, if the particles are fine with a large specific surface area, the rate of dissolution increases and the hardening

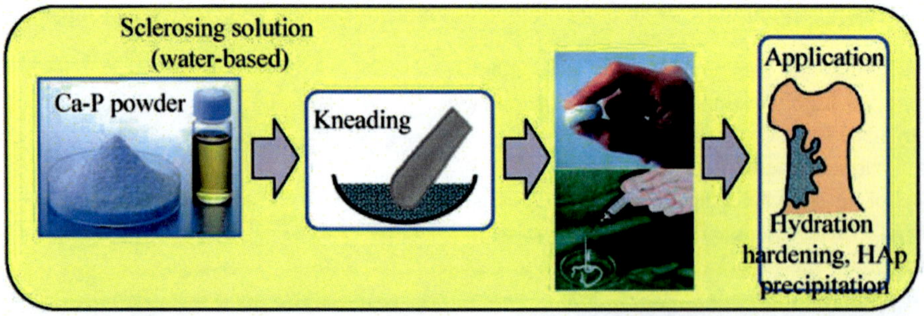

Fig. 20.14 Outline of bioactive bone paste application. The phosphate-based powder and water-based sclerosing solution are kneaded into a paste form, which is applied to areas with bone defects or fractures

Note 20.9 Hard and dense bones that constitute the surface layer of bones.

Note 20.10 Bones that are structured like a sponge and constitute the internal portion of bones.

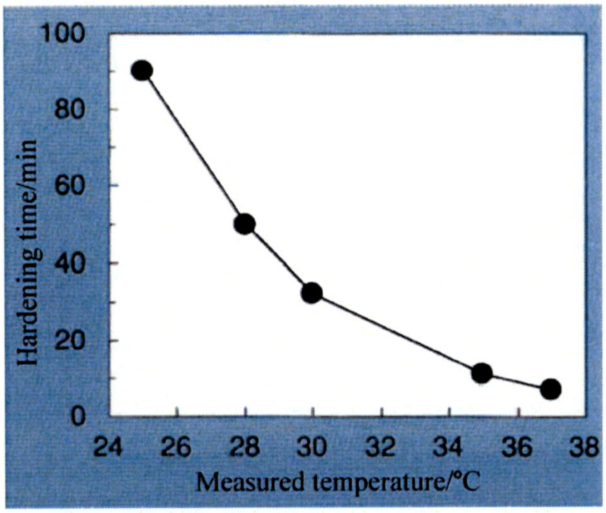

Fig. 20.15 Relationship between the temperature and hardening time. The hardening reaction depends on the temperature. The reaction is promoted as the temperature increases

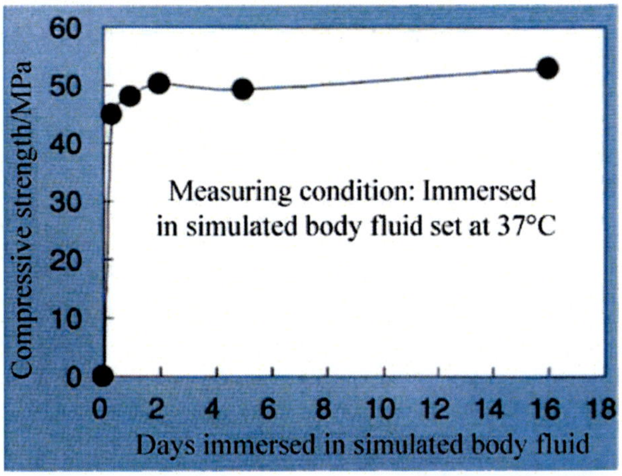

Fig. 20.16 Strength of bioactive bone paste vs. time. After being kneaded, strength increases over time, reaching the highest level of strength within several hours to several days

time decreases, affecting the strength after hardening. The particle size, specific surface area, dispersion state, of the powder are important factors that determine the characteristics of the bone paste. The powder–liquid ratio is also an important factor.

The produced powder and sclerosing solution are, respectively, sterilized and airtight-packed in an appropriate manner. The powder tends to be affected by moisture and thus special attention is needed to prevent exposure to moisture.

20.3.3 Future Prospects

The bioactive bone paste has been widely used for clinical application after its release. Research has continued for several types of new bone paste with improved hardening characteristics and operability. The material is expected to be used as a DDS carrier of bone growth promoting substances and anticancer drugs as well as the cell scaffold in regenerative medicine. The material will be used for various purposes in the future to improve people's QOL.

Ceramic Dental Implant (1984) 20.4

Recovery of the occlusal function (ability to chew one's food) was a long-cherished dream for people who have lost teeth due to accidents or diseases. Various materials have been used as dental implants since long ago, but none of these materials seems to have achieved satisfactory results. Current dental implants originated from the 1952 discovery by Branemark of Sweden, who discovered high biocompatibility of pure titanium with bones. While pure titanium dental implants were widely used throughout the world, ceramic materials with higher biocompatibility with living bodies were developed. In Japan, dental implants made of single crystal or polycrystalline alumina ceramics as well as those made of hydroxyapatite were released and used widely in clinical applications after receiving approval from the Japanese Ministry of Health & Welfare (presently known as the Japanese Ministry of Health, Labour and Welfare).

20.4.1 Background of Development

When teeth are lost due to accidents or diseases, the patients are treated using the bridge method, where the unaffected teeth are used to support a bridge tooth or structure, or by using false teeth if there are no unaffected teeth available to attach a bridge. However, people are not comfortable using false teeth and strongly wish to recover the functions of their lost teeth. The implant method that uses dental implants was introduced responding to these needs.

It is believed that dental implants have a very long history. A mummy, who had green stones (thought to be emeralds) implanted on upper and lower jaw bones, was discovered from the remains of the Inca Empire. Various materials have been tried since that time, but it seems that satisfactory results were difficult to attain.

Following the discovery by Branemark of Sweden, (who discovered high biocompatibility of pure titanium with bones) pure titanium dental implants were widely used throughout the world. Meanwhile, ceramic materials with a higher biocompatibility with living bodies were developed. In Japan, dental implants made of single crystal or polycrystalline alumina ceramics as well as those made of hydroxyapatite (Note 20.11) (hereinafter HAp) were released and used widely in clinical applications. Dental implants made of HAp are explained below.

20.4.2 Characteristics

20.4.2.1 Interaction with Bones

When materials such as plastic and cobalt-chrome alloy are implanted in bones, soft tissues are formed on the surface of the materials and the bones do not bond or contact directly with these materials. When pure titanium or alumina ceramics are used, soft tissues are not formed and the materials contact directly with bones, exhibiting superior biocompatibility with living bodies. It was proved during the process of development that HAp chemically bonds with bones, while having the same properties as the materials mentioned above. Figure 20.17 is a transmission electron microscope image showing the boundary between HAp and bone.

The relationship between pure titanium and alumina ceramics with bones is frequently called bioinert, while the relationship between HAp and bones is called bioactive (Note 20.12).

Note 20.11 Calcium phosphate compound expressed as $Ca_{10}(PO_4)_6(OH)_2$. It is the major component of minerals in vertebrate bones. Artificially synthesized ceramics feature excellent biocompatibility with living bodies and are used widely in living bodies.

Note 20.12 It is also called bioactive. It is the property of a material that bonds directly with bones when it is implanted in bones. In contrast, the materials that contact directly with bones but do not bond with bones are bioinert.

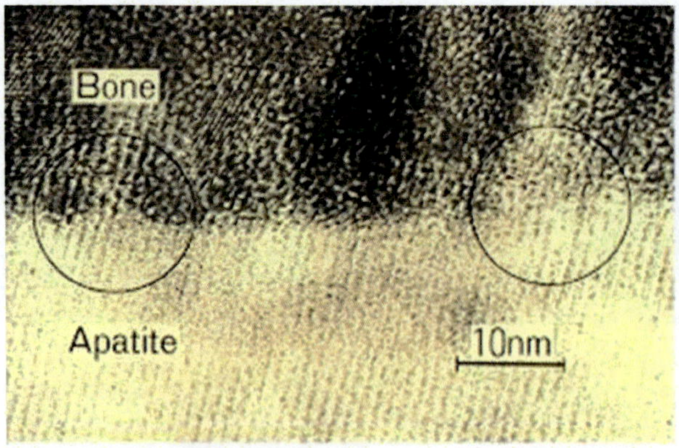

Fig. 20.17 Boundary layer between HAp and bone. HAp is the major component of minerals in bones, and artificially synthesized HAp and the HAp in bones are in direct contact. It is observed that the direction of crystal lattices in both of the HAp are partially aligned, indicating that HAp and bones are chemically bonded (courtesy of Makoto Ogiso at Tokyo Medical and Dental University)

20.4.2.2 Relationship with Gum Tissue

In the case of natural teeth, healthy gums are directly in contact with the tooth enamel, preventing the intrusion of bacteria. Upon investigation of the relationship between HAp dental implants and gums, it was discovered that the tissues at the boundary between HAp dental implants and gums are the same as those at the boundary between gum and enamel, indicating epithelial attachment of gum tissue on HAp. Later, it was reported that a similar attachment is observed between alumina ceramics and gums. Therefore, it is believed that epithelial tissue attachment is attributable to firmly fixing the implants in the jaw bones, not the material that is in contact with the gums. This fact indicated that the area that contacts the gums does not always have to be made of HAp, which led to the development of a two-piece implant.

20.4.2.3 Types and Speci cations

The original dental implant consisted of one piece, where the abutment of the superstructure was exposed in the mouth immediately after implanting (Note 20.13) in the jaw bone. Production of the one-piece dental implant was approved by the Ministry of Health & Welfare (name at that time) in 1983, and products were released during that same year. HAp dental implants directly bond with bones. However, several months are required for the recovery of bone and for complete bonding. Unfortunately, bonding with bones is not achieved if they move during this period. It is difficult to prevent movement of dental implants when one-piece dental implants are used. For this reason, the two-piece implants were developed, made by using HAp for the areas that contact jaw bones and titanium alloy for the other side. In this type of dental implant, a dental implant is implanted under the mucous membrane. It is left undisturbed until it is completely bonds with the bone. Then a titanium alloy component is connected to the dental implant with a screw and the abutment for fixing the superstructure is exposed in the mouth. These types of products are shown in Fig. 20.18.

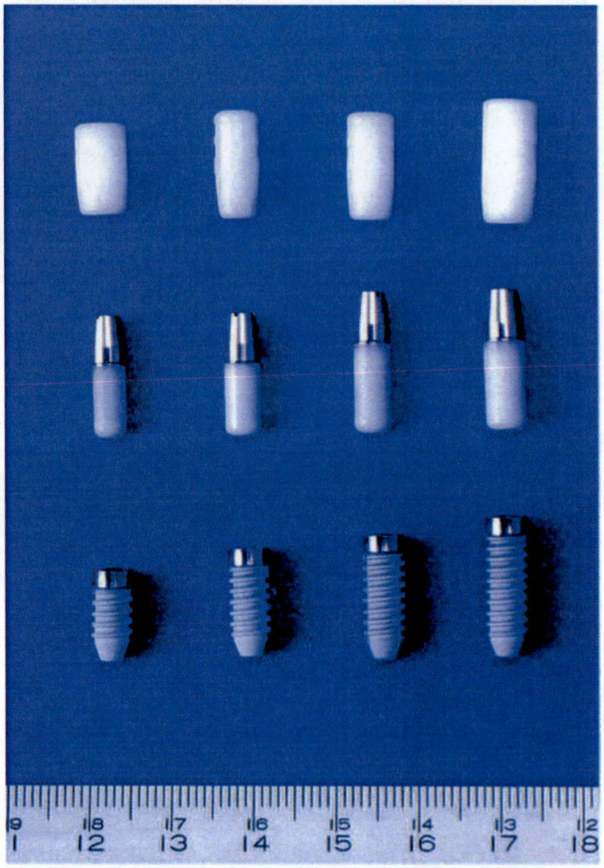

Fig. 20.18 Types of HAp dental implants. The *top row* are one-piece implants, the *middle row* are two-piece and the *bottom row* are prototypes of the two-piece implants made by the thermal spraying method. In respect to the two-piece products in the middle, the external cylinder contacted with bone is made of HAp, while the internal cylinder and the abutment are made of titanium alloy. The external cylinder and the internal cylinder are bonded using a special adhesive

Note 20.13 The term originally means "insertion" or "placement," but it indicates a dental implant in many cases. In particular, "dental implant" is used as a synonym of dental implants.

20.4.2.4 Manufacturing Method

HAp was produced through the dry forming process using isostatic pressing→grinding by a numerically-controlled lathe→sintering. Ceramics are generally made with this same type of process. During the early stage of the development, it was believed that the sintering of HAp was difficult and therefore it was difficult to improve its hardness. The target was set at a bending strength of 200 MPa or higher for the commercialization of dental implants. Development was performed with a focus on the manufacturing method of powders that can be sintered to a density close to theoretical density at low temperature. For that purpose, a series of improvements were made to the synthesis method. It was previously believed that the temperatures need to be 1,200°C or higher for the sintering of HAp. However, through the accumulation of expertise, a technology for producing a dense sintered body with a relative density of 99.2% or higher at 1,050°C was developed, realizing the targeted strength. Figure 20.19 shows the structure of the surface of HAp that was used as dental implants.

Fig. 20.19 Microstructure of HAp surface. HAp is sintered and obtained a dense body with an average grain size of about 0.3 μm

20.4.3 Future Prospects

HAp dental implants were popular originally, but the production was discontinued because the success rate of long-term clinical use was less than 80% due to the severe conditions inside the mouth. However, the superior biocompatibility of HAp with living bodies is widely known and the material is used as bones throughout the world. Information on use of the material in living bodies will continue to be investigated. It is expected that more advanced dental implants will be developed taking advantage of characteristics of ceramics in the future.

Artificial Joint (1982) 20.5

Artificial joint replacement, intended for pain relief and recovery of joint functions that have been lost due to accidents and diseases, is widely accepted as a very effective treatment procedure. However, the service life is limited because of wearout of the material in the replacement joint. Improvement is required for artificial joints to have a longer service life. Joints have two functions, load bearing function and the motor function. The service life of artificial joints depends upon the material's abrasion resistance. Therefore, reduction in the abrasion of joints (bearings) is effective as one of the means for extending their service life. Metal-resin bearings were used conventionally. Artificial joints that incorporate alumina-resin bearings take advantage of the low abrasion properties of ceramics and have been contributing to the reduction in abrasion of artificial joints. Ceramic based artificial joints are used as replacements in hips, knees, shoulders, elbows, feet, etc.

20.5.1 Background of Development

When human mobility or bodily functions have been lost and a patient's pain gets worse at a joint due to accident and disease, artificial material replacement has grown popular, intended for pain relief and recovery of a joint's function. According to statistics, approximately 140,000 cases of artificial joint replacement were performed in Japan during fiscal 2006. Of the total cases, 85,000 cases were performed on hip joints and 50,000 cases were on knee joints. More than 97% of the total cases were performed on the two major lower extremities. The major artificial joints are shown in Fig. 20.20.

Artificial joint replacement has been contributing largely to enhancement in the quality of life (QOL) (Note 20.14), by relieving pain and allowing patients to walk with their own legs. However, the service life is limited due to the artificial material, and issues such as breakage of the joining portion with body tissues including bones have been pointed out. For these reasons, development of artificial joints with a longer service life is needed.

Note 20.14 Quality of life is a term of measurement on how fulfilling and satisfactory one's life is.

20.5.2 Characteristics

20.5.2.1 Characteristics and Speci cations of Products

An artificial hip joint product is shown in Fig. 20.21. The artificial hip joint is composed of a pelvic-side cup made of polyethylene resin, an artificial femoral head assembled to the cup and a metallic supporting rod (stem) used to fix the femoral head on the femur. The femoral head, in combination with the resin cup, constitutes a spherical bearing. Abrasion of the resin by the bearing is the cause of degradation and the factor that determines the service life of artificial joints in many cases. In order to reduce abrasion of the resin, alumina has been proposed and applied for use, instead of metal, as the femoral head material. The alumina used as the femoral head in an artificial hip joint features high strength due to its high purity and fine crystal structure with an average grain size of 1–2 μm. The low abrasion properties were realized due to the high strength and high hardness of the material as well as the precise spherical shape and surface processing. The surface roughness of bearings is specified as <0.02 μRa by an international standard, while the surface of the alumina can be processed to a roughness of 0.01 μRa. The low abrasion property of alumina femoral heads has also been proved by clinical use. The effectiveness of the femoral heads has been reported in many conference presentations. Alumina ceramics have also been used on the surface of artificial knee joints for the reduction in the resin material abrasion.

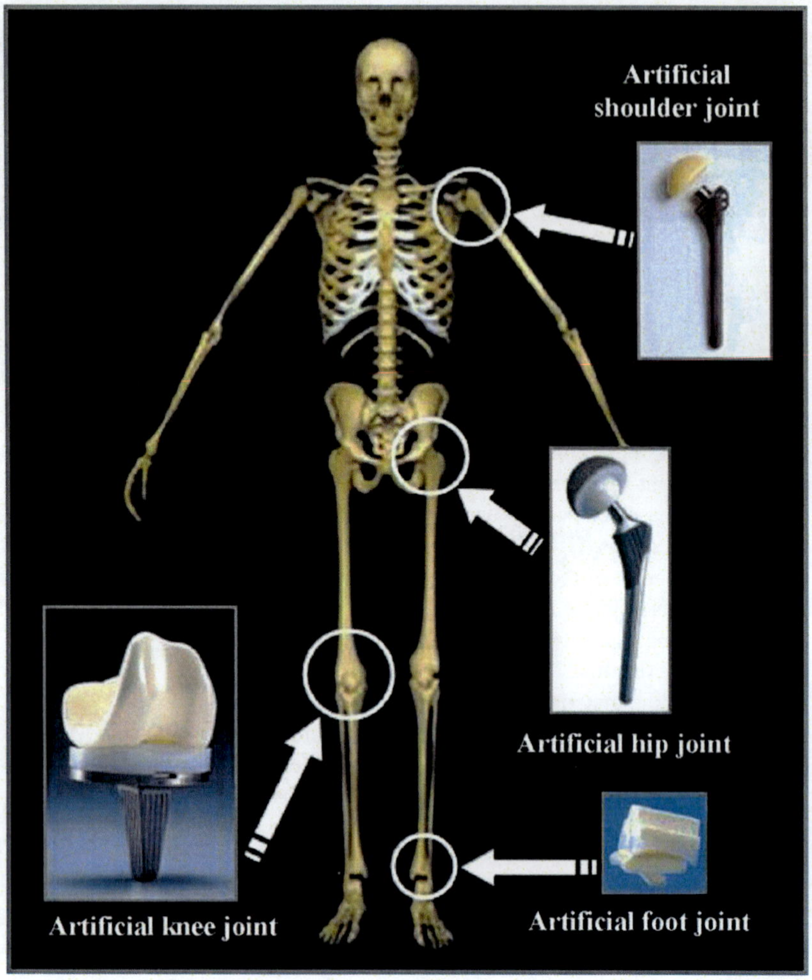

Fig. 20.20 Major artificial joints. Artificial joints are widely used in arms (*upper limbs*) and feet (*lower limbs*), which are called motor organs. Ninety-seven percent of artificial joints are used in hips and knees, which are under heavy loads and degrade with aging

20.5.2.2 Manufacturing Method

Figure 20.22 shows the manufacturing method of ceramic femoral heads. Alumina powder, mixed with a binder for creation of an intended shape, is filled in a special rubber mold. Pressure is applied to produce a cylindrical formed body. The formed body is ground into the intended shape and then is heated in a furnace at high temperature to produce a sintered body. The sintered body is precisely machined and polished before completion of the product.

Alumina femoral heads are intended for medical applications, and therefore, they are sterilized, packed and delivered after a complete inspection.

20.5.3 Future Prospects

Artificial joint replacement is a treatment procedure that can be very effective for the recovery of joint function. In Japan's rapidly aging society, the roles of artificial joints are expected to expand responding to the need to support active post-retirement life. With this background, zirconia with even higher strength and toughness began to be used in artificial joints and its effectiveness has been demonstrated. Ceramic materials have been steadily evolving. Major properties of the materials are listed in Table 20.1.

As part of the ceramic technologies intended for artificial joints, the technology for adding biocompatibility to metal

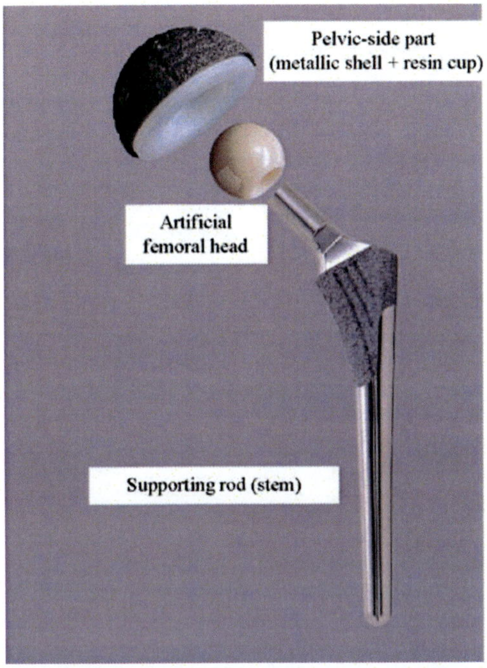

Fig. 20.21 Artificial hip joints. The spherical bearing of artificial hip joints consists of a pelvic-side cup made of resin and an artificial femoral head, which play the role of the caput femoris. The other parts include a metallic shell used to fix the cup on the bone and a supporting rod (stem) used to secure the artificial femoral head, etc.

surfaces (Note 20.15) has also been developed, although it was not introduced in this section. For improvement of durability, stable bonding of body tissues with artificial materials is essential. Currently, stable and long-term bonding, which is realized by coating metal surfaces with calcium phosphate-based ceramic, featuring the properties similar to bones, is in clinical trials. Research and development of ceramics that are harmless to human bodies with demonstrated functionality for use in bodies is expected to increase in the near future.

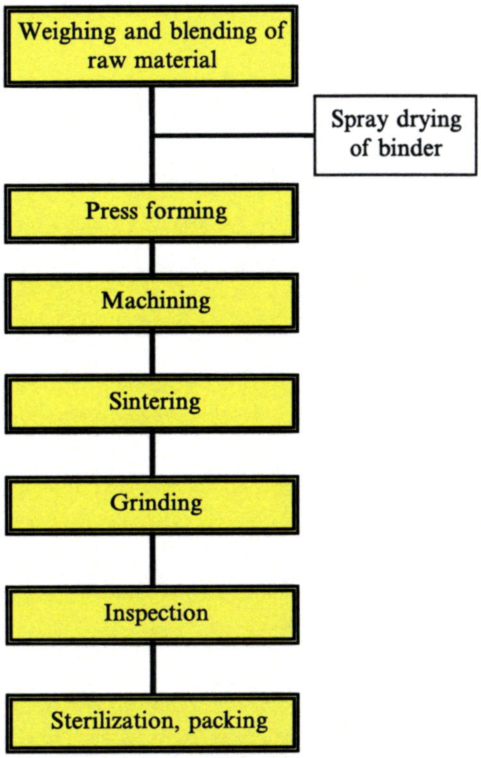

Fig. 20.22 Manufacturing process of ceramic femoral head. Powder of raw material, mixed with a binder for creation of an intended shape, is filled in a special rubber mold. Pressure is applied to produce a cylindrical formed body. The formed body undergoes machining and is sintered, and the surface is then polished to complete the product

Table 20.1 Properties of ceramics used to produce artificial joints

Material	Alumina	Zirconia
Purity	$Al_2O_3 > 99.5\%$	$ZrO_2 + HfO_2 + Y_2O_3 > 99\%$
Density (g/cm^3)	3.97	6.08
Young's modulus (GPa)	400	210
Hardness (Hv)	1,900	1,300
Bending strength (MPa)	640	1,600

Note 20.15 Material properties that are friendly to living bodies and prevent the material from being detected as foreign to the human body. Materials used to produce artificial joints are required to be evaluated for biocompatibility so as to ensure safety.

20.6 Ceramic Teeth (Porcelain Teeth, Porcelain Tooth Crown) (Porcelain Teeth: 1774, Porcelain Crown: 1889)

Our ancestors tried various means to recover the functionality of their teeth after losing them due to decay or accidents. "Post crown" and "false teeth" were invented as replacements for teeth. Artificial teeth used as "post crown" and "false teeth" are made of ceramics, resins (plastics) and metals. Ceramics are chemically stable and hard, maintaining the ability of chewing food for a long period. Furthermore, the material is suitable for reproducing a color tone similar to that of natural teeth and is nearly free from discoloration and staining, providing long-term durability.

20.6.1 Background of Development

Teeth are very important in terms of our diet, pronunciation and appearance as human beings. Artificial teeth are thought to have originated in ancient Egypt. They were made of animal bones, wood, etc. Artificial teeth used for "false teeth" are currently made of ceramics, resins (plastics) and metals. They are also called porcelain teeth, resin teeth and metal teeth. Porcelain teeth made of ceramics are superior to resin teeth in terms of abrasion resistance, resistance to discoloration and staining, capability to use for chewing food and long service life. Figures 20.23 and 20.24, respectively, shows a picture of false teeth and a picture of porcelain teeth. "Post crowns" are also made of ceramics, metals and resins. The majority of them are made of ceramics because of their functions, esthetic properties and biocompatibility they have with living bodies, and ceramic crowns (Note 20.16) are widely used. Porcelain teeth and ceramic crowns made of ceramics are explained below.

Note 20.16 Used to restore crowns on natural teeth. Made of a single ceramic material or ceramic and metallic materials.

20.6.2 Characteristics

20.6.2.1 Porcelain Teeth

In Japan, a potter (ceramic artist) made prototypes of porcelain teeth in 1774, and porcelain false teeth were released in 1788 after research. Artificial teeth made from porcelain were used for a long period, but both porcelain teeth and resin teeth were used after the commercialization of resin teeth made of acrylic resin (methacrylate ester polymer). It is expected that artificial teeth has esthetic properties and physiochemical properties similar to those of natural teeth. However, porcelain teeth and resin teeth both have advantages and disadvantages. Therefore, both of them are used currently.

The raw material used in porcelain teeth consists of feldspar, silica stone, clay, etc., like potteries and porcelains, and pigment made of metal oxides is mixed with the raw material to produce porcelain teeth. Feldspar begins to melt at 1,100°C, turning into clear glass at a higher temperature. Porcelain teeth made of glass is fragile, and therefore, silica is blended to improve its strength. However, if the percentage of silica is too high, the clarity that provides the appearance of natural teeth decreases. To solve this problem, a vacuum firing method was developed to produce porcelain teeth. If the formed porcelain teeth are heat treated with the vacuum firing method, bubble formation inside is reduced substantially, enhancing clarity dramatically. If this method is employed, the content of silica stone can be increased, realizing high esthetic properties and strength at the same time.

Artificial teeth are manufactured based on the theory that contours of a person's central incisors (Note 20.17) follow one's facial contours. They are basically square, convexed and egg shaped, and each of them is made in various sizes and color tones.

Note 20.17 Teeth in the center of upper and lower jaws, four in total.

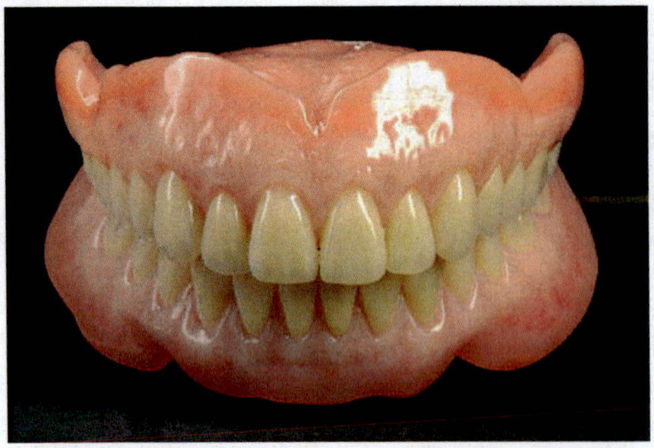

Fig. 20.23 False teeth. False teeth are classified into full dentures and partial dentures. Artificial teeth are made of ceramics or resins, while the gums are made of metals or resins

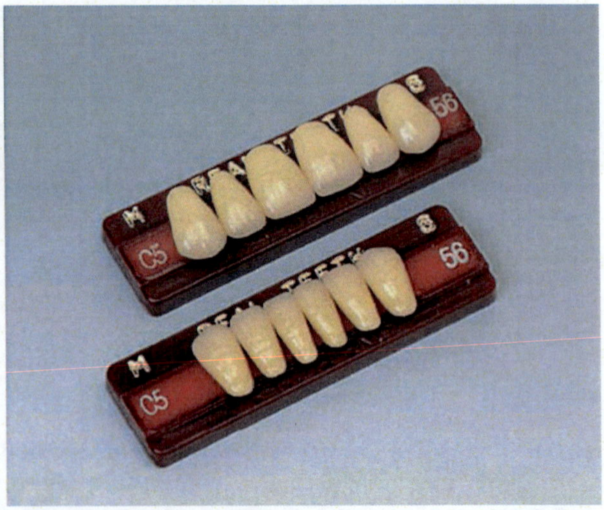

Fig. 20.24 Porcelain teeth. Porcelain teeth are made by adding pigments made of metal oxides to feldspar, silica stone and clay, like potteries and porcelains

20.6.2.2 Ceramic Post Crown

A "post crown" is an artificial tooth that caps or covers a natural tooth that has decayed or has been damaged in accidents. It is fixed over the surface of a tooth that has been ground to reduce its size. The manufacturing method of ceramic crowns made of feldspar (hereinafter ceramic) was introduced in America in 1889. Later, in 1903, manufacturing methods of ceramic post crowns made of platinum foil and ceramics were introduced. However, products caused many issues such as breakage of post crowns in the mouth due to insufficient strength. In order to enhance the strength, the porcelain-fused-to-metal crown, made by means the fusion of metal and ceramic materials, was introduced in 1956 and is still used even today. However, the post crown was made of opaque metal, making it difficult to reproduce the clarity of natural teeth, degrading the esthetic properties. Research and development continued to replace the metal material with ceramic material. In the 1990s, ceramic post crowns made of alumina were released for clinical use. In the 2000s, ceramic post crowns made of zirconia with higher strength and toughness were released for clinical use. In line with the advance in the CAD/CAM technologies (Note 20.18), zirconia began to be used clinically for front teeth as well as for bridges (Note 20.19) for teeth including molars. This is due to their excellent physical and chemical properties. Ceramic post crowns of nonmetallic structures are translucent and similar to natural teeth. They provide esthetic effects and also are friendly to patients who are allergic to metals. A ceramic post crown of a nonmetallic structure is shown in Fig. 20.25.

Note 20.18 Stands for computer-aided design/computer-aided(-assisted)manufacturing. Under the CAD/CAM system, technology information covering the entire processes from designing to manufacturing is processed effectively by computers.

Note 20.19 When multiple teeth are lost, artificial teeth are fixed by using adjacent unaffected teeth as supports in a manner similar to a bridge so as to recover function.

20.6.2.3 Manufacturing Method (Porcelain Teeth)

Feldspar, silica stone and clay are ground into fine powders. Porcelain teeth come in various colors, and pigments are mixed to reproduce color tones of dentine and the enamel of natural teeth. Then, a binder is added to the mixture before kneading. Kneaded material of a dentine color and kneaded material of an enamel color are filled in a die used to form porcelain teeth for press forming. The formed body needs to be larger than the final product in consideration of shrinkage after sintering. After low-temperature drying and calcination, the burrs on edges are removed from the formed body, and then the formed body is vacuum-fired at a high temperature.

20.6.2.4 Manufacturing Method (Ceramic Crowns)

Post crowns made of a ceramic are generally made by incorporating CAD/CAM technologies. The shape of the tooth that

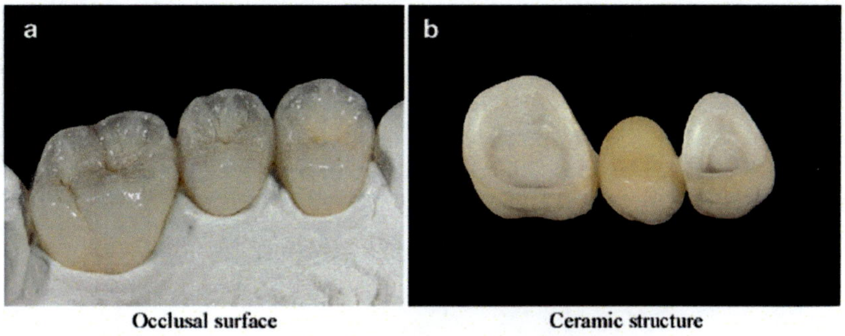

Fig. 20.25 Post crowns (ceramic post crown). Used to restore natural teeth. Made of a single ceramic material or ceramic and metallic materials. Example of (**a**) Occlusal surface (**b**) Ceramic structure

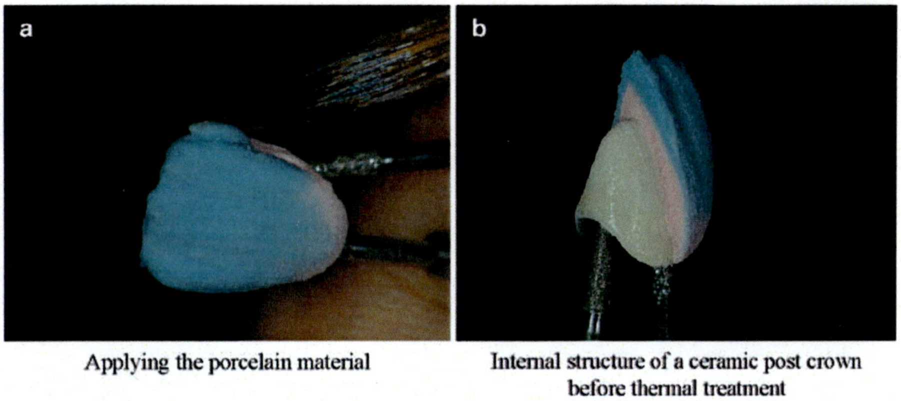

Fig. 20.26 Applying the porcelain material and its internal structure. (**a**) Applying the porcelain. The porcelain powder is kneaded with a special solution and formed on the metallic or ceramic surface using a brush for applying the porcelain to create ceramic post crown. (**b**) Internal structure of a ceramic post crown. The internal structure of the post crown was made by applying porcelain on a ceramic piece

has been ground is measured optically, and a partially-sintered ceramic block is ground by machining based on the measured data and is then sintered. Porcelain that matches the thermal expansion coefficient of the ceramic is fused to the ceramic product. The porcelain is a powder of various color tones including ivory and enamel colors. The porcelain powder is kneaded with a special solution and applied to form the ceramic product using a brush. The procedure is performed by a dental technician (Fig. 20.26). The post crown covered with porcelain is sintered in a vacuum furnace in the same manner as porcelain teeth. It is then trimmed and sintered again for a glossy finish.

20.6.3 Future Prospects

Materials used for dental treatment are classified into organic, inorganic and metallic materials. All of them have their own advantages and disadvantages. Ceramic materials are hard and feature high chemical stability. They are suitable for reproducing the color tones of natural teeth and are nearly free from discoloration and staining. They are excellent materials that provide a long-term cosmetic effect. Demand for porcelain teeth has not been expanding in recent years, but they continue to be used. Meanwhile, demands for an all-ceramic post crown that is used for aesthetic dental treatments have been expanding rapidly, following advances in manufacturing methods incorporating CAD/CAM technologies.

Literature

1. Jacho M (1981) Clin Orthp 157:259–278 **(20.1)**
2. Aoki H, Kato K (1975) Ceramics 10:469–478 **(20.1)**
3. Sakamoto M, Nakasu M et al (2007) J Biomed Mater Res 82A:238–242 **(20.1)**
4. Toriyama M, Kawamura S (1987) Synthesis of high-sinterability mechanochemical β-tricalcium phosphate. J Ceram Soc Jpn 95(7):741–745 **(20.2)**
5. Nikkei Biobusiness (2002) Artificial bones that regenerate bones. December 2002 issue, Tokyo, pp 124–127 (in Japanese) **(20.2)**
6. Tanaka T, Chazono M, Komaki H (2006) Bone prosthetic materials – basic and clinical application of β-TCP. J Jpn Orthop Assoc 80:270–275 **(20.2)**

7. Hakamatsuka Y, Irie H (2006) Developments for commercialization of β-TCP madreporic body. Biomaterial 24(2):100–107 **(20.2)**
8. Monma H, Kanazawa T (1976) J Ceram Soc Jpn 84:209–213 **(20.3)**
9. Hirano M (1993) New Ceram 5 **(20.3)**
10. Sawamura T, Hattori M, Okuyama M (2005) Ceramics 40:831–834 **(20.3)**
11. Ogiso M (1983) J Mouth Dis 50:1 (in Japanese) **(20.4)**
12. Ogiso M (1984) Limits and potential of endosseous implant. Quintessence Publishing Co., Ltd., Tokyo, pp 53–66 (in Japanese) **(20.4)**
13. Medium-Term Forecast of Medical Bionics (Artificial Organ) Market and Thorough Analysis of New Comers, 2007 version. Yano Research Institute Ltd. (2007) (in Japanese) **(20.5)**
14. ISO7206-2 (1996) **(20.5)**
15. Oonishi H (1990) J Biomechanics 23(4):382 **(20.5)**
16. Nobumoto S (1989) Easy-to-understand dental materials. Shofu Inc., Kyoto, pp 205–208, 228 (in Japanese) **(20.6)**
17. Kawakami M (1996) Second-version, new dental materials and equipment. Ishiyaku Publishers Inc., Tokyo, pp 106–114 (in Japanese) **(20.6)**

Part VII

Ceramics for Medical and Sanitation Sector

21 Medical and Health Products

Devices and materials that take advantage of various characteristics of ceramics have been developed and used in medical and physicochemical related products. Introduced in this section are piezoelectric ceramics for ultrasonic probes and virus absorption and decomposition filters, which are ceramic products that have been used in medical and analysis fields.

Medical ultrasonographs are used widely for diagnosis in the medical field because they enable easy and safe observation of tomogram images of organs and the movement of fetuses. High-permittivity piezoelectric ceramics are used as the material for piezoelectric oscillators of ultrasonic probes used in these devices (Sect. 21.1). The array ultrasonic probe, which has been commercialized and used widely, is made by dicing piezoelectric ceramic and acoustic matching layers into a width of 0.1–0.3 mm. A total of more than one million of them are manufactured throughout the world. Virus absorption and decomposition filters are made by immobilizing hydroxyapatite, which is used for artificial bone prosthetic material, on a filter surface. This was first applied for facial masks (Sect. 21.2). Initial products were made by immobilizing hydroxyapatite using adhesives and problems developed such as deterioration due to washing. Following the development of the hot adhesion method, where hydroxyapatite is laminated directly on the mask, virus absorption properties were enhanced, enabling cleaning the filter with water.

Y. Imanaka et al. (eds.), *The Ceramic Society of Japan, Advanced Ceramic Technologies & Products*,
DOI 10.1007/978-4-431-54108-0_21, © Springer Japan 2012

21.1 Piezoelectric Ceramics Used in Probes for Medical Ultrasonography (1975)*

Piezoelectric ceramics (Fig. 21.1) intended for ultrasonic probes (Note 21.1) are used (Fig. 21.2) in medical equipment for creating ultrasonographs (Fig. 21.3), that are used to observe internal organs i.e. the movements of pre-born babies and hearts. This piezoelectric material consists of high-permittivity piezoelectric ceramics made by adding a special additive called a relaxor with a complex perovskite structure containing lead [$Pb(B_1,B_2)O_3$] to lead zirconate titanate (PZT). The piezoelectric ceramic resonators are 8–15 mm in width, 25–75 mm in length and 0.1–0.5 mm in thickness, are ground precisely and are mounted with an electrode on each side.

Currently, an array of ultrasonic probes (Fig. 21.4) are widely used, the piezoelectric ceramic for the acoustic matching layer are diced to a width of 0.1–0.3 mm. More than one million resonators intended for ultrasonic probes have been manufactured since 1975 throughout the world.

21.1.1 Background of Development

The application of medical diagnostic devices (Fig. 21.3) that utilize ultrasound (Note 21.2) was started in the 1970s. The devices, which enable easy and safe observation of cross-sectional images of internal organs and movements of pre-born babies, began to be used in a wide range of fields for diagnosis related to the heart, abdominal organs, pre-born babies, mammary glands, the thyroid gland, blood vessels and urinary organs. In order to improve the resolution of images, channels of the ultrasonic probes (Fig. 21.4) were increased. Responding to this trend, demands for high-performance piezoelectric ceramic materials with high piezoelectric constant, permittivity and uniformity has increased.

21.1.2 Characteristics

21.1.2.1 Characteristics and Specifications

Lead zirconate titanate (PZT) ceramic was developed in the 1960s, but the material featured low sinterability and its permittivity and piezoelectric constant were insufficient. To improve the sinterability of PZT and increase its piezoelectric constant and permittivity, piezoelectric materials consisting of three components (Fig. 21.5) (which are made by adding lead-based complex perovskite called the relaxor to PZT), were developed by a number of companies with Japanese companies as leaders during the period of 1965–1975. The major relaxors include lead magnesium niobate [$Pb(Mg_{1/3}Nb_{2/3})O_3$], lead nickel niobate [$Pb(Ni_{1/3}Nb_{2/3})O_3$] and lead yttrium niobate [$Pb(Y_{1/2}Nb_{1/2})O_3$], and they are used as relaxors even today. These piezoelectric materials feature a piezoelectric constant d_{33} of 400–800 pC/N, relative permittivity of 2,000–6,000 and Curie temperature, Tc of 150–300°C.

21.1.2.2 Manufacturing Method

Piezoelectric materials consisting of three components are normally manufactured by the solid state reaction method. Powders of lead oxide, titanium oxide, zirconium oxide, niobium oxide, magnesium carbonate, nickel oxide, yttrium oxide, etc. are weighed and mixed using the wet method. The mixture is milled and dried, and then goes through calcination at around 900°C. The material is milled again into a fine powder, which is mixed with organic binders and are then press formed. The powders are filled into a ceramic shell and

*The number in parenthesis indicates the year that the product was first commercialized.

Note 21.1 Devices that transmit and receive ultrasonic waves through direct contact with human bodies. Piezoelectric ceramics are used inside the devices.

Note 21.2 Sound waves within a range of 20 kHz to 10 MHz. This frequency range cannot be detected by human ears.

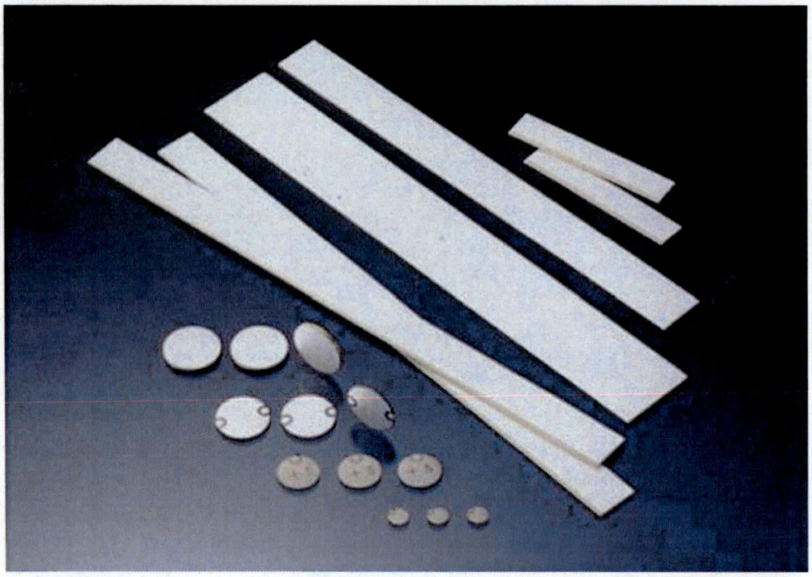

Fig. 21.1 Piezoelectric ceramic resonator for medical ultrasonography (courtesy of Fuji Ceramics Corporation). Standard size piezoelectric ceramic resonators are 8–15 mm in width, 25–75 mm in length and 0.1–0.5 mm in thickness. They are ground precisely and are mounted with an electrode on each side

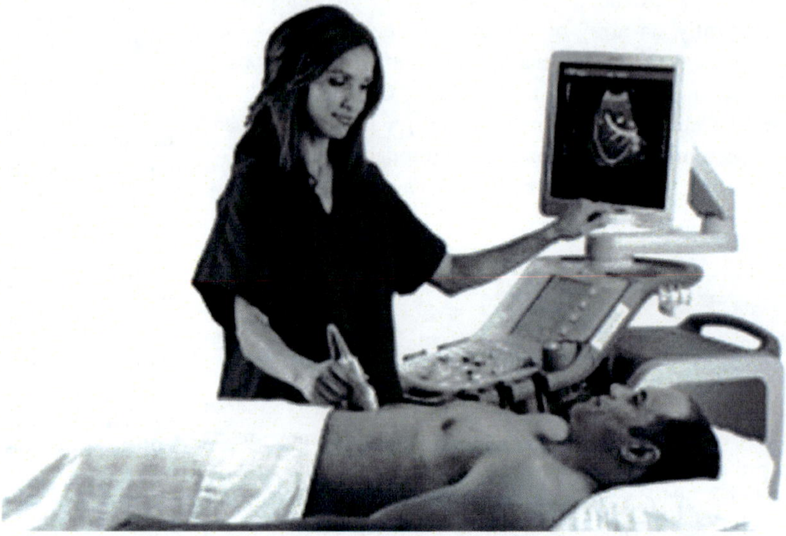

Fig. 21.2 Various types of ultrasonic proves. (*1*) Cardiovascular sector, (*2*) abdominal convex, (*3*) high-frequency linear (excerpt from the catalog of ultrasonographs manufactured by Toshiba Medical Systems Corporation)

sintered at around 1,200°C. The sintered body is ground precisely to a specified thickness, and then silver or gold electrodes are formed on it. The piezoelectric resonator is completed by the application of a DC voltage of about 2 kV/mm. Typical resonators for medical use are sized 70 mm × 15 mm × 0.4 mm.

21.1.3 Future Prospects

Piezoelectric ceramic materials have advantages in commercialization because a large substrate can be produced at low cost, but their piezoelectric characteristics are limited. Relaxor-based piezoelectric single crystals have been

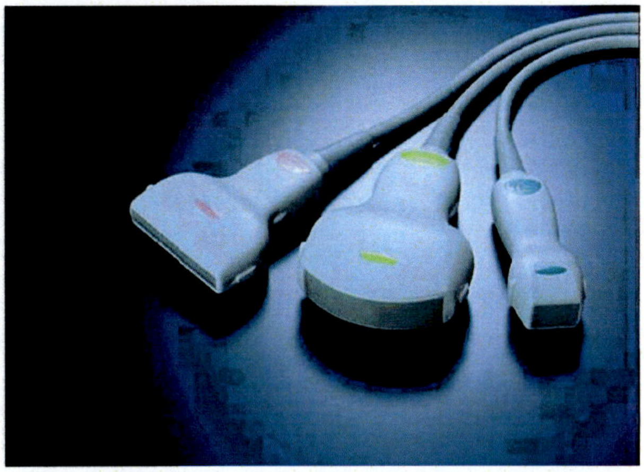

Fig. 21.3 Medical ultrasonograph (manufactured by Toshiba Medical Systems Corporation). A medical ultrasonograph that allows for easy and safe observation of cross-sectional images of internal organs and movements of pre-born babies

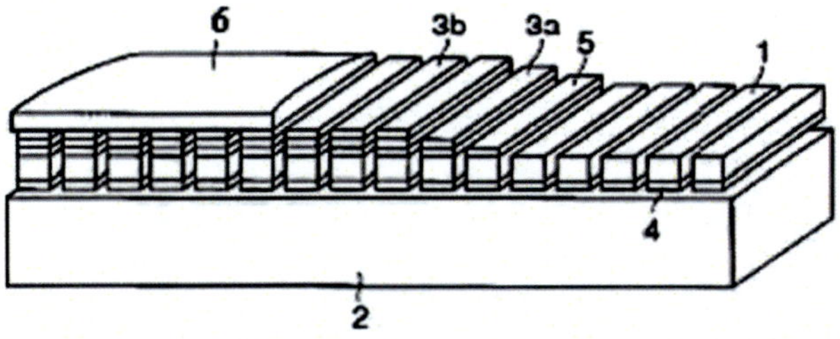

Fig. 21.4 Schematic of an ultrasonic probe. An ultrasonic probe is composed of the backing material, the piezoelectric ceramic material, the acoustic matching layer and the acoustic lens. The piezoelectric ceramic is diced to the width of 0.1–0.3 mm together with the matching layer. (*1*) Piezoelectric ceramic, (2) backing material, (*3a*) matching layer-1, (*3b*) matching layer-2, (*4*) lower electrode, (*5*) upper electrode, (*6*) acoustic lens

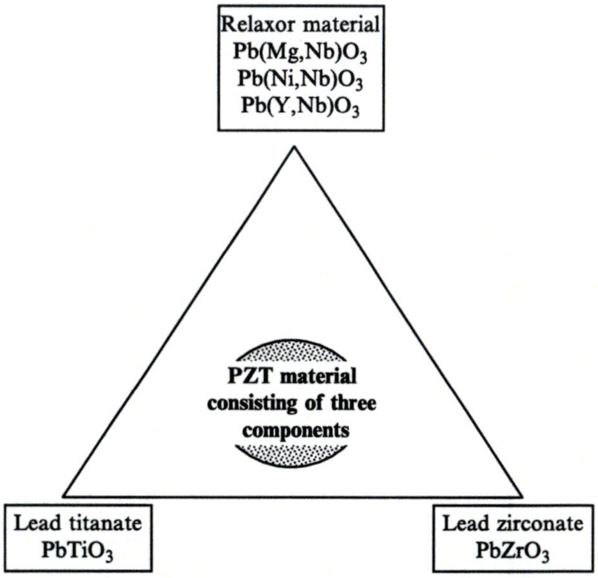

Fig. 21.5 Phase diagram of PZT materials consisting of three components. Materials consisting of three components are superior to PZT materials consisting of two components in piezoelectric constant and permittivity

attracting attention and have been studied throughout the world since 1995. Relaxor-based piezoelectric single crystals feature higher piezoelectric constants $d_{33} > 1{,}500$ pC/N, which is about twice that of PZT ceramics consisting of three components. Furthermore, lead-free piezoelectric materials have also been studied in consideration for the environment, and a barium titanate-based material with a piezoelectric constant of $d_{33} = 500$ pC/N has also been reported. These new materials have yet to replace PZT ceramics consisting of three components because of the issues of cost and lower image resolution, but they are expected to improve in the future.

21.2 Virus Absorbing Air Filter (1991)

Various studies on the application of hydroxyapatite have been performed responding to the introduction of high quality hydroxyapatite in the market following the commercialization of hydroxyapatite artificial bones and establishment of synthesis technologies. The virus absorbing air filter incorporating hydroxyapatite was commercialized in 1991 for use in air filtering face masks. Absorption performance of the original filter was low because the filter was bonded on the base material with an adhesive which blocked air exposure to part of the hydroxyapatite. Later, the absorption performance was improved by bonding it onto nonwoven fabric (Note 21.3), which was made using the span bond method (Note 21.4) and the thermal bond method (Note 21.5), directly using the heating adhesion method without using adhesive. Following this improvement, the filters now can be washed with neutral detergent. The filters used in antiviral face masks and air purifiers are expected to further advance for more effective prevention of airborne transmission of infectious diseases including influenza.

21.2.1 Background of Development

Hydroxyapatite [$Ca_{10}(PO_4)_6(OH)_2$, HAp] absorbs proteins and nucleic acid. It has been used as a filler in the liquid chromatography method for separation and purification of proteins and nucleic acid. It also features high biocompatibility with living bodies and has been used in artificial bones through active research and development. In line with the advance in artificial bones, synthesis methods have been improved, remarkably enhancing the stability in purity and quality and expanding its applications. Furthermore, developments are under way for its use as deodorant since HAp was proved to absorb ammonia and nitrogen oxide. It was also discovered that HAp absorbs viruses having proteins on its surface, such as the influenza virus. Taking advantage of these characteristics, filters coated with HAp have been developed for the removal of viruses (Note 21.6) and odors. They are used in antivirus face masks and air purifiers as shown in Fig. 21.6.

21.2.2 Characteristics

21.2.2.1 Characteristics and Specifications

Virus absorbing HAp is bound homogeneously on the surface of the base material using various methods so that HAp cannot be removed easily.

Nonwoven plastic fabric is normally used as a base material of filters for antiviral masks. For use in gauze face masks, the filters need to be resistant to washing with water or neutral detergent. Meanwhile, disposable masks do not have to be resistant to washing, unlike gauze masks.

Nonwoven fabric made by the same method as for antivirus masks and net-like plastic filters formed by using a mold, are used in air purifiers. HAp particles are bound on the filter surface.

Note 21.3 Fabrics for clothes are generally woven from threads made by twisting fibers. Nonwoven fabrics are made by bonding or entwining fibers by physical or chemical reactions and are "not woven." Wet tissues and felts are typical nonwoven fabrics.

Note 21.4 Thermoplastic polymer such as polyethylene is melted and ejected to form continuous fibers of a fabric. The fiber is double layered, thermoplastic polymer on the surface and thermal resistant polymer in the core.

Note 21.5 Fibers made using the span bond method, etc. are melted at high temperature to bond the fibers to adjacent fibers.

Note 21.6 Smaller than cells (including bacteria). They cannot multiply on their own and infect cells. They have the characteristics of nonliving materials and living matters at the same time. Many scholars today classify viruses as non-cellular organisms or nonliving matter.

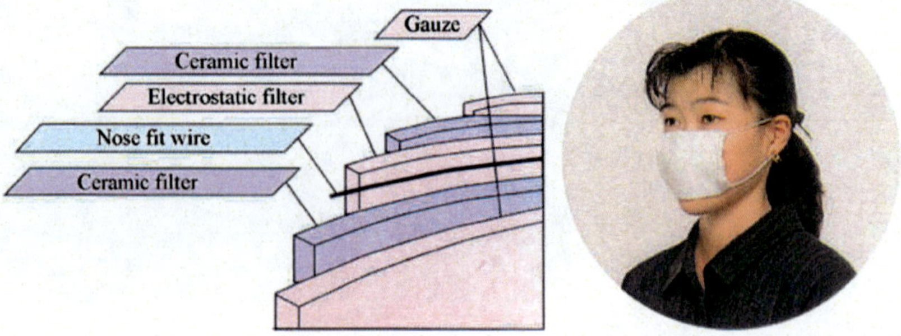

Fig. 21.6 Antivirus air filtering face mask. The illustration shows the configuration of a face mask. The electrostatic filter is sandwiched between two ceramic filters (for virus absorption). The picture shows an antivirus face mask on the market

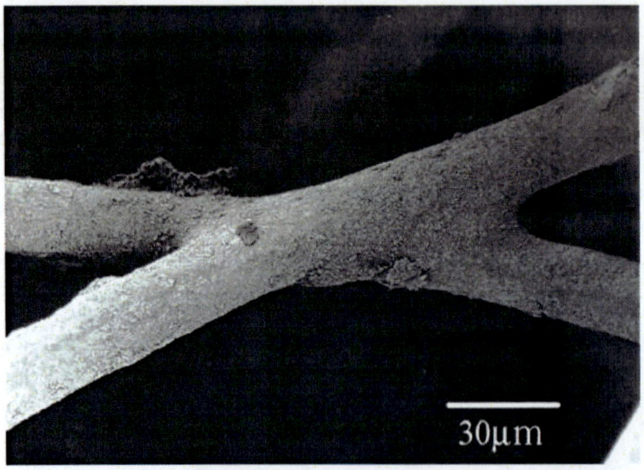

Fig. 21.7 SEM image of a fiber of the unwoven fabric used in ceramic filters. Virus absorbing ceramics are bound on the surface of unwoven fabric using the heating adhesion method. The whole surface of the fabric is covered with ceramic particles

21.2.2.2 Manufacturing Method

In the early years, filter base materials were dipped in a solution of HAp with virus absorption properties, organic adhesive, dispersant and antifoaming agent or sprayed with a solution so as to bind HAp to the surface. In this method, the surface of HAp is partially covered by adhesive, dispersant and antifoam agent, blocking the virus absorption sites and lowering virus absorption ability. The influence of the adhesive was particularly large and care was needed in the selection and amount (concentration) of the adhesive. HAp on this type of filter is removed if it is washed with water. Therefore, they all are disposable filters.

To overcome these disadvantages, improvements in the absorption performance and making them washable, the thermal adhesion method where nonwoven fabric (which is made with a combination of the span bond method and the thermal bond method and has thermoplastic polymer on the surface) is dipped in the HAp slurry. This is followed by the heat treatment for binding of HAp on the fabric surface (Figs. 21.7 and 21.8). Following the adoption of the heating adhesion method, virus absorption ability was improved remarkably as the HAp surface is not covered by adhesive, dispersant or antifoaming agent (Table 21.1). At the same time, the filters can be washed with water or neutral detergent. The majority of current products are manufactured by the heating adhesion method.

Other methods include the HAp particle injection method and the method where HAp is deposited on the surface of the filter material.

21.2.3 Future Prospects

Even in Japan, where the hygienic and medical technology is advanced, the number of deaths caused by influenza virus reaches approximately 1,800 a year (2005, National Institute of Infectious Diseases). During the past few years, a highly toxic avian flu virus emerged in Asia resulting in the deaths of infected people. There is a concern that the virus may mutate to a new type of influenza virus that is transmitted among humans. The new type of influenza is expected to cause high casualties throughout the world if it spreads.

Unlike HEPA filters (Note 21.7), current HAp filters are simplified filters and are not capable of capturing 100% of all airborne viruses. However, they are effective for the prevention of influenza because the incidence of influenza can be reduced if the number of viruses is reduced (Fig. 21.9).

Development of simplified air filters that can capture a larger percentage of influenza viruses for the dramatic reduction of infection rate is under development.

Note 21.7 Filters that remove foreign particles and dusts from the air to produce purified air. According to the Japanese Industrial Standards (JIS), the air filter should be capable of collecting 99.97% or more of the particles with a diameter of 0.3 μm at the rated wind volume and the initial pressure loss should be 245 Pa or less.

Fig. 21.8 Cross-sectional structure of a fiber of the unwoven fabric used in ceramic filters. Thermal resistant polymer fiber (*yellow*), thermoplastic polymer (*orange*) and HAp particles (*light blue*)

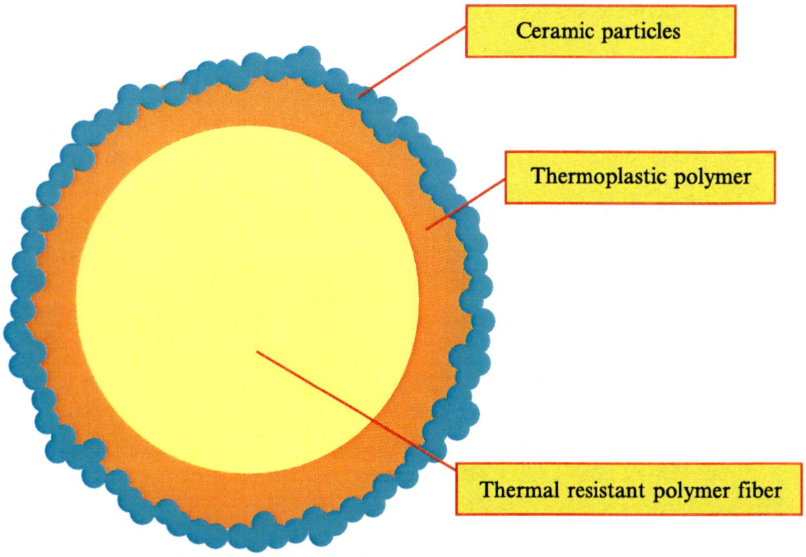

Table 21.1 Influence of washing antiviral face mask with a washing machine on virus absorption ability

	Virus absorption ratio of unwoven fabric prior to washing with water (%)	Virus absorption ratio of unwoven fabric after washing with water (%)
Unwoven fabric filter on which ceramic particles are bound by organic adhesive	88	75
Unwoven fabric filter on which ceramic particles are bound by water-soluble adhesive	93	0
Unwoven fabric filter on which ceramic particles are bound by the heating adhesion method	99	97

An influenza virus solution is filtrated by each type of unwoven fabric to measure virus absorption ability. Drops in absorption ability are measured doing the same after washing the filters with a washing machine

Fig. 21.9 Absorption of influenza viruses (SEM image). The *left picture* shows the surface of a fiber in ceramic filter unwoven fabric before absorption of influenza viruses, and the *right picture* shows the surface after absorption of influenza viruses (*spherical* and *bar* shaped)

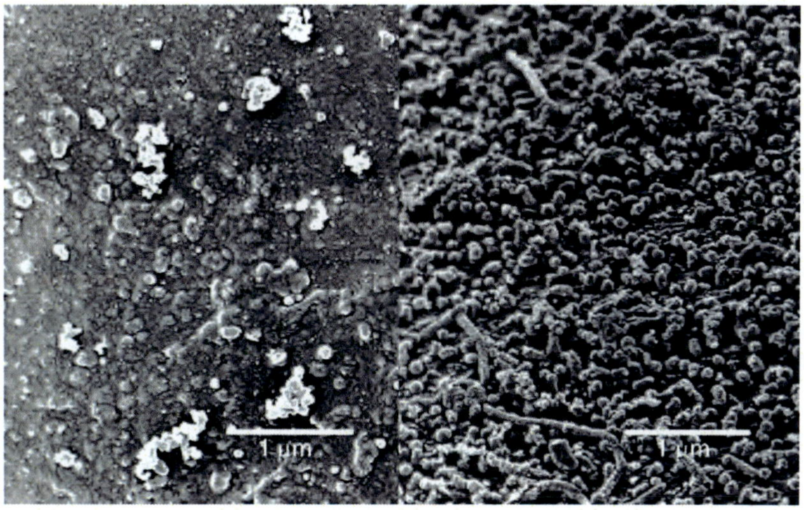

Literature

1. Ouchi H (1965) J Am Ceram Soc 48:630–636 (21.1)
2. Yamashita Y (2004) Jpn J Appl Phys 43:6679–6682 (21.1)
3. Karaki T (2007) Jpn J Appl Phys 46(4):L97–L98 (21.1)
4. Tiselius A et al (1956) Arch Biochem Biophys 65:132–155 (21.2)
5. Aoki H (1971) J Stomatol Soc Jpn 40:277 (in Japanese) (21.2)
6. Driskell TD et al (1973) J Dent Res 52:123 (21.2)
7. Hiraide T (1993) Ceramics 28:642–646 (in Japanese) (21.2)
8. Tsuru S et al (1991) Bio-Mater Eng 1:143–147 (21.2)

Part VIII

Ceramics Raw Commodity Sector of Activity

Everyday Ceramic Items

22

Ceramics have been used since ancient times in a variety of everyday items that need to be thermally resistant. With regard to thermal properties, priority is placed on low thermal expansion characteristics. A variety of scientific and technological achievements in ceramics have been incorporated. Everyday items that have been contributing to society, as well as to products used in the jewelry field, are introduced in this section.

Zirconia is used in kitchen knives, slicers and stationery scissors (Sect. 22.1). These items, unlike metallic items, feature high hardness, sharpness and corrosion resistance and are free from rust and do not leave a metallic taste in the item that has been cut. The grater, incorporating fine ceramic technology, is glazed on the surface projections so as to smooth out the surface and allows for easy removal of stains.

Ceramics are also used in electrical light bulb components. In incandescent lights, soda-lime glass is used in the main bulb portion and lead glass is used in the stem portion (Sect. 22.2). Functional ceramic materials are used in the main tube, front glass, mirror, cap and bond of halogen light bulbs, that contain a small amount of halogen gas. For example, steatite ($MgO{\cdot}SiO_2$) or alumina are used in the cap in consideration for thermal resistance and electrical insulation properties.

The white light of light emitting diodes (LED) is realized by a combination of a blue LED and yellow phosphor (Sect. 22.3). They feature high efficiency, a long service life, high luminance and a wide color variation and are promising lighting devices. Ceramic materials are used in the LED chip substrate, the circuit board and in the phosphor.

Sodium lamps, featuring higher efficiency and luminance than incandescent lamps and fluorescent lamps, are widely used as exterior lights on roads, at sports facilities (Sect. 22.4). Sodium lamps are composed of two layers, the external tube and the internal light emitting tube, and translucent alumina is used in the light emitting tube. It is required to be high in translucency and is characterized by high purity and a dense structure. Applications of translucent alumina have been expanding rapidly for use in high-pressure sodium lamps for exterior illumination and metal halide lamps for interior illumination.

Semiconductor gas sensors are widely used in households for security and disaster prevention as well as for a comfortable and healthy life (Sect. 22.5). They are used to detect the leakage of methane gas and carbon monoxide generated by incomplete combustion. Responding to the issues arising from air tight houses and sick house syndrome, air conditioners are equipped with air purifying and ventilator functions and are mounted with a number of sensors that detect air pollution.

Piezoelectric buzzers (Sect. 22.6) are used in a variety of electronic devices ranging from home appliances such as microwave ovens and microcomputer controlled rice cookers, electric calculators, clocks, game machines as well as office automation (OA) devices. Responding to the needs for small, thin and light products and operation under low-voltage, products that incorporate co-firing technologies of laminated ceramic green sheets and electrodes have also been produced. Piezoelectric speakers, which are similar products, have also begun to be used widely.

Tiles are widely used as building exterior materials and interior materials for bathrooms, toilets and kitchens (Sect. 22.7). In recent years, humidity control, thermal insulation and anti-soiling functions are required for tiles. Ultralight heat resistant tiles are used on the surface of the space shuttle, which incorporates leading-edge space technology.

Ceramics release infrared rays when they are heated. The radiant heat is easily absorbed by the heating object, thereby effectively increasing the temperature. Far-infrared ceramic heaters are also effective for reducing CO_2 generation resulting in increasing applications (Sect. 22.8).

In the jewelry field, ceramics such as alumina and zirconia as well as cermet are used in watch crystals and bands (Sect. 22.9). The characteristics of ceramics, such as high hardness, light weight and diversified color tones and textures, are effective for improving these products. A variety of surface treatments are performed to serve diversified market needs.

Y. Imanaka et al. (eds.), *The Ceramic Society of Japan, Advanced Ceramic Technologies & Products*,
DOI 10.1007/978-4-431-54108-0_22, © Springer Japan 2012

The anti-reflection (AR) coat is a ceramic film formed on the surface of eyeglass lenses so as to reduce light reflectance (Sect. 22.10). The film reduces ghost image phenomena such as double images and flickers, which are caused by the light reflected from the lens surface. Normally, ceramic films are formed on lenses by vacuum deposition.

Synthetic jewels are made by artificially reproducing chemical compositions and structures of natural jewels (Sect. 22.11). Recrystallized emeralds and opals have both been commercialized. The former is produced by growing crystals with the flux method and the latter is made by arranging silicon dioxide grains regularly to produce characteristics of jewels, including opalescence.

Ceramic Knife, Grater, Slicer and Scissors (1984)* 22.1

Fine ceramic kitchen goods such as ceramic knives, slicers and scissors have sharp blades formed by zirconia and are used to cut food. Zirconia is harder than metals, remains sharper for a longer time and is free from metallic smells (odor and taste peculiar to metal products). Ceramic graters are attached to kitchenware or enforced porcelain and protrusions formed on the grating area are applied with glaze. The finished products are recognized by consumers as "ceramics." Since the introduction of these products in 1984, they have gradually penetrated the market and are widely used not only in Japan but also in countries around the world.

22.1.1 Background of Development

The majority of conventional kitchen cutlery and household stationery were made of metals and resins. Today, however, a number of these items are made of ceramic, which have abrasion resistance and corrosion resistance that are not observed in conventional materials. They have penetrated the market and are frequently sold in shops or online these days. Zirconia and alumina are used to produce kitchen cutlery such as knives (Fig. 22.1), slicers (Fig. 22.2) for cutting food and ceramic graters (Fig. 22.3) for grating food as well as stationery items such as scissors (Fig. 22.4).

*The number in parenthesis indicates the year that the product was first commercialized.

22.1.2 Characteristics

22.1.2.1 Characteristics and Specifications of Products

These products have abrasion resistance and corrosion resistance of ceramics. They do not degrade the taste of food; they remain sharp for a long period of time and are free from both a metallic smell and rust. Zirconia ceramics are used as the blades of ceramic knives, slicers and scissors. Resin or metallic handles are attached to the processed ceramic blades.

The edge durability test is performed to obtain a figure of merit that indicates a knife's abrasion resistance. A fixed load is applied to a knife to cut pieces of a specified type of paper. The sharpness of the knife is indicated by the number of pieces of paper that can be cut at one time. This test is repeated over and over again in order to measure the durability of the blade.

The test results are shown in Fig. 22.5. Ceramic knives are harder than stainless or forged steel knives and sharpness remains longer than metallic knives based upon the paper cutting test results.

Graters are usually made by using pottery and enforced porcelain as the base material. Food is grated by the protrusions on the surface. The surface is glazed for smoothness. Therefore, they can be cleaned just by being placed under flowing water and are easy to clean. Silicon rubber is applied on the bottom to prevent slippage.

22.1.2.2 Manufacturing Method

There are many ceramic manufacturing methods and they vary depending on manufacturers. General manufacturing methods are explained below. The common uniaxial press method is mainly employed for production of ceramic knives

Fig. 22.1 Ceramic knives incorporating zirconia. The handles are colored. The blades are made from zirconia. The handles are made from polypropylene

Fig. 22.2 Ceramic slicer. A cabbage is being cut into strips by a ceramic slicer. The blade is made from zirconia and the body is made from ABS resin

Fig. 22.3 Ceramic grater. Silicon rubber is bonded on the back to prevent slipping

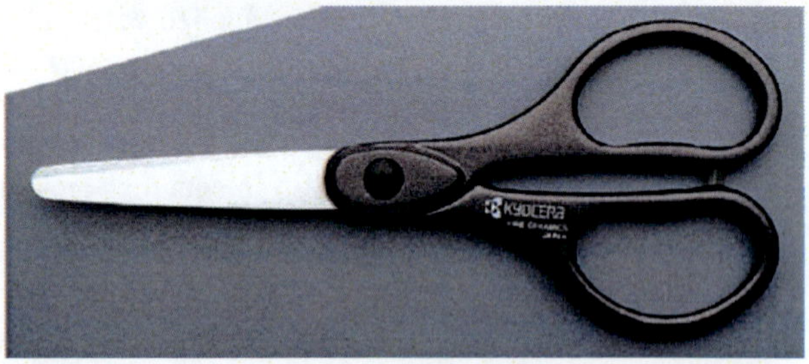

Fig. 22.4 Ceramic scissors. The blades are made from zirconia and the body is made from aluminum

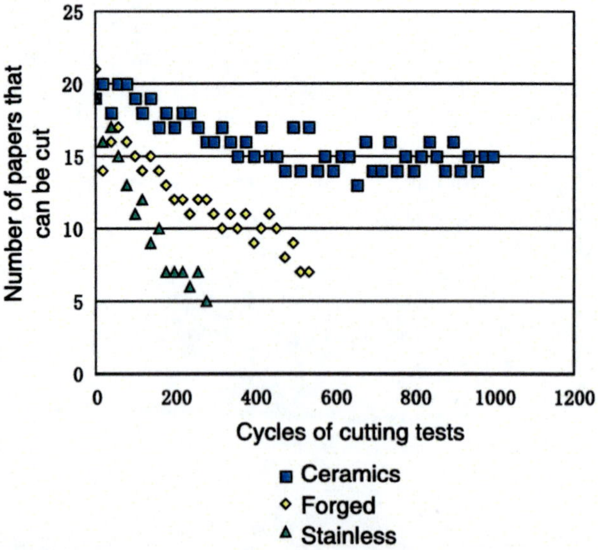

Fig. 22.5 Results of edge durability test. Reproduced from the referenced literature

and scissors. The zirconia material is mixed with binders to create a slurry. The slurry is spray dried by using a spray drier to create granules of a specified size. The granules are poured into a die and formed by applying a specified pressure to the top and bottom. (Injection forming and extrusion forming are sometimes employed.) Then, dewaxing and sintering are performed to create ceramics.

The ceramics are processed into specified shapes using a diamond grinder to create ceramic parts. The ceramics are attached to handles or bodies made of resin, metal or wood by insert molding, screws or by bonding. As for graters, the casting method is usually employed. The slurry made from the raw material is poured into a water-absorbing die made of plaster and pressure is applied during forming. After removal from the die, drying and sintering are performed and glaze is applied on the surface. The product is finished by bonding slip-proof silicon rubber to the bottom.

22.1.3 Future Prospects

It took 25 years before the advantages of ceramic knives and ceramic scissors began to be recognized within the cutlery and stationary market. In recent years, they are used in Japan as well in the North American, European and Asian markets. This market is expected to expand further in the future.

Light Bulbs (1890) 22.2

Light bulbs were put into practical use by Edison in 1879. In Japan, they were commercialized by Hakunetsusha (currently Toshiba) in 1890. Light bulbs are classified into several different categories such as incandescent light bulbs and halogen light bulbs, in which inert gas such as argon and a minor amount of halogen gas are enclosed. Halogen light bulbs are used as spot lights in shops and as the headlights in automobiles because of their brightness and high efficiency. Thermal resistance is required because light is emitted from a heated thin resistant wire by applying a voltage to it. For this reason, ceramics such as the glass, mirrors, caps, adhesives, etc. are used in light bulbs.

22.2.1 Background of Development

In October 1879, Edison succeeded in making the first carbon light bulb using carbonized cotton thread as the filament. Later, in 1910, Coolidge in the U.S. invented the "tungsten light bulb," remarkably extending the service life of light bulbs. In 1913, Langmuir in the U.S. further extended the service life of light bulbs by sealing inert gas inside that does not react chemically with tungsten in the light bulb in order to prevent sublimation due to the high temperature of the tungsten filament. In 1959, Zubler in the U.S developed a light bulb with a longer service life by sealing halogen gas inside the light bulb silica glass tube to prevent the blackening of light bulbs. In Japan, the halogen bulbs were commercialized in 1961.

22.2.2 Characteristics

22.2.2.1 Products

Figure 22.6 shows the appearances and structures of incandescent light bulbs and halogen light bulbs.

Table 22.1 shows major ceramic parts in light bulbs.

22.2.2.2 Characteristics and Specifications

Table 22.2 shows types of glass used to produce tubes and their characteristics. Since many years ago, the main bulb has been made of soda-lime glass, while the stem has been made of lead glass. However, the lead glass is being replaced by non-lead glass in consideration for the environment. The main tube of a halogen light bulb is made of silica glass or high-alumina glass, which has high thermal resistance. The front glass and the mirror are made of borosilicate glass. Multilayer optical films that are dichroic are formed on outer surface of the tube of a single-ended type halogen light bulb without a mirror like surface (JD) and on the inner surface of the mirror of a single-ended type halogen light bulb with a mirrored surface (JDR) in order to control infrared rays.

22.2.2.2.1 Adhesive

Water/alkoxide-based silica or alumina-silica is used as adhesive.

22.2.2.2.2 Caps

Ceramic caps with high heat resistance are used in halogen lamps, as shown in Fig. 22.7. Steatite ($MgO{\cdot}SiO_2$) or alumina (Al_2O_3) is selected in consideration of their radiation performance and cost depending on products.

22.2.2.3 Manufacturing Method

22.2.2.3.1 Glass

The manufacturing method of each type of glass is shown in Table 22.3.

22.2.2.3.2 Adhesive

Water-based adhesives are classified into colloidal silica type adhesives that are dehydrated and condensated in the process of drying and alkoxide sol adhesives in which hydrolysis and

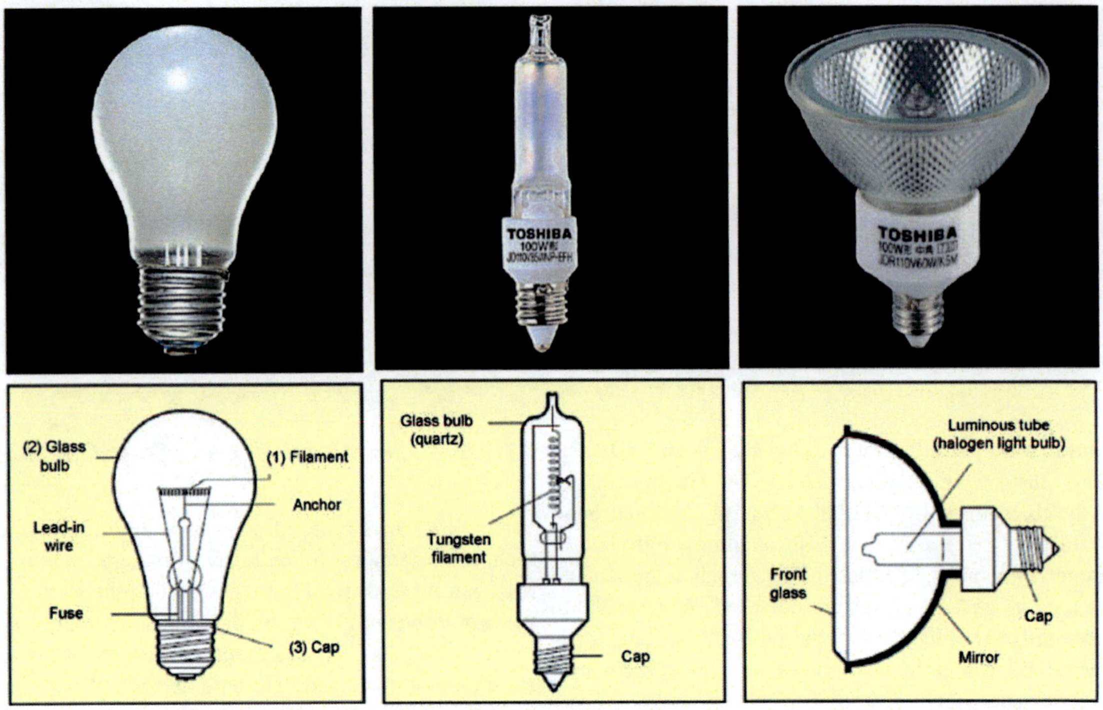

Fig. 22.6 Appearances and structures of incandescent light bulbs and halogen light bulbs. From *left* to *right*: an incandescent light bulb, a single-ended type halogen light bulb without a mirror (JD) and a double-ended type halogen light bulb with a mirror (JDR). Ceramic components are the glass of the incandescent light bulb and the glass, mirrors, caps and adhesives of the halogen light bulbs

Table 22.1 Major ceramic parts

Type	Component
Incandescent light bulbs	Glass
Halogen light bulbs	Glass, mirror, cap and adhesive

Table 22.2 Types of light bulb glasses and their characteristics

	Silica glass	Borosilicate glass	Soda-lime glass	Lead glass
α	5	38	97	92
T_s	1,670	640	692	615
T_g		530	540	434
Component system	SiO_2	B_2O_3–SiO_2	Na_2O–CaO–SiO_2	PbO–SiO_2

α (10^{-7}/°C): thermal expansion coefficient, T_s (°C): softening point, T_g (°C): glass-transition point

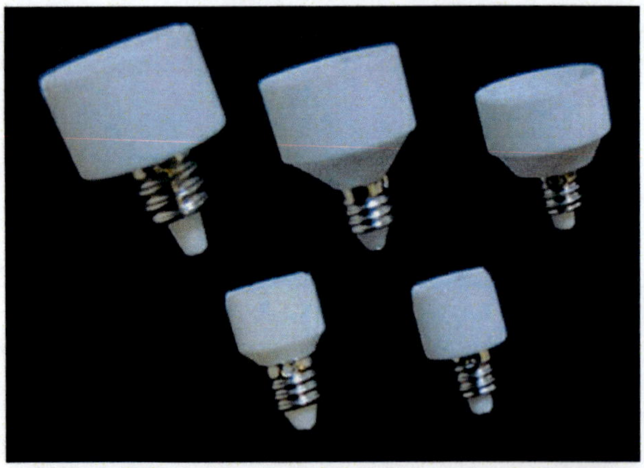

Fig. 22.7 Appearance of caps for halogen light bulbs. Steatite ($MgO \cdot SiO_2$) or alumina (Al_2O_3) is used to produce the ceramic parts

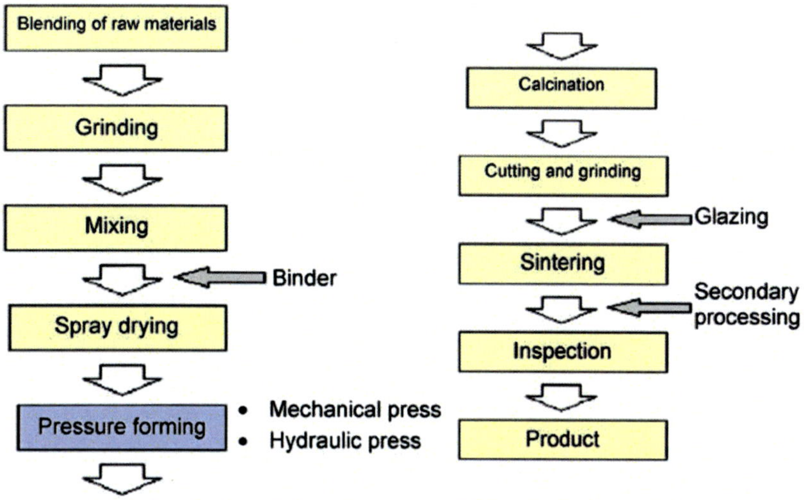

Fig. 22.8 Pressure forming method. Caps are manufactured by the pressure forming method

Table 22.3 Manufacturing method of each type of glass

Type of lamp	Glass	Manufacturing method
Incandescent light bulbs	Bulb	Ribbon machine method (Note 22.1)
	Tube	Danner method (Note 22.2)
Halogen light bulbs	Tube	Down draw method (Note 22.3)
	Mirror	Press method (Note 22.4)

polymerization occurs. Fine particle silica and alumina are used in both types.

22.2.2.3.3 Cap

Caps are manufactured by the pressure forming method, which is shown in Fig. 22.8.

22.2.3 Future Prospects

In consideration of the environment (prevention of global warming, protection of the environment), incandescent light bulbs that consume a large amount of electricity have gradually been replaced by bulb-shaped fluorescent lamps and LED lamps that consume less electricity with a long service life and high efficiency. Meanwhile, light bulbs will continue to be used in the environment where a priority is placed on color rendering properties, in high-temperature high-humidity environments, in agricultural green houses where the heat generated from light bulbs are utilized and in other places.

Note 22.1 A forming method in which the material is heated and melt extruded in the form of a tube, which is held between dies, blowing air inside to create a hollow part.

Note 22.2 The material consisting mainly of carbonate is melted at a temperature of 1,300°C, and the molten glass is wrapped around a hollow ceramic part called a sleeve. Hot air is supplied from behind the sleeve and the sleeve is drawn in the horizontal direction to create a tube.

Note 22.3 The material is melted and drawn toward the bottom. There is also the up draw method.

Note 22.4 Molten glass is flown into a die and press formed by a machine to form a glass product.

LED Light (1996) 22.3

The LED (light emitting diode) is widely used in industry and is incorporated in mobile phones and a variety of display devices. Following the improvement in the luminance efficiency of LEDs, luminaire (lighting) applications have recently seen a rapid increase. For lighting applications, pseudo white is normally produced by a combination of light from a blue LED and yellow (plus red and green, in some cases) phosphor, which is complementary to blue. LEDs are promising as new interior and exterior lights because of such characteristics like energy-efficiency, long-service life, a wide variation of luminescent colors and high luminance. Ceramic components are currently used in chip substrates, substrates, phosphors, because of high radiation properties and thermal resistance. Ceramic parts used in mounting the LEDs are expected to solve the issue of handling the heat generated from chips, which arises following the enhancement of light intensity and efficiency.

22.3.1 Background of Development

Visible light LEDs, which are produced from devices made by growing a GaAs epitaxial film on a GaAs substrate, was realized and used in the displays of calculators and watches for the first time in 1962. The white LED was realized by using three primary colors, the blue LED was commercialized in 1993 and red and green LEDs were commercialized earlier. However, the tri-color system has a complicated circuit and its color rendering properties (Note 22.5), which are important for illumination, are insufficient. These issues were solved in 1996 by means of the introduction of a system made by combining blue GaN-LED with yellow YAG phosphor. The efficiency (Note 22.6) was very low (5 lm/W) at the time of introduction. Materials including the chip and phosphor as well as the structure were improved later, realizing 100 lm/W in products and 150 lm/W in laboratories. It is promising as a new illumination source because of its energy-saving properties, long service life and high light intensity. The LED has a 50-year history, but LED lights have a history of only 12 years and the technologies are expected to be enhanced in the future.

22.3.2 Characteristics

22.3.2.1 Products

Figure 22.9 shows the structure of the LED incorporated in LED bulbs (Fig. 22.10) and LED luminaires (Fig. 22.11). The major ceramic parts are the chip substrate, the substrate and the phosphor.

22.3.2.1.1 Chip Materials

Figure 22.12 shows a cross sectional view of chip. InGaN chips that emit a wide range of light including ultraviolet, blue and green colors are used in white LED chips. The

Note 22.5 The characteristics of light sources such as lamps that affect the visibility of objects are called color rendering properties. The color chart for evaluation is lit by the standard light specified by JIS (Japanese Industrial Standards) and the light to be evaluated, and the deviation in color is quantified and expressed by the color rendering index [average color rendering index (Ra) and special color rendering index (R9–R15)]. Ra of the standard light is 100 and higher values indicate better color visibility.

Note 22.6 Known as luminance efficiency or lamp efficiency and expressed by total flux per unit electric power lm/W (lumen per watt). Efficiencies of major light sources are 40–110 lm/W for fluorescent lamps, 10–18 lm/W for incandescent light bulbs, 20 lm/W for halogen light bulbs and 50–130 lm/W for HID (high intensity discharge) lamps.

semiconductor layer is made by forming an n-type nitride semiconductor layer (n-GaN: Si), a luminous layer (InGaN) and a p-type nitride semiconductor layer (p-GaN: Mg) on a sapphire (Al_2O_3) substrate. The electrodes are composed of a positive electrodes with a transparent electrode (Au/Ni) and a p-side electrode pad formed on the p-type semiconductor layer and a negative electrode having an n-side electrode pad formed on the n-type nitride semiconductor layer. The chip is enclosed by a protective overcoat layer (SiO_2). In addition to sapphire (Al_2O_3) substrates, silicon carbide (SiC) substrates have been commercialized.

22.3.2.1.2 Substrate Materials

Ceramic materials are used to prevent deterioration of luminance efficiency due to heat generated from the chip and to reduce changes in light color due to deterioration during usage. This is done in order to achieve high efficiency and high light intensity. Characteristics required of substrates are as follows:

1. High radiation performance
2. Low deterioration due to light
3. Low wavelength dependency of reflectance
4. High insulation properties

Alumina (Al_2O_3) is used mainly as the material. Silicon (Si) is sometimes used to manufacture submounts. In recent years, ceramics have also been used to manufacture reflectors in addition to substrates.

22.3.2.1.3 Phosphors

The YAG (yttrium aluminum garnet) phosphors are commonly used for blue color excitation. The YAG phosphor is a light emitting oxide made by adding cerium (Ce),

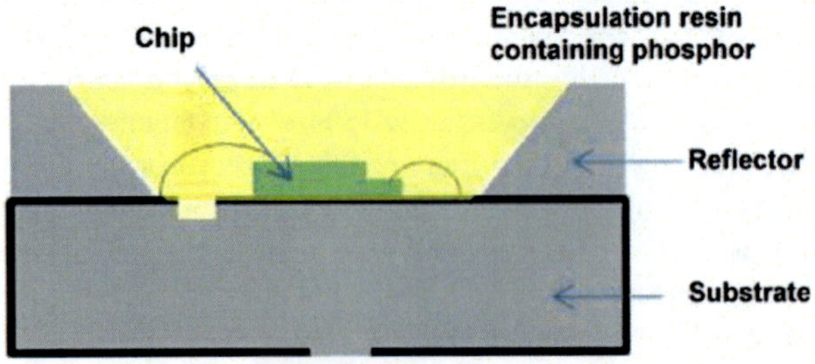

Fig. 22.9 Structure of LED. The white LED system that creates *white color* using a blue light emitting chip and *yellow* phosphor. The chip mounted on the substrate is enclosed by encapsulation resin in which phosphors are dispersed

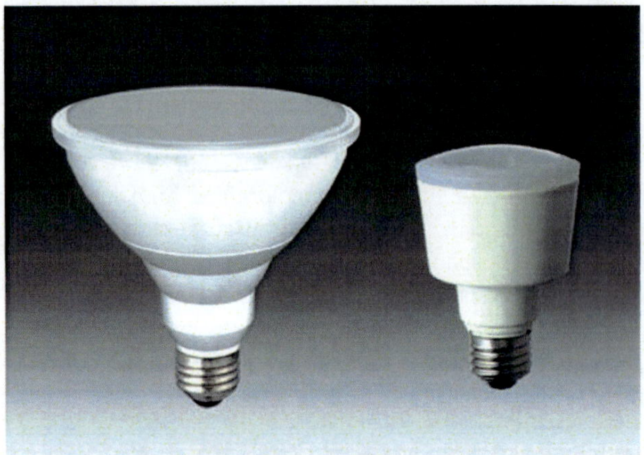

Fig. 22.10 LED bulbs. The electrical connectors of LED bulbs are the same shape as those of incandescent light bulbs, allowing interchangeability between the two types. They use less energy and offer long service life

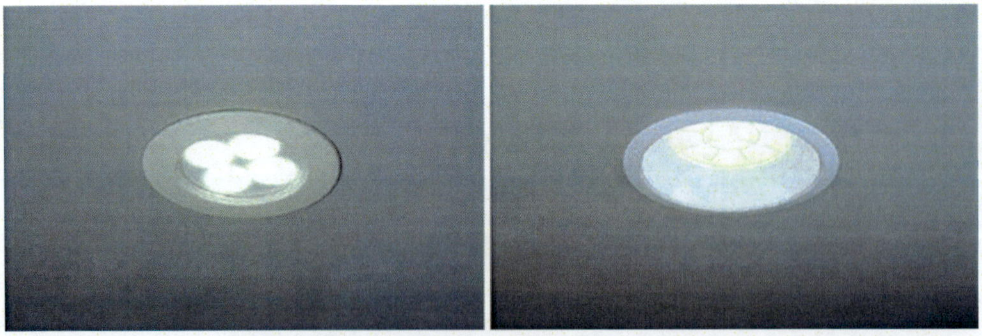

Fig. 22.11 LED luminaires. LED luminaires are currently used mainly as down lights and spot lights. They are in the process of being introduced into base lights

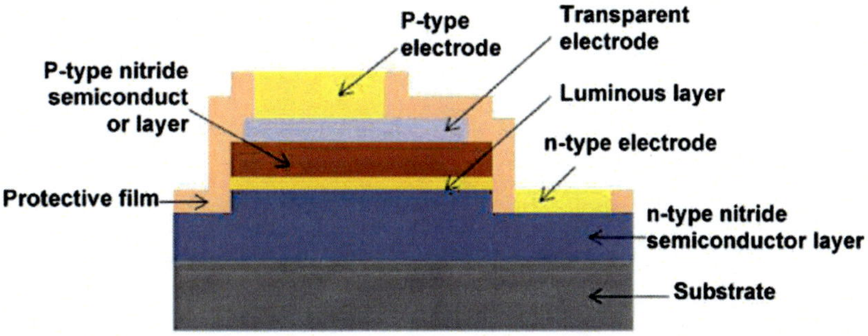

Fig. 22.12 Cross sectional view of a chip. The layers are formed on the sapphire (Al_2O_3) substrate and the silicon carbide (SiC) substrate

an activating agent, to yttrium aluminate ($Y_3Al_5O_{12}$). It emits yellow colored light, which creates pseudo white color in combination with the blue color emitted from the chip. Another example of an oxide phosphor is the silicate phosphor made by adding europium (Eu), an activating agent, to strontium–barium silicate $(Sr, Ba)_2SiO_4$. Emission colors of oxide phosphors can be adjusted. This color range is from green to orange and colors are changed by changing the Sr-Ba composition ratio. Red and green phosphors are sometimes added to improve the color rendering index (Ra), which affects the visibility of colors, for use in luminaire. Red phosphor $CaAlSiN_3$: Eu nitride phosphor is attracting attention. Developments of oxynitride, which has excellent thermal properties, have been accelerated in recent years.

22.3.2.1.4 Chip Materials

Single crystals such as sapphire (Al_2O_3) used as the substrate material are manufactured by the crystal growth method. The crystal growth methods are classified largely into three methods, the gas-phase method, the liquid-phase method and the solid-phase method. The liquid-phase method is generally employed.

22.3.2.1.5 Substrate Materials

Alumina substrates are manufactured by press forming, using the sheet forming method or dies.

22.3.2.1.6 Phosphors

In order to manufacture oxide phosphors such as YAG, the raw material and flux, an auxiliary agent, are mixed using the dry method. The mixture is then calcined at a high temperature under atmospheric pressure, followed by the process of washing and refinement of the grain diameter by means of grinding. Meanwhile, there are a number of issues to be solved regarding nitride phosphors. They need to be calcined under high pressure and in a nitrogen gas ambient. The purity of the nitride substantially affects the characteristics of products. The processing method is expected to be simplified in the future.

22.3.3 Future Prospects

LED lights, which are currently used as LED bulbs in shops and facilities, are expected to be the major luminaire in offices and households in the future. LED lights possess great potential because they respond quickly to input voltage and can be used for visible light communication. It is expected that the light intensity and efficiency will be further enhanced and ceramic parts featuring high radiation performance and insulation properties are expected to play important roles in handling the heat generated from the chips.

22.4 Sodium Lamp (1963)

Sodium lamps utilize an electrical gas discharge in sodium vapor for the purpose of illumination. In the 1960s high-pressure sodium lamps were realized following the commercialization of translucent alumina ceramics that resist reaction with high-temperature sodium vapor. They are widely used for exterior illumination in roads, parks, etc. The translucent alumina ceramics used in the luminous tubes is made by forming high-purity alumina into tubes and sintering them in a reducing atmosphere, that show a total transmittance of 90 % or higher. The translucent alumina features high dimensional accuracy, thermal resistance and alkali resistance, which has expanded its application to metal halide lamps (Note 22.7) and semiconductor apparatus components.

22.4.1 Background of Development

Sodium lamps emit light by utilizing an electrical discharge in the sodium vapor. The lamps are classified into two types, the low-pressure type and the high-pressure type, depending on the pressure of the sodium vapor inside the luminous tube during lighting. Low-pressure sodium lamps have a high lamp efficiency (Note 22.8) (Fig. 22.13) and are used to illuminate tunnels, etc. However, they emit yellow orange colored light and do not allow color discrimination. The color rendering property (Note 22.9) can be improved by increasing the sodium vapor pressure, but the special sodium-resistant glass used as the luminous tube of low-pressure sodium lamps is not resistant to the high temperature and high pressure of sodium vapor. High-pressure sodium lamps were realized following the start of the production of translucent alumina ceramics in the 1960s. The lamp has a dual structure composed of the outer tube and the inner luminous tube. Translucent ceramics are used to manufacture the luminous tubes. Niobium electrodes are sealed by fritted glass on both sides of the luminous tube and sodium and mercury, the luminous materials, are sealed inside together with gas for starting (Fig. 22.14). Sodium lamps are higher in efficiency and luminance when compared with incandescent lamps and fluorescent lamps. They are widely used to illuminate exterior spaces such as roads, streets and public spaces as well as factories and sports facilities. They are also used as light sources in the growing of vegetables (Fig. 22.15).

22.4.2 Characteristics

The translucency of a luminous tube (which is an important characteristic) is attributable to the following characteristics:

1. The amount of impurities is limited and the alumina is of high purity
2. Dense structure with a limited number of residual pores

Impurities and pores cause absorption/scattering of light and need to be minimized (Fig. 22.16). Although the tubes have translucency because of these characteristics, the appearance is not clear but is milky white. It is believed that the color comes from grain boundary scattering due to optical anisotropy attributable to the crystal structure of alumina and residual fine pores that remain. When the grain diameter increases, the clarity (in-line transmittance) is enhanced, while the strength is lowered, leading to breakage during production and later during use. Clarity is not always necessary for lamps for general use, and what is important is the total amount of transmitted light (total transmittance). The temperature of luminous tubes exceeds 1,000°C during the illumination of lamps, and therefore, thermal resistance,

Note 22.7 A type of high-pressure electrical discharge lamp. Light with balanced RGB (high color rendering property) can be emitted by using metal halides featuring a variety of emission spectra from different luminous materials.

Note 22.8 The value obtained by dividing the amount of light (light flux) by consumed power. Expressed in 1 m (lumen: unit of light flux)/W.

Note 22.9 A term used to express the color visibility of objects. When an object is illuminated by a light source with a high color rendering property, the color visibility becomes closer to the color visibility in sunlight.

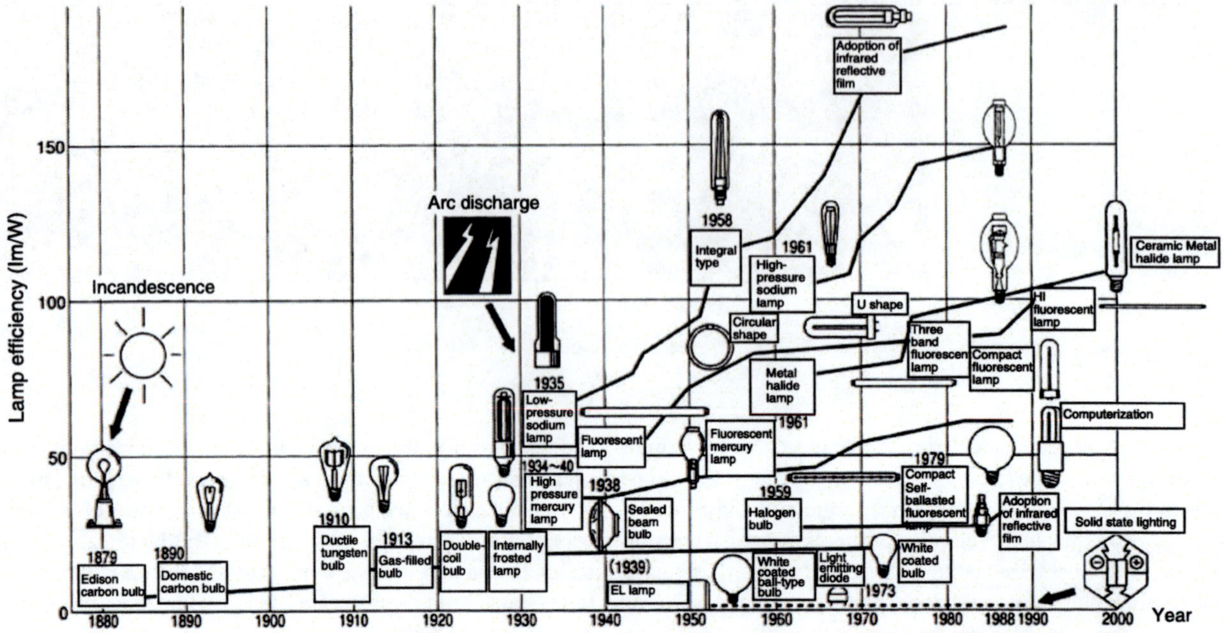

Fig. 22.13 Progress in lamp efficiency. Progress in the lamp efficiency of major light sources. The efficiency of sodium lamps is higher than that of other light sources

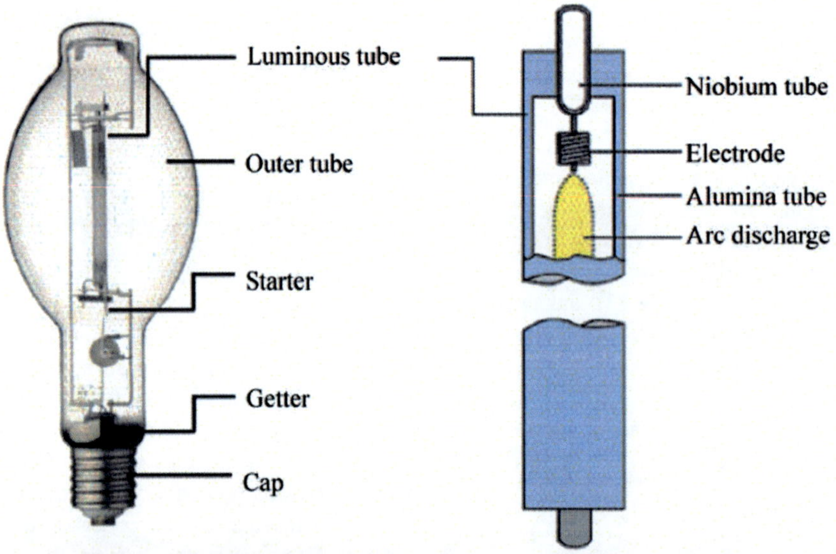

Fig. 22.14 Photograph of a high-pressure sodium lamp and structure of a luminous tube. Appearance of a general high-pressure sodium lamp. The lamp has a dual structure composed of the outer tube and the inner luminous tube. Translucent alumina is used to manufacture the internal luminous tube. Niobium electrodes are sealed by fritted glass on both sides of the luminous tube. Sodium amalgam and gas for starting are sealed inside

Fig. 22.15 Roadside high-pressure sodium lamps in operation

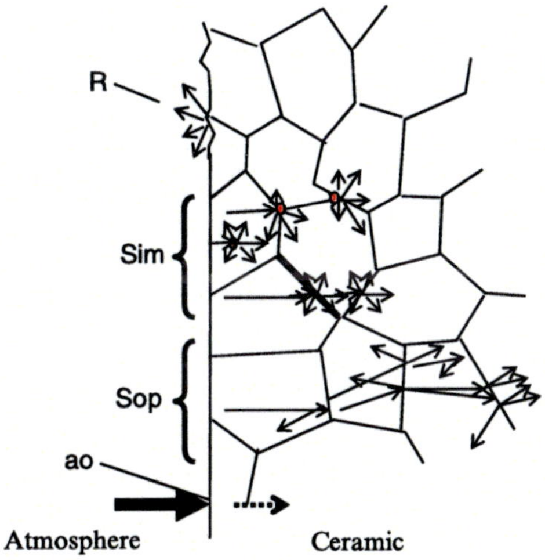

Light permeability I/Io=(1-R)2exp(-bt), b=ao+Sim+Sop

I/Io: Ratio of input and output

t: Thickness of ceramic

R: Influence of surface reflection

Sim: Influence of imperfect structure (bubbles, impurities, grain boundary)

Sop: Influence of interaction due to optical anisotropy of each grain

ao: Influence of absorption

Fig. 22.16 Light permeability model of translucent ceramics. Impurities and pores cause light scattering/absorption and need to be minimized

strength and corrosion resistance to high-temperature sodium vapor are required of the luminous tubes (Table 22.4). The grain diameter and the grain boundary phase are maintained at appropriate values for practical use by adding a small amount of sintering agent (metal oxide) and by controlling the sintering temperature.

22.4.2.1 Manufacturing Method

High-purity alumina (99.99 %) is used as the raw material because impurities deteriorate translucency, as mentioned above. The tube shapes are formed by the extrusion, casting and rubber press methods. The manufacturing process

Table 22.4 General characteristics of translucent (milky white) polycrystalline α-Al_2O_3 used in sodium lamps

Characteristics	Unit	Value
Purity	%	99.9
Grain diameter	μm	35
Total transmittance	%	96[a]
Specific weight		3.99
Water absorption rate	%	0.0
4-Point bending strength	MPa	300[b]
Fracture toughness K_{IC} (SEPB method)	MPa $m^{1/2}$	4.2
Young's modulus	GPa	410
Poisson's ratio		0.24
Vickers hardness	Hv	1,850
Thermal expansion coefficient (40–800°C)	/K	8.1×10^{-6}
Thermal conductivity	W/m K	33
Specific heat	J/kg K	790
Volume electrical resistivity	Ω cm	1×10^{16}
Voltage resistance	kV/mm	20
Permittivity		10

[a]Values based on internal diameter of 8 mm, thickness of 0.75 mm and total length of 105 mm
[b]JIS R1601

utilizes the rubber press method and is explained below. Initially, raw material powders are mixed with water and a small amount of sintering agent is dissoluted by ball milling. It is then mixed with organic binder and is granulated by a spray drier. In the forming process, the granules are press formed by the rubber press method to create tubes. Then, the ends of the tubes are assembled with separately-formed ring-shaped parts, depending on the specifications. In the calcination process, the tubes are temporarily calcined at 1,000–1,500°C to remove organic components such as the binder, and then are sintered in a hydrogen atmosphere at 1,700–1,900°C to create dense sintered bodies without pores and also to bond the ring-shaped parts on the main body.

22.4.3 Future Prospects

The metal halide lamp, a type of high-pressure electrical discharge lamp (similar to the high-pressure sodium lamp) was conventionally made from silica glass. However, ceramic metal halide lamps were commercialized in the 1990s. The performance of lamps has been improved year after year taking advantage of the high dimensional accuracy, thermal resistance and corrosion resistance of ceramics. Ceramic halide lamps, featuring high luminance and high efficiency as well as high color rendering properties and high color stability, are used as spot lights in showcase windows and in lights for shopping malls, etc. which are required to be of high quality and high efficiency. Following this trend, the range of application of translucent ceramics, which were mainly used in high-pressure sodium lamps for exterior illumination, has been expanding to the field of interior illumination at a rapid pace. This trend is expected to continue.

Semiconductor Gas Sensors (1968) 22.5

Ceramic gas sensors that are currently used in households are classified into various types based upon their mode of operation such as the semiconductor type, the catalytic combustion type (Note 22.10), the solid electrolyte type (Note 22.11), etc. The usage of household gas sensors are roughly divided into two different categories. One is the "safety and disaster prevention" sensor, where the sensors detect leakage of methane gas and liquefied petroleum gas (LPG) as well as carbon dioxide produced by incomplete combustion. The other is a monitor for "measuring air quality" where the sensor measures air pollutants and automatically controls ventilation fans and air purifiers to reduce the pollutant concentration. The semiconductor gas sensors (Fig. 22.17) are widely used in the two fields explained above. The semiconductor gas sensors detect gases utilizing the property of electrical resistance, which changes when the sensor is exposed to a combustible gas or a kind of reducing gas. Mass production of sensors made from tin oxide was started in 1968 for the first time in the world.

The housing of gas sensors has an explosion-proof stainless steel (SUS) mesh cap or an activated carbon filter, depending on the purpose of the gas detection and usage.

22.5.1 Background of Development

Usage of semiconductor gas sensors by households is closely related to the improvement in the airtightness of houses. In the "safety and disaster prevention" field, accidents caused by flammable gas leakage have been decreasing due to efforts of governments and the expansion of gas alarms. However, accidents caused by incomplete combustion increased due to an improvement in the airtightness of houses. Responding to the changing trends, semiconductor gas sensors that detect carbon dioxide began to be mounted on gas alarms beginning around 1995.

In the "air quality" field, odors inside houses, generated by pets, garbage and cooking, are becoming more serious due to the improved airtightness of houses. In addition, people are becoming more sensitive to the quality of air because of volatile organic compounds (VOC) (which are emitted from building materials, carpeting and furniture, is the cause of the sick house syndrome) and have become sensitive to pollen resulting in allergic reactions. People are more concerned than ever about keeping indoor air free from pollutants and pollen. Ventilation fans and air purifiers are offered as products that keep indoor air clean. Many of the air conditioners today are equipped with air purification and ventilation functions. A number of semiconductor gas sensors are mounted on these devices to control them by detecting pollution in the air (Fig. 22.18 shows sensors mounted on products).

The multi-function gas leak alarm equipped with an incomplete combustion detecting function is mounted on a gas sensor capable of detecting CH_4 and CO at the same time, while the air conditioner indoor unit equipped with air purification and ventilation functions is mounted with a sensor capable of detecting pollution in the air.

22.5.2 Characteristics

22.5.2.1 Gas Sensing Material

The gas sensing material used in semiconductor gas sensors is tin oxide. The sensor is composed of a gas sensing component and a heater that heats the gas sensor from 300°C to

Note 22.10 When a platinum wire coil coated with a catalyst is heated and brought into contact with a gas, the gas and oxygen react with the catalyst, generating heat. The temperature of the platinum wire coil increases, leading to an increase in the resistance value of the platinum wire. The resistance value is measured and varies in proportion to the detected gas concentration.

Note 22.11 The sensor has an operating electrode and a reference electrode on each side of the solid electrolyte. When gas concentration on the operating electrode rises, electromotive force attributable to eccentrically-located ions develops between the two electrodes. The electromotive force is converted to electrical signals for gas detection.

Fig. 22.17 Appearance of semiconductor gas sensors

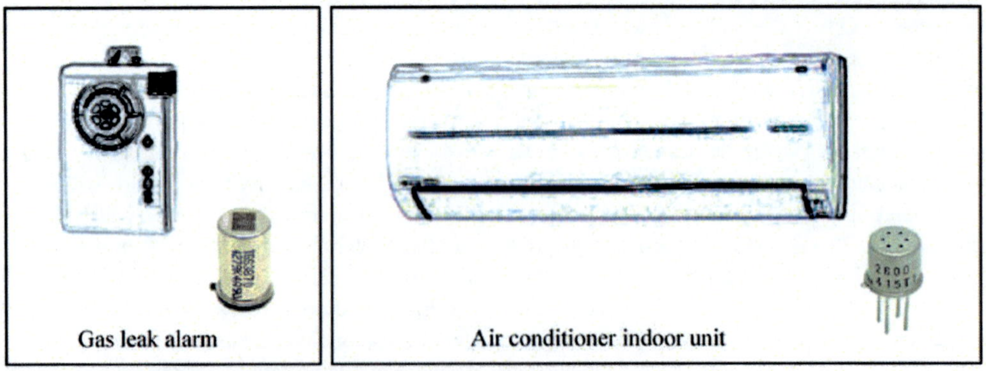

Fig. 22.18 Examples of sensors mounted on products

450°C for use (the structure is shown in Fig. 22.19). When the heated tin oxide is exposed to a combustible gas (H_2 in the example), a type of reducing gas, an oxidation reaction between the gas and absorbed oxygen occurs on the tin oxide surface. As a result, the amount of oxygen that was absorbed within the tin oxide surface decreases, lowering the surface potential barrier and inducing the freer movement of electrons. The electrical resistance is lowered because of this reason. Figure 22.20 shows the mechanism of gas detection. The resistance value of the sensor changes in response to the change in gas concentration. Figure 22.21 shows changes in sensor resistance values responding to changes in concentration of various types of gas. Semiconductor gas sensors are characterized by their compact size, high durability, high mass productivity, comparatively low production cost and a simple sensor drive circuit (an example of the circuit is shown in Fig. 22.22).

The gas sensing material (tin oxide) and the heater material are printed, respectively, on opposite sides of the alumina substrate. The alumina substrate, which is heated to about 400°C electrically, is suspended so as to prevent heat loss.

If a reducing gas atmosphere exists around the sensor, the amount of oxygen absorbed on the surface of tin oxide decreases, resulting in the decrease in grain boundary potential barriers. As a result, electrons move more freely from grain to grain.

Typical sensitivity characteristics of sensors intended for methane gas detection. The vertical axis indicates the sensor resistance ratio (R_S/R_O).

R_S = Sensor resistance value in various concentrations of gas
R_O = Sensor resistance value in 5,000 ppm methane

Sensor resistance (R_S) can be calculated from the voltage on both ends (V_{OUT}) of a resistive load (R_L) series-connected to the sensor.

$$R_S = \left(\left(V_C - V_{OUT}\right) / V_{OUT}\right) \times R_L$$

Fig. 22.19 An example of the gas sensor structure

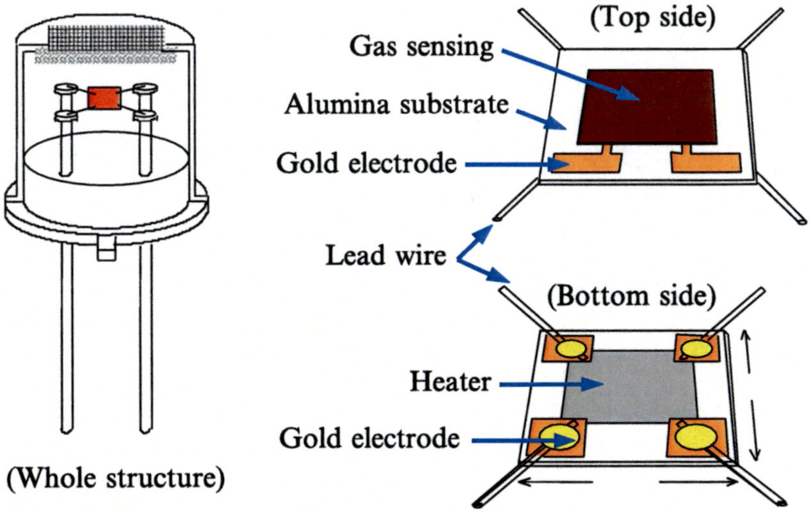

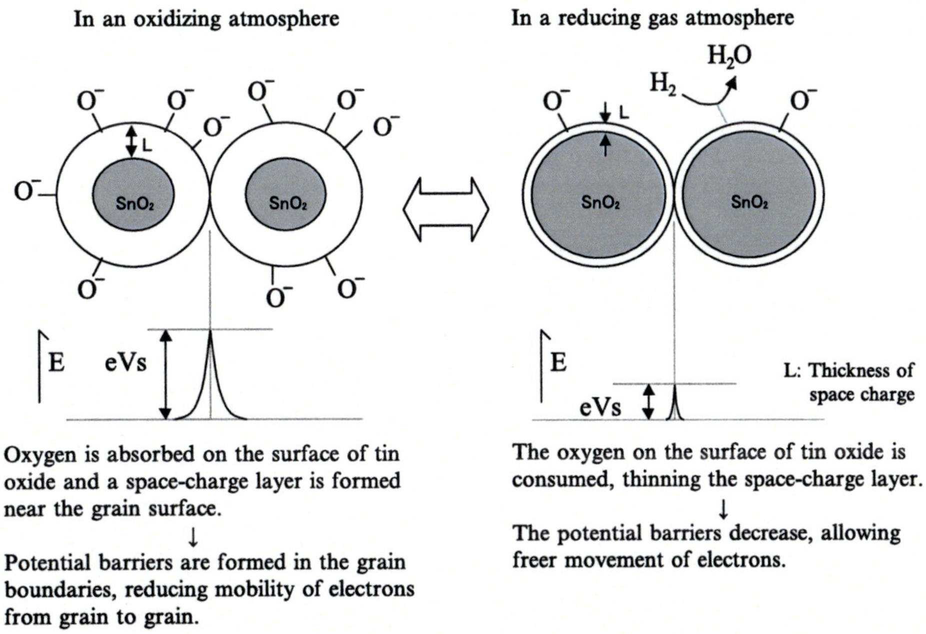

Fig. 22.20 Gas detection mechanism (schematic diagrams of the grain boundary of SnO_2 grains)

22.5.2.2 Manufacturing Method

A catalyst is added to the calcined and sintered tin oxide powder to create the gas sensing material. The type of catalyst is selected depending on the gas to be detected. The gas sensing material is processed into shapes that allow heating and measurement of electrical signals. As for printed sensors such as those shown in Fig. 22.20, gas sensing material is formed by screen-printing on the alumina substrate on which the heater and electrodes are printed.

22.5.3 Future Prospects

Since the start of world's first mass production of semiconductor gas sensors incorporating tin oxide in 1968, the structure and material have been improved. This was in response to the demands for miniaturization and the reduction in power consumption. This also took place in order to address the needs to detect diversified types of gas. Products incorporating the MEMS technologies (Note 22.12) have

Note 22.12 Stands for Micro-Electro-Mechanical Systems. Indicates a micro-electro mechanical elements and technologies for creating them.

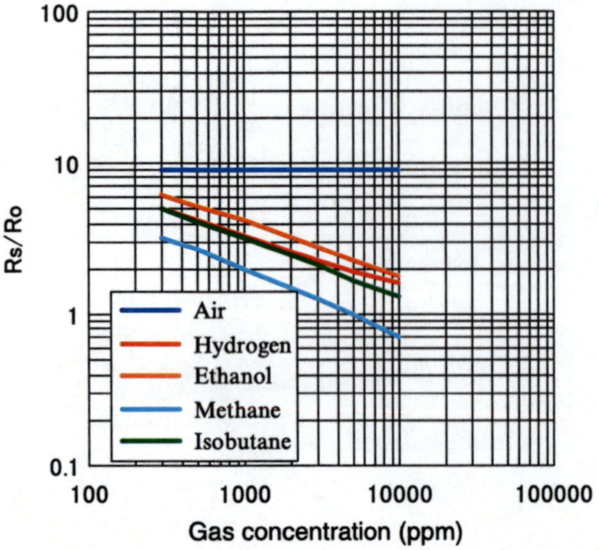

Fig. 22.21 Gas sensitivity characteristic

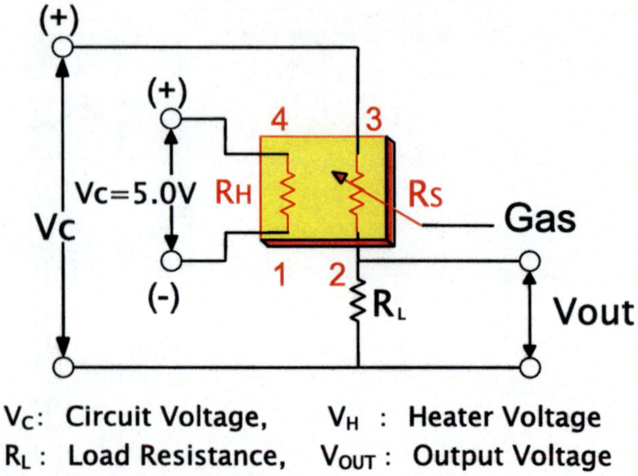

Fig. 22.22 Circuit for sensor basic measurements

been commercialized in various fields in recent years. Introduction of the MEMS technologies have also been promoted for production of semiconductor gas sensors. Introduction of the MEMS technologies will lead to a substantial improvement in the downsizing of sensors and a significant reduction in power consumption. The introduction will lead to the realization of battery-powered alarms that have long been awaited and to the further reduction in power consumption by home appliances. Application of MEMS technologies will also lead to the realization of low-cost multi sensors having multiple gas sensing functions mounted on a single sensor. Through processing of signals from the sensor, the detection of various types of gas will be allowed and the selectivity and accuracy will also be improved.

Piezoelectric Buzzers (1975) 22.6

Compact, thin and low power consumption piezoelectric buzzers rapidly expanded in the market responding to the demands for reduction in the size, thickness and weight for use in mobile devices including desktop calculators, clocks and hand held video games. This occurred together with the incorporation of ICs in electronic devices represented by home appliances such as electronic ovens and micro-computer controlled rice cookers as well as OA devices such as printers. They have also recently been used for music reproduction as speakers in mobile phones. Because piezoelectric buzzers are made by simply laminating a piezoelectric ceramic thin plate with a metal plate, they exhibit low cost and allow a reduction in thickness. Following the progress in the ceramic sheet forming technologies and the lamination technologies incorporating the method to sinter internal electrodes and ceramics simultaneously as one unit in recent years, low-cost and thin piezoelectric speakers with high-performance have been introduced.

22.6.1 Background of Development

Compact, thin and low power consumption piezoelectric buzzers rapidly expanded in the market responding to the demands for reduction in the size, thickness and weight of mobile devices including desktop calculators, clocks and handheld video games. This occurred together with the incorporation of ICs in electronic devices represented by home appliances such as electronic ovens and micro-computer controlled rice cookers as well as OA devices such as printers (Fig. 22.23). Responding to the recent popularity in thin models of mobile phones and mobile audio players, piezoelectric ceramics are currently used in thin buzzers (speakers) that feature a wide range of reproduction frequencies and are capable of reproducing music.

Piezoelectric buzzers of various shapes and packages, including the pin type and the drip-proof type, are manufactured for a variety of applications.

22.6.2 Characteristics

22.6.2.1 Products

Figure 22.24 shows the basic operating principle of a piezoelectric buzzer incorporating piezoelectric ceramics. The piezoelectric buzzer is made by laminating a thin piezoelectric diaphragm on a metal plate. The piezoelectric ceramics expand and contract responding to the application of alternating voltage, developing repeated bending motion to generate sound waves. Piezoelectric buzzers are simple both in the basic structure and the manufacturing process, which enables the manufacturing of thin products at a low cost. Piezoelectric buzzers normally employ a single-plate element made by forming external electrodes on a piezoelectric ceramic element, and the distance between the electrodes are several hundred μm. Therefore, application of a high voltage of several tens of volts is required to generate high volumes (sound pressure). Recently, piezoelectric ceramic devices made by laminating ceramic layers of several tens of μm and electrode layers alternatively (employing multi-layer stacking technologies in which internal electrodes and ceramics are sintered simultaneously as one unit) have been commercialized. This has made it possible to generate high sound pressure by application of several volts. Internal electrodes of laminated piezoelectric ceramic devices are generally formed by silver–palladium alloy. Following the recent technological evolution of piezoelectric ceramics and the development of piezoelectric ceramic materials that can be sintered at a temperature below 1,000°C, the ratio of palladium in the alloy used to form the internal electrodes can now be reduced.

22.6.2.2 Manufacturing Method

In the early days of development, piezoelectric ceramic devices were made by slicing or grinding cylindrical or prismatic ceramic blocks. Therefore, the processing cost was high, making it difficult to lower the price of piezoelectric buzzers. Following the progress in ceramic production technologies, piezoelectric ceramic devices can now be made without the grinding process, by forming a green sheet (Note 22.13) of several hundreds of μm, punching or cutting it and then sintering the cut pieces. This reduces the processing cost.

Laminated piezoelectric buzzers that can be driven by low voltages have been manufactured recently. The production process is shown in Fig. 22.25. Raw materials consisting of lead oxide, zirconium dioxide, titanium dioxide, etc. are blended and mixed, and the mixture of powders is calcinated to produce lead zirconate titanate (PZT), the piezoelectric ceramic material. Solvents such as water, alcohol, etc. and organic binders are added to the powder and a green sheet of several tens of μm in thickness is produced by the doctor-blade method (Note 22.14). The internal electrodes are then screen-printed on the green sheet using silver–palladium alloy paste. Several printed green sheets are then pressed together. The organic binders are removed by heating process, and a laminated sintered body is created by sintering the internal electrodes and piezoelectric ceramics simultaneously at a temperature of 900–1,200°C.

The sintered body is cut into the intended shapes, external electrodes are formed and electrical field of several kV per mm is applied between the electrodes to orient the domains known as poling to produce piezoelectric ceramic elements. The elements are bonded on metal plates and packaged to produce piezoelectric buzzers. Appearance of a piezoelectric buzzer produced by the method described above is shown in Fig. 22.26 and one of its characteristics is shown in Fig. 22.27. It is very thin and compact, with a thickness of 1.2 mm and the area of 19 mm × 13 mm. High sound pressure is attained by application of voltage as low as several volts.

Although it is compact and very thin with an outer dimension of 19 mm × 13 mm and a thickness of 1.2 mm, it is capable of reproducing music at high sound pressure. It is mounted in thin mobile phones and audio players.

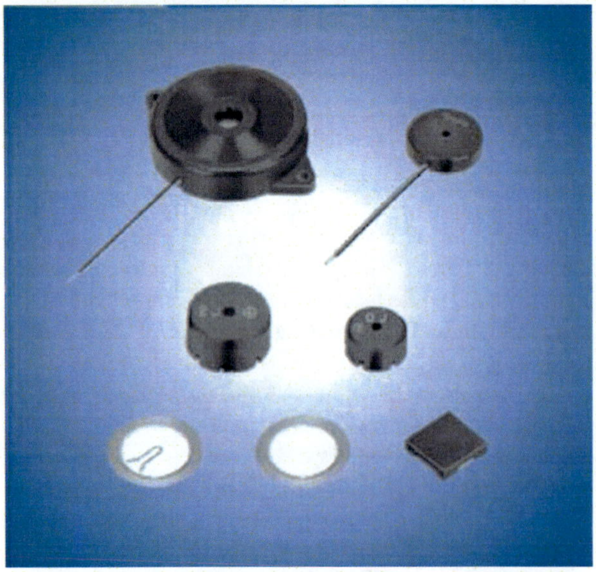

Fig. 22.23 Piezoelectric buzzers

22.6.3 Future Prospects

Mobile phones and mobile audio devices have been downsized and slimmed remarkably while they are equipped with more functions. Following the increase in built-in parts, buzzers that can be mounted in narrow spaces are required. It is expected that applications for piezoelectric buzzers (which are effective for reduction of thickness) to compact mobile devices will continue to increase.

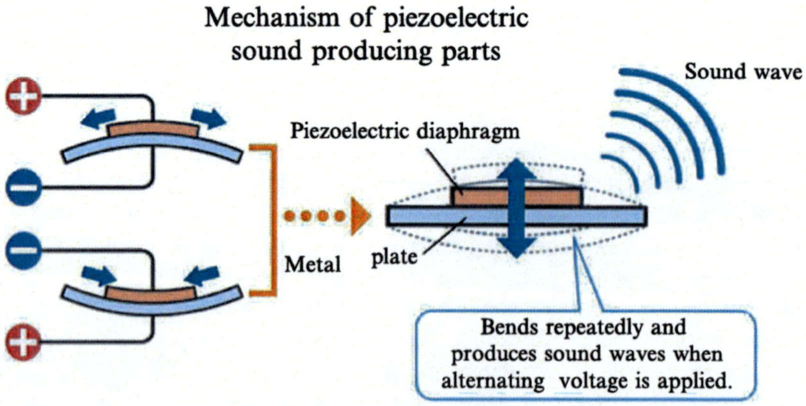

Fig. 22.24 Operating principle of piezoelectric buzzers. The piezoelectric diaphragm expands and contracts depending on the direction of voltage application. When alternating voltage is applied, repeatedly bending produces a vibration resulting in sound waves

Note 22.13 Formed body in the shape of a tape, made from slurry, a mixture of ceramic powder, organic binders, by using a forming machine.

Note 22.14 One of the ceramic forming methods. Slurry, a mixture of ceramic powder and organic binders, is fed through a gap between a film and a blade (doctor blade) to form a tape-shaped ceramic sheet.

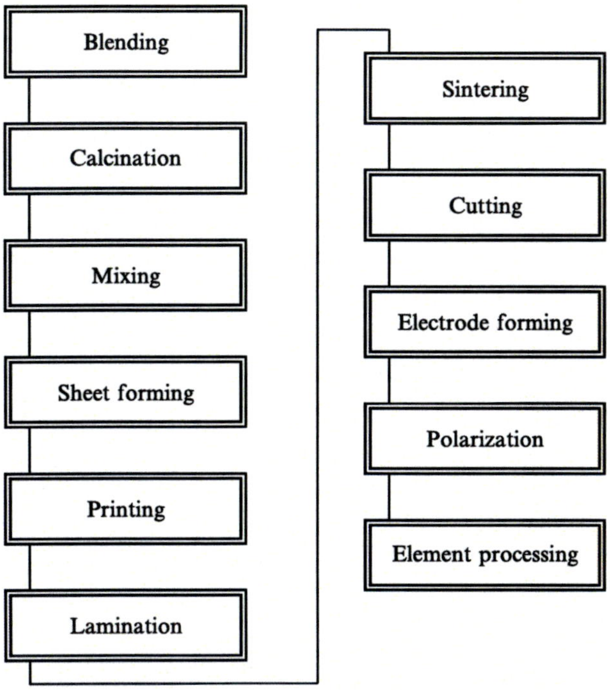

Fig. 22.25 Production process of laminated piezoelectric ceramic elements. Electrodes are formed on the green sheet by the screen printing method, and a multi-layer stack of green sheets is sintered to create a sintered body. Then the laminated layer is cut, electrodes are formed and polarization is performed in a high electrical field to produce laminated piezoelectric ceramic elements

Fig. 22.26 Appearance of a laminated piezoelectric speaker

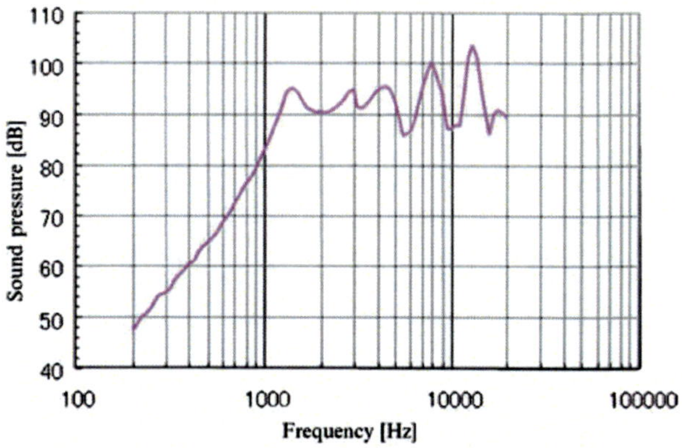

Fig. 22.27 Sound pressure/frequency characteristics of laminated piezoelectric speakers. Flat sound pressure characteristics have been achieved in a wide range of frequencies, which are required for speakers

Tiles (2650 BC) 22.7

The word *tile* is derived from a Latin word meaning "to cover or enclose." Tiles are used both in building interiors and exteriors as well as on the space shuttle. In Japan, they are widely used by the construction industry as a ceramic based surface finishing material. The history of tiles used as construction materials goes back to 2650 BC, when they were used on the walls of the underground passage in the pyramids in ancient Egypt. Tiles need to have properties required of functional materials as well as designs required of surface finishing materials. Therefore, while the manufacturing technologies are advanced, traditional decorating methods are also incorporated. It is important to maintain tradition and artistic technologies for the production of "fired wares" and develop new technologies at the same time.

22.7.1 Background of Development

The word *tile* originated from the Latin word "Tegula," which means "to cover," "to coat" or "to enclose." The ceramic thermal insulation material that covers the space shuttle, shown in Fig. 22.28, implies the original meaning of the word "tile" (Note 22.15). In Japan, tiles have various applications such as for floor tiles, wall tiles, decorated bricks and wall bricks. Ceramic tiles used on walls and floors as construction materials were decided to commonly be called "tiles" in 1922 (11th year of the Taisho era). Since then, the term "tile" has been generally used to refer to such construction tiles.

Note 22.15 When a space shuttle makes its re-entry into the atmosphere, the maximum temperature at the tip of the nose reaches 1,600°C. Special tiles were developed to protect the body from this high temperature. The tiles are made by solidifying silica glass fibers with a specific weight of 0.12 and have a high thermal insulation value and thermal resistance. Some insulation tiles were repaired on the space shuttle in 2005 during a mission by Soichi Noguchi, a Japanese astronaut.

22.7.2 Characteristics

22.7.2.1 History of Tiles (BC–Present)

The oldest tiles in the world are the tiles that cover the underground passage of Step Pyramids constructed in 2650 BC in ancient Egypt (Fig. 22.29). These tiles consist of alkaline components such as Na, K, Ca and Mg and Si, which are similar to glass components, and contain nearly no clay, which was used as the raw material of earthenware and bricks since earlier times. The surface is coated with a turquoise blue glaze, which contains Cu as the pigment. Later, tiles were used widely to decorate the walls of religious buildings in the Persian Empire. They were brought to Europe during the period when Islam expanded to Spain. They were further developed through the fusion with cultures and technologies of European countries. They began to be mass-produced as industrial products during the industrial revolution in the United Kingdom during the eighteenth and nineteenth centuries, expanding to countries throughout the world.

They were brought to Japan in the Meiji era (1869–1911) by Europeans who stayed in Japan on business and used interior tiles to decorate porch floors and furnaces of their western style houses. During the same period, G. Wagener, a German chemist referred to as the father of the modern ceramic technology, visited Japan to introduce the manufacturing technology of modern tiles. Wagener established a tile manufacturing company and sold the tiles under the name of Asahi Ceramics. The ceramic industry of Japan was later led by many of his students who studied under him at Tokyo Vocational School (currently the Tokyo Institute of Technology). However, it took some time before interior tiles began to be used widely in Japanese houses, the majority of which were wooden houses with sliding paper screens. Interior tiles expanded to individual houses following the use of interior tiles in bath tubs and floors of public bath houses and spas during the period from the end of the Taisho era (1912–1926) and the beginning of the Showa era

Fig. 22.28 Exterior insulation tile used by the space shuttle (Gladstone Museum, UK). The insulation tile is made by coating silica glass fibers fused at a specific ratio of 0.12 with glass that contains silicon carbide

Fig. 22.29 The tiles that decorate the basement of ancient Egyptian pyramids (INAX Live Museum, Tokoname City). They are made from quartz and contain alkaline components such as Na, K, Ca and Mg and Si. Copper is used as the pigment of the turquoise *blue glaze*

(1927–1988). Interior tiles expanded to households in line with the growing concern on the sanitation of bathrooms and toilets of Japanese houses. Following the rapid flow of the American culture into Japan after the World War II ended in 1945, houses were westernized and interior tiles began to be used widely in bathrooms, kitchens and toilets of houses, bringing brightness into Japanese houses. However, following the spread of western style toilets and unit baths since the second half of the Showa era, usage and production of interior tiles has been decreasing.

Meanwhile, exterior tiles were developed for various applications in Japan. The tiles were brought to Japan from China at the end of the sixth century and were used not only as roofing materials but also as floor and exterior materials, which were used in walls of temples and castles. The technology to use tiles on exterior walls was applied to buildings at the end of the Meiji era, when exterior tiles were first used in full-scale ferroconcrete buildings. The technology to use tiles as an exterior finishing material of ferroconcrete buildings spread throughout Japan after the Great Kanto Earthquake in the 12th year of the Taisho era. The former main building of the Imperial Hotel, which was designed by Frank Lloyd Wright, a well-known American architect of the twentieth century, triggered the spread of exterior tiles. Wright used bricks as surface finishing materials, not as structural materials (Fig. 22.30). The Great Kanto Earthquake occurred on the day of the opening of the Imperial Hotel. The hotel withstood the earthquake while many of the buildings made of bricks collapsed. After this event, the technology to cover exterior walls of ferroconcrete buildings with exterior tiles that look like bricks expanded. Scratch tiles (Note 22.16) used in the Imperial Hotel became popular at

Note 22.16 The tiles are made by scratching the surface with a jig having nails arranged in a line. In Japan, they were first used by Frank Lloyd Wright on the exterior of the former main building of the Imperial Hotel.

Fig. 22.30 Tiles used on the exterior of the former main building of the Imperial Hotel (INAX Live Museum, Tokoname City). Tiles are used both as concrete formworks and surface finishing materials

Fig. 22.31 Walls composed of three-dimensional tiles and glass panels (INAX Osaka Office Building, Osaka City). A glass panel is placed between triangular tiles that are 60 cm on each side. Units, each containing 12 blocks, were assembled to form the walls

Table 22.5 Manufacturing processes of tiles and variations

Blending of raw materials (types of raw materials)	Forming (forming methods)	Decoration (decoration methods)	Sintering (sintering condition)
Clay	Dry press forming	Spray glazing	Oxidized calcination
Feldspar	Wet extrusion forming	Screen glazing	Reduction firing
Quartz sand	Press forming with gypsum molding die	Centrifugal glazing	
Pyrophyllite	Slip casting	Screen print decoration	(Types of sintering equipment and time)
Lime		Polishing	Tunnel furnace (sintering time: 30–40 h)
Chamotte		Sandblast	Roller hearth furnace (sintering time: 20–100 h)
		Split after drying (split face)	Shuttle furnace (sintering time: 30–40 h)
		Scratch	
		Inlay	

Combinations of raw materials, manufacturing methods and decoration methods are limitless, and the manufacturing method is selected depending on the tile intended function and design

the beginning of the Showa era and were used as exterior materials of government buildings and university buildings throughout Japan. The examples are the Ishikawa Prefectural Office building and Waseda University Osumi Hall. After the war, mosaic tiles with a surface area of 50 cm^2 or less also became popular, in addition to brick effect exterior tiles (Note 22.17). In the Heisei era, large tiles and three-dimensional terracotta louvers (Note 22.18) began to be used as well as glass compound tiles (Fig. 22.31).

Note 22.17 Bricks are ceramic materials that are used as structural materials of masonry buildings, while tiles are ceramic materials used as surface finishing materials and do not function as structural materials. Many of the exterior tiles used today are similar to bricks in appearance.

Note 22.18 Terracotta generally means unglazed sculptures, but in construction terms, they indicate complex shaped and large ceramic construction materials used on exterior walls of buildings. Long cylindrical or board-shaped ceramic materials used as louvers of buildings, etc. are sometimes called terracotta.

22.7.2.2 Manufacturing Method

Manufacturing processes that are currently incorporated for the production of tiles and variations of tiles are listed in Table 22.5. As the table indicates, a number of decoration methods incorporating traditional ceramic technologies are used even today, although the technologies have advanced and production has been computerized.

22.7.3 Future Prospects

Tiles that originated in ancient Egypt about 4,500 years ago still remain with the same glow, proving the durability of ceramic tiles. The major functions of ceramic tiles are surface protection and durability improvement of walls, but the designs are also important because they are intended as decoration materials. New tiles with humidity control, insulation and antifouling performance were developed recently, but these tiles also are required to be superior in terms of design. Louis Sullivan, a US architect, said "Form ever follows function." "Fired wares" appears to be a more appropriate term than "ceramics" to express the functionality of tiles. Tiles need to maintain their aesthetic beauty and texture of traditional fired wares while maintaining advanced functionality.

Far-Infrared Ceramic Heaters (1965) 22.8

Ceramics at elevated temperature release electromagnetic waves (wavelength range of 2.5–30 μm) that are easily absorbed by irradiated substances other than metals (Fig. 22.32). Far-infrared ceramic heaters take advantage of this property and radiate their heat without an intermediate media. Therefore, the heat is absorbed directly by the irradiated substance, making it possible to raise its temperature. Because of these reasons, far-infrared ceramic heaters are used widely in the industrial and consumer products as the heat source in many heating and drying devices. The applications have been increasing because they operate using electricity.

22.8.1 Background of Development

Production of automobile parts (interior parts, exterior parts, springs, brake parts, etc.) includes processes involving drying and heat treatment. Hot gas, metallic heaters, near-infrared ray lamps and far-infrared ceramic heaters have been used as the heat source of these processes. Application of far-infrared ceramic heaters was previously limited to the process of drying paint so as to prevent possible explosions, because paints often contain organic solvents. These organic based paints have been replaced by water-soluble paints recently in consideration for the environment. Preventive measures against explosions during drying are no longer necessary if water-soluble paints are used. Therefore, the application of far-infrared heaters has been increasing because they are clean and enable speedy heating, while contributing to space saving through efficient design considerations.

Meanwhile, interior materials in homes and buildings have been changing responding to the needs for lightweight and soundproof properties. Specifically, carpets are changing from double-layer types to triple-layer types. Therefore, in carpets in addition to the front surface and the back, an intermediate layer may contain heating elements for use during winter. Responding to the change in specifications, methods that incorporate both the traditional heat sources and far-infrared ceramic heaters have been developed and used widely. Furthermore, exterior materials are becoming larger, and the application of far-infrared ceramic heaters is also expanding in this field. Some of the current far-infrared ceramic heaters are sized 2 m × 2 m or larger. The entire area of the heated material needs to be heated evenly, and therefore, the application of far-infrared ceramic heaters, which can heat materials evenly at higher absorption efficiency (Fig. 22.33), has been expanding. As explained above, the application of far-infrared ceramic heaters has been expanding for heating and drying of automobile parts.

Far-infrared rays (Note 22.19) exhibit a higher thermal absorption rate and heating efficiency than other heat sources.

22.8.2 Characteristics

22.8.2.1 Products

The heaters (Fig. 22.34) are made by embedding metallic heating wires within ceramics. Far-infrared rays are emitted evenly from the entire ceramic surface.

22.8.2.2 Advantages of Ceramic Heaters

Far-infrared ceramic heaters have the following four characteristics (Table 22.6).

1. High emissivity: efficient heating, drying is possible.
2. Homogeneous heating: the heating elements are sheet-shaped and feature excellent irradiation distribution, making it possible to heat a wide area evenly and uniformly.

Note 22.19 Infrared rays within the wavelength of 3 μm to 1 mm are defined as far-infrared rays.

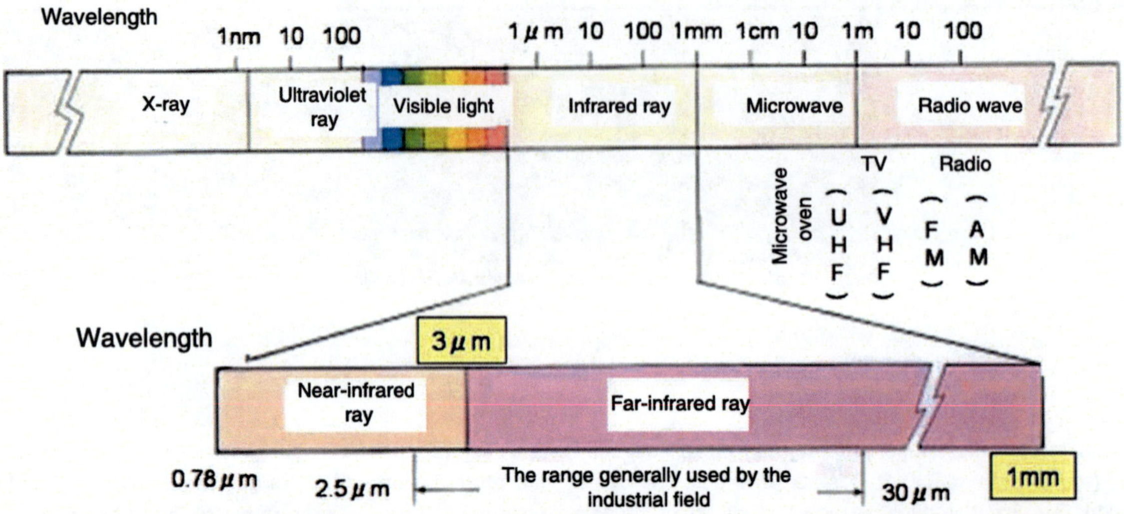

Fig. 22.32 Far infrared rays in the electromagnetic wave spectrum

Fig. 22.33 Absorption efficiency of far-infrared rays

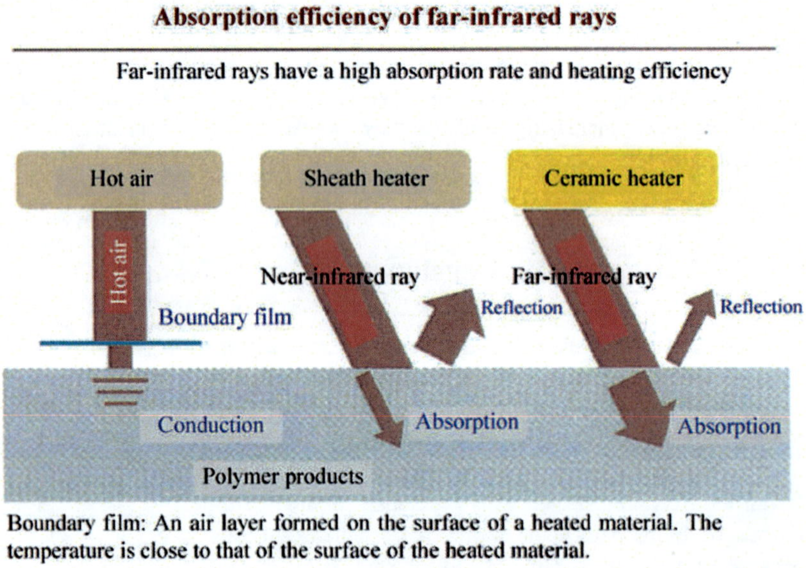

3. High-accuracy temperature control: temperature detecting terminal can be mounted inside the heater, allowing flexible control of the heater surface temperature.
4. Long service life: heating wires are enclosed in ceramics to prevent oxidizing. Various types of heaters are compared below (Table 22.7).

22.8.3 Future Prospects

Heating and drying processes are essential in manufacturing in various industries. Effective and controlled heating directly affect the production cost and the quality of products. Responding to an increase in the quality of heated parts and their composition, companies are required to combine multiple heat sources for the expansion of applications, and they also need to reduce CO_2 emissions as measures to prevent global warming. The applications for far-infrared ceramic heaters will continue to expand in the future.

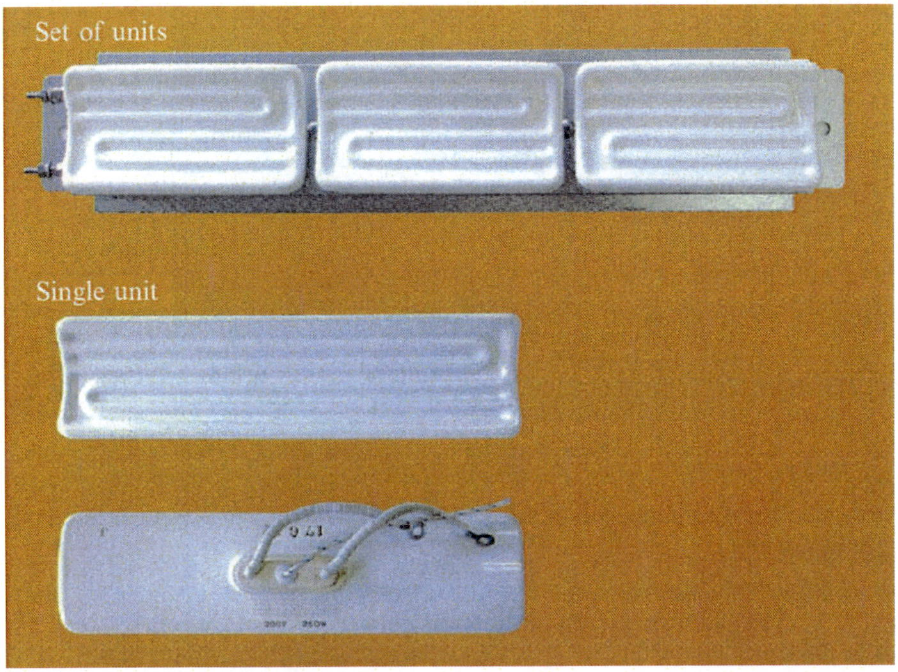

Fig. 22.34 Far-infrared ceramic heaters

Table 22.6 Characteristics of far-infrared ceramic heaters

Items	Values indicating performances
Wavelength of infrared rays (μm)	Mainly 3–7
Emissivity (ε) (Note 22.20)	0.96
Maximum heater temperature	600°C

Table 22.7 Comparison of various heaters

Items	Far-infrared ceramic heaters	Sheathed heaters	Infrared ray lamps	Gas heat source
Heating time[a]	100	150	200	300
Power consumption[a]	100	200	300	270 (converted to electric power)
Installation area	Small	Medium	Medium	Large
Temperature control	High	Coarse	Coarse	Coarse
Maintenance	Periodical check	Continuous check	Continuous check	Continuous check
Service life	Long	Short	Short	Short

[a]The values in this table are relative values compared to the performance of the far-infrared radiation heater which is assumed to be 100

Note 22.20 The ratio between the amount of energy released from the surface of a substance of a certain temperature and the amount of energy released from an imaginary object that absorbs 100% of the energy supplied by a black body at the same temperature.

Decorative Ceramic Watch Cases, Bands (1965) 22.9

Ceramics have been used as watch cases and bands for 40 years, since 1972, and were originally based upon alumina ceramics. Currently, zirconia ceramics and cermet are used. Variations in color tones are possible such as black, white, silver and gold. A variety of surface finishes can be produced varying from mirror finish, translucence finish, hairline finish in order to achieve the desired finish or texture. The materials are processed in a manner to serve a diversified market.

22.9.1 Background of Development

Watch makers have added value to watches by introducing various colors, designs and functions, thereby contributing to the development of the watch industry. Driving systems for automatic time keeping (utilizing the principles of solar batteries and self-winding mechanisms) have also been introduced recently.

Watch makers have also pushed the advance of materials for watch cases and bands. Stainless steel was originally used, followed by adoption of noble metals such as gold and silver for the luxury watch market. In addition, ultra-hard materials began to be used responding to the needs for scratch-resistant watch cases and bands. Ceramics, having original color tones and textures, scratch resistant properties and lightweight began to be used later (Fig. 22.35).

22.9.2 Characteristics and Manufacturing Method

22.9.2.1 Characteristics

Ceramics such as alumina, zirconia and cermet are used to manufacture watch cases and bands. They exhibit the following characteristics:

1. They are resistant to scratches, maintaining the appearance of new products long after purchase because of their superhardness.
2. Their specific weight is lower than ultra-hard metals which helps to reduce the weight of watches.
3. Skin irritation is reduced because they do not contain metals that sometimes cause allergic reactions.
4. They enable creation of color tones unique to each material.
5. Color tones are steadfast even after long-term usage.

22.9.2.2 Products

Ceramic watch cases and bands are manufactured with the following process (Fig. 22.36). First, watch cases and bands are formed into shapes by powder press forming, extrusion forming, injection molding, etc. For production of recent watch cases, injection molding is adopted because complex three-dimensional shapes are common for cases. In this forming process, ceramic materials are molded using specified dies. The molding process is followed by the cutting process for shaping. However, dewaxing and sintering are normally performed after the molding process. The sintering process is very important in producing the characteristics of ceramics. They go through a grinding process, where they are processed into specified shapes and accurate dimensions, thereby matching the shapes required of watch cases and bands. Finally, mirror finishing, translucence finishing, hairline finishing, etc. are performed before completion of watch cases and bands (Fig. 22.36).

22.9.3 Future Prospects

Ceramics are materials with tremendous potential as shown in Figs. 22.37 and 22.38, new colors in addition to existing colors are possible with ceramics, making it possible to modify original color tones to suit watch makers. Ceramics are light and scratch resistant. Watch makers can offer original watch colors for differentiation and customer appeal. This market is expected to grow (Figs. 22.37 and 22.38).

Fig. 22.35 Watches that incorporate ceramics. Ceramics are used in the watch cases and bands. The demand for watches incorporating ceramic parts continue to increase because of their texture and surface characteristics are unique to ceramics

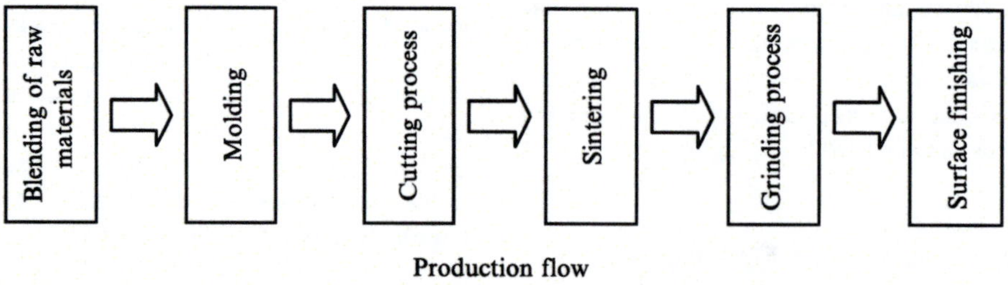

Fig. 22.36 Ceramic production process used for watches and bands

Fig. 22.37 Ceramics of various colors. A variety of colored ceramics are offered, including *reddish* and *bluish* colors, deep and pale colors, gold color ceramics and silver color ceramics that produce metallic luster, responding to needs

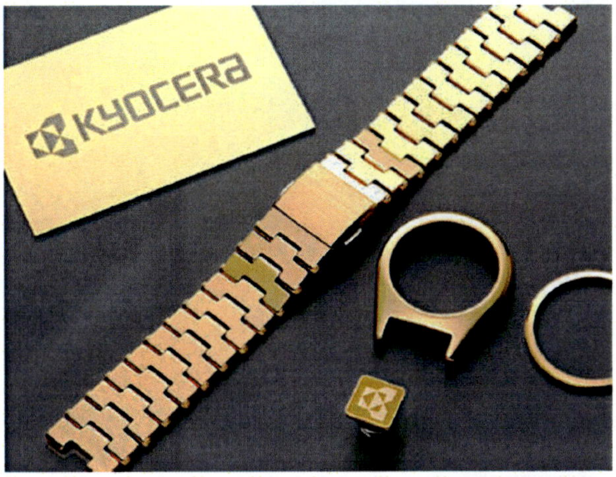

Fig. 22.38 New gold color ceramics. Newly developed gold color ceramics, which is brighter than previous gold colored ceramics. The luxurious texture is similar to that of "18-carat gold"

Eyeglass Lenses (AR Coating) (First Half of 1970s) 22.10

Eyeglasses are normally composed of lenses and a frame. Eyeglass lenses are used mainly for vision correction, ocular correction and eye protection. Optical lenses utilize the refraction and reflection of light to correct vision. The optical system of eyeglass lenses and the optical system of eyes are positioned and configured so that they function as one optical system, by holding the eyeglass lenses in a frame. Eyeglass lenses have a long history. They were first used in Medieval Europe. Lens materials, designing and surface treatment technologies are key technologies for production of eyeglass lenses. Lens materials are made of glass or plastics. In terms of design, they are classified into single focus lenses, multifocal lenses and progressive power lenses. Surface treatment is performed to prevent reflection and improve abrasion resistance. In Japan, anti-reflective (AR) coated eyeglass lenses, which were commercialized in the 1970s, have more than an 80% market share. Additional functionality (water-repellent properties, abrasion resistance, etc.) of the lenses is expected to continue. Requirements for eyeglass lenses will continue to be strict but become more diversified. Demands for low-price lenses are also expected to increase. Eyeglass lenses are optical lenses that will continue to play very important roles in vision care.

22.10.1 Background of Development

Eyeglasses are normally composed of lenses (Note 22.21) and a frame. Eyeglass lenses have been used mainly for vision correction, ocular correction and eye protection, etc. In recent years, however, priorities are also placed on the fashionableness of eyeglasses, resulting in diversified frame designs. Optical lenses that take advantage of refraction and reflection of light are framed to produce eyeglasses. Eyeglass lenses are classified into four types in terms of their function:

1. Vision correction lenses: lenses for refractive error correction (short-sightedness, far-sightedness, distorted vision), lenses for focusing error correction.
2. Ocular correction lenses: lenses for ocular misalignment correction (heterophoria/strabismus prisms).
3. Eye protection lenses: lenses for harmful light removal and dimming, light shielding and anti-glare lenses, visualization lenses, photochromatic lenses, polarizing lenses.
4. Special purpose lenses: lenses for amblyopia treatment, telephoto lenses, Fresnel lenses.

There are a wide variety of eyeglass lenses, as explained above. Variations in lenses, created by combination of frames, designs, materials and surface treatments, are limitless. Eyeglass lenses are products that are customized responding to customer needs.

22.10.2 Characteristics

22.10.2.1 Design

In terms of applications and designs, eyeglass lenses are classified into three types, single focus lenses, multifocal lenses and progressive power lenses. In the progressive power lenses, unlike optical lenses used in other fields, there is no boundary between vision range for far sightedness and the vision range for short sightedness, containing a series of degrees in a single lens. (Classification by design is shown in Fig. 22.39.)

Single focus lenses include lenses for correction of short-sightedness and lenses for correction of far-sightedness and

Note 22.21 Eyeglass lenses are optical lenses that take advantage of refraction and reflection of light. They are used mainly for vision correction, ocular correction and eye protection. The functions are realized with a combination of elemental technologies involving designs, materials, surface treatments (including anti-reflection films), etc. They are shaped to fit into frames, which are manufactured in consideration for fashion, etc. All required performance is contained in each of the lenses that constitute eyeglasses.

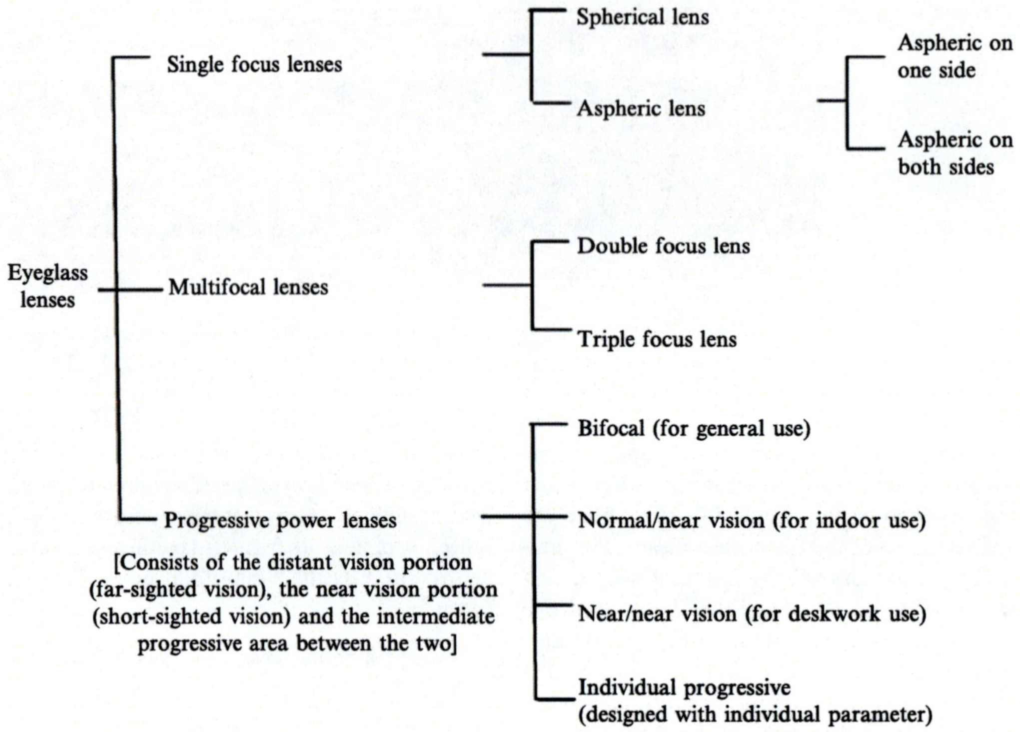

Fig. 22.39 Classification of eyeglass lenses by design

presbyopia. Designs include spherical designs and aspheric designs. Aspheric lenses are classified into lenses that are aspheric on one side and ones that are aspheric on both sides.

Multifocal lenses are classified into double focus lenses and triple focus lenses. The vision range for far-sightedness of the double focus lens contains degrees for the correction of short-sightedness or far-sightedness, while the vision range for short-sightedness contains degrees for correction of presbyopia. Triple focus lens has an additional vision range (the intermediate vision range) which contains intermediate degrees.

The progressive power lenses are the optical lenses that exhibit the highest performance among eyeglass lenses. There is no boundary between vision ranges for far-sightedness and vision ranges for short sightedness, containing a series of degrees that change gradually.

The progressive power lenses are represented by bifocal lenses. In addition to all-purpose bifocal lenses, a variety of progressive power lenses designed with consideration for usage environments have been commercialized. They include the normal/near vision type for indoor use, near/near vision type for deskwork use and individual progressive type optimized for each consumer.

22.10.2.2 Materials

Glass and plastics are normally used as lens materials. It is important that the materials are homogeneous and have high transparency. The materials are required to have balanced refractive index, Abbe number (Note 22.22) and specific weight. They also need to have high weather resistance, durability and workability. Figure 22.40 below shows refractive indexes and Abbe numbers of major eyeglass lens materials. In general, the lens becomes thinner as the refractive index becomes higher. Vision becomes clearer as the Abbe number becomes higher. The refractive index and Abbe's number are reciprocal. Glass lenses have been replaced by plastic lenses in the past several years.

22.10.2.3 Surface Treatment

Surface treatments such as coatings are important to supplement the optical properties of the lens. Coating processes are

Note 22.22 Abbe number is an index for evaluation of color dispersion (change in refraction index responding to the change in wavelength) in transparent bodies.

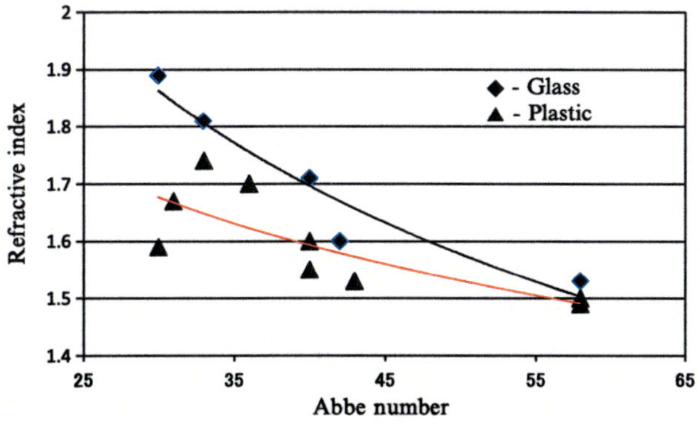

Fig. 22.40 Refractive indexes and Abbe numbers of major eyeglass lenses. Eyeglass lenses are made of glass or plastics as shown in this figure. The Abbe number of eyeglass lens materials becomes lower as the refractive index becomes higher, as is the case for other optical lenses

classified broadly into three types, the hard coating that reinforces surface hardness, the AR coating (Note 22.23) that reduces reflectance of light and functional film coating that add functions such as dimming and blocking polarized light. Staining is performed during the surface treatment process so as to create a variety of colored lenses. Surface treatments including coating processes need to be selected in consideration for compatibility (such as adhesion property) of the treatment with the lens material. A variety of products are offered, featuring various film configurations and film properties (such as abrasion and heat resistance), intended for various purposes and usages. General consumers expect low-cost surface treatment that enhances resistance to stains, scratches and heat. Previously, hard coating, AR coating and water repellent coating were common. However, various types of multifunctional coatings, such as abrasion/scratch resistant coating, shock absorbing coating, anti-frost/stain-proof/anti-static coating, heat resistant coating and organic compound coating were commercialized recently.

22.10.2.4 AR Coating

The majority of the light that enters the lens passes through the lens, but part of the light is reflected by the front and back surfaces. The light that travels straight in the air, as shown in Fig. 22.41, is refracted and reflected at the same time when it passes through the eyeglass lens. When lenses without AR coating are used, the light reflected by the lens surface causes double images called ghosts (virtual images) or flickers, resulting in unclear vision.

The light ray that travels straight in air is refracted and reflected at the same time when it passes through the eyeglass lens. When lenses without AR coating are used, the light reflected by the lens surface causes double images called a ghost (virtual images) or flickers, resulting in unclear vision. In addition, reflections by the lens surface may confuse the user wearing the glasses.

The refractive index (n) and the reflectance (R) of a lens are obtained from the Fresnel formula $R=(n-1/n+1)\ 2$. The principle of the AR coating is based on the "light interference" effect. Figure 22.42 below shows a lens with a refractive index of NL, which is coated with a thin film featuring a refractive index of N1 and a thickness of d. The ray of the light that enters the lens at an incident angle of less than 90° normal to the film surface is reflected by the surface of the thin film is expressed as R1. The ray of the light reflected by the back of the thin film is expressed as R2 (in the figure, the light is inclined for convenience). When the film thickness is 1/4 of the light wavelength λ of the incident light, the reflected rays are separated into wave groups R1 and R2. Phases of R1 and R2 are opposite to each other, balancing and interfering with each other, disappearing when two conditions (phase condition (Note 22.24) $N1d=\lambda/4$ (λ; wavelength) and amplitude condition (Note 22.25) $N1=\sqrt{NL}$) are satisfied. With respect to the AR coating, the monolayer film coating dominated the market in the past. However, the multilayer film coating dominates the current market, and reflectance is reduced by applying inorganic oxide multilayer films. The reflectance of lens materials varies depending on the refractive index. However, it is normally 8–12%. Reflectance is reduced to approximately 0.5–4% by application of the AR coating. The reflectance of AR-coated eyeglass lenses today range from around 0.5%, 1%, 2–4%. Vision through eyeglasses continues to approach the level of that of the naked eye. A surface treatment technology that adds functions other than the antireflection function (such as water repellent and abrasion resistant functions) to the AR coating has also been introduced.

Note 22.23 AR stands for Anti-reflection.

An AR coating reduces reflectance of stray light on the lens surface and increases the transmitted light utilizing the interference principle of light. It also removes trouble such as the generation of double images called ghosts (virtual images), which are caused light reflected by the lens surface and prevent clear vision. The AR coatings can be a monolayer or multilayer film coatings.

Note 22.24 Phase conditions are expressed by $N1d=\lambda/4$ (N1: refractive index of thin film, d: thickness of thin film, λ: wavelength). For prevention of reflection, the optical length of the thin film needs to be 1/4 of the wavelength of the light. This is called the phase condition.

Note 22.25 The amplitude condition is expressed by the expression $N1=\sqrt{NL}$ (N1: refractive index of thin film, $\sqrt{NL}$: square root of refractive index of lens). The amplitude of light reflected by the film surface has to be equivalent to the amplitude of light reflected by the back of the film surface. This is called the amplitude condition.

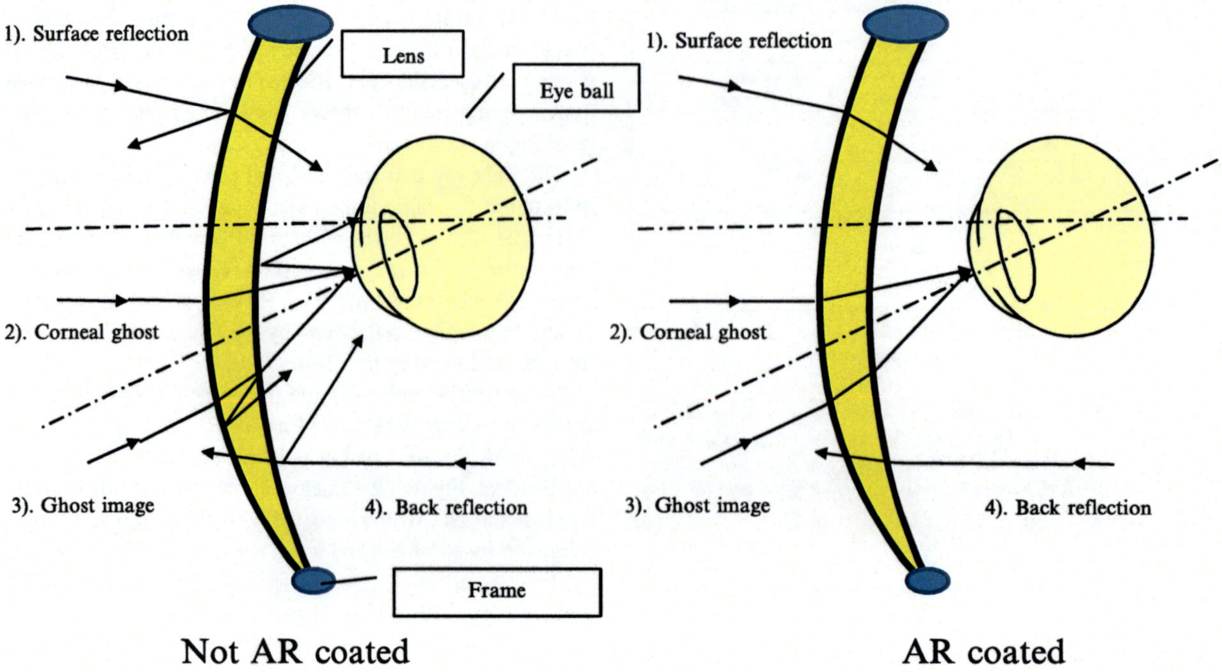

Fig. 22.41 AR coating

22.10.2.5 Manufacturing Method

There are three major processes for eyeglass lenses; the stock lens production process, the customized lens production process and a process for fitting the processed and surface treated lenses in frames for the completion of eyeglasses. Eyeglass lenses are made of glass or plastics. They are produced with elemental technologies involving designing, materials, forming, processing, surface treatments (hard coating, AR coating, dimming, polarizing and tinting). The glass is mainly produced by blending, melting and forming, etc. Thermosetting plastics are produced by the cast polymerization method, while thermoplastics are produced by the injection molding method, etc. Eyeglass lenses are produced through a series of processes (such as cutting, grinding, polishing and shaping) and the surface treatment based on prescriptions received from users. Typical eyeglasses and lenses are shown in Fig. 22.43.

The vacuum deposition method is normally employed for preparing AR coatings, which is performed to reduce the reflectance of stray light. However, the spin coating method is sometimes employed. In the vacuum deposition method, the substance to be deposited is heated and evaporated in a vacuum chamber by the resistance heating method or the electron beam method to form a clear thin film of the evaporated substance on the lens surface. Multilayer film coating (which consists of multiple layers of films with different refractive indexes) is currently popular as it considers the refractive index of materials and functions such as anti-reflection functions.

Eyeglasses are composed of a frame and lenses that are processed to match the frame, as Fig. 22.43 shows. Eyeglass lenses are classified into two types; stock lenses, which are thin and cut into shapes that fit the frames, and semi-finished lenses, which are thick and applied with surface treatment and shaping before fitting in frames to complete eyeglasses. Eyeglass lenses applied with surface treatments for tinting, etc. are also offered, in addition to clear lenses.

22.10.3 Future Prospects

Eyeglasses will attract more attention and prices will continue to drop in the future as the populations in developed countries age and middle-income populations in developing countries rise. In Japan, the U.S., and Europe and in Asia, needs for eyeglass lenses vary depending on the region. In order to respond to the high volume and diversified needs and the needs for low-price products, greater effort is required for innovative technologies as well as improvements in production technologies. Eyeglass lenses are optical lenses that are expected to remain major products that support vision care.

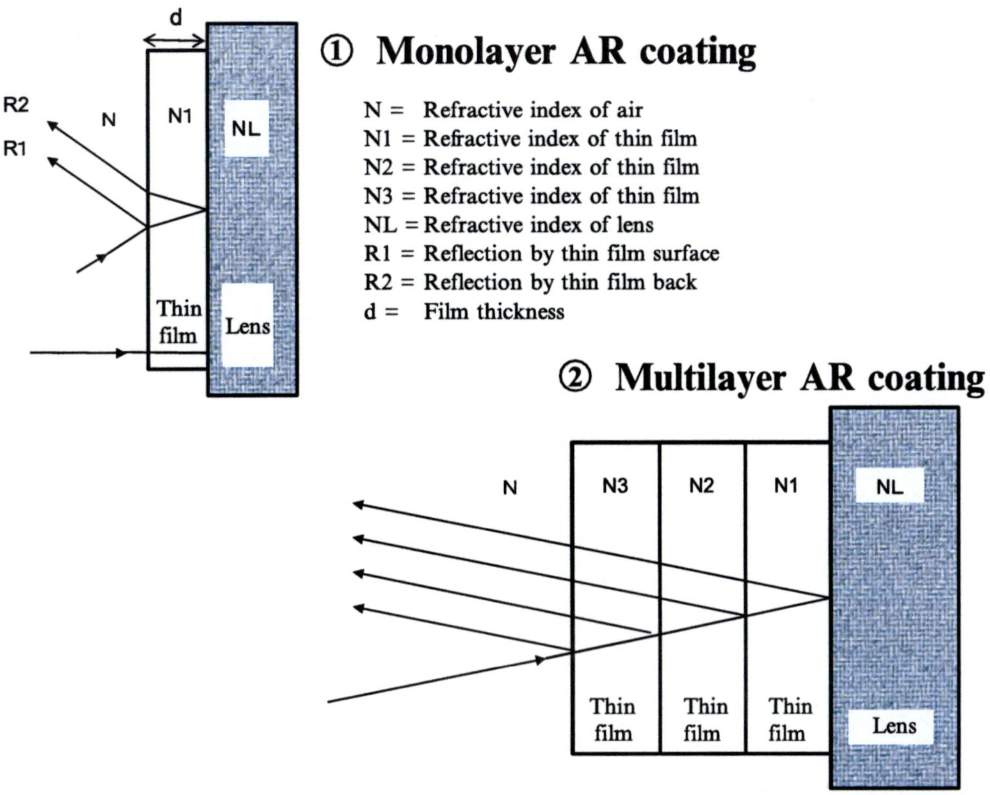

Fig. 22.42 Principle of AR coating. As indicated by ① and ② a lens with a refractive index of NL is coated with a thin film featuring a refractive index of N1 and a thickness of d. Light enters the lens at an incident angle of 90°. The light reflected by each side of the film is expressed as R1 and R2, respectively. The reflected beams are separated into two wave groups R1 and R2. They overlap and interfere with each other, disappearing completely when two conditions (phase condition $N1d=\lambda/4$ (λ: wavelength) and amplitude condition $N1=\sqrt{NL}$) are satisfied. *Note*: In the figure, the light is a little tilted for easy differentiation

Fig. 22.43 Typical eyeglasses and lenses

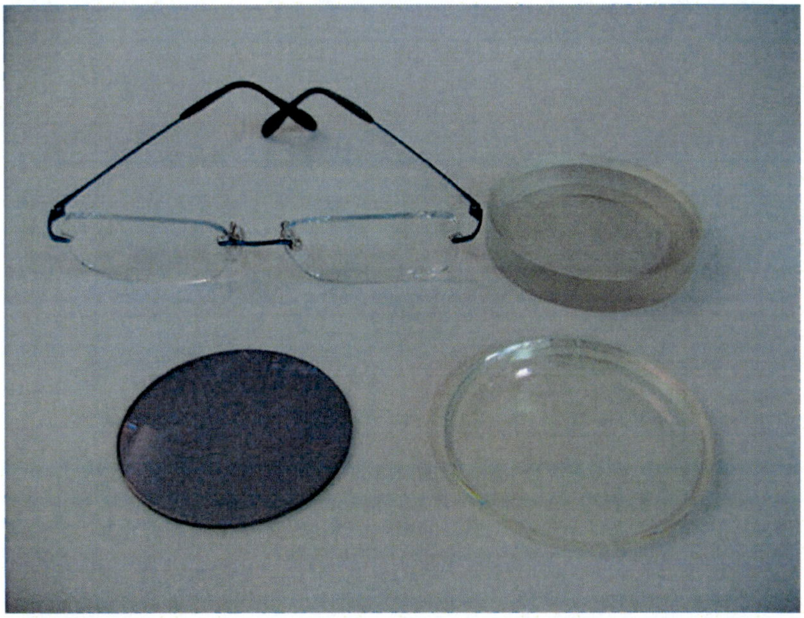

22.11 Synthetic Jewelry (1975)

Jewels have been considered precious since old times. Artificial jewels that look like natural ones have been made since the dawn of history. Efforts to produce synthetic jewels having the same chemical composition and structure of natural jewels, by precisely reproducing the minerals and crystals found on Earth, began in the nineteenth century, when research on properties of jewels was being advanced. Fremy and Feil of France succeeded in the synthesis of ruby for the first time in 1877. The industrial production of ruby was started in 1902 by Verneuil of France (Note 22.26). Since that time, a variety of jewels have been synthesized by various methods. Kyocera a company in Japan has developed various synthetic jewels, including ruby and opal since it succeeded in the development of emerald in 1975.

22.11.1 Background of Development

Jewels are natural resources and have been mined since before recorded history. Depletion of natural resources is progressing worldwide. The Japanese economy improved during the age of high economic growth and people began to have a strong yearning to possess jewels. However, jewels were still too expensive to buy for many people. Under these circumstances, the needs for high-quality, inexpensive jewels increased. Companies and research institutes around the world began to compete in the development of synthetic jewelry in the 1970s.

Note 22.26 Fremy et al. succeeded in synthesis by the flux method, but did not succeed in commercialization, because the synthetic ruby was more expensive than natural ruby due to the expensive cost. Meanwhile, Verneuil developed the Verneuil method (flame-melting method) and succeeded in commercialization.

22.11.2 Characteristics and Manufacturing Method

22.11.2.1 Characteristics

Kyocera's synthetic jewels are classified into two groups by structure. The group of "recrystallized jewels" composed of single crystals and the group of "created jewels" composed of amorphous material. Figure 22.44 shows pictures of man-made recrystallized emerald and created opal, which, respectively, represent the groups. Characteristics of recrystallized emerald and natural emerald are compared in Table 22.8.

Figure 22.45 is a picture of the internal structure of the created opal. Natural opal consists of regular three-dimensional array of fine silica dioxide particles with diameters on the order of 200–300 nm. The unique patterns called "speckles" and the play of color that changes color patterns depending on the viewing angle, which characterize opal, are generated by Bragg diffraction of light (Note 22.27). The chemical composition and structure of natural opal, as well as unique patterns called "speckles" (Fig. 22.46) and the play of color, are reproduced in the created opal.

22.11.2.2 Manufacturing Method

The preparation method of recrystallized jewels is different from that of created jewels and varies depending on the type of jewel. The processing of recrystallized emerald and created opal are explained below as examples.

Note 22.27 Bragg diffraction: diffraction observed when the Brag diffraction condition $n\lambda=2\ d\sin\theta$ is satisfied. Angle of the incident light corresponds to θ and grain diameter corresponds to d. The spectrum of wave length λ and the color tones are changed by adjusting θ and d.

Fig. 22.44 Synthetic gemstones: recrystallized emerald (*left*), created opal (*right*). Color tones of the best quality natural jewels were reproduced, realizing flawless and high-clarity jewels

Table 22.8 Characteristics of recrystallized emerald and natural emerald

	Recrystallized emerald	Natural emerald
Chemical composition	$3BeO{\cdot}Al_2O_3{\cdot}6SiO_2$	$3BeO{\cdot}Al_2O_3{\cdot}6SiO_2$
Crystal system	Hexagonal system	Hexagonal system
Hardness (Mohs)	7.5–8.0	7.5–8.0
Specific weight	2.65–2.70	2.65–2.74
Melting point	1,410°C	1,410°C
Clarity	Clear-translucent	Clear-translucent
Refractive index	1.563–1.568	1.565–1.598
Birefringence	0.005	0.005–0.008
Pleochroism	Green and blue	Green and blue
Inclusion	Liquid phase, solid phase	Liquid phase, solid phase, gas phase

Fig. 22.46 "Speckles," the patterns in the created opal. The patterns are unique to opal and none of the patterns are identical

Fig. 22.45 Internal picture of created opal (16,000×). Silica dioxide spheres of 250 nm are arranged in a regular three-dimensional array. The tone of opal is changed by adjusting the condition for light diffraction through control of the size of silica dioxide spheres (Note 22.27).

The recrystallized emerald is grown by the flux method using a flux. The raw material consisting of emerald elements is melted in a flux with a high melting point to create a saturated condition. Emerald elements in the solution are deposited as crystals when it is cooled down slowly over 6 months to 1 year. Temperature control at high temperature ranges and the cooling speed are important in this process. Achievement of an optimal saturated condition at around 1,400°C and control of the cooling speed at 1°C/day or slower are the key points for growing high quality crystals. Figure 22.47 shows the rough stone of the recrystallized emerald.

The preparation method of the created opal is explained below. Fine silica dioxide particles with the basic structure of an opal are synthesized and dispersed in water. When the liquid is left still standing, the silica dioxide particles settle down and form regularly-arrayed sedimentation, developing the "speckles" that characterize opal. When the particles are completely settled, the supernatant liquid is removed. The sediment is dried into a solid body similar to chalk and is applied

Fig. 22.47 Rough stone of recrystallized emerald. The crystals were grown by the flux method at high temperature followed by cooling to room temperature which takes over 6 months to 1 year

with a special treatment to produce the created opal. Control of the size of silica dioxide spheres is important in this process.

The "speckles" become clearer and the play of color becomes more prominent when uniform sized particle are used. For production of the created opal, 13 months are required in total and special technologies are needed. Kyocera is the only company that has been successful in the commercial production of synthetic opals.

22.11.3 Future Prospects

Production of high-quality natural jewels has been declining and it is likely that these resources will be depleted in the near future. Meanwhile, the needs for jewels are expected to increase, responding to the change in attitude toward jewels from possession as properties to utilization as fashion items. Responding to the increasing demand for jewels, the supply of natural jewels will be maintained at the current level by incorporating advanced technologies to improve the quality of low-grade jewels that were abandoned in the past.

Under this market trend, the needs for high quality and inexpensive synthetic jewels that have distinctive characteristics will continue to grow. In terms of environmental protection, synthetic jewels have the least effect on the environment and can be positioned as environmentally-conscious products that match the needs of society.

Literature

1. Takenouchi K (2003) Tribologist 48:559–563 (in Japanese) (22.1)
2. Promotional Department at the Illuminating Engineering Institute of Japan (ed) (2005) Basic knowledge on illumination, intermediate version (revised edition) (Series: New Illumination Class), pp 16, 30–32 (in Japanese) (22.4)
3. Promotional Department at the Illuminating Engineering Institute of Japan (ed) (2004) Light source (revised edition) (Series: New Illumination Class), pp 2–4, 56–66 (in Japanese) (22.4)
4. Yamaguchi Y (ed) (1984) Ceramic science series 3. Optoceramics. Gihodo Shuppan, Tokyo, pp 40, 48–50, 76–77 (in Japanese) (22.4)
5. Yamazoe N (2007) In: Karube Y (ed) Dictionary of biosensors and chemical sensors. Techno System, pp 475–484 (22.5)
6. Yanashima K (ed) Science of eyeglasses 21 basic edition. HOYA Corporation Vision Care Company (22.10)
7. Itoi M et al (2001) Eyeglasses. Medical-Aoi Publications, Inc. (22.10)
8. Takahashi K et al (2008) Control and measurement of polarization and birefringence in LCD/optical materials and their applications technical information society, pp 195–211 (22.10)
9. Webster R (1980) GEMS. Gemmological Association of All Japan Co., Ltd., Tokyo, pp 335–336 (22.11)
10. Isogami M (2007) Sci Eng Mater 44:89–94 (22.11)

Index

Y. Imanaka et al. (eds.), *The Ceramic Society of Japan, Advanced Ceramic Technologies & Products*,
DOI 10.1007/978-4-431-54108-0, © Springer Japan 2012

Q

R